Instructor's Manual
to accompany Stein/Barcellos

CALCULUS
AND ANALYTIC GEOMETRY

FIFTH EDITION

ANTHONY BARCELLOS
American River College
Sacramento, California

McGraw-Hill, Inc.
New York St. Louis San Francisco Auckland Bogotá
Caracas Lisbon London Madrid Mexico City Milan
Montreal New Delhi San Juan Singapore
Sydney Tokyo Toronto

This book is printed on recycled, acid-free paper containing a minimum of 50% recycled de-inked fiber.

**Instructor's Manual to Accompany
CALCULUS AND ANALYTIC GEOMETRY
Fifth Edition
byAnthony Barcellos**

Copyright ©1992 by McGraw-Hill, Inc. All rights reserved. Printed in the United States of America. The contents, or parts thereof, may be reproduced for use with
**CALCULUS AND ANALYTIC GEOMETRY
Fifth Edition
by Stein/Barcellos**
provided such reproductions bear copyright notice, but may not be reproduced in any form for any other purpose without permission of the publisher.

ISBN 0-07-061189-0

Table of Contents

- **Advice to Instructors** 1

1. **An Overview of Calculus** 31
2. **Functions, Limits, and Continuity** 35
3. **The Derivative** 61
4. **Applications of the Derivative** 79
5. **The Definite Integral** 113
6. **Topics in Differential Calculus** 137
7. **Computing Antiderivatives** 165
8. **Applications of the Definite Integral** 201
9. **Plane Curves and Polar Coordinates** 229
10. **Series** 251
11. **Power Series and Complex Numbers** 275
12. **Vectors** 305
13. **The Derivative of a Vector Function** 327
14. **Partial Derivatives** 341
15. **Definite Integrals over Plane and Solid Regions** 387
16. **Green's Theorem** 417
17. **The Divergence Theorem and Stokes' Theorem** 435

Appendices

- **A** Real Numbers 449
- **B** Graphs and Lines 449
- **C** Topics in Algebra 453
- **D** Exponents 454
- **E** Mathematical Induction 455
- **F** The Converse of a Statement 456
- **G** Conic Sections 457
- **H** Logarithms and Exponentials Defined through Calculus 461
- **I** The Taylor Series for $f(x, y)$ 462
- **J** Theory of Limits 463
- **K** The Interchange of Limits 464
- **L** The Jacobian 465
- **M** Linear Differential Equations with Constant Coefficients 470
- **●** Supplementary Conceptual Problems 471

A Note on Notation

This manual uses the same notation and techniques as the text by Stein and Barcellos. However, for the sake of convenience and to save space, we also take two shortcuts. The symbol $\underset{H}{=}$ represents an equality that follows from an application of l'Hôpital's rule (Sec. 6.8 and thereafter). Also, the text writes out units of measurement in full, such as "meters per second per second," whereas we usually employ the standard abbreviations, such as "m/sec^2." We may occasionally use mag **A** in place of $\|\mathbf{A}\|$ for the magnitude of the vector **A** in those cases where the expression for **A** is complicated.

Preface

This *Instructor's Manual* has two major components: pedagogical notes for the instructor and answers to even-numbered exercises in the fifth edition of *Calculus and Analytic Geometry* by Sherman K. Stein and Anthony Barcellos. Supplementary conceptual exercises are appended at the end of the manual.

Most of the answers in this manual take the form of solution sketches, less detailed than the solutions from the *Student's Solutions Manual* but still sufficient to show the chain of reasoning. Where an answer alone suffices, such as with many of the routine exercises, only an answer is given. Occasionally the solutions include notes to the instructor to expand on the intent of the particular exercise or to discuss details which the instructor may find useful. In addition to even-numbered solutions, we also provide the solutions to the odd-numbered exploration problems which were excluded from the *Student's Solutions Manual*.

All of the solutions have been checked and double-checked in hopes of eliminating all errors. With over three thousand solutions, however, it is quite possible that a few errors have managed to slip through anyway. Please bring any errors that you might find to the attention of Anthony Barcellos at P.O. Box 2249, Davis, CA 95617.

About This Book. Camera-ready copy was produced on a Hewlett-Packard LaserJet III from documents created with WordPerfect 5.1. Most illustrations were created by CoPlot and CoDraw from CoHort Software (P.O. Box 1149, Berkeley, CA 94701, 510/524-9878), which generated .WPG files for direct import into WordPerfect. Several figures were created in the Microsoft Windows version of Mathematica 2.0 from Wolfram Research, Inc. (100 Trade Center Drive, Champaign, IL 61820, 217/398-0700) and translated into WordPerfect format by Hijaak from Inset Systems, Inc. (71 Commerce Drive, Brookfield, CT 06804, 203/740-2400).

Acknowledgments. We wish to thank the editorial staff at McGraw-Hill for their assistance with this manual, including former math editors Robert Weinstein and Richard Wallis. Maggie Lanzillo was a constant source of encouragement even while gently reminding us about deadlines.

Many of the people who participated in preparation of the Stein/Barcellos textbook also contributed in various ways to this manual. We acknowledge in particular the efforts of **Keith Sollers** and **Mallory Austin** who drafted solutions to thousands of exercises over the past two years and were key contributors to this manual. Other significant contributions of solutions came from Dean Hickerson, Duane Kouba, Don Johnson, and Timothy Thayer. Keith, Mallory, and Dean also provided the bulk of the proofreading, which they performed painstakingly, but are not to be held responsible for any errors left by the author. Erik Mans was also of great assistance in proofing the later chapters. In addition to those who worked on the text, Travis Andrews and Kelly Riddle contributed to the answer-checking. Heroic efforts were made by **Richard Kinter, Judith Kinter,** and **Michael Kinter,** who typed the reams of solutions and in the process became among the world's greatest experts in WordPerfect 5.1's equation editor.

—Anthony Barcellos

Advice to Instructors

In discussing each chapter we first comment on the chapter as a whole, then call attention to certain exercises.

1 An Overview of Calculus

This chapter permits a calculus course to begin with calculus, rather than precalculus. Even students whose precalculus is not strong are ready to understand this chapter. Students who have had some calculus and think that the definite integral is "$F(b) - F(a)$" can be set right at this point.

Options (0–3 lectures)
- 0 lectures, leaving the chapter to be read.
- 1 lecture, surveying the first two sections.
- 2 lectures, one on each of the first two, leaving the third as reading.
- 3 lectures, one on each section.

An interesting alternative is covering Secs. 3.1 and 5.1 the first two lectures instead, or right after this chapter.

2 Functions, Limits, and Continuity

There are many ways to treat the 10 sections of this chapter. The particular approach or emphasis will depend on the preparation of the students, the time available, and the instructor's perspective.

Options (5–13 lectures)
- 5 lectures, one each on 2.1, 2.3, 2.4, 2.7, and 2.8.
- 9 lectures, one combining 2.9 and 2.10.
- 13 lectures, 2.4, 2.6, and 2.8 each getting two.

2.1 (Functions) Because of the importance of the function concept, this section offers a variety of perspectives for the instructor to choose from (more than should be covered in a single lecture).

2.2 (Composite functions) This section reinforces the function concept. It is not needed until Sec. 3.6, when composite functions are differentiated.

2.3 (Limits) One lecture.

Notes on Chapter 2

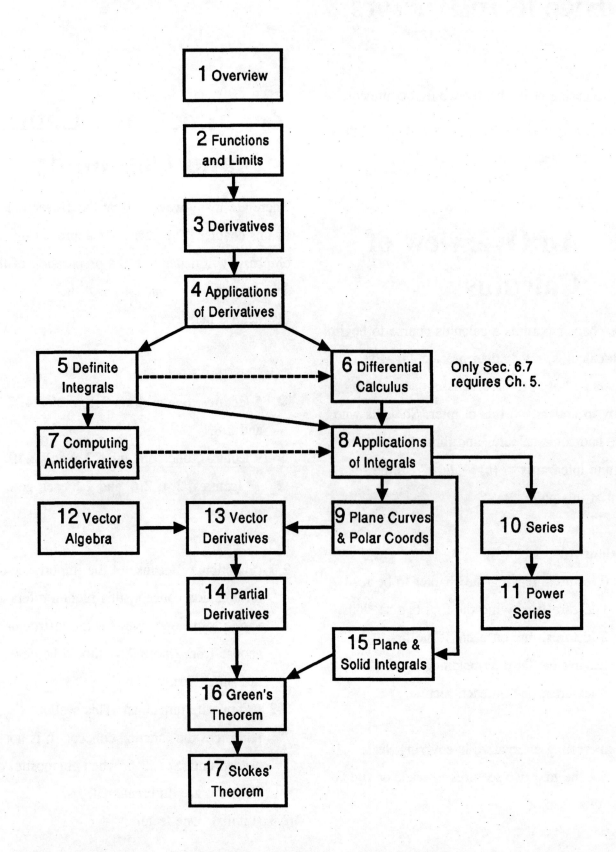

Organizational Options After Chapter 4

The early chapters of the text have been constructed to afford the instructor maximum flexibility in designing the course syllabus. The basic options are outlined below.

- **Early Introduction of the Definite Integral.** The text introduces the definite integral and several of its applications in Chapter 5 (The Definite Integral) immediately after treatment of the main uses of the derivative in Chapter 4 (Applications of the Derivative). This is convenient for a semester system, where the elements of both differential and integral calculus can be discussed during the first term. Further applications then follow in Chapter 6 (Topics in Differential Calculus) and Chapter 8 (Applications of the Definite Integral), with Chapter 7 (Computing Antiderivatives) providing several techniques for formal integration. The main option here is whether to spend much time on Chapter 7, or to cut it short in favor of using integral tables, computers, or other shortcuts for the applications in Chapter 8.
  Sequence of chapters: ④ → ⑤ → ⑥ → ⑦ → ⑧

- **Early Transcendental Functions/Differential Calculus.** Some instructors prefer to complete the treatment of differential calculus before discussing the definite integral. This is especially convenient for schools on a quarter system, where the shorter term prompts a narrower focus for the syllabus. It also provides the earliest possible introduction of all of the transcendental functions. In this case, Chapter 4 (Applications of the Derivative) is followed immediately by Chapter 6 (Topics in Differential Calculus). All of Chapter 6 can be covered except for Sec. 6.7 (The Differential Equation of Natural Growth and Decay), which depends on Chapter 5 and must therefore be left until Chapter 5 is covered.
  Sequence of chapters: ④ → ⑥ → ⑤ → ⑦ → ⑧

- **Early Applications of Definite Integrals.** Most of the applications of integration in Chapter 8 (Applications of the Definite Integral) can be treated immediately after Chapter 5 (The Definite Integral). Section 5.3 shows how to represent area, volume, distance, and mass as definite integrals, an early indication of the many applications of integrals. Sections 8.1 (Computing Area by Parallel Cross Sections), 8.2 (Some Pointers on Drawing), 8.3 (Setting Up a Definite Integral), and 8.4 (Computing Volumes) are a natural sequel to Chapter 5. Most of the antiderivatives do not require Chapter 6 (where log, exp, and inverse trigonometric functions are introduced and differentiated) and those that do are easy to skip in the examples and exercises. If the instructor inserts Sec. 7.1 (Shortcuts, Integral Tables, and Machines) between Chapters 5 and 8, many integrals that would otherwise require the formal techniques of Chapter 7 can be done by integral tables or symbolic math systems. (Chapter 8 problems that use the techniques of Chapter 7 are specifically identified in the exercise sets.)
  Sequence of chapters: ④ → ⑤ → ⑧ → ⑥ → ⑦

2.4 (Computation of limits) Though the essential ideas could be covered in one lecture, the concepts, especially "indeterminate limits" could merit two lectures.

2.5 (Some tools for graphing) If time presses, omit; then introduce the ideas as asides when needed in later lectures.

2.6 (A review of trigonometry) This is precalculus material, but many students need some review, most generally two lectures. However, a lecture covering the definition of sine and cosine, their graphs, and their identities will provide adequate background for Sections 2.7 and 3.5. (The students could refer to this section later, as needed.)

2.7 (The limit of $(\sin \theta)/\theta$ as θ approaches 0) Many students feel that the quotient of two small numbers is near 1. ("0/0 = 1") This section presents the first non-trivial indeterminate limits. The instructor may take a very casual approach, asserting that "clearly chord/arc is near 1." But then Exercise 20 presents another limit which everyone intuitively suspects is also 1 (but turns out to be 3/4). Observe that $(1 - \cos \theta)/\theta$ is treated geometrically in order to convey the idea of "one quantity approaching 0 much faster than another does." The traditional treatment, which involves an algebraic trick, survives as Exercise 15. Note that the argument for $(\sin \theta)/\theta$ differs from the usual one in the areas compared; $(\sin \theta)/\theta$ appears directly, not as the reciprocal of $\theta/(\sin \theta)$.

2.8 (Continuous functions) One lecture can cover the essence of this section: continuity, extreme-value and intermediate-value theorems. The text also covers continuity at an end point, but this can be left for the student to read. Spending an extra day so that students could work on more interesting exercises may pay off in the long run.

2.9 and 2.10 (Precise definitions of limits) There are many ways to deal with these two sections. Some instructors omit them. Some feel that the students should at least be exposed to these definitions, since such exposure will help those students who will take further math. One could cover only Sec. 2.9, where the diagrams and algebra are much simpler than in Sec. 2.10. Also, rather than covering the two sections in two lectures, one may insert them *gradually* throughout the course in short digressions.

EXERCISES

2.1 (Functions) Exercises 54, 55 and 64 concern drunk driving, a topic of special interest to drivers aged 16-24. Many students do not realize that the quotient $f(t)/g(t)$ measures risk. The behavior of this quotient is dramatically time-sensitive.

2.2 (Composite functions) Exercises 35 and 36 present a significant contrast that requires thought. Exercises 31-34 reinforce the idea of a polynomial.

2.3 (The limit of a function) Several exercises present non-trivial limits (the kind that cannot be discovered by an algebraic trick), E23-25, for instance. Note that E26 provides background for e, defined in Chapter 6. E32 can be used to introduce several topics, such as determining a minimum value. E35 is a complete tri-ex exploratory problem (tri-ex = explore, extract, explain). The student may observe that f decreases but is always greater than 1/2. (Many think it approaches 0.)

2.4 (Computations of limits) Exercises 41-43 require the student to provide examples. The arguments in E38-40 also demand that the student "think," not just grind out limits.

2.5 (Some tools for graphing) E29-36 are short writing exercises. E41 could lead to a fruitful discussion. Contrast E42 and 43.

2.6 (A review of trigonometry) In E54 the answers are not unique. (*e.g.*, $\sin \theta = \cos(\pi/2 - \theta)$ or $\sin \theta = \sqrt{1 - \cos^2 \theta}$). Note that E55 introduces Sec. 2.7.

2.7 (The limit of $(\sin \theta)/\theta$ as θ approaches 0) Note E20, whose answer is not 1. E23-28 describe graphs that require the student to understand the behavior of $\sin x$.

2.8 (Continuous functions) E38-50 are especially instructive. One or more might be included in a lecture.

2.9 (Precise definitions of "$\lim_{x \to \infty} f(x) = \infty$" and "$\lim_{x \to \infty} f(x) = L$") E27-30 test understanding.

2.10 (Precise definition of "$\lim_{x \to a} f(x) = L$") Note E19-24.

3 The Derivative

Options (5-8 lectures)
- 5 lectures, one combining 3.1 and 3.2.
- 6 lectures, one on each section.
- 8 lectures, 3.2 and 3.4 each getting two.

Students with a smattering of calculus may think that the derivative is *defined* as slope. That is one reason there are four different motivations (applications) of the derivative in 3.1. Once a couple of them are done in detail, the other two can be sketched quickly. One may include the definition of the derivative from Sec. 3.2 in the lecture on 3.1. (In fact, it is hard to resist.)

One way to treat Sec. 3.2 is to differentiate x^n, n a positive integer, x^{-1}, and assert that $D(x^a) = ax^{a-1}$ for any exponent, which is justified in Chap. 6. We made that assumption now to avoid the non-instructive proof for $D(x^{m/n})$.

Section 3.3 introduces the Δ notation and antiderivative in addition to distinguishing differentiable from continuous. The notion of antiderivatives is mentioned also in Chap. 4, reinforced by exercises in order to be available in Chap. 5.

The proofs in 3.4, by using the Δf, Δg, Δu notation, avoid tricks such as abruptly adding and subtracting an algebraic expression. Note that antiderivatives of x^a are discussed, though the notation $\int f(x)\, dx$ does not appear until Chap. 5.

In 3.5 we differentiate the six trigonometric functions. This is an appropriate time to show why radian measure is better than degrees.

In 3.6 the proof of the chain rule avoids

division by $\Delta u = 0$, but in a manner that is student-friendly. The case $\Delta u = 0$ is covered by Exercise 55.

EXERCISES

3.1 (Four problems with one theme) E31 and 32 illustrate other applications. E46 is a check on the concept of slope.

3.2 (The derivative) E38 and 39 check the student's understanding of slope, while E40 anticipates an extremum test. E41 is a check on the definition of derivative.

3.3 (The derivative and continuity) E23, 24, 29-32 concern antiderivatives. (Practice to prepare students for Chaps. 4, 5, 6, 7.) E25-28 and 34 are easy tests of understanding.

3.4 (The derivatives of the sum, difference, product, and quotient) This is a section where drill exercises (E1-46) should be emphasized. E47,48 reinforce antiderivatives.

3.5 (The derivatives of the trigonometric functions) E44,45 might be used to anticipate the chain rule.

3.6 (The derivative of a composite function) E1-44 are drill exercises to offer adequate practice and E45-48 reinforce antiderivatives.

3.S E53, 54, 55 (economics) E56 (biology), E72-74 (use a calculator to check and develop understanding of the derivative.)

4 Applications of the Derivative

Options (7–13 lectures)
- 7 lectures, omit 4.4, 4.6, and 4.10.
- 9 lectures, omit 4.4.
- 13 lectures, with 4.1, 4.7, and 4.9 each getting two.

4.1 (Three theorems about the derivative) $f' = 0$ at extremum, Rolle's theorem, and mean-value theorem. Each is amply motivated; if time is a factor, the motivations—without the proofs—would suffice. If there is time, this section merits 2 lectures.

4.2 (The first derivative and graphing) Here are the definitions of relative and global extrema and critical number and critical point. (Note that at a critical point $f' = 0$. We do not include the case when f' does not exist.) The section ends with maximizing f on $[a, b]$. Though there are several definitions, the section may be treated in 1 lecture.

4.3 (Motion and the second derivative) This

section also defines the higher derivatives. It treats only constant acceleration, and is used in Sec. 4.10, growth of a function.

4.4 (Related rates) If time is a factor, this traditional topic may be skipped. Since subsequent sections do not depend on it, 4.4 can also be postponed till later.

4.5 (The second derivative and graphing) This also gives the second-derivative test for extrema.

4.6 (Newton's method for solving an equation) This is another application that can be skipped if time presses. Calculators reduce its urgency.

4.7 (Applied maximum and minimum problems) This crucial section is one of the high points of calculus. It could be covered in 1 lecture, with its exercises scattered through several assignments. There is enough material to merit two lectures, should the instructor wish to emphasize applications.

4.8 (Implicit differentiation) is referred to in Sec. 14.10, Lagrange multipliers. Since implicit differentiation depends only on the chain rule, it may be treated earlier if instructors wish to have this technique available for extremum problems or related rates.

4.9 (The differential and linearization) presents two ideas used later. The differential is used formally in integration (Chap. 7). The differential approximation is examined in Sec. 4.10, which is background for the analysis of the error in Taylor series in Sec. 11.3. It is generalized to 2 variables and the tangent plane in Sec. 14.8. The use of the differential in estimates (*e.g.*, $\sqrt{65}$) may be deemphasized in the age of the calculator.

4.10 (The second derivative and growth of a function) This is one of the new sections in this edition. It shows the average student what f'' really does. A student who understands the assignment, which is couched in terms of cars, will not be surprised by Lagrange's formula for the error in Sec. 11.2. Of course the section could be delayed and covered between Secs. 11.1 and 11.2. (The choice may be influenced by time constraints.)

EXERCISES

4.1 (Three theorems about the derivative) E19-28 (short conceptual), E33-36 (ask for graph of f based on graph of f') E47-48 (antiderivatives).

4.2 (The first derivative and graphing) E9-16 (short conceptual).

4.3 (Motion and the second derivative) E24,

30 (the more challenging).

4.4 (Related rates) E23, 24 (the more interesting). If you assign E22, use Example 3 to note that $dV/dt = \pi r^2 (dh/dt)$.

4.5 (The second derivative and graphing) E21-28, 31, 32, 34 (short conceptual).

4.6 (Newton's method for solving an equation) E20-22, 28, 29 (the most interesting).

4.7 (Applied maximum and minimum problems) There are ample exercises. Consider including one of them in each of several assignments on subsequent topics.

4.8 (Implicit differentiation) Note E27-29.

4.9 (The differential and linearization) E47-49 (of special interest), especially E48.

4.10 (The second derivative and growth of a function) The heuristic approach in the text is supplemented by E14,15 which are purely mathematical. Note E16, background for Chap. 11. Also E17-19 (conceptual).

5 The Definite Integral

Options (5-8 lectures)
- 5 lectures, omit 5.4 and 5.6.
- 7 lectures, one on each section.
- 8 lectures, with two on 5.1.

5.1 (Estimates in four problems) presents 4 problems that lead to the same idea, a limit of sums. Students who think the definite integral is simply "area" will be alerted. If students are familiar with summation notation, the definite integral could be defined in this lecture.

5.2 (Summation notation and approximating sums) has summation notation, develops formulas for $\Sigma\, i$ and $\Sigma\, i^2$ using telescoping sums, and gives the summation notation for Riemann sums. If the students are well prepared, the lecture on 5.2 could cover part of 5.3.

5.3 (The definite integral) defines the definite integral, states its relation to area, distance, volume, and mass, and evaluates $\int_a^b x^2\, dx$. Students with a smattering of calculus may be thinking, "Why all the fuss; it's just $F(b) - F(a)$." For this reason you may discuss $\int_0^1 \sqrt{1 + x^3}\, dx$, for example.

5.4 (Estimating the definite integral) Presents the trapezoidal and Simpson's methods. In the 4th. edition this was in Chap. 7. It is now in Chap. 5 to help give the idea of the definite integral time to sink in before it gets mixed up with $F(b) - F(a)$. (It also is

Advice to Instructors

the basis for the new Sec. 5.6).

5.5 (Properties of the antiderivative and the definite integral) Presents basic properties of $\int_a^b f(x)\,dx$ and $\int f(x)\,dx$. It is here that the notation for an antiderivative is introduced.

5.6 (Background for the fundamental theorem of calculus) Gives the students a chance to anticipate the fundamental theorem of calculus, in particular, to realize that $G(x) = \int_a^x f(t)\,dt$ is a function and its derivative "seems to be" $f(x)$. If the students grasp this down-to-earth section, then the instructor is in a position to treat the culminating Sec. 5.7 as an anticlimax, a simple consequence of Sec. 5.6.

5.7 (The fundamental theorem of calculus) presents the Fundamental Theorem of Calculus. The fact that $G'(x) = f(x)$ is now called the first fundamental theorem, since it is proved first and is the basis of $F(b) - F(a) = \int_a^b f(x)\,dx$, which is now called the second fundamental theorem. In later chapters the second is usually referred to as the "The fundamental theorem of calculus." Note Example 5, which shows that the definite integral should not be thought of as "$F(b) - F(a)$." Note the use of a paintbrush on p. 307; one could be used in the lecture. The chapter summary includes a list of common functions whose antiderivatives are not elementary.

EXERCISES

5.1 (Estimates in four problems) E20 and 33 (for those who think $F(b) - F(a)$ is the definition of the definite integral), E29 (business), E34 (kinetic energy), E36 (writing), E38 (drill on differentiation), E39 (antiderivatives), E40 (drawing).

5.2 (Summation notation and approximating sums) E29 (computer science), E34 (geometric series by telescoping sums).

5.3 (The definite integral) E19 (obtains $\int_0^b x^3\,dx$), E20 (obtains $\int_0^b x^5\,dx$), E26 (volume of a cone), E28–31 (writing), E32 (anticipates fundamental theorem).

5.4 (Estimating the definite integral) E17 (surprising), E21-26 (justify Simpson's method), E30 (writing), E33, 34 (short conceptual).

5.5 (Properties of the antiderivative and the definite integral) E39 (writing, another approach to the average).

5.6 (Background for the fundamental theorem

of calculus) E14-16 (students who do these are ready for Sec. 5.7. Indeed, Sec. 5.7 is almost redundant.)

5.7 (The fundamental theorem of calculus) E35, 36 (reinforcing the distinction between elementary and non-elementary) E44 (volume of ball), E46 (why the area under the velocity curve is the distance).

5.S (Summary) E45 (a different motivation of the second fundamental theorem), E53-54 (reinforce idea of non-elementary antiderivative), E61,62 (challenging exercises that depend on first fundamental theorem), E64 (operations research), E65 (direct proof of second fundamental theorem), E67-69 (challenging), E70,71 (short and conceptual, these involve several ideas).

6 Topics in Differential Calculus

Options (4–12 lectures)
- 4 lectures, one on 6.1 and 6.2 combined, omit 6.4, 6.7, 6.8, and 6.9.
- 9 lectures, one on each section.
- 12 lectures, with 6.3, 6.7, and 6.8 each getting two.

6.1 (Logarithms) Many students need this review. It is pre-calculus material.

6.2 (The number e) If time is a pressure, just define e and bypass the motivation by compound interest. Then 6.2 could be treated with 6.3.

6.3 (The derivative of a logarithmic function) This includes logarithmic differentiation,
$$\int \frac{dx}{x}, \int \frac{f'(x)}{f(x)} \, dx.$$

6.4 (One-to-one functions and their inverses) Precalculus material, but many students benefit from a review. Sec. 6.5 could be started during this lecture.

6.5 (The derivative of b^x) We also obtain $D(x^a) = ax^{a-1}$.

6.6 (The derivatives of the inverse trigonometric functions) There would be time to review the graph of $\tan^{-1} x$ and $\sin^{-1} x$.

6.7 (The differential equation of natural growth and decay) If time presses, this section could be omitted. However, it is an important application. See the sidebars on radioactive half-life and the AIDS epidemic on p. 374. (E42 describes another model of the spread of AIDS. It might be pointed out that the data on this epidemic are not reliable.)

6.8 (l'Hôpital's rule) This section could be

Advice to Instructors

omitted if one prefers to use power series, as discussed in Sec. 11.5.

6.9 (The hyperbolic functions and their inverses) If time presses, this could be omitted or assigned as reading.

EXERCISES

6.1 (Logarithms) E21–24, 31, 32 (short conceptual), E34 and E36 (challenging).

6.2 (The number e) E9-15 (short conceptual), E16 (challenging).

6.3 (The derivative of a logarithmic function) E27-30 (practice differentiating), E41, 42 (challenging), E44 (a different approach to e), E45 (a well-known paradox).

6.4 (One-to-one functions and their inverses) E19-25 (short conceptual).

6.5 (The derivative of b^x) E46 (important graph), E50 (treats $D(x^a)$ when $x < 0$).

6.6 (The derivatives of the inverse trigonometric functions) E51, 52 (offer contrasts; background for Chap. 7).

6.7 (The differential equation of natural growth and decay) E28 (Doomsday), E29 (this "paradox" raises a fundamental question about modeling natural growth), E30 (one model for three different applications), E31 (harvesting), E34 (penetration of sunlight), E35 (economics), E38 (drug), E39-41 (inhibited growth), E42 (AIDS).

6.8 (l'Hôpital's rule) E49 (failure of intuition), E52 (short conceptual).

6.9 (The hyperbolic functions and their inverses) E37 (application to motion).

6.S (Summary) E139 (a review and writing), E141, 142 (short conceptual), E144 (review by proving), E152 (challenging), E158, 159 (proof of l'Hôpital's rule), E160, 161 (short conceptual), E162 (challenging), E163, 164 (a review through short conceptual).

7 Computing Antiderivatives

There is no general agreement on how much time should be spent on techniques of integration. While it is true that computers and some calculators can do many formal integrations, it is also true that for many years integral tables could do them too, yet integration techniques were taught.

Options (6–9 lectures)
- 6 lectures, omit 7.7.
- 7 lectures, one on each section.
- 9 lectures, with 7.2 and 7.7 each getting two.

There is a consensus that we must teach substitution (Sec. 7.2) and integration by parts (Sec. 7.3). Partial fractions (Secs. 7.4 and 7.5) are used in differential equations and in combinatorics (generating functions). Sec. 7.6 (special techniques) treats such important integrands as $\sin mx \sin nx$, $\sin^2 x$, $\sec x$, $\tan x$, the substitution $u = \sqrt[n]{ax + b}$, and trig substitutions. Sec. 7.7 gives the student an opportunity to practice. With this chapter as background, a student should feel comfortable with integration software, be aware that an antiderivative need not be elementary, and recognize integrals that are easily done by hand.

If time presses, Sec. 7.7 could be left to be read. An instructor's point of view will determine how much attention should be given to Secs. 7.4–7.6.

EXERCISES

7.1 (Shortcuts, integral tables, and machines) The exercises are all straightforward.

7.2 (The substitution method) E47-49 (provocative).

7.3 (Integration by parts) E27 (writing), E28 (a formula for $\int P(x)e^x \, dx$, $P(x)$ a polynomial; note that this can be used to justify "tabular integration" for instructors who like to introduce this shortcut; see the solution), E37, 38 (reinforce notion of elementary f), E47 (paradox), E48 (shows theoretical use of integration by parts).

7.4 (How to integrate certain rational functions) E39-43 strengthen algebraic background.

7.5 (Integration of rational functions by partial fractions) E55 (rational root theorem), E56 (factoring a cubic).

7.6 (Special techniques) E53-55 ($u = \tan(\theta/2)$), E60, ($\int \sec \theta \, d\theta$ and Mercator map).

7.7 (What to do in the face of an integral) E60-62 (short conceptual).

7.S (Summary) E164-173 (valuable background, reinforcing the limited utility of $F(b) - F(a)$), E182-190 (further practice with non-elementary integrals).

8 Applications of the Definite Integral

The main purpose of this chapter is to reinforce the concept of the definite integral and develop skill in applying it. The central idea of 8.1 and 8.4 (area and volume by cross sections) is already developed in Chap. 5. The only section in this chapter that is absolutely essential to later development is Sec. 8.8 (improper integrals), which is the basis of the integral test for convergence of a series (Sec. 10.4).

Options (4–9 lectures)
- 4 lectures, omit 8.5, 8.6, and 8.7; assign 8.2 as reading.
- 8 lectures, one on each section.
- 9 lectures, with two on 8.8.

8.1 (Computing area by parallel cross sections) Practice applying the idea developed in Chap. 5.

8.2 (Some pointers on drawing) develops the drawing skill necessary to set up area and volume problems. It is worth a part or all of a lecture, though it could be assigned to be read.

8.3 (Setting up a definite integral) describes how integrals are set up in the "real world."

8.4 (Computing volumes) Practice applying the idea developed in Chap. 5.

8.5 (The shell method) is optional.

8.6 (The centroid of a plane region) is related to Sec. 15.3 (centers of mass in the plane and space, with variable density).

8.7 (Work) is optional. The concept of work is needed in Sec. 16.3, but it can be defined then. The lecture might start with the "launch of a rocket" example on p. 501.

8.8 (Improper integrals) The lecture might start with the "launch to infinity" example on p. 511.

EXERCISES

8.1 (Computing area by parallel cross sections) E30 (uses first fundamental theorem), E32 (reviews Newton's method and estimating a definite integral), E37, 38 (tri-ex).

8.2 (Some pointers on drawing) E13 (If at least one student has access to a dowel and saw, this is especially instructive.)

8.3 (Setting up a definite integral) E10 (surface area of sphere), E30 (makes an important point).

8.4 (Computing volumes) E20 (important general result for "cones")

8.5 (The shell method) E21 (short conceptual), E23 (reinforces idea of non-elementary integral).

8.6 (The centroid of a plane region) E17 (a common misconception), E18 (combines several ideas), E19 (at first glance, counterintuitive, combines several ideas), E21 (surely students can find cardboard and scissors; in Sec. 15.3 it is shown in an exercise that the balancing lines are indeed concurrent, a fact needed to show that the centroid is independent of choice of coordinates).

8.7 (Work) E8-12 (short conceptual).

8.8 (Improper integrals) E33 (a test of intuition), E34 (a nice application of integration by parts), E41, 42 (non-numerical computations).

9 Plane Curves and Polar Coordinates

Sections 9.5 (area of surface of revolution) and 9.7 (reflection properties of conics) are not needed in later chapters. Sec. 9.6 (curvature) is used in Sec. 13.4 to describe the normal component of the acceleration vector. One could wait till then to define curvature.

Options (4–7 lectures)

- 4 lectures, omit 9.5, 9.6, and 9.7.
- 5 lectures, omit 9.5 and 9.7.
- 7 lectures, one on each section.

9.1 (Polar coordinates) Precalculus material.

9.2 (Area in polar coordinates) Though the Riemann sums approach is included at the end of the section, the informal approach focuses on the little triangle that is the basis of the method.

9.3 (Parametric equations) If there is time, the rotary engine might merit a lecture.

9.4 (Arc length and speed on a curve) The formal approach is again included, but the informal approach does reveal the essence. As Exercise 29 shows, seldom can arc length be computed in closed form. Speed is the key concept in this section.

9.5 (The area of a surface of revolution) Exercise 27 in Sec. 9.S shows that it is unusual to be able to find the surface area in closed form. That exercise could be an interesting part of the lecture.

9.6 (Curvature) Only the definition of curvature, $|d\phi/ds|$ is needed later.

9.7 (The reflection properties of the conic sections) Though not used later, it describes a variety of applications in many fields. It's a morale builder.

9.S Note the special Guide Quiz on Chaps. 3 to 9.

EXERCISES

9.1 (Polar coordinates) E33-35 (reviews finding a maximum), E40 (reviews Newton's method).

9.2 (Area in polar coordinates) E18 (reviews

first fundamental theorem), E24-26 (conceptual tri-ex's).

9.3 (Parametric equations) E33 (a geometric view of l'Hôpital's rule), E36 conceptual, background for Sec. 12.2 (Projections)).

9.4 (Arc length and speed on a curve) E24 (short, conceptual), E27 (reviews several ideas), E28 (conceptual), E29 (shows arc length can seldom be computed in closed form), E30 (a little geometry).

9.5 (The area of a surface of revolution) E17 (a moving bit of history), E18 (conceptual), E27 (a case where the area but not the arc length can be given in closed form), E29, 30 (theoretical basis of this section), E31 (short conceptual).

9.6 (Curvature) E16-18 (applications of curvature), E27 (a perplexing question for discussion).

9.7 (The reflection properties of the conic sections) E14 (polar coordinates), E20 (depends on E14, a test of intuition).

9.S Note the Guide Quiz on Chaps. 3-9, which emphasizes concepts. In Review Exercises: E20, 21 (short conceptual), E22 (a good review with a surprisingly simple answer, after substantial computations), E25 (generalized mean-value theorem), E27 (surface area not usually computable in closed form).

10 Series

This chapter is concerned primarily with series with constant terms.

Options (7-8 lectures)
- 7 lectures, one on each section.
- 8 lectures, with two on 10.7.

10.1 (An informal introduction to series) This survey of Chaps. 10 and 11 is important. It lets the student know the motivation behind the convergence tests, control of errors, and use of power series.

10.2 (Sequences) Note that this covers the important sequences $\{r^n\}$ and $\{k^n/n!\}$. The relation of k^n to the atom bomb is described at the end of the section. (It makes a vivid introduction to sequences.)

10.3 (Series) It should be emphasized that neither a calculator nor a computer can tell whether a series converges simply by adding up terms. (Compare $\Sigma\, n^{-1}$ and $\Sigma\, n^{-1.0001}$.)

10.4 (The integral test) This is a convenient place to emphasize the error and controlling it. The staircase pictures display the whole theory.

10.5 (Comparison tests) Again a few staircase

diagrams show the idea. It might be pointed out that the test can sometimes be used to control the error.

10.6 (Ratio tests) Emphasize that up through this section the tests applied to positive series (or negative series).

10.7 (Tests for series with both positive and negative terms) This section merits 2 lectures. Note that the proof of the absolute-convergence test is far more student-friendly than the traditional $a_n = (a_n + |a_n|) - |a_n|$ proof, which was in the 4th Edition, and is still available as E38. The proof now used shows that the "positive part" and the "negative part" separately converge. The diagrams help make this clear.

10.S (Summary) has a map that shows the dependencies between the various tests. Some mathematicians say, "Just prove the absolute ratio test, save time." However, as the map shows, the ratio test does not stand on its own feet. Nor does it suffice for determining convergence at the limit of convergence of a power series, in Chap. 11.

EXERCISES

10.1 (An informal introduction to series) E12–14 obtain special series ($1/(1 - x)$, $\ln(1 + x)$, and $\tan^{-1} x$).

10.2 (Sequences) E22 (sequences and Newton's method), E23, 24 (sequences and Riemann sums), E27–30 (using the precise definition of the limit of a sequence)

10.3 (Series) E23–25 (repeating decimals), E27 (geometric series in economics), E28 (geometric series in medicine), E29 (geometric series and the creation of money), E30 (geometric series and the multiplier effect), E32 ($\sum n2^{-n}$ and gambling), E33 (ancient treatment of $\sum n2^{-n}$)

10.4 (The integral test) E15–22 (working with the error), E30, 31 (infinite products), E32 (infinity of primes)

10.5 (Comparison tests) E40 (decimal representation of e)

10.6 (Ratio tests) E29 (proof of root test)

10.7 (Tests for series with both positive and negative terms) E37 (a seeming paradox), E38 (the standard proof of the absolute-convergence test)

10.S (Summary) E42–47 (Stirling's approximation for $n!$ and Wallis's formula for $\pi/2$), E48 (fascinating exploration of certain sequences given recursively), E49 (good review exercise)

11 Power Series and Complex Numbers

This chapter has already been introduced in Sec. 10.1. Before Sec. 11.2, one could review Sec. 4.10, which gives a heuristic argument to show that the error in using the differential is controlled by the second derivative. (This would help prepare the student for the form of the error in Sec. 11.2 and for the proof in Sec. 11.3.)

Options (4–8 lectures)
- 4 lectures, omit 11.3, 11.6, and 11.7.
- 7 lectures, omit 11.3, with two on 11.6.
- 7 lectures, one on each section.
- 8 lectures, with two on 11.6.

Since Chapter 11 is not a prerequisite to vector analysis, partial derivatives, or multiple integrals, there is a wide choice of which sections to cover in this chapter. Sections 11.1–3 form a unit on Taylor series. Secs. 11.4-5 concern power series in general. Finally, Secs. 11.6-7 present the complex numbers (which many students need in sophomore engineering, physics, and math courses, but usually do not understand) and Euler's theorem on $e^{i\theta}$. These two sections can serve as the climax of calculus of a single variable.

If one lecture on 11.6 would not be enough for your class, we would recommend omitting 11.3 to free the necessary time rather than short-change complex numbers.

EXERCISES

11.1 (Taylor series) E31-32 (short conceptual).

11.2 (The error in Taylor series) E17-18 (not computational, short), E19 (good review), E21-25 (together show that a Taylor series need not represent the function), E26 (a theoretical application of the error formula).

11.3 (Why the error in Taylor series is controlled by a derivative), E1-4 (checking the proof), E7-10 (show why errors in numerical integration are controlled by higher derivatives).

11.4 (Power series and radius of convergence) E13-16, 33-34, 36 (short conceptual).

11.5 (Manipulating power series) E3 (getting $\tan^{-1} x$ series as in E14 in Sec. 10.1, p. 574), E21 (short conceptual), E22 (efficient calculation of logs), E24 (good review and writing exercise), E26 (another view of l'Hôpital's rule).

11.6 (Complex numbers) E3-10 (plotting complex numbers, making them geometric and down to earth), E17 (some writing), E24, 26, 27 (short conceptual), E31-33 (reviews parametric equations), E34 (reviews polar equations), E38 (a new view of partial fractions: in praise of complex numbers), E39 (factoring real polynomials: in praise of complex numbers).

11.7 (The relation between the exponential and the trigonometric functions) E15-20 (short conceptual), E27-28 (log of complex number and definition of complex exponential), E34 (a short tri-ex).

11.S (Summary) Review Exercises: E1 (nice review of series), E34 (writing, a good review), E45 (writing), E53 (short conceptual), E58 (theory), E59 (a better estimate of $f'(x)$), E60 (a paradox to be resolved), E61 (theoretical), E63, 64 (theoretical), E65-71 (obtain $\sum n^{-2} = \pi^2/6$).

12 Vectors

This chapter, as its title indicates, contains no calculus. Sec. 12.5 (determinants of orders 2 and 3) is precalculus material, which may be omitted or assigned as reading. There are two sections of geometric applications, Secs. 12.4 and 12.7. The first uses the dot product to obtain equations of lines and planes. The second, using also the cross product, illustrates several geometric applications. (If time presses, this section may be omitted.)

Options (5-10 lectures)
- 5 lectures, omit 12.5 and 12.7.
- 8 lectures, with two on 12.4.
- 10 lectures, with 12.2, 12.3, and 12.4 each getting two.

12.1 (The algebra of vectors) Though there are many definitions, they are simple and can be covered in one lecture.

12.2 (Projections) The notion of projection is fundamental in theory and in applications. For that reason we emphasize the concept by assembling in one section the types of projection scattered through several sections in the previous edition. A lecture could present a couple of types

Advice to Instructors

thoroughly, leaving the others to be read. Two lectures, if time does not press.

12.3 (The dot product of two vectors) This section would benefit from taking two lectures. Note that the formula for the dot product in terms of components is obtained vectorially. We feel that it is more instructive and natural than the "law of cosines" proof, which is now in Exercise 51.

12.4 (Lines and planes) For the most part, this may be presented as an application of the dot product.

12.5 (Determinants) This reviews the precalculus material needed for cross products.

12.6 (The cross product of two vectors) In contrast to the definition of the dot product, we use components to define cross product. Exercises 39-41 outline the geometric approach, which we feel requires diagrams too involved for a lecture.

12.7 (More on lines and planes) This applies both the dot and cross product to seven geometric problems. Good background for computer graphics. Even if it is not covered in a lecture, its exercises could be assigned.

EXERCISES

12.1 (The algebra of vectors) E28 (short conceptual), E33, 34, 35 (a good check on understanding), E39, 40 (obtain an inequality).

12.2 (Projections) E1-6 (these combine drawing with concepts), E7 (important reinforcement), E11, 12, (drawing and conceptual), E21, 22, 23, 24 (short, conceptual), E25 (background for next section), E26 (challenging).

12.3 (The dot product of two vectors) E44 (sounds harder than it is), E45 (emphasizes meaning of $\mathbf{A} \cdot \mathbf{u}$), E46 (short conceptual), E47-49 (business application of dot product), E51 (the standard "law of cosines" way to express dot product in components), E54, 55 (for computer science students).

12.4 (Lines and planes), E35, 39 (good review), E42 (conceptual), E43 (reviews finding minimum by differentiation), E45 (conceptual).

12.5 (Determinants) E22-24 (conceptual), E25-26 (for computer science students).

12.6 (The cross product of two vectors) E29, 30, 34 (short conceptual), E35 (amusing), E36 (crystals), E37 (the error appears in a physics text), E38 (conceptual; we still don't have an easy way to show $z = 1$), E39-41 (a geometric approach to the cross product), E42-44 (conceptual, motivated by computer graphics).

12.7 (More on lines and planes) E26 (related to computer graphics), E27 (computer graphics), E28 (solar energy), E29 (fabricating steel hoppers), E30 (conceptual).

12.S (Summary) The Guide Quiz emphasizes concepts. Review Exercises: E32 (related to "hidden line" in computer graphics), E35, 36 (short conceptual), E39 (Gram-Schmidt), E42, 43 (challenging, good review), E45 (short, but challenging), E46 (short conceptual), E47 (challenging), E48 (good review), E49 (good review).

13 The Derivative of a Vector Function

Only Secs. 13.1 and 13.2 are needed in later chapters (namely in Secs. 14.7, 14.8, and 14.10 in Chapter 14, and in Chapters 16 and 17).

Options (2–6 lectures)
- 2 lectures, omit 13.3, 13.4, and 13.5.
- 5 lectures, one on each section.
- 6 lectures, with two on 13.5.

EXERCISES

13.1 (The derivative of a vector function) E3, 4 (review definition of a derivative), E28 (short conceptual, reviewing derivative), E29 (conceptual), E30-36 (practice in integration).

13.2 (Properties of the derivative of a vector function) E19 (short conceptual), E20 (practice in differentiation).

13.3 (The acceleration vector) E17-21 (satellites), E21 (good review), E24-26 (practice in integration), E27 (reviews concept of derivative), E28 (obtains v^2/r informally), E29 (reviews basics).

13.4 (The components of acceleration) E23-24 (short conceptual), E37-42 (generalizations to curves in space).

13.5 (Newton's law implies Kepler's three laws) The exercises obtain Kepler's laws from Newton's.

13.S (Summary) The Guide Quiz is mainly conceptual. Review Exercises: E15, 16 (short conceptual).

14 Partial Derivatives

Options (7–14 lectures)
- 7 lectures, omit 14.10 and 14.11; assign 14.2 and 14.4 as reading.
- 13 lectures, with 14.7 and 14.9 each getting two.
- 14 lectures, with 14.6, 14.7, and 14.9 each getting two.

Sections 14.1-14.3 acclimate the student to surfaces and level curves, in particular quadric surfaces. They contain no calculus. Sec. 14.4 generalizes limits and continuity to functions of two variables. If time presses, only the highlights of these need be covered, perhaps in 2 lectures, leaving the rest for reference. Sec. 14.5 introduces partial derivatives. At the urging of engineers we emphasize the notation $\partial f/\partial x$ rather than f_x. (Engineers reserve subscript notation for components of a vector.)

Sec. 14.6 on the chain rule has a companion, Sec. 14.11, on the same subject. The latter treats the delicate, but important, case when a middle variable is also a terminal (bottom) variable. Note that the mnemonic diagrams are slightly simpler than the traditional ones. Section 14.7 (Directional derivatives and the gradient) and Sec. 14.8 (Normals and the tangent plane) are essential for Chaps. 16 and 17, while Secs. 14.9 (Critical points and extrema) and Sec. 14.10 (Lagrange multipliers) are not.

EXERCISES

14.1 (Graphs) E1-34 are drawing exercises. The students need practice drawing surfaces, without worrying about calculus concepts. E35 combines drawing with calculus.

14.2 (Quadric surfaces) E1-8 and 13-28 (drawing), E33-36 (short conceptual), E38 (nice with wire and string).

14.3 (Functions and their level curves) E1-19 (drawing), E20, 21 (short conceptual).

14.4 (Limits and continuity) E21-22 (ϵ, δ exercises), E24 (an important point).

14.5 (Partial derivatives) E1-26 (computation), E29-31 (check on definition of partial derivative), E33 (short conceptual), E36 (conceptual review), E40 (short conceptual).

14.6 (The chain rule) E1-6 (verifying chain rule), E11-21 (verifying that certain types of functions satisfy certain partial differential equations), E24 (a proof), E25 (extending proof to more variables), E26 (paradox).

- 14.7 (Directional derivatives and the gradient) E13-18 (very short conceptual), E31, 32 (good preparation for Chap. 16), E34 (short conceptual), E36 (convincing), E41 (a short check on understanding).
- 14.8 (Normals and the tangent plane) E23 (implicit in proof of Theorem 2), E24 (short conceptual), E26–27 (orthogonal curves), E28 (review), E32 (a general writing review).
- 14.9 (Critical points and extrema) E1-10 (computations), E11-16 (short conceptual), E42 (when discriminant is 0), E43 (regression line), E45 (surprising).
- 14.10 (Lagrange multipliers) E19 (the sailboat example completed), E20, 21 (review of vector algebra), E29-32 (obtaining general inequalities), E33 (economics), E35 (computer science).
- 14.11 (The chain rule revisited) E5-8 (thermodynamics), E11 (mentioned in the preface for students).
- 14.S (Summary) E27 (no one-to-one correspondence between plane and space with continuous partial derivatives), E31-36 (include practice differentiating), E56, 57 (review first fundamental theorem of calculus), E74 (challenging review), E75-77 (gradient in polar coordinates).

15 Definite Integrals Over Plane and Solid Regions

If time presses combine 15.1 and 15.2, also 15.6 and 15.7. (Note that we use the coordinate-free notations $\int_R f(P)\, dA$ and $\int_R f(P)\, dV$ rather than $\iint_R f(x,y)\, dx\, dy$ and $\iiint_R f(x,y,z)\, dx\, dy\, dz$.)

Options (3–7 lectures)
- 3 lectures, one on 15.1 and 15.2 combined; omit 15.3, 15.6, and 15.7.
- 5 lectures, one on 15.1 and 15.2 combined, and one on 15.6 and 15.7 combined.
- 7 lectures, one on each section.

EXERCISES

15.1 (The definite integral of a function over a region in the plane) E1-10 (computations reinforce the definition of $\int_R f(P)\, dA$, E12-13 (short conceptual), E15 (reviews level curves), E16, 17 (Monte Carlo method), E18 (conceptual), E19 (contains an unsolved problem).

15.2 (Computing $\int_R f(P)\, dA$ using rectangular coordinates) E43, 44 (integrating an odd function), E45 (reviews concept of derivative).

15.3 (Moments and centers of mass) E15-17 (short conceptual), E18 (applies E17), E19 (shows that the center of gravity is well defined, a fact usually tacitly assumed).

15.4 (Computing $\int_R f(P)\, dA$ using polar coordinates) E39 (note the preamble before the exercise: could also make a nice lecture), E40 (the normal curve), E42 (find the error), E43 (epidemics), E44-49 (transportation: location of warehouses), E50 (a more detailed explanation of r in the integrand).

15.5 (The definite integral of a function over a region in space) E36 (short conceptual), E37 (parallel axis theorem), E38 (a shortcut), E39 (short conceptual: geology), E40 (intuitive argument for the repeated integral).

15.6 (Computing $\int_R f(P)\, dV$ using cylindrical coordinates) E39 (short conceptual), E40 (good review of distance from a point to a plane).

15.7 (Computing $\int_R f(P)\, dV$ using spherical coordinates) E39 (interesting contrast), E40 (short conceptual), E41, 42 (shortcuts), E43 (contrast with example in text), E44 (contrast with E43), E47 (find error), E48 (improper integral over a solid), E49 (useful).

15.S (Summary) Review Exercises: E34 (gravity paradox), E36 (reviews "elementary function"), E37 (the Schwarz inequality), E40 (obtains both Taylor series and its remainder in integral form), E42 (an inequality involving centroids), E53 (reviews gradient).

16 Green's Theorem

A thorough, unhurried treatment of this chapter, which concerns mainly vector fields in the plane, provides a sound basis for Chap. 17, which covers Stokes' theorem and the divergence theorem.

Options (6–10 lectures)
- 6 lectures, one on each section.
- 9 lectures, with 16.2, 16.4, and 16.5 each getting two.
- 10 lectures, with 16.2, 16.4, 16.5, and 16.6 each getting two.

16.1 (Vector and scalar fields) This section defines a variety of scalar and vector fields needed in Chaps. 16 and 17, and introduces the unit vector $\hat{r}$. All of these concepts will be given physical meaning in subsequent sections.

16.2 (Line integrals) Note the attention given to special integrals such as $\oint_C y\, dx$.

This is done for two reasons: to give substance to a line integral and to develop the key to one of the two proofs of Green's theorem. Note that

$$\int_C f(P)\, ds$$

is defined differently from the other three line integrals.

16.3 (Four applications of line integrals) The three physical applications: work, circulation, and flux will be referred to several times. The relation of a line integral to subtended angle is background for the steradian example in Chap. 17, which provides geometric insight into Gauss's theorem.

16.4 (Green's theorem) The first proof involves the cancellation principle (useful in physics) and provides a new use of the differential. The second proof, which is more common, reviews repeated integrals. Note that the proofs go in "opposite directions."

16.5 (Applications of Green's theorem) Note that Example 1 is related to subtended angle (See E17). Emphasize that $\hat{r}/r$ in the plane is divergence-free and $\hat{r}/r^2$ in space is divergence-free.

16.6 (Conservative vector fields) An appropriate climax to a calculus course.

EXERCISES

There is a healthy balance of computational and conceptual exercises. Engineers and physicists urge us to emphasize the latter, for the integral theorems of vector analysis are used as often to develop a physical principle as to evaluate some quantity, and the mathematical concepts are used to define or describe physical concepts.

16.1 (Vector and scalar fields) E21, 22 (a significant contrast), E23, 24 (a similar contrast), E31, 32 (similar contrast), E34 (uses separable differential equations), E34, 36 (significant contrast).

16.2 (Line integrals) E15, 16 (emphasize: these line integrals are needed later in the first proof of Green's theorem), E32-35, 39, 40 (examples of $\int_C f\, ds$), E36 (short conceptual).

16.3 (Four applications of line integrals) E13, 14 (short conceptual), E15-18 (these writing exercises reinforce important concepts), E24 (short conceptual), E25, 26 (short conceptual, background for Gauss's law), E27 (conceptual, good review of derivative), E28, 29 (short conceptual), E30 (practice differentiating).

16.4 (Green's theorem) E15, 16 (proofs are a way to review concepts), E17-19 (short conceptual), E29 (long conceptual but good background for physics), E30 (a simple check on basics).

16.5 (Applications of Green's theorem) E12-16 (check on understanding), E18-20 (the zero-integral theorem reinforced), E23 (this is what the engineering students will see later), E26 (longer writing project), E28 (short conceptual).

16.6 (Conservative vector fields) E11-14 (short checks on understanding), E17 (reinforces several ideas), E20 (thermodynamics), E22 (reinforces fundamentals), E23 (short conceptual).

16.S (Summary) Review Exercises: E24-26 (paddle wheels and circular motion), E29 (a challenging use of line integrals), E30-33 (a basic property of harmonic functions), E35-36 (apply the preceding E30-33), E37 (another way to determine the divergence-free central fields in the plane), E38 (short conceptual).

17 The Divergence Theorem and Stokes' Theorem

The student who understands Chap. 16 is well prepared for Chap. 17. If the instructor has emphasized proofs in Chap. 16, then the proofs may be de-emphasized here, and attention focused on the statement and use of the divergence theorem and Stokes' theorem.

Options (4–6 lectures)
- 4 lectures, one on each section.
- 5 lectures, with two on 17.2.
- 6 lectures, with 17.1 and 17.2 each getting two.

17.1 (Surface integrals) We develop surface

integrals in a way that emphasizes projections, because the latter concept is used often in applications. Note that Example 2 is critical to the proof of the divergence theorem in Sec. 17.2. If you don't plan to prove the divergence theorem, then advise students to skip Example 2.

17.2 (Divergence theorem) The theorem and its corollaries have analogs in Chap. 16. Engineering and physics majors will be happy to see Examples 3–7.

17.3 (Stokes' theorem) Chap. 16 mentioned Stokes' theorem in the plane. This would be a good starting point.

17.4 (Applications of Stokes' theorem) We obtain a coordinate-free definition of curl.

EXERCISES

Keep in mind that most students will use the theorems in this chapter conceptually and to simplify calculations rather than carry out a straightforward evaluation of an integral.

17.1 (Surface integrals) E9-12 (help set basics for proof in next section), E29, 30 (important surface integrals), E31-33 (integrating over a cone), E36 (short conceptual), E37 (important review).

17.2 (The divergence theorem) E25-26 (short, important conceptual applications), E27-28 (reinforce proof), E30-32 (short conceptual), E33 (reinforces steradians and improper integrals), E34-35 (basic for steradians), E36 (coordinate-free definition of divergence), E37 (conceptual, reinforces several ideas).

17.3 (Stokes' theorem) E13-16 (short conceptual questions), E25 (another way to show central fields are conservative), E28 (the cancellation principle reviewed), E29 (playing with a Möbius band), E30 (the relation between orientability and the cancellation principle), E31 (another way to show central fields are conservative), E32 (a "paradox").

17.4 (Applications of Stokes' theorem) E1, 2 (very short conceptual), E3 (good review, writing), E4 (reviews several concepts), E5-12 (important applications).

17.S (Summary) Guide Quiz is primarily conceptual. Review Exercises: E12 (a check on fundamentals), E21, 25 (major applications), E39 (a review), E47 (hydrostatics), E50-51 (a key property of harmonic functions), E52-57 (integrating a vector field rather than a scalar field).

Appendices

A Real Numbers

This section treats division by 0, inequalities, defines the various intervals, and reviews absolute value. E32 ($3.1416 \neq \pi$), E33 (rationals and repeating decimals).

B Graphs and Lines

B.1 (Coordinate systems and graphs) This covers the xy coordinate system, distance formula, and graph of an equation.

B.2 (Lines and their slopes) This covers slope of a line, tests for parallel and perpendicular lines, the slope-intercept form, point-slope form, (and in Exercise 30, intercept form). E36 (obtains the test for perpendicularity).

C Topics in Algebra

This appendix reviews rationalizing, the quadratic formula, the binomial theorem, finite geometric series, and the factor theorem. These topics may be covered when and if needed.

D Exponents

This appendix begins with the definition of 2^x when x is a positive integer, states the basic law of exponents, $b^{x+y} = b^x b^y$, and develops the definition of b^x for rational x from this law.

E Mathematical Induction

The text mentions induction in a few exercises. E1-6 refer to precalculus topics, but E7-10 refer to calculus.

F The Converse of a Statement

When students fail to distinguish an implication from its converse, refer them to this section. E1-12 are precalculus examples and E13-24 are calculus examples.

G Conic Sections

This is a full treatment of conic sections, translation and rotation of axes, and conic sections in polar coordinates. (The last is used in Sec. 13.5, which obtains Kepler's laws from Newton's.

H Logarithms and Exponentials Defined through Calculus

This is a self-contained development of $\ln x$ and b^x through calculus. It is independent of the development in Secs. 6.3-6.5.

The following three appendices treat topics usually treated in advanced calculus courses. However, many students who do not take such a course will need to refer to them.

I The Taylor Series for $f(x, y)$

Appendix I develops Taylor's series for $f(x, y)$ and applies it to obtain the test for an extremum stated in Sec. 14.9.

J Theory of Limits

Appendix J uses the (ϵ, δ) definition of a limit to prove that a polynomial is continuous.

K The Interchange of Limits

Appendix K obtains the equality of the mixed partials and the derivative of $\int_a^b f(x, y) \, dx$ with respect to y. In both cases the hypotheses are made strong enough to permit the proofs to be accessible to freshmen. The third section introduces a central theme in advanced calculus: "When can you interchange the order of taking limits?" (The two topics mentioned above are shown to be special cases.)

L The Jacobian

This appendix develops the Jacobian as a magnification and its use in changes of coordinates.

M Linear Differential Equations with Constant Coefficients

Second-order linear differential equations with constant coefficients are used often in engineering and physics courses. This appendix develops their solutions.

1 An Overview of Calculus

1.1 The Derivative

2 (a)

x	$\sqrt[3]{x^3 + x}$	$\sqrt[3]{x^3 + x} - x$
1	1.2599210	0.2599210
10	10.0332228	0.0332228
100	100.0033332	0.0033332
1000	1000.0003333	0.0003333

(b) The difference appears to approach 0.

4 (a)

x	$x^3 - 1$	$x^2 - 1$	$(x^3 - 1)/(x^2 - 1)$
2	7	3	2.3333333
1.5	2.375	1.25	1.9
1.1	0.331	0.21	1.5761905
1.01	0.030301	0.0201	1.5075124
1.001	0.0030030	0.002001	1.5007501
0.9	−0.271	−0.19	1.4263158
0.99	−0.029701	−0.0199	1.4925126

(b) The ratio appears to approach 1.5.

6 (a) $16(2.01)^2 - 16(2)^2 = 0.6416$ ft

(b) $\dfrac{0.6416}{0.01} = 64.16$ ft/sec

(c) $\dfrac{16(2.001^2 - 2^2)}{2.001 - 2} = 64.016$ ft/sec

(d) $\dfrac{16(2^2 - 1.999^2)}{2 - 1.999} = 63.984$ ft/sec

(e) 64 ft/sec

8 (a)

x	$\sin x$	$(\sin x)/x$
30°	0.5000000	0.0166667
10°	0.1736482	0.0173648
5°	0.0871557	0.0174311
1°	0.0174524	0.0174524
0.1°	0.0017453	0.0174533

(b) The ratio appears to approach some number near 0.01745.

10 (a) $\dfrac{1 - \cos 1}{1^2} \approx 0.4596977$, $\dfrac{1 - \cos 0.1}{0.1^2} \approx 0.4995835$, $\dfrac{1 - \cos 0.01}{0.01^2} \approx 0.4999958$

(b) The ratio appears to approach 1/2.

1.2 The Integral

2. (a)

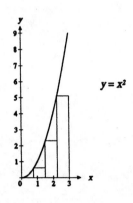

(b)

Rectangle	Height	Width	Area
First	0	3/4	0
Second	9/16	3/4	27/64
Third	9/4	3/4	27/16
Fourth	81/16	3/4	243/64

Total area $= 0 + \dfrac{27}{64} + \dfrac{27}{16} + \dfrac{243}{64} = \dfrac{189}{32}$

4. (a)

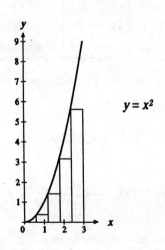

(b)

Rectangle	Height	Width	Area
First	0	3/5	0
Second	9/25	3/5	27/125
Third	36/25	3/5	108/125
Fourth	81/25	3/5	243/125
Fifth	144/25	3/5	432/125

Total area $= 0 + \dfrac{27}{125} + \dfrac{108}{125} + \dfrac{243}{125} + \dfrac{432}{125}$

$= \dfrac{162}{25} = 6.48$

6. (a)

Rectangle	Height	Width	Area
First	36/25	1/5	36/125
Second	49/25	1/5	49/125
Third	64/25	1/5	64/125
Fourth	81/25	1/5	81/125
Fifth	4	1/5	4/5

The total area of the five overestimating rectangles

is $\dfrac{36}{125} + \dfrac{49}{125} + \dfrac{64}{125} + \dfrac{81}{125} + \dfrac{4}{5} = \dfrac{330}{125}$

$= \dfrac{66}{25} = 2.64.$

(b)

Rectangle	Height	Width	Area
First	1	1/5	1/5
Second	36/25	1/5	36/125
Third	49/25	1/5	49/125
Fourth	64/25	1/5	64/125
Fifth	81/25	1/5	81/125

The total area of the five underestimating rectangles

is $\dfrac{1}{5} + \dfrac{36}{125} + \dfrac{49}{125} + \dfrac{64}{125} + \dfrac{81}{125} = \dfrac{255}{125}$

$= \dfrac{51}{25} = 2.04.$

8 (a)

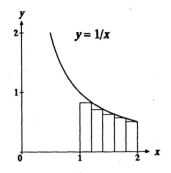

Rectangle	Height	Width	Area
First	5/6	1/5	1/6
Second	5/7	1/5	1/7
Third	5/8	1/5	1/8
Fourth	5/9	1/5	1/9
Fifth	1/2	1/5	1/10

The area of the five underestimating rectangles is $\dfrac{1}{6} + \dfrac{1}{7} + \dfrac{1}{8} + \dfrac{1}{9} + \dfrac{1}{10} =$

$\dfrac{1627}{2520} \approx 0.6456349.$

(b) The area of the ten underestimating rectangles is $\dfrac{1}{11} + \dfrac{1}{12} + \dfrac{1}{13} + \dfrac{1}{14} + \dfrac{1}{15} + \dfrac{1}{16} +$

$\dfrac{1}{17} + \dfrac{1}{18} + \dfrac{1}{19} + \dfrac{1}{20} \approx 0.6687714.$

(c) The area of the ten overestimating rectangles is $\dfrac{1}{10} + \dfrac{1}{11} + \dfrac{1}{12} + \dfrac{1}{13} + \dfrac{1}{14} + \dfrac{1}{15} +$

$\dfrac{1}{16} + \dfrac{1}{17} + \dfrac{1}{18} + \dfrac{1}{19} \approx 0.7187714.$

(d) The area under the curve must be between 0.66877 and 0.71878.

10 (a) The area of the four overestimating rectangles is $\dfrac{1}{1^3+1}\cdot\dfrac{1}{2} + \dfrac{1}{(3/2)^3+1}\cdot\dfrac{1}{2} + \dfrac{1}{2^3+1}\cdot\dfrac{1}{2}$

$+ \dfrac{1}{(5/2)^3+1}\cdot\dfrac{1}{2} \approx 0.4499165.$

(b) The area of the four "midpoint" rectangles is

$\dfrac{1}{(5/4)^3+1}\cdot\dfrac{1}{2} + \dfrac{1}{(7/4)^3+1}\cdot\dfrac{1}{2} +$

$\dfrac{1}{(9/4)^3+1}\cdot\dfrac{1}{2} + \dfrac{1}{(11/4)^3+1}\cdot\dfrac{1}{2}$

$\approx 0.3112284.$

1.3 Survey of the Text

2 (a) If the angle whose tangent equals x is

$x - \dfrac{x^3}{3} + \dfrac{x^5}{5} - \dfrac{x^7}{7} + \cdots$, then the angle whose tangent equals 1 must be

$1 - \dfrac{1^3}{3} + \dfrac{1^5}{5} - \dfrac{1^7}{7} + \cdots =$

$1 - \dfrac{1}{3} + \dfrac{1}{5} - \dfrac{1}{7} + \cdots$. But the angle whose tangent equals 1 is $\pi/4$, so we have found an unending numerical expression for $\pi/4$.

(b) Since $\dfrac{\pi}{4} = 1 - \dfrac{1}{3} + \dfrac{1}{5} - \dfrac{1}{7} + \cdots$, it follows that $\pi = 4 - \dfrac{4}{3} + \dfrac{4}{5} - \dfrac{4}{7} + \cdots$. We obtain the following successive approximations for π:

$4 - \dfrac{4}{3} \approx 2.6667,\; 4 - \dfrac{4}{3} + \dfrac{4}{5} \approx 3.4667,$

$$4 - \frac{4}{3} + \frac{4}{5} - \frac{4}{7} \approx 2.8952,$$

$$4 - \frac{4}{3} + \frac{4}{5} - \frac{4}{7} + \frac{4}{9} \approx 3.3397.$$ (The results do not appear to be homing in on π very quickly.)

(c) The angle whose tangent is $1/\sqrt{3}$ is $\pi/6$.

Hence $\dfrac{\pi}{6} =$

$$\frac{1}{\sqrt{3}} - \frac{(1/\sqrt{3})^3}{3} + \frac{(1/\sqrt{3})^5}{5} - \frac{(1/\sqrt{3})^7}{7} + \cdots$$

$$= \frac{1}{3^{1/2}} - \frac{1}{3^{5/2}} + \frac{1}{5 \cdot 3^{5/2}} - \frac{1}{7 \cdot 3^{7/2}} + \cdots.$$

(d) This is not a good approach.

(e) Using $x = 1/2$, we have

$$\frac{1}{2} - \frac{(1/2)^3}{3} + \frac{(1/2)^5}{5} - \frac{(1/2)^7}{7} + \cdots$$

≈ 0.4634673. Using the arctangent key on a calculator yields 0.4636476.

2 Functions, Limits, and Continuity

2.1 Functions

2

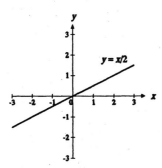

$f(x) = x/2$ is the equation of a line with slope 1/2 and y intercept 0.

4

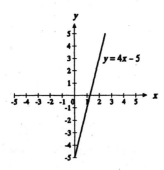

$f(x) = 4x - 5$ is the equation of a line with slope 4 and y intercept -5.

6

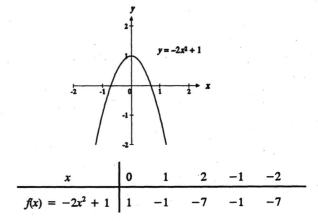

x	0	1	2	-1	-2
$f(x) = -2x^2 + 1$	1	-1	-7	-1	-7

$f(x) = -2x^2 + 1$ is the equation of a parabola with a maximum value at (0, 1).

8

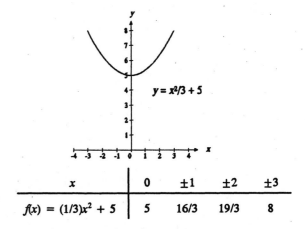

x	0	± 1	± 2	± 3
$f(x) = (1/3)x^2 + 5$	5	16/3	19/3	8

$f(x) = \dfrac{1}{3}x^2 + 5$ is the equation of a parabola with a minimum value at (0, 5).

10

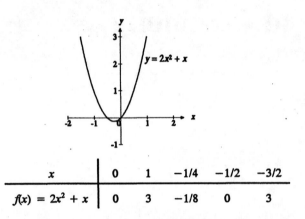

x	0	1	−1/4	−1/2	−3/2
$f(x) = 2x^2 + x$	0	3	−1/8	0	3

$f(x) = 2x^2 + x$ is the equation of a parabola with a minimum value at $(-1/4, -1/8)$.

12

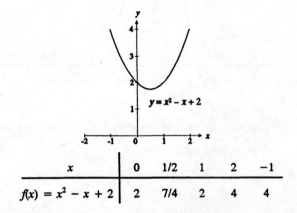

x	0	1/2	1	2	−1
$f(x) = x^2 - x + 2$	2	7/4	2	4	4

$f(x) = x^2 - x + 2$ is the equation of a parabola with a minimum value at $(1/2, 7/4)$.

14

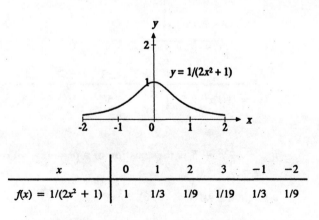

x	0	1	2	3	−1	−2
$f(x) = 1/(2x^2 + 1)$	1	1/3	1/9	1/19	1/3	1/9

16

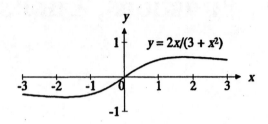

x	0	1	2	3	−1	−2
$f(x) = 2x/(3 + x^2)$	0	1/2	4/7	1/2	−1/2	−4/7

18 The domain consists of all nonnegative real numbers; the interval is $[0, \infty)$. The range consists of all nonnegative real numbers; the interval is $[0, \infty)$.

20 In order for $\sqrt{x - 1}$ to make sense, we need $x - 1 \geq 0$, or $x \geq 1$. Thus, the domain is the interval $[1, \infty)$. The range is the interval $[0, \infty)$.

22 We require $x^2 - 4 \geq 0$, so $x^2 \geq 4$ or $|x| \geq 2$. Thus the domain is the intervals $(-\infty, -2]$ and $[2, \infty)$. Since the output becomes arbitrarily large as $|x|$ gets large, the range is the interval $[0, \infty)$.

24 The function is defined for all $x \neq 0$ (since division by zero is taboo). Thus the domain is all non-zero real numbers. Solve $b = 2/x$ for x to find the range: $x = 2/b$. For meaningful x, b must be nonzero. Thus the range is all real numbers except zero.

26 In order for $f(x)$ to be defined, we need $x - 2 \neq 0$ or $x \neq 2$. Therefore, the domain is all real numbers except 2. Solve $b = 1/(x - 2)$ for x to find the range: $x - 2 = 1/b$, so $x = 1/b + 2$. For meaningful x, b must be nonzero. Thus the range is all real numbers except zero.

28 We need $2 + x^2 \neq 0$ or $x^2 \neq -2$ which is true for all real x. Thus, the domain is the interval $(-\infty, \infty)$. Solve $b = 1/(2 + x^2)$ for x to find the range. Immediately, it can be seen that b is positive since $2 + x^2 > 0$ for all x. We have $2 + x^2 = 1/b$, so $x^2 = 1/b - 2$, and thus $x = \pm\sqrt{1/b - 2}$. For real x, we must have $1/b - 2 \geq 0$ or $1/b \geq 2$. It follows that $b \leq 1/2$. In summary, the range is the interval $(0, 1/2]$.

30 (a) $f(-3) = \dfrac{1}{1 + (-3)} = -\dfrac{1}{2} = -0.5$

  (b) $f(3) = \dfrac{1}{1 + 3} = \dfrac{1}{4} = 0.25$

  (c) $f(9) = \dfrac{1}{1 + 9} = \dfrac{1}{10} = 0.1$

  (d) $f(99) = \dfrac{1}{1 + 99} = \dfrac{1}{100} = 0.01$

32 (a) $f(5 - 3) = f(2) = 1/2^2 = 1/4 = 0.25$

  (b) $f(4 - 6) = f(-2) = 1/(-2)^2 = 1/4 = 0.25$

34 Note that $\dfrac{f(2 + h) - f(2)}{h} = \dfrac{(2 + h)^3 - 2^3}{h} =$

$\dfrac{8 + 12h + 6h^2 + h^3 - 8}{h} = \dfrac{h^3 + 6h^2 + 12h}{h}$

$= h^2 + 6h + 12$.

  (a) For $h = 1$, $h^2 + 6h + 12 = 1 + 6 + 12 = 19$.

  (b) For $h = 0.05$, $h^2 + 6h + 12 = (0.05)^2 + 6(0.05) + 12 = 12.3025$.

  (c) For $h = 0.0001$, $h^2 + 6h + 12 = (0.0001)^2 + 6(0.0001) + 12 \approx 12.0006$.

  (d) For $h = -0.001$, $h^2 + 6h + 12 = (-0.001)^2 + 6(-0.001) + 12 \approx 11.9940$.

  (e) As h approaches 0, $h^2 + 6h + 12$

$= \dfrac{f(2 + h) - f(2)}{h}$ approaches 12.

36 $f(a + h) - f(a) = \dfrac{1}{(a + h)} - \dfrac{1}{a} = \dfrac{a - (a + h)}{a(a + h)}$

$= \dfrac{-h}{a(a + h)}$

38 $\dfrac{f(x + h) - f(x)}{h} = \dfrac{1}{h}\left(\dfrac{1}{2x + 2h + 1} - \dfrac{1}{2x + 1}\right)$

$= \dfrac{1}{h} \cdot \dfrac{(2x + 1) - (2x + 2h + 1)}{(2x + 2h + 1)(2x + 1)}$

$= \dfrac{1}{h} \cdot \dfrac{-2h}{(2x + 2h + 1)(2x + 1)}$

$= -\dfrac{2}{(2x + 2h + 1)(2x + 1)}$

40 (a) is the graph of a function; (b) and (c) are not.

42 The area of a circle with radius x is πx^2. Thus $f(x) = \pi x^2$. The domain is all nonnegative real numbers. (To exclude the degenerate circle of radius 0, choose the domain to be all positive real numbers.)

44 The volume of a cube of side x is x^3. The domain is all nonnegative real numbers. (To exclude the degenerate cube of side 0, choose the domain to be all positive real numbers.)

46 The length of the hypotenuse is given by the Pythagorean theorem. So $f(x) = \sqrt{3^2 + x^2} = \sqrt{9 + x^2}$. The domain is all $x \geq 0$.

48 $f(x) = \overline{AB} + \overline{BC} = \sqrt{4^2 + x^2} + \sqrt{3^2 + (4 - x)^2}$

$= \sqrt{x^2 + 16} + \sqrt{x^2 - 8x + 25}$. The domain is $0 \leq x \leq 4$.

50 $f(x) = \dfrac{\overline{AP}}{1.5} + \dfrac{\overline{PB}}{4}$

$= \dfrac{2}{3}\sqrt{x^2 + 2^2} + \dfrac{1}{4}\sqrt{(6-x)^2 + 5^2}$

$= \dfrac{2}{3}\sqrt{x^2 + 4} + \dfrac{1}{4}\sqrt{x^2 - 12x + 61}$ hours

52 (a) $f(0) = 0, f(a) = b$

(b) $f(a/2) = b/2$

(c) Triangles ABE and CDE are similar. This implies that $\dfrac{\overline{AB}}{x}$

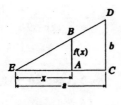

$= \dfrac{b}{a}$. So $\overline{AB} = f(x) = \dfrac{b}{a}x$. This makes sense because $f(0) = 0$, $f(a) = b$, and $f(a/2) = b/2$, and this checks out with (a) and (b).

54 (a) From Fig. 6, the maximum occurs at 3 P.M. and the minimum at 3:30 A.M.

(b) From Fig. 6, the maximum occurs at 4:30 P.M. and 11:00 P.M. and the minimum at 4 A.M.

56

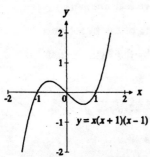

(a) If $f(x) = 0$, then $0 = x(x+1)(x-1)$ and we have $x = 0$ or $x = -1$ or $x = 1$.

(b) $x = 0, 1, -1$

(c) $f(x)$ crosses the y axis when $x = 0$. Therefore,

the y intercept is $f(0) = 0$.

58 (a) $f(1+1) = 4 \neq 2 = f(1) + f(1)$, so $f(a+b)$ does not always equal $f(a) + f(b)$.

(b) $f(a) + f(b) = 3a + 3b = 3(a+b) = f(a+b)$, so $f(a+b) = f(a) + f(b)$.

(c) $f(a) + f(b) = -4a - 4b = -4(a+b) = f(a+b)$, so $f(a+b) = f(a) + f(b)$.

(d) $f(9+16) = 5 \neq 3 + 4 = f(9) + f(16)$, so $f(a+b)$ does not always equal $f(a) + f(b)$.

(e) $f(a) + f(b) = (2a+1) + (2b+1) = 2(a+b) + 2 \neq 2(a+b) + 1 = f(a+b)$, so $f(a+b) \neq f(a) + f(b)$.

60 $f(x) = 1; f(x) = 1729^x; f(x) = \begin{cases} 7, & x < 0 \\ -1, & x = 0 \\ 1/7, & x > 0 \end{cases}$

62 $f(x) = 0; f(x) = x^7; f(x) = x^2|x|$

64 (a) 4 P.M.

(b) 4 P.M.

(c) The maximum risk is when accidents/driver is at a maximum. This occurs at 1 A.M.

(d) The minimum risk is when accidents/driver is at a minimum. This occurs at 5 P.M.

(e) At 1 A.M. the risk is approximately 0.0962. At 5 P.M. the risk is approximately 0.0175. The most risky hour is about 5.5 times as risky as the least risky hour.

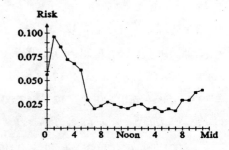

2.1 Functions

Time	Accidents/driver
M	0.0566
1	0.0962
2	0.0857
3	0.0725
4	0.0679
5	0.0615
6	0.0295
7	0.0204
8	0.0235
9	0.0276
10	0.0250
11	0.0222

Time	Accidents/driver
N	0.0208
1	0.0247
2	0.0260
3	0.0203
4	0.0218
5	0.0175
6	0.0205
7	0.0186
8	0.0295
9	0.0296
10	0.0379
11	0.0405

16 $u = (2x)^2 - 1 = 4x^2 - 1$, so $y = (4x^2 - 1)^2$.

18 $u = \sin(x + x^2)$, so $y = \sin^3(x + x^2)$.

20 Let $u = \sqrt{x} + 1$. Then say $y = u^{10}$. (Other answers are possible.)

22 Let $u = 1 + x^2$. Then say $y = \sqrt[3]{u}$. (Other answers are possible.)

24 Let $u = \sin x$. Then $y = u^2$. (Other answers are possible.)

26 Let $w = 1 + 2x$. Let $v = w^{50}$. Then $y = \sqrt[3]{v}$. (Other answers are possible.)

28 (a) $f(g(0.6)) \approx f(0.2) \approx 0.3$
    (b) $g(f(0.3)) \approx g(0.5) \approx 0.25$
    (c) $f(f(0.5)) \approx f(0.7) \approx 0.8$

30 All $x \neq 1$; all x except 0 and 1; all x except 0 and 1. $(f \circ f \circ f)(x) = f(f(f(x))) = f\left(f\left(\dfrac{1}{1-x}\right)\right) =$

$f\left(\dfrac{1}{1 - \dfrac{1}{1-x}}\right) = f\left(\dfrac{1-x}{-x}\right) = \dfrac{1}{1 - \dfrac{1-x}{-x}}$

$= \dfrac{x}{x + 1 - x} = x.$

2.2 Composite Functions

Note: The answers to Exercises 1 through 6 depend on the type of calculator used. In addition, there are often several different correct answers.

2 To evaluate $(\sin x)^2$ at $x = 3$, first set your calculator to radian mode, then push $\boxed{3}$ $\boxed{\text{SIN}}$ $\boxed{x^2}$.

4 To evaluate $(\sqrt{1 + x})^3$ at $x = 4$ push $\boxed{1}$ $\boxed{+}$ $\boxed{4}$ $\boxed{=}$ $\boxed{\sqrt{x}}$ $\boxed{y^x}$ $\boxed{3}$ $\boxed{=}$.

6 To evaluate $\left(\cos \sqrt[3]{x}\right)^2$ at $x = 2$, first set your calculator to radian mode, then push $\boxed{2}$ $\boxed{y^x}$ $\boxed{3}$ $\boxed{1/x}$ $\boxed{=}$ $\boxed{\text{COS}}$ $\boxed{x^2}$.

8 $y = (2x + 1)^2$

10 $y = \dfrac{1}{2x^2 - 3}$

12 $y = \left(\sqrt{x}\right)^2 = x$, for $x \geq 0$.

14 $y = \sin \sqrt{x}$

32 We seek f such that $f(g(x)) = g(f(x))$, that is $f(x^2) = g(ax^2 + bx + c)$. Thus
$ax^4 + bx^2 + c = (ax^2 + bx + c)^2$
$= a^2x^4 + 2abx^3 + (2ac + b^2)x^2 + 2bcx + c^2$.
Since this must be true for all x, we have $a = a^2$, $2ab = 0$, $2ac + b^2 = b$, $2bc = 0$, and $c = c^2$.
Since $a \neq 0$ this leads to $a = 1$, $b = 0$, and $c = 0$. Thus $f(x) = x^2$.

34 None.
$f(g(x)) = 2ax^2 + 2bx + (2c + 3)$.
$g(f(x)) = a(2x + 3)^2 + b(2x + 3) + c$

$= 4ax^2 + (12a + 2b)x + (9a + 3b + c)$. Hence if $f \circ g = g \circ f$, then $2a = 4a$, violating $a \neq 0$.

36 No. $(f \circ g)(-1) = f(g(-1)) = [g(-1)]^4 \geq 0$, so $(f \circ g)(-1) \neq -1$.

2.3 The Limit of a Function

2 $\lim\limits_{x \to 1} (4x - 2) = 4 \cdot 1 - 2 = 2$

4 $\lim\limits_{x \to 3} \dfrac{x^2 - 9}{x - 3} = \lim\limits_{x \to 3} (x + 3) = 3 + 3 = 6$

6 $\lim\limits_{x \to 1} \dfrac{x^6 - 1}{x^3 - 1} = \lim\limits_{x \to 1} (x^3 + 1) = 1^3 + 1 = 2$

8 $\lim\limits_{x \to 5} \dfrac{3x + 5}{4x} = \dfrac{3 \cdot 5 + 5}{4 \cdot 5} = 1$

10 $\lim\limits_{x \to 3} \pi^2 = \pi^2$

12 $\lim\limits_{x \to 1^+} \sqrt{4x - 4} = \sqrt{4 \cdot 1 - 4} = 0$

14 For $x < 1$, $|x - 1| = -(x - 1)$, so
$\lim\limits_{x \to 1^-} \dfrac{x - 1}{|x - 1|} = \lim\limits_{x \to 1^-} \dfrac{x - 1}{-(x - 1)} = \lim\limits_{x \to 1^-} (-1) = -1.$

16 $\lim\limits_{h \to 0} \dfrac{(1 + h)^2 - 1}{h} = \lim\limits_{h \to 0} (2 + h) = 2$

18 $\lim\limits_{x \to 3} \dfrac{1/x - 1/2}{x - 2} = \dfrac{1/3 - 1/2}{3 - 2} = -\dfrac{1}{6}$

20 $\lim\limits_{x \to 1} \dfrac{3^x - 3}{2^x} = \dfrac{3^1 - 3}{2^1} = 0$

22 (a) 2
    (b) 2
    (c) 1
    (d) 2

24

x	1	0.1	0.01
$3^x - 2^x$	1	0.0443	0.004091
$(3^x - 2^x)/x$	1	0.4435	0.4091

x	-0.01	-0.1	-1
$3^x - 2^x$	-0.0040	-0.0371	-0.1667
$(3^x - 2^x)/x$	0.4018	0.3707	0.1667

$\dfrac{3^x - 2^x}{x}$ appears to approach a number near 0.40 as $x \to 0$. (The limit exists.)

26 (a),(b)

x	1	0.1	0.01	0.001
$(1 + x)^{1/x}$	2	2.5937	2.7048	2.7169

x	-0.1	-0.01	-0.001
$(1 + x)^{1/x}$	2.8680	2.7320	2.7196

(c) $(1 + x)^{1/x}$ appears to approach a number near 2.72 as $x \to 0$. (The limit exists.)

28 (a)

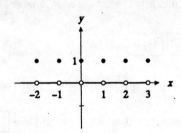

(b) Yes. $\lim\limits_{x \to 3} f(x) = 0$.

(c) $\lim\limits_{x \to 3.5} f(x) = 0$.

(d) For all numbers a, $\lim\limits_{x \to a} f(x) = 0$.

30 (a)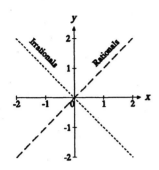

(b) No. (See (e).)

(c) No. (See (e).)

(d) Yes. If x is near 0 then $f(x) = \pm x$ is near 0. Hence $\lim_{x \to 0} f(x) = 0$.

(e) Only for $a = 0$. For any a, there are both rationals and irrationals near a, so $f(x)$ can be near either a or $-a$. For $a \neq 0$, $a \neq -a$, so $\lim_{x \to a} f(x)$ does not exist.

32 (a)

x	1.0	0.5	0.4	0.3	0.2	0.1	0.01
x^x	1	0.7071	0.6931	0.6968	0.7248	0.7943	0.9550

(b) It appears to be about 0.69. (In fact, it is $e^{-1/e} \approx 0.6922$.)

(c) Yes. It appears to be 1.

34 (a) For $a = 0$, $\lim_{x \to 0} \dfrac{f(x) - f(0)}{x - 0} = \lim_{x \to 0} \dfrac{2^x - 2^0}{x}$

$= \lim_{x \to 0} \dfrac{2^x - 1}{x}$, which from Example 4 is approximately 0.69.

For $a = 1$, $\lim_{x \to 1} \dfrac{f(x) - f(1)}{x - 1} = \lim_{x \to 1} \dfrac{2^x - 2^1}{x - 1}$

$= \lim_{x \to 1} \dfrac{2^x - 2}{x - 1}$. This limit is approximately $1.38 = 2(0.69)$.

For $a = 2$, $\lim_{x \to 2} \dfrac{f(x) - f(2)}{x - 2} = \lim_{x \to 2} \dfrac{2^x - 2^2}{x - 2}$

$= \lim_{x \to 2} \dfrac{2^x - 4}{x - 2}$. This limit is approximately $2.76 = 4(0.69)$.

For $a = 3$, $\lim_{x \to 3} \dfrac{f(x) - f(3)}{x - 3} = \lim_{x \to 3} \dfrac{2^x - 2^3}{x - 3}$

$= \lim_{x \to 3} \dfrac{2^x - 8}{x - 3}$. This limit is approximately $5.52 = 8(0.69)$.

(b) It seems that, when $a = 4$, the limit is approximately $11.04 = 2^4(0.69)$.

(c) For any number a, $\lim_{x \to a} \dfrac{2^x - 2^a}{x - a} \approx 2^a(0.69)$.

(d) Rewrite $\dfrac{2^x - 2^a}{x - a}$ as $\dfrac{2^a(2^{x-a} - 1)}{x - a}$. Then,

$\lim_{x \to a} \dfrac{2^x - 2^a}{x - a} = 2^a \lim_{x \to a} \dfrac{2^{x-a} - 1}{x - a}$. Let $u = x - a$. As $x \to a$, $u \to 0$, so by substitution we have $2^a \lim_{x \to a} \dfrac{2^{x-a} - 1}{x - a} = 2^a \lim_{u \to 0} \dfrac{2^u - 1}{u}$. But

$\lim_{u \to 0} \dfrac{2^u - 1}{u} \approx 0.69$ (by Example 4), so we

conclude that $\lim_{x \to a} \dfrac{2^x - 2^a}{x - a} \approx 2^a(0.69)$.

2.4 Computations of Limits

2 $\lim_{x \to 4} (2x^2 - 5) = 2 \cdot 4^2 - 5 = 27$

4 $\lim_{x \to 3} \dfrac{(7x^2 - 10)}{x - 1} = \dfrac{7 \cdot 3^2 - 10}{3 - 1} = \dfrac{53}{2}$

6 $\lim\limits_{x \to 5} [(x^2 - x)(2x - 7)] = (5^2 - 5)(2 \cdot 5 - 7)$

$= 20 \cdot 3 = 60$

8 $\lim\limits_{x \to \infty} (5x - 9) = \lim\limits_{x \to \infty} x\left(5 - \dfrac{9}{x}\right) = \infty$, since $\lim\limits_{x \to \infty} x$

$= \infty$ and $\lim\limits_{x \to \infty}\left(5 - \dfrac{9}{x}\right) = 5$.

10 $\lim\limits_{x \to \infty} (3x^2 - 7x + 2) = \lim\limits_{x \to \infty} x^2\left(3 - \dfrac{7}{x} + \dfrac{2}{x^2}\right) =$

∞, since $\lim\limits_{x \to \infty} x^2 = \infty$ and $\lim\limits_{x \to \infty}\left(3 - \dfrac{7}{x} + \dfrac{2}{x^2}\right)$

$= 3$.

12 $-\infty$

14 ∞

16 $-\infty$

18 0

20 2

22 $\lim\limits_{x \to -\infty} \dfrac{5x^3 + 2x}{x^2 + x + 7} = \lim\limits_{x \to -\infty} \dfrac{5x^3}{x^2} = \lim\limits_{x \to -\infty} 5x = -\infty$

24 $-\infty$

26 ∞

28 (a) ∞

(b) ∞

(c) ∞

30 $\lim\limits_{x \to -\infty} \dfrac{\sqrt{9x^2 + x + 3}}{6x} = \lim\limits_{x \to -\infty} \dfrac{\sqrt{9x^2 + x + 3}}{-\sqrt{36x^2}}$

$= -\lim\limits_{x \to -\infty} \sqrt{\dfrac{9x^2 + x + 3}{36x^2}} = -\sqrt{\dfrac{9}{36}} = -\dfrac{1}{2}$

32 $\lim\limits_{x \to -\infty} \dfrac{\sqrt{x^2 + 3x + 1}}{\sqrt{16x^2 + x + 2}} = \lim\limits_{x \to -\infty} \sqrt{\dfrac{x^2 + 3x + 1}{16x^2 + x + 2}}$

$= \sqrt{\dfrac{1}{16}} = \dfrac{1}{4}$

34 (a) The product of two very large positive numbers is also very large and positive.

(b) The quotient can be any number in $(0, \infty)$.

(c) The difference can be any number in $(-\infty, \infty)$.

36 Note that $\dfrac{2^x}{4^x} = \left(\dfrac{2}{4}\right)^x = \left(\dfrac{1}{2}\right)^x$.

Now, $\lim\limits_{x \to \infty} \dfrac{2^x}{4^x} = \lim\limits_{x \to \infty} \left(\dfrac{1}{2}\right)^x = 0$, $\lim\limits_{x \to -\infty} \dfrac{2^x}{4^x} =$

$\lim\limits_{x \to -\infty} \left(\dfrac{1}{2}\right)^x = \infty$, and $\lim\limits_{x \to 0} \dfrac{2^x}{4^x} = \lim\limits_{x \to 0} \left(\dfrac{1}{2}\right)^x = 1$.

38 $\lim\limits_{x \to \infty} \left(\dfrac{3x^2 + 2x}{x + 5} - 3x\right)$

$= \lim\limits_{x \to \infty} \left(\dfrac{3x^2 + 2x - 3x(x + 5)}{x + 5}\right)$

$= \lim\limits_{x \to \infty} \dfrac{3x^2 + 2x - 3x^2 - 15x}{x + 5} = \lim\limits_{x \to \infty} \dfrac{-13x}{x + 5}$

$= \lim\limits_{x \to \infty} \dfrac{-13x}{x} = \lim\limits_{x \to \infty} (-13) = -13$. Thus, neither

citizen is correct. If we let $f(x) = \dfrac{3x^2 + 2x}{x + 5}$ and

$g(x) = 3x$, then $\lim\limits_{x \to \infty} f(x) = \infty$ and $\lim\limits_{x \to \infty} g(x) = \infty$,

so we must be careful in determining

$\lim\limits_{x \to \infty} (f(x) - g(x))$.

40 Since $\lim\limits_{x \to 0^+} f(x) = 1$ and $\lim\limits_{x \to 0} g(x) = \infty$, more

information is needed about $f(x)$ and $g(x)$ to

2.4 Computations of Limits

determine $\lim_{x \to 0^+} f(x)^{g(x)}$. If $f(x) = 2^x$ and $g(x) = 1/x^2$, then $\lim_{x \to 0^+} f(x)^{g(x)} = \lim_{x \to 0^+} 2^{1/x} = \infty$. But if $f(x) = 1$ and $g(x) = 1/x$, $\lim_{x \to 0^+} f(x)^{g(x)} = \lim_{x \to 0^+} (1)^{1/x} = 1$. Finally, if $f(x) = 2^x$ and $g(x) = 1/x$, then $\lim_{x \to 0^+} f(x)^{g(x)} = \lim_{x \to 0^+} (2^x)^{1/x} = 2$. So, depending on our choice of $f(x)$ and $g(x)$, either citizen could be right or both could be incorrect.

42 (a) $\lim_{x \to \infty} (f(x) + g(x)) = \infty$, since both $f(x)$ and $g(x)$ become arbitrarily large as $x \to \infty$.

   (b) The limit cannot be determined. For example, if $f(x) = 2x^2$ and $g(x) = x^2$, we have
   $\lim_{x \to \infty} (f(x) - g(x)) = \lim_{x \to \infty} (2x^2 - x^2) = \lim_{x \to \infty} x^2 = \infty$. However, if $f(x) = x^2$ and $g(x) = 2x^2$, then $\lim_{x \to \infty} (f(x) - g(x)) = \lim_{x \to \infty} (x^2 - 2x^2) = \lim_{x \to \infty} -x^2 = -\infty$. (See also Exercise 38.)

   (c) $\lim_{x \to \infty} f(x)g(x) = \infty$, since both $f(x)$ and $g(x)$ become arbitrarily large as $x \to \infty$.

   (d) The limit cannot be determined. For example, if $f(x) = x^4$ and $g(x) = x^2$, we have $\lim_{x \to \infty} \frac{g(x)}{f(x)} = \lim_{x \to \infty} \frac{x^2}{x^4} = \lim_{x \to \infty} \frac{1}{x^2} = 0$. However, if $f(x) = x$ and $g(x) = 2x$, $\lim_{x \to \infty} \frac{g(x)}{f(x)} = \lim_{x \to \infty} \frac{2x}{x}$
   $= \lim_{x \to \infty} 2 = 2$.

44 $\lim_{x \to \infty} \frac{P(x)}{Q(x)} = \lim_{x \to \infty} \frac{ax^n}{bx^m} = \lim_{x \to \infty} \frac{a}{b} x^{n-m}$

   (a) If $m = n$, then $n - m = 0$, so $\lim_{x \to \infty} \frac{P(x)}{Q(x)}$
   $= \lim_{x \to \infty} \frac{a}{b} x^0 = \lim_{x \to \infty} \frac{a}{b} = \frac{a}{b}$.

   (b) If $m < n$, then $n - m > 0$, so $\lim_{x \to \infty} \frac{P(x)}{Q(x)} =$
   $\lim_{x \to \infty} \frac{a}{b} x^{n-m} = \infty$.

   (c) If $m > n$, then $n - m < 0$, so $\lim_{x \to \infty} \frac{P(x)}{Q(x)} =$
   $\lim_{x \to \infty} \frac{a}{b} x^{n-m} = 0$.

46 (a)

x	100	1000	10,000	100,000
$\sqrt{x^2 + 100} - x$	0.4988	0.0500	0.0050	0.0005

   (c) $\lim_{x \to \infty} \left(\sqrt{x^2 + 100} - x\right)$
   $= \lim_{x \to \infty} \frac{\left(\sqrt{x^2 + 100} - x\right)\left(\sqrt{x^2 + 100} + x\right)}{\sqrt{x^2 + 100} + x}$
   $= \lim_{x \to \infty} \frac{x^2 + 100 - x^2}{\sqrt{x^2 + 100} + x}$
   $= \lim_{x \to \infty} \frac{100}{\sqrt{x^2 + 100} + x} = 0$ since
   $\frac{1}{\sqrt{x^2 + 100} + x} \to 0$ as $x \to \infty$.

2.5 Some Tools for Graphing

2. (a) Since $f(x) = \sqrt{1 - x^2} = \sqrt{1 - (-x)^2} = f(-x)$, $f(x)$ is even.

   (b) Since $f(x) = 5x^4 - x^2 = 5(-x)^4 - (-x)^2 = f(-x)$, $f(x)$ is even.

   (c) Since $f(x) = 7/x^6 = 7/(-x)^6 = f(-x)$, $f(x)$ is even.

4. (a) Since $f(x) = 2x - x^3$, and $f(-x) = 2(-x) - (-x)^3 = -2x + x^3 = -(2x - x^3) = -f(x)$, $f(x)$ is odd.

   (b) Since $f(x) = x^3/(1 + x^2)$, and $f(-x) = (-x)^3/(1 + (-x)^2) = -x^3/(1 + x^2) = -f(x)$, $f(x)$ is odd.

   (c) Since $f(x) = \sqrt[5]{x}$, and $f(-x) = \sqrt[5]{-x} = \sqrt[5]{-1}\sqrt[5]{x} = -\sqrt[5]{x} = -f(x)$, $f(x)$ is odd.

6. (a) $f(x) = 2x - 1$, $f(-x) = -2x - 1 \neq -f(x) = -2x + 1$, in general; and $f(-x) \neq f(x) = 2x + 1$, in general, so $f(x)$ is neither even nor odd.

   (b) $f(x) = x^3 + x^2$, $f(-x) = -x^3 + (-x)^2 = -x^3 + x^2 \neq -f(x) = -x^3 - x^2$, in general; and $f(-x) \neq f(x) = x^3 + x^2$, in general, so $f(x)$ is neither even nor odd.

   (c) $f(x) = x^2 + 1/x$, $f(-x) = (-x)^2 + 1/(-x) = x^2 - 1/x \neq -f(x) = -x^2 - 1/x$, in general; and $f(-x) \neq f(x) = x^2 + 1/x$, in general, so $f(x)$ is neither even nor odd.

8. (a) $f(x) = (1 + x)/(1 - x)$ and $f(-x) = (1 - x)/(1 + x)$. Since $f(-x) \neq f(x)$ and $f(-x) \neq -f(x)$, in general, $f(x)$ is neither odd nor even.

   (b) $f(x) = (4x^2 - 3x^4)/x^3$ and $f(-x) = [4(-x)^2 - 3(-x)^4]/(-x)^3 = -(4x^2 - 3x^4)/x^3$

$= -f(x)$. So $f(x)$ is odd.

   (c) $f(x) = x^3\sqrt[3]{x}$ and $f(-x) = (-x)^3\sqrt[3]{-x} = x^3\sqrt[3]{x} = f(x)$. So $f(x)$ is even.

10. x intercept: $0 = 3x - 7$, so $x = \frac{7}{3}$.

    y intercept: $y = 3 \cdot 0 - 7 = -7$

12. x intercepts: $0 = 2x^2 + 5x + 3 = (2x + 3)(x + 1)$, so $x = -1$ or $-3/2$.

    y intercept: $y = 2 \cdot 0^2 + 5 \cdot 0 + 3 = 3$

14. x intercept: $0 = x^2 + x + 1$, so $x = \frac{-1 \pm \sqrt{1^2 - 4 \cdot 1 \cdot 1}}{2 \cdot 1} = \frac{-1 \pm \sqrt{-3}}{2}$. Since x is not real at this value, $y = x^2 + x + 1$ has no real x intercept.

    y intercept: $y = 0^2 + 0 + 1 = 1$

16. Since $\lim_{x \to \infty} y = 0$ and $\lim_{x \to -\infty} y = 0$, $y = 0$ is the only horizontal asymptote. Since $y = \frac{x - 2}{x^2 - 9}$ is undefined for $x^2 - 9 = 0$ or $x = \pm 3$, $x = \pm 3$ are the only vertical asymptotes.

18. Note that $\lim_{x \to \infty} y = 0$ and $\lim_{x \to -\infty} y = 0$. So $y = 0$ is the only horizontal asymptote. Since the denominator of y is never zero, there are no vertical asymptotes.

20. We see that $\lim_{x \to \infty} y = 0$ and $\lim_{x \to -\infty} y = 0$ as well. Thus, $y = 0$ is the only horizontal asymptote. Note that $x^2 + 2x + 1 = (x + 1)^2 = 0$ when $x = -1$, so the only vertical asymptote is $x = -1$.

2.5 Some Tools for Graphing

22

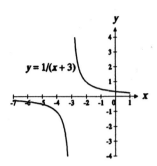

24

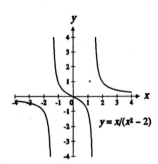

26

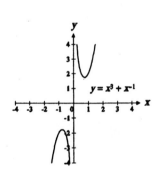

28

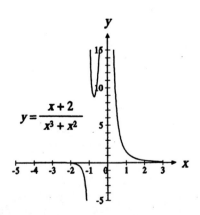

30 (a) Their sum is also even.

(b) Their product is also even.

(c) Their quotient is also even.

32 (a) If f is an even function, then $f(0)$ can take on any real value or be undefined.

(b) If f is an odd function, $f(0)$ must either equal 0 or be undefined.

34 Any even polynomial can be expressed as $a_0 + a_2 x^2 + a_4 x^4 + \cdots = p_e(x)$. Clearly $p_e(x) = p_e(-x)$, and the addition of any x^n of odd degree to $p_e(x)$ would destroy this property.

36 If $P(x) = ax^m +$ (lower terms) and $Q(x) = bx^n +$ (lower terms) then, for large $|x|$, $\dfrac{P(x)}{Q(x)} \approx \dfrac{a}{b} x^{m-n}$.

For $m < n$, $\lim\limits_{|x| \to \infty} \dfrac{P(x)}{Q(x)} = 0$. For $m = n$, the limit is a/b. So for $m \leq n$, there is a horizontal asymptote. For $m > n$, $\lim\limits_{|x| \to \infty} \left|\dfrac{P(x)}{Q(x)}\right| = \infty$ so there is no horizontal asymptote.

38

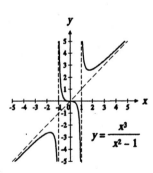

40

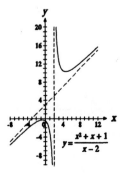

42 Let $f(x) = a^{p(x)}$, where a is a positive real number and $p(x)$ is any odd function. Then $f(x)$ satisfies the condition $f(-x) = 1/f(x)$, and there are infinitely many such functions.

43 Yes. The function $f(x) = 0$ satisfies this condition. To see that it's the only such function, note that, for all x, $f(x) = f(-(-x)) = 2f(-x) = 2 \cdot 2f(x) = 4f(x)$, so $f(x) = 0$.

44 For $f(x)$ to be odd it must satisfy $f(-x) = -f(x)$ for all x in its domain. Then $f(-x) = \dfrac{1}{3^{-x} - 1} + k =$

$-f(x) = \dfrac{-1}{3^x - 1} - k$. Then $\dfrac{1}{1/3^x - 1} + k =$

$\dfrac{-1 - k(3^x - 1)}{3^x - 1}$. Then $\dfrac{3^x}{1 - 3^x} + k =$

$\dfrac{1 - k + 3^x k}{1 - 3^x}$. Hence $\dfrac{3^x + k - 3^x k}{1 - 3^x} =$

$\dfrac{1 - k + 3^x k}{1 - 3^x}$, so it follows that $3^x + k - 3^x k = 1 - k + 3^x k$, so $3^x(1 - k) + k = 3^x k + (1 - k)$. For this to be true for all x we must have $1 - k = k$ which implies $k = 1/2$. Conversely we find that the function $f(x) = \dfrac{1}{3^x - 1} + \dfrac{1}{2}$ is odd; that is,

$f(-x) = \dfrac{1}{3^{-x} - 1} + \dfrac{1}{2} = \dfrac{3^x}{1 - 3^x} + \dfrac{1}{2} =$

$\dfrac{1}{1 - 3^x} - 1 + \dfrac{1}{2} = -\left(\dfrac{1}{3^x - 1} + \dfrac{1}{2}\right) = -f(x).$

For $f(x)$ to be even, it must satisfy $f(-x) = f(x)$ for all x. In particular, $f(-1) = f(1)$. But $f(-1) = k - \dfrac{3}{2}$ and $f(1) = k + \dfrac{1}{2}$ are not equal for any k.

2.6 A Review of Trigonometry

2 (a) π (b) $\pi/3$ (c) $\pi/4$ (d) 0

4 (a) 45° (b) 30° (c) 360° (d) 180°

6 $(0.5)(3) = 1.5$ inches

8 (a) $\dfrac{3}{\pi} = \dfrac{x}{180}$, so 3 radians $= \dfrac{540}{\pi} \approx 171.89°$.

(b) $\dfrac{1}{180} = \dfrac{x}{\pi}$, so 1 degree $= \dfrac{\pi}{180}$ radians ≈ 0.0175 radians.

10 (a) 2 radians $= \dfrac{360}{\pi}$ degrees $\approx 114.59°$, so draw an angle of 114.59°.

(b) Using the string, draw a circle with radius equal to the length of the string. Then mark a point A on the circle and measure 2 lengths of the string along the circle to a point B. If the center of the circle is O, then the angle AOB will measure 2 radians.

12 (a)

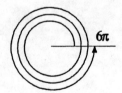

(b)

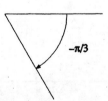

14 $\cos\dfrac{\pi}{6} = \sin\left(\dfrac{\pi}{2} - \dfrac{\pi}{6}\right) = \sin\dfrac{\pi}{3} = \dfrac{\sqrt{3}}{2}$

2.6 A Review of Trigonometry

16 (a) $\dfrac{\sqrt{3}}{2}$

　　(b) $-1/2$

　　(c) $-\dfrac{\sqrt{3}}{2}$

　　(d) $1/2$

　　(e) $\dfrac{\sqrt{3}}{2}$

　　(f) $1/2$

18 $\sin\dfrac{\pi}{4}\cos\dfrac{\pi}{4} + \cos\dfrac{\pi}{4}\sin\dfrac{\pi}{4} =$

$\dfrac{\sqrt{2}}{2}\cdot\dfrac{\sqrt{2}}{2} + \dfrac{\sqrt{2}}{2}\cdot\dfrac{\sqrt{2}}{2} = 1 = \sin\dfrac{\pi}{2} = \sin\left(\dfrac{\pi}{4}+\dfrac{\pi}{4}\right)$

20 $\sin\dfrac{\pi}{12} = \sin\left(\dfrac{\pi}{4} - \dfrac{\pi}{6}\right) =$

$\sin\dfrac{\pi}{4}\cos\dfrac{\pi}{6} - \cos\dfrac{\pi}{4}\sin\dfrac{\pi}{6} = \dfrac{1}{\sqrt{2}}\cdot\dfrac{\sqrt{3}}{2} - \dfrac{1}{\sqrt{2}}\cdot\dfrac{1}{2}$

$= \dfrac{\sqrt{6}}{4} - \dfrac{\sqrt{2}}{4} = \dfrac{\sqrt{6}-\sqrt{2}}{4}$

22 $\sin 2\theta = \sin(\theta+\theta) = \sin\theta\cos\theta + \cos\theta\sin\theta$
　　$= 2\sin\theta\cos\theta$

24 $\cos 2\theta = \cos^2\theta - \sin^2\theta = (1-\sin^2\theta) - \sin^2\theta$
　　$= 1 - 2\sin^2\theta$

26 $\cos 2\theta = 1 - 2\sin^2\theta$
　　$\cos 2\theta - 1 = -2\sin^2\theta$
　　$\sin^2\theta = \dfrac{1-\cos 2\theta}{2}$

28 (a) $+, +, +$
　　(b) $+, -, -$

30 (a) $\tan\dfrac{\pi}{6} = \dfrac{\sin(\pi/6)}{\cos(\pi/6)} = \dfrac{1}{\sqrt{3}}$

　　(b) $\tan\dfrac{\pi}{3} = \dfrac{\sin(\pi/3)}{\cos(\pi/3)} = \sqrt{3}$

32 (a) $\tan^{-1}(-1) = -\pi/4$, so the angle of inclination is $\pi - \pi/4 = 3\pi/4 = 135°$.

　　(b) $\tan^{-1} 1 = \pi/4 = 45°$

　　(c) $\tan^{-1} 2 \approx 1.1071 \approx 63.43°$

　　(d) $\tan^{-1} 3 \approx 1.2490 \approx 71.57°$

　　(e) $\tan^{-1}(-3) \approx -1.2490$, so the angle of inclination is approximately $\pi - 1.2490 \approx 1.8925 \approx 108.43°$.

34 (a) $\tan^{-1}\dfrac{1}{10} \approx 0.0997 \approx 5.71°$

　　(b) $\tan^{-1} 10 \approx 1.4711 \approx 84.29°$

36 (a) $b = c\sin\theta$

　　(b) $a = c\cos\theta$

　　(c) $b = a\tan\theta$

38 $\cos\dfrac{\pi}{3} = \dfrac{1}{2}, \sin\dfrac{\pi}{3} = \dfrac{\sqrt{3}}{2}, \tan\dfrac{\pi}{3} = \sqrt{3}$

40 (a) $\cos\dfrac{\pi}{3} = \dfrac{1}{2} = \dfrac{x}{3}$, so $x = \dfrac{3}{2}$.

　　(b) $\cos\dfrac{\pi}{4} = \dfrac{1}{\sqrt{2}} = \dfrac{x}{5}$, so $x = \dfrac{5}{\sqrt{2}}$.

　　(c) $\tan\dfrac{\pi}{3} = \sqrt{3} = \dfrac{x}{4}$, so $x = 4\sqrt{3}$.

42 (a) $\csc\theta = \dfrac{1}{\sin\theta}$, so $\csc\dfrac{\pi}{6} = \dfrac{1}{1/2} = 2$,

$\csc\dfrac{\pi}{4} = \dfrac{1}{1/\sqrt{2}} = \sqrt{2} \approx 1.4$, $\csc\dfrac{\pi}{3} = \dfrac{1}{\sqrt{3}/2} = \dfrac{2}{\sqrt{3}} \approx 1.15$, and $\csc\dfrac{\pi}{2} = \dfrac{1}{1} = 1$.

(b)

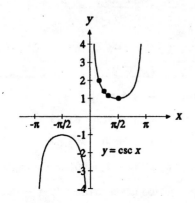

44 $\cos(A + (-B)) = \cos A \cos(-B) - \sin A \sin(-B)$
    $= \cos A \cos B + \sin A \sin B$

46 Let $u = \pi/2 - \theta$. Then $\theta = \pi/2 - u$ and
    $\sin(\pi/2 - u) = \cos u$.

48 $\sin(A + (-B)) = \sin A \cos(-B) + \cos A \sin(-B)$
    $= \sin A \cos B - \cos A \sin B$

50 (a) $\cos 2\theta = 1 - 2 \sin^2 \theta$, so $2 \sin^2 \theta = 1 - \cos 2\theta$ and thus $\sin^2 \theta = \frac{1}{2}(1 - \cos 2\theta)$.

It follows that $\sin \theta = \pm \sqrt{\frac{1}{2}(1 - \cos 2\theta)}$.

(b) $\sin \frac{\pi}{4} = \sqrt{\frac{1}{2}\left(1 - \cos \frac{\pi}{2}\right)} = \frac{1}{\sqrt{2}}$

(c) $\sin\left(-\frac{\pi}{4}\right) = -\sqrt{\frac{1}{2}\left(1 - \cos \frac{\pi}{2}\right)} = -\frac{1}{\sqrt{2}}$

52 (a) $1 = \cos^2 \theta + \sin^2 \theta$
        $(\cos \theta)^{-2} = 1 + \sin^2 \theta (\cos \theta)^{-2}$
        $\sec^2 \theta = 1 + \tan^2 \theta$

    (b) $1 = \sin^2 \theta + \cos^2 \theta$
        $(\sin \theta)^{-2} = 1 + \cos^2 \theta (\sin \theta)^{-2}$
        $\csc^2 \theta = 1 + \cot^2 \theta$

54 $\sin \theta = \cos(\pi/2 - \theta)$
    $\tan \theta = \cos(\pi/2 - \theta)/(\cos \theta)$

$\cot \theta = (\cos \theta)/\cos(\pi/2 - \theta)$
$\sec \theta = 1/\cos \theta$
$\csc \theta = 1/\cos(\pi/2 - \theta)$

56 The even trigonometric functions are $\cos \theta$ and $\sec \theta$, while the odd trigonometric functions are $\sin \theta$, $\csc \theta$, $\tan \theta$, and $\cot \theta$.

2.7 The Limit of $(\sin \theta)/\theta$ as θ Approaches 0

2 (a) $\frac{1}{2} \theta r^2 = \frac{1}{2} \cdot \frac{\pi}{4} \cdot 6^2 = \frac{9\pi}{2}$ in^2

   (b) $\frac{1}{2} \cdot 3 \cdot 6^2 = 54$ in^2

   (c) $45° = \pi/4$ radians. By (a), Area $= \frac{9\pi}{2}$ in^2.

4 2

6 2/3

8 1

10 $\lim_{\theta \to 0} \theta \cot \theta = \lim_{\theta \to 0} \left(\frac{\tan \theta}{\theta}\right)^{-1} = 1$

12 Note that $\tan \theta \approx \theta$ for small θ (in radians), so $\tan 0.04 \approx 0.04$. The calculator shows that $\tan 0.04 \approx 0.040021$.

14 Let x be the radian measure of $9°$. Then $\frac{9°}{360°} = \frac{x}{2\pi}$, so $x = \frac{9}{360} \cdot 2\pi = \frac{\pi}{20}$. Thus, $\tan 9° = \tan \frac{\pi}{20} \approx \frac{\pi}{20} \approx 0.15708$. But, thanks to the calculator, we all know that $\tan 9°$ is *really* about 0.15838.

16 Multiply $\frac{1 - \cos \theta}{\theta^2}$ by $\frac{1 + \cos \theta}{1 + \cos \theta}$ and simplify:

$$\frac{1-\cos\theta}{\theta^2} \cdot \frac{1+\cos\theta}{1+\cos\theta} = \frac{1-\cos^2\theta}{\theta^2(1+\cos\theta)}$$

$$= \frac{\sin^2\theta}{\theta^2(1+\cos\theta)}.$$

So, $\lim_{\theta\to 0} \frac{1-\cos\theta}{\theta^2} = \lim_{\theta\to 0} \frac{\sin^2\theta}{\theta^2(1+\cos\theta)}$

$$= \lim_{\theta\to 0} \frac{\sin^2\theta}{\theta^2} \cdot \frac{1}{1+\cos\theta}$$

$$= \lim_{\theta\to 0} \left(\frac{\sin\theta}{\theta}\right)^2 \cdot \lim_{\theta\to 0} \frac{1}{1+\cos\theta} = 1^2 \cdot \frac{1}{1+1}$$

$$= \frac{1}{2}.$$

18 $\lim_{\theta\to 0} \frac{\theta - \sin\theta}{\theta^3} = \frac{1}{6}$, so calculator computations will give values near 0.1667 until θ is too small for calculator accuracy.

20 (b) Let the radius be 1. Then Area of ABC =
$\frac{1}{2}\overline{AB}\cdot\overline{AC} = \frac{1}{2}(1-\cos\theta)\sin\theta$ and Area of
shaded region = (Area of sector) − (Area of
OAC) = $\frac{\theta}{2} - \frac{1}{2}\overline{OA}\cdot\overline{AC}$ =
$\frac{1}{2}(\theta - \cos\theta\sin\theta)$. Hence their quotient is
$\frac{(1-\cos\theta)\sin\theta}{\theta - \cos\theta\sin\theta}$. Letting $\theta = 0.01$ implies
that the value of the quotient is about 0.74999625.

22 (a) All nonzero x

(b) $g(-x) = \frac{1-\cos(-x)}{-x} = \frac{1-\cos x}{-x}$

$= -\frac{1-\cos x}{x} = -g(x)$

(c) $0 \leq g(x) \leq \frac{2}{x}$ for $x > 0$, so $\lim_{x\to\infty} g(x) = 0$ by the squeeze principle.

(d) $g(x) = 0$ if and only if $\cos x = 1$ and $x \neq 0$, which occurs if and only if $x = 2n\pi$ for a nonzero integer n.

(e)

x	0.1	$\pi/2$	$3\pi/2$	2π	3π
$1-\cos x$	0.00	1.00	1.00	0.00	2.00
$g(x)$	0.05	0.64	0.21	0.00	0.21

(f),(g)

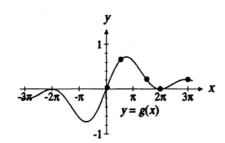

(h) 0

24 (a) $f\left(\frac{1}{\left(2n+\frac{3}{2}\right)\pi}\right) = \sin\left(2n+\frac{3}{2}\right)\pi = \sin\frac{3\pi}{2}$

$= -1$

(b)

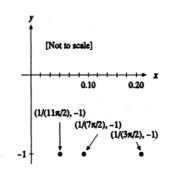

26 $x \to \infty$ so $1/x \to 0$ and $\sin 1/x \to 0$.

28 (a) $x \sin x = 0$ if and only if either $x = 0$ or $\sin x = 0$. But this requires that $x = n\pi$ for some integer n). Hence $x = n\pi$ for some integer n.

   (b) $f(x) = x$ implies that $x(\sin x - 1) = 0$, which occurs only when either $x = 0$ or $\sin x = 1$. Hence $x = 0$ or $x = \pi/2 + 2n\pi$ for some integer n.

   (c) $f(x) = -x$ implies that $x(\sin x + 1) = 0$, which occurs only when either $x = 0$ or $\sin x = -1$. Hence $x = 0$ or $x = 3\pi/2 + 2n\pi$ for some integer n.

   (d),(e)

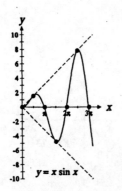

   (f) Yes. It equals 0.
   (g) No

2.8 Continuous Functions

2 $f(1/2) = 1/2$, so $1/2$ is in the domain of $f(x)$, so condition 1 holds. $\lim_{x \to 1/2} f(x) = 1$, so condition 2 holds. $\lim_{x \to 1/2} f(x) = 1 \ne 1/2 = f(1/2)$, so condition 3 fails. Thus $f(x)$ is not continuous at $a = 1/2$.

4 $f(1) = 1/2$, so 1 is in the domain of $f(x)$; hence condition 1 holds. $\lim_{x \to 1^-} f(x) = \infty$, so condition 2 fails. Therefore $f(x)$ is not continuous (from the left) at $a = 1$.

6 $f(0) = 0$, so 0 is in the domain of $f(x)$; hence condition 1 holds. $\lim_{x \to 0^+} x \sin 1/x = \lim_{1/x \to \infty} \dfrac{\sin 1/x}{1/x}$ $= 0$, so condition 2 holds. $f(0) = 0 = \lim_{x \to 0^+} x \sin(1/x)$, so condition 3 holds. Thus $f(x)$ is continuous (from the right) at $a = 0$.

8 $f(1/2) = 1/2$, so $1/2$ is in the domain of $f(x)$ and condition 1 holds. $\lim_{x \to 1/2} f(x) = \infty$, so condition 2 fails. Hence $f(x)$ is not continuous at $a = 1/2$.

10 $f(0) = 3/2$, so 0 is in the domain of $f(x)$ and condition 1 holds. Note that $\lim_{x \to 0^-} f(x) = 3/2 \ne 1/2$ $= \lim_{x \to 0^+} f(x)$, so $\lim_{x \to 0} f(x)$ does not exist, condition 2 fails, and $f(x)$ is not continuous at $a = 0$. However, considering continuity from the left only, $\lim_{x \to 0^-} f(x)$ $= f(0)$, conditions 1, 2, and 3 hold, and $f(x)$ is continuous *from the left* at $a = 0$.

12 (a)

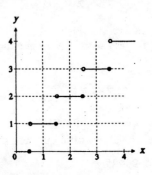

   (b) Yes. 3
   (c) Yes. 4
   (d) No
   (e) No

2.8 Continuous Functions

(f) Everywhere except $n + 1/2$ for integers n.

(g) $n + 1/2$ for integers n.

14 Yes. Since $\lim\limits_{x \to 1} \dfrac{x^3 - 1}{x - 1} = \lim\limits_{x \to 1} (x^2 + x + 1) = 3$,

defining $f(1) = 3$ will work.

16 Since $\lim\limits_{x \to 1} \dfrac{x - 1}{|x - 1|}$ does not exist, no choice of $f(1)$

will satisfy $\lim\limits_{x \to 1} \dfrac{x - 1}{|x - 1|} = f(1)$.

18 (a) The maximum value is 1, at $x = 0$.

(b) The maximum value is 1, at $x = 0, 2\pi$.

20 Yes in both cases, since the function is continuous and the interval is closed.

22 (a) Yes, at $x = 0$.

(b) Yes, at $x = 0$.

(c) No.

(d) Yes, at $x = 0$.

24 (a) No.

(b) Yes, at $x = 0$.

26 $f(1) = -2 < 0 < 15 = f(2)$, so $f(x) = 0$ for some x in $[1, 2]$.

28 $f(-1) = 3 < 5 < 8 = f(4)$, so the intermediate-value theorem applies. $f(c) = 5$ leads to

$c^2 - 2c - 5 = 0$, so $c = \dfrac{2 \pm \sqrt{2^2 - 4 \cdot 1 \cdot (-5)}}{2} =$

$1 \pm \sqrt{6}$. Only $1 + \sqrt{6}$ is in the interval.

30 $c = \dfrac{\pi}{6}, \dfrac{11\pi}{6}, \dfrac{13\pi}{6}, \dfrac{23\pi}{6}, \dfrac{25\pi}{6}$.

32 $f(0) = 1 > 0 > -1 = f(1)$ so, by the intermediate-value theorem, $f(x) = 0$ for some x in $[0, 1]$.

34 Yes. Let $f(x) = 2^x - x^3$. Then $f(1) = 1 > 0 > -4 = f(2)$, so $f(c) = 0$ for some c in $[1, 2]$. (In fact, there are two real solutions: $c \approx 1.37347$ and $c \approx 9.93954$.)

36 No. No, the conditions of the intermediate-value theorem are not met: f is not continuous at $x = 0$, since $f(0)$ is not defined.

38 (a) $1/x \to 0$ as $x \to \infty$, so $\lim\limits_{x \to \infty} 2^{1/x} = 1$.

(b) $1/x \to 0$ as $x \to -\infty$, so $\lim\limits_{x \to -\infty} 2^{1/x} = 1$.

(c) $1/x \to \infty$ as $x \to 0^+$ so $\lim\limits_{x \to 0^+} 2^{1/x} = \infty$; hence $\lim\limits_{x \to 0^+} f(x)$ does not exist.

(d) $1/x \to -\infty$ as $x \to 0^-$, so $\lim\limits_{x \to 0^-} 2^{1/x} = 0$.

(e)

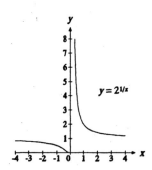

(f) $\lim\limits_{x \to 0^+} f(x)$ does not exist, so $\lim\limits_{x \to 0} f(x)$ does not exist, and we therefore cannot define $f(0)$ so that f is continuous at 0.

40 (a) Let $P(x) = a_n x^n + a_{n-1} x^{n-1} + \cdots + a_0$ be a polynomial of odd degree n. If $a_n > 0$, Exercise 39 tells us that there is a number r such that $P(r) = 0$. If $a_n < 0$, then $-P(x) = -a_n x^n - a_{n-1} x^{n-1} - \cdots - a_0$ is a polynomial of odd degree n, with positive lead coefficient, so Exercise 39 tells us that there is a number r such that $-P(r) = 0$, so $P(r) = 0$. Thus, $P(x)$ has a root r, so by the factor theorem $x - r$ is

a factor of $P(x)$. The polynomial $x - r$ has degree one, so $P(x)$ has a factor of degree one.

(b) For all x, $x^2 + 1 \geq 1$, $x^4 + 1 \geq 1$, and $x^{100} + 1 \geq 1$, so none of these polynomials has a root r. By the factor theorem, there is no r such that $x - r$ is a factor of one of these polynomials. A factor of first degree is of the form $ax + b = a(x + b/a) = a(x - (-b/a)) = a(x - r)$, and $x - r$ is a factor if and only if $a(x - r)$ is a factor; but none of these polynomials has a root so none has a factor of first degree.

(c) $(x^2 + \sqrt{2}x + 1)(x^2 - \sqrt{2}x + 1)$
$= x^2 x^2 + x^2(-\sqrt{2}x) + x^2 \cdot 1 + \sqrt{2}x \cdot x^2 +$
$\quad \sqrt{2}x(-\sqrt{2}x) + \sqrt{2}x \cdot 1 + x^2 - \sqrt{2}x + 1$
$= x^4 - \sqrt{2}x^3 + x^2 + \sqrt{2}x^3 - 2x^2 + \sqrt{2}x$
$\quad + x^2 - \sqrt{2}x + 1$
$= x^4 + 1$

42 Letting $f(x) = A(x) - B(x)$, we see that $f(a) = A(a) - B(a) = 0 - A = -A$; furthermore $f(b) = A(b) - B(b) = A - 0 = A$. By assumption, $f(x)$ is continuous on $[a, b]$. Since $f(a) = -A \leq 0 \leq A = f(b)$, the intermediate-value theorem implies that $f(c) = 0$ for some number c in $[a, b]$. For this number we have $A(c) = B(c)$, so the line L_c is the one we want.

44 Yes. Introduce an xy coordinate system so that K_1 and K_2 lie in the first quadrant, as shown. Consider a line L that makes an angle θ with the x axis. By Exercise 41 there are lines L_1 and L_2 parallel to L, which cut in half the areas of K_1 and K_2, respectively. Let $f(\theta) = d_1 - d_2$ where d_1 and d_2

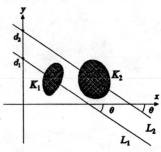

are the y intercepts of L_1 and L_2, respectively. For θ greater than but sufficiently close to $-\pi/2$, we see from the first figure that $d_1 < d_2$ and $f(\theta) < 0$.

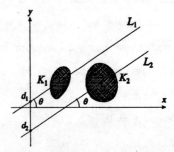

For θ less than but sufficiently close to $\pi/2$ we have $d_1 > d_2$, as shown in the second figure, so $f(\theta) > 0$. By the intermediate-value theorem there is a number γ, $-\pi/2 < \gamma < \pi/2$, such that $f(\gamma) = 0$. This means that L_1 and L_2 have the same y intercept for $\theta = \gamma$. Since their slopes are equal, they must coincide.

46

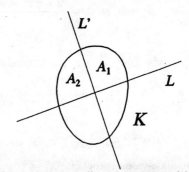

Introduce an xy coordinate system and let L be the line of angle of inclination θ which cuts K into two pieces of equal area. By Exercise 41 we know that L exists. Similarly, let L' be the line of inclination

2.8 Continuous Functions

$\theta + \pi/2$ that cuts K in half. Suppose the area of K is A and let $f(\theta) = A_1$, as shown in the figure. Then $f(\theta + \pi/2) = A_2$ and $A_1 + A_2 = f(\theta) + f(\theta + \pi/2) = A/2$. If $f(\theta) = f(\theta + \pi/2) = A/4$ we are done; otherwise one of $f(\theta)$ and $f(\theta + \pi/2)$ is less than $A/4$ and the other is greater than $A/4$. It follows that there is some intermediate value γ such that $f(\gamma) = A/4$. Then the line L of inclination $\theta = \gamma$ and the corresponding L' cut K into four parts of equal area.

48 (a) $f(2) = f(1 + 1) = f(1) + f(1) = c + c = 2c$

(b) $f(0) = f(0 + 0) = f(0) + f(0) = 2f(0)$; $f(0) = 2f(0)$, so $f(0) = 0$.

(c) $0 = f(0) = f(1 + (-1)) = f(1) + f(-1) = c + f(-1)$, so $f(-1) = -c$.

(d) $f(n) = f(n - 1 + 1) = f(n - 1) + f(1) = f(n - 2 + 1) + f(1) = f(n - 2) + f(1) + f(1) = \cdots = f(1) + \cdots + f(1) = nf(1) = nc = cn$.

(e) If n is a negative integer, then $-n$ is a positive integer. Thus $0 = f(0) = f(n + (-n)) = f(n) + f(-n) = f(n) + c(-n) = f(n) - cn$, so $f(n) = cn$.

(f) $c = f(1) = f\left(\frac{1}{2}\right) + f\left(\frac{1}{2}\right) = 2f\left(\frac{1}{2}\right)$, so $f\left(\frac{1}{2}\right) = c/2$.

(g) Let n be a positive integer. Then

$c = f(1) = f\left(\frac{n}{n}\right) = f\left(\frac{n-1}{n} + \frac{1}{n}\right) =$

$f\left(\frac{n-1}{n}\right) + f\left(\frac{1}{n}\right) = f\left(\frac{n-2}{n} + \frac{1}{n}\right) + f\left(\frac{1}{n}\right)$

$= f\left(\frac{n-2}{n}\right) + f\left(\frac{1}{n}\right) + f\left(\frac{1}{n}\right) = \cdots =$

$f\left(\frac{1}{n}\right) + f\left(\frac{1}{n}\right) + \cdots + f\left(\frac{1}{n}\right) = nf\left(\frac{1}{n}\right)$. Thus $nf(1/n) = c$, so $f(1/n) = c/n$. Now, let n be a negative integer. Then $-n$ is a positive integer, and $f\left(-\frac{1}{n}\right) = f\left(\frac{1}{-n}\right) = \frac{c}{-n} = -\frac{c}{n}$.

Thus $0 = f(0) = f\left(\frac{1}{n} + \left(-\frac{1}{n}\right)\right) =$

$f\left(\frac{1}{n}\right) + f\left(-\frac{1}{n}\right) = f\left(\frac{1}{n}\right) - \frac{c}{n}$, so $f\left(\frac{1}{n}\right) = \frac{c}{n}$.

Thus, for any nonzero integer n, $f\left(\frac{1}{n}\right) = \frac{c}{n}$.

(h) Suppose m is a positive integer. $f\left(\frac{m}{n}\right) =$

$f\left(\frac{m-1}{n} + \frac{1}{n}\right) = f\left(\frac{m-1}{n}\right) + f\left(\frac{1}{n}\right) =$

$f\left(\frac{m-2}{n} + \frac{1}{n}\right) + f\left(\frac{1}{n}\right) =$

$f\left(\frac{m-2}{n}\right) + f\left(\frac{1}{n}\right) + f\left(\frac{1}{n}\right) = \cdots =$

$f\left(\frac{1}{n}\right) + \cdots + f\left(\frac{1}{n}\right) = mf\left(\frac{1}{n}\right) = m \cdot \frac{c}{n} =$

$c\left(\frac{m}{n}\right)$. If m is a negative integer, then $-m$ is a positive integer, so $f\left(\frac{-m}{n}\right) = c\left(\frac{-m}{n}\right) =$

$-c \cdot \frac{m}{n}$. Now $0 = f(0) = f\left(\frac{m}{n} + \frac{-m}{n}\right) =$

$f\left(\frac{m}{n}\right) + f\left(\frac{-m}{n}\right) = f\left(\frac{m}{n}\right) - c\left(\frac{m}{n}\right)$, so $f\left(\frac{m}{n}\right)$

$= c\left(\frac{m}{n}\right)$. If $m = 0$, then $m/n = 0$ and $f\left(\frac{m}{n}\right)$

$= f(0) = 0 = c \cdot \dfrac{0}{n} = c \cdot \dfrac{m}{n}$. Thus, if m is an integer, and n is a positive integer, we have $f\left(\dfrac{m}{n}\right) = c\left(\dfrac{m}{n}\right)$.

(i) There are rational numbers arbitrarily close to x. (Consider, for example, the decimal expansion of x.) As $r \to x$ through rational values, $f(r) = cr \to cx$. By continuity, $f(x) = cx$.

50 (a) No. Consider $f(x) = x + 1$.

(b) Yes. Consider the continuous function $g(x) = f(x) - x$. Since $g(0) = 1$ and $g(1) = -1$, the intermediate-value theorem implies that $g(x) = 0$ for some x in $[0, 1]$. For this x, $f(x) = x$.

2.9 Precise Definitions of "$\lim_{x \to \infty} f(x) = \infty$" and "$\lim_{x \to \infty} f(x) = L$"

2 (c) $D = \dfrac{1000}{4} = 250$

4 (a) Any $D \geq \dfrac{1200}{6} = 200$ works.

(b) Any $D \geq \dfrac{1800}{6} = 300$ works.

6 Let $D = E/4$. For $x > D$, $4x > 4D = E$.

8 Let $D = E + 600$.

10 Let $D = \dfrac{E + 1200}{3}$.

12 Let $D = \dfrac{E + 300}{2}$. For $x > D$, $\cos x \leq 1$, so $2x - 300 \cos x \geq 2x - 300 > 2D - 300 = E$.

14 Let $D = E^{1/3}$.

16 (c) For $x > 0$, $|f(x) - 0| = 2/x < 1/100$ if and only if $x > 200$. $D = 200$.

(d) Let $D = 2/\epsilon$.

18 Let $D = 1/\epsilon$. For $x > D$, $\left|\dfrac{x + \cos x}{x} - 1\right| = \left|\dfrac{\cos x}{x}\right| \leq \dfrac{1}{x} < \epsilon$.

20 Let $D = 3/\epsilon$. For $x > D$, $\left|\dfrac{2x + 3}{x} - 2\right| = \left|\dfrac{3}{x}\right| < \epsilon$.

22 Let $D = 5/3 + \dfrac{40}{9\epsilon}$. For $x > D$, $\left|\dfrac{2x + 10}{3x - 5} - \dfrac{2}{3}\right| = \dfrac{40}{3(3x - 5)} < \epsilon$.

24 Let $\epsilon = 1$. For integers n, $\left|\sin\left(2\pi n + \dfrac{3\pi}{2}\right) - \dfrac{1}{2}\right| = \left|-1 - \dfrac{1}{2}\right| = 3/2 \geq \epsilon$, so there is no D such that $|\sin x - 1/2| < \epsilon$ for $x > D$.

26 Let $\epsilon = 1$. For all $x \geq \dfrac{L + 1}{2}$, $|2x - L| \geq \epsilon$, so there is no D such that $|2x - L| < \epsilon$ for $x > D$.

28

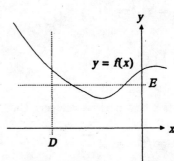

For each number E there is a number D such that $f(x) > E$ for all $x < D$.

30

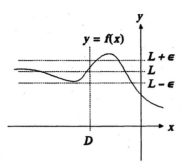

For each positive number ϵ there is a number D such that $|f(x) - L| < \epsilon$ for all $x < D$.

32 No, the argument is incorrect. Because D is not supposed to be a function of x, D should not contain $\cos x$.

2.10 Precise Definition of "$\lim_{x \to a} f(x) = L$"

2 In this case, $a = 3$ and $L = 11$. The precise definition requires that $0 < |x - 3| < \delta$ should lead to $|(4x - 1) - 11| < \epsilon$. This latter inequality simplifies to $|4x - 12| = 4|x - 3| < \epsilon$, so we find that $\delta = \epsilon/4$ will serve.

4 This time $a = 5$ and $L = 7$. Hence $0 < |x - 5| < \delta$ should imply that $|(2x - 3) - 7| < \epsilon$. This latter inequality simplifies to $|2x - 10| = 2|x - 5| < \epsilon$, so we see that $\delta = \epsilon/2$ suffices.

6 We have $\epsilon = 0.04$ and $\lim_{x \to 2} f(x) = 1$. Note that $a = 2$ and $L = 1$ in this case. We need $0 < |x - 2| < \delta$ to imply that $\left|\left(-\frac{x}{3} + \frac{5}{3}\right) - 1\right| < 0.04 = \epsilon$. But this latter inequality simplifies to $\left|-\frac{x}{3} + \frac{2}{3}\right| = \left|-\frac{1}{3}(x - 2)\right| = \frac{1}{3}|x - 2| < 0.04$, so $|x - 2| < 0.12$. Hence $\delta = 0.12$ (or any smaller positive value) will suffice.

8 Let $\delta = \sqrt{\frac{\epsilon}{4}}$.

10 Let $\delta = \frac{4\epsilon}{5}$. For $|x - 1| < \delta$, $\left|\frac{5x + 3}{4} - 2\right| = \frac{5}{4}|x - 1| < \epsilon$.

12 Let $\delta = 1/8$. For $|x - 1| < \delta$, $|x + 2| \leq |3| + |x - 1| < 3 + 1/8 < 4$, so $|x^2 + x - 2| = |x + 2| \cdot |x - 1| < 4\delta = 1/2$. (In fact, any positive $\delta \leq \frac{\sqrt{11} - 3}{2}$ will work.)

14 We have $\sqrt{x} = 0.98$ when $x = 0.9604$ and $\sqrt{x} = 1.02$ when $x = 1.0404$. Hence $0.98 < \sqrt{x} < 1.02$ when $0.9604 < x < 1.0404$. We can rewrite the latter inequality as $-0.0396 < x - 1 < 0.0404$, showing that our interval is not quite symmetric. Choosing $\delta = 0.0396$ (the smaller of the two in absolute value), we have $|\sqrt{x} - 1| < 0.02$ whenever $|x - 1| < \delta$, as desired.

16 (a) If $|x - 4| < \delta < 1$ then $x > 3$, so $|\sqrt{x} + 2| > \sqrt{3} + 2$ and $|\sqrt{x} - 2| = \left|\frac{x - 4}{\sqrt{x} + 2}\right| < \frac{\delta}{\sqrt{3} + 2}$.

(b) Let $\delta = (\sqrt{3} + 2)\epsilon$.

18 (a) If $|x - 2| < \delta < 1$ then $x > 1$ so $\left|\frac{1}{x} - \frac{1}{2}\right| = \left|\frac{x - 2}{2x}\right| < \frac{\delta}{2 \cdot 1} = \frac{\delta}{2}$.

(b) Let $\delta = 2\epsilon$.

20 For each positive number ϵ there is a positive number δ such that for all x that satisfy the inequality $a - \delta < x < a$ it is true that $|f(x) - L| < \epsilon$.

22 For each number E there is a positive number δ such that for all x that satisfy the inequality $0 < |x - a| < \delta$ it is true that $f(x) < E$.

24 For each number E there is a positive number δ such that for all x that satisfy the inequality $a - \delta < x < a$ it is true that $f(x) > E$.

26 (a) Let $\delta = 1/10$.

(b) Let $\delta = \sqrt[3]{\epsilon}$.

28 Let $\epsilon = 1/2$. If $\lim\limits_{x \to 2} x^2 = 3$, there must be a number δ such that $0 < |x - 2| < \delta$ implies that $|x^2 - 3| < 1/2$. In particular, $x = 2 + \delta/2$ satisfies $0 < |x - 2| < \delta$. But for this x,
$|x^2 - 3| = (2 + \delta/2)^2 - 3 > 2^2 - 3 = 1 > 1/2$, so the desired δ does not exist.

2.S Review Exercises

2 $\dfrac{f(x + h) - f(x)}{h}$

$= \dfrac{1}{h}\left[2(x+h)^2 - \dfrac{1}{x+h} - \left(2x^2 - \dfrac{1}{x}\right)\right]$

$= \dfrac{1}{h}\left(4xh + 2h^2 + \dfrac{h}{x(x+h)}\right)$

$= 4x + 2h + \dfrac{1}{x(x+h)}$

4 $x^{1/6}$. Domain: all $x \geq 0$; range: $[0, \infty)$.

6 $\tan x$. Domain: all $x \neq n\pi + \pi/2$ for integer n; range: $(-\infty, \infty)$.

8 $\sqrt{4 - x^2}$. Domain: $[-2, 2]$; range: $[0, 2]$.

10 -0.39

12 $\dfrac{f(2 + h) - f(2)}{h}$

$= \dfrac{3 + 2(2 + h) + (2 + h)^2 - (3 + 2\cdot 2 + 2^2)}{h}$

$= \dfrac{1}{h}(6h + h^2) = 6 + h$

14 0

16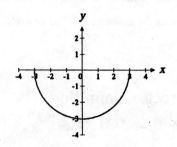

18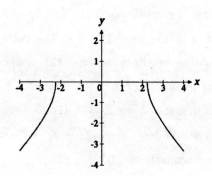

20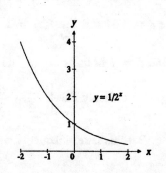

22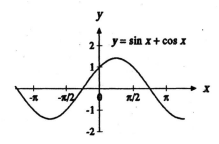

24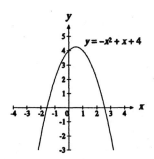

26 (b), (d), and (e)

28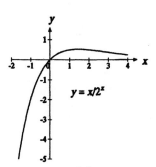

30 (a) $2(xy + xz + yz)$ in^2

(b) xyz in^3

(c) $4(x + y + z)$ in

32

x	1	0.5	0.4	0.3	0.1	0.001
$f(x)$	1	0.707	0.693	0.697	0.794	0.993

(b)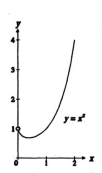

34 $\lim\limits_{x \to 1} \dfrac{x^3 - 1}{x^2 - 1} = \lim\limits_{x \to 1} \dfrac{x^2 + x + 1}{x + 1} = \dfrac{3}{2}$

36 $\lim\limits_{x \to 0} \dfrac{x^4 - 16}{x^3 - 8} = \dfrac{-16}{-8} = 2$

38 $\lim\limits_{x \to -\infty} \dfrac{x^9 + 6x + 3}{x^{10} - x - 1} = \lim\limits_{x \to -\infty} \dfrac{1}{x} \cdot \dfrac{1 + 6/x^8 + 3/x^9}{1 - 1/x^9 - 1/x^{10}}$

$= 0$

40 $\lim\limits_{x \to -\infty} \dfrac{x^4 + x^2 + 1}{3x^2 + 4} = \lim\limits_{x \to -\infty} x^2 \cdot \dfrac{1 + 1/x^2 + 1/x^4}{3 + 4/x^2} = \infty$

42 $\lim\limits_{x \to 81} \dfrac{x - 81}{\sqrt{x} - 9} = \lim\limits_{x \to 81} (\sqrt{x} + 9) = 18$

44 $\lim\limits_{x \to -\infty} (\sqrt{2x^2} - \sqrt{2x^2 - 6x})$

$= \lim\limits_{x \to -\infty} \dfrac{6x}{\sqrt{2x^2} + \sqrt{2x^2 - 6x}} = \lim\limits_{x \to -\infty} \dfrac{6}{\sqrt{2} + \sqrt{2 - 6/x}}$

$= \dfrac{6}{2\sqrt{2}} = \dfrac{3}{\sqrt{2}}$

46 $\lim\limits_{x \to 4^-} \dfrac{1}{x - 4} = -\infty$

48 $\lim\limits_{x \to 3^+} [2x] = 6$

50 As $x \to 0^-$, $1/x \to -\infty$, so $2^{1/x} \to 0$.

52 As $x \to -\infty$, $1/x \to 0$, so $2^{1/x} \to 1$.

54 $\lim_{x \to -\infty} \dfrac{(x+1)^{100}}{(2x+50)^{100}} = \lim_{x \to -\infty} \dfrac{(1+1/x)^{100}}{2^{100}(1+25/x)^{100}}$

$= 2^{-100}$

56 $\dfrac{\cos x}{1-\sin x} = \dfrac{1+\sin x}{\cos x}$. As $x \to \pi/2^-$, $1 + \sin x$

$\to 1$ and $\cos x \to 0^+$ so $\dfrac{1+\sin x}{\cos x} \to \infty$. As

$x \to \pi/2^+$, $1 + \sin x \to 1$ and $\cos x \to 0^-$ so

$\dfrac{1+\sin x}{\cos x} \to -\infty$. Hence the desired limit does

not exist, and is neither ∞ nor $-\infty$.

58 $\lim_{x \to \infty} \dfrac{\sin x}{3x} = 0$, since $\left|\dfrac{\sin x}{3x}\right| < \dfrac{1}{3x}$.

60 As $x \to \pi/2^+$, $\cos x \to 0^-$, so $\sec x = \dfrac{1}{\cos x} \to$

$-\infty$.

62 As $x \to 0^-$, $\sin x \to 0^-$, so $\csc x = \dfrac{1}{\sin x} \to -\infty$.

64 $\lim_{x \to \infty} x \sin \dfrac{1}{x} = \lim_{y \to 0} \dfrac{1}{y} \sin y = 1$

66 The graph oscillates between $y = x^2$ and $y = -x^2$. The limit does not exist and is neither ∞ nor $-\infty$.

68 $\cos^2 \theta - \sin^2 \theta = \cos 2\theta$ oscillates between $y = 1$ and $y = -1$. The limit does not exist.

70 $f(x) = \dfrac{20}{x}$, $g(x) = x$

72 $f(x) = x + 3$, $g(x) = x$

74 $f(x) = x^2$, $g(x) = x$

76 (a) Yes, since the interval is closed.
    (b) No. $x^3 + x + 1 \to -\infty$ as $x \to -\infty$.

78 (a) Yes, at $x = 2$.
    (b) Yes, at $x = 2$.

(c) Yes, at $x = 100$.
(d) No. $1/x^3 \to 0^+$ as $x \to \infty$.

80 Let $f(x) = x^3 - 2x^2 - 3x + 1$. Since $\lim_{x \to -\infty} f(x) = -\infty$, $f(0) = 1 > 0$, $f(2) = -5 < 0$, and $\lim_{x \to \infty} f(x) = \infty$, the intermediate-value theorem gives roots in the intervals mentioned.

82 (a) $\lim_{x \to 3} [f(x) - g(x)] = 0 - 0 = 0$

    (b) $\lim_{x \to 3} \sin f(x) = \sin 0 = 0$

    (c) $\lim_{x \to 3} \cos f(x) = \cos 0 = 1$

    (d) $\lim_{x \to 3} f(x)g(x) = 0 \cdot 0 = 0$

    (e) The limit may be any real number, ∞, $-\infty$, or may not exist. The table shows several examples.

$f(x)$	$g(x)$	$\lim_{x \to 3} [f(x)]^3/g(x)$
$x - 3$	$(x-3)^3/r$	r ($r \neq 0$)
$x - 3$	$(x-3)^2$	0
$x - 3$	$(x-3)^5$	∞
$x - 3$	$-(x-3)^5$	$-\infty$
$x - 3$	$(x-3)^4$	Does not exist

84 $\sin 0 = 0 < 1/2 < 1 = \sin \dfrac{9\pi}{2}$, so the intermediate-value theorem applies. Now $\sin c = 1/2$ if and only if $c = \dfrac{\pi}{6} + 2n\pi$ or $\dfrac{5\pi}{6} + 2n\pi$

for some integer n. In $[0, 9\pi/2]$ the solutions are

$c = \dfrac{\pi}{6}, \dfrac{5\pi}{6}, \dfrac{13\pi}{6}, \dfrac{17\pi}{6}, \dfrac{25\pi}{6}$.

2.5 Summary: Review Exercises

86 $\lim\limits_{x \to 0} \dfrac{\tan x - \sin x}{x^2} = \lim\limits_{x \to 0} \dfrac{\dfrac{\sin x}{\cos x} - \sin x}{x^2}$

$= \lim\limits_{x \to 0} \dfrac{(\sin x)\left(\dfrac{1}{\cos x} - 1\right)}{x^2}$

$= \lim\limits_{x \to 0} \dfrac{\sin x}{x^2} \cdot \left(\dfrac{1 - \cos x}{\cos x}\right)$

$= \lim\limits_{x \to 0} \dfrac{\sin x}{x} \cdot \dfrac{1 - \cos x}{x} \cdot \dfrac{1}{\cos x} = 1 \cdot 0 \cdot 1 = 0$

88 Yes. Let $f(0) = 0$.

90 (a) For $x > 0$, $(4^x + 3^x)^{1/x} > 2^{1/x}$, so
$$\lim\limits_{x \to 0^+} (4^x + 3^x)^{1/x} = \infty.$$

(b) Let $x = -y$. As $x \to 0^-$, $y \to 0^+$ so $4^{-y} + 3^{-y} \to 2$ and $(4^{-y} + 3^{-y})^{1/y} \to \infty$. Hence
$$(4^x + 3^x)^{1/x} = \dfrac{1}{(4^{-y} + 3^{-y})^{1/y}} \to 0.$$

(c) $(4^x + 3^x)^{1/x} = 4(1 + (3/4)^x)^{1/x}$ so, for $x > 0$, $4 < (4^x + 3^x)^{1/x} < 4 \cdot 2^{1/x}$. Since $2^{1/x} \to 1$ as $x \to \infty$, $(4^x + 3^x)^{1/x} \to 4$ (by the squeeze principle).

91 (a) No. For example, let K be a disk and let P be its center. Then every chord through P is cut in half by P.

(b) Yes. For any angle θ, consider the chord through P at angle θ counterclockwise from the x axis. This chord meets the boundary at two points A and B, where A is at angle θ and B is at angle $\theta + \pi$. Let $f(\theta) = \dfrac{\overline{PA}}{\overline{PB}}$. Clearly f is continuous. Also, $f(\theta + \pi) = \dfrac{\overline{PB}}{\overline{PA}} = \dfrac{1}{f(\theta)}$.

In particular, $f(\pi) = \dfrac{1}{f(0)}$, so one of $f(0)$ and $f(\pi)$ is at least 1, while the other is at most 1. Either way, 1 is between $f(0)$ and $f(\pi)$ so, by the intermediate-value theorem, $f(\theta) = 1$ for some θ in $[0, \pi]$. Thus the chord at this angle θ is divided into two pieces of equal length.

3 The Derivative

3.1 Four Problems with One Theme

2. Since θ is in the second quadrant, while the arctangent function gives values from the fourth quadrant, we must add π radians (180°) to our results.

   (a) Since $\tan^{-1} m = \tan^{-1}(-1) = -\pi/4 = -45°$, we have $\theta = 3\pi/4 = 135°$.

   (b) Since $\tan^{-1} m = \tan^{-1}(-2) \approx -1.107 \approx -63.4°$, we have $\theta \approx 2.034 \approx 116.6°$.

4. (a),(b)

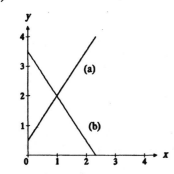

6. $\lim\limits_{h \to 0} \dfrac{\left(\frac{1}{2} + h\right)^2 - \left(\frac{1}{2}\right)^2}{h} = \lim\limits_{h \to 0} (1 + h) = 1$

8. $\lim\limits_{h \to 0} \dfrac{(1 + h)^2 - 1^2}{h} = \lim\limits_{h \to 0} (2 + h) = 2$

10. $\lim\limits_{h \to 0} \dfrac{(1 + h)^3 - 1^3}{h} = \lim\limits_{h \to 0} (3 + 3h + h^2) = 3$

12. (a) $\lim\limits_{h \to 0} \dfrac{(0 + h)^3 - 0^3}{h} = \lim\limits_{h \to 0} h^2 = 0$

    (b)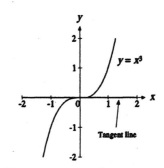

14. $\lim\limits_{h \to 0} \dfrac{16\left[\left(\frac{1}{2} + h\right)^2 - \left(\frac{1}{2}\right)^2\right]}{h} = \lim\limits_{h \to 0} 16(1 + h)$

    $= 16$ ft/sec

16. $\lim\limits_{h \to 0} \dfrac{16\left[\left(\frac{1}{4} + h\right)^2 - \left(\frac{1}{4}\right)^2\right]}{h} = \lim\limits_{h \to 0} 16\left(\frac{1}{2} + h\right)$

    $= 8$ ft/sec

18. (a) $\dfrac{(3.01)^3 - 3^3}{3.01 - 3} = 27.0901$ ft/sec

    (b) $\dfrac{(3 + h)^3 - 3^3}{h} = 27 + 9h + h^2$ ft/sec

    (c) $\lim\limits_{h \to 0^+} (27 + 9h + h^2) = 27$ ft/sec

20. (a) $\dfrac{1.9^3 - 2^3}{1.9 - 2} = 11.41$

    (b) $\dfrac{(2 + h)^3 - 2^3}{h} = 12 + 6h + h^2$

22 (a) We know slope of tangent line at $(-1, 1)$

equals $\lim\limits_{h \to 0} \dfrac{(-1+h)^2 - (-1)^2}{(-1+h) - (-1)} =$

$\lim\limits_{h \to 0} (-2 + h) = -2.$

(b)
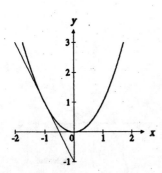

24 (a) $\dfrac{3.1^2 - 3^2}{3.1 - 3} = 6.1$

(b) $\dfrac{3.01^2 - 3^2}{3.01 - 3} = 6.01$

(c) $\dfrac{3.001^2 - 3^2}{3.001 - 3} = 6.001$

(d) $\lim\limits_{h \to 0} \dfrac{(3+h)^2 - 3^2}{h} = \lim\limits_{h \to 0} (6 + h) = 6$

26 (a) $\dfrac{1.5^2 - 1.49^2}{1.5 - 1.49} = 2.99$

(b) $\dfrac{1.5^2 - 1.499^2}{1.5 - 1.499} = 2.999$

(c) $\lim\limits_{h \to 0^-} \dfrac{1.5^2 - (1.5 + h)^2}{-h} = \lim\limits_{h \to 0^-} (3 + h) = 3$

28 (a) $2.01^2 - 2^2 = 0.0401$ g

(b) Density $\approx \dfrac{0.0401}{0.01} = 4.01$ g/cm

(c) Density $\approx \dfrac{2^2 - 1.99^2}{2 - 1.99} = 3.99$ g/cm

(d) $\lim\limits_{h \to 0^+} \dfrac{(2+h)^2 - 2^2}{h} = \lim\limits_{h \to 0^+} (4 + h)$

$= 4$ g/cm

(e) $\lim\limits_{h \to 0^-} \dfrac{2^2 - (2+h)^2}{h} = \lim\limits_{h \to 0^-} (4 + h)$

$= 4$ g/cm

30 (a) Slope of tangent line at $(-1, -1)$

$= \lim\limits_{h \to 0} \dfrac{(-1+h)^3 - (-1)^3}{(-1+h) - (-1)}$

$= \lim\limits_{h \to 0} \dfrac{3h - 3h^2 + h^3}{h} =$

$\lim\limits_{h \to 0} (3 - 3h + h^2) = 3.$

(b)
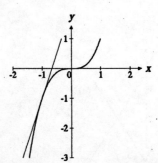

It resembles the line $y = 3x + 2.$

32 (a) $\dfrac{3.1^2 - 3^2}{3.1 - 3} = 6.1$ million dollars per year

(b) $\dfrac{3.01^2 - 3^2}{3.01 - 3} = 6.01$ million dollars per year

(c) $\lim\limits_{h \to 0} \dfrac{(3+h)^2 - 3^2}{h} = \lim\limits_{h \to 0} (6 + h) = 6$

million dollars per year

34 (a) $\dfrac{[2(1+h)^2 + (1+h)] - [2 \cdot 1^2 + 1]}{h}$

$= 5 + h$ ft/sec

3.1 Four Problems with One Theme

$= 5 + h$ ft/sec

(b) $\lim_{h \to 0^+} (5 + h) = 5$ ft/sec

36 $\lim_{h \to 0} \dfrac{16[(t + h)^2 - t^2]}{h} = \lim_{h \to 0} 16(2t + h)$

$= 32t$ ft/sec

38 $\lim_{h \to 0^+} \dfrac{(x + h)^2 - x^2}{h} = \lim_{h \to 0^+} (2x + h) = 2x$ g/cm

40 (a)

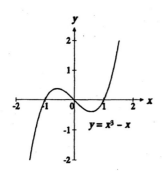

(b) $\lim_{h \to 0} \dfrac{[(x + h)^3 - (x + h)] - [x^3 - x]}{h} =$

$\lim_{h \to 0} (3x^2 - 1 + 3xh + h^2) = 3x^2 - 1$

(c) $3x^2 - 1 = 0$ if and only if $x = \pm \dfrac{1}{\sqrt{3}}$. The

points are $\pm \left(\dfrac{1}{\sqrt{3}}, -\dfrac{2}{3\sqrt{3}} \right)$.

(d) $3x^2 - 1 = 1$ if and only if $x = \pm \sqrt{\dfrac{2}{3}}$. The

points are $\pm \left(\sqrt{\dfrac{2}{3}}, -\dfrac{1}{3}\sqrt{\dfrac{2}{3}} \right)$.

42 The tangent at (a, a^2) has slope $2a$ and equation
$y - a^2 = 2a(x - a)$; that is $y = 2ax - a^2$.

(a) If $9 = 2a \cdot 4 - a^2$ then $a^2 - 8a + 9 = 0$, so
$a = 4 \pm \sqrt{7}$. Since $4 + \sqrt{7} > 4$, she should

shut off her engine at the point $(4 - \sqrt{7},$
$(4 - \sqrt{7})^2) = (4 - \sqrt{7}, 23 - 8\sqrt{7})$.

(b) If $-9 = 2a \cdot 4 - a^2$ then $a^2 - 8a - 9 = 0$,
so $a = -1$ or 9. She should shut off the
engine at $(-1, 1)$.

44 Find the slope of the line through $(1, 2)$ and

$(1 + h, 2^{1+h})$. We have slope $= \dfrac{2^{(1+h)} - 2^1}{(1 + h) - 1} =$

$\dfrac{2(2^h - 1)}{h} = m$.

(a) Here $h = 0.1$. Thus, $m \approx 1.435$.

(b) Here $h = 0.01$. Thus, $m \approx 1.391$.

(c) Here $h = -0.1$. Thus, $m \approx 1.339$.

(d) Here $h = -0.01$. Thus, $m \approx 1.382$.

46 (a),(b)

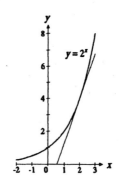

(c) The slope is about 2.77.

3.2 The Derivative

2 $(5x)' = \lim_{h \to 0} \dfrac{5(x + h) - 5x}{h} = \lim_{h \to 0} 5 = 5$

4 $(3x - 1)' = \lim_{h \to 0} \dfrac{3(x + h) - 1 - (3x - 1)}{h}$

$= \lim_{h \to 0} 3 = 3$

6 $(-x^2)' = \lim_{h \to 0} \dfrac{-(x + h)^2 - (-x^2)}{h} =$

$$\lim_{h\to 0} \frac{-x^2 - 2xh - h^2 + x^2}{h} = \lim_{h\to 0} (-2x - h)$$

$$= -2x$$

8 $(3x^2 - x)' =$

$$\lim_{h\to 0} \frac{3(x+h)^2 - (x+h) - (3x^2 - x)}{h}$$

$$= \lim_{h\to 0} \frac{3x^2 + 6xh + 3h^2 - x - h - 3x^2 + x}{h}$$

$$= \lim_{h\to 0} (6x - 1 + 3h) = 6x - 1$$

10 $\left(\frac{1}{2}\sqrt{x}\right)' = \lim_{h\to 0} \dfrac{\frac{1}{2}\sqrt{x+h} - \frac{1}{2}\sqrt{x}}{h}$

$$= \frac{1}{2} \lim_{h\to 0} \left(\frac{\sqrt{x+h} - \sqrt{x}}{h} \cdot \frac{\sqrt{x+h} + \sqrt{x}}{\sqrt{x+h} + \sqrt{x}}\right)$$

$$= \frac{1}{2} \lim_{h\to 0} \frac{(x+h) - x}{h(\sqrt{x+h} + \sqrt{x})}$$

$$= \frac{1}{2} \lim_{h\to 0} \frac{1}{\sqrt{h+x} + \sqrt{x}} = \frac{1}{2} \cdot \frac{1}{2\sqrt{x}} = \frac{1}{4\sqrt{x}}$$

12 $(7x^2 - \sqrt{x})'$

$$= \lim_{h\to 0} \frac{7(x+h)^2 - \sqrt{x+h} - (7x^2 - \sqrt{x})}{h}$$

$$= \lim_{h\to 0} \left(\frac{7(x+h)^2 - 7x^2}{h} - \frac{\sqrt{x+h} - \sqrt{x}}{h}\right)$$

$$= \lim_{h\to 0} \left(14x + 7h - \frac{1}{\sqrt{x+h} + \sqrt{x}}\right) = 14x - \frac{1}{2\sqrt{x}}$$

14 $(5x - x^3)'$

$$= \lim_{h\to 0} \frac{5(x+h) - (x+h)^3 - (5x - x^3)}{h}$$

$$= \lim_{h\to 0} \frac{5x + 5h - x^3 - 3x^2 h - 3xh^2 - h^3 - 5x + x^3}{h}$$

$$= \lim_{h\to 0} (5 - 3x^2 - 3xh - h^2) = 5 - 3x^2$$

16 $\left(\dfrac{1}{x^2} + x\right)'$

$$= \lim_{h\to 0} \left[\frac{1/(x+h)^2 + x + h - (1/x^2 + x)}{h}\right]$$

$$= \lim_{h\to 0} \left[1 + \frac{x^2 - (x+h)^2}{h(x+h)^2 x^2}\right]$$

$$= \lim_{h\to 0} \left[1 - \frac{2x}{(x+h)^2 x^2} - \frac{h}{(x+h)^2 x^2}\right]$$

$$= -\frac{2}{x^3} + 1$$

18 $(x^4)' = 4x^3 = 4\left(\dfrac{1}{2}\right)^3 = \dfrac{1}{2}$

20 $(x^5)' = 5x^4 = 5(\sqrt{2})^4 = 20$

22 $(\sqrt[3]{t})' = \dfrac{1}{3}t^{-2/3} = \dfrac{1}{3}(-1)^{-2/3} = \dfrac{1}{3}$

24 $(x^{\sqrt{2}})' = \sqrt{2}x^{\sqrt{2}-1} = \sqrt{2} \cdot 3^{\sqrt{2}-1}$

26 $(x^{2/3})' = \dfrac{2}{3}x^{-1/3} = \dfrac{2}{3}[(-27)^{-1/3}] = -\dfrac{2}{9}$

28 $(x^{4/5})' = \dfrac{4}{5}x^{-1/5} = \dfrac{4}{5}\left[\left(\dfrac{1}{32}\right)^{-1/5}\right] = \dfrac{8}{5}$

30 Note that $f(x) = x^{-1/3}$. By Theorem 2, $f'(x) =$

$-\dfrac{1}{3}x^{-4/3}$ and $f'(8) = -\dfrac{1}{3}(8^{-4/3}) = -\dfrac{1}{48}$.

32 $(x^{\sqrt{2}})' = \sqrt{2}x^{\sqrt{2}-1} = \sqrt{2} \cdot 3^{\sqrt{2}-1} \approx 2.229$

34 (a) $\dfrac{2.01^4 - 2^4}{2.01 - 2} = 32.240801$ ft/sec

3.2 The Derivative

(b) $\dfrac{2^4 - 1.99^4}{2 - 1.99} = 31.760799$ ft/sec

(c) $(t^4)' = 4t^3 = 4 \cdot 2^3 = 32$ ft/sec

36 (a) $\dfrac{2.01^3 - 2^3}{2.01 - 1} = 12.0601$ g/cm

(b) $\dfrac{2.01^3 - 1.99^3}{2.01 - 1.99} = 12.0001$ g/cm

(c) $(x^3)' = 3x^2 = 3 \cdot 2^2 = 12$ g/cm

38 (a) 1

(b) -1.6

(c) 0

(d) 2

(e)

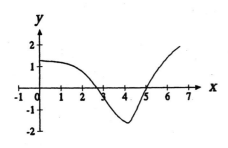

40 (a) $(x + x^2)' =$

$\lim\limits_{h \to 0} \dfrac{x + h + (x + h)^2 - (x + x^2)}{h}$

$= \lim\limits_{h \to 0} \dfrac{h + 2xh + h^2}{h} = 1 + 2x$

(b)

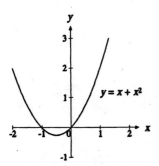

(c) $0 = 1 + 2x$ when $x = -1/2$

(d) $(-1/2, -1/4)$

(e) $-1/4$

42 (a) Domain of $x^{3/5}$: $(-\infty, \infty)$. All we need is for $x^{3/5}$ to be real. Since we can always take the fifth root of a real number (whether positive, negative, or zero), no problem arises.

Domain of $x^{3/4}$: $[0, \infty)$. Clearly, $x^{3/4}$ is real only for nonnegative x.

Domain of $x^{4/3}$: $(-\infty, \infty)$. Again, $x^{4/3}$ is real for all choices of real x.

(b) The table shows the domain for each case of $x^{m/n}$. Note the symmetry caused by the fact that $m/n = (-m)/(-n)$.

m/n in lowest terms	$m < 0$	$m = 0$	$m > 0$
$n < 0$, even	$[0, \infty)$	N/A	$(0, \infty)$
$n < 0$, odd	$(-\infty, \infty)$	$(-\infty, 0), (0, \infty)$	$(-\infty, 0), (0, \infty)$
$n > 0$, odd	$(-\infty, 0), (0, \infty)$	$(-\infty, 0), (0, \infty)$	$(-\infty, \infty)$
$n > 0$, even	$(0, \infty)$	N/A	$[0, \infty)$

44 Let $P(n) = (x + h)^n$; since $P(0) = 1$ and $P(1) = x + h$ are trivial cases for what we wish to prove, we begin with $P(2) = (x + h)^2 = x^2 + 2xh + Q$. We know that $Q = h^2$ has h^2 as a factor, and that $P(2)$ has the desired form. We assume $P(n) = x^n + nx^{n+1}h + Q$, where Q has h^2 as a factor.

Now, $P(n + 1) = (x + h)^{n+1} = (x + h)(x + h)^n$

$= (x + h)(P(n)) = (x + h)(x^n + nx^{n-1}h + Q)$

$= x^{n+1} + nx^nh + xQ + hx^n + nx^{n-1}h^2 + hQ$

$= x^{n+1} + (n + 1)x^nh + xQ + nx^{n-1}h^2 + hQ$

$= x^{n+1} + (n + 1)x^nh + Q'$

(where $Q' = xQ + nx^{n-1}h^2 + hQ$ has h^2 as a

factor). We have shown that $P(n)$ implies $P(n + 1)$. Since $P(2)$ holds, it follows that $P(n)$ holds for all $n > 2$.

3.3 The Derivative and Continuity

2 $\lim\limits_{\Delta x \to 0} \dfrac{(x + \Delta x)^4 - x^4}{\Delta x} =$

$\lim\limits_{\Delta x \to 0} \dfrac{x^4 + 4x^3 \Delta x + 6x^2(\Delta x)^2 + 4x(\Delta x)^3 + (\Delta x)^4 - x^4}{\Delta x}$

$= \lim\limits_{\Delta x \to 0} \left(4x^3 + 6x^2 \Delta x + 4x(\Delta x)^2 + (\Delta x)^3\right) = 4x^3$

4 $\lim\limits_{\Delta x \to 0} \dfrac{3\sqrt{x + \Delta x} - 3\sqrt{x}}{\Delta x}$

$= \lim\limits_{\Delta x \to 0} \dfrac{3\sqrt{x + \Delta x} - 3\sqrt{x}}{\Delta x} \cdot \dfrac{\sqrt{x + \Delta x} + \sqrt{x}}{\sqrt{x + \Delta x} + \sqrt{x}}$

$= \lim\limits_{\Delta x \to 0} \dfrac{3}{\sqrt{x + \Delta x} + \sqrt{x}} = \dfrac{3}{2\sqrt{x}}$

6 $\lim\limits_{\Delta x \to 0} \dfrac{[5(x + \Delta x)^2 + 3(x + \Delta x) + 2] - [5x^2 + 3x + 2]}{\Delta x}$

$= \lim\limits_{\Delta x \to 0} (10x + 5\Delta x + 3) = 10x + 3$

8 $\lim\limits_{\Delta x \to 0} \dfrac{\dfrac{5}{(x + \Delta x)^2} - \dfrac{5}{x^2}}{\Delta x} = \lim\limits_{\Delta x \to 0} \dfrac{5(-2x - \Delta x)}{x^2(x + \Delta x)^2}$

$= -\dfrac{10}{x^3}$

10 $\lim\limits_{\Delta x \to 0} \dfrac{[(x + \Delta x)^3 - 5(x + \Delta x) + 1992] - [x^3 - 5x + 1992]}{\Delta x}$

$= \lim\limits_{\Delta x \to 0} (3x^2 + 3x\Delta x + \Delta x^2 - 5) = 3x^2 - 5$

12 (a) $\Delta f = \dfrac{1}{2 + \dfrac{1}{5}} - \dfrac{1}{2} = -\dfrac{1}{22}$

(b) $\Delta f = \dfrac{1}{2 - \dfrac{1}{8}} - \dfrac{1}{2} = \dfrac{1}{30}$

14 (a) $\Delta y = \Delta f = f(2.8) - f(3) = 0.3$

(b)

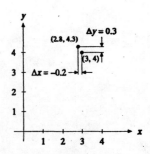

16 (a) For $a = 0$, $\lim\limits_{x \to a} f(x) = 2 \neq 1 = f(a)$.

(b) For $a = 1$ and 3, $\lim\limits_{x \to a} f(x) = f(a)$, but

$\lim\limits_{x \to a^-} \dfrac{f(x) - f(a)}{x - a} \neq \lim\limits_{x \to a^+} \dfrac{f(x) - f(a)}{x - a}$.

18 (a) f is continuous for all a.

(b) For $a = 0$ and 4, $\lim\limits_{x \to a^-} \dfrac{f(x) - f(a)}{x - a} \neq$

$\lim\limits_{x \to a^+} \dfrac{f(x) - f(a)}{x - a}$. For $a = 2$,

$\lim\limits_{x \to a} \dfrac{f(x) - f(a)}{x - a} = \infty$. Hence f is not

differentiable for $a = 0$, 2, or 4.

20 $D\left(\dfrac{1}{3x + 5}\right) = \lim\limits_{h \to 0} \dfrac{\dfrac{1}{3(x + h) + 5} - \dfrac{1}{3x + 5}}{h}$

$= \lim\limits_{h \to 0} \dfrac{[3x + 5] - [3(x + h) + 5]}{h(3x + 5)[3(x + h) + 5]}$

3.3 The Derivative and Continuity

$$= \lim_{h \to 0} \frac{-3}{(3x+5)(3x+3h+5)} = -\frac{3}{(3x+5)^2}$$

22 (a)

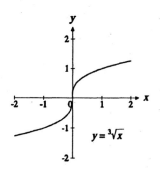

(b)

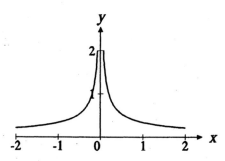

(c) Yes. See the graph in (a).

(d) No, f is not differentiable at 0. (See the graph in (b).)

24 (a) Any $1/x^2 + C$, where C is a constant.

(b) Any $\sqrt{x} + C$, where C is a constant.

26

28

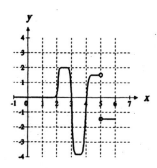

30 (a) $x^2 + C$

(b) $\sqrt[3]{x} + C$

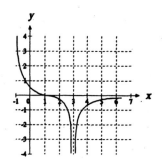

32 (a) $\frac{3}{4}x^{4/3} + C$

(b) $\frac{2}{3}x^{3/2} + C$

34 We know $f'(x) \approx \dfrac{f(x + \Delta x) - f(x)}{\Delta x}$ for small Δx.

Thus $\Delta x f'(x) \approx f(x + \Delta x) - f(x)$ and $f(x + \Delta x) \approx f(x) + \Delta x f'(x)$. So, $f(3 + 0.07) \approx f(3) + 0.07 \cdot f'(3)$, $f(3.07) \approx 3.965$.

36 The derivative is 2.

3.4 The Derivatives of the Sum, Difference, Product, and Quotient

2 $D(2x^2 - x) = D(2x^2) - D(x) = 4x - 1$

4 $D(2x^3 - x^2 + x) = D(2x^3) - D(x^2) + D(x)$
$= 6x^2 - 2x + 1$

6 $D(x^6 + 5x^2 + 2) = D(x^6) + D(5x^2) + D(2)$
$= 6x^5 + 10x + 0 = 6x^5 + 10x$

8 $D(2x^3 + 3\sqrt{x}) = D(2x^3) + D(3\sqrt{x})$
$= 6x^2 + \frac{3}{2}x^{-1/2}$

10 $D(5/x^2 + \sqrt[4]{x}) = D(5x^{-2}) + D(x^{1/4})$
$= -10x^{-3} + \frac{1}{4}x^{-3/4}$

12 $D[(3x - 4)(5x + 1)]$

$= (3x - 4)D(5x + 1) + (5x + 1)D(3x - 4)$

$= (3x - 4)5 + (5x + 1)3 = 15x - 20 + 15x + 3$

$= 30x - 17$

14. $D[(2x^2 - 1)(x^2 - 3)]$

$= (2x^2 - 1)D(x^2 - 3) + (x^2 - 3)D(2x^2 - 1)$

$= (2x^2 - 1)(2x) + (x^2 - 3)(4x)$

$= 4x^3 - 2x + 4x^3 - 12x = 8x^3 - 14x$

16. $D[(5x^5 + x)(2x + 1)]$

$= (5x^5 + x)D(2x + 1) + (2x + 1)D(5x^5 + x)$

$= (5x^5 + x)(2) + (2x + 1)(25x^4 + 1)$

$= 10x^5 + 2x + 50x^5 + 2x + 25x^4 + 1$

$= 60x^5 + 25x^4 + 4x + 1$

18. $D[(x + 3\sqrt{x})(2x + 1)]$

$= (x + 3\sqrt{x})D(2x + 1) + (2x + 1)D(x + 3\sqrt{x})$

$= (x + 3\sqrt{x})(2) + (2x + 1)\left(1 + \frac{3}{2\sqrt{x}}\right)$

$= 2x + 6\sqrt{x} + 2x + 3\sqrt{x} + 1 + \frac{3}{2\sqrt{x}}$

$= 4x + 9\sqrt{x} + 1 + \frac{3}{2\sqrt{x}}$

20. $\dfrac{d}{dx}\left(\dfrac{7x - \sqrt{x^3}}{6}\right) = \dfrac{1}{6}\dfrac{d}{dx}(7x - x^{3/2})$

$= \dfrac{1}{6}\left(7 - \dfrac{3}{2}x^{1/2}\right) = \dfrac{7}{6} - \dfrac{1}{4}x^{1/2}$

22. $\left(\dfrac{1 + x}{2 - x}\right)' = \dfrac{(2 - x)(1 + x)' - (1 + x)(2 - x)'}{(2 - x)^2}$

$= \dfrac{(2 - x)\cdot 1 - (1 + x)(-1)}{(2 - x)^2} = \dfrac{3}{(2 - x)^2}$

24. $\dfrac{d}{dx}\left(\dfrac{2x^2 - 7}{3x + 1}\right)$

$= \dfrac{(3x + 1)(2x^2 - 7)' - (2x^2 - 7)(3x + 1)'}{(3x + 1)^2}$

$= \dfrac{(3x + 1)(4x) - (2x^2 - 7)(3)}{(3x + 1)^2}$

$= \dfrac{12x^2 + 4x - 6x^2 + 21}{(3x + 1)^2} = \dfrac{6x^2 + 4x + 21}{(3x + 1)^2}$

26. $\dfrac{d}{ds}\left(\dfrac{s^2 - s}{5s^2 + s + 2}\right)$

$= \dfrac{(5s^2 + s + 2)(s^2 - s)' - (s^2 - s)(5s^2 + s + 2)'}{(5s^2 + s + 2)^2}$

$= \dfrac{(5s^2 + s + 2)(2s - 1) - (s^2 - s)(10s + 1)}{(5s^2 + s + 2)^2} =$

$\dfrac{10s^3 + 2s^2 + 4s - 5s^2 - s - 2 - 10s^3 + 10s^2 - s^2 + s}{(5x^2 + s + 2)^2}$

$= \dfrac{6s^2 + 4s - 2}{(5s^2 + s + 2)^2}$

28. $\dfrac{d}{dw}\left[\dfrac{2w + \sqrt{w^3}}{3w}\right] = \dfrac{d}{dw}\left[\dfrac{2}{3} + \dfrac{w^{3/2}}{3w}\right]$

$= \dfrac{d}{dw}\left[\dfrac{2}{3} + \dfrac{w^{1/2}}{3}\right] = 0 + \dfrac{1}{3}\cdot\dfrac{1}{2}w^{-1/2} = \dfrac{1}{6}w^{-1/2}$

30. $\left(\dfrac{1}{x + \sqrt{x}}\right)' = \dfrac{-(x + \sqrt{x})'}{(x + \sqrt{x})^2} = -\dfrac{1 + \dfrac{1}{2\sqrt{x}}}{(x + \sqrt{x})^2}$

$= -\dfrac{1 + 2\sqrt{x}}{2\sqrt{x}(x + \sqrt{x})^2}$

32. $\dfrac{d}{dx}\left[\dfrac{1}{2x^2 - x + 5}\right] = -\dfrac{(2x^2 - x + 5)'}{(2x^2 - x + 5)^2}$

3.4 The Derivative of the Sum, Difference, Product, and Quotient

$$= -\frac{4x - 1}{(2x^2 - x + 5)^2}$$

34 $\frac{d}{dx}\left[\frac{(2x+9)(3x^2-x)}{x^2}\right] = \frac{d}{dx}[6x + 25 - 9x^{-1}]$

$$= 6 + 0 + 9x^{-2} = 6 + \frac{9}{x^2}$$

36 $\left[\left(x + \frac{2}{x}\right)(x^2 + 6x + 1)\right]'$

$$= (x^3 + 6x^2 + 3x + 12 + 2x^{-1})'$$

$$= 3x^2 + 12x + 3 - 2x^{-2}$$

38 $\left(\frac{1}{2x+1}\right)' = -\frac{2}{(2x+1)^2}$, which equals $-\frac{2}{25}$

when $x = 2$. Thus the tangent line is $y - \frac{1}{5} = -\frac{2}{25}(x - 2)$.

40 $\left(\frac{x+1}{x+2}\right)' = \frac{(x+2)(1) - (x+1)(1)}{(x+2)^2} = \frac{1}{(x+2)^2}$, which equals 1 when $x = -1$. Hence the tangent line is $y - 0 = 1(x - (-1))$; that is, $y = x + 1$.

42 $D(5\sqrt{t}) = 5 D(\sqrt{t}) = 5 \cdot \frac{1}{2\sqrt{t}}$; when $t = 9$, we have

$5 \cdot \frac{1}{2\sqrt{9}} = 5 \cdot \frac{1}{2 \cdot 3} = \frac{5}{6}$, so velocity equals 5/6.

44 $(4x^2 + x + 2)' = 8x + 1 = 8 \cdot 2 + 1 = 17$

46 $\left(\frac{x}{x+1}\right)' = \frac{(x+1)x' - x(x+1)'}{(x+1)^2} = \frac{1}{(x+1)^2}$

$$= \frac{1}{(2+1)^2} = \frac{1}{9}$$

48 Pick any two C's to obtain two antiderivatives.

(a) $x^5/5 + C$

(b) $\frac{2}{3}x^{3/2} + C$

(c) $-\frac{1}{2}x^{-2} + C$

(d) $\frac{3}{2}x^2 - \frac{5}{x} + C$

50 (a) $D(x^{-4}) = D\left(\frac{1}{x^4}\right) = \frac{-4x^3}{(x^4)^2} = -4x^{-5}$

(b) $D(x^{-m}) = D\left(\frac{1}{x^m}\right) = -\frac{mx^{m-1}}{(x^m)^2} = -\frac{mx^{m-1}}{x^{2m}}$

$$= -mx^{-m-1}$$

52 The proof is by induction. Let $P(n)$ be the statement that $D(1/x^n) = -n/x^{n+1}$, and assume $P(n)$ is true. Then $P(n + 1)$ is the corresponding statement for $D(1/x^{n+1})$. Note that $D(1/x^{n+1}) =$

$$D\left(\frac{1}{x} \cdot \frac{1}{x^n}\right) = \frac{1}{x} \cdot D\left(\frac{1}{x^n}\right) + \frac{1}{x^n} \cdot D\left(\frac{1}{x}\right)$$

$$= \frac{1}{x} \cdot \frac{-n}{x^{n+1}} + \frac{1}{x^n} \cdot \frac{-1}{x^2} = \frac{-n}{x^{n+2}} - \frac{1}{x^{n+2}}$$

$$= \frac{-(n+1)}{x^{(n+1)+1}}.$$ Hence $P(n)$ implies $P(n+1)$. Since $P(1)$ is true (because $D(1/x) = -1/x^2$), it follows that $P(n)$ is also true for all $n > 1$.

54 $(f - g)'(x) = \lim_{h \to 0} \frac{(f-g)(x+h) - (f-g)(x)}{h}$

$$= \lim_{h \to 0}\left(\frac{f(x+h) - f(x)}{h} - \frac{g(x+h) - g(x)}{h}\right)$$

$$= f'(x) - g'(x)$$

56 (a) Usually negative. (For small x it may be positive, since some people still think "If it's

that cheap, there must be something wrong with it, so I won't buy it.")

(b) Gallons per cent

(c) Since Δy and y have the same units (amounts) and Δx and x have the same units (prices), both ratios $\Delta y/y$ and $\Delta x/x$ are dimensionless. Hence $(\Delta y/y)/(\Delta x/x)$ is dimensionless.

(d) $\epsilon = \lim_{\Delta x \to 0} \frac{\Delta y/y}{\Delta x/x} = \lim_{\Delta x \to 0} \left(\frac{x}{y} \cdot \frac{\Delta y}{\Delta x}\right)$

$= \frac{x}{y} \lim_{\Delta x \to 0} \frac{\Delta y}{\Delta x} = \left(\frac{x}{y}\right) y'$

(e) For $\Delta x = \frac{x}{100}$, $\Delta y = -\frac{2y}{100}$, so $\frac{\Delta y/y}{\Delta x/x} = -2$ and $\epsilon \approx -2$.

(f) $\epsilon \approx \frac{-1/100}{2/100} = -\frac{1}{2}$

(g) The quantity ϵ is negative in standard economic applications, so one typically discusses the value of $|\epsilon|$. If $|\epsilon| > 1$, then profit decreases with price. In this case—see (e)—the percentage change in price is exceeded by the percentage change in demand; the impact of the price change is "stretched" and the situation is called *elastic*. If $|\epsilon| < 1$, profit increases with price. In this case—see (f)—the percentage change in price is greater than the resultant percentage change in demand; since this is the opposite of the previous situation, it is called *inelastic*.

(h) $\epsilon = \frac{x}{y} y' = \frac{x}{x^{-3}} (x^{-3})' = x^4(-3)x^{-4} = -3$

58 Yes. Suppose $y = mx + b$ is tangent to $y = x^2$ at (x_0, y_0) and to $y = -x^2 + 2x - 2$ at (x_1, y_1). Then m equals the derivative of x^2 at x_0; that is, $m =$

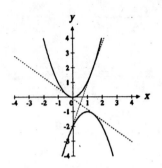

$2x_0$. Also, $x_0^2 = y_0 = mx_0 + b$, so $b = x_0^2 - mx_0 = -x_0^2 = -\frac{m^2}{4}$. Similarly, m equals the derivative of $-x^2 + 2x - 2$ at x_1; that is, $m = -2x_1 + 2$. Also $-x_1^2 + 2x_1 - 2 = y_1 = mx_1 + b$, so $b = -x_1^2 + 2x_1 - 2 - mx_1 = -x_1^2 + 2x_1 - 2 + 2x_1^2 - 2x_1 = x_1^2 - 2 = \left(\frac{2-m}{2}\right)^2 - 2 = \frac{m^2 - 4m - 4}{4}$. Hence $-\frac{m^2}{4} = \frac{m^2 - 4m - 4}{4}$, $m^2 - 2m - 2 = 0$, and $m = 1 \pm \sqrt{3}$. So $b = -\frac{m^2}{4} = -\frac{2 \pm \sqrt{3}}{2}$, and there are two lines: $y = (1 + \sqrt{3})x - \frac{2 + \sqrt{3}}{2}$ and $y = (1 - \sqrt{3})x - \frac{2 - \sqrt{3}}{2}$.

60 (a) The slope is $3a^2$, so the equation of the line through (a, a^3) is $y - a^3 = 3a^2(x - a)$; that is, $y = 3a^2x - 2a^3$.

(b) No. If (x, y) is an intersection then $3a^2x - 2a^3 = y = x^3$, so $0 = x^3 - 3a^2x + 2a^3 = (x - a)^2(x + 2a)$. Intersections are (a, a^3) and $(-2a, -8a^3)$. Thus, there is another intersection if and only if $a \neq 0$.

3.5 The Derivatives of the Trigonometric Functions

2. $D(7 \cos x) = 7 D(\cos x) = -7 \sin x$

4. $D(-6 \cot x) = -6 D(\cot x) = -6(-\csc^2 x)$
   $= 6 \csc^2 x$

6. $D(5 \csc x) = 5 D(\csc x) = 5(-\csc x \cot x)$
   $= -5 \csc x \cot x$

8. $(x^3 \cos x)' = (x^3)(\cos x)' + (x^3)'(\cos x)$
   $= -x^3 \sin x + 3x^2 \cos x$

10. $\left(\dfrac{1 - \sin x}{\cos x}\right)' = (\sec x - \tan x)'$
    $= \sec x \tan x - \sec^2 x$

12. $(x^3 \sec x)' = x^3 \sec x \tan x + 3x^2 \sec x$

14. $(3 \csc x + 2 \tan x)' = -3 \csc x \cot x + 2 \sec^2 x$

16. $(x^2 \cos x \cot x)' = (x^2)' \cos x \cot x +$
    $x^2 (\cos x)' \cot x + x^2 \cos x (\cot x)'$
    $= 2x \cos x \cot x - x^2 \sin x \cot x - x^2 \cos x \csc^2 x$
    $= 2x \cos x \cot x - x^2 \cos x - x^2 \cos x \csc^2 x$
    $= x(\cos x)(2 \cot x - x - x \csc^2 x)$

18. $\left(\dfrac{x}{1 + \sec x}\right)' = \dfrac{(1 + \sec x)(x)' - (x)(1 + \sec x)'}{(1 + \sec x)^2}$
    $= \dfrac{1 + \sec x - x \sec x \tan x}{(1 + \sec x)^2}$

20. $(\cos x + x \sin x)' = x \cos x$

22. $(3x^2 \sin x - 6 \sin x - x^3 \cos x + 6x \cos x)'$
    $= (3x^2 \cos x + 6x \sin x) - 6 \cos x -$
    $(x^3(-\sin x) + 3x^2 \cos x) + 6(x(-\sin x) + \cos x)$
    $= x^3 \sin x$

24. $(-\cot x - x)' = \csc^2 x - 1 = \cot^2 x$

26. $[(3x^2 - 6) \cos x + (x^3 - 6x) \sin x]'$
    $= [(3x^2 - 6)(-\sin x) + 6x \cos x]$
    $+ [(x^3 - 6x) \cos x + (3x^2 - 6) \sin x]$
    $= x^3 \cos x$

28. $D(\cos x) = -\sin x$
    (a) $-\sin \pi/4 = -\dfrac{1}{\sqrt{2}}$
    (b) $-\sin\left(-\dfrac{2\pi}{3}\right) = \dfrac{\sqrt{3}}{2}$
    (c) $-\sin 2 \approx -0.909$

30. $D(\sec x) = \sec x \tan x$
    (a) $\sec \pi/4 \tan \pi/4 = \sqrt{2}$
    (b) $\sec 3\pi/4 \tan 3\pi/4 = \sqrt{2}$
    (c) $\sec 0.5 \tan 0.5 \approx 0.623$

32. $D(\cot \theta) = D\left(\dfrac{\cos \theta}{\sin \theta}\right)$
    $= \dfrac{(\sin \theta)(-\sin \theta) - (\cos \theta)(\cos \theta)}{\sin^2 \theta} = \dfrac{-1}{\sin^2 \theta}$
    $= -\csc^2 \theta$

34.   (a)   Rate of rising $= (2 \sin t)' = 2 \cos t$ ft/hr
    (b)   At low tide, $\sin t = -1$ so $\cos t = 0$ and thus the rate is 0. At sea level, $\sin t = 0$ so $\cos t = \pm 1$ and the rate is ± 2 ft/hr. The surface rises (or falls) most rapidly at mean sea level.

36.   (a)   Slope $= (\tan x)' = \sec^2 x = \sec^2 \dfrac{\pi}{4} = (\sqrt{2})^2$
    $= 2$
    (b) $\tan^{-1} 2 \approx 63.435° \approx 1.10715$ radians

38. $(\cos x)' = -\sin x$, so the angle is $= \tan^{-1}\left(-\sin \dfrac{\pi}{3}\right)$
    $= \tan^{-1}\left(-\dfrac{\sqrt{3}}{2}\right) \approx -40.89° \approx -0.71372$ radians.

40.   (a) $5 \tan x$
    (b) $-6 \cot x$

42.   (a) $-5 \cos x - 6 \sin x$
    (b) $3 \tan x$

44 $(\sin 7x)' = \lim_{h \to 0} \dfrac{\sin 7(x+h) - \sin 7x}{h}$

$= \lim_{h \to 0} \dfrac{(\sin 7x)(\cos 7h - 1) + (\cos 7x)(\sin 7h)}{h} =$

$(-7 \sin 7x) \lim_{h \to 0} \dfrac{1 - \cos 7h}{7h} + (7 \cos 7x) \lim_{h \to 0} \dfrac{\sin 7h}{7h}$

$= (-7 \sin 7x)0 + (7 \cos 7x)1 = 7 \cos 7x$

46 (a)

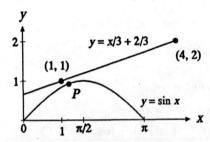

(c) The actual slope of the tangent at P is given by $(\sin x)' = \cos x = 1/3$. Solving, we see that $x = \cos^{-1} 1/3$, $y = \sin(\cos^{-1} 1/3)$ or $x \approx 1.231$, $y \approx 0.943$.

3.6 The Derivative of a Composite Function

2 Let $y = u^{20}$, where $u = 2 - 7x$. Then $\dfrac{dy}{dx} =$

$\dfrac{dy}{du} \cdot \dfrac{du}{dx} = 20u^{19}(-7) = -140(2 - 7x)^{19}$.

4 Let $y = u^{100}$, where $u = x^4 - 6$. Then $\dfrac{dy}{dx} =$

$\dfrac{dy}{du} \cdot \dfrac{du}{dx} = 100u^{99} \cdot 4x^3 = 100(x^4 - 6)^{99} \cdot 4x^3$

$= 400x^3(x^4 - 6)^{99}$.

6 Let $y = 5 \cos u$, where $u = 2x$. Then $\dfrac{dy}{dx} =$

$\dfrac{dy}{du} \cdot \dfrac{du}{dx} = -5 \sin u \cdot (2) = -10 \sin 2x$.

8 Let $y = u^3$, where $u = \sin v$ and $v = 2x$. Then $\dfrac{dy}{dx}$

$= \dfrac{dy}{du} \cdot \dfrac{du}{dv} \cdot \dfrac{dv}{dx} = 3u^2 \cdot \cos v \cdot (2) = 6 \sin^2 v \cos v$

$= 6 \sin^2 2x \cos 2x$.

10 Let $y = u^3$, where $u = \cos v$ and $v = 4x$. Then $dy/dx = 3u^2(-\sin v)(4) = -12 \cos^2 v \sin v$

$= -12 \cos^2 4x \sin 4x$.

12 Let $y = u^5$, where $u = \sec v$ and $v = 2x$. Then $dy/dx = 5u^4 \cdot (\sec v \tan v) \cdot 2 = 10 \sec^5 v \tan v$

$= 10 \sec^5 2x \tan 2x$.

14 Let $y = \sqrt{u}$, where $u = \csc x$. Then $\dfrac{dy}{dx} =$

$\dfrac{1}{2\sqrt{u}} \cdot (-\csc x \cot x) = \dfrac{-\csc x \cot x}{2\sqrt{\csc x}}$

$= -\dfrac{1}{2} \cot x \sqrt{\csc x}$.

16 $\dfrac{dy}{dx} = \dfrac{1}{3}(x^3 + 8)^{-2/3} \cdot 3x^2 = x^2(x^3 + 8)^{-2/3}$

18 $\dfrac{dy}{dx} = 5 \sin^4(3x + 2) \cdot \cos(3x + 2) \cdot 3$

$= 15 \sin^4(3x + 2) \cos(3x + 2)$

20 $\dfrac{dy}{dx} = \sqrt{1 + x^2} \cdot 3 \sec^2 5x (\sec 5x \tan 5x) 5 +$

$(\sec^3 5x) \dfrac{1}{2\sqrt{1 + x^2}} (2x)$

$= 15\sqrt{1 + x^2} \sec^3 5x \tan 5x + \dfrac{x \sec^3 5x}{\sqrt{1 + x^2}}$

22 $[x^5(x^2 + 2)^3 \cos^3 5x]'$

$= [(x^5)'(x^2 + 2)^3 \cos^3 5x + x^5 [(x^2 + 2)^3]' \cos^3 5x$

$+ x^5(x^2 + 2)^3[(\cos 5x)^3]'$

3.6 The Derivative of a Composite Function

$= 5x^4(x^2 + 2)^3 \cos^3 5x + x^5 \cdot 3(x^2 + 2)^2 \cdot 2x \cdot \cos^3 5x$
$\quad + x^5(x^2 + 2)^3 \cdot 3(\cos 5x)^2(-\sin 5x) \cdot 5$

$= x^4(x^2 + 2)^2 \cos^2 5x \, [5(x^2 + 2) \cos 5x +$
$\quad 6x^2 \cos 5x - 15x(x^2 + 2) \sin 5x]$

$= x^4(x^2 + 2)^2 \cos^2 5x \, [(11x^2 + 10) \cos 5x -$
$\quad (15x^3 + 30x) \sin 5x]$

24 $(x^3 \cos^2 3x \sin^2 2x)'$

$= 3x^2 \cos^2 3x \sin^2 2x +$
$\quad x^3 \cdot 2 \cos 3x \cdot (-\sin 3x) \cdot 3 \cdot \sin^2 2x +$
$\quad x^3 \cos^2 3x \cdot 2 \sin 2x \cdot \cos 2x \cdot 2$

$= x^2 \cos 3x \sin 2x \, [3 \cos 3x \sin 2x -$
$\quad 6x \sin 3x \sin 2x + 4x \cos 3x \cos 2x]$

26 $[(3x - 5)^{-7}]' = -7(3x - 5)^{-8} \cdot 3 = -\dfrac{21}{(3x - 5)^8}$

28 $\left[\dfrac{(x^3 - 1)^5 \cot^3 x}{x}\right]' =$

$\dfrac{x[(x^3 - 1)^5 \cot^3 x]' - (x^3 - 1)^5 \cot^3 x \, [x]'}{x^2} =$

$\dfrac{x[(x^3 - 1)^5 3 \cot^2 x (-\csc^2 x) + (\cot^3 x) 5(x^3 - 1)^4 \cdot 3x^2] - (x^3 - 1)^5 \cot^3 x}{x^2}$

$= \dfrac{(x^3 - 1)^4 \cot^2 x \, [x[-3(x^3 - 1) \csc^2 x + 15x^2 \cot x] - (x^3 - 1) \cot x]}{x^2}$

$= \dfrac{(x^3 - 1)^4 \cot^2 x}{x^2}[(-3x^4 + 3x) \csc^2 x + (14x^3 + 1) \cot x]$

30 $\left[\left(\dfrac{1 + 2x}{1 + 3x}\right)^4\right]' =$

$4\left(\dfrac{1 + 2x}{1 + 3x}\right)^3 \cdot \dfrac{(1 + 3x) \cdot 2 - (1 + 2x) \cdot 3}{(1 + 3x)^2} =$

$4\left(\dfrac{1 + 2x}{1 + 3x}\right)^3 \cdot \dfrac{2 + 6x - 3 - 6x}{(1 + 3x)^2} = \dfrac{-4(1 + 2x)^3}{(1 + 3x)^5}$

32 $\dfrac{d}{dx}(\tan^2 \sqrt{x}) = 2 \tan \sqrt{x} \, (\sec^2 \sqrt{x}) \dfrac{1}{2\sqrt{x}}$

$= \dfrac{\tan \sqrt{x} \sec^2 \sqrt{x}}{\sqrt{x}}$

34 $\dfrac{d}{dx}((1 - x^2)^{-1/2}) = -\dfrac{1}{2}(1 - x^2)^{-3/2}(-2x)$

$= \dfrac{x}{(1 - x^2)^{3/2}}$

36 $\dfrac{d}{dx}\left(\dfrac{(2x + 1)^3}{(3x + 1)^4}\right)$

$= \dfrac{(3x + 1)^4 \cdot 3(2x + 1)^2 \cdot 2 - (2x + 1)^3 \cdot 4(3x + 1)^3 \cdot 3}{(3x + 1)^8}$

$= -\dfrac{6(2x + 1)^2(x + 1)}{(3x + 1)^5}$

38 $(\sec 2x \tan 5x)' = \left[\sqrt[3]{(2x + 1)^2 + x}\right]'$

$= \dfrac{1}{3}((2x + 1)^2 + x)^{-2/3}(8x + 5)$

40 $\sec 2x \cdot \sec^2 5x \cdot 5 + \sec 2x \tan 2x \cdot 2 \cdot \tan 5x$

$= 5 \sec 2x \sec^2 5x + 2 \sec 2x \tan 2x \tan 5x$

42 $\dfrac{d}{dx}\left(-\dfrac{1}{3} \cos 3x + \dfrac{1}{9} \cos^3 3x\right)$

$= -\dfrac{1}{3}(-\sin 3x) \cdot 3 + \dfrac{1}{9} \cdot 3 \, (\cos^2 3x) \cdot (-\sin 3x) \cdot 3$

$= \sin 3x - \sin 3x \cos^2 3x = \sin^3 3x$

44 $\dfrac{d}{dx}\left[\dfrac{1}{20}(5 + 2x)^5 - \dfrac{5}{16}(5 + 2x)^4\right]$

$= \dfrac{1}{20} \cdot 5(5 + 2x)^4 \cdot 2 - \dfrac{5}{16} \cdot 4(5 + 2x)^3 \cdot 2$

$= \dfrac{1}{2}(5 + 2x)^4 - \dfrac{5}{2}(5 + 2x)^3$

$= \dfrac{1}{2}(5 + 2x)^3[5 + 2x - 5] = x(5 + 2x)^3$

46 $\sin^3 x + C$

48 $-\frac{1}{4}\cos^4 x + C$

50 (a)

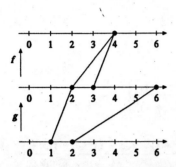

(b) In order to compute $h'(x) = [f(g(x))]' = f'(g(x)) \cdot g'(x)$, we need to know $g(x)$, $g'(x)$, and $f'(g(x))$. Since we know all these values only at $x = 1$, we can find only $h'(1)$. Thus, $h'(1) = f'(g(1)) \cdot g'(1) = f'(2) \cdot g'(1) = 5 \cdot 3 = 15$.

52 Let $h(x) = g(u)$, where $u = x^2$. Then $\frac{dh}{dx} =$

$\frac{dh}{du} \cdot \frac{du}{dx} = \frac{dg}{du} \cdot \frac{du}{dx}$. We know $\frac{dg}{du} = \frac{1}{u^3 + 1}$, so

$\frac{dh}{dx} = \frac{1}{u^3 + 1} \cdot 2x = \frac{2x}{x^6 + 1}$.

54 Let $V(r)$ denote the volume at any radius r. Then

$\frac{dV}{dt} = \frac{dV}{dr} \cdot \frac{dr}{dt}$. Now, $\frac{dV}{dt} = 1 \frac{\text{ft}^3}{\text{sec}}$ and, when r

$= 3$ ft, $\frac{dV}{dr} = \frac{d}{dr}\left(\frac{4}{3}\pi r^3\right) = 4\pi r^2 = 36\pi$ ft^2.

Solving for dr/dt in the equation $1 \frac{\text{ft}^3}{\text{sec}} =$

36π ft$^2 \frac{dr}{dt}$ yields $\frac{dr}{dt} = \frac{1}{36\pi} \frac{\text{ft}}{\text{sec}}$

≈ 0.008842 ft/sec.

3.S Review Exercises

2 $(\sqrt{3x})' = \lim_{h \to 0} \frac{\sqrt{3(x+h)} - \sqrt{3x}}{h} =$

$\lim_{h \to 0} \frac{3}{\sqrt{3(x+h)} + \sqrt{3x}} = \frac{3}{2\sqrt{3x}} = \frac{\sqrt{3}}{2\sqrt{x}}$

4 $[(2x + 1)^2]' = \lim_{h \to 0} \frac{(2(x+h) + 1)^2 - (2x + 1)^2}{h}$

$= \lim_{h \to 0} (8x + 4h + 4) = 8x + 4$

6 $(\sin 5x)' = \lim_{h \to 0} \frac{\sin 5(x+h) - \sin 5x}{h} =$

$(-5 \sin 5x) \lim_{h \to 0} \frac{1 - \cos 5h}{5h} + (5 \cos 5x) \lim_{h \to 0} \frac{\sin 5h}{5h}$

$= 5 \cos 5x$

8 $(t^4 - 5t^2 + 2)' = 4t^3 - 10t$

10 $\left[\frac{(3x + 1)^4}{(2x - 1)^2}\right]'$

$= \frac{(2x - 1)^2 \cdot 4(3x + 1)^3 \cdot 3 - (3x + 1)^4 \cdot 2(2x - 1) \cdot 2}{(2x - 1)^4}$

$= \frac{4(3x - 4)(3x + 1)^3}{(2x - 1)^3}$

12 $\left(\sqrt{5x^2 - x}\right)' = \frac{10x - 1}{2\sqrt{5x^2 - x}}$

14 $[(t^2 + 1)^{3/4}]' = \frac{3}{4}(t^2 + 1)^{-1/4} \cdot 2t = \frac{3}{2}t(t^2 + 1)^{-1/4}$

16 $(\cos^3 7x)' = 3\cos^2 7x \cdot (-\sin 7x) \cdot 7$

$= -21 \cos^2 7x \sin 7x$

3.S Summary: Review Exercises

18 $\left[\dfrac{(3x-2)^{-5}}{11}\right]' = \dfrac{1}{11}(-5)(3x-2)^{-6}\cdot 3$

$= -\dfrac{15}{11}(3x-2)^{-6}$

20 $(x^2 \cos 4x)' = x^2(-\sin 4x)\cdot 4 + 2x \cos 4x$

$= 2x \cos 4x - 4x^2 \sin 4x$

22 $\left[\left(\dfrac{\sin 2x}{1+\tan 3x}\right)^3\right]' =$

$3\left(\dfrac{\sin 2x}{1+\tan 3x}\right)^2 \dfrac{(1+\tan 3x)\cdot 2\cos 2x - \sin 2x \cdot 3\sec^2 3x}{(1+\tan 3x)^2}$

$= \dfrac{3\sin^2 2x}{(1+\tan 3x)^4}[2\cos 2x(1+\tan 3x) - 3\sin 2x \sec^2 3x]$

24 $\left[\dfrac{x^2 \sin 5x}{(2x+1)^3}\right]' =$

$\dfrac{(2x+1)^3[x^2 \cos 5x \cdot 5 + 2x \sin 5x] - x^2 \sin 5x \cdot 3(2x+1)^2 \cdot 2}{(2x+1)^6}$

$= (2x+1)^{-4}[5x^2(2x+1)\cos 5x + 2x(1-x)\sin 5x]$

26 $\dfrac{d}{dx}(\sqrt[4]{x^3 + \sqrt{x+\sin x}}) = \dfrac{d}{dx}[(x^3 + \sqrt{x+\sin x})^{1/4}]$

$= \dfrac{1}{4}(x^3 + \sqrt{x+\sin x})^{-3/4}\left[\dfrac{d}{dx}(x^3) + \dfrac{d}{dx}(\sqrt{x+\sin x})\right]$

$= \dfrac{1}{4(x^3+\sqrt{x+\sin x})^{3/4}}\left[3x^2 + \dfrac{1+\cos x}{2\sqrt{x+\sin x}}\right]$

$= \dfrac{6x^2\sqrt{x+\sin x} + 1 + \cos x}{8\sqrt{x+\sin x}(x^3+\sqrt{x+\sin x})^{3/4}}$

28 $[\sqrt[5]{\csc^{11} 3x}]' = \dfrac{11}{5}(\csc 3x)^{6/5}(-\csc 3x \cot 3x)\cdot 3$

$= -\dfrac{33}{5}(\csc 3x)^{11/5}\cot 3x$

30 $(\sqrt{5x-1})' = \dfrac{5}{2\sqrt{5x-1}}$

32 $\left(\dfrac{x^3+1}{x^2}\right)' = 1 - \dfrac{2}{x^3}$

34 $\left[\left(\dfrac{3x+1}{2x+1}\right)^4\right]'$

$= 4\left(\dfrac{3x+1}{2x+1}\right)^3 \cdot \dfrac{(2x+1)\cdot 3 - (3x+1)\cdot 2}{(2x+1)^2}$

$= \dfrac{4(3x+1)^3}{(2x+1)^5}$

36 $\left[\dfrac{\sin 4x}{x\sqrt{2x+1}}\right]'$

$= \dfrac{x\sqrt{2x+1}(\sin 4x)' - \sin 4x (x\sqrt{2x+1})'}{(x\sqrt{2x+1})^2} =$

$\dfrac{x\sqrt{2x+1}\, 4\cos 4x - (\sin 4x)(x(2x+1)^{-1/2} + \sqrt{2x+1})}{x^2(2x+1)}$

$= \dfrac{(8x^2+4x)\cos 4x - (3x+1)\sin 4x}{x^2(2x+1)^{3/2}}$

38 $\left(\dfrac{x^2}{4} - \dfrac{x\sin 2ax}{4a} - \dfrac{\cos 2ax}{8a^2}\right)' =$

$\dfrac{x}{2} - \dfrac{1}{4a}(x\cdot 2a\cos 2ax + \sin 2ax) - \dfrac{1}{8a^2}(-2a\sin 2ax)$

$= \dfrac{x}{2}(1 - \cos 2ax) = x \sin^2 ax$

40 $\left(-\dfrac{1}{a}\cos ax + \dfrac{1}{3a}\cos^3 ax\right)'$

$= -\dfrac{1}{a}(-a\sin ax) + \dfrac{1}{3a}\cdot 3\cos^2 ax\cdot(-\sin ax)\cdot a$

$= (\sin ax)(1 - \cos^2 ax) = \sin^3 ax$

42 (a) Rate of seepage
    (b) Seepage rate approaches zero.
44 Slope = $8x^3 - 12x$, which equals 40 for $x = 2$. Hence the equation of the tangent line is $y - 16 = 40(x - 2)$; that is, $y = 40x - 64$.
46 Density at x is $\dfrac{1}{2\sqrt{x}}$.
    (a) 1
    (b) 1/2
    (c) No
48 (a) Other lines, such as the y axis, also meet the graph just once.
    (b) The tangent line through (1, 1) also meets the curve at $(-2, -8)$.
    (c) The tangent line is the line with the same slope as the curve at the given point.
50 Rate = $3t^2$ g/hr
    (a) 0 g/hr
    (b) 3 g/hr
    (c) 12 g/hr
52 (a) 16.4
    (b) Average rate of profit between 2 and 2.1 years after the start.
    (c) Slope of the line through $(2, 4 \cdot 2^2)$ and $(2.1, 4 \cdot 2.1^2)$.
    (d) Average velocity between 2 and 2.1 seconds after the start.
54 (a) Rate at which the car's value increases.
    (b) When the value is decreasing. When the value is increasing. Derivative is usually negative.
56 (a) $4^2 - 3^2 = 7$ grams
    (b) $3.01^2 - 3^2 = 0.0601$ grams
    (c) $t^2 - 3^2 = t^2 - 9$ grams
    (d) Average = $\dfrac{t^2 - 9}{t - 3} = t + 3$ g/hr

58 (a) Rate at which cloud top rises.
    (b) Decreases to 0.
    (c) Appears to approach approximately 12.5.
    (d) Drawing the tangent line at (0, 0) shows that the cloud is rising at about 6 mi/min.
    (e) Drawing the tangent line at (1, 5) shows that the cloud is rising at about 4 mi/min.
60 (a) Yes. Both equal $2x + 3x^2$.
    (b) No. $5x^4 \neq 2x \cdot 3x^2$
    (c) Yes. Both equal $2x - 3x^2$.
    (d) No. $-\dfrac{1}{x^2} \neq \dfrac{2x}{3x^2}$.
62 (a) $\dfrac{2xz}{y}$
    (b) $-\dfrac{x^2 z}{y^2}$
    (c) $\dfrac{x^2}{y}$
    (d) $-3 \sin 3x \sin 4y$
    (e) $4 \cos 3x \cos 4y$
64 (a) $f'(0) = \lim\limits_{\Delta x \to 0} \dfrac{f(\Delta x) - f(0)}{\Delta x} =$
    $\lim\limits_{\Delta x \to 0} \left(\Delta x \sin \dfrac{1}{\Delta x} \right) = 0$, since $\Delta x \to 0$ and $\left| \sin \dfrac{1}{\Delta x} \right| \leq 1$.
    (b) $x \neq 0$, so $f'(x) =$
    $x^2 \left(\cos \dfrac{1}{x} \right)\left(-\dfrac{1}{x^2} \right) + 2x \sin \dfrac{1}{x}$
    $= 2x \sin \dfrac{1}{x} - \cos \dfrac{1}{x}$.
    (c) For integers n, $f'\left(\dfrac{1}{2\pi n} \right) =$

3.S Summary: Review Exercises

$$2 \cdot \frac{1}{2\pi n} \sin 2\pi n - \cos 2\pi n = -1, \text{ which}$$

doesn't approach $f'(0) = 0$ as $n \to \infty$.

66 $f'(2)$, where $f(w) = (1 + w^2)^3$. $f'(w) = 3(1 + w^2)^2 \cdot 2w$, so $f'(2) = 300$.

68 $f(x) = 5$ for all x. *Proof*: If not, then $f(x) \geq 6$ or $f(x) \leq 4$ for some x. By the intermediate-value theorem, $f(c) = 5.5$ or $f(c) = 4.5$ for some c between 0 and x, contradicting the fact that $f(c)$ is an integer.

70 (a) $\dfrac{(fg)'}{fg} = \dfrac{f'g + fg'}{fg} = \dfrac{f'}{f} + \dfrac{g'}{g}$

(b) $\dfrac{(f/g)'}{f/g} = \dfrac{\frac{gf' - fg'}{g^2}}{f/g} = \dfrac{gf' - fg'}{fg}$

$= \dfrac{f'}{f} - \dfrac{g'}{g}$

(c) $\dfrac{(fgh)'}{fgh} = \dfrac{f'}{f} + \dfrac{g'}{g} + \dfrac{h'}{h}$

72 (a) $f'(1) = 1$

(b) $f'(2) = 1/2$

(c) $f'(1/2) = 2$, $f'(3) = 1/3$, $f'(4) = 1/4$

Filling in the table:

x	1/2	1	2	3	4
$f'(x)$	2	1	1/2	1/3	1/4

(d) $f'(x) = 1/x$

# 4 Applications of the Derivative

4.1 Three Theorems About the Derivative

2 (a)

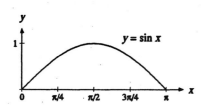

(b) Yes, at $x = \pi/2$.

(c) No. Minima occur at $x = 0$ and $x = \pi$, where $f'(x) = \pm 1$.

4 (a)

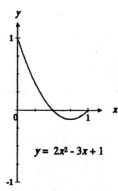

(b) The maximum occurs at (0, 1).

(d) $\dfrac{dy}{dx} = 4x - 3 = 0$ when $x = 3/4$.

(e) The minimum occurs at the point $(3/4, -1/8)$.

6 (a)

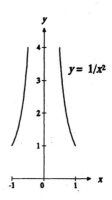

(b) $f(-1) = 1 = f(1)$

(c) No. $f'(x) = -2/x^3$ is never 0.

(d) $f(0)$ does not exist.

8 $f(x) = x^3 - x$ is continuous and differentiable on $[-1, 1]$ and $f(-1) = 0 = f(1)$. Thus $f'(c) = 3c^2 - 1 = 0$ when $c = \pm 1/\sqrt{3}$.

10 $f(0) = 1 = f(4\pi)$, while $f'(x) = \cos x - \sin x = 0$ when $\cos x = \sin x$, or $\tan x = 1$. Hence $c = \pi/4$, $5\pi/4$, $9\pi/4$, and $13\pi/4$.

12 Let $f(x) = 2x^2 + x + 1$. Then $f(-2) = 7$ and $f(3) = 22$. By the mean-value theorem, $f'(c) = \dfrac{f(3) - f(-2)}{3 - (-2)} = \dfrac{22 - 7}{5} = 3$. Hence $4c + 1 = 3$ and $c = 1/2$.

14 Let $f(x) = 5x - 7$. Then $f(0) = -7$ and $f(4) = 13$. By the mean-value theorem, $f'(c) = \dfrac{f(4) - f(0)}{4 - 0} = \dfrac{13 - (-7)}{4} = 5$. But $f'(x) = 5$, so any value of c in $(0, 4)$ will serve.

16 (a),(b),(c)

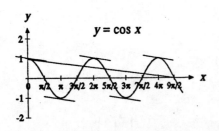

(d) From the figure, we see that there are five points in $[0, 9\pi/2]$ where $f'(c) = -\sin c = \dfrac{f(9\pi/2) - f(0)}{9\pi/2 - 0} = -\dfrac{2}{9\pi}$.

(e) We want those points in $[0, 9\pi/2]$ where $\sin c = 2/(9\pi)$. With the help of the graph (and a calculator) we find that $c \approx 0.0708, 3.0708, 6.3540, 9.3540,$ or 12.6372.

18 (a) $(\csc^2 x)' = 2\csc x\,(-\csc x \cot x)$
$ = -2\csc^2 x \cot x$
$ (\cot^2 x)' = 2\cot x\,(-\csc^2 x)$
$ = -2\csc^2 x \cot x$

(b) $C = \csc^2 x - \cot^2 x = \dfrac{1}{\sin^2 x} - \dfrac{\cos^2 x}{\sin^2 x} = \dfrac{1 - \cos^2 x}{\sin^2 x} = 1$

20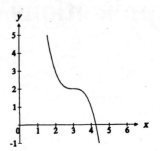

22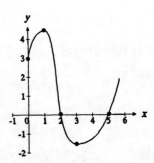

24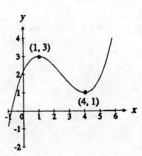

26 If $f'(x) \geq 0$ for all x, then f is a function that never decreases. But $f(2) = 5$ and $f(3) = -1$, a contradiction.

28 If $f(x) = 2$ at $x = 1$ and 3, then Rolle's theorem implies that $f'(x)$ must be 0 somewhere between 1 and 3. But we are told that $f'(x)$ is never 0 between 1 and 3, a contradiction.

30 (a) $(5x - 4\cos x)' = 5 + 4\sin x$, which is always positive, so $5x - 4\cos x$ is always increasing. Hence $5x - 4\cos x = 0$ has at most one solution. At $x = 0$, $5x - 4\cos x = -4 < 0$, while at $x = 1$, $5x - 4\cos x =$

$5 - 4\cos 1 > 0$, so the intermediate-value theorem implies that $5x - 4\cos x$ must equal 0 somewhere between 0 and 1.

(b) See (a).

(c) The solution is about 0.64113.

32 (a) If $f'(x) > 0$ for all x, then the function is always increasing. The equation $f(x) = 3$ can have at most one solution.

(b) If $f'(x) > 0$ for $x < 7$ and $f'(x) < 0$ for $x > 7$, then $f(x) = 3$ can have at most one solution less than 7 and at most one solution greater than 7.

34

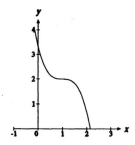

36

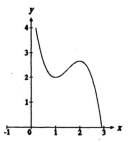

38 $f(x) = x^3 + ax^2 + c$, so $f'(x) = 3x^2 + 2ax$. Since $f(0) = c > 0$ and $\lim_{x \to -\infty} f(x) = -\infty$, we know there is at least one negative root. Since $a < 0$, we have $f'(x) > 0$ for $x < 0$, so $f(x)$ is an increasing function on $(-\infty, 0)$. The negative root is unique.

40 (a) The unique root of $a_0 + a_1 x = 0$ is $x = -a_0/a_1$, which always exists if $a_1 \neq 0$.

(b) The roots of $a_0 + a_1 x + a_2 x^2$ are $x_1 = \dfrac{-a_1 - \sqrt{a_1^2 - 4a_0 a_2}}{2a_2}$ and $x_2 = \dfrac{-a_1 + \sqrt{a_1^2 - 4a_0 a_2}}{2a_2}$, which are distinct provided that $\sqrt{a_1^2 - 4a_0 a_2} \neq 0$.

(c) Suppose there are *three* distinct roots: $x_1 < x_2 < x_3$. Then $f(x_1) = f(x_2) = f(x_3) = 0$, so Rolle's theorem implies that there exist numbers c_1 and c_2 such that $x_1 < c_1 < x_2 < c_2 < x_3$ and $f'(c_1) = f'(c_2) = 0$. But $f'(x) = a_1 + 2a_2 x$, which is a linear equation possessing only a single root, whereas we appear to have found two: c_1 and c_2. Hence $f(x)$ cannot have three distinct roots.

(d) Since $f(x) = a_0 + a_1 x + a_2 x^2 + a_3 x^3$ is a polynomial of degree 3, $f'(x)$ is a polynomial of degree 2 and possesses at most two distinct roots. Rolle's theorem implies the existence of a root of $f'(x)$ between each pair of consecutive roots of $f(x)$, so if $f(x)$ had more than three distinct roots, it would contradict the known number of possible roots of $f'(x)$.

(f) We know from (a), (c), and (d) that polynomials of degrees 1, 2, and 3 have at most 1, 2, or 3 distinct roots, respectively. Assume that a polynomial of degree n possesses at most n distinct roots. If $f(x)$ has degree $n + 1$, then $f'(x)$ has degree n and, by assumption, has at most n distinct roots. Rolle's theorem implies the existence of a root of $f'(x)$ between each pair of consecutive roots of $f(x)$, so if $f(x)$ had more than $n + 1$ distinct roots, it would contradict the known number of possible roots of $f'(x)$. Hence the number of distinct roots a polynomial can have is no greater than its degree.

42 If the mass of the left x units of string is $f(x)$, then the average density of the string between $x = a$ and $x = b$ is $\frac{f(b) - f(a)}{b - a}$. The mean-value theorem states that at some point c between a and b, the density at c equals the average density over the interval $[a, b]$; that is, that $f'(c) = \frac{f(b) - f(a)}{b - a}$.

44 (a) Corollary 2. The fixed distance is some constant C.
   (b) Corollary 1. The derivative is always 0 because all tangents are horizontal.

46 Yes. The derivative could occasionally be zero. (For example, $f(x) = x^3$ is increasing, but $f'(x) = 3x^2$ is 0 for $x = 0$.)

48 (a) $2x^4 + C$
   (b) $-\frac{1}{2} \cos 2x + C$
   (c) $-\frac{1}{x} + C$
   (d) $\frac{3}{4} x^{4/3} + C$

50 (a) Since $y = g(x)$ is the equation of a line passing through $(a, f(a))$ and $(b, f(b))$, the slope is $\frac{f(b) - f(a)}{b - a}$ and the point-slope formula yields $g(x) - f(a) = \frac{f(b) - f(a)}{b - a}(x - a)$, the desired result.

   (b) The coefficient of x in the formula $g(x) = f(a) + \frac{f(b) - f(a)}{b - a} x - \frac{f(b) - f(a)}{b - a} a$ is $\frac{f(b) - f(a)}{b - a}$, so that is $g'(x)$.

4.2 The First Derivative and Graphing

2 Minimum at 0.

4 Minimum at 1.

6 $f'(x) = 6x(x + 1)$ changes from $+$ to $-$ at -1 and from $-$ to $+$ at 0. Maximum at -1; minimum at 0.

8 $f'(x) = -x \sin x$. Critical points are $x = n\pi$ for integers n. For n positive and even or negative and odd, $n\pi$ is a maximum. For n positive and odd or negative and even, $n\pi$ is a minimum. 0 is neither a maximum nor a minimum.

10 If $(1, 2)$ is a critical point, $f'(1) = 0$. In addition, $f'(x) < 0$ for all x (except $x = 1$) implies that the sign of $f'(x)$ doesn't change at $(1, 2)$. Thus, $(1, 2)$ is neither a maximum nor a minimum.

12 Given $\lim_{x \to -\infty} f(x) = 4$, we know that $y = 4$ is a horizontal asymptote. Since $4 > 3 > 2$, $f(x)$ is decreasing over $(-\infty, 1)$. $f(x)$ must be increasing thereafter because $\lim_{x \to \infty} f(x) = \infty$.

4.2 The First Derivative and Graphing

14 $f(x)$ is decreasing on $(-\infty, 1)$, since $f(x) \to \infty$ as $x \to -\infty$. $f(x)$ is increasing on $(1, 4)$, since there is

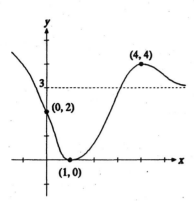

no critical point in $(1, 4)$. $f(x)$ is decreasing on $(4, \infty)$, since $f(x) \to 3$ as $x \to \infty$. Given these facts, $x = 1$ is a minimum and $x = 4$ is a maximum. Also the y intercept is 2, while the x intercept coincides with the critical point.

16 $f'(x)$ never equals zero, so $f(x)$ is either always increasing or always decreasing. Since $f(x) \to 0$ as

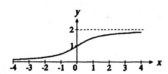

$x \to -\infty$ while $f(x) \to 2$ as $x \to \infty$, we know $f(x)$ is always increasing and passes through $(0, 1)$.

18

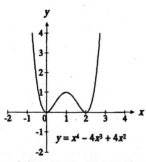

There are minima at $(0, 0)$ and $(2, 0)$; a maximum occurs at $(1, 1)$.

20

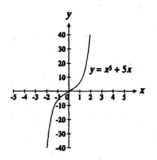

22

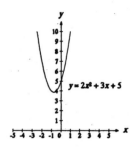

There is a global minimum at $(-3/4, 31/8)$.

24

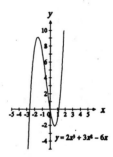

There is a local minimum at
$\left(\frac{1}{2}(\sqrt{5} - 1), -\frac{1}{2}(5\sqrt{5} - 7)\right)$ and a local maximum at
$\left(-\frac{1}{2}(\sqrt{5} + 1), \frac{1}{2}(5\sqrt{5} + 7)\right)$.

26

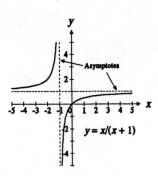

28

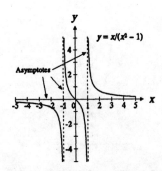

30

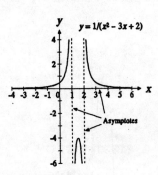

There is a local maximum at (3/2, −4).

32

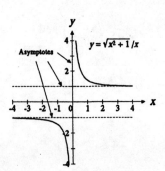

34 $f'(x) = 4 - 2x$ changes from + to − at $x = 2$.

Maximum: $f(2) = 4$; the minimum is the lesser of $f(0)$ and $f(5)$, namely, $f(5) = -5$.

36 $f'(x) = 4x - 5 < 0$ for x in $[-1, 1]$. Maximum: $f(-1) = 7$; minimum: $f(1) = -3$.

38 $f'(x) = \dfrac{1 - x^2}{(x^2 + 1)^2}$ changes from + to − at $x = 1$.

Maximum: $f(1) = 1/2$; the minimum is the lesser of $f(0)$ and $f(3)$, namely, $f(0) = 0$.

40 $f'(x) = \dfrac{1 - x}{(x^2 + 1)^{3/2}}$ changes from + to − at $x = 1$. Maximum: $f(1) = \sqrt{2}$; the minimum is the lesser of $f(0)$ and $f(3)$, namely, $f(0) = 1$.

42 $f'(x) = \cos x + \sin x = 0$ when $x = \dfrac{3\pi}{4}$. This is a maximum value on $[0, \pi]$ since $f'(x) > 0$ for $x < \dfrac{3\pi}{4}$ and $f'(x) < 0$ for $x > \dfrac{3\pi}{4}$. Now, $f(0) = -1$ and $f(\pi) = 1$. Thus, there is a maximum of $\sqrt{2}$ at $x = \dfrac{3\pi}{4}$ and a minimum of -1 at $x = 0$.

44

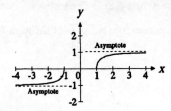

46

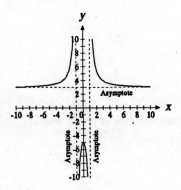

48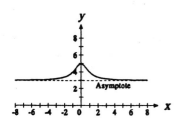

50 (a) Since the degree of g is odd, g has a real root, at which f/g has a vertical asymptote.

(b) If $f(x) = a_n x^n + \cdots + a_0$ and $g(x) = b_n x^n + \cdots + b_0$ with $a_n \neq 0$ and $b_n \neq 0$, then $\lim_{x \to \infty} \frac{f(x)}{g(x)} = \frac{a_n}{b_n}$, so f/g has $y = \frac{a_n}{b_n}$ as a horizontal asymptote.

(c) Since the degree of f is less than the degree of g, $\lim_{x \to \infty} \frac{f(x)}{g(x)} = 0$, so f/g has the x axis as an asymptote.

52 (a) $f'(x) > 0$ for $x > 1$, $f'(x) < 0$ for $x < 1$ implies $f'(x) = 0$ at $x = 1$ and $f(x)$ takes on a global minimum at $x = 1$. Thus the minimum value is 1 on [0, 2].

(b) We know $f(x)$ does not have a maximum in (0, 2). Since $f(0) = 3 > 2 = f(2)$, the maximum value of $f(x)$ is 3 on [0, 2].

4.3 Motion and the Second Derivative

2 $\frac{dy}{dx} = 5$, $\frac{d^2y}{dx^2} = 0$

4 $\frac{dy}{dx} = 6x^5$, $\frac{d^2y}{dx^2} = 30x^4$

6 $\frac{dy}{dx} = 12x^2 - 2x + 1$, $\frac{d^2y}{dx^2} = 24x - 2$

8 $\frac{dy}{dx} = \frac{(x-1) \cdot 2x - x^2 \cdot 1}{(x-1)^2} = \frac{x^2 - 2x}{(x-1)^2}$,

$\frac{d^2y}{dx^2} = \frac{(x-1)^2 \cdot (2x-2) - (x^2-2x) \cdot 2(x-1)}{(x-1)^4}$

$= \frac{2(x-1)^3 - 2x(x-2)(x-1)}{(x-1)^4} = \frac{2}{(x-1)^3}$

10 $\frac{dy}{dx} = \sec x \tan x$, $\frac{d^2y}{dx^2} = (\sec x)(\sec^2 x + \tan^2 x)$

12 $\frac{dy}{dx} = \frac{\tan x - x \cdot \sec^2 x}{\tan^2 x} = \cot x - x \csc^2 x$,

$\frac{d^2y}{dx^2} = -2 \csc^2 x + 2x \csc^2 x \cot x$

$= 2 \csc^2 x \, (x \cot x - 1)$.

14 $\frac{dy}{dx} = 3(x+1)^2$, $\frac{d^2y}{dx^2} = 6(x+1)$

16 $\frac{dy}{dx} = 2x \sec^2 x^2$, $\frac{d^2y}{dx^2} =$

$2 \sec^2 x^2 + 4x \cdot \sec^2 x^2 \tan x^2 \cdot 2x$

$= 2 \sec^2 x^2 \left[4x^2 \tan x^2 + 1 \right]$

18 $G'(t) > 0$, but $G''(t) < 0$.

20 $y = -16t^2 + 96$. $y = 0$ when $t = \sqrt{6}$, so the ball falls for $\sqrt{6}$ seconds.

22 (a) In this case, $v_0 = 0$, $a = -32$ ft/s^2, and $y_0 = 555 + 55/120 \approx 555.4583$ ft. Thus, $y \approx -16t^2 + 555.4583$. When $y = 0$, $16t^2 \approx 555.4583$ and $t \approx \pm 5.89$ sec. The negative root makes no sense, so $t \approx 5.89$ sec.

(b) In this case, $v_0 = 0$, $a = -32$ ft/sec^2, and $y_0 = 984.25$ ft. So $y = -16t^2 + 984.25$. When $y = 0$, $16t^2 = 984.25$ and $t \approx \pm 7.84$ sec. It takes about 7.84 sec.

24. Since $(y' - at)' = y'' - a = 0$, $y' - at$ is a constant. At $t = 0$, $y' - at = v_0$, so $y' = at + v_0$ for all t. Since $\left(y - \frac{a}{2}t^2 - v_0 t\right)' = y' - at - v_0 = 0$, $y - \frac{a}{2}t^2 - v_0 t$ is a constant. At $t = 0$, it equals y_0, so $y = \frac{a}{2}t^2 + v_0 t + y_0$ for all t.

26. (a) By Exercise 24, $y = -4t^2 + 8t + 10$ feet after t seconds.

    (b) $y' = -8t + 8$ is zero at $t = 1$. Maximum is $-4 \cdot 1^2 + 8 \cdot 1 + 10 = 14$ ft.

28. (a) $y = -16t^2 + 6t - 4$ feet after t seconds.

    (b) $y' = -32t + 6$ is zero at $t = 3/16$. Maximum is $-16\left(\frac{3}{16}\right)^2 + 6\left(\frac{3}{16}\right) - 4 = -\frac{55}{16}$ ft.

30. If the initial velocity is v_0 then, by Exercise 24, $y = -16t^2 + v_0 t$ and $y' = -32t + v_0$. Maximum height is reached when $t = \frac{v_0}{32}$. The ball hits the ground when $t = \frac{v_0}{16}$. So it takes as long to fall as to rise. The velocity when it hits is $-32\left(\frac{v_0}{16}\right) + v_0 = -v_0$, the negative of its initial velocity. The speeds are equal.

4.4 Related Rates

2. Since $x \cdot \frac{dx}{dt} = s \cdot \frac{ds}{dt}$ and $\frac{dx}{dt} = 5$, we have $\frac{ds}{dt} = \frac{x}{s} \cdot 5 = \frac{5x}{\sqrt{x^2 + 30^2}}$.

    (a) $\frac{5 \cdot 1}{\sqrt{1 + 900}} = \frac{5}{\sqrt{901}}$ ft/sec

    (b) $\frac{5 \cdot 100}{\sqrt{10000 + 900}} = \frac{50}{\sqrt{109}}$ ft/sec

4. (a)

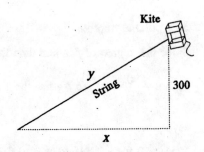

    (b) We first see that $x^2 + 300^2 = y^2$ is an equation involving the variables of interest. Differentiating with respect to time, we get $2x\frac{dx}{dt} = 2y\frac{dy}{dt}$ or $\frac{dx}{dt} = \frac{y}{x}\frac{dy}{dt}$. We are given $y = 500$ ft and $dy/dt = 20$ ft/sec. Now, $x = \sqrt{y^2 - 300^2} = \sqrt{500^2 - 300^2} = 400$ ft, so $\frac{dx}{dt} = \frac{500}{400} \cdot 20 = 25$ ft/sec, and this is the kite's velocity.

4.4 Related Rates

6

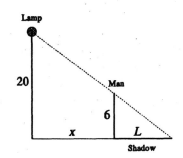

$\dfrac{L}{6} = \dfrac{L+x}{20}$, so $L = \dfrac{3}{7}x$ and $\dfrac{dL}{dt} = \dfrac{3}{7} \cdot \dfrac{dx}{dt}$.

Since $dx/dt = 5$ ft/sec, $\dfrac{dL}{dt} = \dfrac{15}{7}$ ft/sec, no matter how far the man is from the lamp.

8 $V = \dfrac{4}{3}\pi r^3$, so $\dfrac{dV}{dt} = \dfrac{4\pi}{3} \cdot 3r^2 \cdot \dfrac{dr}{dt} = 4\pi r^2 \cdot \dfrac{dr}{dt}$,

and $\dfrac{dr}{dt} = \dfrac{1}{4\pi r^2} \cdot \dfrac{dV}{dt}$.

(a) When $r = 2$, $\dfrac{dr}{dt} = \dfrac{1}{4\pi \cdot 2^2}(-1)$

$= -\dfrac{1}{16\pi}$ in/sec.

(b) When $r = 1$, $\dfrac{dr}{dt} = \dfrac{1}{4\pi \cdot 1^2}(-1)$

$= -\dfrac{1}{4\pi}$ in/sec.

10 $h^2 = x^2 + y^2$, so $h \cdot \dfrac{dh}{dt} = x \cdot \dfrac{dx}{dt} + y \cdot \dfrac{dy}{dt}$. When $x = 3$, $y = 4$, $dx/dt = 5$, and $dy/dt = -6$, $h = 5$ so $\dfrac{dh}{dt} = \dfrac{1}{5}(3 \cdot 5 + 4(-6)) = -\dfrac{9}{5}$. The hypotenuse is decreasing at 9/5 ft/sec.

12 Let l be the length of the rectangle and w be the width, then $l = 12$ ft, $w = 5$ ft, $dl/dt = 7$ ft/sec, and $dw/dt = -3$ ft/sec.

(a) The area is $A = lw$, so $\dfrac{dA}{dt} = \dfrac{d}{dt}(lw)$

$= l \cdot \dfrac{dw}{dt} + w \cdot \dfrac{dl}{dt} = 12(-3) + 5 \cdot 7$

$= -36 + 35 = -1$ ft²/sec.

(b) The perimeter is $P = 2l + 2w$, so $\dfrac{dP}{dt}$

$= \dfrac{d}{dt}(2l + 2w) = 2 \cdot \dfrac{dl}{dt} + 2 \cdot \dfrac{dw}{dt}$

$= 2 \cdot 7 + 2(-3) = 14 - 6 = 8$ ft/sec.

(c) The diagonal has length $D = \sqrt{l^2 + w^2}$, so

$\dfrac{dD}{dt} = \dfrac{d}{dt}\left(\sqrt{l^2 + w^2}\right) = \dfrac{l \cdot \dfrac{dl}{dt} + w \cdot \dfrac{dw}{dt}}{\sqrt{l^2 + w^2}}$

$= \dfrac{12 \cdot 7 + 5(-3)}{\sqrt{5^2 + 12^2}} = \dfrac{69}{13}$ ft/sec.

14 Since $\dfrac{d\theta}{dt} = \dfrac{7}{x^2 + 49}\dot{x}$, $\ddot{\theta}$

$= \dfrac{7}{x^2 + 49}\ddot{x} - \dfrac{14x}{(x^2 + 49)^2}\dot{x}^2$. But $\dot{x} = -10$, so

$\ddot{x} = 0$ and $\ddot{\theta} = -\dfrac{1400x}{(x^2 + 49)^2}$.

(a) When $x = 7$, $\ddot{\theta} = -\dfrac{50}{49}$ radians/min².

(b) When $x = 1$, $\ddot{\theta} = -\dfrac{14}{25}$ radians/min².

16 (a) $\tan\theta = y/x$, so $y = x\tan\theta$.

(b) $\dot{y} = x(\sec^2\theta)\dot{\theta} + (\tan\theta)\dot{x}$

(c) $\ddot{y} = x(\sec^2\theta)\ddot{\theta} + 2(\sec^2\theta)\dot{x}\dot{\theta} + (\tan\theta)\ddot{x}$

$+ 2x(\sec^2 \theta \tan \theta) \dot{\theta}^2$.

[Other answers are possible; for example, in (b)
$\dfrac{d\theta}{dt} = \dfrac{x\dot{y} - y\dot{x}}{x^2 + y^2}$.]

18 Let P equal pressure and x equal altitude. We are given that $dP/dx = -128(0.88)^x$ mb/km and $dx/dt = 5$ km/sec. Therefore, $\dfrac{dP}{dt} = \dfrac{dP}{dx} \dfrac{dx}{dt} = -640(0.88)^x$ mb/sec.

(a) At $x = 1$ km, $dP/dt = -640(0.88) = -563.2$ millibars/sec.

(b) At $x = 50$ km, $dP/dt = -640(0.88)^{50} \approx -1.07$ millibars/sec.

20 $V = \dfrac{4}{3}\pi r^3$ and $A = 4\pi r^2$. If $dV/dt = kA$, then
$4\pi r^2 \cdot \dfrac{dr}{dt} = \dfrac{dV}{dt} = kA = 4\pi r^2 k$, so $dr/dt = k$ is constant.

22 Let the surface of the water have radius r and area A. Let V be the volume of the water and h its

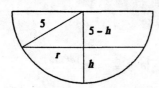

depth. (Compare Example 3.) Then $\dot{V} = A\dot{h} = \pi r^2 \dot{h}$. We have $5^2 = r^2 + (5 - h)^2$, so $r^2 = 10h - h^2$ and $\dot{V} = \pi(10h - h^2)\dot{h}$. Since $\dot{V} = 1$,
$\dot{h} = \dfrac{1}{\pi(10h - h^2)}$.

(a) When $h = 3$, $\dot{h} = \dfrac{1}{21\pi}$ ft/min.

When $h = 4$, $\dot{h} = \dfrac{1}{24\pi}$ ft/min.

When $h = 5$, $\dot{h} = \dfrac{1}{25\pi}$ ft/min.

(b) Since $\dot{h} = \dfrac{1}{\pi(10h - h^2)}$, $\ddot{h} = $
$\dfrac{1}{\pi} \dfrac{-1}{(10h - h^2)^2}(10\dot{h} - 2h\dot{h}) = \dfrac{(2h - 10)\dot{h}}{\pi(10h - h^2)^2}$
$= \dfrac{2h - 10}{\pi^2(10h - h^2)^3}$.

When $h = 3$, $\ddot{h} = -\dfrac{4}{9261\pi^2}$ ft/min².

When $h = 4$, $\ddot{h} = -\dfrac{1}{6912\pi^2}$ ft/min².

When $h = 5$, $\ddot{h} = 0$ ft/min².

24 We have $V = \dfrac{4}{3}\pi r^3$, so $r = \sqrt[3]{\dfrac{3V}{4\pi}}$.

(a) When $V = 0.4$ cc, $r \approx 0.4571$ cm. When $V = 1.6$ cc, $r \approx 0.7256$ cm. Since dr/dt is constant over the 74 hours, we have $dr/dt \approx$
$\dfrac{0.7256 - 0.4571}{74} = \dfrac{0.2685}{74} \approx$
0.003628 cm/hr.

(b) We have $\dfrac{dV}{dt} = 4\pi r^2 \dfrac{dr}{dt}$, so $\dfrac{dV}{dt} =$
$\dfrac{4\pi r^3}{3} \dfrac{3}{r} \dfrac{dr}{dt} = \dfrac{3V}{r} \dfrac{dr}{dt} = 3V \sqrt[3]{\dfrac{4\pi}{3V}} \dfrac{dr}{dt} =$
$\sqrt[3]{36\pi V^2} \dfrac{dr}{dt} \approx 0.01755 V^{2/3}$ cc/hr.

(c) When $V = 1$ cc, $dV/dt \approx 0.01755$ cc/hr.

4.5 The Second Derivative and Graphing

2 $D(D(x^3 - 6x^2 + 1)) = D(3x^2 - 12x) = 6x - 12$
$= 6(x - 2)$, which is greater than 0 when $x > 2$; hence $x^3 - 6x^2 + 1$ is concave up on $(2, \infty)$. Furthermore, $6(x - 2) < 0$ when $x < 2$, so $x^3 - 6x^2 + 1$ is concave down on $(-\infty, 2)$. Since $x^3 - 6x^2 + 1$ changes concavity at $x = 2$, $(2, -15)$ is an inflection point.

4   Let $f(x) = 2x^2 - 5x$. Then $f'(x) = 4x - 5$ and $f''(x) = 4$. Since $f''(x) > 0$ for all x, f is concave up for all x.

6   Let $f(x) = x^5$. Then $f'(x) = 5x^4$ and $f''(x) = 20x^3$. Now, $f''(x) > 0$ for $x > 0$ and $f''(x) < 0$ for $x < 0$. So f is concave up on $(0, \infty)$. On $(-\infty, 0)$, f is concave down. At $x = 0$, f changes concavity, so $(0, 0)$ is an inflection point.

8   Let $f(x) = 3x^5 - 5x^4$. Then $f'(x) = 15x^4 - 20x^3$ and $f''(x) = 60x^3 - 60x^2 = 60x^2(x - 1)$. Now, $f''(x) > 0$ for $x > 1$, so f is concave up on $(1, \infty)$; $f''(x) \leq 0$ for $x < 1$, so f is concave down on $(-\infty, 1)$. Hence f changes concavity at $x = 1$, so f has an inflection point at $(1, -2)$. (Although $f''(0) = 0$, the concavity does not change there, so $(0, 0)$ is not an inflection point.)

10   Let $f(x) = \dfrac{1}{1 + x^4}$. Then $f'(x) = \dfrac{-4x^3}{(1 + x^4)^2}$ and

$f''(x) = \dfrac{(1 + x^4)^2(-12x^2) - (-4x^3) \cdot 2(1 + x^4) \cdot 4x^3}{(1 + x^4)^4}$

$= \dfrac{-12x^2(1 + x^4) + 32x^6}{(1 + x^4)^3} = \dfrac{4x^2(5x^4 - 3)}{(1 + x^4)^3}$. Now

$f''(x) > 0$ when $5x^4 - 3 > 0$; that is, for $|x| > \sqrt[4]{3/5}$. Hence f is concave up for $|x| > \sqrt[4]{3/5}$ and concave down for $|x| < \sqrt[4]{3/5}$; the inflection points are $\left(\sqrt[4]{3/5}, 5/8\right)$ and $\left(-\sqrt[4]{3/5}, 5/8\right)$.

12   Let $f(x) = x^2/2 + 1/x$. Then $f'(x) = x - 1/x^2$ and $f''(x) = 1 + 2/x^3$. Now, $f''(x) = 0$ when $1 + 2/x^3 = 0$, or $x = -\sqrt[3]{2}$. Since $f''(x) > 0$ for $x < -\sqrt[3]{2}$, f is concave up on $(-\infty, -\sqrt[3]{2})$. Subsequently, f is concave down on $(-\sqrt[3]{2}, 0)$. For $x > 0$, $f''(x) > 0$, and f is concave up on $(0, \infty)$. Since f changes concavity at $x = -\sqrt[3]{2}$, $(-\sqrt[3]{2}, 0)$ is an inflection point. Note that there is no point of inflection at $x = 0$ because f is not continuous at 0.

14   Let $f(x) = \sin x + \sqrt{3} \cos x$. Then $f'(x) = \cos x - \sqrt{3} \sin x$ and $f''(x) = -\sin x - \sqrt{3} \cos x$. Now, $f''(x) = 0$ when $-\sin x = \sqrt{3} \cos x$ or when $\tan x = -\sqrt{3}$. This occurs when $x = -\pi/3 + k\pi$, where k is an integer. When $x = -\pi/3 + 2k\pi$, $f''(x)$ goes from $+$ to $-$. When $x = -\pi/3 + (2k + 1)\pi$, $f''(x)$ goes from $-$ to $+$. In summary, f has inflection points at $(-\pi/3 + k\pi, 0)$, is concave up on $\left(-\dfrac{\pi}{3} + 2(k-1)\pi, -\dfrac{\pi}{3} + 2k\pi\right)$, and is concave down on $\left(-\dfrac{\pi}{3} + 2k\pi, -\dfrac{\pi}{3} + (2k+1)\pi\right)$.

16   Let $f(x) = \cos x + \sin x$. Then $f'(x) = -\sin x + \cos x$ and $f''(x) = -\cos x - \sin x$. Therefore $f''(x) = 0$ when $\tan x = -1$, or when $x = -\pi/4 + k\pi$, where k is an integer. When $x = -\pi/4 + 2k\pi$, $f''(x)$ goes from $+$ to $-$. When $x = -\pi/4 + (2k + 1)\pi$, $f''(x)$ goes from $-$ to $+$. So f has inflection points at $(-\pi/4 + k\pi, 0)$, is concave up on $\left(-\frac{\pi}{4} + (2k-1)\pi, -\frac{\pi}{4} + 2k\pi\right)$, and is concave down on $\left(-\frac{\pi}{4} + 2k\pi, -\frac{\pi}{4} + (2k+1)\pi\right)$.

18 Let $f(x) = 2x^3 + 9x^2$. Then $f(x) = x^2(2x + 9) = 0$ when $x = 0$ or $-9/2$. Now $f'(x) = 6x^2 + 18x = 6x(x + 3) = 0$ when $x = 0$ or -3. Also, $f''(x) = 12x + 18 = 6(2x + 3) = 0$ when $x = -3/2$.

20 $f(x) = x^4 + 4x^3 + 6x^2 - 2$. We have $f'(x) = 4x^3 + 12x^2 + 12x = 4x(x^2 + 3x + 3)$, which changes from $-$ to $+$ at $x = 0$; thus the only critical point is $(0, -2)$, which is also the y intercept. $f''(x) = 12x^2 + 24x + 12 = 12(x + 1)^2 \geq 0$ for all x, so there are no inflection points. In plotting points for the graph we find that the x intercepts are near 0.5 and -0.75. (These are not exact values; with a calculator we can find more decimal places: 0.493359 and -0.749008.)

22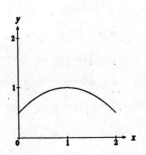

24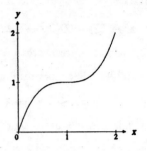

26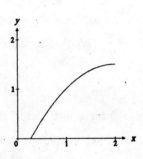

28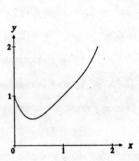

30 $f(x) = x + (x + 1)^{1/3}$. Since $f(0) = 0 + (0 + 1)^{1/3} = 1$, $(0, 1)$ is the y intercept. We have $y' = (x + (x + 1)^{1/3})' = 1 + \frac{1}{3}(x + 1)^{-2/3}$. Note that $y' > 0$ for all x. Hence y is increasing for all x. Since $\lim_{x \to -\infty} y = -\infty$ and $\lim_{x \to \infty} y = \infty$, we see that y has exactly one root (by the intermediate-value theorem and the fact that y is continuous on $(-\infty, \infty)$). We can approximate this root using the bisection method. It is about -0.682328. Now, let

4.5 The Second Derivative and Graphing

us examine $y'' = -\frac{1}{3} \cdot \frac{2}{3}(x + 1)^{-5/3}$. Note that $y'' > 0$ for $x < -1$ and $y'' < 0$ for $x > -1$; y'' is undefined at $x = -1$. Thus $(-1, -1)$ is an inflection point. The graph follows.

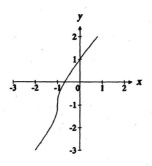

32

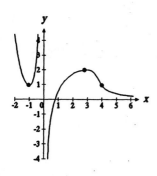

34 (a) Increasing: $-3 < x < -1$, $0 < x < 1$, $x > 4$

Decreasing: $x < -3$, $-1 < x < 0$, $1 < x < 4$

Concave upward: $x < -2$, $x > 3$

Concave downward: $-2 < x < 0$, $0 < x < 3$

Maxima: $-1, 1$; minima: $-3, 4$

(b)

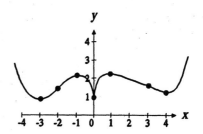

(c)

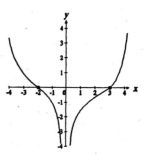

36 The second derivative of a polynomial of degree six is a polynomial of degree four. To find inflection points, we need to identify the sign changes of this fourth-degree polynomial.

(a) Since a fourth-degree polynomial can have no real roots (take $f(x) = x^4 + 1$, for example), it is possible for a polynomial of degree six to have no inflection points.

(b) No. If a fourth-degree polynomial changed sign exactly once, then it would approach ∞ as $x \to -\infty$ and $-\infty$ as $x \to \infty$ or vice versa, but that's impossible for a polynomial of even degree.

38 $\frac{dy}{dt} = ky(M - y) = kMy - ky^2$, so $\frac{d^2y}{dt^2} =$

$kM \cdot \frac{dy}{dt} - 2ky \cdot \frac{dy}{dt} = k(M - 2y)\frac{dy}{dt}$, which equals

0 if $y = M/2$. By assumption, $dy/dt > 0$ (because

y is *growing*), so the sign of $\frac{d^2y}{dt^2}$ depends on that of $M - 2y$. Now $M - 2y < 0$ for $y > M/2$ and $M - 2y > 0$ for $y < M/2$, so $\frac{d^2y}{dt^2}$ changes sign at $y = M/2$. It has an inflection point, as claimed.

40 If $y' = 0$ then $y'' = \frac{d}{dx}(y') = \frac{d}{dx}(\sin y + 2y + x)$

$= (\cos y)y' + 2y' + 1 = 1$. Since $y'' > 0$, the function has a local minimum.

42 (a) Since $f''(a) = \lim\limits_{\Delta x \to 0} \frac{f'(a + \Delta x) - f'(a)}{\Delta x}$ and it is given that $f''(a)$ is negative,

$\lim\limits_{\Delta x \to 0} \frac{f'(a + \Delta x) - f'(a)}{\Delta x} < 0$.

(b) Since a is a critical number, $f'(a) = 0$. Thus

$\lim\limits_{\Delta x \to 0^+} \frac{f'(a + \Delta x)}{\Delta x} < 0$, implying that

$f'(a + \Delta x) < 0$.

(c) Similarly, $\lim\limits_{\Delta x \to 0^-} \frac{f'(a + \Delta x)}{\Delta x} < 0$, implying

that $f'(a + \Delta x) > 0$ (since $\Delta x < 0$).

(d) Here, $f'(a + \Delta x) > 0$ for $\Delta x < 0$ and $f'(a + \Delta x) < 0$ for $\Delta x > 0$. So $f'(x)$ goes from + to − at $x = a$. Therefore, $f(x)$ has a relative maximum at $x = a$.

4.6 Newton's Method for Solving an Equation

2 Here, $x_2 = x_1 - \frac{f(x_1)}{f'(x_1)} = 3 - \frac{0.06}{0.3} = 2.8$.

4 We want a function for which $x = \sqrt[3]{a}$ is a root, so let $f(x) = x^3 - a$. Then $x_{i+1} = x_i - \frac{f(x_i)}{f'(x_i)} = x_i -$

$\frac{x_i^3 - a}{3x_i^2} = x_i - \frac{x_i}{3} + \frac{a}{3x_i^2} = \frac{2}{3}x_i + \frac{a}{3x_i^2}$.

6 Letting $a = 19$ and $x_1 = 4$ in Exercise 3, $x_2 =$

$\frac{1}{2}\left(4 + \frac{19}{4}\right) = \frac{35}{8} = 4.375$ and $x_3 =$

$\frac{1}{2}\left(\frac{35}{8} + 19 \cdot \frac{8}{35}\right) \approx 4.359$.

8 Letting $a = 25$ and $x_1 = 3$ in Exercise 4, we have

$x_2 = \frac{2}{3} \cdot 3 + \frac{25}{3 \cdot 3^2} = 2 + \frac{25}{27} = \frac{79}{27} \approx 2.926$

and $x_3 = \frac{2}{3} \cdot \frac{79}{27} + \frac{25}{3 \cdot (79/27)^2} =$

$\frac{158}{81} + \frac{25 \cdot 27^2}{3 \cdot 79^2} = \frac{158}{81} + \frac{6075}{6241} \approx 2.924$.

10 Note: The result in this exercise is so accurate that we write all intermediate results to 12 decimal places in order to compute the error in (b).

(a) Here, $a = 2$ and $x_1 = 1$, so $x_2 =$

$\frac{2}{3} \cdot 1 + \frac{2}{3 \cdot 1^2} = \frac{4}{3} \approx 1.333333333333$, $x_3 =$

$\frac{2}{3} \cdot \frac{4}{3} + \frac{2}{3}\left(\frac{3}{4}\right)^2 = \frac{8}{9} + \frac{3}{8} = \frac{91}{72} \approx$

4.6 Newton's Method for Solving an Equation

1.263888888889, $x_4 = \dfrac{2}{3} \cdot \dfrac{91}{72} + \dfrac{2}{3} \cdot \left(\dfrac{72}{91}\right)^2 \approx$

1.2599335493450, and $x_5 = \dfrac{2}{3} \cdot x_4 + \dfrac{2}{3} \cdot \left(\dfrac{1}{x_4}\right)^2$

≈ 1.259921050018. Since x_4 and x_5 agree to four decimal places, we are done.

(b) From the calculator, $\sqrt[3]{2} \approx 1.259921049895$. Hence $|\sqrt[3]{2} - x_5| \approx 1.23 \times 10^{-10}$, and x_5 and $\sqrt[3]{2}$ agree to nine decimal places.

12 (a) $f(2) = -1 < 0 < 65 = f(3)$

(b) $x_{i+1} = x_i - \dfrac{x_i^4 + x_i - 19}{4x_i^3 + 1}$ so $x_2 = \dfrac{67}{33} \approx$

2.0303 and $x_3 \approx 2.0297$.

14 (a) $f(1) = -1 < 0 < 10 = f(2)$, so there's at least one root. Note that $f'(x) = 6x^2 - 2x = 2x(3x-1) > 0$ for $1 < x < 2$, so there's at most one root.

(b) $x_2 = x_1 - \dfrac{2x_1^3 - x_1^2 - 2}{6x_1^2 - 2x_1} = \dfrac{53}{42} \approx 1.2619$

16 (a)

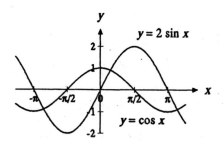

(b) $x_1 = 0.5$ is approximately a root of $f(x) = 2 \sin x - \cos x$.

(c) $x_2 = x_1 - \dfrac{2 \sin x_1 - \cos x_1}{2 \cos x_1 + \sin x_1} \approx 0.4636316$,

$x_3 \approx 0.4636476$. Since x_2 and x_3 agree to four places, we are done.

18 (a) $x_2 = x_1 - \dfrac{x_1^2 + 1}{2x_1} = \dfrac{x_1^2 - 1}{2x_1} = 0.75$,

$x_3 = -0.29$, $x_4 \approx 1.57$, $x_5 \approx 0.47$

(b)

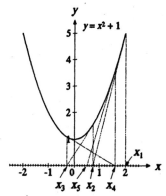

(c) $x_2 = -\sqrt{3}/3$, $x_3 = \sqrt{3}/3$. x_n oscillates between $\sqrt{3}/3$ and $-\sqrt{3}/3$ forever.

(d) $x_2 = 0$, so $x_3 = \dfrac{x_2^2 - 1}{2x_2}$ is not defined.

20 $2\theta = 2 \cos \theta$, so $\theta - \cos \theta = 0$. Since $(\theta - \cos \theta)' = 1 + \sin \theta > 0$ for $0 < \theta < \pi/2$, there's only one solution. Letting $\theta_1 = 1$, we have

$\theta_2 = \theta_1 - \dfrac{\theta_1 - \cos \theta_1}{1 + \sin \theta_1} = \dfrac{\theta_1 \sin \theta_1 + \cos \theta_1}{1 + \sin \theta_1} \approx$

0.7504, $\theta_3 \approx \theta_4 \approx 0.7391$. To two places, $\theta \approx 0.74$. (Of course, the fact that θ_3 and θ_4 agree to four places does not guarantee that the actual root will agree to four, or even two, places, but it usually does.)

22 $2 \sin \theta = \dfrac{3}{4} \cdot 2\theta$, so $f(\theta) = 3\theta - 4 \sin \theta = 0$. Note that $f'(\theta) = 3 - 4 \cos \theta$ is negative for $0 < \theta <$

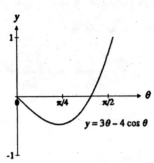

$\cos^{-1}\frac{3}{4}$ and positive for $\cos^{-1}\frac{3}{4} < \theta < \frac{\pi}{2}$. Since

$f(0) = 0$, $f(\theta) < 0$ for $0 < \theta < \cos^{-1}\frac{3}{4}$. Hence in

the interval $\left[\cos^{-1}\frac{3}{4}, \frac{\pi}{2}\right]$, $f(\theta)$ increases from

$f\left(\cos^{-1}\frac{3}{4}\right) < 0$ to $f\left(\frac{\pi}{2}\right) = \frac{3\pi}{2} - 4 > 0$. So f has

exactly one root in $(0, \pi/2)$. For $\theta_1 = 1$, we have

$\theta_2 = \theta_1 - \dfrac{3\theta_1 - 4\sin\theta_1}{3 - 4\cos\theta_1} = \dfrac{4(\sin\theta_1 - \theta_1\cos\theta_1)}{3 - 4\cos\theta_1}$

≈ 1.4362, $\theta_3 \approx 1.2962$, $\theta_4 \approx 1.2761$, $\theta_5 \approx \theta_6 \approx$ 1.2757. To two places, $\theta \approx 1.28$.

24 (a) $\left(\dfrac{\sin x}{x}\right)' = \dfrac{x\cdot\cos x - \sin x}{x^2} = 0$ implies $\sin x$

$= x\cos x$, so $x = \tan x$.

(b) $\dfrac{\sin(-x)}{-x} = \dfrac{-\sin x}{-x} = \dfrac{\sin x}{x}$

(c)

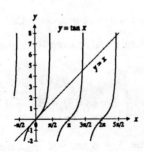

They cross once in each of $\left[\dfrac{\pi}{2}, \dfrac{3\pi}{2}\right]$ and

$\left[\dfrac{3\pi}{2}, \dfrac{5\pi}{2}\right]$.

(d) $(\tan x - x)' = \sec^2 x - 1 \geq 0$ for x in

$\left[\dfrac{\pi}{2}, \dfrac{3\pi}{2}\right]$. Thus $\tan x - x$ is an increasing

function on $\left[\dfrac{\pi}{2}, \dfrac{3\pi}{2}\right]$. Hence there is at most

one root in $[\pi/2, 3\pi/2]$.

(e) Infinitely many, since there are infinitely many solutions to $\tan x - x = 0$.

(f) Here $x_2 = x_1 - \dfrac{\tan x_1 - x_1}{\sec^2 x_1 - 1} =$

$x_1 - \dfrac{\sin x_1 \cos x_1 - x_1 \cos^2 x_1}{1 - \cos^2 x_1} =$

$\dfrac{x_1 - \sin x_1 \cos x_1}{\sin^2 x_1}$. Choose $x_1 = \dfrac{7\pi}{5} \approx$

4.3982. Then $x_2 \approx 4.5376$, $x_3 \approx 4.5025$, $x_4 \approx 4.4938$, and $x_5 \approx x_6 \approx 4.4934$. So the

critical number in $\left[\dfrac{\pi}{2}, \dfrac{3\pi}{2}\right]$ is approximately

$4.493 \approx 1.43\pi$.

26 Let $f(x) = \sin x - x$. Now $f'(x) = \cos x - 1 \leq 0$ for all x. Thus f has at most one real root. But $f(0) = 0$, so f has exactly one root.

28 We are given that $\lim\limits_{n\to\infty} x_n = L$, and that $f'(L) \neq 0$.

Now, taking the limit as $n \to \infty$ of both sides of Newton's formula, we have $\lim\limits_{n\to\infty} x_{n+1} =$

$\lim\limits_{n\to\infty}\left(x_n - \dfrac{f(x_n)}{f'(x_n)}\right) = \lim\limits_{n\to\infty} x_n - \lim\limits_{n\to\infty}\dfrac{f(x_n)}{f'(x_n)}$. Thus,

4.6 Newton's Method for Solving an Equation

since $\lim_{n\to\infty} x_{n+1} = \lim_{n\to\infty} x_n = L$, our result reduces to $L = L - \frac{f(L)}{f'(L)}$ (since $f'(L) \neq 0$). We conclude that $f(L) = 0$.

4.7 Applied Maximum and Minimum Problems

2 We must maximize $A(x) = 80x - \frac{1}{2}x^2$ for $0 \leq x \leq 160$. $A'(x) = 80 - x = 0$ when $x = 80$, so $x = 80$ and $y = \frac{1}{2}(160 - x) = 40$ gives the maximum area.

4 We must maximize $V(x) = x(5 - 2x)(7 - 2x) = 4x^3 - 24x^2 + 35x$ for $0 \leq x \leq 5/2$. $V'(x) = 12x^2 - 48x + 35$, so the maximum occurs at $x = \frac{48 \pm \sqrt{624}}{24} = 2 \pm \frac{1}{6}\sqrt{39}$. Since x cannot exceed $5/2$, the optimal cut has length $2 - \frac{1}{6}\sqrt{39}$ inches.

6 We must maximize $V(x) = x(6 - 2x)(10 - 2x) = 4(x^3 - 8x^2 + 15x)$ for $0 \leq x \leq 3$. $V'(x) = 4(3x^2 - 16x + 15)$, so the maximum occurs at $x = \frac{16 \pm \sqrt{76}}{6} = \frac{8 \pm \sqrt{19}}{3}$. But $x \leq 3$, so the optimal cut length is $\frac{8 - \sqrt{19}}{3}$ inches.

8 We must maximize $S = \pi r^2 + 2\pi r \cdot \frac{100}{\pi r^2} = \pi r^2 + \frac{200}{r}$ for $r > 0$. $\frac{dS}{dr} = 2\pi r - \frac{200}{r^2}$, so $r = \sqrt[3]{\frac{100}{\pi}}$ at the maximum. The height is $h = \frac{100}{\pi r^2} = \sqrt[3]{\frac{100}{\pi}}$.

10 Let the base be b and the height be h. Then the amount of material is $M = b^2 + 4bh$. b and h satisfy $b^2 h = 1000$, so $M = b^2 + 4b \cdot \frac{1000}{b^2} = b^2 + \frac{4000}{b}$ and $\frac{dM}{db} = 2b - \frac{4000}{b^2}$. Hence M is minimal for $b = \sqrt[3]{2000} = 10\sqrt[3]{2}$ and $h = \frac{1000}{b^2} = 5\sqrt[3]{2}$.

12

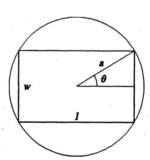

From the figure, we see that $P = 2l + 2w = 2(2a\cos\theta) + 2(2a\sin\theta) = 4a\cos\theta + 4a\sin\theta$. Now, $\frac{dP}{d\theta} = -4a\sin\theta + 4a\cos\theta = 0$ when $\sin\theta = \cos\theta$, or when $\theta = \pi/4$. Since $\frac{d^2P}{d\theta^2} = -4a\cos\theta - 4a\sin\theta < 0$ at $\theta = \pi/4$, this gives a maximum perimeter. Hence, $l = w = \sqrt{2}a$ and the rectangle is a square.

14 $h = A/w$, so the perimeter is $P(w) = 2(w + h) = $

$2(w + A/w)$. $P'(w) = 2\left(1 - \dfrac{A}{w^2}\right)$ changes from $-$ to $+$ when $w = h = \sqrt{A}$.

16 Area $= A(x) = xy - \dfrac{1}{2}x^2 = x(1-x) - \dfrac{1}{2}x^2 = x - \dfrac{3}{2}x^2$. This must be maximized for $0 \le x \le 1/2$. $A'(x) = 1 - 3x$ changes from $+$ to $-$ at $x = 1/3$. Thus $x = 1/3$ mi, $y = 2/3$ mi.

18 Let the numbers be s and c. Then $s = 1 - c$, so the product is $P(c) = s^2c^3 = c^3 - 2c^4 + c^5$. $P'(c) = 3c^2 - 8c^3 + 5c^4 = c^2(1-c)(3-5c)$, which changes from $+$ to $-$ at $c = 3/5$. The numbers are $2/5$ and $3/5$.

20

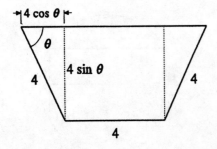

(a) From the figure, we see that Area $= A(\theta) = 16 \sin\theta + 16 \sin\theta \cos\theta$. Furthermore, we have $A'(\theta) = 16(2\cos^2\theta + \cos\theta - 1) = 16(1 + \cos\theta)(2\cos\theta - 1)$, which changes sign from $+$ to $-$ when $\cos\theta = 1/2$; that is, for $\theta = \pi/3$.

(b) Yes. $x = 4 + 8\cos\theta = 4 + 8 \cdot 1/2 = 8$.

22 Let the cylinder have height h and radius r. Then $108 = h + 2\pi r$, so $h = 108 - 2\pi r$ and the area is $A(r) = 2\pi r^2 + 2\pi rh = 2\pi(108r - (2\pi - 1)r^2)$. $A'(r) = 2\pi(108 - 2(2\pi - 1)r)$ changes from $+$ to $-$ at $r = \dfrac{54}{2\pi - 1}$. Hence $r = \dfrac{54}{2\pi - 1}$ inches, $h = \dfrac{108(\pi - 1)}{2\pi - 1}$ inches.

24 Let the box have height h and base $b \times b$. Then $108 = h + 4b$, so $h = 108 - 4b$ and the area is $A(b) = 2b^2 + 4bh = 432b - 14b^2$. $A'(b) = 432 - 28b$, which changes from $+$ to $-$ at $b = \dfrac{108}{7}$. The box is $\dfrac{108}{7}$ in by $\dfrac{108}{7}$ in by $\dfrac{324}{7}$ in.

26 (a) Cost is proportional to $2x^2 + 4xy \cdot 2 = 2x^2 + 8xy$.

(b) Since $x^2y = 100$, $y = \dfrac{100}{x^2}$ and the cost is proportional to $C(x) = 2x^2 + \dfrac{800}{x}$. $C'(x) = 4x - \dfrac{800}{x^2} = \dfrac{4}{x^2}(x^3 - 200)$ changes from $-$ to $+$ at $x = \sqrt[3]{200} = 2\sqrt[3]{25}$. Hence $x = 2\sqrt[3]{25}$ inches and $y = \sqrt[3]{25}$ inches.

28 (a) Small

(b) Large

(c) At s mi/hr, the trip takes $\dfrac{600}{s}$ hours, so the cost (in cents) is $C(s) = 600(20 + s/2) + 1800 \cdot \dfrac{600}{s} = 600\left(20 + \dfrac{s}{2} + \dfrac{1800}{s}\right)$. $C'(s) = 600\left(\dfrac{1}{2} - \dfrac{1800}{s^2}\right) = \dfrac{300}{s^2}(s^2 - 3600)$ changes from $-$ to $+$ at $x = 60$. Optimal speed is 60 mi/hr. (This neglects the cost of speeding tickets.)

(d) No

30 (a) Directly from the power plant to the factory. At an angle across the river and then along the river.

(c) Cost is proportional to $C(x) = 5\sqrt{1 + x^2} + 3(s - x)$, for $0 \le x \le s$. $C'(x) = \dfrac{5x}{\sqrt{1 + x^2}} - 3 = \dfrac{5}{\sqrt{1 + 1/x^2}} - 3$ changes sign from $-$ to $+$ when $\dfrac{5}{\sqrt{1 + 1/x^2}} = 3$; that is, for $x = 3/4$. For $s \le 3/4$, the cheapest route has $x = s$; that is, the cable is straight. For $s \ge 3/4$, $x = 3/4$; that is, the cable runs $5/4$ miles underwater and $s - 3/4$ miles along the bank.

32 (a) $\dfrac{dR}{d\theta} = \dfrac{2v_0^2}{g} \cos 2\theta$ changes from $+$ to $-$ at $\theta = \pi/4 = 45°$.

(b) R is maximal when $\sin 2\theta = 1$; that is, for $\theta = 45°$.

34 (a)

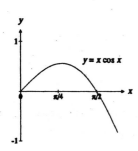

(b) We have $y' = -x \sin x + \cos x$ and $y'' = -x \cos x - 2 \sin x$. For x in $(0, \pi/2)$, $\sin x > 0$ and $\cos x > 0$, so $y''(x) < 0$. Now since $y'(0) = 1$ and $y'(\pi/2) = -\pi/2$, we know that $y'(c) = 0$ for exactly one c in $(0, \pi/2)$. Since $y''(c) > 0$, $x = c$ gives a relative maximum of y.

(c) $y' = 0$ when $-x \sin x + \cos x = \cos x - x \sin x = 0$

(d) $x_2 = x_1 - \dfrac{\cos x_1 - x_1 \sin x_1}{-2 \sin x_1 - x_1 \cos x_1} = \dfrac{\pi}{4} + \dfrac{1 - \pi/4}{+ 2 + \pi/4} \approx 0.8624$

(e) $x_3 \approx 0.8603$

36 To find the critical points of $x^3 + \cos x$, we must find the roots of $f'(x) = 3x^2 - \sin x$. By Rolle's theorem, between any two roots of $f'(x)$ there must be a root of $f''(x) = 0$. But $f''(x) = 6x - \cos x$ is an increasing function (since $f'''(x) = 6 + \sin x \ge 5$). Hence $f''(x) = 0$ has at most one root and $f'(x)$ has at most two roots. $x = 0$ is one such root. Since $f'(0.1) \approx -0.07 < 0$ and $f'(1) = 3 - \sin 1 > 0$, there is one other root. By Newton's method we find that $f'(x) = 0$ for $x \approx 0.3274$. Comparing the value of f at this point and at the endpoints of $[0, \pi/2]$ shows that the minimum is about $f(0.3274) \approx 0.9820$.

38 Since $\left(\dfrac{A}{\sqrt{A^2 + B^2}}\right)^2 + \left(\dfrac{B}{\sqrt{A^2 + B^2}}\right)^2 = 1$, there is an angle α such that $\dfrac{A}{\sqrt{A^2 + B^2}} = \cos \alpha$ and $\dfrac{B}{\sqrt{A^2 + B^2}} = \sin \alpha$. Then $A \cos t + B \sin t = \sqrt{A^2 + B^2}(\cos \alpha \cos t + \sin \alpha \sin t) = \sqrt{A^2 + B^2} \cos(t - \alpha)$. Hence the maximum and minimum are $\sqrt{A^2 + B^2}$ and $-\sqrt{A^2 + B^2}$.

40 From the figure, $x = b\cos\theta - a\cot\theta$. This must be maximized for $\sin^{-1}\frac{a}{b} \leq \theta \leq \frac{\pi}{2}$. Now $\frac{dx}{d\theta} =$

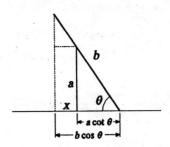

$-b\sin\theta + a\csc^2\theta = (a - b\sin^3\theta)\csc^2\theta$

changes from $+$ to $-$ at $\theta = \sin^{-1}\sqrt[3]{a/b}$. For this θ, $\cos\theta = \frac{\sqrt{b^{2/3} - a^{2/3}}}{b^{1/3}}$ and $\cot\theta =$

$\frac{\sqrt{b^{2/3} - a^{2/3}}}{a^{1/3}}$, so $x = (b^{2/3} - a^{2/3})^{3/2}$.

42 $U(r) = U_0\left(r_0^{12}r^{-12} - 2r_0^6 r^{-6}\right)$, so

$U'(r) = U_0\left(-12r_0^{12}r^{-13} + 12r_0^6 r^{-7}\right) =$

$12U_0 r_0^6 r^{-13}(r^6 - r_0^6)$, which changes from $-$ to $+$ at $r = r_0$. Hence $U(r)$ is minimal for $r = r_0$.

44

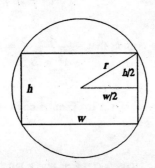

The stiffness is proportional to $S = h^3 w =$ $h^3\sqrt{4r^2 - h^2}$, which must be maximized on $0 \leq h$

$\leq 2a$. $\frac{dS}{dh} = \frac{4h^2(3r^2 - h^2)}{\sqrt{4r^2 - h^2}}$ changes from $+$ to $-$

at $h = \sqrt{3}r$. Then $w = \sqrt{4r^2 - h^2} = r$, so the height should be $\sqrt{3}$ times the width.

46 If $P = (x, x^2)$ then $f(x) = \overline{PA}^2 + \overline{PB}^2 = [(x+1)^2 + (x^2 - 1)^2] + [(x-1)^2 + (x^2 - 1)^2] = 2(x^4 - x^2 + 2)$. The derivative is $4x(2x^2 - 1)$, which equals 0 for $x = 0$ or $\pm\frac{1}{\sqrt{2}}$. Considering these points and the endpoints $x = \pm 1$, we find:

x	0	$\pm 1/\sqrt{2}$	± 1
$f(x)$	4	7/2	4

(a) The minimum occurs at $\left(\pm\frac{1}{\sqrt{2}}, \frac{1}{2}\right)$.

(b) The maximum occurs at $(0, 0)$ and $(\pm 1, 1)$.

48 $\frac{dy}{dS} = \frac{C}{4} - \frac{R}{VS^2} = \frac{CVS^2 - 4R}{4VS^2}$, which changes from $-$ to $+$ at $S = \sqrt{\frac{4R}{CV}}$, so this value of S gives the minimum cost. But then $CVS^2 = 4R$ so $\frac{CS}{4} = \frac{R}{VS}$.

50 This is equivalent to Exercise 40, so the maximal y is $(b^{2/3} - a^{2/3})^{3/2}$.

52

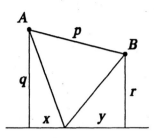

(a) From the figure, the length of the road is $L = \sqrt{q^2 + x^2} + \sqrt{r^2 + y^2}$, where $y = \sqrt{p^2 - (q-r)^2} - x$. Clearly the minimum occurs where $\dfrac{dL}{dx} = 0$. $\dfrac{dL}{dx} = $

$\dfrac{x}{\sqrt{q^2+x^2}} + \dfrac{y}{\sqrt{r^2+y^2}} \cdot \dfrac{dy}{dx} = $

$\dfrac{x}{\sqrt{q^2+x^2}} - \dfrac{y}{\sqrt{r^2+y^2}}$. Setting this equal to 0, we find that $xr = qy = q\left(\sqrt{p^2 - (q-r)^2} - x\right)$, so $x = \dfrac{q\sqrt{p^2 - (q-r)^2}}{q+r}$, $y = \dfrac{r\sqrt{p^2 - (q-r)^2}}{q+r}$,

and $L = \sqrt{p^2 + 4qr}$.

(b)

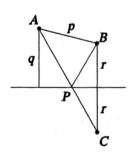

Let P be the point where the path meets the road. Then $\overline{AP} + \overline{PB} = \overline{AP} + \overline{PC}$ is minimal when A, P, and C are collinear. Hence

$\overline{AP} + \overline{PC} = \overline{AC} = $

$\left[\left(\sqrt{p^2 - (q-r)^2}\right)^2 + (q+r)^2\right]^{1/2} = \sqrt{p^2 + 4qr}$.

54 If $P = (x, 0)$, the slope of PA is $-2/x$; the slope of

PB is $\dfrac{1}{1-x}$. Hence $\tan(APB) = \left|\dfrac{-\dfrac{2}{x} - \dfrac{1}{1-x}}{1 + \left(-\dfrac{2}{x}\right)\dfrac{1}{1-x}}\right|$

$= \left|\dfrac{x-2}{x^2-x+2}\right|$. Let $f(x) = \dfrac{x-2}{x^2-x+2}$. Then $f'(x)$

$= \dfrac{x(4-x)}{(x^2-x+2)^2}$ changes from $-$ to $+$ at $x = 0$

and back to $-$ at $x = 4$. Since $\lim\limits_{x \to \pm \infty} f(x) = 0$, the

minimum and maximum of f are $f(0) = -1$ and $f(4) = 1/7$, respectively. Hence APB is maximal when $x = 0$ and $P = (0, 0)$.

56

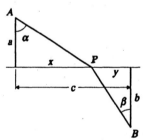

The travel time is $T = \dfrac{1}{v_1}\sqrt{a^2 + x^2} + \dfrac{1}{v_2}\sqrt{b^2 + y^2}$,

where $y = c - x$. Hence $\dfrac{dT}{dx} = \dfrac{1}{v_1} \cdot \dfrac{x}{\sqrt{a^2+x^2}} + $

$\dfrac{1}{v_2} \cdot \dfrac{y}{\sqrt{b^2+y^2}} \cdot \dfrac{dy}{dx} = \dfrac{\sin \alpha}{v_1} - \dfrac{\sin \beta}{v_2}$. So if $\dfrac{dT}{dx} = 0$, then $\dfrac{\sin \alpha}{v_1} = \dfrac{\sin \beta}{v_2}$.

4.8 Implicit Differentiation

2 Explicit: $y = \sqrt{x^2 - 3}$, $\dfrac{dy}{dx} = \dfrac{2}{\sqrt{x^2 - 3}}$, which equals 2 when $x = 2$.

Implicit: $2x - 2y \cdot \dfrac{dy}{dx} = 0$, $\dfrac{dy}{dx} = \dfrac{x}{y}$, which equals 2 at $(2, 1)$.

4 Explicit: $y = -\sqrt{100 - x^2}$, $\dfrac{dy}{dx} = \dfrac{x}{\sqrt{100 - x^2}}$, which equals $\dfrac{3}{4}$ when $x = 6$.

Implicit: $2x + 2y \cdot \dfrac{dy}{dx} = 0$, $\dfrac{dy}{dx} = -\dfrac{x}{y}$, which equals $\dfrac{3}{4}$ at $(6, -8)$.

6 $6y^2 \cdot \dfrac{dy}{dx} + 4(x \cdot \dfrac{dy}{dx} + y) + 2x = 0$, $\dfrac{dy}{dx} = -\dfrac{x + 2y}{2x + 3y^2}$, which equals $-\dfrac{3}{5}$ at $(1, 1)$.

8 $1 + (\sec^2 xy)(x \cdot \dfrac{dy}{dx} + y) = 0$, $\dfrac{dy}{dx} = -\dfrac{y + \cos^2 xy}{x}$, which equals $-\dfrac{1}{2} - \dfrac{\pi}{4}$ at $(1, \pi/4)$.

10 $S = 2\pi r^2 + 2\pi rh$, $0 = 4\pi r + 2\pi\left(r\dfrac{dh}{dr} + h\right)$, $r\dfrac{dh}{dr} = -2r - h$

$V = \pi r^2 h$, $\dfrac{dV}{dr} = \pi\left(r^2 \dfrac{dh}{dr} + 2rh\right) = \pi(r(-2r - h) + 2rh) = \pi r(h - 2r)$. When $\dfrac{dV}{dr} = 0$, $h/r = 2$. (That is, height equals diameter.)

12

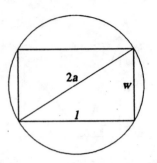

The constraint is $w^2 + l^2 = (2a)^2 = 4a^2$ (from the Pythagorean theorem), so differentiating implicitly with respect to w yields $\dfrac{d}{dw}(w^2 + l^2) =$

$2w + 2l\dfrac{dl}{dw} = 0$. So $\dfrac{dl}{dw} = -\dfrac{w}{l}$. Now, if we denote the perimeter by p, we have $p = 2w + 2l$. Again, differentiating implicitly with respect to w gives $\dfrac{dp}{dw} = 2 + 2 \cdot \dfrac{dl}{dw} = 2\left(1 - \dfrac{w}{l}\right)$, which equals 0 when $w = l$. Thus $4a^2 = w^2 + l^2 = w^2 + w^2 = 2w^2$ and $w = l = \sqrt{2}a$. The rectangle of largest perimeter is a square, and this agrees with Exercise 12 of Sec. 4.7.

14 Let the sides be x and y. Since $xy = A$, $\dfrac{dy}{dx}x + y = 0$, so $\dfrac{dy}{dx} = -y/x$. The perimeter is $P = 2(x + y)$, so $\dfrac{dP}{dx} = 2(1 + \dfrac{dy}{dx}) = \dfrac{2(x - y)}{x}$.

When $\dfrac{dP}{dx} = 0$, $x = y$.

16 Let base be b and height be h. Then $108 = h + 4b$, so $0 = \dfrac{dh}{db} + 4$ and $\dfrac{dh}{db} = -4$. The area

4.8 Implicit Differentiation

is $A = 2b^2 + 4bh$, so we have $\dfrac{dA}{db} = 4b + 4h +$

$4b\dfrac{dh}{db} = 4(h - 3b)$. When $\dfrac{dA}{db} = 0$, $h = 3b$.

Hence $108 = 3b + 4b = 7b$, so $b = \dfrac{108}{7}$ and

$h = \dfrac{324}{7}$.

18 $\sec(x + 2y)\tan(x + 2y)\cdot\left(1 + 2\dfrac{dy}{dx}\right) -$

$\sin(x - 2y)\cdot\left(1 - 2\dfrac{dy}{dx}\right) + \dfrac{dy}{dx} = 0,\ \dfrac{dy}{dx} =$

$\dfrac{\sin(x - 2y) - \sec(x + 2y)\tan(x + 2y)}{2\sec(x + 2y)\tan(x + 2y) + 2\sin(x - 2y) + 1}$

20 $3\sin^2(xy)\cos(xy)\cdot\left(x\dfrac{dy}{dx} + y\right) -$

$\sin(x + y)\cdot\left(1 + \dfrac{dy}{dx}\right) + 1 = 0$

$\dfrac{dy}{dx} = \dfrac{\sin(x + y) - 1 - 3y\sin^2(xy)\cos(xy)}{3x\sin^2(xy)\cos(xy) - \sin(x + y)}$

22 $5x^4 + xy' + y + 5y^4y' = 0,\ y' = -\dfrac{5x^4 + y}{x + 5y^4} =$

$-\dfrac{7}{81},\ 20x^3 + xy'' + y' + y' + 20y^3(y')^2 + 5y^4y''$,

$y'' = -\dfrac{20x^3 + 2y' + 20y^3(y')^2}{x + 5y^4} = -\dfrac{137{,}926}{531{,}441}$

24 $3x^2 + x^2y' + 2xy - 3xy^2y' - y^3 = 0,\ y' =$

$\dfrac{3x^2 + 2xy - y^3}{3xy^2 - x^2} = \dfrac{15}{2}$,

$6x + x^2y'' + 2xy' + 2xy' + 2y - 3xy^2y'' -$
$6xy(y')^2 - 3y^2y' - 3y^2y' = 0$,

$y'' = \dfrac{6x + 2y + 4xy' - 6y^2y' - 6xy(y')^2}{3xy^2 - x^2}$

$= -323$

26 $3x^2 + 2xyy' + y^2 + 5x^3y^4y' + 3x^2y^5 = 0,\ y' =$

$-\dfrac{3x^2 + y^2 + 3x^2y^5}{2xy + 5x^3y^4} = -1$. The slope of line

through $(1, 1)$ and $(-2, 3)$ is $-2/3 \neq -1$. Hence $(-2, 3)$ is not on the line.

28 $ny^{n-1}y' = mx^{m-1},\ y' = \dfrac{m}{n}\cdot\dfrac{x^{m-1}}{y^{n-1}} = \dfrac{m}{n}\cdot\dfrac{x^{m-1}}{x^{m-m/n}}$

$= \dfrac{m}{n}x^{(m/n)-1}$

30 (a) $\dfrac{d}{dx}\left(\sqrt{4 - 9x^2}\right) = \dfrac{-18x}{2\sqrt{4 - 9x^2}} = \dfrac{-9x}{\sqrt{4 - 9x^2}}$

(b) $\dfrac{d}{dx}(x^{1/3}\cos^3 x) =$

$x^{1/3}\cdot 3\cos^2 x\,(-\sin x) + \cos^3 x\left(\dfrac{1}{3}x^{-2/3}\right)$

$= x^{1/3}\cos^2 x\left[\dfrac{\cos x}{3x} - 3\sin x\right]$

$= \dfrac{1}{3}x^{-2/3}\cos^2 x\,(\cos x - 9x\sin x)$

(c) $\dfrac{d}{dx}\left(\dfrac{\sec 2x}{1 + \tan 2x}\right)$

$= \dfrac{(1 + \tan 2x)(\sec 2x \tan 2x)\cdot 2 - (\sec 2x)(\sec^2 2x)\cdot 2}{(1 + \tan 2x)^2}$

$= \dfrac{2\sec 2x \tan 2x + 2\sec 2x \tan^2 2x - 2\sec^3 2x}{(1 + \tan 2x)^2}$

$= \dfrac{2\sec 2x\,(\tan^2 2x + \tan 2x - \sec^2 2x)}{(1 + \tan 2x)^2}$

$$= \frac{2\sec 2x\,(\tan 2x - 1)}{(\tan 2x + 1)^2} \text{ or } \frac{d}{dx}\left(\frac{\sec 2x}{1 + \tan 2x}\right)$$

$$= \frac{d}{dx}\left(\frac{1}{\cos 2x + \sin 2x}\right)$$

$$= -\frac{-2\sin 2x + 2\cos 2x}{(\cos 2x + \sin 2x)^2}$$

$$= \frac{2(\sin 2x - \cos 2x)}{(\sin 2x + \cos 2x)^2}$$

4.9 The Differential and Linearization

2

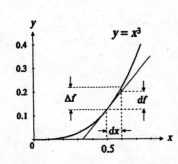

$df = 3x^2\,dx = 0.075$, $\Delta f = (1/2 + 0.1)^3 - (1/2)^3$
$= 0.091$

4 $df = \frac{1}{3}x^{-2/3}\,dx = -4/27 \approx -0.148$, $\Delta f =$

$\sqrt[3]{27 - 4} - \sqrt[3]{27} = \sqrt[3]{23} - 3 \approx -0.156$

6

$df = (\cos x)\,dx = -\pi/24 \approx -0.131$, $\Delta f =$

$\sin\left(\frac{\pi}{3} - \frac{\pi}{12}\right) - \sin\frac{\pi}{3} = \frac{1}{\sqrt{2}} - \frac{\sqrt{3}}{2}$

$= \frac{1}{2}(\sqrt{2} - \sqrt{3}) \approx -0.159$

8 (a) $f(x) = \sqrt{x}$

(b) $x = 25$

(c) $f(26.1) \approx f(25) + f'(25)(26.1 - 25)$

$= 5 + \frac{1}{2 \cdot 5} \cdot 1.1 = 5.11$

10 $f(x) = x^{1/2}$, $f'(x) = \frac{1}{2}x^{-1/2}$, $\sqrt{103} = f(100 + 3) \approx$

$f(100) + f'(100) \cdot 3 = 10.15$

12 $f(x) = x^{1/3}$, $f'(x) = 1/3 x^{-2/3}$, $\sqrt[3]{28} = f(27 + 1)$

$\approx f(27) + f'(27) \cdot 1 = 3 + 1/27 \approx 3.037$

14 $\tan(\pi/4 + 0.01) \approx \tan \pi/4 + (\sec^2 \pi/4) \cdot 0.01$

≈ 1.02

16 $\sin(\pi/3 + 0.02) \approx \pi/3 + (\cos \pi/3) \cdot 0.02$

$= \frac{\sqrt{3}}{2} + 0.01 \approx 0.876$

18 $\sin(0 + 0.3) \approx \sin 0 + (\cos 0) \cdot 0.3 = 0.3$

20 $(16 - 0.3)^{1/2} \approx 16^{1/2} + (1/2 \cdot 16^{-1/2})(-0.3)$
$= 3.9625$

22 $\sin(0 + \pi/90) \approx \sin 0 + (\cos 0)\pi/90 = \pi/90$

≈ 0.03491

24 $\tan(0 + \pi/60) \approx \tan 0 + (\sec^2 0)\pi/60 = \pi/60$

≈ 0.05236

26 $f(x) = x^{10}$, $f'(x) = 10x^9$, $(1 + h)^{10} = f(1 + h)$

$\approx f(1) + f'(1)h = 1 + 10h$

28 $f(x) = x^{1/3}$, $f'(x) = \frac{1}{3}x^{-2/3}$, $\sqrt[3]{1 + h} = f(1 + h) \approx$

$f(1) + f'(1)h = 1 + h/3$

30 $\dfrac{x\,dx}{\sqrt{1+x^2}}$

32 $-5 \sin 5x\,dx$

34 $3x^2 \sec^2 x^3\,dx$

36 $x^3 \cdot 2 \sec 5x \cdot \sec 5x \tan 5x \cdot 5\,dx + 3x^2 \cdot \sec^2 5x\,dx$
    $= x^2(\sec^2 5x)(10x \tan 5x + 3)\,dx$

38 (a) $f(x) = x^{-2}$ and $f'(x) = 2x^{-3}$, so $f(a) = f(2) =$
    $\dfrac{1}{4}$ and $f'(a) = f'(2) = -\dfrac{1}{4}$. Thus $p(x) =$

    $f(a) + f'(a)(x-a) = \dfrac{1}{4} - \dfrac{1}{4}(x-2)$

    $= -\dfrac{1}{4}x + \dfrac{3}{4}$.

    (b) In this case, $f(a) = f(3) = 1/9$ and $f'(a) =$
    $f'(3) = -2/27$. Hence $p(x) =$

    $\dfrac{1}{9} - \dfrac{2}{27}(x-3) = -\dfrac{2}{27}x + \dfrac{2}{9} + \dfrac{1}{9} =$

    $-\dfrac{2}{27}x + \dfrac{1}{3}$.

40

x	$f(x)$	$p(x)$	$f(x) - p(x)$
1.5	0.666667	0.625	0.041667
1.9	0.526316	0.525	0.001316
1.95	0.512821	0.5125	0.000321
1.99	0.502513	0.5025	0.000013
1.999	0.500250	0.50025	0.00000013

Here, $f(x) = 1/x$, $f'(x) = -1/x^2$, and $f(2) = 1/2$ while $f'(2) = -1/4$. Thus, $p(x) = \dfrac{1}{2} - \dfrac{1}{4}(x-2)$

$= -\dfrac{1}{4}x + 1$.

42 Let each side be x and area be A. Then $\dfrac{dA}{A} =$

$\dfrac{d(x^2)}{x^2} = \dfrac{2x\,dx}{x^2} = 2 \cdot \dfrac{dx}{x}$. Since $\dfrac{|dx|}{x} \leq 0.05$,

$\dfrac{|dA|}{A} \leq 0.1$; the error in the area is at most about

10%. (It may in fact be as much as 10.25%.)

44 (a) $df = 3x^2\,\Delta x$, $\Delta f = (x + \Delta x)^3 - x^3$
    $= 3x^2\,\Delta x + 3x\,\Delta x^2 + \Delta x^3$

    (b),(c)

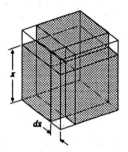

The differential df consists of the three square slabs shaded in the figure above. The actual change in volume Δf consists of everything *except* the cube of side x shaded in the figure below.

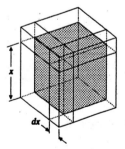

46 (a) $(\sin \theta)' = \cos \theta$ so $\sin \theta = \sin(0 + \theta)$
    $\approx \sin 0 + (\cos 0)\theta = \theta$
    $(\cos \theta)' = -\sin \theta$ so $\cos \theta = \cos(0 + \theta)$
    $\approx \cos 0 + (-\sin 0)\theta = 1$
    $(\tan \theta)' = \sec^2 \theta$ so $\tan \theta = \tan(0 + \theta)$
    $\approx \tan 0 + (\sec^2 0)\theta = \theta$

(b) Note that $\dfrac{1 - \cos\theta}{\theta^2} =$

$$\dfrac{(1 - \cos\theta)(1 + \cos\theta)}{\theta^2(1 + \cos\theta)} =$$

$\dfrac{1}{1 + \cos\theta}\left(\dfrac{\sin\theta}{\theta}\right)^2$. Hence $\lim_{\theta\to 0}\dfrac{1 - \cos\theta}{\theta^2} =$

$\left(\lim_{\theta\to 0}\dfrac{1}{1 - \cos\theta}\right)\left(\lim_{\theta\to 0}\dfrac{\sin\theta}{\theta}\right)^2 = \dfrac{1}{2}\cdot 1^2 = \dfrac{1}{2}$.

So the difference between $\cos\theta$ and 1 is about $\dfrac{1}{2}\theta^2$. But the difference between $\cos\theta$ and $1 - \dfrac{1}{2}\theta^2$ is much smaller; in fact,

$$\lim_{\theta\to 0}\dfrac{\cos\theta - \left(1 - \dfrac{1}{2}\theta^2\right)}{\theta^2} =$$

$\lim_{\theta\to 0}\left(\dfrac{1}{2} - \dfrac{1 - \cos\theta}{\theta^2}\right) = \dfrac{1}{2} - \dfrac{1}{2} = 0.$

47 (b) $\lim_{\Delta x\to\infty}\dfrac{\Delta f}{df} = 1$

(c) $\lim_{\Delta x\to 0}\dfrac{\Delta f}{df} = \lim_{\Delta x\to 0}\dfrac{f(a + \Delta x) - f(a)}{f'(a)\,\Delta x} =$

$\dfrac{1}{f'(a)}\lim_{\Delta x\to 0}\dfrac{f(a + \Delta x) - f(a)}{\Delta x}$, since $f'(a)$ is a nonzero constant. However,

$\lim_{\Delta x\to 0}\dfrac{f(a + \Delta x) - f(a)}{\Delta x} = f'(a)$ by definition,

so $\lim_{\Delta x\to 0}\dfrac{\Delta f}{df} = \dfrac{1}{f'(a)}\lim_{\Delta x\to 0}\dfrac{f(a + \Delta x) - f(a)}{\Delta x}$

$= \dfrac{1}{f'(a)}\cdot f'(a) = 1$. We are done.

48 (a) Since the differential approximation is only close to Δf for Δx small, we must find $f(1.6)$ in steps by first finding $f(1.1)$, then $f(1.2)$, etc. Thus

$f(1.1) \approx f(1) + f'(1)\cdot 0.1$
$= 3 + (0.7)(0.1) = 3.07$
$f(1.2) \approx f(1.1) + f'(1.1)\cdot 0.1$
$= 3.07 + (0.5)(0.1) = 3.12$
$f(1.3) \approx f(1.2) + f'(1.2)\cdot 0.1$
$= 3.12 + (0.4)(0.1) = 3.16$
$f(1.4) \approx f(1.3) + f'(1.3)\cdot 0.1$
$= 3.16 + (0.2)(0.1) = 3.18$
$f(1.5) \approx f(1.4) + f'(1.4)\cdot 0.1$
$= 3.18 + (0.3)(0.1) = 3.21$
$f(1.6) \approx f(1.5) + f'(1.5)\cdot 0.1$
$= 3.21 + (0.4)(0.1) = 3.25$

So $f(1.6) \approx 3.25$.

(b) The distance traveled during $[1, 1.6]$ is
$f(1.6) - f(1) \approx 3.25 - 3 \approx 0.25$.

4.10 The Second Derivative and Growth of a Function

2 Here $v_0 = y_0 = 0$ and $y(t) \le \dfrac{1}{2}Mt^2$. Thus $y(100)$

$\le \dfrac{1}{2}(10)\cdot(100)^2 = 50{,}000$ feet.

4 In this case $m = 4.1$, $M = 15.5$, and $t = 6$. So $1/2\cdot 4.1\cdot 6^2 \le y(6) \le 1/2\cdot 15.5\cdot 6^2$, or 73.8 mi $\le y(6) \le 279$ mi.

6 Applying the growth theorem with $t_0 = 2$, $t = 5$, and $M = 2$, we have $f(5) \le \dfrac{1}{2}\cdot 2(5 - 2)^2 = 9$.

8 Applying the growth theorem with $t_0 = 1$, $t = 4$, $m = 2.9$, and $M = 3.1$, we have

4.10 The Second Derivative and Growth of a Function

$$\frac{1}{2} \cdot 2.9(4-1)^2 \leq f(4) \leq \frac{1}{2} \cdot 3.1(4-1)^2$$

$$13.05 \leq f(4) \leq 13.95$$

10 (a) Here $f'(x) = \frac{1}{2}x^{-1/2}$, so $f(3) = \sqrt{3}$ and $f'(3)$

$= \frac{1}{2\sqrt{3}}$. Thus $p(x) = f(3) + f'(3)(x-3) =$

$\sqrt{3} + \frac{1}{2\sqrt{3}}(x-3) = \frac{1}{2\sqrt{3}}x + \frac{1}{2}\sqrt{3}$

$= \frac{1}{2}\left(\frac{x}{\sqrt{3}} + \sqrt{3}\right).$

(b)

x	$E(x) = f(x) - p(x)$	$(x-3)^2$	$E(x)/(x-3)^2$
4	-0.0207259	1	-0.0207259
3.1	-0.0002366	0.01	-0.0236635
3.01	-0.0000024	0.0001	-0.0240163
2.99	-0.0000024	0.0001	-0.0240964

(c) $f''(x) = -\frac{1}{4}x^{-3/2}$, so $\frac{1}{2}f''(3) = -\frac{1}{8} \cdot 3^{-3/2} \approx$

-0.024056. Hence $\frac{f''(3)}{2}$ does agree with the

fourth column in (b).

12 (a) Since $E(x)$ decreases as $(\Delta x)^2$, and $E(3.1) = 0.02$ when $\Delta x = 0.1$, we would expect

$E(3.01)$ to be close to $0.02 \cdot (\frac{1}{10})^2 = 0.0002$

when $\Delta x = 0.01$.

(b) For Δx small, $\frac{E(x + \Delta x)}{(\Delta x)^2} \approx \frac{f''(x)}{2}$, so

$$\frac{2 E(3.1)}{(0.1)^2} = \frac{2 \cdot 0.02}{0.01} = 4.$$

14 (a) Under the assumption that $f'(0) = 0$, $g(0) = f'(0) - M \cdot 0 = 0 - 0 = 0$.

Under the assumption that $f''(t) \leq M$, $g'(t) = f''(t) - M \leq 0$.

(b) Since g has a negative derivative on $[0, t]$, Corollary 3 in Sec. 4.1 implies that g is decreasing on $[0, t]$. Hence $g(t) \leq g(0) = 0$.

(c) Because $f'(t) = Mt + g(t)$ and $g(t) \leq 0$, we conclude that $f'(t) \leq Mt$.

16 Introduce the error function $E(b) = f(b) - p(b)$, where $p(x)$ is the differential approximation, and let m and M be the minimum and maximum, respectively, of $f''(x)$ for x in $[a, b]$. By Theorem 2,

$$\frac{m(b-a)^2}{2} \leq E(b) \leq \frac{M(b-a)^2}{2}.$$

Since $f''(x) = m$ and $f''(y) = M$ for some x and y in $[a, b]$ and f'' is continuous, the intermediate-value theorem implies that $E(b) = \frac{f''(c)(b-a)^2}{2}$ for some c in $[a, b]$. Hence $f(b) = p(b) + E(b) = f(a) + f'(a)(b-a) + \frac{1}{2}f''(c)(b-a)^2$.

18 (a) Let $D(x) = f(a) + f'(a)(x - a) + \frac{f''(a)}{2}(x-a)^2$.

Now $D(a) = f(a)$, $D'(a) = f'(a)$, and $D''(a) = f''(a)$. So $D(x)$ more closely resembles $f(x)$ at a because their second derivatives are equal at a, whereas with $p(x)$ this is not necessarily the case.

4.S Review Exercises

2 Let $f(x) = 3\tan x + x^3$, so $f'(x) = 3\sec^2 x + 3x^2 > 0$ for x in $[0, \pi/4]$. Hence f is increasing on $[0, \pi/4]$. Since $f(0) = 0 < 2$ and $f(\pi/4) = 3 + (\pi/4)^3 > 2$, $f(x) = 2$ for exactly one x in $[0, \pi/4]$.

4 Let x be the length, in miles, along the side of the road and y the length of the perpendicular sides. Thus $y = 1/x$. The cost is proportional to $C = 5x + 3(x + 2y) = 8x + 6/x$. This must be minimized for $x > 0$. $\dfrac{dC}{dx} = 8 - \dfrac{6}{x^2} = \dfrac{8x^2 - 6}{x^2}$ changes sign from $-$ to $+$ at $\dfrac{\sqrt{3}}{2}$. The pasture is $\dfrac{\sqrt{3}}{2}$ miles (along the road) by $\dfrac{2}{\sqrt{3}}$ miles (perpendicular to the road).

6 $P' > 0$, but $P'' < 0$, where P is the overall wholesale price index.

8 Slope, velocity, magnification, population growth rate, density, acceleration.

10 Let the rectangle have width w and height h. Let P be the prescribed perimeter. Then $P = 3w + 2h$, so $h = \dfrac{1}{2}(P - 3w)$. The area is $A = wh + \dfrac{\sqrt{3}}{4}w^2 = \dfrac{P}{2}w - \dfrac{6 - \sqrt{3}}{4}w^2$. Then $\dfrac{dA}{dw} = \dfrac{P}{2} - \dfrac{6 - \sqrt{3}}{2}w$ changes sign from $+$ to $-$ at $w = \dfrac{P}{6 - \sqrt{3}}$. Hence $h = \dfrac{(3 - \sqrt{3})P}{2(6 - \sqrt{3})} = \dfrac{5 - \sqrt{3}}{22}P$.

12 Let the sphere have radius r and the cube have side s; then $1 = 4\pi r^2 + 6s^2$ so $0 = 8\pi r + 12s\dfrac{ds}{dr}$ and $\dfrac{ds}{dr} = -\dfrac{2\pi r}{3s}$. The total volume is $V = \dfrac{4}{3}\pi r^3 + s^3$. Its extreme values must be found for $0 \le r \le \dfrac{1}{\sqrt{4\pi}}$. $\dfrac{dV}{dr} = 4\pi r^2 + 3s^2 \cdot \dfrac{ds}{dr} = 4\pi r^2 - 2\pi rs = 2\pi r(2r - s)$, which changes sign from $-$ to $+$ when $2r = s$.

(a) Solving $2r = s$ and $1 = 4\pi r^2 + 6s^2$ gives $r = \dfrac{1}{2\sqrt{\pi + 6}}$ and $s = \dfrac{1}{\sqrt{\pi + 6}}$. The amount used for the sphere is $4\pi r^2 = \dfrac{\pi}{\pi + 6}$ ft^2; the amount used for the cube is $6s^2 = \dfrac{6}{\pi + 6}$ ft^2.

(b) $V(r)$ is a maximal either for $r = 0$, $s = \dfrac{1}{\sqrt{6}}$, or for $r = \dfrac{1}{\sqrt{4\pi}}$, $s = 0$. Since $V(0) = \dfrac{1}{6\sqrt{6}} < \dfrac{1}{6\sqrt{\pi}} = V\left(\dfrac{1}{\sqrt{4\pi}}\right)$, all of the glass should be used for the sphere.

14 $f(x) = x^{1/3}$, $f'(x) = \dfrac{1}{3}x^{-2/3}$, $\sqrt[3]{8 + h} = f(8 + h) \approx f(8) + f'(8)h = 2 + h/12$.

16 (a) Let $f(x) = \dfrac{1}{\sqrt{x}}$, so $f'(x) = -\dfrac{1}{2}x^{-3/2}$. For

4.S Summary: Review Exercises

small x, $f(1 - x^2) \approx f(1) + f'(1)(-x^2)$

$= 1 + \dfrac{1}{2}x^2$

(b) Here $x = 0.2$, so $f(1 - (0.2)^2) \approx$

$1 + \dfrac{1}{2}(0.2)^2 = 1.02$. Exactly, $f(1 - (0.2)^2)$

$= f(0.96) = \dfrac{1}{\sqrt{0.96}} \approx 1.0206$. The estimate

is quite good, off by only about

$|1.02 - 1.0206| = 0.0006$.

18 (a) Let $f(x) = \dfrac{1}{\sqrt{x}}$, so $f'(x) = -\dfrac{1}{2}x^{-3/2}$. Thus, for

small x, $f(2 + 7x) \approx f(2) + f'(2)(7x)$

$= \dfrac{1}{\sqrt{2}} - \dfrac{7x}{4\sqrt{2}}$.

(b) Here $x = 0.1$, so $f(2 + 7 \cdot 0.1) = f(2.7) =$

$\dfrac{1}{\sqrt{2.7}} \approx 0.6086$. The estimate from (a) is

$\dfrac{1}{\sqrt{2}} - \dfrac{7 \cdot 0.1}{4\sqrt{2}} \approx 0.5834$. The estimate is not

very close because though $x = 0.1$ is small
relative to 2, $7x = 0.7$ is not, and the
differential approximation fails.

20 (a) Here $dr = 1.1 - 1 = 0.1$, and $dV = 4\pi r^2\, dr$

$\approx 4\pi \cdot 1^2 \cdot 0.1 = 0.4\pi$ in^3.

(b) Since $\dfrac{dV}{V} = 3\dfrac{dr}{r}$, $\left|\dfrac{dV}{V}\right| = 3 \cdot \left|\dfrac{0.1}{1}\right| = 0.3$ or

a 30% change in volume.

(c) Here $dr = 2.1 - 2 = 0.1$, and $dV = 4\pi r^2\, dr$

$\approx 4\pi \cdot 2^2 \cdot 0.1 = 1.6\pi$ in^3.

(d) Now $\left|\dfrac{dV}{V}\right| = 3\left|\dfrac{0.1}{2}\right| = 0.15$, or a 15%

change in volume.

22 Since the error in height h is negligible, we use the
differential of the volume $V = \pi r^2 h$ with respect to

the radius r. Then $\dfrac{dV}{V} = \dfrac{2\pi r h\, dr}{\pi r^2 h} = 2\dfrac{dr}{r}$. Since

$\left|\dfrac{dV}{V}\right| \leq 0.03$, we need $\left|\dfrac{dr}{r}\right| \leq 0.015$. Thus, a

1.5% error is the most that can be permitted when

measuring the radius. Finally, we note that $\left|\dfrac{dC}{C}\right| =$

$\left|\dfrac{2\pi\, dr}{2\pi r}\right| = \left|\dfrac{dr}{r}\right|$, so a 1.5% error is the most that

can be tolerated in the measurement of the
circumference as well.

24 $f(x) = x^3/6 + ax + b$ for constants a and b.

26 Note that $p(x) = f(a) + f'(a)(x - a) =$

$\dfrac{1}{(1 + 2)^2} + \dfrac{-2}{(1 + 2)^3} \cdot (x - 2) = \dfrac{1}{9} - \dfrac{2}{27}(x - 2)$,

and $df = f'(a)\, dx = -\dfrac{2}{27}\, dx = -\dfrac{2}{27}\Delta x$.

Δx	$f(a+\Delta x) - p(a+\Delta x)$	Δf	df	$\Delta f - df$
0.2	0.0013600	−0.0134549	−0.0148148	0.0013600
0.1	0.0003546	−0.0070528	−0.0074074	0.0003546
0.01	0.0000037	−0.0007371	−0.0007407	0.0000037
−0.01	0.0000037	0.0007445	0.0007407	0.0000037

28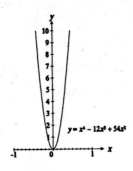

30 (a) $|x-1|$
   (b) $x^{1/3}$
   (c) $(x-2)^3$
   (d) $(x-2)^4$

32 Let height be h and base be x. Then the volume is $V = x^2h$ and cost is $C = 2ax^2 + 4bxh$.

   (a) With C fixed, $h = \dfrac{C - 2ax^2}{4bx}$, so $V = \dfrac{1}{4b}(Cx - 2ax^3)$ and $\dfrac{dV}{dx} = \dfrac{1}{4b}(C - 6ax^2)$

   changes sign from $+$ to $-$ at $x = \sqrt{\dfrac{C}{6a}}$.

   Then $h = \dfrac{\sqrt{aC}}{b\sqrt{6}}$ and $\dfrac{h}{x} = \dfrac{a}{b}$.

   (b) With V fixed, $h = \dfrac{V}{x^2}$ so $C = 2ax^2 + \dfrac{4bV}{x}$

   and $\dfrac{dC}{dx} = 4ax - \dfrac{4bV}{x^2} = \dfrac{4}{x^2}(ax^3 - bV)$

   changes sign from $-$ to $+$ at $x = \sqrt[3]{\dfrac{bV}{a}}$.

   Then $h = \sqrt[3]{\dfrac{a^2V}{b^2}}$ and $\dfrac{h}{x} = \dfrac{a}{b}$.

34 (a) $\sec(\pi/3 + h) \approx \sec \pi/3 + (\sec \pi/3 \tan \pi/3)h$
       $= 2 + 2\sqrt{3}h$

   (b) $\sqrt[3]{1 + h^2} = (1 + h^2)^{1/3} \approx 1^{1/3} + \dfrac{1}{3} \cdot 1^{-2/3} \cdot h^2$
       $= 1 + \dfrac{1}{3}h^2$

   (c) $\dfrac{1}{(1-h)^2} = (1-h)^{-2} \approx$
       $1^{-2} + (-2) \cdot 1^{-3}(-h) = 1 + 2h$

36 $\sec^2\left(\dfrac{\pi}{4}xy\right) \cdot \dfrac{\pi}{4}\left(x\dfrac{dy}{dx} + y\right) + 3y^2 \cdot \dfrac{dy}{dx} + 1 = 0$

   $\dfrac{dy}{dx} = -\dfrac{\dfrac{\pi}{4}y\sec^2\left(\dfrac{\pi}{4}xy\right) + 1}{\dfrac{\pi}{4}x\sec^2\left(\dfrac{\pi}{4}xy\right) + 3y^2} = -\dfrac{\pi + 2}{\pi + 6}$

38 (a) $5(2.1 - 2) = 0.5$ feet
   (b) 0.5
   (c) 0.5 feet

40 If $dy/dx = ax^2$ then $\dfrac{d^2y}{dx^2} = \dfrac{d}{dx}(ax^2) = 2ax$.

42 $\sin\left(\dfrac{\pi}{6} + h\right) \approx \sin\dfrac{\pi}{6} + \left(\cos\dfrac{\pi}{6}\right)h = \dfrac{1}{2} + \dfrac{\sqrt{3}}{2}h$

44 Let $f(\theta) = \sin^2\theta \cos\theta$. Then $f'(\theta) = (2\cos^2\theta - \sin^2\theta)\sin\theta$ is 0 when $\sin\theta = 0$ or $\sin^2\theta = 2\cos^2\theta$. Since $\sin^2\theta + \cos^2\theta = 1$, the latter gives $\sin^2\theta = 2/3$ and $\cos^2\theta = 1/3$. The maximum is $\dfrac{2}{3}\dfrac{\sqrt{1}}{3} = \dfrac{2}{3\sqrt{3}}$.

46 Let x be the length of a side of the inscribed square, and suppose that its corners touch the sides of the outer square a distance y from the outer square's corners. Then the area, which we consider

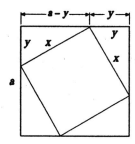

as a function of y for $0 \leq y \leq a$, is $A(y) = x^2 = y^2 + (a - y)^2 = y^2 + a^2 - 2ay + y^2 = 2y^2 - 2ay + a^2$, so $A'(y) = 4y - 2a$, which changes from $-$ to $+$ when $4y = 2a$; that is, $y = a/2$. Hence the smallest square has $y = a/2$ and sides of length $x = \dfrac{a}{\sqrt{2}}$.

48 Let the point A be given by (x_0, y_0). Consider the distance from A to an arbitrary point $P = (x, f(x))$ on the curve. Let us minimize the square of the distance (to avoid square roots) and thereby minimize the distance itself: $\overline{AP}^2 = (x - x_0)^2 + (f(x) - y_0)^2$, $\dfrac{d}{dx}(\overline{AP}^2) = 2(x - x_0) + 2(f(x) - y_0) \cdot f'(x)$. This equals 0 when $2(x - x_0) = -2(f(x) - y_0) \cdot f'(x)$, so $\dfrac{f(x) - y_0}{x - x_0} \cdot f'(x) = -1$. Since $\dfrac{f(x) - y_0}{x - x_0}$ is the slope of the line AP and $f'(x)$ is the slope of the tangent line at $(x, f(x))$, we conclude that the lines are perpendicular.

50 (a) Let $f(x) = 2x^5 - 10x + 5$; then $f'(x) = 10x^4 - 10$, which is positive for $x < -1$ and $x > 1$, negative for $-1 < x < 1$. Since $\lim_{x \to -\infty} f(x) = -\infty$, $f(-1) = 13 > 0$, $f(1) =$

$-3 < 0$, and $\lim_{x \to \infty} f(x) = \infty$, it follows that f has three roots, one in each of the intervals $(-\infty, -1), (-1, 1),$ and $(1, \infty)$.

(b) The roots are about -1.601, 0.507, and 1.329.

52 $f(x) = x^3 + x - 6$. Then $f'(x) = 3x^2 + 1$ is positive for all x. Since $\lim_{x \to -\infty} f(x) = -\infty$ and $\lim_{x \to \infty} f(x) = \infty$, there is exactly one root. Applying Newton's method with $x_1 = 1$ yields $x_2 = 2$, $x_3 \approx 1.692$, $x_4 \approx 1.636$, $x_5 \approx x_6 \approx 1.634$. The root is approximately 1.634.

54 Let x be the horizontal distance from the worker to the pulley, and let $y = \overline{BC}$. Now $\dfrac{dx}{dt} = 5$ and $x^2 + 10^2 = y^2$, so $2x\dfrac{dx}{dt} = 2y\dfrac{dy}{dt}$, hence $\dfrac{dy}{dt} = \dfrac{x}{y} \cdot \dfrac{dx}{dt} = 5\dfrac{x}{y}$. The load rises at the rate dy/dt, since the load rises as fast as the rope is pulled.

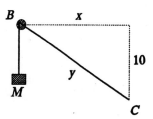

(a) When $y = 15$, $x = \sqrt{15^2 - 10^2} = 5\sqrt{5}$, so $\dfrac{dy}{dt} = 5 \cdot \dfrac{x}{y} = \dfrac{5 \cdot 5\sqrt{5}}{15} = \dfrac{5\sqrt{5}}{3}$ ft/sec.

(b) When $y = 100$, $x = \sqrt{100^2 - 10^2} = 30\sqrt{11}$, so $\frac{dy}{dt} = 5 \cdot \frac{x}{y} = \frac{5 \cdot 30\sqrt{11}}{100} = \frac{3\sqrt{11}}{2}$ ft/sec.

56 If the radius is r and the height is h, then $V = \frac{\pi}{3}r^2 h$, so $h = \frac{3V}{\pi}r^{-2}$. Also, $l = \sqrt{r^2 + h^2} = \left(r^2 + \frac{9V^2}{\pi^2}r^{-4}\right)^{1/2}$, so the area is $A = \pi r^2 + \pi r l = \pi\left(r^2 + \sqrt{r^4 + 9V^2\pi^{-2}r^{-2}}\right)$. Hence $\frac{dA}{dr} = \pi\left[2r + (2r^3 - 9V^2\pi^{-2}r^{-3})(r^4 + 9V^2\pi^{-2}r^{-2})^{-1/2}\right]$. Note that $A \to \infty$ as $r \to 0^+$ or $r \to \infty$. Hence there is a minimum, at which $\frac{dA}{dr} = 0$. Setting $\frac{dA}{dr} = 0$ yields $2\sqrt{r^6 + 9V^2\pi^{-2}} = 9V^2\pi^{-2}r^{-3} - 2r^3$,

$4r^6 + 36V^2\pi^{-2} = 81V^4\pi^{-4}r^{-6} - 36V^2\pi^{-2} + 4r^6$,

$81V^4\pi^{-4}r^{-6} = 72V^2\pi^{-2}$, and $r = \frac{1}{\sqrt{2}}\sqrt[3]{\frac{3V}{\pi}}$.

Hence $h = \frac{3V}{\pi}r^{-2} = 2\sqrt[3]{\frac{3V}{\pi}}$.

58 Let S be the square of the distance from $(x, 3x+7)$ to $(0,0)$. Thus $S = x^2 + (3x+7)^2 = 10x^2 + 42x + 49$. Hence $\frac{dS}{dx} = 20x + 42$ changes sign from $-$ to $+$ at $x = -\frac{21}{10}$. The point is $\left(-\frac{21}{10}, \frac{7}{10}\right)$.

60 (a) $f'(x) = (x-1)^n + n(x-1)^{n-1}(x-2)$, so $f'(1) = 0$.

(b) For small $\epsilon > 0$, $f(1 + \epsilon) = \epsilon^n(-1 + \epsilon) < 0$

and $f(1 - \epsilon) = (-\epsilon)^n(-1 - \epsilon) = (-1)^{n+1}\epsilon^n(1 + \epsilon)$. So $x = 1$ gives a maximum if n is odd, neither otherwise.

62 (a),(b)

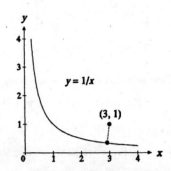

(c) Let S be the square of the distance from $(x, 1/x)$ to $(3, 1)$, so $S = (x-3)^2 + \left(\frac{1}{x} - 1\right)^2$

$= x^2 - 6x + 10 - 2x^{-1} + x^{-2}$. Hence $\frac{dS}{dx} = 2x - 6 + 2x^{-2} - 2x^{-3}$. At the nearest point, $\frac{dS}{dx} = 0$, so $x^4 - 3x^3 + x - 1 = 0$.

(d) Let $f(x) = x^4 - 3x^3 + x - 1$. Then $f(2.9) = -0.5389$ and $f(3) = 2$, so $f(x) = 0$ for some x in $(2.9, 3)$.

64 Recall from geometry that an angle inscribed in a semicircle is a right angle; furthermore, the

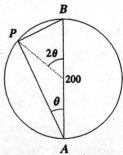

measure of a central angle is twice that of the corresponding inscribed angle. Hence in the figure we have the right triangle APB and the angles θ and

2θ. Then $\overline{AP} = 200 \cos \theta$ and $\widehat{BP} = 200\theta$, so the travel time is $T(\theta) = \dfrac{\overline{AP}}{100} + \dfrac{\widehat{BP}}{200} = 2 \cos \theta + \theta$.

This must be minimized for $0 \leq \theta \leq \pi/2$. $\dfrac{dT}{d\theta} = 1 - 2 \sin \theta$ changes sign from $+$ to $-$ at $\theta = \pi/6$. Hence the minimum occurs at either $\theta = 0$ or $\theta = \pi/2$. Since $T(\pi/2) = \pi/2 < 2 = T(0)$, the minimum occurs at $\theta = \pi/2$. The swimmer should walk the entire distance.

66 (a) $\Delta A \approx dA = d(\pi r^2) = 2\pi r \, dr$

(b) If we cut and straighten the region between concentric circles with radii r and $r + dr$, we obtain, approximately, a $2\pi r$ by dr rectangle.

68 If $xy = 2$, then $y = 2x^{-1}$ so $S = x^2 + y^2 = x^2 + 4x^{-2}$ and $dS/dx = 2x - 8x^{-3} = 2x^{-3}(x^4 - 4)$ changes sign from $-$ to $+$ at $x = \sqrt{2}$. Hence both numbers should be $\sqrt{2}$.

70 Note: The exercise asks only for *sketches* of the required graphs. This solution goes further by providing approximate equations for the necessary figures, which may be useful for the instructor but it is *not* expected that students should be able to come up with these on their own—especially in last figure of (b), which is based on an antiderivative.

(a) In the case of Figure 5, $v(t)$ is a constant, so $a(t) = v'(t) = 0$ and the graph is just a straight line coinciding with the t axis. In the case of Figure 6, however, $v(t) = 5t$, so $a(t) = 5$.

In Figure 7, the graph of $v(t)$ appears to be a parabola with equation $v(t) = 10t - 5t^2$, in which case $a(t) = 10 - 10t$. Even if not exact, this gives a good guide for the shape of the curve.

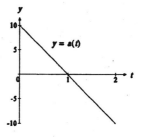

Finally, Figure 8 resembles a sine wave, $v(t) \approx 5 \sin(\pi t/2)$. Hence $a(t) \approx \dfrac{5\pi}{2} \cos \dfrac{\pi t}{2}$.

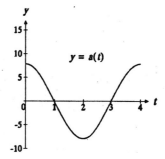

(b) In the case of Figure 5, $s(t) = 10t$.

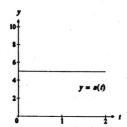

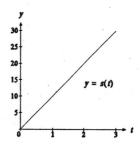

Using the information obtained in (a), it appears in the case of Figure 6 that $s(t) = 5t^2/2$.

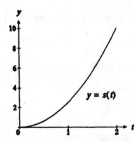

Using the information obtained in (a), it appears in the case of Figure 7 that $s(t)$ is close to $5t^2 - \dfrac{5}{3}t^3$.

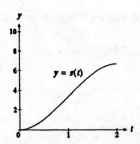

Using the information obtained in (a), it appears in the case of Figure 8 that $s(t)$ is close to $\dfrac{10}{\pi}\left(1 - \cos\dfrac{\pi t}{2}\right)$.

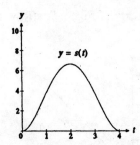

5 The Definite Integral

5.1 Estimates in Four Problems

2 (a)

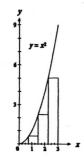

(c) The total area of the four rectangles is

$$\frac{3}{4}\left(\frac{0}{4}\right)^2 + \frac{3}{4}\left(\frac{3}{4}\right)^2 + \frac{3}{4}\left(\frac{6}{4}\right)^2 + \frac{3}{4}\left(\frac{9}{4}\right)^2$$

$$= \frac{27}{64}(0^2 + 1^2 + 2^2 + 3^2) = \frac{27}{64} \cdot 14 = \frac{378}{64}$$

$$= 5.90625.$$

4 The area of the estimate is $\frac{3}{5}\left(\frac{3}{10}\right)^2 + \frac{3}{5}\left(\frac{9}{10}\right)^2 +$

$$\frac{3}{5}\left(\frac{15}{10}\right)^2 + \frac{3}{5}\left(\frac{21}{10}\right)^2 + \frac{3}{5}\left(\frac{27}{10}\right)^2$$

$$= \frac{3}{5}\left(\frac{3}{10}\right)^2(1^2 + 3^2 + 5^2 + 7^2 + 9^2) = \frac{4455}{500}$$

$$= 8.91.$$

6 The area of the underestimate is $\frac{3}{15}\left(\frac{0}{15}\right)^2 +$

$$\frac{3}{15}\left(\frac{3}{15}\right)^2 + \frac{3}{15}\left(\frac{6}{15}\right)^2 + \cdots + \frac{3}{15}\left(\frac{42}{15}\right)^2$$

$$= \left(\frac{3}{15}\right)^3(0^2 + 1^2 + 2^2 + \cdots + 14^2) = \frac{27405}{3375}$$

$$= 8.12.$$

8 The area of the underestimate equals $\frac{1}{5}\left(\frac{5}{5}\right)^2 +$

$$\frac{1}{5}\left(\frac{6}{5}\right)^2 + \frac{1}{5}\left(\frac{7}{5}\right)^2 + \frac{1}{5}\left(\frac{8}{5}\right)^2 + \frac{1}{5}\left(\frac{9}{5}\right)^2$$

$$= \frac{1}{5^3}(5^2 + 6^2 + 7^2 + 8^2 + 9^2) = \frac{255}{125} = 2.04.$$

10 The area of the estimate is $\frac{1}{10}\left(\frac{21}{20}\right)^2 + \frac{1}{10}\left(\frac{23}{20}\right)^2$

$$+ \frac{1}{10}\left(\frac{25}{20}\right)^2 + \cdots + \frac{1}{10}\left(\frac{39}{20}\right)^2$$

$$= \frac{1}{10 \cdot 20^2}(21^2 + 23^2 + 25^2 + \cdots + 39^2) = \frac{9330}{4000}$$

$$= 2.3325.$$

12 The midpoints of [0, 1], [1, 5/3], [5/3, 11/4], and [11/4, 3] are 1/2, 4/3, 53/24, and 23/8, respectively. Thus the approximating area is

$$(1 - 0)\left(\frac{1}{2}\right)^2 + \left(\frac{5}{3} - 1\right)\left(\frac{4}{3}\right)^2 + \left(\frac{11}{4} - \frac{5}{3}\right)\left(\frac{53}{24}\right)^2$$

$$+ \left(3 - \frac{11}{4}\right)\left(\frac{23}{8}\right)^2 = \frac{1}{4} + \frac{2}{3} \cdot \frac{16}{9} + \frac{13}{12} \cdot \frac{2809}{576}$$

$$+ \frac{1}{4} \cdot \frac{529}{64} = \frac{1}{4} + \frac{32}{27} + \frac{36517}{6912} + \frac{529}{256} = \frac{1265}{144}$$

$$= 8.7847\overline{2}.$$

14 (c) The total area equals $\frac{1}{5}\left(\frac{5}{6}\right) + \frac{1}{5}\left(\frac{5}{7}\right) + \frac{1}{5}\left(\frac{5}{8}\right)$

$+ \frac{1}{5}\left(\frac{5}{9}\right) + \frac{1}{5}\left(\frac{5}{10}\right) = \frac{1}{6} + \frac{1}{7} + \frac{1}{8} + \frac{1}{9}$

$+ \frac{1}{10} = \frac{1627}{2520} \approx 0.6456.$

16 (a),(b)

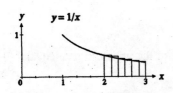

(d) The total area equals $\frac{1}{6}\left(\frac{6}{12}\right) + \frac{1}{6}\left(\frac{6}{13}\right) +$

$\frac{1}{6}\left(\frac{6}{14}\right) + \frac{1}{6}\left(\frac{6}{15}\right) + \frac{1}{6}\left(\frac{6}{16}\right) + \frac{1}{6}\left(\frac{6}{17}\right)$

$= \frac{1}{12} + \frac{1}{13} + \frac{1}{14} + \frac{1}{15} + \frac{1}{16} + \frac{1}{17}$

$= \frac{51939}{123760} \approx 0.4197.$

18 (a),(b)

(d) The total area equals $\frac{1}{2}\sqrt{\frac{3}{2}} + \frac{1}{2}\sqrt{\frac{4}{2}} +$

$\frac{1}{2}\sqrt{\frac{5}{2}} + \frac{1}{2}\sqrt{\frac{6}{2}} + \frac{1}{2}\sqrt{\frac{7}{2}} + \frac{1}{2}\sqrt{\frac{8}{2}} = \sqrt{\frac{3}{8}}$

$+ \sqrt{\frac{4}{8}} + \sqrt{\frac{5}{8}} + \sqrt{\frac{6}{8}} + \sqrt{\frac{7}{8}} + \sqrt{\frac{8}{8}} \approx$

$4.9115.$

20 (a),(b)

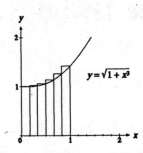

(d) The total area equals $\frac{1}{6}\sqrt{1 + \left(\frac{1}{6}\right)^3} +$

$\frac{1}{6}\sqrt{1 + \left(\frac{2}{6}\right)^3} + \frac{1}{6}\sqrt{1 + \left(\frac{3}{6}\right)^3} +$

$\frac{1}{6}\sqrt{1 + \left(\frac{4}{6}\right)^3} + \frac{1}{6}\sqrt{1 + \left(\frac{5}{6}\right)^3} +$

$\frac{1}{6}\sqrt{1 + \left(\frac{6}{6}\right)^3} = \frac{1}{36\sqrt{6}}(\sqrt{217} + \sqrt{224} +$

$\sqrt{243} + \sqrt{280} + \sqrt{341} + \sqrt{432}) \approx 1.1484.$

22 (a) The total area equals $2 \cdot \frac{1}{2^2} + 2 \cdot \frac{1}{4^2} + 2 \cdot \frac{1}{6^2}$

$= \frac{49}{72} = 0.680\overline{5}.$

(b) The total area equals $1 \cdot \frac{1}{(3/2)^2} + 1 \cdot \frac{1}{(5/2)^2} +$

$1 \cdot \frac{1}{(7/2)^2} + 1 \cdot \frac{1}{(9/2)^2} + 1 \cdot \frac{1}{(11/2)^2} +$

$1 \cdot \frac{1}{(13/2)^2} \approx 0.7922.$

(c) The total area equals $1 \cdot \frac{1}{1^2} + 1 \cdot \frac{1}{2^2} + 1 \cdot \frac{1}{3^2}$

$+ 1 \cdot \frac{1}{4^2} + 1 \cdot \frac{1}{5^2} + 1 \cdot \frac{1}{6^2} \approx 1.4914.$

5.1 Estimates in Four Problems

(d) The total area equals $1 \cdot \frac{1}{2^2} + 1 \cdot \frac{1}{3^2} + 1 \cdot \frac{1}{4^2}$

$+ 1 \cdot \frac{1}{5^2} + 1 \cdot \frac{1}{6^2} + 1 \cdot \frac{1}{7^2} \approx 0.5118.$

24 (a) The total area equals $\frac{1}{50} \cdot \frac{1}{51/50} +$

$\frac{1}{50} \cdot \frac{1}{52/50} + \cdots + \frac{1}{50} \cdot \frac{1}{100/50} = \frac{1}{51} +$

$\frac{1}{52} + \cdots + \frac{1}{100} \approx 0.6882.$

(b) The total area equals $\frac{1}{50} \cdot \frac{1}{50/50} +$

$\frac{1}{50} \cdot \frac{1}{51/50} + \cdots + \frac{1}{50} \cdot \frac{1}{99/50} = \frac{1}{50} + \frac{1}{51}$

$+ \cdots + \frac{1}{99} \approx 0.6982.$

26 (a) The mass is approximately $\frac{3}{10}\left(\frac{3}{20}\right)^2 +$

$\frac{3}{10}\left(\frac{9}{20}\right)^2 + \cdots + \frac{3}{10}\left(\frac{57}{20}\right)^2 = \frac{3591}{400}$

$= 8.9775$ grams.

(b) The mass is approximately $\frac{3}{10}\left(\frac{3}{10}\right)^2 +$

$\frac{3}{10}\left(\frac{6}{10}\right)^2 + \cdots + \frac{3}{10}\left(\frac{30}{10}\right)^2 = \frac{2079}{200}$

$= 10.395$ grams.

(c) The mass is approximately $\frac{3}{10}\left(\frac{0}{10}\right)^2 +$

$\frac{3}{10}\left(\frac{3}{10}\right)^2 + \cdots + \frac{3}{10}\left(\frac{27}{10}\right)^2 = \frac{1539}{200}$

$= 7.695$ grams.

(d) The mass in Problem 2 is less than 10.395 grams but larger than 7.695 grams.

28 (b) Distance $< \frac{1}{2}(2^{1/2} - 1) + \frac{1}{2}(2^1 - 1) + \cdots$

$+ \frac{1}{2}(2^4 - 1) = \frac{1}{2}(\sqrt{2} + 2 + 2\sqrt{2} + 4 +$

$4\sqrt{2} + 8 + 8\sqrt{2} + 16 - 8)$

$= \frac{1}{2}(22 + 15\sqrt{2}) \approx 21.6066$ km

(c) Distance $> \frac{1}{2}(2^0 - 1) + \frac{1}{2}(2^{1/2} - 1) + \cdots$

$+ \frac{1}{2}(2^{7/2} - 1) = \frac{1}{2}(1 + \sqrt{2} + 2 + 2\sqrt{2} + 4$

$+ 4\sqrt{2} + 8 + 8\sqrt{2} - 8) = \frac{1}{2}(7 + 15\sqrt{2})$

≈ 14.1066 km

30 Using similar triangles, we see that the radius of the circular cross section x feet down from the vertex of the cone is x feet.

(a) The partition of Problem 4 leads to cylindrical slabs to estimate the volume, as shown in the figure.

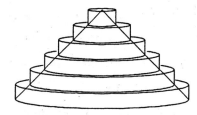

(b) The volume of a cylindrical slab is $\pi r^2 h$, where r is its radius and h is its height. Here all cylinders are of equal height, 1/2. Since the radius of a cross section of the cone equals its distance from the vertex, the estimate is

$$\pi\left(\frac{1}{2}\right)^2\left(\frac{1}{2}\right) + \pi\left(\frac{2}{2}\right)^2\left(\frac{1}{2}\right) + \pi\left(\frac{3}{2}\right)^2\left(\frac{1}{2}\right) +$$

$$\pi\left(\frac{4}{2}\right)^2\left(\frac{1}{2}\right) + \pi\left(\frac{5}{2}\right)^2\left(\frac{1}{2}\right) + \pi\left(\frac{6}{2}\right)^2\left(\frac{1}{2}\right)$$

$$= \frac{\pi}{8}(1^2 + 2^2 + 3^2 + 4^2 + 5^2 + 6^2) = \frac{91\pi}{8}$$

$$\approx 35.74 \text{ ft}^3.$$

(c)

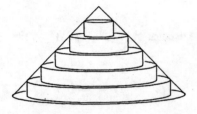

(d) Volume $\approx \pi\left(\frac{0}{2}\right)^2\left(\frac{1}{2}\right) + \pi\left(\frac{1}{2}\right)^2\left(\frac{1}{2}\right) +$

$$\pi\left(\frac{2}{2}\right)^2\left(\frac{1}{2}\right) + \pi\left(\frac{3}{2}\right)^2\left(\frac{1}{2}\right) + \pi\left(\frac{4}{2}\right)^2\left(\frac{1}{2}\right) +$$

$$\pi\left(\frac{5}{2}\right)^2\left(\frac{1}{2}\right)$$

$$= \frac{\pi}{8}(0^2 + 1^2 + 2^2 + 3^2 + 4^2 + 5^2) = \frac{55\pi}{8}$$

$$\approx 21.60 \text{ ft}^3$$

32 (a) $0.2[\sin 0 + \sin(0.2) + \cdots + \sin(1.2)] +$
$(\pi/2 - 1.4) \sin(1.4) \approx 0.8970$

(b) $0.2[\sin(0.2) + \sin(0.4) + \cdots + \sin(1.4)] +$
$(\pi/2 - 1.4) \sin(\pi/2) \approx 1.0966$

34 Speed at distance x centimeters from end $= 2\pi x$

cm/sec. Mass of each section $= \frac{32}{6} = \frac{16}{3}$ grams.

Energy of section from x to $x + 1/2$ centimeters

$$\approx \frac{1}{2} \cdot \frac{16}{3}\left[2\pi\left(x + \frac{1}{4}\right)\right]^2 = \frac{32\pi^2}{3}\left(x + \frac{1}{4}\right)^2 \text{ ergs.}$$

Total energy $\approx \dfrac{32\pi^2}{3}\left[\left(\dfrac{1}{4}\right)^2 + \left(\dfrac{3}{4}\right)^2 + \cdots +$

$\left(\dfrac{11}{4}\right)^2\right] = \dfrac{572\pi^2}{3}$ ergs.

36

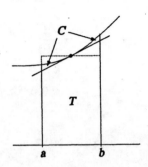

We have $f''(x) \geq 0$ on [1, 4], so f is concave up on [1, 4]. The figure shows an arbitrary section, with endpoints a and b, $1 \leq a \leq b < 4$, and length $b - a < 3$. The midpoint of the section is $x = \dfrac{a + b}{2}$. Since f is concave up on [1, 4], the curve lies above its tangent line at $x = \dfrac{a + b}{2}$. Let A equal the area under $y = f(x)$ on $[a, b]$, let T be the area under the tangent line and above $[a, b]$, and let C be the area above the tangent line and below the curve, above $[a, b]$. From the figure, $C > 0$, so $A = T + C > T$. Hence T is an underestimate of A. The area of the estimating rectangle (indicated by the dashed line) is equal to T, the area of the trapezoid. Hence the rectangles give an underestimate of the area.

5.1 Estimates in Four Problems

38 (a) $\dfrac{4}{3}(1+x^2)^{1/3} \cdot 2x = \dfrac{8}{3}x(1+x^2)^{1/3}$

(b) $\dfrac{1}{\left(\sqrt[3]{5x}\right)^2}\left[\sqrt[3]{5x}[(1+x^3)\cdot 3\cos 3x + 3x^2 \sin 3x]\right.$

$\left. - (1+x^3)\sin 3x \cdot \left(\dfrac{1}{3}\right)\sqrt[3]{5x}^{-2/3}\right]$

$= \dfrac{9x(1+x^3)\cos 3x + (8x^3 - 1)\sin 3x}{3\sqrt[3]{5x}^{4/3}}$

(c) $\dfrac{3}{8} + \dfrac{3}{32}(\sin 4x + 4x\cos 4x) +$

$\dfrac{1}{8}[\cos^3 2x \cdot 2 \cos 2x +$

$\sin 2x \cdot 3 \cos^2 2x \, (-\sin 2x)2]$

$= \dfrac{3}{8} + \dfrac{3}{32}\sin 4x + \dfrac{3}{8}x\cos 4x +$

$\dfrac{1}{4}\cos^4 2x - \dfrac{3}{4}\cos^2 2x \sin^2 2x$

(d) $\dfrac{3}{8}(-2)(2x+3)^{-3}\cdot 2 - \dfrac{1}{4}(-1)(2x+3)^{-2}\cdot 2$

$= \dfrac{1}{2}(2x+3)^{-2} - \dfrac{3}{2}(2x+3)^{-3}$

$= \dfrac{x}{(2x+3)^3}$

(e) $\dfrac{1}{6}\cdot 3 \cos^2 2x \cdot(-\sin 2x)\cdot 2 - \dfrac{1}{2}(-\sin 2x)\cdot 2$

$= \sin 2x - \sin 2x \cos^2 2x = \sin^3 2x$

(f) $3x^2\sqrt{x^2-1}\,\tan 5x + \dfrac{x^4}{\sqrt{x^2-1}}\tan 5x +$

$5x^3\sqrt{x^2-1}\,\sec^2 5x$

40 (a)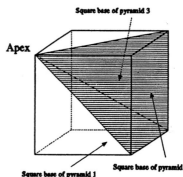

(b) Let the side of the cube equal 3 feet. Then the volume of the cube is 27 ft³ and the volume of one of the tents (equivalent to the one in Problem 4) is $1/3 \cdot 27$ ft³ $= 9$ ft³.

(c) Since all four problems are equivalent, we can conclude that the area in Problem 1 is 9, the mass in Problem 2 is 9 grams, and the distance traveled in Problem 3 is $8\cdot 9 = 72$ miles.

5.2 Summation Notation and Approximating Sums

2 (a) $2^2 + 3^2 + 4^2 = 29$

(b) 29

(c) $(1^2 + 1) + (2^2 + 2) + (3^2 + 3) = 20$

4 (a) $\dfrac{1}{3} + \dfrac{1}{4} + \dfrac{1}{5} = \dfrac{47}{60}$

(b) $1 + 1 + 1 + 1 + 1 = 5$

(c) $\dfrac{1}{2} + \dfrac{1}{4} + \dfrac{1}{8} = \dfrac{7}{8}$

6 (a) $\displaystyle\sum_{i=2}^{100}\dfrac{1}{i}$

(b) $\displaystyle\sum_{i=1}^{5}\dfrac{1}{2i+1}$

(c) $\sum_{i=0}^{50} \frac{1}{(2i+1)^2}$

8 (a) $\sum_{i=0}^{99} 8t_i^2(t_{i+1} - t_i)$

(b) $\sum_{i=1}^{n} 8t_i^2(t_i - t_{i-1})$

10 (a) $\frac{1}{3}(x_{100}^3 - x_0^3)$

(b) $\frac{1}{x_{70}} - \frac{1}{x_4}$

12 (a) $(a_1 - b_1) + (a_2 - b_2) + (a_3 - b_3)$
$= (a_1 + a_2 + a_3) - (b_1 + b_2 + b_3)$

(b) Let $a_1 = 0, a_2 = 1, b_1 = 1, b_2 = 0$.

(c) $(b_1 + b_2 + b_3)a_1 + (b_1 + b_2 + b_3)a_2 + (b_1 + b_2 + b_3)a_3$
$= (a_1 + a_2 + a_3)b_1 + (a_1 + a_2 + a_3)b_2 + (a_1 + a_2 + a_3)b_3$

14 $x_i - x_{i-1} = \frac{3-1}{4} = \frac{1}{2}, \sum_{i=1}^{n} f(c_i)(x_i - x_{i-1})$

$= \sum_{i=1}^{4} 3x_{i-1} \cdot \frac{1}{2} = \frac{3}{2}\left(1 + \frac{3}{2} + 2 + \frac{5}{2}\right) = 10.5$

16 $\sum_{i=1}^{4} (x_i^3 + x_i) \cdot \frac{1}{2} = \frac{1}{2}\left(\left[\left(\frac{3}{2}\right)^3 + \frac{3}{2}\right] + [2^3 + 2] + \left[\left(\frac{5}{2}\right)^3 + \frac{5}{2}\right] + [3^3 + 3]\right) = 31.5$

18 $\sum_{i=1}^{4} 2^{x_{i-1}} \cdot \frac{1}{2} = \frac{1}{2}(2^1 + 2^{3/2} + 2^2 + 2^{5/2})$
$= 3 + 3\sqrt{2} \approx 7.24$

20 (a) $\frac{100 \cdot 101 \cdot 201}{6} = 338{,}350$

(b) $2\frac{n(n+1)(2n+1)}{6} + 3\frac{n(n+1)}{2} + 6n$
$= \frac{1}{6}(4n^3 + 15n^2 + 47n)$

22 $(n+1)^5 - 1 = \sum_{i=1}^{n} [(i+1)^5 - i^5]$

$= \sum_{i=1}^{n} (5i^4 + 10i^3 + 10i^2 + 5i + 1) = 5\sum_{i=1}^{n} i^4$

$+ 10\left(\frac{n^4}{4} + \frac{n^3}{2} + \frac{n^2}{4}\right) + 10\left(\frac{n^3}{3} + \frac{n^2}{2} + \frac{n}{6}\right)$

$+ 5\left(\frac{n^2}{2} + \frac{n}{2}\right) + n$

$= 5\sum_{i=1}^{n} i^4 + \frac{5}{2}n^4 + \frac{25}{3}n^3 + 10n^2 + \frac{31}{6}n$, so

$\sum_{i=1}^{n} i^4 = \frac{1}{5}\bigg[(n+1)^5 - 1 - \frac{5}{2}n^4 - \frac{25}{3}n^3 -$
$10n^2 - \frac{31}{6}n\bigg] = \frac{1}{5}n^5 + \frac{1}{2}n^4 + \frac{1}{3}n^3 - \frac{1}{30}n$.

24 In the following, $x_i = 1 + (6i)/n$, where n is the number of sections.

(a) $\sum_{i=1}^{6} x_{i-1}^3(x_i - x_{i-1})$

(b) $\sum_{i=1}^{6} x_i^3(x_i - x_{i-1})$

(c) $\sum_{i=1}^{6} \left(\frac{x_i + x_{i-1}}{2}\right)^3 (x_i - x_{i-1})$

26 $\sqrt[3]{0}(1-0) + \sqrt[3]{1}(4-1) + \sqrt[3]{8}(10-4) = 15$

28 $\sum_{i=0}^{n-1} c^{n-1-i}d^i$ and $\sum_{i=0}^{n-1} c^i d^{n-1-i}$ both work.

5.2 Summation Notation and Approximating Sums

30 We have $x_i = \pi i/n$, so $\sum_{i=1}^{n} f(c_i)(x_i - x_{i-1}) =$

$$\sum_{i=1}^{n} \left(\sin \frac{\pi(i-1)}{n}\right) \frac{\pi}{n}.$$

32 (a) The appropriate Riemann sum is

$$\sum_{i=1}^{100} f(c_i)(x_i - x_{i-1}) = \sum_{i=1}^{100} 2^{-x_i} \cdot \left(\frac{1}{100}\right), \text{ where } x_i$$

$= i/100$. Now, $\sum_{i=1}^{100} 2^{-x_i} \cdot \left(\frac{1}{100}\right) =$

$$\frac{1}{100} \sum_{i=1}^{100} 2^{-x_i}$$

$$= \frac{1}{100}\left(2^{-\frac{1}{100}} + 2^{-\frac{2}{100}} + 2^{-\frac{3}{100}} + \cdots + 2^{-\frac{100}{100}}\right)$$

≈ 0.7189.

(b) The appropriate Riemann sum is

$$\sum_{i=1}^{100} f(c_i)(x_i - x_{i-1}) = \sum_{i=1}^{100} 2^{-x_{i-1}} \cdot \left(\frac{1}{100}\right). \text{ Now}$$

$$\sum_{i=1}^{100} 2^{-x_{i-1}} \cdot \left(\frac{1}{100}\right) = \frac{1}{100} \sum_{i=1}^{100} 2^{-x_{i-1}}$$

$$= \frac{1}{100}\left(2^{-\frac{0}{100}} + 2^{-\frac{1}{100}} + 2^{-\frac{2}{100}} + \cdots + 2^{-\frac{99}{100}}\right)$$

≈ 0.7239.

34 $(1 + r + r^2 + \cdots + r^{n-1})(1 - r) = \sum_{i=0}^{n-1} b_i(1 - r)$

$$= \sum_{i=0}^{n-1} (r^i - r^{i+1}) = r^0 - r^{(n-1)+1} = 1 - r^n, \text{ so } 1$$

$+ r + r^2 + \cdots + r^{n-1} = \dfrac{1 - r^n}{1 - r}$ if $r \neq 1$.

5.3 The Definite Integral

2 (a) 2

    (b) 3

    (c) 1

4 Shaded area $= \int_0^b x^2\, dx = \dfrac{b^3}{3}$

Area of $ABCD = b \cdot b^2 = b^3$

Ratio $= 1/3$

6 (a) a

    (b) $a + \dfrac{b-a}{n}$

    (c) $a + \dfrac{2(b-a)}{n},\ a + \dfrac{i(b-a)}{n}$

    (d) $\sum_{i=1}^{n} x_{i-1}^2 (x_i - x_{i-1})$

$$= \sum_{i=1}^{n} \left(a + \frac{(i-1)(b-a)}{n}\right)^2 \cdot \frac{b-a}{n}$$

$$= \frac{b-a}{n} \sum_{i=0}^{n-1} \left(a + \frac{i(b-a)}{n}\right)^2$$

$$= \frac{b-a}{n}\left(\sum_{i=0}^{n-1} a^2 + \sum_{i=0}^{n-1} \frac{2a(b-a)}{n} i + \sum_{i=0}^{n-1} \frac{(b-a)^2}{n^2} i^2\right)$$

$$= \frac{b-a}{n}\left(a^2 n + \frac{2a(b-a)}{n} \cdot \frac{(n-1)[(n-1)+1]}{2} + \frac{(b-a)^2}{n^2} \cdot \frac{(n-1)[(n-1)+1][2(n-1)+1]}{6}\right)$$

$$= \frac{b-a}{n}\left(a^2 n + a(b-a)(n-1) + \frac{(b-a)^2(n-1)(2n-1)}{6n}\right)$$

$$= \frac{b-a}{6}\left(2(a^2 + ab + b^2) + \frac{3(a^2 - b^2)}{n} + \frac{(b-a)^2}{n^2}\right)$$

(e) As $n \to \infty$, the expression in (d) approaches
$$\frac{b-a}{6} \cdot 2(a^2 + ab + b^2) = \frac{b^3 - a^3}{3}.$$

8 (a) Length of ith section of string
   (b) Density at a point of the ith section
   (c) Approximate mass of the ith section
   (d) Approximate mass of the string
   (e) Actual mass of the string

10 Since $\int_a^b f(x)\, dx = \lim_{\text{mesh} \to 0} \sum_{i=1}^n f(c_i)(x_i - x_{i-1})$,

$\int_1^\pi \sin x \, dx = \lim_{\text{mesh} \to 0} \sum_{i=1}^n (\sin c_i)(x_i - x_{i-1})$, where $x_0 = 1$ and $x_n = \pi$. $(x_i - x_{i-1})$ represents the length of one of the n sections in the partition of $[1, \pi]$, and the sampling point c_i is contained in $[x_{i-1}, x_i]$.

12 $\int_2^5 t^2 \, dt = \frac{5^3 - 2^3}{3} = 39$ ft

14 All sections have length 1.
   (a) $2^0 \cdot 1 + 2^1 \cdot 1 + 2^2 \cdot 1 = 7$
   (b) $2^1 \cdot 1 + 2^2 \cdot 1 + 2^3 \cdot 1 = 14$

16 $\frac{1}{10}(0^3 + 0.1^3 + 0.2^3 + \cdots + 0.9^3) = \frac{81}{400}$
    $= 0.2025$

18 (a) $\frac{1}{10}(\sqrt{0} + \sqrt{0.1} + \cdots + \sqrt{0.9}) \approx 0.6105$
   (b) $\frac{1}{10}(\sqrt{0.1} + \sqrt{0.2} + \cdots + \sqrt{1}) \approx 0.7105$
   (c) $\frac{1}{10}(\sqrt{0.05} + \sqrt{0.15} + \cdots + \sqrt{0.95}) \approx 0.6684$

20 (a) $(n+1)^6 - 1 = \sum_{i=1}^n ((i+1)^6 - i^6)$

$= 6 \sum_{i=1}^n i^5 + 15 \sum_{i=1}^n i^4 + 20 \sum_{i=1}^n i^3 +$
$15 \sum_{i=1}^n i^2 + 6 \sum_{i=1}^n i + \sum_{i=1}^n 1 = 6S +$
$15\left(\frac{1}{5}n^5 + \frac{1}{2}n^4 + \frac{1}{3}n^3 - \frac{1}{30}n\right) +$
$20\left(\frac{1}{4}n^4 + \frac{1}{2}n^3 + \frac{1}{4}n^2\right) +$
$15\left(\frac{1}{3}n^3 + \frac{1}{2}n^2 + \frac{1}{6}n\right) + 6\left(\frac{1}{2}n^2 + \frac{1}{2}n\right) + n$
$= 6S + 3n^5 + \frac{25}{2}n^4 + 20n^3 + \frac{31}{2}n^2 + 6n,$
$6S = n^6 + 6n^5 + 15n^4 + 20n^3 + 15n^2 + 6n - (3n^5 + \cdots + 6n)$
$= n^6 + 3n^5 + \frac{5}{2}n^4 - \frac{1}{2}n^2, \; S = \sum_{i=1}^n i^5$
$= \frac{1}{6}n^6 + \frac{1}{2}n^5 + \frac{5}{12}n^4 - \frac{1}{12}n^2$
$= \frac{1}{12}n^2(n+1)^2(2n^2 + 2n - 1)$

(b) $\int_0^b x^5 \, dx = \lim_{n \to \infty} \sum_{i=1}^n \left(\frac{ib}{n}\right)^5 \cdot \frac{b}{n}$
$= \lim_{n \to \infty} \frac{b^6}{n^6}\left(\frac{1}{6}n^6 + \frac{1}{2}n^5 + \frac{5}{12}n^4 - \frac{1}{12}n^2\right)$
$= \frac{b^6}{6}$

22 $(2 \cdot 2 + 1) + (2 \cdot 3 + 1) + (2 \cdot 4 + 1) = 21$

24 (a) $\int_0^5 \pi(\sqrt{x/\pi})^2 \, dx = \int_0^5 x \, dx$
   (b) $\int_0^5 x \, dx = \frac{5^2}{2} - \frac{0^2}{2} = 12.5$

26 $\int_0^h \pi \left(\frac{ax}{h}\right)^2 dx = \frac{\pi a^2}{h^2} \cdot \frac{h^3}{3} = \frac{\pi a^2 h}{3}$

28 (a) $\sum_{i=1}^{n} \frac{1}{x_{i-1}}(x_i - x_{i-1})$

(b) $\sum_{i=1}^{n} \frac{1}{3x_{i-1}}(3x_i - 3x_{i-1}) = \sum_{i=1}^{n} \frac{1}{x_{i-1}}(x_i - x_{i-1})$

(c) Take the limit of the result in (b) as the mesh m of the partition of [1, 2] approaches 0. Note that the mesh of the partition [3, 6] is $3m$, so it also approaches 0.

30 We know that $\sum_{i=1}^{n} f(c_i)(x_i - x_{i-1})$ (with $x_0 = 1$ and $x_n = 5$) approximates $\int_1^5 f(x)\, dx$. However, it is also true that $\sum_{i=1}^{n} f(c_i)(x_i - x_{i-1}) \leq$

$\sum_{i=1}^{n} (-3)(x_i - x_{i-1}) = (-3)(x_n - x_0) = -12$, since $f(x) \leq -3$ for all x in [1, 5]. We may take the limit as the mesh $\to 0$ of both sides of the inequality, yielding $\int_1^5 f(x)\, dx \leq -12$.

32 $\int_0^x t^2\, dt = \frac{x^3}{3}$, so $G(x) = \frac{x^3}{3}$. Thus $\frac{dG}{dx} = x^2$.

34 (a) Each section has length $\frac{5-1}{n} = \frac{4}{n}$. Hence the mesh is $\frac{4}{n}$.

(b) The mesh is at least $4/n$ and at most 4.

5.4 Estimating a Definite Integral

2 $a = 0, b = 2, n = 4, h = (2 - 0)/4 = 1/2$:

$\int_0^2 \frac{dx}{1 + x^2} \approx \frac{1/2}{2}\left[\frac{1}{1 + 0^2} + 2 \cdot \frac{1}{1 + (1/2)^2} + 2 \cdot \frac{1}{1 + (2/2)^2} + 2 \cdot \frac{1}{1 + (3/2)^2} + \frac{1}{1 + (4/2)^2}\right]$

$= \frac{1}{4}\left[1 + \frac{8}{5} + 1 + \frac{8}{13} + \frac{1}{5}\right] = \frac{1}{4} \cdot \frac{287}{65} = \frac{287}{260}$

≈ 1.1038.

4 $a = 0, b = 2, n = 3, h = \frac{2 - 0}{3} = \frac{2}{3}$:

$\int_0^2 \sin \sqrt{x}\, dx \approx$

$\frac{2/3}{2}\left[\sin \sqrt{0} + 2 \sin \sqrt{\frac{2}{3}} + 2 \sin \sqrt{\frac{4}{3}} + \sin \sqrt{\frac{6}{3}}\right]$

$\approx \frac{1}{3}[0 + 1.4575 + 1.8293 + 0.9878] \approx 1.4249$.

6 $a = 1, b = 3, n = 6, h = \frac{3-1}{6} = 1/3$:

$\int_1^3 \frac{2^x}{x}\, dx \approx \frac{1/3}{2}\left[\frac{2^1}{1} + 2 \cdot \frac{2^{4/3}}{4/3} + 2 \cdot \frac{2^{5/3}}{5/3} + 2 \cdot \frac{2^2}{2} + 2 \cdot \frac{2^{7/3}}{7/3} + 2 \cdot \frac{2^{8/3}}{8/3} + \frac{2^3}{3}\right]$

$= \frac{1}{6}\left[2 + \frac{3}{2} \cdot 2^{4/3} + \frac{6}{5} \cdot 2^{5/3} + 4 + \frac{6}{7} \cdot 2^{7/3} + \frac{3}{4} \cdot 2^{8/3} + \frac{8}{3}\right]$

≈ 4.2230.

8 $a = 1, b = 3, n = 2, h = (3 - 1)/2 = 1$:

$\int_1^3 \cos x^2\, dx \approx \frac{1}{2}[\cos 1^2 + 2 \cos 2^2 + \cos 3^2]$

$\approx \frac{1}{2}[0.5403 + (-1.3073) + (-0.9111)]$

≈ -0.8391.

10 $a = 0$, $b = 1$, $n = 4$, $h = (1 - 0)/4 = 1/4$:

$$\int_0^1 \frac{dx}{x^3 + 1} \approx \frac{1/4}{3}\left[\frac{1}{0^3 + 1} + 4\cdot\frac{1}{(1/4)^3 + 1} + \right.$$

$$\left. 2\cdot\frac{1}{(2/4)^3 + 1} + 4\cdot\frac{1}{(3/4)^3 + 1} + \frac{1}{1^3 + 1}\right]$$

$$= \frac{1}{12}\cdot\frac{82141}{8190} = \frac{82141}{98280} \approx 0.8358.$$

12 $a = 0$, $b = 1$, $n = 2$, $h = (1 - 0)/2 = 1/2$:

$$\int_0^1 \frac{dx}{x^4 + 1} \approx \frac{1/2}{3}\left[\frac{1}{0^4 + 1} + 4\cdot\frac{1}{(1/2)^4 + 1} + \frac{1}{1^4 + 1}\right]$$

$$= \frac{1}{6}\left[1 + \frac{64}{17} + \frac{1}{2}\right] = \frac{1}{6}\cdot\frac{179}{34} = \frac{179}{204} \approx 0.8775.$$

14 Let $A(x)$ denote the cross-sectional area of the ship's hull as a function of x.

(a) By the trapezoidal method with $a = 0$, $b = 120$, and $h = 20$, the volume of the hull is

$$\int_0^{120} A(x)\, dx \approx \frac{20}{2}[f(0) + 2f(20) + 2f(40)$$

$$+ 2f(60) + 2f(80) + 2f(100) + f(120)]$$

$$= 10[0 + 2\cdot 200 + 2\cdot 400 + 2\cdot 450 + 2\cdot 420$$

$$+ 2\cdot 300 + 150] = 36{,}900 \text{ ft}^3.$$

(b) By Simpson's method, with $a = 0$, $b = 120$, and $h = 20$, the volume of the hull is

$$\int_0^{120} A(x)\, dx \approx \frac{20}{3}[f(0) + 4f(20) + 2f(40)$$

$$+ 4f(60) + 2f(80) + 4f(100) + f(120)]$$

$$= \frac{20}{3}[0 + 4\cdot 200 + 2\cdot 400 + 4\cdot 450 + 2\cdot 420$$

$$+ 4\cdot 300 + 150] \approx 37{,}266.6667 \text{ ft}^3.$$

16 Volume $\approx \frac{2}{3}[4.0 + 4(3.7) + 2(3.5) + 4(3.1) +$

$2(3.1) + 4(2.9) + 2(2.7) + 4(2.4) + 2.1]$

$\approx 48.7333 \text{ ft}^3$

18 For a given n (and therefore a given h), let L be the left-point estimate and R be the right-point estimate. Then $\dfrac{L + R}{2} =$

$$\frac{h[f(x_0) + f(x_1) + \cdots + f(x_{n-1})] + h[f(x_1) + f(x_2) + \cdots + f(x_n)]}{2}$$

$$= \frac{h}{2}[f(x_0) + 2f(x_1) + 2f(x_2) + \cdots + 2f(x_{n-1}) + f(x_n)].$$

This is equivalent to the trapezoidal estimate.

20 Using the result of Exercise 22 of Sec. 5.2, the actual value of $\displaystyle\int_0^b x^4\, dx = \lim_{n\to\infty}\sum_{i=1}^n \left(\frac{ib}{n}\right)^4\cdot\frac{b}{n}$

$$= \lim_{n\to\infty} \frac{b^5}{n^5}\left(\frac{n^5}{5} + \frac{n^4}{2} + \frac{n^3}{3} - \frac{n}{30}\right) = \frac{b^5}{5}, \text{ so}$$

$\displaystyle\int_0^1 x^4\, dx = \frac{1}{5}$. By Simpson's method, with $n = 2$, we have $\dfrac{1/2}{3}[0^4 + 4(1/2)^4 + 1^4] = \dfrac{5}{24}$, so the error is $\dfrac{5}{24} - \dfrac{1}{5} = \dfrac{1}{120}$. The maximum value of

$\dfrac{d^4(x^4)}{dx^4}$ for x in $[0, 1]$ is $M_4 = 24$ (the constant value of the fourth derivative of x^4), so $(b - a)M_4 h^4/180 = (1 - 0)(24)(1/2)^4/180 \approx 1/120$, which is the difference between our estimate and the actual result.

22 The constants A, B, and C must satisfy $A(-h)^2 + B(-h) + C = f(-h)$, $A\cdot 0^2 + B\cdot 0 + C = f(0)$, $Ah^2 + Bh + C = f(h)$. The middle equation yields $C = f(0)$. The first and last become

$Ah^2 - Bh = f(-h) - f(0)$, $Ah^2 + Bh = f(h) - f(0)$. Adding these two equations, $2Ah^2 = f(h) + f(-h) - 2f(0)$. Hence $A = \dfrac{f(h) + f(-h) - 2f(0)}{2h^2}$. Subtracting gives $2Bh = f(h) - f(-h)$ so $B = \dfrac{f(h) - f(-h)}{2h}$. Hence, the three points lie on the parabola $y = \left[\dfrac{f(h) + f(-h) - 2f(0)}{2h^2}\right]x^2 + \left[\dfrac{f(h) - f(-h)}{2h}\right]x + f(0)$.

24 As suggested, let $x = c + t$ and $dx = dt$. Then $t = x - c$ varies from $-h$ to h. Hence $\int_{c-h}^{c+h} f(x)\, dx = \int_{-h}^{h} f(t + c)\, dt = \int_{-h}^{h} g(t)\, dt$, where $g(t) = f(t + c) = A(t + c)^2 + B(t + c) + C$ still represents a parabola. By Exercise 21, $\int_{-h}^{h} g(t)\, dt = \dfrac{h}{3}[g(-h) + 4g(0) + g(h)] = \dfrac{h}{3}[f(c - h) + 4f(c) + f(c + h)]$, as desired.

26 (a) If $f(x) = x^3$, then $\int_{-h}^{h} f(x)\, dx = 0$ (since $f(x)$ is an odd function). Further, $\dfrac{h}{3}[f(-h) + 4f(0) + f(h)] = \dfrac{h}{3}[-h^3 + 0 + h^3] = 0$. Therefore $\int_{-h}^{h} f(x)\, dx = 0 = \dfrac{h}{3}[f(-h) + 4f(0) + f(h)]$.

(b) Let $f(x)$ be a cubic polynomial. If Simpson's rule is exact on each pair of adjacent intervals, then it is exact on the entire range of integration. We therefore consider only the case $n = 2$. For $n = 2$, we have $h = \dfrac{b-a}{2}$, $x_0 = a$, $x_1 = a + h$, and $x_2 = a + 2h = b$. Let $c = x_1$. Then the integral is $\int_{a}^{b} f(x)\, dx = \int_{c-h}^{c+h} f(x)\, dx$, while Simpson's estimate is $\dfrac{h}{3}[f(c - h) + 4f(c) + f(c + h)]$. Letting $x = c + t$, we have $dx = dt$ and $\int_{c-h}^{c+h} f(x)\, dx = \int_{-h}^{h} f(c + t)\, dt$. Since $f(x)$ is a cubic, so is $f(c + t)$, say $f(c + t) = Eg(t) + q(t)$, where $g(t) = t^3$ and $q(t)$ is a quadratic polynomial. Then $\int_{-h}^{h} f(c + t)\, dt = \int_{-h}^{h} [Eg(t) + q(t)]\, dt = E\int_{-h}^{h} g(t)\, dt + \int_{-h}^{h} q(t)\, dt = E \cdot \dfrac{h}{3}[g(-h) + 4g(0) + g(h)] + \dfrac{h}{3}[q(-h) + 4q(0) + q(h)] = \dfrac{h}{3}[(Eg(-h) + q(-h)) + 4(Eg(0) + q(0)) + (Eg(h) + q(h))] = \dfrac{h}{3}[f(c - h) + 4f(c) + f(c + h)]$, which is Simpson's estimate. Thus Simpson's estimate is exact for cubics.

28 (a) $\dfrac{(5 - 1)10h^2}{12} \leq 0.01$, so $h \leq \sqrt{\dfrac{3}{1000}}$. Since $h = 4/n$, we must have $n \geq 4\sqrt{\dfrac{1000}{3}} \approx 73.03$;

hence $n \geq 74$ and $h \leq 2/37$.

(b) $\dfrac{(5-1)50h^4}{180} \leq 0.01$, so $h \leq \sqrt[4]{\dfrac{9}{1000}}$. Since $h = 4/n$, we must have $n \geq 4\sqrt[4]{\dfrac{1000}{9}} \approx 12.99$; since n must be even, we have $n \geq 14$ and $h \leq 2/7$.

30 (a) Since $R = h(f(x_1) + f(x_2) + \cdots + f(x_{n-1}) + f(b))$ and $L = h(f(a) + f(x_1) + \cdots + f(x_{n-2}) + f(x_{n-1}))$, $R - L = h(f(x_1) + f(x_2) + \cdots + f(x_{n-1}) + f(b) - f(a) - f(x_1) - \cdots - f(x_{n-2}) - f(x_{n-1})) = h(f(b) - f(a))$.

(b) Let $[x_i, x_{i+1}]$ denote an arbitrary section in the partition of $[a, b]$, with sampling point c_{i+1} and the right and left endpoint estimates included in the figure. 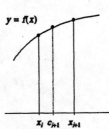 We see that for any c_{i+1} in $[x_i, x_{i+1}]$, $f(x_i) < f(c_{i+1}) < f(x_{i+1})$ and $hf(x_i) < hf(c_{i+1}) < hf(x_{i+1})$. Hence, any arbitrary sampling point estimate lies between the left and right endpoint estimates on that section. So this is true for all sections in the partition of $[a, b]$. Let S denote the arbitrary sampling point estimate of the area under $y = f(x)$ and above $[a, b]$. We conclude that $R > S$ and $L < S$. We also know that $R > A$ and $L < A$, since f is increasing on $[a, b]$. Thus $R - L > S - L > S - A$, so $R - L > S - A$. From (a), we see that $h(f(b) - f(a)) > S - A$. So S differs from A by less than $h(f(b) - f(a))$.

(c) From the mean-value theorem, there exists a c in $[a, b]$ such that $f'(c)(b - a) = f(b) - f(a)$.

Using (b), we conclude that $h(f(b) - f(a)) = h \cdot f'(c)(b - a) > S - A$; also, $hM_1(b-a) \geq hf'(c)(b - a)$, by definition. Therefore $M_1(b - a)h > S - A$ and we are done.

(d) Let $[x_i, x_{i+1}]$ denote an arbitrary section in the partition, and let N_i denote the maximum value of $|f'(x)|$ for x in $[x_i, x_{i+1}]$. Then from (c), the error is less than $\sum_{i=0}^{n-1} N_i(x_{i+1} - x_i)^2 \leq$

$\sum_{i=0}^{n-1} M_i h(x_{i+1} - x_i) = M_1 h(x_n - x_0) =$

$M_1(b - a)h$, as before.

32 (a),(b),(d)

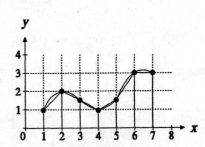

(c) $\int_1^7 f(x)\, dx \approx \dfrac{h}{2}[f(1) + 2f(2) + 2f(3) + 2f(4) + 2f(5) + 2f(6) + f(7)] = \dfrac{1}{2}[1 + 2 \cdot 2 + 2 \cdot 1.5 + 2 \cdot 1 + 2 \cdot 1.5 + 2 \cdot 3 + 3]$
$= 11$

(e) $\int_1^7 f(x)\, dx \approx \dfrac{h}{3}[f(1) + 4f(2) + 2f(3) + 4f(4) + 2f(5) + 4f(6) + f(7)] = \dfrac{1}{3}[1 + 4 \cdot 2 + 2 \cdot 1.5 + 4 \cdot 1 + 2 \cdot 1.5 + 4 \cdot 3 + 3]$
≈ 11.3333

34 (a)

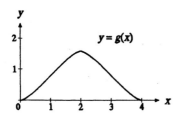

(b)

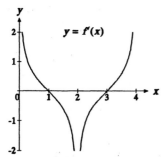

(c) Since $y = f(x)$ looks like two semicircular arcs of radius 1, one above the x axis and the other below, $g(4)$ should equal 0, $g(2)$ should equal $\pi/2$ (or close to it), and $g(0) = 0$; $g(x)$ should be symmetric about $x = 2$. The derivative $f'(x)$ should also be symmetric about $x = 2$ and appears to be very large (or even undefined) at $x = 0, 2,$ and 4. Also, it appears that $f'(1) = f'(3) = 0$.

5.5 Properties of the Antiderivative and the Definite Integral

2 $\int \dfrac{7}{x^2}\, dx = 7 \int x^{-2}\, dx = 7\left(\dfrac{x^{-1}}{-1}\right) + C$

$= -\dfrac{7}{x} + C.$ Check: $\left(-\dfrac{7}{x} + C\right)' = -7 \cdot \dfrac{-1}{x^2} + 0$

$= \dfrac{7}{x^2}.$

4 $\int \left(6x^2 + \dfrac{1}{\sqrt{x}}\right) dx = 6 \int x^2\, dx + \int x^{-1/2}\, dx =$

$6 \cdot \dfrac{x^3}{3} + \dfrac{x^{1/2}}{1/2} + C = 2x^3 + 2\sqrt{x} + C.$ Check:

$(2x^3 + 2\sqrt{x} + C)' = 6x^2 + \dfrac{1}{\sqrt{x}}.$

6 (a) $\int \sin x\, dx = -\cos x + C.$ Check:

$(-\cos x + C)' = -(-\sin x) = \sin x$

(b) $\int \sin 3x\, dx = -\dfrac{\cos 3x}{3} + C.$ Check:

$\left(-\dfrac{1}{3} \cos 3x + C\right)' = -\dfrac{1}{3}(-3 \sin 3x)$

$= \sin 3x.$

8 $\int \sec x \tan x\, dx = \sec x + C,$ since $(\sec x + C)'$ $= \sec x \tan x.$

10 $\int \csc^2 x\, dx = -\cot x + C,$ since $(-\cot x + C)'$ $= \csc^2 x.$

12 (b) True, by definition of antiderivative.

14 (a) $\int_1^2 x\, dx = \dfrac{2^2}{2} - \dfrac{1^2}{2} = \dfrac{3}{2}$

(b) $\int_2^1 x\, dx = -\int_1^2 x\, dx = -\dfrac{3}{2}$

(c) $\int_3^3 x\, dx = 0$

16 (a) $\int 3x^2\, dx = x^3 + C$

(b) $\int_1^4 3x^2\, dx = 4^3 - 1^3 = 63$

18 Applying Property 7 with $m = -1, M = 4, a = -2,$ and $b = 7,$ we have

$m(b - a) \leq \int_a^b f(x)\, dx \leq M(b - a)$

$$-1(7-(-2)) \le \int_{-2}^{7} f(x)\, dx \le 4(7-(-2))$$

$$-9 \le \int_{-2}^{7} f(x)\, dx \le 36.$$

Hence $\int_{-2}^{7} f(x)\, dx$ lies in $[-9, 36]$.

20 $\int_{1}^{2} (5x + 2)\, dx = \int_{1}^{2} 5x\, dx + 2\int_{1}^{2} dx =$

$\dfrac{5}{2}\cdot 2^2 - \dfrac{5}{2}\cdot 1^2 + 2\cdot(2-1) = \dfrac{19}{2}$, and $\dfrac{19}{2} =$

$f(c)(2-1)$. So $f(c) = \dfrac{19}{2}$, and $f(c) = 5c + 2 =$

$\dfrac{19}{2}$ reduces to $c = \dfrac{3}{2}$.

22 $\int_{1}^{4} (x^2 + x)\, dx = \int_{1}^{4} x^2\, dx + \int_{1}^{4} x\, dx =$

$\dfrac{4^3}{3} - \dfrac{1^3}{3} + \dfrac{4^2}{2} - \dfrac{1^2}{2} = \dfrac{63}{3} + \dfrac{15}{2} = \dfrac{57}{2}$, and

$\dfrac{57}{2} = f(c)(4-1)$. So $f(c) = \dfrac{19}{2}$, and $f(c) =$

$c^2 + c = \dfrac{19}{2}$. This becomes $2c^2 + 2c - 19 = 0$.

Thus $c = \dfrac{-2 \pm \sqrt{2^2 + 4\cdot 2\cdot 19}}{2\cdot 2} = -\dfrac{1}{2} \pm \dfrac{\sqrt{39}}{2}$. We

choose $c = -\dfrac{1}{2} + \dfrac{\sqrt{39}}{2}$ because it lies in $[1, 4]$.

24 (a) $\int_{1}^{3} (2f(x) + 6g(x))\, dx = 2\int_{1}^{3} f(x)\, dx +$

$6\int_{1}^{3} g(x)\, dx = 2\cdot 4 + 6\cdot 5 = 38$

(b) $\int_{3}^{1} (f(x) - g(x))\, dx$

$= \int_{3}^{1} f(x)\, dx - \int_{3}^{1} g(x)\, dx$

$= -\int_{1}^{3} f(x)\, dx + \int_{1}^{3} g(x)\, dx = -4 + 5$

$= 1$

26 The average is $\dfrac{\int_{a}^{b} f(x)\, dx}{b-a} = \dfrac{\int_{a}^{b} c\, dx}{b-a} =$

$\dfrac{c\int_{a}^{b} dx}{b-a} = \dfrac{c(b-a)}{b-a} = c$ (as expected).

28 $f'(x) = 1 > 0$, so $f(x) = x$ takes on a minimum on $[1, 4]$ at $f(1) = 1$ and a maximum on $[1, 4]$ at $f(4) = 4$. The average value of $f(x)$ on $[1, 4]$ is given

by $\dfrac{\int_{1}^{4} x\, dx}{4-1} = \dfrac{1}{3}\left(\dfrac{4^2}{2} - \dfrac{1^2}{2}\right) = \dfrac{5}{2}$.

30 $f'(x) = 2x > 0$ for x in $(0, 5)$, so $f(x) = x^2$ takes on a minimum on $[0, 5]$ at $f(0) = 0$ and a maximum on $[0, 5]$ at $f(5) = 25$. The average

value of $f(x)$ on $[0, 5]$ is given by $\dfrac{\int_{0}^{5} x^2\, dx}{5-0} =$

$\dfrac{1}{5}\left(\dfrac{5^3}{3}\right) = \dfrac{25}{3}$.

32 Average value of $f(x)$ on $[0, 8] = \dfrac{\int_{0}^{8} f(x)\, dx}{8-0} \approx$

$\dfrac{1}{8}\cdot\dfrac{1}{3}[5 + 4\cdot 1 + 2\cdot 2 + 4\cdot 4 + 2\cdot 3 + 4\cdot 3 + 2\cdot 2$

$+ 4\cdot 6 + 1] = \dfrac{1}{24}\cdot 76 = \dfrac{19}{6}$.

34 The statement is true from some functions (for example, $f(x) = g(x) = 0$) but false for most functions (for example, $f(x) = g(x) = 1$).

5.5 Properties of the Antiderivative and the Definite Integral

36 $\left(\dfrac{2x}{a^2}\sin ax + \dfrac{2}{a^3}\cos ax - \dfrac{x^2}{a}\cos ax + C\right)' =$

$\dfrac{2}{a^2}[x(a\cos ax) + \sin ax] + \dfrac{2}{a^3}(-a\sin ax) -$

$\dfrac{1}{a}[x^2(-a\sin ax) + 2x\cos ax] = \dfrac{2x\cos ax}{a} +$

$\dfrac{2\sin ax}{a^2} - \dfrac{2\sin ax}{a^2} + x^2\sin ax - \dfrac{2x\cos ax}{a}$

$= x^2\sin ax$

38 We are given that $4 = \dfrac{\int_1^3 f(x)\,dx}{3 - 1}$, since 4 is the

average value of $f(x)$ on $[1, 3]$, and that $5 =$

$\dfrac{\int_3^6 f(x)\,dx}{6 - 3}$ because 5 is the average value of $f(x)$

on $[3, 6]$. Thus $\int_1^3 f(x)\,dx = 4(3 - 1) = 8$ and

$\int_3^6 f(x)\,dx = 5(6 - 3) = 15$. Hence $\int_1^6 f(x)\,dx$

$= \int_1^3 f(x)\,dx + \int_3^6 f(x)\,dx = 8 + 15 = 23$. So

the average value of $f(x)$ on $[1, 6]$ is $\dfrac{\int_1^6 f(x)\,dx}{6 - 1}$

$= \dfrac{23}{5}.$

5.6 Background for the Fundamental Theorems of Calculus

2 (a)

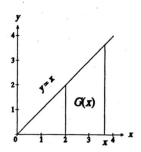

(b) $G(x) = \dfrac{1}{2}(2 + x)(x - 2) = \dfrac{1}{2}(x^2 - 4)$

$= \dfrac{1}{2}x^2 - 2$

(c) $G'(x) = x = f(x)$

4 (a) $G(4) \approx 1.3863$, $G(4.1) \approx 1.4110$

(b)

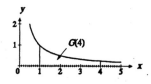

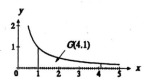

(c) $G'(4) \approx \dfrac{G(4.1) - G(4)}{0.1} \approx$

$\dfrac{1.4110 - 1.3863}{0.1} = \dfrac{0.0247}{0.1} = 0.247$

(d) $f(4) = \dfrac{1}{4} = 0.25$, which is about equal to

the value in (c).

6 (a) $G(1) \approx 0.8356$, $G(1.01) \approx 0.8406$

(c) $G'(1) \approx \dfrac{G(1.01) - G(1)}{0.01}$

$\approx \dfrac{0.8406 - 0.8356}{0.01} = \dfrac{0.005}{0.01}$

$= 0.5$

(d) $f(1) = 1/2$

8 (a) $G(2) \approx 0.3607$, $G(1.99) \approx 0.3582$

(c) $G'(2) \approx \dfrac{G(1.99) - G(2)}{-0.01} \approx$

$\dfrac{0.3582 - 0.3607}{-0.01} = \dfrac{-0.0025}{-0.01} = 0.25$

(d) $f(2) = 1/4$

10 (a) $G(x) = \int_1^x t^2\, dt = \dfrac{x^3}{3} - \dfrac{1^3}{3} = \dfrac{1}{3}x^3 - \dfrac{1}{3}$

(b) $G'(x) = x^2$, so $G'(7) = 7^2 = 49$.

(c) Yes, since $G'(7) = 49 = f(7)$.

(d) Yes, since $G'(8) = 64 = f(8)$.

12 (a)

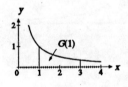

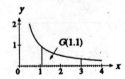

(b) $G(1) > G(1.1)$

(c) G is a decreasing function.

(d) G' is negative since G is decreasing.

(e) Here $G(1.01) \approx 1.0887$ while $G(1) \approx 1.0986$.

Thus $G'(1) \approx \dfrac{G(1.01) - G(1)}{0.01} \approx$

$\dfrac{1.0887 - 1.0986}{0.01} = \dfrac{-0.0099}{0.01} = -0.99$.

(f) Since $f(1) = 1.0$, $G'(1)$ seems to be the negative of $f(1)$.

14 (a)

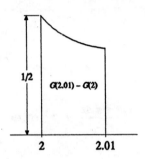

(b) $\dfrac{G(2.01) - G(2)}{0.01}$ is the average value of $f(t)$

on $[2, 2.01]$, which should be roughly $f(2) = 1/2$ since f varies by only $f(2) - f(2.01) \approx 0.0025$ over $[2, 2.01]$.

16 (a) Since $G(x + \Delta x) - G(x) > 0$, $G(x)$ is increasing and $G'(x)$ should be positive.

(b) Since $G(x + \Delta x) - G(x) < 0$, $G(x)$ is decreasing and $G'(x)$ should be negative.

5.7 The Fundamental Theorems of Calculus

2 (a) $(x + \sec x)\big|_0^{\pi/4} = \dfrac{\pi}{4} + \sec\dfrac{\pi}{4} - 0 - \sec 0$

$= \dfrac{\pi}{4} + \sqrt{2} - 1 = 1.1996$

(b) $\dfrac{1}{x}\bigg|_2^3 = \dfrac{1}{3} - \dfrac{1}{2} = -\dfrac{1}{6}$

(c) $\sqrt{x - 1}\big|_5^{10} = \sqrt{9} - \sqrt{4} = 1$

4 The definite integral of f from a to b is equal to the difference of the antiderivative of f evaluated at b and at a (in that order).

5.7 The Fundamental Theorems of Calculus

6. $\int_{-1}^{3} 2x^4 \, dx = \frac{2x^5}{5}\Big|_{-1}^{3} = \frac{2}{5}(243 + 1) = \frac{488}{5}$

8. $\int_{-1}^{2} (6x - 3x^2) \, dx = (3x^2 - x^3)\Big|_{-1}^{2}$

   $= 3 \cdot 4 - 8 - (3 \cdot 1 - (-1)) = 4 - 4 = 0$

10. $\int_{\pi/4}^{3\pi/4} 3 \sin x \, dx = -3 \cos x \Big|_{\pi/4}^{3\pi/4}$

    $= -3\left(-\frac{\sqrt{2}}{2} - \frac{\sqrt{2}}{2}\right) = 3\sqrt{2}$

12. $\int_{0}^{\pi/6} \cos 3x \, dx = \frac{1}{3} \sin 3x \Big|_{0}^{\pi/6}$

    $= \frac{1}{3}\left(\sin \frac{\pi}{2} - \sin 0\right) = \frac{1}{3}$

14. $\int_{1}^{9} \frac{1}{\sqrt{x}} \, dx = \int_{1}^{9} x^{-1/2} \, dx = \frac{x^{1/2}}{1/2}\Big|_{1}^{9}$

    $= 2(\sqrt{9} - \sqrt{1}) = 4$

16. $\int_{2}^{4} \frac{4}{x^3} \, dx = \frac{-2}{x^2}\Big|_{2}^{4} = \frac{-2}{16} - \frac{-2}{4} = -\frac{1}{8} + \frac{1}{2}$

    $= \frac{3}{8}$

18. $\frac{1}{2-1} \int_{1}^{2} x^4 \, dx = \frac{x^5}{5}\Big|_{1}^{2} = \frac{1}{5}(32 - 1) = \frac{31}{5}$

20. $\frac{1}{\pi/2 - 0} \int_{0}^{\pi/2} \cos x \, dx = \frac{2}{\pi} \sin x \Big|_{0}^{\pi/2} = \frac{2}{\pi}$

22. $\frac{1}{\frac{\pi}{6} - \frac{\pi}{8}} \int_{\pi/8}^{\pi/6} \sec 2x \tan 2x \, dx = \frac{24}{\pi} \cdot \frac{1}{2} \sec 2x \Big|_{\pi/8}^{\pi/6}$

    $= \frac{12}{\pi}\left(\sec \frac{\pi}{3} - \sec \frac{\pi}{4}\right) = \frac{12}{\pi}(2 - \sqrt{2})$

24. $\int_{2}^{3} \frac{1}{x^2} \, dx = \frac{-1}{x}\Big|_{2}^{3} = -\frac{1}{3} + \frac{1}{2} = \frac{1}{6}$

26. $\int_{25}^{36} \sqrt{x} \, dx = \frac{x^{3/2}}{3/2}\Big|_{25}^{36} = \frac{2}{3}(36^{3/2} - 25^{3/2})$

    $= \frac{2}{3}(216 - 125) = \frac{182}{3}$

28. $\int_{1}^{8} 7 \sqrt[3]{t} \, dt = 7 \frac{t^{4/3}}{4/3}\Big|_{1}^{8} = \frac{21}{4}(8^{4/3} - 1^{4/3})$

    $= \frac{21}{4}(16 - 1) = \frac{315}{4}$ feet

30. $\int_{1/4}^{1} 4\sqrt{x} \, dx = 4 \frac{x^{3/2}}{3/2}\Big|_{1/4}^{1} = \frac{8}{3}\left(1^{3/2} - \left(\frac{1}{4}\right)^{3/2}\right)$

    $= \frac{8}{3}\left(1 - \frac{1}{8}\right) = \frac{7}{3}$ grams

32. $\int_{1}^{5} \frac{1}{x^3} \, dx = \frac{x^{-2}}{-2}\Big|_{1}^{5} = -\frac{1}{2}\left(\frac{1}{5^2} - \frac{1}{1^2}\right) = \frac{12}{25}$

34. (a) $\int_{1}^{3} x^2 \, dx$ is a definite integral, which is defined as a limit of sums. The expression $\int x^2 \, dx \Big|_{1}^{3}$ is the difference of two values of an antiderivative of x^2.

    (b) The second fundamental theorem says that the two quantities are equal.

36. By the first fundamental theorem, $G(x) = \int_{0}^{x} \sin t^2 \, dt$ is an antiderivative of $\sin x^2$; however, there is no way to express $G(x)$ as a simple formula involving only elementary functions (and no integrals). Hence $\sin x^2$ does not have an *elementary* antiderivative.

(a) False

(b) True

38 (a) $\frac{d}{dx}\left(\int_1^x t^4 \, dt\right) = x^4$

(b) $\frac{d}{dx}\left(\int_x^1 t^4 \, dt\right) = \frac{d}{dx}\left(-\int_1^x t^4 \, dt\right) =$

$-\frac{d}{dx}\left(\int_1^x t^4 \, dt\right) = -x^4$

40 $\frac{d}{dx}\left(\int_{-1}^x 3^{-t} \, dt\right) = 3^{-x}$

42

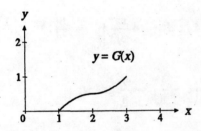

The function f is linear over [1, 2] and [2, 3], so its integrals are easy to compute. For $1 \le x \le 2$,

we have $G(t) = \int_1^x (2-t) \, dt = \left(2t - \frac{t^2}{2}\right)\Big|_1^x =$

$-\frac{1}{2}(x^2 - 4x + 3)$ and $G(2) = 1/2$, as expected.

Then for $2 \le x \le 3$, we have $G(t) =$

$\frac{1}{2} + \int_2^x (t-2) \, dt = \frac{1}{2} + \left(\frac{t^2}{2} - 2t\right)\Big|_2^x =$

$\frac{1}{2}(x^2 - 4x + 5)$, and $G(2) = 1$, as expected.

44

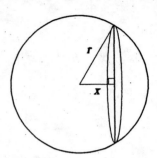

(a) From the figure we see that the radius of the disk is $\sqrt{r^2 - x^2}$.

(b) The area of the disk is $\pi\left(\sqrt{r^2 - x^2}\right)^2 = \pi r^2 - \pi x^2$.

(c) The volume of the sphere is

$\int_{-r}^r (\pi r^2 - \pi x^2) \, dx = 2 \int_0^r (\pi r^2 - \pi x^2) \, dx$

$= 2\left(\pi r^2 x - \pi \frac{x^3}{3}\right)\Big|_0^r = 2\left(\frac{2}{3}\pi r^3\right) = \frac{4}{3}\pi r^3,$

as expected.

46 (a) Recalling that $s'(t) = v(t)$, where $s(t)$ is the position function (see Sec. 4.3), we see from the fundamental theorem of calculus that

$\int_a^b v(t) \, dt = s(b) - s(a)$. In other words, the definite integral of the velocity function equals the change in position.

(b) If we graph $y = v(t)$, then $v'(t)$ gives the slope of the graph at each point; but we already know that $v'(t)$ is the acceleration function, so the slope of the graph equals acceleration.

48 (a) $\frac{1}{T}\left[A + \int_0^T f(t) \, dt\right]$ is the average over the period T of overhaul cost and total depreciation. We want this average to be as small as possible.

5.7 The Fundamental Theorems of Calculus

(b) $\dfrac{dg}{dT} = \dfrac{Tf(T) - A - \int_0^T f(t)\, dt}{T^2}$

(c) When $\dfrac{dg}{dT} = 0$, then $Tf(T) = A + \int_0^T f(t)\, dt$

$= Tg(T)$, so $f(T) = g(T)$.

(d) (c) says that the best you can do is to make your overall average cost equal your rate of depreciation.

50 (e) $H(x + \Delta x) - H(x) =$

$\int_{x+\Delta x}^b f(t)\, dt - \int_x^b f(t)\, dt = -\int_x^{x+\Delta x} f(t)\, dt$

$\approx -f(x)\,\Delta x$

(f) $H'(x) = \lim\limits_{\Delta x \to 0} \dfrac{H(x + \Delta x) - H(x)}{\Delta x}$

$= \lim\limits_{\Delta x \to 0} \dfrac{\int_{x+\Delta x}^b f(t)\, dt - \int_x^b f(t)\, dt}{\Delta x}$

$= \lim\limits_{\Delta x \to 0} \dfrac{\left[\int_{x+\Delta x}^b f(t)\, dt - \int_x^{x+\Delta x} f(t)\, dt - \int_{x+\Delta x}^b f(t)\, dt\right]}{\Delta x}$

$= \lim\limits_{\Delta x \to 0} \dfrac{-\int_x^{x+\Delta x} f(t)\, dt}{\Delta x} = \lim\limits_{\Delta x \to 0} \dfrac{-f(c)\,\Delta x}{\Delta x}$, for

some c in $[x, x + \Delta x]$, which in turn reduces to $\lim\limits_{\Delta x \to 0}(-f(c)) = -f(x)$.

5.S Review Exercises

2 $\int_{-1}^{2}(6x^2 + 10x^4)\, dx = (2x^3 + 2x^5)\big|_{-1}^{2} = 84$

4 $\int_{\pi/6}^{\pi/4} 3\cos 2x\, dx = \dfrac{3}{2}\sin 2x\Big|_{\pi/6}^{\pi/4} = \dfrac{3}{2}\left(1 - \dfrac{\sqrt{3}}{2}\right)$

6 $\int_0^1 (x+1)^{-4}\, dx = -\dfrac{1}{3}(x+1)^{-3}\Big|_0^1 = \dfrac{7}{24}$

8 $\dfrac{1}{3}\tan 3x + C$

10 $\dfrac{5}{3}\sec 3x + C$

12 $\dfrac{1}{5}\csc 5x + C$

14 $\int \left(x + \dfrac{1}{x}\right)^2 dx = \int (x^2 + 2 + x^{-2})\, dx$

$= \dfrac{1}{3}x^3 + 2x - x^{-1} + C$

16 $-12x^{-2} + C$

18 (a) $3x^2 \cos x^3$

(b) $\dfrac{1}{3}\sin x^3 + C$

20 (a) $(2^1 - 2^0) + (2^2 - 2^1) + \cdots + (2^{100} - 2^{99})$

$= 2^{100} - 1$

(b) $(2^1 - 2^0) + (2^2 - 2^1) + \cdots + (2^{101} - 2^{100})$

$= 2^{101} - 1$

(c) $\left(\dfrac{1}{1} - \dfrac{1}{2}\right) + \left(\dfrac{1}{2} - \dfrac{1}{3}\right) + \cdots + \left(\dfrac{1}{100} - \dfrac{1}{101}\right)$

$= \dfrac{100}{101}$

22 Partition $[1, 2]$ into n equal sections and use right endpoints as sampling points.

$\int_1^2 (5 + x^2)\, dx = \lim\limits_{\text{mesh}\to 0} \sum_{i=1}^{n} f(c_i)(x_i - x_{i-1})$

$= \lim\limits_{n\to\infty} \sum_{i=1}^{n}\left[5 + \left(1 + \dfrac{i}{n}\right)^2\right]\cdot\dfrac{1}{n}$

$$= \lim_{n\to\infty} \sum_{i=1}^{n} \left[6 + \frac{2i}{n} + \frac{i^2}{n^2}\right]\frac{1}{n}$$

$$= \lim_{n\to\infty} \frac{1}{n}\left[\sum_{i=1}^{n} 6 + \frac{2}{n}\sum_{i=1}^{n} i + \frac{1}{n^2}\sum_{i=1}^{n} i^2\right]$$

$$= \lim_{n\to\infty} \frac{1}{n}\left[6n + \frac{2}{n}\left(\frac{n^2}{2} + \frac{n}{2}\right) + \frac{1}{n^2}\left(\frac{n^3}{3} + \frac{n^2}{2} + \frac{n}{6}\right)\right]$$

$$= \lim_{n\to\infty}\left[6 + 1 + \frac{1}{n} + \frac{1}{3} + \frac{1}{2n} + \frac{1}{6n^2}\right] = \frac{22}{3}$$

24. $\int_{\pi/4}^{3\pi/4} 4\sin x\, dx = -4\int_{\pi/4}^{3\pi/4}(-\sin x)\, dx$

$= -4\cos x\Big|_{\pi/4}^{3\pi/4} = -4\left[\cos\frac{3\pi}{4} - \cos\frac{\pi}{4}\right] = 4\sqrt{2}$

26. $\int_{3}^{12}\sqrt{3x}\, dx = \sqrt{3}\int_{3}^{12} x^{1/2}\, dx = \sqrt{3}\cdot\frac{2}{3}x^{3/2}\Big|_{3}^{12}$

$= \frac{2\sqrt{3}}{3}(12^{3/2} - 3^{3/2}) = 42$

28. Let $u = 4x$. Then $\frac{d}{dx}\left(\int_{3}^{4x}\sin^3 t\, dt\right) =$

$\frac{d}{du}\left(\int_{3}^{u}\sin^3 t\, dt\right)\cdot\frac{du}{dx} = (\sin^3 u)\cdot 4 = 4\sin^3 4x$.

30. Let $u = 3x$ and $v = 4x$. Then $\frac{d}{dx}\left(\int_{3x}^{4x}\frac{1}{1+t^2}\, dt\right)$

$= \frac{d}{dx}\left(\int_{3x}^{0}\frac{1}{1+t^2}\, dt + \int_{0}^{4x}\frac{1}{1+t^2}\, dt\right) =$

$\frac{d}{du}\left(\int_{u}^{0}\frac{1}{1+t^2}\, dt\right)\cdot\frac{du}{dx} + \frac{d}{dv}\left(\int_{0}^{v}\frac{1}{1+t^2}\, dt\right)\cdot\frac{dv}{dx}$

$= -3\cdot\frac{1}{1+9x^2} + 4\cdot\frac{1}{1+16x^2}$

$= \frac{1 - 12x^2}{(1+9x^2)(1+16x^2)}.$

32. Let $u = \sin 3x$ and $v = \cos 3x$. Then

$\frac{d}{dx}\left(\int_{\sin 3x}^{\cos 3x} 2^t\, dt\right) = \frac{d}{dx}\left(\int_{\sin 3x}^{0} 2^t\, dt + \int_{0}^{\cos 3x} 2^t\, dt\right)$

$= \frac{d}{du}\left(\int_{u}^{0} 2^t\, dt\right)\cdot\frac{du}{dx} + \frac{d}{dv}\left(\int_{0}^{v} 2^t\, dt\right)\cdot\frac{dv}{dx}$

$= -2^u\cdot(3\cos 3x) + 2^v(-3\sin 3x)$

$= -3[2^{\sin 3x}\cos 3x + 2^{\cos 3x}\sin 3x]$.

34. $\int_{1}^{0} x^2\, dx = -\int_{0}^{1} x^2\, dx = -\frac{1}{3}$

36. $\int_{3}^{2}(x+1)^{-2}\, dx = -\int_{2}^{3}(x+1)^{-2}\, dx$

$= -[-(x+1)^{-1}]\Big|_{2}^{3} = 4^{-1} - 3^{-1} = -\frac{1}{12}$

38. First $f'(x) = 2\cos 2x$, $f''(x) = -4\sin 2x$, $f^{(3)}(x) = -8\cos 2x$, and $f^{(4)}(x) = 16\sin 2x$. Now let M_n be the maximum value of $|f^{(n)}(x)|$ over $[1, 2]$. To find M_2, we must identify all extrema of $f''(x) = -4\sin 2x$ over $[1, 2]$. Thus $f^{(3)}(x) = -8\cos 2x = 0$ when $x = (2n+1)\pi/4$, for any integer n. None of these values occur in $[1, 2]$ so the extrema occurs at the endpoints. Since $4\sin 2 \approx 3.6372$ while $4\sin 4 \approx -3.0272$, we see that $M_2 = 4\sin 2$. Similarly, $M_4 = 16\sin 2$.

(a) We desire $\frac{1}{12}(2-1)M_2 h^2 < \frac{1}{1000}$, so $h <$

$\left(\frac{12}{4000\sin 2}\right)^{1/2} \approx 0.0574$. We need at least

$\frac{2-1}{0.0574} \approx 17.4$ sections, so the number of

sections must be at least 18 and $h \le 1/18$.

(b) We desire $\frac{1}{180}(2-1)M_4 h^4 < \frac{1}{1000}$, so

$h < \left(\frac{180}{16000 \sin 2}\right)^{1/4} \approx 0.3335$. We need at

least $\frac{2-1}{0.3335} \approx 2.998$ sections. The number

of sections must be even, so it must be at least 4 and $h \leq 1/4$.

40 (a) With $h = 1/2$: $\int_1^2 x^4 \, dx \approx$

$\frac{1/2}{2}\left[1^4 + 2\cdot\left(\frac{3}{2}\right)^4 + 2^4\right] = \frac{1}{4}\cdot\frac{217}{8} = \frac{217}{32}$.

With $h = 1/4$: $\int_1^2 x^4 \, dx \approx$

$\frac{1/4}{2}\left[1^4 + 2\left(\frac{5}{4}\right)^4 + 2\left(\frac{3}{2}\right)^4 + 2\left(\frac{7}{4}\right)^4 + 2^4\right]$

$= \frac{1}{8}\cdot\frac{3249}{64} = \frac{3249}{512}$.

(b) Let $E(h)$ denote the error when sections of length h are used. Then $E(1/2) =$

$\left|\frac{31}{5} - \frac{217}{32}\right| = \frac{93}{160}$ and $E(1/4) =$

$\left|\frac{31}{5} - \frac{3249}{512}\right| = \frac{373}{2560}$.

(c) $\frac{E(1/4)}{E(1/2)} = \frac{373/2560}{93/160} = \frac{373}{1488}$

(d) With $h = 1/20$: $\int_1^2 x^4 \, dx \approx \frac{9,929,333}{1,600,000}$, so

$E(1/20) = \left|\frac{31}{5} - \frac{9,929,333}{1,600,000}\right| = \frac{9,333}{1,600,000}$,

while $E(1/2) = 93/160$. Thus $\frac{E(1/2)}{E(1/20)} =$

$\frac{93/160}{9,333/1,600,000} = \frac{310,000}{3,111} \approx 99.65$

$\approx \left(\frac{1/2}{1/20}\right)^2$.

42 (a) With $f(t) = \int_2^t \sin x^2 \, dx$, we have

$\lim_{\Delta x \to 0} \frac{\int_2^{5+\Delta x} \sin x^2 \, dx - \int_2^5 \sin x^2 \, dx}{\Delta x} =$

$\lim_{\Delta x \to 0} \frac{f(5+\Delta x) - f(5)}{\Delta x}$, which is the definition

of $f'(5)$. Using the first fundamental theorem of calculus, $f'(t) = \sin t^2$, so $f'(5) = \sin 25$.

(b) First, compute $\frac{d}{dx}\left(\left(\int_0^{x^2} \frac{dt}{\sqrt{1-5t^3}}\right)^2\right)$. By the

chain rule, $\frac{d}{dx}\left(\left(\int_0^{x^2} \frac{dt}{\sqrt{1-5t^3}}\right)^2\right) =$

$2\left(\int_0^{x^2} \frac{dt}{\sqrt{1-5t^3}}\right)\cdot\frac{d}{dx}\left(\int_0^{x^2} \frac{dt}{\sqrt{1-5t^3}}\right)$, where

$\frac{d}{dx}\left(\int_0^{x^2} \frac{dt}{\sqrt{1-5t^3}}\right) = \frac{d}{du}\left(\int_0^u \frac{dt}{\sqrt{1-5t^3}}\right)\frac{du}{dx}$

$= \left(\frac{1}{\sqrt{1-5u^3}}\right)(2x) = \frac{2x}{\sqrt{1-5x^6}}$, also by the

chain rule with $u = x^2$ (and the first fundamental theorem of calculus). Combining

these results, $\frac{d}{dx}\left(\left(\int_0^{x^2} \frac{dt}{\sqrt{1-5t^3}}\right)^2\right) =$

$\frac{4x}{\sqrt{1-5x^6}}\int_0^{x^2} \frac{dt}{\sqrt{1-5t^3}}$. Now, compute the

derivative of this expression by the product

rule: $\dfrac{d^2}{dx^2}\left(\left(\int_0^{x^2}\dfrac{dt}{\sqrt{1-5t^3}}\right)^2\right) =$

$\left(\dfrac{4x}{\sqrt{1-5t^6}}\right)\left(\int_0^{x^2}\dfrac{dt}{\sqrt{1-5t^3}}\right)' +$

$\left(\dfrac{4x}{\sqrt{1-5x^6}}\right)'\left(\int_0^{x^2}\dfrac{dt}{\sqrt{1-5t^3}}\right) =$

$\dfrac{4x}{\sqrt{1-5x^6}}\dfrac{2x}{\sqrt{1-5x^6}} +$

$\left[\dfrac{\sqrt{1-5x^6}(4) - 4x(1/2)(1-5x^6)^{-1/2}(-30x^5)}{1 - 5x^6}\right]\left(\int_0^{x^2}\dfrac{dt}{\sqrt{1-5t^3}}\right)$

$= \dfrac{8x^2}{1-5x^6} + \dfrac{4(1-5x^6)+60x^6}{(1-5x^6)^{3/2}}\int_0^{x^2}\dfrac{dt}{\sqrt{1-5t^3}} =$

$\dfrac{8x^2}{1-5x^6} + \dfrac{40x^6 + 4}{(1-5x^6)^{3/2}}\int_0^{x^2}\dfrac{dt}{\sqrt{1-5t^3}}.$

44 The area under a curve $y = f(t)$ for $a \le t \le b$ is given by $\int_a^b f(t)\,dt$. If the curve is a graph of speed as a function of time, then the integral gives the distance traveled, so the area represents distance.

46 (a)

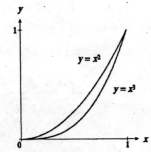

(b) From the figure, $c(x) = x^2 - x^3$.

(c) Area $= \int_0^1 c(x)\,dx = \int_0^1 (x^2 - x^3)\,dx$

(d) Area $= \int_0^1 (x^2 - x^3)\,dx = \left(\dfrac{x^3}{3} - \dfrac{x^4}{4}\right)\Big|_0^1$

$= \left(\dfrac{1}{3} - \dfrac{1}{4}\right) - (0 - 0) = \dfrac{1}{12}$

48 Volume $= \int_0^3 \pi(\sqrt[3]{x})^2\,dx = \pi\int_0^3 x^{2/3}\,dx =$

$\pi\dfrac{3}{5}x^{5/3}\Big|_0^3 = \dfrac{3\pi}{5}\cdot 3^{5/3} = \dfrac{3^{8/3}\pi}{5}$

50 In $[0, \pi/2]$, $\sin x = \cos x$ when $x = \pi/4$. Also, $c(x) = \cos x - \sin x$ is the cross-sectional length. So the area is $\int_0^{\pi/4} c(x)\,dx =$

$\int_0^{\pi/4}(\cos x - \sin x)\,dx = (\sin x + \cos x)\Big|_0^{\pi/4}$

$= \sin\dfrac{\pi}{4} + \cos\dfrac{\pi}{4} - \sin 0 - \cos 0 = \sqrt{2} - 1.$

52 To find the limits of integration, find where $x = x^2 - 4x + 5$; that is, where $x^2 - 5x + 5 = 0$.

Thus $x = \dfrac{5}{2} \pm \dfrac{\sqrt{5}}{2}$. The cross-sectional length is

$c(x) = x - (x^2 - 4x + 5) = -x^2 + 5x - 5$. Thus the area is given by $\int_{\frac{1}{2}(5-\sqrt{5})}^{\frac{1}{2}(5+\sqrt{5})} c(x)\,dx =$

$\int_{\frac{1}{2}(5-\sqrt{5})}^{\frac{1}{2}(5+\sqrt{5})} (-x^2 + 5x - 5)\,dx =$

$\left(-\dfrac{x^3}{3} + \dfrac{5}{2}x^2 - 5x\right)\Big|_{\frac{1}{2}(5-\sqrt{5})}^{\frac{1}{2}(5+\sqrt{5})} = \dfrac{1}{3}\left[\left(\dfrac{1}{2}(5-\sqrt{5})\right)^3 - \left(\dfrac{1}{2}(5+\sqrt{5})\right)^3\right] + \dfrac{5}{2}\left[\left(\dfrac{1}{2}(5+\sqrt{5})\right)^2 - \left(\dfrac{1}{2}(5-\sqrt{5})\right)^2\right]$

$+ 5\left[\left(\dfrac{1}{2}(5-\sqrt{5})\right) - \left(\dfrac{1}{2}(5+\sqrt{5})\right)\right] =$

$\frac{1}{24}(-160\sqrt{5}) + \frac{5}{8}(20\sqrt{5}) + \frac{5}{2}(-2\sqrt{5}) = \frac{5\sqrt{5}}{6}$.

54 (a) $\int_1^2 \frac{2^x}{x^2} dx \approx \frac{1/4}{3}[f(1) + 4f(5/4) + 2f(3/2)$

$+ 4f(7/4) + f(2)] = \frac{1}{12}[2 + 4 \cdot \frac{16}{25} \cdot 2 \cdot 2^{1/4} +$

$2 \cdot 4/9 \cdot 2 \cdot 2^{1/2} + 4 \cdot \frac{16}{49} \cdot 2 \cdot 2^{3/4} + 1] \approx 1.3330$

(b) $\int_1^2 \frac{1}{x^2} dx = \int_1^2 x^{-2} dx = -x^{-1}\big|_1^2$

$= -2^{-1} - (-1)^{-1} = \frac{1}{2}$

(c) $\int_1^2 \frac{x^3 + 1}{x^2} dx = \int_1^2 \left(x + \frac{1}{x^2}\right) dx$

$= \left(\frac{x^2}{2} - \frac{1}{x}\right)\bigg|_1^2 = 2 - \frac{1}{2} - \frac{1}{2} + 1 = 2$

56 It is the approximating sum with $n = 100$ and $x_i = c_i = \frac{i}{100}$.

58 (a) It is the approximating sum with $x_i = c_i = i/n$.

(b) $\frac{1}{4^3}(1^2 + 2^2 + 3^2 + 4^2) = \frac{15}{32}$

(c) $\int_0^1 x^2 dx = \frac{1}{3}x^3\big|_0^1 = \frac{1}{3}$

(d) By (a) and (c), it equals 1/3.

60 (a) $0 = \int_0^0 f(t) dt = [f(0)]^2$, so $f(0) = 0$.

(b) $f(x) = \frac{d}{dx}\left[\int_0^x f(t) dt\right] = \frac{d}{dx}(f(x)^2) =$

$2f(x)f'(x)$, so $f'(x) = 1/2$ for $x > 0$. Hence

$f(x) = \int_0^x f'(t) dt = x/2$.

62 We have $\frac{1}{b-a}\int_a^b v(t) dt = \frac{1}{2}[v(a) + v(b)]$, so

$\int_a^b v(t) dt = \frac{1}{2}(b - a)[v(a) + v(b)]$.

Differentiating with respect to b gives

$v(b) = \frac{1}{2}[(b - a)v'(b) + v(a) + v(b)]$, so

$v(a) = v'(b)a + [v(b) - bv'(b)]$. Let $c = v'(b)$, $d = v(b) - bv'(b)$, and $a = t$.

64 Since $\frac{d}{dN}\left(\int_0^N R(t) dt\right) = R(N)$, the quotient rule gives the result.

66 (a) Let all c_i's be non-dyadic.

(b) Let all c_i's be dyadic.

(c) No matter how small the mesh, the approximating sums can be 0 or 3, so they don't approach a limit.

68 Following the hint, $\frac{d}{da}\left(\int_1^a f(x) dx\right) =$

$\frac{d}{da}\left(\int_b^{ab} f(x) dx\right)$ and $f(a) = bf(ab)$. Setting $a = 1$

gives $1 = bf(b)$, so $f(b) = 1/b$ for all positive b. Hence $f(x) = 1/x$ is the only function that satisfies the conditions.

6 Topics in Differential Calculus

6.1 Logarithms

2. (a) $2/3 = \log_8 4$
   (b) $3 = \log_{10} 1000$
   (c) $-4 = \log_{10} 0.0001$
   (d) $0 = \log_3 1$
   (e) $\frac{1}{2} = \log_{10} \sqrt{10}$
   (f) $-2 = \log_{1/2} 4$

4. (a)

x	1/16	1/4	1	2	4	8	16
$\log_4 x$	-2	-1	0	1/2	1	3/2	2

   (b)

   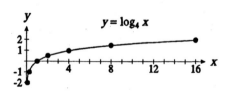

6. (a) $10^3 = 1000$
   (b) $5^{-2} = 1/25$
   (c) $(1/2)^2 = 1/4$
   (d) $64^{7/6} = 128$

8. (a) 100
   (b) 0.01
   (c) 7
   (d) p

10. (a) 4.2
    (b) -2.1
    (c) 4.1

12. $\log_{6/5} 3$

14. $\log_{900} 5$

16. $1/a$, by Exercise 15.

18. (a) $\frac{1}{2} \log_{10} 2 \approx 0.15$
    (b) $-\log_{10} 2 \approx -0.30$
    (c) $\log_{10} 2 - \log_{10} 3 \approx -0.18$
    (d) $\frac{1}{3} \log_{10} 3 \approx 0.16$
    (e) $\log_{10} 2 + 2 \log_{10} 3 \approx 1.26$
    (f) $2 \log_{10} 2 + \log_{10} 3 \approx 1.08$
    (g) $\log_{10} 3 - 2 \log_{10} 2 \approx -0.12$
    (h) $1 + \log_{10} 0.75 \approx 0.88$
    (i) $-\log_{10} 7.5 \approx -0.88$
    (j) $-1 + \log_{10} 0.75 = -1.12$
    (k) $1 + \frac{5}{3} \log_{10} 2 + \log_{10} 3 \approx 1.98$
    (l) $2 \log_{10} 3 - 5 \log_{10} 2 \approx -0.54$

20. (a) $\dfrac{\log_{10} 3}{\log_{10} 1/2} \approx -1.58$

    (b) $\dfrac{\log_{10} 1/2}{\log_{10} 7} \approx -0.36$

22. (a) Reflected across x axis
    (b) Shrink the graph of $y = \log_b x$ vertically by a factor of 2 to get the graph of $\log_{b^2} x$.

24. $\log_b \dfrac{c}{d} = \log_b \left(\dfrac{b^{\log_b c}}{b^{\log_b d}} \right) = \log_b \left(b^{\log_b c - \log_b d} \right)$

26 $\frac{5}{2} \log_{10}(1 + x^2) + 2 \log_{10}(3 + x) + \frac{1}{4} \log_{10}(1 + 2x)$

28 $\frac{1}{2} \log_{10} x + \frac{1}{2} \log_{10}(1 + x) - \frac{3}{4} \log_{10}(1 + 2x)$

= $\log_b c - \log_b d$

30 Yes, whenever $cd = c + d$. For example, let $c = d = 2$.

32 (c) In adding lengths together, one is actually adding the logarithms of the numbers marked on the scale, rather than the numbers themselves. Addition of logarithms corresponds to multiplication, so the result is the desired product.

(d) To divide one simply subtracts logarithms. Therefore one can divide on the slide rule by subtracting lengths instead of adding them.

34 Suppose $\log_3 2 = m/n$ for integers m and n. Since $\log_3 2$ is positive, we may further assume that both m and n are positive. We then have $3^m = (3^{m/n})^n = (3^{\log_3 2})^n = 2^n$. But 3^m is an odd integer, while 2^n is even. This is a contradiction, so $\log_3 2$ must be irrational.

36 (a) First, note that 10^{n-1} is the smallest number with n digits, and 10^n is the smallest number with $n + 1$ digits. Also note that $\log_{10} 10^{n-1} = n - 1$ and $\log_{10} 10^n = n$. So $n - 1 \le \log_{10} a < n$ when $10^{n-1} \le a < 10^n$, since $y = \log_{10} x$ is an increasing function. In other words, the logarithm of an n-digit number to the base 10 (with numbers to the right of the decimal point truncated) is $n - 1$. So to find the number of digits a number has, take the logarithm of it to the base 10, save the number to the left of the decimal point, and

add 1 to it. Now let $k = 391581 \cdot 2^{216193} - 1$. Since $k + 1$ is not a power of 10, $k + 1$ has the same number of digits as k. Let $[x]$ be the greatest integer less than or equal to x. Then the number of digits in k is $[\log_{10} k] + 1$

$= [\log_{10} 391581 \cdot 2^{216193}] + 1$

$= [\log_{10} 391581 + 216193 \log_{10} 2] + 1$

$= [65086.171] + 1 = 65087.$

(b) Since 65087 digits $\cdot$ 1 page/6400 digits $\approx$ 10.1698 pages, it would take 11 pages to print it out.

6.2 The Number e

2

x	-0.1	-0.01	-0.001	-0.0005
$(1 + x)^{1/x}$	2.86797	2.73200	2.71964	2.71896

4 ∞

6 $\lim_{h \to 0} (1 - h)^{1/h} = \left[\lim_{h \to 0} (1 + (-h))^{1/(-h)} \right]^{-1} = e^{-1}$

8 $\lim_{n \to \infty} \left(1 + \frac{3}{n}\right)^{n/2} = \left[\lim_{n \to \infty} \left(1 + \frac{3}{n}\right)^{n/3} \right]^{3/2} = e^{3/2}$

9

x	-0.9	-0.99	-0.999
$(1 + x)^{1/x}$	12.915	104.76	1006.9

$(1 + x)^{1/x}$ seems to approach ∞.

10 It approaches 1.

12 (a) $1000 \cdot 1.08 = 1080$ dollars

(b) $1000 \cdot 1.04^2 = 1081.60$ dollars

(c) $1000\left(1 + \frac{0.08}{12}\right)^{12} \approx 1083.00$ dollars

(d) $1000\left(1 + \dfrac{0.08}{365}\right)^{365} \approx 1083.28$ dollars

(e) $1000e^{0.08} \approx 1083.29$ dollars

14

x	-3	-2	-1	0	1	2	3
e^x	0.0498	0.1353	0.3679	1.0000	2.7183	7.3891	20.0855

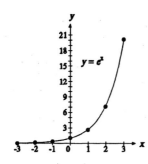

16 (a)

n	-2	-10	-100
$(1 + 1/n)^n$	4	2.8680	2.7320

(b) e

(c) Let $n = -1 - m$. Then $\left(1 + \dfrac{1}{n}\right)^n$

$= \left(1 - \dfrac{1}{m+1}\right)^{-m-1} = \left(\dfrac{m}{m+1}\right)^{-m-1}$

$= \left(\dfrac{m+1}{m}\right)^{m+1} = \left(1 + \dfrac{1}{m}\right)\left(1 + \dfrac{1}{m}\right)^m$. Hence

$\lim\limits_{n \to -\infty} \left(1 + \dfrac{1}{n}\right)^n$

$= \lim\limits_{m \to \infty} \left(1 + \dfrac{1}{m}\right) \cdot \lim\limits_{m \to \infty} \left(1 + \dfrac{1}{m}\right)^m$

$= 1 \cdot \lim\limits_{m \to \infty} \left(1 + \dfrac{1}{m}\right)^m = \lim\limits_{n \to \infty} \left(1 + \dfrac{1}{n}\right)^n$.

6.3 The Derivatives of the Logarithmic Functions

2   $(\ln(1 + x^3))' = \dfrac{1}{1 + x^3} \cdot (1 + x^3)' = \dfrac{3x^2}{1 + x^3}$

4   $(x \ln x^2)' = x(\ln x^2)' + (\ln x^2)(x)'$

$= x \cdot \dfrac{1}{x^2} \cdot 2x + (\ln x^2) \cdot (1) = 2 + \ln x^2$

6   $[(\ln x)^3]' = \dfrac{3}{x} (\ln x)^2$

8   $(\sec 5x \ln 2x)' = (\sec 5x)(\ln 2x)' + (\ln 2x)(\sec 5x)'$

$= (\sec 5x) \cdot \dfrac{1}{2x} \cdot 2 + (\ln 2x)(\sec 5x \tan 5x) \cdot (5)$

$= \dfrac{\sec 5x}{x} + 5 \sec 5x \tan 5x \ln 2x$

$= (\sec 5x)\left(\dfrac{1}{x} + 5 \tan 5x \ln 2x\right)$

10   $[\cos(\ln x)]' = -\sin(\ln x) \cdot \dfrac{1}{x} = -\dfrac{1}{x} \sin(\ln x)$

12   $\left[\dfrac{x}{3} - \dfrac{1}{9} \ln(3x + 1)\right]' = \dfrac{1}{3} - \dfrac{1}{9} \cdot \dfrac{3}{3x + 1} = \dfrac{x}{3x + 1}$

14   $\left[x + 3 - 6 \ln(x + 3) - \dfrac{9}{x + 3}\right]'$

$= 1 - \dfrac{6}{x + 3} + \dfrac{9}{(x + 3)^2} = \dfrac{x^2}{(x + 3)^2}$

16   $[\ln(x + \sqrt{x^2 + 1})]' = \dfrac{1 + \dfrac{x}{\sqrt{x^2 + 1}}}{x + \sqrt{x^2 + 1}} = \dfrac{1}{\sqrt{x^2 + 1}}$

18   $\dfrac{d}{dx}\left(-\dfrac{1}{3x} + \dfrac{5}{9} \ln(3x + 5) - \dfrac{5}{9} \ln x\right)$

$= \dfrac{1}{3x^2} + \dfrac{5}{3(3x + 5)} - \dfrac{5}{9x} = \dfrac{15 - 16x}{9x^2(3x + 5)}$

20 $\dfrac{d}{dx}\left[\sqrt{x^2+1}\left(\ln\left(\sqrt{x^2+1}-1\right)-\ln x\right)\right] =$

$\sqrt{x^2+1}\left(\dfrac{\frac{x}{\sqrt{x^2+1}}}{\sqrt{x^2+1}-1}-\dfrac{1}{x}\right)+\dfrac{x}{\sqrt{x^2+1}}\left(\ln\left(\sqrt{x^2+1}-1\right)-\ln x\right)$

$= \dfrac{1}{x}+\dfrac{x}{\sqrt{x^2+1}}\left(\ln\left(\sqrt{x^2+1}-1\right)-\ln x\right)$

22 $\dfrac{d}{dx}\left[\dfrac{1}{2}\ln(2x+1)+\dfrac{1}{3}\ln(3x+2)-5\ln(x^2+1)\right]$

$= \dfrac{1}{2x+1}+\dfrac{1}{3x+2}-\dfrac{10x}{x^2+1}$

24 $\dfrac{d}{dx}\left[\log_2((x^2+1)^3 \sin 3x)\right]$

$= \dfrac{d}{dx}(3\log_2(x^2+1))+\dfrac{d}{dx}(\log_2(\sin 3x))$

$= 3\cdot\dfrac{\log_2 e}{x^2+1}\cdot 2x+\dfrac{\log_2 e}{\sin 3x}\cdot 3\cos 3x$

$= 3\log_2 e\left[\dfrac{2x}{x^2+1}+\cot 3x\right]$

26 $y'=\dfrac{1-3\ln x}{x^4}$; maximum at $x=e^{1/3}$.

$y''=\dfrac{12\ln x-7}{x^5}$; inflection point at $x=e^{7/12}$.

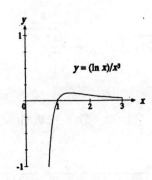
$y=(\ln x)/x^3$

28 $y=\sqrt{1+x^2}\,\sqrt[3]{(1+\cos 3x)^5}$,

$\ln y = \dfrac{1}{2}\ln(1+x^2)+\dfrac{5}{3}\ln(1+\cos 3x)$,

$\dfrac{1}{y}\cdot\dfrac{dy}{dx}=\dfrac{x}{1+x^2}-\dfrac{5\sin 3x}{1+\cos 3x}$,

$\dfrac{dy}{dx}=$

$\sqrt{1+x^2}\,\sqrt[3]{(1+\cos 3x)^5}\left(\dfrac{x}{1+x^2}-\dfrac{5\sin 3x}{1+\cos 3x}\right)$

30 $y=\dfrac{\cot^3 x}{\sqrt[3]{x}\,\sqrt{(x^3+2)^5}}$,

$\ln y = 3\ln\cot x-\dfrac{1}{3}\ln x-\dfrac{5}{2}\ln(x^3+2)$,

$\dfrac{1}{y}y'=\dfrac{-3\csc^2 x}{\cot x}-\dfrac{1}{3x}-\dfrac{15x^2}{2(x^3+2)}$

$= -3\sec x\csc x-\dfrac{1}{3x}-\dfrac{15x^2}{2(x^3+2)}$,

$y'=$

$\dfrac{-\cot^3 x}{\sqrt[3]{x}\sqrt{(x^3+2)^5}}\left(3\sec x\csc x+\dfrac{1}{3x}+\dfrac{15x^2}{2(x^3+2)}\right)$

32 Since $|\ln 2-0.693|=0.0001472$ and $|\ln 2-0.694|=0.0008528$, h should be chosen so that the error in the estimate is less than 0.0008. Now M_4 is the maximum value of $|f^{(4)}(x)|$ over the interval $[1,2]$, and the maximum of $24/x^5$ over this interval is 24. Since $a=1$ and $b=2$, we want

Error $= \dfrac{1}{180}(2-1)h^4\cdot 25 = \dfrac{5}{36}h^4 < 0.0008$.

Thus, choose $h<\sqrt[4]{\dfrac{18}{3125}}\approx 0.2755$, say $h=$

1/4. Then, by Simpson's method, $\int_1^2 \frac{dx}{x} \approx$

$\frac{1/4}{3}\left[\frac{1}{4/4} + 4 \cdot \frac{1}{5/4} + 2 \cdot \frac{1}{6/4} + 4 \cdot \frac{1}{7/4} + \frac{1}{8/4}\right] =$

$\frac{1}{12} \cdot \frac{1747}{210} = \frac{1747}{2520} \approx 0.69325$, which agrees with

$\ln 2 \approx 0.6931$ to three decimal places.

34 (a)

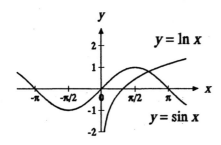

(c) 2.2191

36 $\ln(1 + h) \approx \ln(1) + (\ln x)'(1) \cdot h \approx h$

38 (a) $\frac{1}{3} \ln|3x + 2| + C$

(b) $-\ln|\cos x| + C$

(c) $\ln(3x^2 + x + 5) + C$

(d) Since $[(\ln x)^2]' = 2(\ln x) \cdot \frac{1}{x}$, $\int \ln x \cdot \frac{1}{x} dx$

$= \frac{1}{2} \int 2 \ln x \cdot \frac{1}{x} dx = \frac{1}{2} (\ln x)^2 + C$

40 $\int_2^5 \frac{x}{1+x^2} dx = \frac{1}{2} \ln(1+x^2)\Big|_2^5 = \frac{1}{2} \ln \frac{26}{5}$

42 (a) From the binomial theorem, we know that

$\left(1 + \frac{1}{n}\right)^n = \sum_{i=0}^{n} \frac{n!}{i!(n-i)!}(1^{n-i})\left(\frac{1}{n}\right)^i =$

$\frac{1}{0!} \cdot 1^n \cdot \left(\frac{1}{n}\right)^0 + \frac{n}{1!} \cdot 1^{n-1} \cdot \left(\frac{1}{n}\right)^1 +$

$\frac{n(n-1)}{2!} \cdot 1^{n-2}\left(\frac{1}{n}\right)^2 + \frac{n(n-1)(n-2)}{3!} 1^{n-3}\left(\frac{1}{n}\right)^3$

$+ \cdots + \frac{n(n-1)(n-2)\cdots 1}{n!} \cdot \left(\frac{1}{n}\right)^n$

$= \frac{1}{0!} + \frac{1}{1!} \cdot \left(\frac{n}{n}\right) + \frac{1}{2!} \cdot \left(\frac{n}{n} \cdot \frac{n-1}{n}\right)$

$+ \frac{1}{3!} \cdot \left(\frac{n}{n} \cdot \frac{n-1}{n} \cdot \frac{n-2}{n}\right) + \cdots$

$+ \frac{1}{n!} \cdot \left(\frac{n}{n} \cdot \frac{n-1}{n} \cdot \frac{n-2}{n} \cdots \frac{1}{n}\right).$

Now, for $n > 1$, the coefficients of all the terms involving $1/i!$, $2 \leq i \leq n$, are less than 1. Hence we conclude that, for $n > 1$,

$1 + \left(\frac{1}{n}\right)^n < \frac{1}{0!} + \frac{1}{1!} + \frac{1}{2!} + \cdots + \frac{1}{n!}.$

(b) $\frac{1}{0!} + \frac{1}{1!} + \frac{1}{2!} + \cdots + \frac{1}{6!} \approx 2.7181 \approx e$

44 (a)

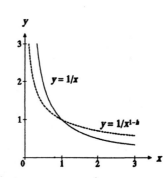

(b) $\int_1^{A(h)} \frac{dx}{x^{1-h}} = \int_1^{A(h)} x^{h-1} dx = \frac{x^h}{h}\Big|_1^{A(h)} =$

$\frac{1}{h}(A(h)^h - 1) = 1$, when $A(h) = (1 + h)^{1/h}$.

(c) For x in [2, 3], $\int_1^x \frac{dt}{t}$ is continuous so, by

the intermediate-value theorem, since $\int_1^2 \frac{dx}{x}$

<1 and $\int_1^3 \frac{dx}{x} > 1$, there must exist some B in $[2, 3]$ such that $\int_1^B \frac{dx}{x} = 1$.

(d) Since for all $x > 1$, $\int_1^x \frac{dt}{t^{1-h}} > \int_1^x \frac{dt}{t}$, we can only have $\int_1^{A(h)} \frac{dx}{x^{1-h}} = \int_1^B \frac{dx}{x}$ if $A(h) = (1 + h)^{1/h} < B$.

(e) As $h \to 0$, the curve $1/x^{1-h}$ approaches the curve $1/x$. Thus we expect $\int_1^{A(h)} \frac{dx}{x^{1-h}} \to \int_1^B \frac{dx}{x}$ as $h \to 0$.

6.4 One-to-One Functions and Their Inverse Functions

2 (a) No, since $f(0) = f(2)$.

   (b) Yes. $x = \sqrt{y} + 1$

4 (a) Yes. $x = \sqrt[5]{y} - 1$

   (b) Yes. $x = \sqrt[5]{y} - 1$

6 (a) No, since $\sqrt[4]{1 + (-1)^2} = \sqrt[4]{1 + (1)^2}$.

   (b) Yes. Its inverse is $x = \sqrt{y^4 - 1}$ for y in $[1, \infty)$.

8 (a) No, since $(-1)^{2/3} = (1)^{2/3}$.

   (b) Yes. Its inverse is $x = y^{3/2}$ for y in $[0, \infty)$.

10 No; there are horizontal lines that meet the graph three times.

12 The points on the graph of f are in solid black, those on the graph of the inverse function g are open.

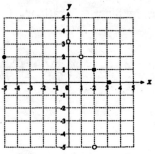

14 The inverse of $y = \log_2 x$ is $x = 2^y$ for y in $(-\infty, \infty)$.

16 (a) Since $\frac{dy}{dx} = -\sin x < 0$ for all x in $(0, \pi)$, $y = \cos x$ is decreasing on $[0, \pi]$. So $y = \cos x$ is one-to-one on $[0, \pi/2]$ and $[0, \pi]$.

   (b)

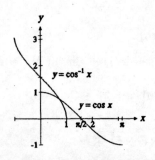

18 (a) Since $\frac{dy}{dx} = \frac{1}{x} > 0$ for x in $(0, \infty)$, $y = \ln x$ is increasing on $(0, \infty)$. Hence it is one-to-one on $(0, \infty)$.

   (b)

20 (a) $a \neq 0$

   (b) $f^{-1}(y) = \frac{y - b}{a}$

6.4 One-to-One Functions and Their Inverse Functions

22 Increasing. If $x_1 < x_2$ then $g(x_1) < g(x_2)$, so
$(f \circ g)(x_1) = f(g(x_1)) < f(g(x_2)) = (f \circ g)(x_2)$.

24 Decreasing. If $x_1 < x_2$ then $g(x_1) > g(x_2)$, that is, $g(x_2) < g(x_1)$. Hence $f(g(x_2)) < f(g(x_1))$; that is, $(f \circ g)(x_1) > (f \circ g)(x_2)$.

6.5 The Derivative of b^x

2 $x \cdot (-4)e^{-4x} + e^{-4x} = (1 - 4x)e^{-4x}$

4 $(\sin 2x) \cdot (\cos e^{-x}) \cdot (-e^{-x}) + (2 \cos 2x)(\sin e^{-x})$
$= 2 \cos 2x \sin e^{-x} - e^{-x} \sin 2x \cos e^{-x}$

6 $\dfrac{(\ln 3)3^{\sqrt{x}}}{2\sqrt{x}}$

8 $\dfrac{d}{dx}[(2 + \cos x)^{\sin x}] = \dfrac{d}{dx}[e^{(\sin x)\ln(2 + \cos x)}]$

$= e^{\sin x \ln(2 + \cos x)}\left[(\sin x)\dfrac{-\sin x}{2 + \cos x} + (\cos x)\ln(2 + \cos x)\right]$

$= (2 + \cos x)^{\sin x}\left[(\cos x)\ln(2 + \cos x) - \dfrac{\sin^2 x}{2 + \cos x}\right]$

10 $[(\tan \sqrt{x})e^{-x}]' = (\tan \sqrt{x})(-e^{-x}) + e^{-x}(\sec^2 \sqrt{x})\dfrac{1}{2\sqrt{x}}$

$= e^{-x}\left[\dfrac{\sec^2 \sqrt{x}}{2\sqrt{x}} - \tan \sqrt{x}\right]$

12 Let $y = \dfrac{10^{x^2}}{\ln(1 + x^2)}$. It follows that $\ln y =$

$(\ln 10)x^2 - \ln(\ln(1 + x^2))$, so $\dfrac{1}{y} \cdot \dfrac{dy}{dx} =$

$2(\ln 10)x - \dfrac{\dfrac{2x}{1 + x^2}}{\ln(1 + x^2)}, \dfrac{dy}{dx} =$

$= \dfrac{10^{x^2}}{\ln(1 + x^2)}\left[2(\ln 10)x - \dfrac{2x}{(1 + x^2)\ln(1 + x^2)}\right]$.

14 Let $y = \dfrac{x^\pi \tan e^x}{2^x}$. Then we have $\ln y =$

$\pi \ln x + \ln \tan e^x - (\ln 2)x$, $\dfrac{1}{y} \cdot \dfrac{dy}{dx} =$

$\dfrac{\pi}{x} + \dfrac{e^x \sec^2 e^x}{\tan e^x} - \ln 2$, and $\dfrac{dy}{dx} =$

$\dfrac{x^\pi \tan e^x}{2^x}\left[\dfrac{\pi}{x} + e^x \sec e^x \csc e^x - \ln 2\right]$.

16 $\dfrac{d}{d\theta}(\ln e^{\cos 3\theta}) = \dfrac{d}{d\theta}(\cos 3\theta) = -3 \sin 3\theta$

18 $\left[\dfrac{xb^{ax}}{a \ln b} - \dfrac{b^{ax}}{a^2(\ln b)^2}\right]'$

$= \dfrac{1}{a \ln b}(x \cdot a(\ln b)b^{ax} + b^{ax}) - \dfrac{1}{a^2(\ln b)^2} \cdot a(\ln b)b^{ax}$

$= xb^{ax}$

20 $\left[\dfrac{1}{ac}\ln(b + ce^{ax})\right]' = \dfrac{1}{ac} \cdot \dfrac{ace^{ax}}{b + ce^{ax}} = \dfrac{e^{ax}}{b + ce^{ax}}$

22 $e^{1.1} \approx e^1 + e^1(0.1) = 1.1e \approx (1.1)(2.72) = 2.992$

24 $10^{1.1} \approx 10^1 + (\ln 10)10^1(0.1) = 10 + \ln 10$
≈ 12.30

26 $\ln 1.1 \approx \ln 1 + \dfrac{1}{1}(0.1) = 0.1$

28 $\log_{10} 0.98 \approx \log_{10} 1 + \dfrac{\log_{10} e}{1}(-0.02)$

$\approx (0.43)(-0.02) = -0.0086$

30 $f(x) = x^2 e^{-x}$, $f'(x) = (2x - x^2)e^{-x}$, and $f''(x) = (2 - 4x + x^2)e^{-x}$.

(a) (0, 0)

(b) See (c).

(c) $f'(x) = x(2-x)e^{-x}$ changes from $-$ to $+$ at $x = 0$ and from $+$ to $-$ at $x = 2$. $(0, 0)$ is a minimum; $(2, 4e^{-2})$ is a maximum.

(d) $f''(x)$ changes sign at $2 \pm \sqrt{2}$.

(e) $y = 0$ is an asymptote as $x \to \infty$.

(f)

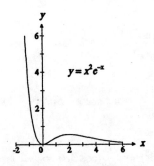

32 $f(x) = x2^{-x}$, $f'(x) = 2^{-x}(1 - x \ln 2)$, $f''(x) = \ln 2 \cdot 2^{-x}(x \ln 2 - 2)$

(a) $(0, 0)$

(b) See (c).

(c) $f'(x) = 2^{-x}(1 - x \ln 2)$ changes from $+$ to $-$ at $x = (\ln 2)^{-1}$. Hence $\left(\dfrac{1}{\ln 2}, \dfrac{1}{\ln 2} \cdot 2^{-\frac{1}{\ln 2}}\right)$ is a maximum.

(d) $f''(x)$ changes sign at $x = \dfrac{2}{\ln 2}$.

(e) $y = 0$ is an asymptote as $x \to \infty$.

(f)

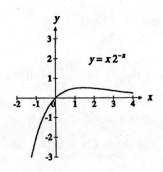

34 $f(x) = xe^x$, $f'(x) = (x + 1)e^x$, $f''(x) = (x + 2)e^x$

(a) $(0, 0)$

(b) See (c).

(c) $f'(x)$ changes from $-$ to $+$ at $x = -1$, so $(-1, -e^{-1})$ is a minimum.

(d) $f''(x)$ changes from $-$ to $+$ at $x = -2$, so $(-2, -2e^{-2})$ is an inflection point.

(e) $y = 0$ is a horizontal asymptote as $x \to -\infty$.

(f)

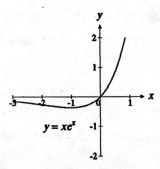

36 $\int_0^{\ln 2} 5e^{-2x}\, dx = -\dfrac{5}{2} e^{-2x} \Big|_0^{\ln 2} = \dfrac{15}{8}$

38 $\int_{-4}^{1} 2^{-x}\, dx = -\dfrac{1}{\ln 2} 2^{-x} \Big|_{-4}^{1} = \dfrac{31}{2 \ln 2}$

40 (a) $f(0) = f(1) = 5$

(b) $f'(c) = (1 - c - c^2)e^c = 0$ when $c^2 + c - 1 = 0$; that is, for $c = \dfrac{-1 \pm \sqrt{5}}{2}$. Of these, only $\dfrac{-1 + \sqrt{5}}{2}$ is in the interval $(0, 1)$.

42 (a) $f(0) = -1 < 0 < (3 - e) + \sin 1 = f(1)$

(b) $x_2 = x_1 - \dfrac{3x_1 + \sin x_1 - e^{x_1}}{3 + \cos x_1 - e^{x_1}} \approx 0.3516$,

$x_3 \approx 0.3604$.

44 $(Ae^{kx})' = kAe^{kx}$, so the derivative of Ae^{kx} is proportional to Ae^{kx}.

46 (a) For $t \geq 0$, $y = e^{-t} \sin t = 0$ whenever $\sin t = 0$, that is, at $t = n\pi$, where n is a nonnegative integer.

(b)
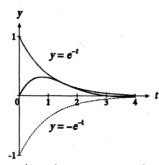

Since $|\sin t| \leq 1$ for all t, $|e^{-t} \sin t| = e^{-t} |\sin t| \leq e^{-t}$ for all $t \geq 0$. Thus $-e^{-t} \leq e^{-t} \sin t \leq e^{-t}$ for all $t \geq 0$.

(c) $y = e^{-t} \sin t = e^{-t}$ whenever $\sin t = 1$, that is, at $t = \dfrac{(4n + 1)\pi}{2}$, where n is a nonnegative integer. So at $t = \dfrac{(4n + 1)\pi}{2}$, $y = e^{-t} \sin t$ intersects the curve $y = e^{-t}$.

(d) $y = e^{-t} \sin t = e^{-t}$ whenever $\sin t = -1$, that is, at $t = \dfrac{(4n + 3)\pi}{2}$, where n is a nonnegative integer. So at $t = \dfrac{(4n + 3)\pi}{2}$, $y = e^{-t} \sin t$ intersects the curve $y = -e^{-t}$. The graph above reflects this information.

(e) $y' = -e^{-t} \sin t + e^{-t} \cos t = e^{-t}(\cos t - \sin t) = 0$ when $\tan t = 1$, that is, at $t = \dfrac{(2n + 1)\pi}{4}$, where n is a nonnegative integer, less than or equal to 7.

6.6 The Derivatives of the Inverse Trigonometric Functions

2 (a)
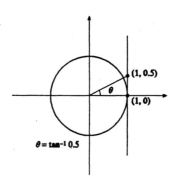

(b)
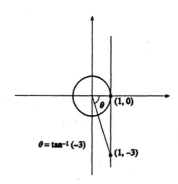

(c)
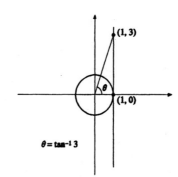

4 (a) $\lim\limits_{x \to \infty} \tan^{-1} x = \dfrac{\pi}{2}$

(b) $\lim\limits_{x \to -\infty} \tan^{-1} x = -\dfrac{\pi}{2}$

6 (a)

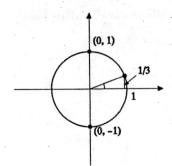

(b)

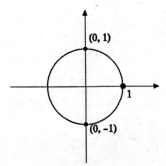

(c)

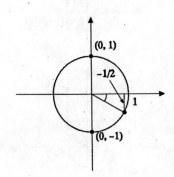

8

x	−1.0	−0.8	−0.6	−0.4	−0.2
$\cos^{-1} x$	3.142	2.498	2.214	1.982	1.772

x	0	0.2	0.4	0.6	0.8	1.0
$\cos^{-1} x$	1.571	1.369	1.159	0.927	0.664	0.0

10

x	1	2	3	4
$\csc^{-1} x$	1.571	0.524	0.340	0.253

x	−1	−2	−3	−4
$\csc^{-1} x$	−1.571	−0.524	−0.340	−0.253

12 (a), (d), and (e) are meaningless

14 $\tan(\sec^{-1} 2) = \tan \pi/6 = \sqrt{3}$

16 $\tan[\sin^{-1}(\sqrt{3}/2)] = \tan \pi/3 = \sqrt{3}$

18 $\sin(\tan^{-1} 0) = \sin 0 = 0$

20 $\cos^{-1}[\cos(-\pi/4)] = \cos^{-1}(\sqrt{2}/2) = \pi/4$

22 $(\tan^{-1} 3x)' = \dfrac{3}{1 + 9x^2}$

24 $(\sin^{-1} e^{-x})' = \dfrac{1}{\sqrt{1 - (e^{-x})^2}}(-e^{-x}) = -\dfrac{1}{\sqrt{e^{2x} - 1}}$

26 $\left(-\dfrac{1}{3} \sin^{-1} \dfrac{3}{x}\right)' = -\dfrac{1}{3} \cdot \dfrac{1}{\sqrt{1 - (3/x)^2}}\left(-\dfrac{3}{x^2}\right)$

$= \dfrac{1}{x^2\sqrt{(x^2 - 9)/x^2}} = \dfrac{1}{|x|\sqrt{x^2 - 9}}$

28 $\left(\dfrac{1}{\sin^{-1} 2x}\right)' = -\dfrac{1}{(\sin^{-1} 2x)^2} \dfrac{2}{\sqrt{1 - (2x)^2}}$

$= -\dfrac{2}{(\sin^{-1} 2x)^2 \sqrt{1 - 4x^2}}$

30 $(x^3 \tan^{-1} 2x)' = x^3 \cdot \dfrac{1}{1 + (2x)^2} \cdot 2 + 3x^2 \tan^{-1} 2x$

$= \dfrac{2x^3}{1 + 4x^2} + 3x^2 \tan^{-1} 2x$

6.6 The Derivatives of the Inverse Trigonometric Functions

32 $[\arcsin(2x-3)]' = \dfrac{1}{\sqrt{1-(2x-3)^2}} \cdot 2$

$= \dfrac{2}{\sqrt{-4x^2+12x-8}} = \dfrac{1}{\sqrt{-x^2+3x-2}}$

34 $(\text{arcsec}\sqrt{x})' = \dfrac{1}{|\sqrt{x}|\sqrt{(\sqrt{x})^2-1}} \cdot \dfrac{1}{2\sqrt{x}} = \dfrac{1}{2x\sqrt{x-1}}$

36 $[\ln(\sin^{-1}5x)^2]'$

$= \dfrac{1}{(\sin^{-1}5x)^2} \cdot 2(\sin^{-1}5x) \cdot \dfrac{1}{\sqrt{1-(5x)^2}} \cdot 5$

$= \dfrac{10}{(\sin^{-1}5x)\sqrt{1-25x^2}}$

38 $\left(10^{\sec^{-1}2x}\right)' = (\ln 10)10^{\sec^{-1}2x} \cdot \dfrac{1}{|2x|\sqrt{(2x)^2-1}} \cdot 2$

$= \dfrac{(\ln 10)10^{\sec^{-1}2x}}{|x|\sqrt{4x^2-1}}$

40 Let $y = 2^x(\log_3 x)(\sec 3x)$. Then $\ln y = (\ln 2)x +$

$\ln(\log_3 x) + \ln(\sec 3x)$, so $\dfrac{1}{y} \cdot \dfrac{dy}{dx} =$

$\ln 2 + \dfrac{1}{\log_3 x} \cdot \dfrac{\log_3 e}{x} + \dfrac{3 \sec 3x \tan 3x}{\sec 3x} =$

$\ln 2 + \dfrac{1}{x \ln x} + 3\tan 3x$ and $\dfrac{dy}{dx} =$

$2^x(\log_3 x)(\sec 3x)\left[\ln 2 + \dfrac{1}{x\ln x} + 3\tan 3x\right]$. (Note that we used the identity $\log_3 x = (\ln x)/(\ln 3)$.)

42 $[(\sin^{-1}\sqrt{x-1})^4]'$

$= 4(\sin^{-1}\sqrt{x-1})^3 \cdot \dfrac{1}{\sqrt{1-(\sqrt{x-1})^2}} \cdot \dfrac{1}{2\sqrt{x-1}}$

$= \dfrac{2(\sin^{-1}\sqrt{x-1})^3}{\sqrt{x-1}\sqrt{2-x}}$

44 $\left(\sqrt{3x^2-1} - \tan^{-1}\sqrt{3x^2-1}\right)'$

$= \dfrac{3x}{\sqrt{3x^2-1}} - \dfrac{1}{1+(\sqrt{3x^2-1})^2} \cdot \dfrac{3x}{\sqrt{3x^2-1}}$

$= \dfrac{3x}{\sqrt{3x^2-1}}\left(1 - \dfrac{1}{3x^2}\right) = \dfrac{\sqrt{3x^2-1}}{x}$

46 $\left(\dfrac{1}{2}\left[(x-3)\sqrt{6x-x^2} + 9\sin^{-1}\dfrac{x-3}{3}\right]\right)'$

$= \dfrac{1}{2}\left[(x-3) \cdot \dfrac{6-2x}{2\sqrt{6x-x^2}} + \sqrt{6x-x^2} + \right.$

$\left. 9\dfrac{1}{\sqrt{1-(x-3)^2/9}} \cdot \dfrac{1}{3}\right]$

$= \dfrac{1}{2}\left[\dfrac{-x^2+6x-9}{\sqrt{6x-x^2}} + \dfrac{6x-x^2}{\sqrt{6x-x^2}} + \dfrac{9}{\sqrt{6x-x^2}}\right]$

$= \dfrac{6x-x^2}{\sqrt{6x-x^2}} = \sqrt{6x-x^2}$

48 $\left(x\sin^{-1}3x + \dfrac{1}{3}\sqrt{1-9x^2}\right)'$

$= x\dfrac{3}{\sqrt{1-(3x)^2}} + \sin^{-1}3x + \dfrac{1}{3} \cdot \dfrac{-18x}{2\sqrt{1-9x^2}}$

$= \sin^{-1}3x$

50 $\left[x\tan^{-1}5x - \dfrac{1}{10}\ln(1+25x^2)\right]'$

$= x\dfrac{5}{1+(5x)^2} + \tan^{-1}5x - \dfrac{1}{10} \cdot \dfrac{50x}{1+25x^2}$

$= \tan^{-1}5x$

52 (a) $\dfrac{d}{dx}\left(-\dfrac{1}{5}\ln\dfrac{5+\sqrt{25-x^2}}{x}\right)$

$= \dfrac{1}{5}\cdot\dfrac{d}{dx}(\ln x) - \dfrac{1}{5}\cdot\dfrac{d}{dx}\left(\ln\left(5+\sqrt{25-x^2}\right)\right)$

$= \dfrac{1}{5x} - \dfrac{1}{5+\sqrt{25-x^2}}\cdot\dfrac{-x}{\sqrt{25-x^2}}$

$= \dfrac{1}{x\sqrt{25-x^2}}$

(b) $\left(-\dfrac{1}{5}\sin^{-1}\dfrac{5}{x}\right)' = -\dfrac{1}{5}\cdot\dfrac{1}{\sqrt{1-(5/x)^2}}\left(-\dfrac{5}{x^2}\right)$

$= \dfrac{1}{|x|\sqrt{x^2-25}}$

54 (a) $d(\tan^{-1} x) = \dfrac{1}{1+x^2}\,dx$

(b) Let $f(x) = \tan^{-1} x$. Then $f(1.1) = f(1+0.1)$

$\approx f(1) + f'(1)\cdot 0.1 = \tan^{-1} 1 + \dfrac{1}{1+1^2}(0.1)$

$= \dfrac{\pi}{4} + 0.05 \approx \dfrac{3.14}{4} + 0.05 = 0.835.$

56 (a) $d(\sec^{-1} x) = \dfrac{1}{|x|\sqrt{x^2-1}}\,dx$

(b) Let $f(x) = \sec^{-1} x$. Then $f(2.08) =$

$f(2+0.08) \approx f(2) + f'(2)\cdot 0.08$

$= \sec^{-1} 2 + \dfrac{1}{|2|\sqrt{2^2-1}}\cdot 0.08 = \dfrac{\pi}{3} + \dfrac{0.04}{\sqrt{3}}$

$\approx 1.070.$

58 (a) $\dfrac{d}{dx}\left(\dfrac{1}{a}\sin^{-1} ax + C\right) = \dfrac{1}{a}\cdot\dfrac{a}{\sqrt{1-(ax)^2}}$

$= \dfrac{1}{\sqrt{1-a^2x^2}}$

(b) $a = 5;\ \dfrac{1}{5}\sin^{-1} 5x + C$

(c) $a = \sqrt{3}:\ \dfrac{1}{\sqrt{3}}\sin^{-1}\sqrt{3}x + C$

60 (a) $\displaystyle\int_{-1}^{1}\dfrac{1}{1+x^2}\,dx = \tan^{-1} x\Big|_{-1}^{1}$

$= \tan^{-1} 1 - \tan^{-1}(-1) = \dfrac{\pi}{4} - \left(-\dfrac{\pi}{4}\right) = \dfrac{\pi}{2}$

(b) $\displaystyle\int_{0}^{\sqrt{3}}\dfrac{1}{1+x^2}\,dx = \tan^{-1} x\Big|_{0}^{\sqrt{3}}$

$= \tan^{-1}\sqrt{3} - \tan^{-1} 0 = \dfrac{\pi}{3} - 0 = \dfrac{\pi}{3}$

(c) $\displaystyle\int_{0}^{10}\dfrac{1}{1+x^2}\,dx = \tan^{-1} x\Big|_{0}^{10}$

$= \tan^{-1} 10 - \tan^{-1} 0 = \tan^{-1} 10$

62 (a) Let $y = \sqrt{1-x^2}$. Then $y^2 = 1-x^2$ and $x^2 + y^2 = 1$. Since the graph of $x^2 + y^2 = 1$

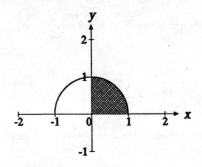

is a unit circle, the graph of $y = \sqrt{1-x^2}$ is the upper half of that circle. (See the graph.) Hence $\displaystyle\int_{0}^{1}\sqrt{1-x^2}\,dx$ corresponds to the shaded area, a quarter of the area of a circle with radius 1. Thus $\int_0^1 \sqrt{1-x^2}\, dx = \frac{1}{4}\cdot\pi\cdot 1^2 = \frac{\pi}{4}$.

(b) $\int_0^1 \sqrt{1-x^2}\, dx \approx$

$\frac{1/6}{3}\left[\sqrt{1-0^2} + 4\sqrt{1-\left(\frac{1}{6}\right)^2} + 2\sqrt{1-\left(\frac{2}{6}\right)^2}\right.$

$+ 4\sqrt{1-\left(\frac{3}{6}\right)^2} + 2\cdot\sqrt{1-\left(\frac{4}{6}\right)^2} +$

$\left. 4\sqrt{1-\left(\frac{5}{6}\right)^2} + \sqrt{1-1^2}\right] = \frac{1}{18}\left[1 + \frac{4}{6}\sqrt{35}\right.$

$\left. + \frac{2}{6}\sqrt{32} + \frac{4}{6}\cdot\sqrt{27} + \frac{2}{6}\cdot\sqrt{20} + \frac{4}{6}\cdot\sqrt{11}\right]$

≈ 0.7775316.

64 (a) $(\tan^{-1} ax)' = \dfrac{a}{1+(ax)^2} = \dfrac{a}{1+a^2x^2}$

(b) By (a), we have $(\tan^{-1}\sqrt{3}x)' = \dfrac{\sqrt{3}}{1+3x^2}$, so the area under $y = \dfrac{1}{1+3x^2}$ and above $[0, 1]$

is $\int_0^1 \dfrac{dx}{1+3x^2} = \dfrac{1}{\sqrt{3}}(\tan^{-1}\sqrt{3}x)\Big|_0^1 =$

$\dfrac{1}{\sqrt{3}}(\tan^{-1}\sqrt{3} - \tan^{-1} 0) = \dfrac{1}{\sqrt{3}}\left(\dfrac{\pi}{3} - 0\right)$

$= \dfrac{\pi}{3\sqrt{3}}$.

6.7 The Differential Equation of Natural Growth and Decay

2    $\dfrac{dy}{dx} = \dfrac{x^2}{y+3}$, so $(y+3)\, dy = x^2\, dx$,

$\int (y+3)\, dy = \int x^2\, dx + C$, and $\dfrac{y^2}{2} + 3y$

$= \dfrac{x^3}{3} + C$.

4    $\dfrac{dy}{dx} = \dfrac{y^2+1}{2x}$, so $\dfrac{dy}{y^2+1} = \dfrac{dx}{2x}$, $\int \dfrac{dy}{y^2+1} =$

$\dfrac{1}{2}\int \dfrac{dx}{x} + C$, and $\tan^{-1} y = \dfrac{1}{2}\ln|x| + C$.

6    $\dfrac{dy}{dx} = \dfrac{\sec^2 x}{\sin y}$, so $\sin y\, dy = \sec^2 x\, dx$,

$\int \sin y\, dy = \int \sec^2 x\, dx + C$, and $-\cos y =$

$\tan x + C$.

8    $\dfrac{dy}{dx} = \dfrac{5+y}{e^x}$, so $\dfrac{dy}{5+y} = e^{-x}\, dx$, $\int \dfrac{dy}{5+y} =$

$\int e^{-x}\, dx + C$, and $\ln|5+y| = -e^{-x} + C$.

10 (a) When $t = 10$, the amount is $2A$. So $2A = Ae^{k\cdot 10}$, $\ln 2 = k\cdot 10$, and $k = \dfrac{\ln 2}{10} \approx 0.0693$.

Hence $A(t) \approx Ae^{0.0693t}$.

(b) Here $2A = Ab^{10}$, so $b^{10} = 2$, and $b = 2^{1/10} \approx 1.072$. Thus $A(t) \approx A(1.072)^t$.

12 (a) $f(t) = 3\cdot 2^t = 12$ when $2^t = 4$, so $t = 2$.

(b) $f(t) = 3\cdot 2^t = 5$ when $2^t = 5/3$, so $t = \dfrac{\ln 5/3}{\ln 2}$.

(c) $f(t) = 3\cdot 2^t = 3e^{kt}$ when $e^k = 2$, so $k = \ln 2 \approx 0.693$.

14. Since $k = \dfrac{\ln 2}{t_2} \approx \dfrac{0.693}{27} \approx 0.0257$, the percent it grows per year is approximately $100 \cdot k \approx 2.57$ percent.

16. Let $f(t)$ be the amount of bacterial culture at time t. By the law of natural growth, $f(t) = 100e^{kt}$ since $f(0) = 100$ grams. Also, since $f(10) = 100e^{k \cdot 10} = 400$, we know that $k = \dfrac{1}{5} \ln 2 \approx 0.139$.

    (a) $f(3) \approx 100e^{0.139 \cdot 3} \approx 151.74$ grams.

    (b) The doubling time is $t_2 = \dfrac{\ln 2}{k} = 5$ hours.

    The quadrupling time is twice the doubling time, or 10 hours. To find the tripling time, we let $f(t) = 3 \cdot 100 = 300$ and solve for t_3. Thus, $300 = 100e^{kt_3}$, so $\ln 3 = kt_3$ and $t_3 = \dfrac{\ln 3}{k} = 5 \dfrac{\ln 3}{\ln 2} \approx 7.925$ hours.

18. Let $A(t)$ be the amount of the substance left after t days; then $A(0) = 12$ grams and $A(1) = 11$ grams. We have $A(t) = A(0)e^{kt} = 12e^{kt}$ and $A(1) = 11 = 12e^{k \cdot 1}$, so $k = \ln 11/12 \approx -0.087$. Thus, $t_{1/2} = -\dfrac{\ln 2}{\ln 11/12} \approx 7.966$ days.

20. Let $I(t)$ be the population (in millions of people) of India after t years, and let $I(0) = 689$ while $I(10) = 832$. Then $I(10) = 832 = 689e^{k \cdot 10}$ and $k = \dfrac{1}{10} \ln \dfrac{832}{689} \approx 0.019$. Thus, $t_2 = \dfrac{\ln 2}{k} \approx \dfrac{0.693}{0.019} = 36.47$ years.

22. (a) No. The correct figure is $30{,}000(1.07)^{200} \approx 22.6$ billion dollars.

    (b) If $30{,}000\left(1 + \dfrac{P}{100}\right)^{200} = 14 \times 10^9$ then $P = 100\left[\left(\dfrac{14 \cdot 10^9}{30{,}000}\right)^{1/200} - 1\right] \approx 6.744$ percent.

    (c) If $30{,}000 e^{\frac{P}{100} \cdot 200} = 14 \times 10^9$, then $P = \dfrac{1}{2} \ln \dfrac{14 \cdot 10^9}{30{,}000} \approx 6.527$ percent.

24. Let $U(t) = U(0)e^{kt}$ be the population (in millions of people) in the United States after t years, with $U(0) = 1.3$ and $t = 0$ corresponding to the year 1751.

    (a) Benjamin Franklin said that $t_2 = 20$ years, so $k = \dfrac{\ln 2}{t_2} = \dfrac{\ln 2}{20} \approx 0.03466$. So in 1990, $t = 1990 - 1751 = 239$ and $U(239) = 1.3e^{0.03466 \cdot 239} \approx 5146.67$. According to Franklin, the population of the U.S. in 1990 would be roughly 5.1 billion!

    (b) Here $U(239) = 250 = 1.3e^{k \cdot 239}$ and $k = \dfrac{1}{239} \ln \dfrac{250}{1.3} \approx 0.022$. Thus $t_2 = \dfrac{\ln 2}{k} \approx \dfrac{0.693}{0.022} = 31.5$ years, about 1.5 times that of Franklin's estimate.

26. (a) $\dfrac{dy}{dt} = k(y - A)$, so $\dfrac{dy}{y - A} = k\,dt$,

    $\displaystyle\int \dfrac{dy}{y - A} = \int k\,dt + C$, and $\ln |y - A| = kt + C$; hence $|y - A| = e^{kt + C}$ and $y = A \pm e^{kt + C} = A \pm e^C e^{kt} = A \pm Be^{kt}$, where $B = e^C$.

    (b) The step is legal because $\dfrac{dy}{dt} = \dfrac{d}{dt}(y - A)$.

6.7 The Differential Equation of Natural Growth and Decay

Now, we know $\frac{d}{dt}(y - A) = k(y - A)$ has the exponential solution $|y - A| = Be^{kt}$, where $B = e^C$. Again, we have $y = A \pm Be^{kt}$.

28 Separation of variables gives $\frac{dP}{P^{1.01}} = k\,dt$, so

$\int \frac{dP}{P^{1.01}} = \int k\,dt + C$ and finally $\frac{P^{-0.01}}{-0.01} = kt + C$. Hence $P^{-0.01} = -0.01(kt + C)$ and $P(t) = [(-0.01(kt + C)]^{-100} = \left[\frac{-100}{kt + C}\right]^{100}$. This holds only if $t < -C/k$ since $P^{-0.01} > 0$. Letting $A = -C$, we see that $t < A/k$. Now $\lim_{t \to A/k^-} P(t) =$

$\lim_{t \to A/k^-} \left[\frac{100}{A - kt}\right]^{100} = \infty$, since $A - kt \to 0^+$. $P(t)$ does not increase forever, since it is defined only for t in $[0, A/k]$.

30 (a) The amount lost in a short interval of time is proportional to both the amount present and the length of the interval, so $\Delta P \approx kP(t)\,\Delta t$, where k is negative.

(b) As $\Delta t \to 0$, we get $\frac{dP}{dt} = kP(t)$, so $P(t) = Ae^{kt}$.

(c) See (a).

32 Let $f(t) = Ab^t$ and $g(t) = Cd^t$, with $A, C > 0$ and $b, d > 1$. Now let $p(t) = f(t)g(t) = AC(bd)^t$, and their product grows exponentially. Also, let $q(t)$

$= \frac{f(t)}{g(t)} = \frac{A}{C}\left(\frac{b}{d}\right)^t$. If $b > d$, then their quotient grows exponentially; if $b \le d$ then it shrinks exponentially. The sum and difference, however, is not exponential. For example, try $f(t) = 2^t$ and $g(t) = 3^t$. If $2^t + 3^t = Ab^t$, then $2^0 + 3^0 = Ab^0$ implies that $A = 2$ and $2^1 + 3^1 = 2b^1$ implies that $b = 2.5$. But then $2^2 + 3^2 = 13$ while $2 \cdot 2.5^2 = 12.5$, so the sum of two exponentials is not exponential. The case of a difference is similar.

34 (a) A photon has a certain probability of passing through Δx meters of water. This probability is independent of the depth the photon has already reached. Hence, we expect a certain fraction, say $0 < f(\Delta x) < 1$, of the photons that reach a depth of x meters, to reach $x + \Delta x$ meters; that is, $I(x + \Delta x) = f(\Delta x)I(x)$. Hence $\Delta I = I(x + \Delta x) - I(x) = (f(\Delta x) - 1)I(x)$. For small Δx, we can approximate $f(\Delta x)$ by $f(0) + \Delta f = 1 + f'(0)\Delta x$. (Observe that $f(0) = 1$ because no light is blocked by a zero thickness of water.) Letting $k = f'(0)$, we have, for small Δx, $\Delta I \approx kI(x)\Delta x$.

(b) $\Delta I \approx kI(x)\Delta x$, so $\frac{\Delta I}{\Delta x} \approx kI(x)$; hence $\frac{dI}{dx} = \lim_{\Delta x \to 0} \frac{\Delta I}{\Delta x} = \lim_{\Delta x \to 0} kI(x) = kI(x)$. Thus $I(x) = Ae^{kx}$, where A is the initial amount of light, $I(0)$; hence $I(x) = I(0)e^{kx}$.

36 (a) The percent increase per unit time

(b) $\frac{P(t + 1) - P(t)}{P(t)} = \frac{Ae^{k(t+1)}}{Ae^{kt}} - 1 = e^k - 1$

(c) Let $f(x) = e^x - 1$. Then $f(k) = f(0 + k) \approx f(0) + f'(0) \cdot k = k$. So $e^k - 1 \approx k$.

38 (b) $\int \frac{dv}{v} = \int (-k)\,dt + C$, so $\ln|v| = -kt + C$, and $v = \pm e^C e^{-kt} = Ae^{-kt}$.

(c) $x = \int v\, dt = \int Ae^{-kt}\, dt = -\frac{1}{k}Ae^{-kt} + B$

(d) $\lim_{t \to \infty} x(t)\big|_0^t$

$= \lim_{t \to \infty}\left(-\frac{1}{k}Ae^{-kt} + B\right) - \left(-\frac{1}{k}A + B\right) = \frac{A}{k}$

40 (a) Separation of variables gives $\frac{dP}{P(M - P)} = k\, dt$, and $\frac{1}{M}\left(\frac{dP}{P} + \frac{dP}{M - P}\right) = k\, dt$.

Integrating yields $\frac{1}{M}[\ln(P) - \ln(M - P)]$

$= kt + C_1$ and $\ln\left(\frac{P}{M - P}\right) = Mkt + C$,

where $C = MC_1$. Thus $\frac{P}{M - P} = e^{Mkt+C} =$

$a_1 e^{Mkt}$, where $a_1 = e^C$. So $P = (M - P)a_1 e^{Mkt}$, $P(t)(1 + a_1 e^{Mkt}) = Ma_1 e^{Mkt}$,

and $P(t) = \frac{Ma_1 e^{Mkt}}{1 + a_1 e^{Mkt}} = \frac{M}{1 + ae^{-Mkt}}$,

where $a = 1/a_1$.

(b) $\lim_{t \to \infty} P(t) = M$, which is reasonable.

(c) $P(0) = \frac{M}{1 + a}$, so $a = \frac{M}{P(0)} - 1$.

42 (a) At $t = 0$, $A(0) = A = 434$. At $t = 3$, $A(3) = 434 \cdot e^{k \cdot 3} = 6242$, so $k = \frac{1}{3}\ln\frac{6242}{434}$

≈ 0.8887.

(b)
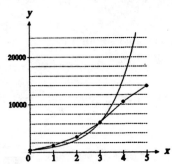

(c) At $t = 6$, $A(6) \approx 434e^{0.8887 \cdot 6} \approx 89{,}791$ deaths.

6.8 l'Hôpital's Rule

2 $\lim_{x \to 1}\dfrac{x^7 - 1}{x^3 - 1} \underset{H}{=} \lim_{x \to 1}\dfrac{7x^6}{3x^2} = \dfrac{7}{3}$

4 $\lim_{x \to 0}\dfrac{\sin x^2}{(\sin x)^2} \underset{H}{=} \lim_{x \to 0}\dfrac{2x \cos x^2}{2 \sin x \cos x}$

$= \lim_{x \to 0}\left(\dfrac{x}{\sin x} \cdot \dfrac{\cos x^2}{\cos x}\right) = 1 \cdot 1 = 1$

6 $\lim_{x \to \infty}\dfrac{x^5}{3^x} \underset{H}{=} \lim_{x \to \infty}\dfrac{5x^4}{(\ln 3)3^x} \underset{H}{=} \lim_{x \to \infty}\dfrac{20x^3}{(\ln 3)^2 3^x} \underset{H}{=}$

$\lim_{x \to \infty}\dfrac{60x^2}{(\ln 3)^3 3^x} \underset{H}{=} \lim_{x \to \infty}\dfrac{120x}{(\ln 3)^4 3^x} \underset{H}{=} \lim_{x \to \infty}\dfrac{120}{(\ln 3)^5 3^x}$

$= 0$

8 $\lim_{x \to 0}\dfrac{\sin x - x}{(\sin x)^3} \underset{H}{=} \lim_{x \to 0}\dfrac{\cos x - 1}{3 \sin^2 x \cos x}$

$\underset{H}{=} \lim_{x \to 0}\dfrac{-\sin x}{6 \cos^2 x \sin x - 3 \sin^3 x}$

$= -\lim_{x \to 0}\dfrac{1}{6 \cos^2 x - 3 \sin^2 x} = -\dfrac{1}{6}$

10 $\lim_{x \to 1}\dfrac{\cos(\pi x/2)}{\ln x} \underset{H}{=} \lim_{x \to 1}\dfrac{-\dfrac{\pi}{2}\sin(\pi x/2)}{1/x} = -\dfrac{\pi}{2}$

12. $\lim_{x \to 0} \dfrac{\sin^{-1} x}{e^{2x} - 1} \underset{H}{=} \lim_{x \to 0} \dfrac{(1 - x^2)^{-1/2}}{2e^{2x}} = \dfrac{1}{2}$

14. Let $y = (1 + \sin 2x)^{\csc x}$. Then $\lim_{x \to 0} \ln y =$

$\lim_{x \to 0} \dfrac{\ln(1 + \sin 2x)}{\sin x} \underset{H}{=} \lim_{x \to 0} \dfrac{\dfrac{2 \cos 2x}{1 + \sin 2x}}{\cos x} = 2$, so

$\lim_{x \to 0} y = e^2$.

16. $\lim_{x \to 0^+} x^2 \ln x = \lim_{x \to 0^+} \dfrac{\ln x}{x^{-2}} \underset{H}{=} \lim_{x \to 0^+} \dfrac{1/x}{-2x^{-3}}$

$= \lim_{x \to 0^+} \left(-\dfrac{x^2}{2}\right) = 0$

18. $\lim_{x \to 0^+} (e^x - 1)(\ln x) = \lim_{x \to 0^+} \dfrac{\ln x}{(e^x - 1)^{-1}} \underset{H}{=}$

$\lim_{x \to 0^+} \dfrac{1/x}{-(e^x - 1)^{-2} e^x} = \lim_{x \to 0^+} \dfrac{-(e^x - 1)^2}{x} \cdot \lim_{x \to 0^+} \dfrac{1}{e^x}$

$= \lim_{x \to 0^+} \dfrac{-(e^x - 1)^2}{x} \underset{H}{=} \lim_{x \to 0^+} \dfrac{-2(e^x - 1)e^x}{1} = 0$

20. $\lim_{x \to \infty} \dfrac{2^x + x}{3^x} \underset{H}{=} \lim_{x \to \infty} \dfrac{(\ln 2) 2^x + 1}{(\ln 3) 3^x} \underset{H}{=} \lim_{x \to \infty} \dfrac{(\ln 2)^2 2^x}{(\ln 3)^2 3^x}$

$= \left(\dfrac{\ln 2}{\ln 3}\right)^2 \lim_{x \to \infty} \left(\dfrac{2}{3}\right)^x = 0$

22. $\dfrac{\log_2 x}{\log_3 x} = \dfrac{(\ln x)/(\ln 2)}{(\ln x)/(\ln 3)} = \dfrac{\ln 3}{\ln 2}$ for $x \neq 1$, so the

limit is $\dfrac{\ln 3}{\ln 2}$.

24. $\lim_{x \to \infty} \left(\sqrt{x^2 + 3} - \sqrt{x^2 + 4x}\right)$

$= \lim_{x \to \infty} \dfrac{(x^2 + 3) - (x^2 + 4x)}{\sqrt{x^2 + 3} + \sqrt{x^2 + 4x}}$

$= \lim_{x \to \infty} \dfrac{(-4 + 3/x)}{\left(\sqrt{1 + 3/x^2} + \sqrt{1 + 4/x}\right)} = \dfrac{-4}{2} = -2$

26. $\lim_{x \to \infty} \dfrac{e^x - 1/x}{e^x + 1/x} = \lim_{x \to \infty} \dfrac{1 - \dfrac{1}{xe^x}}{1 + \dfrac{1}{xe^x}} = 1$

28. $\dfrac{3}{5}$

30. Does not exist.

32. As $x \to \dfrac{\pi}{2}^-$, $\tan x \to \infty$ and $x - \dfrac{\pi}{2} \to 0^-$, so

$\dfrac{\tan x}{x - \pi/2} \to -\infty$. As $x \to \pi/2^+$, $\tan x \to -\infty$ and

$x - \pi/2 \to 0^+$, so $\dfrac{\tan x}{x - \pi/2} \to -\infty$. Hence the limit

is $-\infty$.

34. Let $y = x^{1/x}$. Then $\ln y = \dfrac{\ln x}{x}$. As $x \to 0^+$, $\ln x \to$

$-\infty$ and $x \to 0^+$, so $\ln y \to -\infty$. Hence $y \to 0$.

36. $\lim_{x \to 1} \dfrac{x^2 - 1}{x^3 - 1} \underset{H}{=} \lim_{x \to 1} \dfrac{2x}{3x^2} = \dfrac{2}{3}$

38. $\lim_{x \to 0} \dfrac{xe^x \cos^2 6x}{e^{2x} - 1} = \lim_{x \to 0} (e^x \cos^2 6x) \cdot \lim_{x \to 0} \dfrac{x}{e^{2x} - 1}$

$= \lim_{x \to 0} \dfrac{x}{e^{2x} - 1} \underset{H}{=} \lim_{x \to 0} \dfrac{1}{2e^{2x}} = \dfrac{1}{2}$

40. $\lim_{x \to 0} \dfrac{\csc x - \cot x}{\sin x} = \lim_{x \to 0} \dfrac{1 - \cos x}{\sin^2 x} =$

$\lim_{x \to 0} \dfrac{1 - \cos x}{1 - \cos^2 x} = \lim_{x \to 0} \dfrac{1}{1 + \cos x} = \dfrac{1}{2}$

42. $\lim_{x \to 0} \dfrac{\tan^5 x - \tan^3 x}{1 - \cos x} \underset{H}{=}$

$$\lim_{x \to 0} \frac{(5\tan^4 x - 3\tan^2 x)\sec^2 x}{\sin x}$$

$$= \lim_{x \to 0} (\sec^2 x) \cdot \lim_{x \to 0} \frac{5\tan^4 x - 3\tan^2 x}{\sin x}$$

$$= \lim_{x \to 0} \frac{5\tan^4 x - 3\tan^2 x}{\sin x} \underset{H}{=}$$

$$\lim_{x \to 0} \frac{(20\tan^3 x - 6\tan x)\sec^2 x}{\cos x} = 0$$

44 $\lim_{x \to \pi/4} \frac{\sin 5x}{\sin 3x} = \frac{\sin(5\pi/4)}{\sin(3\pi/4)} = \frac{-1/\sqrt{2}}{1/\sqrt{2}} = -1$

46 $\lim_{x \to 0} \frac{\sin^{-1} x}{\tan^{-1} 2x} \underset{H}{=} \lim_{x \to 0} \frac{(1-x^2)^{-1/2}}{2(1+4x^2)^{-1}} = \frac{1}{2}$

48 Let θ be the angle POB, where O is the origin. Then $P = (\cos\theta, \sin\theta)$ and $Q = (1, \theta)$, so the line

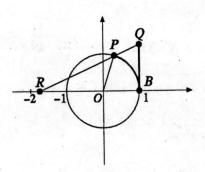

through P and Q is $\frac{y - \theta}{x - 1} = \frac{\theta - \sin\theta}{1 - \cos\theta}$. At R,

$y = 0$ so $x = 1 - \frac{\theta(1 - \cos\theta)}{\theta - \sin\theta} =$

$\frac{\theta\cos\theta - \sin\theta}{\theta - \sin\theta}$. But $\lim_{\theta \to 0} \frac{\theta\cos\theta - \sin\theta}{\theta - \sin\theta}$

$\underset{H}{=} \lim_{\theta \to 0} \frac{-\theta\sin\theta}{1 - \cos\theta} \underset{H}{=} \lim_{\theta \to 0} \frac{-\sin\theta - \theta\cos\theta}{\sin\theta}$

$= -1 - \lim_{\theta \to 0} (\cos\theta)\frac{\theta}{\sin\theta} = -1 - 1 \cdot 1 = -2$.

50 $\lim_{x \to 0} \frac{x^3}{\int_{-x}^{x}(x^2 - u^2)\,du} = \lim_{x \to 0} \frac{x^3}{4x^3/3} = \frac{3}{4}$

52 There exist c_0 and c_1 in (a, x) such that

$$\frac{f(x) - f(a)}{x - a} = f'(c_0) \text{ and } \frac{g(x) - g(a)}{x - a} = g'(c_1),$$

but c_0 and c_1 need not be equal.

54 With $\hbar$, k, and T fixed, $\lim_{\omega \to 0} \frac{\hbar\omega}{e^{\hbar\omega/kT} - 1} \underset{H}{=}$

$$\lim_{\omega \to 0} \frac{\hbar}{\frac{\hbar}{kT}e^{\hbar\omega/kT}} = kT.$$

56

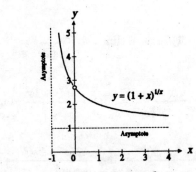

(a) $\ln y = \frac{\ln(1+x)}{x}$ so $\frac{d}{dx}(\ln y) =$

$$\frac{\frac{x}{1+x} - \ln(1+x)}{x^2}. \text{ Let } f(x) =$$

$\frac{x}{1+x} - \ln(1+x)$ for $x > -1$. Then $f'(x) =$

$-\frac{x}{(1+x)^2}$ changes from $+$ to $-$ at $x = 0$.

Hence $f(x)$ has a global maximum at 0, so $f(x) < f(0) = 0$ for all $x \neq 0$. Hence

$\frac{d}{dx}(\ln y) < 0$ for $x > -1$, $x \neq 0$, so y is

decreasing for all such x.

(b) As $x \to -1^+$, $\ln(1 + x) \to -\infty$, so $\ln y \to \infty$ and $y \to \infty$. Thus $x = -1$ is an asymptote.

Also, $\lim\limits_{x \to \infty} \dfrac{\ln(1 + x)}{x} \underset{H}{=} \lim\limits_{x \to \infty} \dfrac{\frac{1}{1+x}}{1} = 0$, so $\lim\limits_{x \to \infty} y = 1$. Thus $y = 1$ is an asymptote.

(c) $\lim\limits_{x \to 0} \dfrac{\ln(1+x)}{x} \underset{H}{=} \lim\limits_{x \to 0} \dfrac{\frac{1}{1+x}}{1} = 1$, so $\lim\limits_{x \to 0} y = e$.

58 $y = x^2 \ln x$ is only defined for $x > 0$. Now $x^2 \ln x = 0$ when $x = 1$. Also, from Example 5, we know that $\lim\limits_{x \to 0^+} x^2 \ln x = \lim\limits_{x \to 0^+} x \cdot \lim\limits_{x \to 0^+} x \ln x = 0$.

In addition $\lim\limits_{x \to \infty} x^2 \ln x = \infty$. Here $\dfrac{dy}{dx} = x^2 \cdot \dfrac{1}{x} + 2x \ln x = x(1 + 2 \ln x)$, which changes sign from $-$ to $+$ at $x = e^{-1/2}$. ($x = 0$ is not in the domain.) Furthermore, $\dfrac{d^2y}{dx^2} = 1 + 2 \ln x + x \cdot \dfrac{2}{x} = 3 + 2 \ln x$ changes sign from $-$ to $+$ when $x = e^{-2/3}$, so $x = e^{-2/3}$ is an inflection point and $x = e^{-1/2}$ is a minimum.

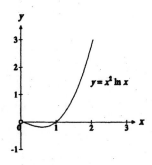

6.9 The Hyperbolic Functions and Their Inverses

2 $\cosh t - \sinh t = \dfrac{e^t + e^{-t}}{2} - \dfrac{e^t - e^{-t}}{2} = \dfrac{2e^{-t}}{2} = e^{-t}$

4 $(\tanh t)' = \dfrac{(\cosh t)(\sinh t)' - (\sinh t)(\cosh t)'}{(\cosh t)^2}$

$= \dfrac{(\cosh t)^2 - (\sinh t)^2}{\cosh^2 t} = \dfrac{1}{\cosh^2 t} = \text{sech}^2 t$

6 $(\coth t)' = \dfrac{(\sinh t)(\cosh t)' - (\cosh t)(\sinh t)'}{(\sinh t)^2} =$

$\dfrac{(\sinh t)^2 - (\cosh t)^2}{\sinh^2 t} = \dfrac{-1}{\sinh^2 t} = -\text{csch}^2 t$

8 $5 \cosh 5x$

10 $\text{sech}(\ln x) = \dfrac{2}{e^{\ln x} + e^{-\ln x}} = \dfrac{2}{x + 1/x} = \dfrac{2x}{x^2 + 1}$,

so the derivative is $\dfrac{(x^2 + 1) \cdot 2 - (2x)(2x)}{(x^2 + 1)^2} = \dfrac{2 - 2x^2}{(x^2 + 1)^2}$.

12 Let $y = \dfrac{(\tanh 2x)(\text{sech } 3x)}{\sqrt{1 + 2x}}$. Then $\ln y = \ln(\tanh 2x) + \ln(\text{sech } 3x) - \dfrac{1}{2} \ln(1 + 2x)$, so $\dfrac{1}{y} y' = 2 \text{ sech } 2x \text{ csch } 2x - 3 \tanh 3x - \dfrac{1}{1 + 2x}$ and

$y' = \dfrac{(\tanh 2x)(\text{sech } 2x)}{\sqrt{1 + 2x}} \left(2 \text{ sech } 2x \text{ csch } 2x - 3 \tanh 3x - \dfrac{1}{1 + 2x} \right)$.

14 $\dfrac{y'}{y} = \dfrac{5}{2} \cdot \dfrac{-5\,\text{csch}^2 5x}{\coth 5x} + \dfrac{1}{3} \cdot 3\,\dfrac{\text{sech}^2 3x}{\tanh 3x} -$

$\dfrac{3 - 4\sin x}{3x + 4\cos x} = -\dfrac{25}{2}\,\text{sech}\,5x\,\text{csch}\,5x +$

$\text{sech}\,3x\,\text{csch}\,3x - \dfrac{3 - 4\sin x}{3x + 4\cos x}$, so $y' =$

$\dfrac{(\coth 5x)^{5/2}(\tanh 3x)^{1/3}}{3x + 4\cos x}\left(-\dfrac{25}{2}\,\text{sech}\,5x\,\text{csch}\,5x +\right.$

$\left.\text{sech}\,3x\,\text{csch}\,3x - \dfrac{3 - 4\sin x}{3x + 4\cos x}\right).$

16 If we let $x = \text{sech}\,t = \dfrac{2}{e^t + e^{-t}} = \dfrac{2}{e^t + 1/e^t}$,

then $e^{2t} + 1 = \dfrac{2}{x}e^t$ and $e^{2t} - \dfrac{2}{x}e^t + 1 = 0$.

Invoking the quadratic formula with unknown e^t

gives $e^t = \dfrac{2/x \pm \sqrt{4/x^2 - 4}}{2} = \dfrac{1}{x} \pm \dfrac{\sqrt{1 - x^2}}{x}$.

Recall that $\text{sech}\,t$ is an even function, so consider $t \geq 0$ only. Then $e^t \geq 1$. Now

$\left[\dfrac{1 + \sqrt{1 - x^2}}{x}\right]\left[\dfrac{1 - \sqrt{1 - x^2}}{x}\right] = 1$, so $\dfrac{1 + \sqrt{1 - x^2}}{x}$

> 1 for $0 < x \leq 1$. So we choose the plus sign,

and $e^t = \dfrac{1 + \sqrt{1 - x^2}}{x}$. Hence $t = \text{sech}^{-1} x =$

$\ln\left(\dfrac{1 + \sqrt{1 - x^2}}{x}\right).$

18 If we let $x = \text{csch}\,t = \dfrac{2}{e^t - e^{-t}} = \dfrac{2}{e^t - 1/e^t}$,

then $e^{2t} - 1 = \dfrac{2}{x}e^t$ and $e^{2t} - \dfrac{2}{x}e^t - 1 = 0$.

Applying the quadratic formula with unknown e^t

gives $e^t = \dfrac{2/x \pm \sqrt{4/x^2 + 4}}{2} = \dfrac{1}{x} \pm \sqrt{1 + \dfrac{1}{x^2}}$.

Now $(\text{csch}\,t)' = -\dfrac{\cosh t}{\sinh^2 t} < 0$ for all t, so $\text{csch}\,t$

is always decreasing. Hence we consider all $t \neq 0$. Since $e^t > 0$ for all t,

$\left(\dfrac{1}{x} + \sqrt{1 + \dfrac{1}{x^2}}\right)\left(\dfrac{1}{x} - \sqrt{1 + \dfrac{1}{x^2}}\right) = -1$, and therefore

$\dfrac{1}{x} + \sqrt{1 + \dfrac{1}{x^2}} > 0$, we select the plus sign. So

$e^t = \dfrac{1}{x} + \sqrt{1 + \dfrac{1}{x^2}}$ and $t = \text{csch}^{-1} x =$

$\ln\left(\dfrac{1}{x} + \sqrt{1 + \dfrac{1}{x^2}}\right),\ x \neq 0.$

20 $(\sinh^{-1} x + C)' = \left[\ln\left(x + \sqrt{x^2 + 1}\right) + C\right]' =$

$\dfrac{1 + \dfrac{x}{\sqrt{x^2 + 1}}}{x + \sqrt{x^2 + 1}} = \dfrac{1}{\sqrt{x^2 + 1}}$, so $\int \dfrac{dx}{\sqrt{x^2 + 1}}$

$= \sinh^{-1} x + C.$

22 $(-\text{csch}^{-1} x + C)' = \left[-\ln\left(\dfrac{1}{x} + \sqrt{1 + \dfrac{1}{x^2}}\right) + C\right]'$

$= \dfrac{\dfrac{1}{x^2} + \dfrac{1/x^3}{\sqrt{1 + 1/x^2}}}{\dfrac{1}{x} + \sqrt{1 + 1/x^2}} = \dfrac{x + \dfrac{1}{\sqrt{1 + 1/x^2}}}{x^2 + x^2\sqrt{1 + x^2}}$

6.9 The Hyperbolic Functions and Their Inverses

$$= \frac{1 + \frac{1}{\sqrt{1+x^2}}}{x + x\sqrt{1+x^2}} = \frac{1}{x\sqrt{1+x^2}}, \text{ so } \int \frac{dx}{x\sqrt{1+x^2}}$$

$$= -\operatorname{csch}^{-1} x + C, \; x > 0.$$

24. $\dfrac{\tanh x + \tanh y}{1 + \tanh x \tanh y}$

$$= \frac{\frac{e^x - e^{-x}}{e^x + e^{-x}} + \frac{e^y - e^{-y}}{e^y + e^{-y}}}{\frac{(e^x + e^{-x})(e^y + e^{-y}) + (e^x - e^{-x})(e^y - e^{-y})}{(e^x + e^{-x})(e^y + e^{-y})}}$$

$$= \frac{(e^y + e^{-y})(e^x - e^{-x}) + (e^x + e^{-x})(e^y - e^{-y})}{(e^x + e^{-x})(e^y + e^{-y}) + (e^x - e^{-x})(e^y - e^{-y})}$$

$$= \frac{2e^{(x+y)} - 2e^{-(x+y)}}{2e^{(x+y)} + 2e^{-(x+y)}} = \tanh(x+y)$$

26. (a) $\cosh^2 x + \sinh^2 x = \left(\dfrac{e^x + e^{-x}}{2}\right)^2 + \left(\dfrac{e^x - e^{-x}}{2}\right)^2$

$$= \frac{2e^{2x} + 2e^{-2x}}{4} = \cosh 2x$$

(b) $2(\sinh x)(\cosh x) = 2\left(\dfrac{e^x - e^{-x}}{2}\right)\left(\dfrac{e^x + e^{-x}}{2}\right)$

$$= \frac{2e^{2x} - 2e^{-2x}}{4} = \sinh 2x$$

28. $\operatorname{sech}^2 x + \tanh^2 x = \left(\dfrac{2}{e^x + e^{-x}}\right)^2 + \left(\dfrac{e^x - e^{-x}}{e^x + e^{-x}}\right)^2 =$

$$\frac{4 + e^{2x} - 2 + e^{-2x}}{(e^x + e^{-x})^2} = \frac{(e^x + e^{-x})^2}{(e^x + e^{-x})^2} = 1$$

30. $y'(0) = \cosh 0 = 1$, so the slope is 1 and the angle is $\pi/4$.

32. (a)

t	0	1	2	3	4
$\cosh t$	1	1.54	3.76	10.07	27.31
$e^t/2$	0.5	1.36	3.69	10.04	27.30

(b)

34. For the case when $a > 0$, $c < 0$, let $y = \left(\dfrac{-c(ax+b)}{a(cx+d)}\right)^{1/2}$. Then $\dfrac{dy}{dx}$

$$= \left(-\frac{c}{a}\right)^{1/2} \cdot \frac{1}{2}\left(\frac{ax+b}{cx+d}\right)^{-1/2} \cdot \frac{a(cx+d) - c(ax+b)}{(cx+d)^2} =$$

$$\frac{1}{2}\left(-\frac{c}{a}\right)^{1/2} \cdot \frac{ad - bc}{(ax+b)^{1/2}(cx+d)^{3/2}}. \text{ Now}$$

$$\frac{d}{dx}\left(\frac{2}{\sqrt{-ac}} \tan^{-1} y\right) = \frac{2}{(-ac)^{1/2}} \cdot \frac{1}{1+y^2} \cdot \frac{dy}{dx} =$$

$$\frac{2}{(-ac)^{1/2}} \cdot \frac{1}{1 + \left(\frac{-c(ax+b)}{a(cx+d)}\right)} \cdot \frac{1}{2}\left(-\frac{c}{a}\right)^{1/2} \cdot \frac{ad - bc}{(ax+b)^{1/2} \cdot (cx+d)^{3/2}}$$

$$= \frac{1}{a} \cdot \frac{a(cx+d)}{a(cx+d) - c(ax+b)} \cdot \frac{ad - bc}{(ax+b)^{1/2}(cx+d)^{3/2}} =$$

$$\frac{1}{(ax+b)^{1/2}(cx+d)^{1/2}}, \text{ so } \int \frac{dx}{\sqrt{ax+b}\sqrt{cx+d}}$$

$$= \frac{2}{\sqrt{-ac}} \tan^{-1}\sqrt{\frac{-c(ax+b)}{a(cx+d)}}, \text{ as claimed. For the}$$

case when $a, c > 0$, let $z = \left(\dfrac{c(ax+b)}{a(cx+d)}\right)^{1/2}$. Then,

from above, $\dfrac{dz}{dx} = \dfrac{1}{2}\left(\dfrac{c}{a}\right)^{1/2} \dfrac{ad - bc}{(ax+b)^{1/2}(cx+d)^{3/2}}$.

Now $\dfrac{d}{dx}\left(\dfrac{2}{\sqrt{ac}}\tanh^{-1} z\right) = \dfrac{2}{(ac)^{1/2}} \cdot \dfrac{1}{1+z^2} \cdot \dfrac{dz}{dx} =$

$\dfrac{2}{(ac)^{1/2}} \cdot \dfrac{1}{1+z^2} \cdot \dfrac{dz}{dx} =$

$\dfrac{2}{(ac)^{1/2}} \cdot \dfrac{1}{1 + \left(\dfrac{c(ax+b)}{a(cx+d)}\right)} \cdot \dfrac{1}{2}\left(\dfrac{c}{a}\right)^{1/2} \dfrac{ad-bc}{(ax+b)^{1/2}(cx+d)^{3/2}}$

$= \dfrac{1}{(ax+b)^{1/2}(cx+d)^{1/2}}$, so, again, $\displaystyle\int \dfrac{dx}{\sqrt{ax+b}\sqrt{cx+d}}$

$= \dfrac{2}{\sqrt{ac}}\tanh^{-1}\sqrt{\dfrac{c(ax+b)}{a(cx+d)}}$, as claimed.

36 $\dfrac{d^2}{dx^2}(\sinh x) = \sinh x$, which changes from $-$ to $+$ only at $x = 0$, so $(0, 0)$ is an inflection point.

Also, $\dfrac{d^2}{dx^2}(\sinh^{-1} x) = -\dfrac{x}{(x^2+1)^{3/2}}$, which changes from $+$ to $-$ only at $x = 0$, so $(0, 0)$ is again an inflection point. This makes sense since $y = \sinh x$ is an odd function.

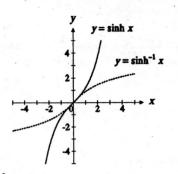

38 (a) $\dfrac{d^2}{dt^2}(A\cosh\sqrt{k}t + B\sinh\sqrt{k}t)$

$= \dfrac{d}{dt}(A\sqrt{k}\sinh\sqrt{k}t + B\sqrt{k}\cosh\sqrt{k}t)$

$= k(A\cosh\sqrt{k}t + B\sinh\sqrt{k}t) = kx$

(b) At $t = 0$, $x = a = A\cosh 0 + B\sin 0 = A$,

and $\dfrac{dx}{dt} = 0 = A\sqrt{k}\sinh 0 + B\sqrt{k}\cosh 0 = B\sqrt{k}$, so $A = a$ and $B = 0$.

6.S Review Exercises

2 $\quad 2(\ln 5)x \cdot 5^{x^2}$

4 $\quad -\dfrac{1}{2}x^{-3/2}$

6 $\quad \dfrac{3}{\sqrt{1-9x^2}}$

8 $\quad \dfrac{1}{3} \cdot \dfrac{1}{1+(x/3)^2} \cdot \dfrac{1}{3} = \dfrac{1}{9+x^2}$

10 $\quad \dfrac{x^2(-5\sin 5x) - (\cos 5x)2x}{x^4}$

$= -\dfrac{5x\sin 5x + 2\cos 5x}{x^3}$

12 $\quad 3x^2\cos 3x + 2x\sin 3x$

14 $\dfrac{e^{\sqrt{x}}}{2\sqrt{x}}$

16 $4 \sin 2x \cos 2x = 2 \sin 4x$

18 $\dfrac{1}{25 + x^2}$

20 $e^{-x}[(\sec^2 x^2)\cdot 2x] + (-e^{-x}) \tan x^2$
   $= e^{-x}(2x \sec^2 x^2 - \tan x^2)$

22 $-15 \csc^2 5x - 15 \csc 3x \cot 3x$

24 $\dfrac{2 \cos 2x}{\sin 2x} = 2 \cot 2x$

26 $x(-5 \sin 5x) + \cos 5x + 5 \cos 5x$
   $= 6 \cos 5x - 5x \sin 5x$

28 $\dfrac{x \cdot \dfrac{-9x}{\sqrt{4 - 9x^2}} - \sqrt{4 - 9x^2}}{x^2} = -\dfrac{4}{x^2\sqrt{4 - 9x^2}}$

30 $-\sin[\log_{10}(3x + 1)] \cdot \dfrac{\log_{10} e}{3x + 1} \cdot 3$

   $= -3(\log_{10} e) \cdot \dfrac{\sin[\log_{10}(3x + 1)]}{3x + 1}$

   $= -\dfrac{3}{\ln 10} \cdot \dfrac{\sin[\log_{10}(3x + 1)]}{3x + 1}$ (Recall that $\log_{10} e = 1/(\ln 10)$.)

32 $\dfrac{(x^2 + 1)^5 \cdot 3 \cos 3x - (\sin 3x) \cdot 5(x^2 + 1)^4 \cdot 2x}{(x^2 + 1)^{10}}$

   $= \dfrac{3(x^2 + 1) \cos 3x - 10x \sin 3x}{(x^2 + 1)^6}$

34 $\dfrac{\sec^2 \sqrt{x}}{2\sqrt{x}}$

36 $-\dfrac{1}{2}(1 - x^2)^{-3/2}(-2x) = \dfrac{x}{(1 - x^2)^{3/2}}$

38 $4\left(\dfrac{1 + 2x}{1 + 3x}\right)^3 \cdot \dfrac{(1 + 3x)\cdot 2 - (1 + 2x)\cdot 3}{(1 + 3x)^2}$

   $= -\dfrac{4(1 + 2x)^3}{(1 + 3x)^5}$

40 $\dfrac{6x}{x^2 + 1}[\ln(x^2 + 1)]^2$

42 $\dfrac{6(x^2 + x) \sin^2 2x \cos 2x - (2x + 1) \sin^3 2x}{(x^2 + x)^2}$

44 $\dfrac{2}{\ln 10} \cdot \dfrac{x}{x^2 + 1}$

46 $\dfrac{1}{\sqrt{1 - (x/5)^2}} \cdot \dfrac{1}{5} = \dfrac{1}{\sqrt{25 - x^2}}$

48 $2(x^2 + 1) \cos 2x + 2x \sin 2x$

50 $5x^4 - 2 + \dfrac{2}{2x + 3}$

52 $6x \sin^2(1 + x^2) \cos(1 + x^2)$

54 $\left[\dfrac{1}{2} \ln(1 + x^2) - \dfrac{1}{2} \ln(1 + x^3)\right]'$

   $= \dfrac{x}{1 + x^2} - \dfrac{3x^2}{2(1 + x^3)}$

56 $\left[\dfrac{1}{2} \ln(4 + x) + \dfrac{1}{3} \ln(x^2 + 1)\right]'$

   $= \dfrac{1}{2(4 + x)} + \dfrac{2x}{3(x^2 + 1)}$

58 $\dfrac{1}{8}\left[\dfrac{2}{2x + 1} - \dfrac{4}{(2x + 1)^2} + \dfrac{2}{(2x + 1)^3}\right] = \dfrac{x^2}{(2x + 1)^3}$

60 $\dfrac{1}{8}\left[2 - \dfrac{12}{2x + 3} + \dfrac{18}{(2x + 3)^2}\right] = \dfrac{x^2}{(2x + 3)^2}$

62 $\dfrac{1}{\sqrt{6}} \cdot \dfrac{1}{1 + \left(x\sqrt{\frac{2}{3}}\right)^2} \cdot \sqrt{\dfrac{2}{3}} = \dfrac{1}{3 + 2x^2}$

64 $\dfrac{1}{5}[e^x(2\cos 2x + 4\sin 2x) + e^x(\sin 2x - 2\cos 2x)]$

$= e^x \sin 2x$

66 $\dfrac{x}{1+x^2} + \tan^{-1} x - \dfrac{1}{2} \cdot \dfrac{2x}{1+x^2} = \tan^{-1} x$

68 $\dfrac{1}{6} \cdot (2 \tan 3x \sec^2 3x) \cdot 3 + \dfrac{1}{3} \cdot \dfrac{-3 \sin 3x}{\cos 3x}$

$= \tan 3x \sec^2 3x - \tan 3x = \tan^3 3x$

70 $\dfrac{1}{8}[e^{-2x}(8x + 4) - 2e^{-2x}(4x^2 + 4x + 2)] = -x^2 e^{-2x}$

72 $\dfrac{1}{2\sqrt{1+\sqrt[3]{x}}} \cdot \dfrac{1}{3} x^{-2/3} = \dfrac{1}{6x^{2/3}\sqrt{1+\sqrt[3]{x}}}$

74 $-\dfrac{1}{(1 + \csc 5x)^2}(-5 \csc 5x \cot 5x)$

$= \dfrac{5 \csc 5x \cot 5x}{(1 + \csc 5x)^2}$

76 Let $y = 2^x 5^{x^2} 7^{x^3}$. Then $\ln y = x \ln 2 + x^2 \ln 5 + x^3 \ln 7$, so $\dfrac{y'}{y} = \ln 2 + 2x \ln 5 + 3x^2 \ln 7$ and

$y' = 2^x 5^{x^2} 7^{x^3} (\ln 2 + 2x \ln 5 + 3x^2 \ln 7)$.

78 $x^3 \cdot 3 \cot^2 \sqrt{x^3} \cdot (-\csc^2 \sqrt{x^3}) \cdot \dfrac{3}{2}\sqrt{x} + 3x^2 \cot^3 \sqrt{x^3}$

$= 3x^2 \cot^3 \sqrt{x^3} - \dfrac{9}{2} x^{7/2} \cot^2 \sqrt{x^3} \csc^2 \sqrt{x^3}$

80 $[x \ln ax - x + C]' = x \cdot \dfrac{a}{ax} + \ln ax - 1 = \ln ax$

82 $\left[\dfrac{x^3}{3} \ln ax - \dfrac{x^3}{9} + C\right]' = \dfrac{x^3}{3} \cdot \dfrac{a}{ax} + x^2 \ln ax - \dfrac{x^2}{3}$

$= x^2 \ln ax$

84 $[\ln |\csc x - \cot x| + C]' = \dfrac{-\cot x \csc x + \csc^2 x}{\csc x - \cot x}$

$= \csc x$

86 $\left[\dfrac{1}{2a} \tan^2 ax + \dfrac{1}{a} \ln |\cos ax| + C\right]'$

$= \dfrac{1}{2a} \cdot 2a \tan ax \sec^2 ax + \dfrac{1}{a} \cdot \dfrac{-a \sin ax}{\cos ax}$

$= \tan ax (\sec^2 ax - 1) = \tan^3 ax$

88 $\left[\dfrac{1}{\sqrt{a}} \ln |x\sqrt{a} + \sqrt{ax^2 + b}| + C\right]'$

$= \dfrac{1}{\sqrt{a}} \cdot \dfrac{\sqrt{a} + \dfrac{ax}{\sqrt{ax^2+b}}}{x\sqrt{a} + \sqrt{ax^2+b}}$

$= \dfrac{1}{\sqrt{a}} \cdot \dfrac{\sqrt{a(ax^2+b)} + ax}{\sqrt{ax^2+b}(x\sqrt{a} + \sqrt{ax^2+b})} = \dfrac{1}{\sqrt{ax^2+b}}$

90 $[x(\ln ax)^2 - 2x \ln ax + 2x + C]' = 2x \ln ax \cdot \dfrac{a}{ax}$

$+ (\ln ax)^2 - \left[2x \cdot \dfrac{a}{ax} + 2 \ln ax\right] + 2 = (\ln ax)^2$

92 $\displaystyle\int_1^2 \dfrac{x^3}{x^4 + 5} dx = \dfrac{1}{4} \ln(x^4 + 5) \Big|_1^2$

$= \dfrac{1}{4} \ln 21 - \dfrac{1}{4} \ln 6 = \dfrac{1}{4} \ln \dfrac{7}{2}$

94 $\left(\dfrac{\log_b x}{\log_c x}\right)' = \dfrac{(\log_c x)\dfrac{\log_b e}{x} - (\log_b x)\dfrac{\log_c e}{x}}{(\log_c x)^2} = 0$

96 (a) $-\frac{1}{3}(3x + 2)^{-1} + C$

(b) $\frac{2}{3}(3x + 2)^{1/2} + C$

(c) $\frac{1}{3}\ln|3x + 2| + C$

98 The tangent line has slope $1/x_0$, so its equation is $y = y_0 + \frac{1}{x_0}(x - x_0)$. When $x = 0$, $y = y_0 - 1$, so $B = (0, y_0 - 1)$. But $A = (0, y_0)$, so $\overline{AB} = 1$.

100 $H'(p) = -\left[p \cdot \frac{1}{p} + \ln p\right] - \left[(1-p)\frac{-1}{1-p} + (-1)\ln(1-p)\right] = \ln\left(\frac{1}{p} - 1\right)$

changes from $+$ to $-$ at $p = 1/2$.

102 Let $f(x) = \ln(1 + 2x)$. Then $f'(x) = \frac{2}{1 + 2x}$ and the limit in question is $f'(3) = 2/7$.

104 Let $f(x) = \ln(1 + x)$. Then $f'(x) = \frac{1}{1 + x}$ and the limit in question is $f'(0) = 1$.

106 Let $P = (x, y)$. The slope of the tangent line at P is $(\ln x)' = \frac{1}{x}$. The slope of the line through $(0, 0)$ and P is $\frac{y - 0}{x - 0} = \frac{y}{x}$. These slopes must be equal, so $\frac{1}{x} = \frac{y}{x}$, $1 = y$, and $x = e^y = e$. Thus $P = (e, 1)$.

108 $\lim\limits_{x \to \pi/4} \frac{\sin x - \sqrt{2}/2}{x - \pi/4} \underset{H}{=} \lim\limits_{x \to \pi/4} \frac{\cos x}{1} = \frac{1}{\sqrt{2}}$

110 $\lim\limits_{x \to 1} \frac{\sin \pi x}{x - 1} \underset{H}{=} \lim\limits_{x \to 1} \frac{\pi \cos \pi x}{1} = -\pi$

112 $\lim\limits_{x \to \infty} 2^x e^{-x} = \lim\limits_{x \to \infty} \left(\frac{2}{e}\right)^x = 0$, since $\left|\frac{2}{e}\right| < 1$.

114 $\lim\limits_{x \to 0} (1 + 3x)^{1/x} = \lim\limits_{x \to 0} [(1 + 3x)^{1/(3x)}]^3 = e^3$

116 $\lim\limits_{x \to \infty} \frac{x^2 + 5}{2x^2 + 6x} = \lim\limits_{x \to \infty} \frac{x^2}{2x^2} = \frac{1}{2}$

118 Let $y = \sin x$. Then $\lim\limits_{x \to 0^+} (\sin x)^{\sin x} = \lim\limits_{y \to 0^+} y^y = 1$, by Example 6 in Sec. 6.8.

120 $\lim\limits_{x \to \infty} \frac{(x^2 + 1)^5}{e^x} = \lim\limits_{x \to \infty} \frac{(x^2 + 1)^5}{x^{10}} \cdot \lim\limits_{x \to \infty} \frac{x^{10}}{e^x} = 1 \cdot 0 = 0$. (See Exercises 48 and 49 in Sec. 6.5.)

122 $\lim\limits_{x \to 0} \frac{xe^x}{e^x - 1} \underset{H}{=} \lim\limits_{x \to 0} \frac{(x + 1)e^x}{e^x} = 1$

124 $\lim\limits_{x \to 0} \frac{\sin x^2}{x \sin x} = \lim\limits_{x \to 0} \frac{\sin x^2}{x^2} \cdot \lim\limits_{x \to 0} \frac{x}{\sin x} = 1 \cdot 1 = 1$

126 $\frac{e + 1}{e - 1}$

128 $\lim\limits_{x \to 0^+} \frac{1 - \cos x}{x - \tan x} \underset{H}{=} \lim\limits_{x \to 0^+} \frac{\sin x}{1 - \sec^2 x}$
$\underset{H}{=} \lim\limits_{x \to 0^+} \frac{\cos x}{-2 \sec^2 x \tan x} = -\infty$

130 $\frac{5^2 + 3^2}{2} = 17$

132 $\lim\limits_{x \to \infty} \frac{e^{-x}}{x^2} = \lim\limits_{x \to \infty} \frac{1}{x^2 e^x} = 0$

134 $\lim\limits_{x \to \infty} \frac{\ln x}{e^x} \underset{H}{=} \lim\limits_{x \to \infty} \frac{1}{xe^x} = 0$

136 Let $y = \left(\dfrac{\sin x}{x}\right)^{1/x}$. Then $\ln y = \dfrac{1}{x} \ln\left(\dfrac{\sin x}{x}\right)$.

Since $\lim\limits_{x \to 0} \dfrac{\sin x}{x} = 1$, $\lim\limits_{x \to 0} \ln\left(\dfrac{\sin x}{x}\right) = 0$ and

l'Hôpital's rule applies: $\lim\limits_{x \to 0} \ln y =$

$\lim\limits_{x \to 0} \dfrac{\ln(\sin x) - \ln x}{x} \underset{H}{=} \lim\limits_{x \to 0} \dfrac{\cot x - 1/x}{1} =$

$\lim\limits_{x \to 0} \left(\dfrac{1}{\tan x} - \dfrac{1}{x}\right) = \lim\limits_{x \to 0} \dfrac{x - \tan x}{x \tan x} \underset{H}{=}$

$\lim\limits_{x \to 0} \dfrac{1 - \sec^2 x}{\tan x + x \sec^2 x} = \lim\limits_{x \to 0} \dfrac{\cos^2 x - 1}{\sin x \cos x + x}$

$= \lim\limits_{x \to 0} \dfrac{\cos 2x - 1}{\sin 2x + 2x} \underset{H}{=} \lim\limits_{x \to 0} \dfrac{-2 \sin 2x}{2 \cos 2x + 2} = 0.$

Thus $\lim\limits_{x \to 0} \left(\dfrac{\sin x}{x}\right)^{1/x} = 1$.

138 $y' = \dfrac{(1 - x - x^2)e^{-x}}{(x + 1)^2}$ changes from $-$ to $+$ at

$x = \dfrac{-1 - \sqrt{5}}{2}$ (so there is a local minimum there)

and from $+$ to $-$ at $x = \dfrac{-1 + \sqrt{5}}{2}$ (so there is a

local maximum there).

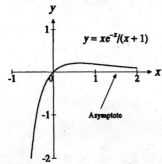

140 $3^{80} = e^{80 \ln 3}$

142 In (a) and (b), the limit is 0. In (c) and (d) it cannot be determined.

144 (a) See Theorem 2 in Sec. 6.6.
(b) See Theorem 3 in Sec. 6.5.
(c) See Theorem 1 in Sec. 6.6.

146 (a) $f(x) = 6^3(5^7 6^8)^x$, so $f'(x) = 6^3 (\ln 5^7 6^8)(5^7 6^8)^x$
$= (\ln 5^7 6^8) f(x)$.
(b) No. $A = 6^3$ and $k = \ln 5^7 6^8$.

148 (a) $\ln y = \dfrac{3}{5} \ln x + 4 \ln(1 + 2x) +$

$3 \ln(\sin 2x) - 2 \ln(\tan 5x)$, so $\dfrac{y'}{y} = \dfrac{3}{5} \cdot \dfrac{1}{x}$

$+ 4 \cdot \dfrac{2}{1 + 2x} + 3 \cdot \dfrac{2 \cos 2x}{\sin 2x} - 2 \cdot \dfrac{5 \sec^2 5x}{\tan 5x}$

$= \dfrac{3}{5x} + \dfrac{8}{1 + 2x} + 6 \cot 2x -$

$10 \sec 5x \csc 5x$; thus $y' =$

$\dfrac{x^{3/5}(1 + 2x)^4 \sin^3 2x}{\tan^2 5x}\left(\dfrac{3}{5x} + \dfrac{8}{1 + 2x} + \right.$

$\left. 6 \cot 2x - 10 \sec 5x \csc 5x\right]$.

(b) $\ln y = 3 \ln x - \dfrac{1}{3} \ln(x^3 + x^2) - \ln(\cos 4x)$,

$\dfrac{y'}{y} = 3 \cdot \dfrac{1}{x} - \dfrac{1}{3} \cdot \dfrac{3x^2 + 2x}{x^3 + x^2} - \dfrac{-4 \sin 4x}{\cos 4x}$

$= \dfrac{6x + 7}{3x(x + 1)} + 4 \tan 4x$, $y' =$

$\dfrac{x^3}{\sqrt[3]{x^3 + x^2} \cos 4x}\left(\dfrac{6x + 7}{3x(x + 1)} + 4 \tan 4x\right)$.

150 (a) $y' = 4x^3 - \dfrac{4}{x} = \dfrac{4}{x}(x^4 - 1)$ changes from

$-$ to $+$ at $x = 1$, so there is a minimum at $(1, 1)$. There is no maximum.

(b) See (a).

(c) $y'' = 12x^2 + \dfrac{4}{x^2} = \dfrac{4}{x^2}(3x^4 + 1) > 0$ for $x > 0$, so there are no inflection points.

152 If $x^y = y^x$ then $f(x) = f(y)$, where $f(x) = \dfrac{\ln x}{x}$.

Note that $f'(x) = \dfrac{1 - \ln x}{x^2}$ is $+$ for $0 < x < e$ and $-$ for $x > e$. So if $f(x) = f(y)$ and $x < y$, then x must be 1 or 2 and y must be 3 or greater. Since $f(y) > 0 = f(1)$, the only solution is $x = 2$, $y = 4$.

154 $\ln 2 = 2 \ln 1.2 - \ln 0.8 - \ln 0.9$

$\ln 3 = 3 \ln 1.2 - 2 \ln 0.8 - \ln 0.9$

$\ln 5 = 4 \ln 1.2 - 3 \ln 0.8 - 2 \ln 0.9$

156 The lilies double each day, and on the 30th day the pond was finally covered, so one day earlier the pond must have been half covered. Thus the answer is the 29th day.

158 (a) Let the cars start at time a and end at time b. Let $f(x)$ be the position of the first car at time x and $g(x)$ be the position of the second car at time x. The distance traveled by the first car is $f(b) - f(a) = 2(g(b) - g(a))$, since the second car travels only half as far. Thus $\dfrac{f(b) - f(a)}{g(b) - g(a)} = 2$. By the generalized mean-value theorem there exists c such that $\dfrac{f'(c)}{g'(c)} = 2$ and $a < c < b$. $f'(x)$ is the speed of the first car and $g'(x)$ is the speed of the second car. Thus, at time c, the first car is traveling twice as fast as the second car, since $f'(c) = 2g'(c)$.

(b) Let $h(x) = f(x) - f(a) - \dfrac{f(b) - f(a)}{g(b) - g(a)}[g(x) - g(a)]$. Then $h(b) = f(b) - f(a) - \dfrac{f(b) - f(a)}{g(b) - g(a)}[g(b) - g(a)] = f(b) - f(a) - (f(b) - f(a)) = 0$, and $h(a) = f(a) - f(a) - \dfrac{f(b) - f(a)}{g(b) - g(a)}[g(a) - g(a)] = 0 - 0 = 0$. Since $h(a) = 0 = h(b)$, Rolle's theorem says that there is a number c such that $a < c < b$ and $h'(c) = 0$. But $h'(x) = f'(x) - \dfrac{f(b) - f(a)}{g(b) - g(a)} g'(x)$, so $f'(c) - \dfrac{f(b) - f(a)}{g(b) - g(a)} g'(c) = 0$. Hence $f'(c) = \dfrac{f(b) - f(a)}{g(b) - g(a)} g'(c)$, so $\dfrac{f'(c)}{g'(c)} = \dfrac{f(b) - f(a)}{g(b) - g(a)}$, as desired.

160 No. Let $f(t) = e^t$ and $g(t) = 2e^t$.

162 Let $f(x) = 1 + x^2$ and $g(x) = \dfrac{\ln 2}{\ln(1 + x^2)}$ for $x \neq 0$.

164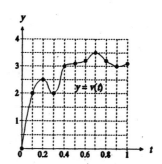

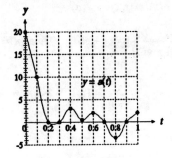

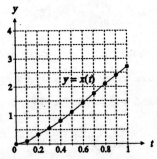

7 Computing Antiderivatives

7.1 Shortcuts, Integral Tables, and Machines

2. $\int (8 + 11x)\, dx = \int 8\, dx + \int 11x\, dx$

   $= 8 \int dx + 11 \int x\, dx = 8x + \dfrac{11}{2}x^2 + C$

4. $\int x^{2/3}\, dx = \dfrac{x^{5/3}}{5/3} + C = \dfrac{3}{5}x^{5/3} + C$

6. $\int \dfrac{dx}{x^3} = \int x^{-3}\, dx = -\dfrac{1}{2}x^{-2} + C = -\dfrac{1}{2x^2} + C$

8. $5 \tan^{-1} x + C$

10. $5 \sin^{-1} x + C$

12. $\ln(1 + e^x) + C$, since $e^x = (1 + e^x)'$

14. $\dfrac{1}{3} \int \dfrac{3}{1 + 3x}\, dx = \dfrac{1}{3} \ln|1 + 3x| + C$

16. $\int \dfrac{1}{1 + x^2}\, dx + \int \dfrac{2x}{1 + x^2}\, dx$

    $= \tan^{-1} x + \ln(1 + x^2) + C$

18. $\int (1 + 2e^x + e^{2x})\, dx = x + 2e^x + \dfrac{1}{2}e^{2x} + C$

20. $\int \dfrac{1 + \sqrt{x}}{x}\, dx = \int \left(\dfrac{1}{x} + \dfrac{1}{\sqrt{x}}\right) dx$

    $= \int \dfrac{dx}{x} + \int x^{-1/2}\, dx = \ln|x| + \dfrac{x^{-1/2 + 1}}{-1/2 + 1} + C$

    $= \ln|x| + 2x^{1/2} + C = \ln|x| + 2\sqrt{x} + C$

22. Since $\sin 3x$ is odd and $\cos 5x$ is even, their product is odd. Hence Shortcut 1 applies:

    $\int_{-\pi/2}^{\pi/2} \sin 3x \cos 5x\, dx = 0$.

24. Since $\sin^3 x = -\sin^3(-x)$, $\sin^3 x$ is an odd function and Shortcut 1 applies: $\int_{-\pi}^{\pi} \sin^3 x\, dx = 0$.

26. $\int_{-3}^{3} \left(x^3 \sqrt{9 - x^2} + 10\sqrt{9 - x^2}\right) dx$

    $= \int_{-3}^{3} x^3 \sqrt{9 - x^2}\, dx + 10 \int_{-3}^{3} \sqrt{9 - x^2}\, dx$, and

    $x^3 \sqrt{9 - x^2}$ is odd, so $\int_{-3}^{3} x^3 \sqrt{9 - x^2}\, dx = 0$ by

    Shortcut 1. By Exercise 25, $\int_{-3}^{3} \sqrt{9 - x^2}\, dx =$

    $\dfrac{9\pi}{2}$. Hence $\int_{-3}^{3} \left(x^3 \sqrt{9 - x^2} + 10\sqrt{9 - x^2}\right) dx =$

    $10 \cdot \dfrac{9\pi}{2} = 45\pi$.

(All formulas used in Exercises 28–32 are given in the covers of the text.)

28. (a) Using Formula 21 with $a = 3$ and $b = 4$ gives $\int \dfrac{dx}{x\sqrt{3x + 4}} = \dfrac{1}{2} \ln \left| \dfrac{\sqrt{3x + 4} - 2}{\sqrt{3x + 4} + 2} \right| + C$.

    (b) Using Formula 23 with $a = 3$ and $b = 4$ gives $\int \dfrac{dx}{x^2 \sqrt{3x + 4}} =$

$$-\frac{\sqrt{3x+4}}{4x} - \frac{3}{2\cdot 4}\int \frac{dx}{x\sqrt{3x+4}}, \text{ which from}$$

(a) reduces to $\int \frac{dx}{x^2\sqrt{3x+4}}$

$$= -\frac{\sqrt{3x+4}}{4x} - \frac{3}{16}\ln\left|\frac{\sqrt{3x+4}-2}{\sqrt{3x+4}+2}\right| + C.$$

30 (a) Using Formula 26 with $a = 4$ and $c = 9$ gives $\int \frac{dx}{4x^2+9} = \frac{1}{\sqrt{4\cdot 9}}\tan^{-1}x\sqrt{\frac{4}{9}} + C$

$$= \frac{1}{6}\tan^{-1}\frac{2}{3}x + C.$$

(b) Using the $a > 0$, $c < 0$ form of Formula 27 with $a = 4$ and $c = -9$ gives $\int \frac{dx}{4x^2-9}$

$$= \frac{1}{2\sqrt{-4(-9)}}\ln\left|\frac{x\sqrt{4} - \sqrt{-(-9)}}{x\sqrt{4} + \sqrt{-(-9)}}\right| + C$$

$$= \frac{1}{12}\ln\left|\frac{2x-3}{2x+3}\right| + C.$$

32 (a) Using Formula 10 with $a = \sqrt{11}$ gives

$$\int \frac{dx}{\sqrt{11-x^2}} = \sin^{-1}\frac{x}{\sqrt{11}} + C.$$

(b) Using Formula 33 with $p = \sqrt{11}$ gives

$$\int \frac{dx}{\sqrt{11+x^2}} = \ln\left|x + \sqrt{11+x^2}\right| + C.$$

34 Formula 42 gives $\int \frac{x\,dx}{\sqrt{2x^2-x+5}} =$

$$\frac{\sqrt{2x^2-x+5}}{2} - \frac{1}{2\cdot 2}\int \frac{dx}{\sqrt{2x^2-x+5}}, \text{ which with}$$

the aid of Formula 40 ($a > 0$) reduces to

$$\int \frac{x\,dx}{\sqrt{2x^2-x+5}} =$$

$$\frac{\sqrt{2x^2-x+5}}{2} - \frac{1}{4\sqrt{2}}\ln\left|4x+1+2\sqrt{2}\sqrt{2x^2-x+5}\right| + C.$$

7.2 The Substitution Method

2 $\int e^u \, du = e^u + C = e^{\sin\theta} + C$

4 $\int_9^{16} \sqrt{u}\,\frac{1}{2}\,du = \frac{1}{2}\cdot\frac{u^{3/2}}{3/2}\Big|_9^{16} = \frac{37}{3}$

6 $\int \frac{du/2}{u^2} = \int \frac{1}{2}u^{-2}\,du = \frac{1}{2}\cdot\frac{u^{-1}}{-1} + C$

$$= -\frac{1}{2u} + C = -\frac{1}{2(1+e^{2x})} + C$$

8 $\int_{1/2}^{1/3} e^u\,(-du) = -e^u\Big|_{1/2}^{1/3} = e^{1/2} - e^{1/3}$

10 $\int \frac{-du/10}{\sqrt{u}} = -\frac{1}{10}\int u^{-1/2}\,du = -\frac{1}{10}\cdot\frac{u^{1/2}}{1/2} + C$

$$= -\frac{\sqrt{u}}{5} + C = -\frac{1}{5}\sqrt{2-5t^2} + C$$

12 $\int_{\pi/4}^{\pi/2} (\sin u)\,2\,du = -2\cos u\Big|_{\pi/4}^{\pi/2} = \sqrt{2}$

14 $\int \sin u\,du = -\cos u + C = -\cos(\ln x) + C$

16 $u = x^2 + 1$. $\int \frac{x\,dx}{(x^2+1)^3} = \int \frac{du/2}{u^3}$

$$= -\frac{1}{4u^2} + C = -\frac{1}{4(x^2+1)^2} + C$$

7.2 The Substitution Method

18 $u = \cos\theta$. $\int \dfrac{\sin\theta}{\cos^2\theta}\,d\theta = \int \dfrac{-du}{u^2} = \dfrac{1}{u} + C$

$= \sec\theta + C$

20 $u = e^x$. $\int e^x \sin e^x\,dx = \int \sin u\,du$

$= -\cos u + C = -\cos e^x + C$

22 $u = 2x+5$. $\int \dfrac{dx}{\sqrt{2x+5}} = \int \dfrac{du/2}{\sqrt{u}} = u^{1/2} + C$

$= \sqrt{2x+5} + C$

24 $u = 4x+3$. $\int \dfrac{dx}{(4x+3)^3} = \int \dfrac{du/4}{u^3} =$

$\dfrac{1}{4}\cdot\dfrac{u^{-2}}{-2} + C = -\dfrac{1}{8(4x+3)^2} + C$

26 $u = x^2 + 3x + 5$. $\int \dfrac{2x+3}{(x^2+3x+5)^4}\,dx = \int \dfrac{du}{u^4}$

$= -\dfrac{1}{3u^3} + C = -\dfrac{1}{3(x^2+3x+5)^3} + C$

28 $u = 1 + \sqrt{x}$. $\int \dfrac{dx}{\sqrt{x}(1+\sqrt{x})^3} = \int \dfrac{2\,du}{u^3}$

$= -\dfrac{1}{u^2} + C = -\dfrac{1}{(1+\sqrt{x})^2} + C$

30 $u = \ln x$. $\int \dfrac{\cos(\ln x)\,dx}{x} = \int \cos u\,du$

$= \sin u + C = \sin(\ln x) + C$

32 $u = 1 + x^4$. $\int \dfrac{x^3}{1+x^4}\,dx = \int \dfrac{du/4}{u} =$

$\dfrac{1}{4}\ln|u| + C = \dfrac{1}{4}\ln(1+x^4) + C$

34 $u = 1 + x$. $\int \dfrac{(u-1)^2\,du}{u^3}$

$= \int (u^{-1} - 2u^{-2} + u^{-3})\,du$

$= \ln|u| + 2u^{-1} - \dfrac{1}{2}u^{-2} + C$

$= \ln|1+x| + \dfrac{2}{1+x} - \dfrac{1}{2(1+x)^2} + C$

36 $u = \ln|x|$. $\int \dfrac{\ln x^2\,dx}{x} = \int \dfrac{2\ln|x|\,dx}{x}$

$= \int 2u\,du = u^2 + C = (\ln|x|)^2 + C$

38 $u = \sin\theta$. $\int_0^{\pi/2} \sin^3\theta \cos\theta\,d\theta = \int_0^1 u^3\,du$

$= \dfrac{u^4}{4}\Big|_0^1 = \dfrac{1}{4}$

40 $u = 3x+1$. $\int_1^2 \dfrac{x^2 - x}{(3x+1)^2}\,dx$

$= \int_4^7 \dfrac{\left(\dfrac{u-1}{3}\right)^2 - \left(\dfrac{u-1}{3}\right)}{u^2} \dfrac{du}{3}$

$= \dfrac{1}{27}\int_4^7 (1 - 5u^{-1} + 4u^{-2})\,du$

$= \dfrac{1}{27}(u - 5\ln|u| - 4u^{-1})\Big|_4^7 = \dfrac{8}{63} + \dfrac{5}{27}\ln\dfrac{4}{7}$

42 $u = \tan\theta$. $\int_0^{\pi/3} \tan^5\theta \sec^2\theta\,d\theta = \int_0^{\sqrt{3}} u^5\,du$

$= \dfrac{u^6}{6}\Big|_0^{\sqrt{3}} = \dfrac{9}{2}$

44 $u = ax + b$. $\int \dfrac{\dfrac{u-b}{a}\cdot\dfrac{du}{a}}{u^2} = \dfrac{1}{a^2}\int\left(\dfrac{1}{u} - \dfrac{b}{u^2}\right)du$

$= \dfrac{1}{a^2}\left(\ln|u| + \dfrac{b}{u}\right) + C$

$$= \frac{1}{a^2}\left(\frac{b}{ax+b} + \ln|ax+b|\right) + C$$

46 (a) For $n = -1$, we must find $\int x(ax+b)^n\, dx$

$$= \int \frac{x}{ax+b}\, dx.$$ Let $u = ax + b$, so $du = a\, dx$ and $x = \frac{u-b}{a}$. Thus $\int \frac{x}{ax+b}\, dx =$

$$\int \frac{(u-b)/a \cdot du/a}{u} = \frac{1}{a^2}\int \frac{u-b}{u}\, du =$$

$$\frac{1}{a^2}\int\left(1 - \frac{b}{u}\right) du = \frac{1}{a^2}(u - b\ln|u|) + K$$

$$= \frac{1}{a^2}(ax + b - b\ln|ax+b|) + K$$

$$= \frac{1}{a^2}(ax - b\ln|ax+b|) + C, \text{ where } C$$

$$= \frac{b}{a^2} + K.$$

(b) For $n = -2$, we must find $\int x(ax+b)^n\, dx =$

$$\int \frac{x}{(ax+b)^2}\, dx.$$ From Exercise 44, it equals

$$\frac{1}{a^2}\left(\frac{b}{ax+b} + \ln|ax+b|\right) + C.$$

(c) Let $u = ax + b$, so from (a) we see that

$$\int x(ax+b)^n\, dx = \int \left(\frac{u-b}{a}\right) \cdot u^n\, \frac{du}{a}$$

$$= \frac{1}{a^2}\int (u^{n+1} - bu^n)\, du$$

$$= \frac{1}{a^2}\left[\frac{u^{n+2}}{n+2} - \frac{bu^{n+1}}{n+1}\right] + C =$$

$$\frac{1}{a^2}\left[\frac{(ax+b)^{n+2}}{n+2} - \frac{b(ax+b)^{n+1}}{n+1}\right] + C, \text{ with}$$

$n \neq -1, -2$.

48 (a) Jill is right. $\cos\theta \neq \sqrt{1-u^2}$ for $\pi/2 < \theta \leq \pi$.

(b) $\int_0^\pi \frac{1 + \cos 2\theta}{2}\, d\theta = \left(\frac{\theta}{2} + \frac{1}{4}\sin 2\theta\right)\Big|_0^\pi$

$$= \frac{\pi}{2}$$

50 Let $u = -x$, so $du = -dx$. Then $\int_{-a}^0 f(x)\, dx =$

$$\int_a^0 f(-u)(-du) = -\int_0^a [-f(u)](-du) =$$

$$-\int_0^a f(u)\, du = -\int_0^a f(x)\, dx, \text{ since } f \text{ is an odd}$$

function and u is a dummy variable. It follows that

$$\int_{-a}^a f(x)\, dx = \int_{-a}^0 f(x)\, dx + \int_0^a f(x)\, dx = 0.$$

52 (a) $y = (\ln x)/x = 0$ when $x = 1$ so $(1, 0)$ is the only x intercept. Since $(\ln x)/x$ is not defined for $x \leq 0$, there is no y intercept. $y' = \frac{1 - \ln x}{x^2}$ changes from $+$ to $-$ at $x = e$, so

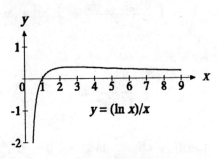

(e, e^{-1}) is a global maximum. Since

$y'' = \dfrac{2\ln x - 3}{x^3}$ changes from $-$ to $+$ at x

$= e^{3/2}$, $(e^{3/2}, \frac{3}{2}e^{-3/2})$ is an inflection point.

Finally, $\lim_{x \to 0^+} \frac{\ln x}{x} = -\infty$ and $\lim_{x \to \infty} \frac{\ln x}{x} \stackrel{H}{=}$

$\lim_{x \to \infty} \frac{1/x}{1} = 0.$

(b) Let $u = \ln x$. Then $\int_e^{e^2} \frac{\ln x}{x} dx = \int_1^2 u \, du$

$= \frac{u^2}{2} \Big|_1^2 = \frac{3}{2}.$

54 The distance traveled from $t = 0$ to $t = 1$ sec is

$\int_0^1 e^{-t} \sin \pi t \, dt.$

(a) From Formula 64 with $x = t$, $a = -1$, and $b = \pi$, we obtain $\int_0^1 e^{-t} \sin \pi t \, dt$

$= \frac{e^{-t}}{1 + \pi^2}(-\sin \pi t - \pi \cos \pi t) \Big|_0^1$

$= \frac{e^{-1}}{1 + \pi^2}(0 + \pi) - \frac{1}{1 + \pi^2}(0 - \pi)$

$= \frac{\pi}{1 + \pi^2}(1 + e^{-1}) \approx 0.395352.$

(b) Using Simpson's method with $n = 4$, we have

$\int_0^1 e^{-t} \sin \pi t \, dt \approx \frac{1/4}{3}[e^{-0} \sin 0 +$

$4e^{-1/4} \sin \frac{\pi}{4} + 2e^{-1/2} \sin \frac{\pi}{2} +$

$4e^{-3/4} \sin \frac{3\pi}{4} + e^{-1} \sin \pi] \approx 0.3960.$

7.3 Integration by Parts

2    $u = x + 3$, $dv = e^{-x} dx$. $\int (x + 3)e^{-x} dx$

$= (x + 3)(-e^{-x}) - \int (-e^{-x}) \, dx$

$= -(x + 3)e^{-x} - e^{-x} + C = -(x + 4)e^{-x} + C$

4    $u = x + 3$, $dv = \cos 2x \, dx$. $\int (x + 3) \cos 2x \, dx$

$= (x + 3) \cdot \frac{1}{2} \sin 2x - \int \frac{1}{2} \sin 2x \, dx$

$= \frac{1}{2}(x + 3) \sin 2x + \frac{1}{4} \cos 2x + C$

6    $u = \ln x$, $dv = (2x + 1) dx$. $\int (2x + 1) \ln x \, dx$

$= (\ln x)(x^2 + x) - \int (x^2 + x) \cdot \frac{1}{x} \, dx$

$= (x^2 + x) \ln x - x^2/2 - x + C$

8    $u = x^2$, $dv = e^{2x} dx$; $\int_0^1 x^2 e^{2x} \, dx =$

$x^2 \cdot \frac{1}{2} e^{2x} \Big|_0^1 - \int_0^1 \frac{1}{2} e^{2x} \, 2x \, dx =$

$\frac{1}{2} e^2 - \int_0^1 x e^{2x} \, dx$. $U = x$, $dV = e^{2x} dx$;

$\frac{1}{2} e^2 - x \cdot \frac{1}{2} e^{2x} \Big|_0^1 + \int_0^1 \frac{1}{2} e^{2x} \, dx =$

$\frac{1}{2} e^2 - \frac{1}{2} e^2 + \frac{1}{4} e^{2x} \Big|_0^1 = \frac{1}{4}(e^2 - 1)$

10    $u = \tan^{-1} 2x$, $dv = dx$. $\int_0^{1/2} \tan^{-1} 2x \, dx =$

$(\tan^{-1} 2x) x \Big|_0^{1/2} - \int_0^{1/2} x \cdot \frac{2}{1 + 4x^2} \, dx =$

$\frac{1}{2} \tan^{-1} 1 - \frac{1}{4} \ln(1 + 4x^2) \Big|_0^{1/2} = \frac{1}{2} \cdot \frac{\pi}{4} - \frac{1}{4} \ln 2$

$= \frac{1}{8}(\pi - 2\ln 2)$

12 $u = \ln x, dv = x^3 dx.$ $\int x^3 \ln x \, dx$

$= (\ln x) \cdot \frac{1}{4}x^4 - \int \frac{1}{4}x^4 \frac{dx}{x}$

$= \frac{1}{4}x^4 \ln x - \int \frac{1}{4}x^3 \, dx$

$= \frac{1}{4}x^4 \ln x - \frac{1}{16}x^4 + C$

14 $u = (\ln x)^n, dv = dx.$ $\int (\ln x)^n \, dx =$

$(\ln x)^n x - \int x \cdot n(\ln x)^{n-1} \cdot \frac{1}{x} \, dx =$

$x(\ln x)^n - n \int (\ln x)^{n-1} \, dx.$ Hence $\int (\ln x)^3 \, dx =$

$x(\ln x)^3 - 3x(\ln x)^2 + 6x(\ln x) - 6x + C$ and

$\int_2^3 (\ln x)^3 \, dx = 3(\ln 3)^3 - 9(\ln 3)^2 + 18(\ln 3) -$

$2(\ln 2)^3 + 6(\ln 2)^2 - 12(\ln 2) - 6$

16 $u = \ln x, dv = x^{-3} dx.$ $\int_e^{e^2} (\ln x)x^{-3} \, dx$

$= (\ln x)\left(-\frac{1}{2}x^{-2}\right)\Big|_e^{e^2} - \int_e^{e^2} \left(-\frac{1}{2}x^{-2}\right) \cdot \frac{1}{x} \, dx$

$= \frac{1}{2e^2} - \frac{1}{e^4} + \frac{1}{2}\int_e^{e^2} x^{-3} \, dx$

$= \frac{1}{2e^2} - \frac{1}{e^4} - \frac{1}{4}x^{-2}\Big|_e^{e^2} = \frac{3e^2 - 5}{4e^4}$

18 $u = e^{-2x}, dv = \sin 3x \, dx;$ $\int e^{-2x} \sin 3x \, dx$

$= e^{-2x}\left(-\frac{1}{3}\cos 3x\right) - \int \left(-\frac{1}{3}\cos 3x\right)(-2e^{-2x} dx)$

$= -\frac{1}{3}e^{-2x} \cos 3x - \frac{2}{3}\int e^{-2x} \cos 3x \, dx.$ Now let

$U = e^{-2x}, dV = \cos 3x \, dx$ for the remaining

integral and continue. $-\frac{1}{3}e^{-2x} \cos 3x -$

$\frac{2}{3}\left(e^{-2x} \cdot \frac{1}{3}\sin 3x - \int \left(\frac{1}{3}\sin 3x\right)(-2e^{-2x} dx)\right)$

$= -\frac{1}{3}e^{-2x} \cos 3x - \frac{2}{9}e^{-2x} \sin 3x +$

$\frac{4}{9}\int e^{-2x} \sin 3x \, dx.$ Hence $\int e^{-2x} \sin 3x \, dx$

$= -\frac{1}{13}e^{-2x}(3\cos 3x + 2\sin 3x) + C.$

20 $u = \ln|x|, dv = 2x \, dx.$ $\int x \ln(x^2) \, dx =$

$\int (\ln|x|) \cdot 2x \, dx = (\ln|x|)x^2 - \int x^2 \cdot \frac{1}{x} \, dx$

$= x^2 \ln|x| - \frac{1}{2}x^2 + C$

22 (a) Let $u = 2x + 7.$ $\int \frac{x \, dx}{\sqrt{2x + 7}} =$

$\int \frac{\frac{u-7}{2} \cdot \frac{1}{2} \, du}{\sqrt{u}} = \int \left(\frac{1}{4}u^{1/2} - \frac{7}{4}u^{-1/2}\right) du$

$= \frac{1}{6}u^{3/2} - \frac{7}{2}u^{1/2} + C$

$= \frac{1}{6}(2x + 7)^{3/2} - \frac{7}{2}(2x + 7)^{1/2} + C$

$= \frac{1}{3}(x - 7)\sqrt{2x + 7} + C$

(b) Let $u = x$ and $dv = \frac{dx}{\sqrt{2x + 7}}.$

24 (a) Let $u = ax + b.$ Then $\int \frac{x \, dx}{\sqrt[3]{ax + b}} =$

7.3 Integration by Parts

$$\int \frac{\frac{u-b}{a} \cdot \frac{1}{a} \, du}{\sqrt[3]{u}} = \int \frac{1}{a^2}(u^{2/3} - bu^{-1/3}) \, du$$

$$= \frac{1}{a^2}\left(\frac{3}{5}u^{5/3} - \frac{3}{2}bu^{2/3}\right) + C$$

$$= \frac{3}{10a^2}u^{2/3}(2u - 5b) + C$$

$$= \frac{3}{10a^2}(ax + b)^{2/3}(2ax - 3b) + C.$$

(b) Use $u = x$ and $dv = (ax + b)^{-1/3} \, dx$.

26 (a) $\int \sin^3 x \, dx$

$$= -\frac{1}{3}\sin^2 x \cos x + \frac{2}{3}\int \sin x \, dx$$

$$= -\frac{1}{3}\sin^2 x \cos x - \frac{2}{3}\cos x + C$$

(b) $\int \sin^5 x \, dx$

$$= -\frac{1}{5}\sin^4 x \cos x + \frac{4}{5}\int \sin^3 x \, dx$$

$$= -\frac{1}{5}\sin^4 x \cos x - \frac{4}{15}\sin^2 x \cos x -$$

$$\frac{8}{15}\cos x + C$$

28 (a) $[(P(x) - P'(x) + P''(x) - \cdots)e^x]'$
$= e^x[P(x) - P'(x) + P''(x) - \cdots] + e^x[P'(x) - P''(x) + P'''(x) - \cdots] = e^x P(x)$

(b) Here $P(x) = 3x^3 - 2x - 2$, $P'(x) = 9x^2 - 2$, $P''(x) = 18x$, $P'''(x) = 18$, and $P^{(n)}(x) = 0$ for $n \geq 4$. From (a), we know $\int P(x)e^x \, dx = [P(x) - P'(e) + P''(x) - \cdots]e^x$ so

$\int (3x^3 - 2x - 2)e^x \, dx = [(3x^3 - 2x - 2) -$
$(9x^2 - 2) + 18x - 18 + 0 - \cdots]e^x + C =$
$(3x^3 - 9x^2 + 16x - 18)e^x + C.$

(c) Let $u = P(x)$ and $dv = e^x \, dx$, so $du = P'(x) \, dx$ and $v = e^x$. Then $\int P(x)e^x \, dx = P(x)e^x - \int P'(x)e^x \, dx$. Now let $U = P'(x)$ and $dV = e^x \, dx$, so $dU = P''(x) \, dx$ and $V = e^x$. Thus $\int P(x)e^x \, dx = P(x)e^x - \left[P'(x)e^x - \int P''(x)e^x \, dx\right] = P(x)e^x - P'(x)e^x + \int P''(x)e^x \, dx$. Continuing the recursion gives the formula in (a).

Note: This technique can be generalized to the shortcut known as "tabular integration." If we momentarily adopt the notation $D^{-1}g$ to represent an antiderivative of g (and $D^{-2}g$ an antiderivative of an antiderivative of g, and so on), the integral $\int f(x)g(x) \, dx$ becomes $f \cdot D^{-1}g - Df \cdot D^{-2}g + D^2f \cdot D^{-3}g - D^3f \cdot D^{-4}g + \cdots$ by repeated application of integration by parts, terminating only if $D^n f = 0$ for some n. In this exercise we examined the case where $f(x)$ is a polynomial and $g(x)$ is e^x. In actual use, one may write down two columns: one of successive derivatives of f and the other of successive antiderivatives of g, then combining factors with alternating signs in the above pattern for the desired result.

30 (a) $y = e^{-x}\sin x = 0$ when $x = 0$ and $x = \pi$, so $(0, 0)$ is the y intercept and $(0, 0)$ and $(\pi, 0)$ are the x intercepts. $y' = e^{-x}(\cos x - \sin x)$ changes from $+$ to $-$ when $\tan x = 1$, or when $x = \pi/4 + k\pi$, where k is an integer, so $\left(\frac{\pi}{4}, \frac{\sqrt{2}}{2}e^{-\pi/4}\right)$ is a maximum on $[0, \pi]$. y''
$= -2e^{-x}\cos x$ changes from $-$ to $+$ at $x = \pi/2$, so $(\pi/2, e^{-\pi/2})$ is an inflection point.

(b) The area is $\int_0^\pi e^{-x}\sin x \, dx$. Let $u = \sin x$

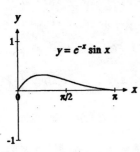

and $dv = e^{-x} dx$, so $du = \cos x \, dx$ and $v = -e^{-x}$. Thus $\int e^{-x} \sin x \, dx = -e^{-x} \sin x - \int (-e^{-x})(\cos x \, dx) = -e^{-x} \sin x + \int e^{-x} \cos x \, dx$. Now let $U = \cos x$ and $dV = e^{-x} dx$, so $dU = -\sin x \, dx$ and $V = -e^{-x}$. Then $\int e^{-x} \sin x \, dx = -e^{-x} \sin x + \left[-e^{-x} \cos x - \int (-e^{-x})(-\sin x \, dx) \right]$ and it follows that $\int e^{-x} \sin x \, dx = -\frac{1}{2} e^{-x}(\sin x + \cos x)$. Finally,

$$\int_0^\pi e^{-x} \sin x \, dx = -\frac{1}{2} e^{-x}(\sin x + \cos x) \Big|_0^\pi$$

$$= \frac{1}{2}(e^{-\pi} + 1).$$

32 The volume is $\int_1^e \pi (\ln x)^2 \, dx$ since the cross-sections are circles of radius $\ln x$. From Exercise 13, we know $\int (\ln x)^2 \, dx = x(\ln x)^2 - 2x \ln x + 2x + C$, so $\pi \int_1^e (\ln x)^2 \, dx = \pi [x(\ln x)^2 - 2x \ln x + 2x]\Big|_1^e = \pi(e - 2)$.

34 Let $y = \sqrt[3]{x}$. Then, by repeated use of parts,

$\int \sin \sqrt[3]{x} \, dx = \int (\sin y) \, 3y^2 \, dy =$

$3 \int y^2 \sin y \, dy$

$= 3\left[-y^2 \cos y + 2y \sin y - 2 \int \sin y \, dy \right]$

$= -3y^2 \cos y + 6y \sin y + 6 \cos y + C =$

$-3\sqrt[3]{x^2} \cos \sqrt[3]{x} + 6\sqrt[3]{x} \sin \sqrt[3]{x} + 6 \cos \sqrt[3]{x} + C.$

36 Let $y = \sqrt[3]{x}$. Then $\int \exp(\sqrt[3]{x}) \, dx = \int e^y \, 3y^2 \, dy$

$= 3 \int y^2 e^y \, dy = 3(y^2 e^y - 2 \int y e^y \, dy)$

$= 3y^2 e^y - 6(y e^y - \int e^y \, dy)$

$= (3y^2 - 6y + 6)e^y + C$

$= (3\sqrt[3]{x^2} - 6\sqrt[3]{x} + 6) \exp(\sqrt[3]{x}) + C.$

38 $u = \tan x$, $dv = x \, dx$. $\int x \tan x \, dx =$

$(\tan x) \cdot \frac{1}{2} x^2 - \int \frac{1}{2} x^2 \sec^2 x \, dx =$

$\frac{1}{2} x^2 \tan x - \frac{1}{2} \int \left(\frac{x}{\cos x} \right)^2 dx$. If $\int \left(\frac{x}{\cos x} \right)^2 dx$

were elementary, then $\int x \tan x \, dx$ would be also.

40 (a) Let $u = \ln(x + 1)$, $dv = dx$, and $v = x$, so

$du = \frac{dx}{x + 1}$, and $\int \ln(x + 1) \, dx =$

$x \ln(x + 1) - \int \frac{x}{x + 1} \, dx$. This last integral

can be evaluated by using substitution ($w = x + 1$), by polynomial division, or by the following device (which is equivalent to polynomial division): $\int \frac{x}{x + 1} \, dx =$

$\int \frac{(x + 1) - 1}{x + 1} \, dx = \int \left(1 - \frac{1}{x + 1} \right) dx =$

$x - \ln(x+1) + K$. Hence $\int \ln(x+1)\,dx =$
$x\ln(x+1) - x + \ln(x+1) + C =$
$(x+1)\ln(x+1) - x + C$, where $C = -K$.

(b) With $u = \ln(x+1)$, $dv = dx$, and $v = x+1$, we have $du = \dfrac{dx}{x+1}$, so

$\int \ln(x+1)\,dx = (x+1)\ln(x+1) - \int \dfrac{x+1}{x+1}\,dx = (x+1)\ln(x+1) - \int dx$

$= (x+1)\ln(x+1) - x + C$.

42 Let $u = (\ln x)^n$ and $dv = dx$. Then $du = n(\ln x)^{n-1} \cdot \dfrac{1}{x}\,dx$ and $v = x$. Hence $\int (\ln x)^n\,dx$

$= (\ln x)^n \cdot x - \int x \cdot n(\ln x)^{n-1} \cdot \dfrac{dx}{x}$

$= x(\ln x)^n - n \int (\ln x)^{n-1}\,dx$.

44 Rewrite $\cos^n ax$ as $\cos^{n-1} ax \cos ax$. Now let $u = \cos^{n-1} ax$ and $dv = \cos ax\,dx$. Then $\int \cos^n ax\,dx$

$= \int \cos^{n-1} ax \cos ax\,dx = (\cos^{n-1} ax)\left(\dfrac{1}{a}\sin ax\right)$

$- \int \left(\dfrac{1}{a}\sin ax\right)(n-1)\cos^{n-2} ax(-\sin ax)(a)\,dx$

$= \dfrac{\cos^{n-1} ax \sin ax}{a} -$

$(n-1) \int (\cos^2 ax - 1)\cos^{n-2} ax\,dx$

$= \dfrac{\cos^{n-1} ax \sin ax}{a} -$

$(n-1) \int (\cos^n ax - \cos^{n-2} ax)\,dx$. So

$n\int \cos^n ax\,dx = \dfrac{\cos^{n-1} ax \sin ax}{a} +$

$(n-1)\int \cos^{n-2} ax\,dx$, and $\int \cos^n ax\,dx$

$= \dfrac{\cos^{n-1} ax \sin ax}{na} + \dfrac{n-1}{n}\int \cos^{n-2} ax\,dx$.

46 Since $(\tan x)(1 + \cos x)^{3/2}$ is an odd function,
$\displaystyle\int_{-\pi/4}^{\pi/4} (\tan x)(1 + \cos x)^{3/2}\,dx = 0$.

48 (a) The period of $\cos ax$ is $\dfrac{2\pi}{a}$. As a gets larger, the period gets smaller, allowing more

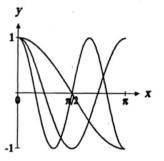

oscillations of $\cos ax$ (between -1 and 1) to fit in a given interval. Pictorially, $\cos ax$ appears more and more like a compressed spring as $a \to \infty$. The graph shows the cases $a = 1, 2,$ and 3.

(b) See the discussion in (c).

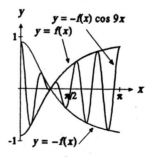

(c) As $a \to \infty$, the oscillations of $f(x)\cos ax$ over $[0, 1]$ and between $y = f(x)$ and $y = -f(x)$ become more compressed. Intuitively, the areas above and below the x axis should become more equal, so $\lim_{a \to \infty} \int_0^1 f(x) \cos ax \, dx$ seems to be 0.

(d) Let $u = f(x)$ and $dv = \cos ax \, dx$. Then

$$\int_0^1 f(x) \cos ax \, dx = \frac{1}{a} \sin ax \, f(x) \Big|_0^1 -$$

$$\int_0^1 \frac{1}{a} \sin ax \, f'(x) \, dx =$$

$$\frac{f(1) \sin a}{a} - \frac{1}{a} \int_0^1 \sin ax \, f'(x) \, dx. \text{ Since } f'(x)$$

is continuous, let M_1 denote the maximum value of $f'(x)$ over $[0, 1]$. Then $-M_1 \leq$

$\int_0^1 \sin ax \, f'(x) \, dx \leq M_1$ because $|\sin ax| \leq$

1 for all a. Thus, $\lim_{a \to \infty} \int_0^1 f(x) \cos ax \, dx =$

$\lim_{a \to \infty} \frac{1}{a} \left(f(1) \sin a - \int_0^1 \sin ax \, f'(x) \, dx \right) = 0,$

since $f(1) \sin a - \int_0^1 \sin ax \, f'(x) \, dx$ is finite.

7.4 How to Integrate Certain Rational Functions

2 Let $u = 3x + 6$. $\int \frac{2 \, dx}{3x + 6} = \int \frac{2 \cdot du/3}{u} =$

$\frac{2}{3} \ln|u| + C = \frac{2}{3} \ln|3x + 6| + C$

4 Let $u = 4x + 1$. $\int \frac{dx}{(4x + 1)^3} = \int u^{-3} \cdot \frac{1}{4} \, du$

$= -\frac{1}{8} u^{-2} + C = -\frac{1}{8(4x + 1)^2} + C.$

6 Let $u = 3x$. $\int \frac{dx}{9x^2 + 1} = \int \frac{du/3}{u^2 + 1}$

$= \frac{1}{3} \tan^{-1} u + C = \frac{1}{3} \tan^{-1} 3x + C$

8 $\int_0^1 \frac{x \, dx}{x^2 + 2} = \frac{1}{2} \ln(x^2 + 2) \Big|_0^1 = \frac{1}{2} \ln \frac{3}{2}$

10 $\int \frac{3x - 5}{x^2 + 9} \, dx = \frac{3}{2} \ln(x^2 + 9) - 5 \int \frac{dx}{x^2 + 9}$

$= \frac{3}{2} \ln(x^2 + 9) - \frac{5}{3} \tan^{-1} \frac{x}{3} + C$

12 $\int \frac{dx}{9x^2 + 4} = \frac{1}{6} \tan^{-1} \frac{3x}{2} + C$

14 $\frac{1}{18} \ln(9x^2 + 4) + C$

16 By Exercises 12 and 14, $\int \frac{2x - 1}{9x^2 + 4} \, dx$

$= \frac{1}{9} \ln(9x^2 + 4) - \frac{1}{6} \tan^{-1} \frac{3x}{2} + C.$

18 $\frac{1}{4} \ln(2x^2 + 3) + C$

20 $\frac{1}{2} \tan^{-1} \frac{x + 1}{2} + C$

22 $\int \frac{x \, dx}{x^2 - 2x + 3}$

$= \int \frac{(x - 1) \, dx}{x^2 - 2x + 3} + \int \frac{dx}{x^2 - 2x + 3}$

$= \frac{1}{2} \ln(x^2 - 2x + 3) + \frac{1}{\sqrt{2}} \tan^{-1} \frac{x - 1}{\sqrt{2}} + C$

24 $\int \frac{dx}{3x^2 - 12x + 13} = \int \frac{dx}{3(x - 2)^2 + 1}$

$= \frac{1}{\sqrt{3}} \tan^{-1}(\sqrt{3}(x - 2)) + C$

7.4 How to Integrate Certain Rational Functions

26 $\displaystyle\int \frac{dx}{x^2 + 4x + 9} = \int \frac{dx}{(x+2)^2 + 5}$

$\displaystyle = \frac{1}{\sqrt{5}} \tan^{-1} \frac{x+2}{\sqrt{5}} + C$

28 $\displaystyle\int \frac{dx}{2x^2 + 6x + 5} = \int \frac{dx}{2\left(x + \frac{3}{2}\right)^2 + \frac{1}{2}}$

$= \tan^{-1}(2x + 3) + C$

30 $\displaystyle\int \frac{2x\,dx}{x^2 + 2x + 5}$

$\displaystyle = \int \frac{2x+2}{x^2+2x+5}\,dx - \int \frac{2\,dx}{(x+1)^2 + 4}$

$\displaystyle = \ln(x^2 + 2x + 5) - \tan^{-1}\frac{x+1}{2} + C$

32 $\displaystyle\int \frac{x\,dx}{5x^2 - 3x + 2} = \frac{1}{10}\int \frac{(10x-3)+3}{5x^2 - 3x + 2}\,dx$

$\displaystyle = \frac{1}{10}\ln(5x^2 - 3x + 2) + \int \frac{6\,dx}{(10x-3)^2 + 31} =$

$\displaystyle \frac{1}{10}\ln(5x^2 - 3x + 2) + \frac{3}{5\sqrt{31}}\tan^{-1}\frac{10x-3}{\sqrt{31}} + C$

34 $\displaystyle\int \frac{x+3}{x^2 + x + 1}\,dx = \frac{1}{2}\int \frac{(2x+1)+5}{x^2 + x + 1}\,dx$

$\displaystyle = \frac{1}{2}\ln(x^2 + x + 1) + 10\int \frac{dx}{(2x+1)^2 + 3}$

$\displaystyle = \frac{1}{2}\ln(x^2 + x + 1) + \frac{5}{\sqrt{3}}\tan^{-1}\frac{2x+1}{\sqrt{3}} + C$

36 $\displaystyle\int \frac{(x+5)\,dx}{2x^2 + 3x + 5} = \frac{1}{4}\int \frac{(4x+3)+17}{2x^2 + 3x + 5}\,dx$

$\displaystyle = \frac{1}{4}\ln(2x^2 + 3x + 5) + 34\int \frac{dx}{(4x+3)^2 + 31}$

$\displaystyle = \frac{1}{4}\ln(2x^2 + 3x + 5) + \frac{17}{2\sqrt{31}}\tan^{-1}\frac{4x+3}{\sqrt{31}} + C$

38 $\displaystyle\int \frac{dx}{2x^2 + x + 1} = \frac{1}{2}\int \frac{dx}{\left(x+\frac{1}{4}\right)^2 + \frac{7}{16}}$

$\displaystyle = \frac{2}{\sqrt{7}}\tan^{-1}\frac{4x+1}{\sqrt{7}} + C$, so $\displaystyle\int_1^2 \frac{dx}{2x^2 + x + 1}$

$\displaystyle = \left(\frac{2}{\sqrt{7}}\tan^{-1}\frac{4x+1}{\sqrt{7}}\right)\bigg|_1^2$

$\displaystyle = \frac{2}{\sqrt{7}}\left(\tan^{-1}\frac{9}{\sqrt{7}} - \tan^{-1}\frac{5}{\sqrt{7}}\right).$

40 (a) We know that $\displaystyle r_1 = \frac{-b + \sqrt{b^2 - 4ac}}{2a}$ and r_2

$\displaystyle = \frac{-b - \sqrt{b^2 - 4ac}}{2a}$. So $a(x - r_1)(x - r_2)$

$= a(x^2 - (r_1 + r_2)x + r_1 r_2)$

$\displaystyle = a\left(x^2 + \frac{b}{a}x + \frac{c}{a}\right) = ax^2 + bx + c.$

(b) We know $r = -b/(2a)$. So $a(x - r)(x - r)$

$\displaystyle = a(x^2 - 2rx + r^2) = a\left(x^2 + \frac{b}{a}x + \frac{b^2}{4a^2}\right) =$

$ax^2 + bx + c$ since $b^2 - 4ac = 0$ implies $b^2/(4a) = c.$

42 (a) $x^2 - 9 = (x+3)(x-3)$

(b) $x^2 - 5 = (x + \sqrt{5})(x - \sqrt{5})$

(c) Irreducible

(d) $x^2 + 3x + 2 = (x+1)(x+2)$

(e) $x^2 + 6x + 9 = (x+3)^2$

(f) Irreducible

(g) The roots of $x^2 + 5x + 2$ are $r_1 =$

$\dfrac{-5+\sqrt{17}}{2}$, $r_2 = \dfrac{-5-\sqrt{17}}{2}$, so $x^2 + 5x + 2$

$= (x - r_1)(x - r_2)$

$= \left(x + \dfrac{5-\sqrt{17}}{2}\right)\left(x + \dfrac{5+\sqrt{17}}{2}\right).$

44 Using Formula 28 with $a = 4$, $c = 1$, and $n = 2$ gives $\displaystyle\int \dfrac{dx}{(4x^2+1)^2} = \dfrac{1}{2}\dfrac{x}{4x^2+1} + \dfrac{1}{2}\int\dfrac{dx}{4x^2+1}.$

Applying Formula 26 yields $\displaystyle\int \dfrac{dx}{(4x^2+1)^2}$

$= \dfrac{x}{2(4x^2+1)} + \dfrac{1}{4}\tan^{-1}2x + C.$

46 (a) Let $u = (x^2+p)^{-n}$ and $dv = dx$. Then

$\displaystyle\int \dfrac{dx}{(x^2+p)^n}$

$= \dfrac{x}{(x^2+p)^n} - \displaystyle\int x(-2nx)(x^2+p)^{-(n+1)}\,dx$

$= \dfrac{x}{(x^2+p)^n} + 2n\displaystyle\int \dfrac{x^2}{(x^2+p)^{n+1}}\,dx.$

(b) Now $\displaystyle\int \dfrac{dx}{(x^2+p)^n} = \dfrac{x}{(x^2+p)^n} +$

$2n\left[\displaystyle\int \dfrac{dx}{(x^2+p)^n} - p\int\dfrac{dx}{(x^2+p)^{n+1}}\right]$, so

$\displaystyle\int \dfrac{dx}{(x^2+p)^{n+1}} =$

$\dfrac{x}{2np(x^2+p)^n} + \dfrac{2n-1}{2np}\displaystyle\int \dfrac{dx}{(x^2+p)^n}$, which is

the recursion given by Formula 38 in the special case $a = 1$, $b = 0$, and $c = p$.

7.5 Integration of Rational Functions by Partial Fractions

2 $\dfrac{5}{(x-1)(x+3)} = \dfrac{c_1}{x-1} + \dfrac{c_2}{x+3}$

4 $\dfrac{3}{(x+1)(x+5)(x+4)}$

$= \dfrac{c_1}{x+1} + \dfrac{c_2}{x+5} + \dfrac{c_3}{x+4}$

6 $\dfrac{x^3}{(x+1)^2(x+2)^2}$

$= \dfrac{c_1}{x+1} + \dfrac{c_2}{(x+1)^2} + \dfrac{c_3}{x+2} + \dfrac{c_4}{(x+2)^2}$

8 $\dfrac{x^3+2}{(x-1)^2(x^2+2x+3)^2} =$

$\dfrac{c_1}{x-1} + \dfrac{c_2}{(x-1)^2} + \dfrac{c_3 x + c_4}{x^2+2x+3} + \dfrac{c_5 x + c_6}{(x^2+2x+3)^2}$

10 $\dfrac{x^2+2}{(x+1)^2(2x+2)(3x+3)} = \dfrac{x^2+2}{6(x+1)^4}$

$= \dfrac{c_1}{x+1} + \dfrac{c_2}{(x+1)^2} + \dfrac{c_3}{(x+1)^3} + \dfrac{c_4}{(x+1)^4}$

12 $\dfrac{x^3}{(x+1)(x+2)} = \dfrac{x^3}{x^2+3x+2}$, which, by long

division, yields $\dfrac{x^3}{(x+1)(x+2)} =$

$x - 3 + \dfrac{7x+6}{(x+1)(x+2)}.$

14 $\dfrac{x^5+x}{(x+1)^2(x-2)} = \dfrac{x^5+x}{x^3-3x-2}$, which, by long

division, yields $\dfrac{x^5+x}{(x+1)^2(x-2)}$

7.5 Integration of Rational Functions by Partial Fractions

$= x^2 + 3 + \dfrac{2x^2 + 10x + 6}{(x+1)^2(x-2)}$.

16 $\dfrac{x+2}{x(x+1)} = \dfrac{2}{x} - \dfrac{1}{x+1}$

18 $\dfrac{2x^2 + 4x + 3}{x(x+1)^2} = \dfrac{3}{x} - \dfrac{1}{x+1} - \dfrac{1}{(x+1)^2}$

20 $\dfrac{x+4}{(x-1)(x+3)} = \dfrac{5/4}{x-1} - \dfrac{1/4}{x+3}$

22 $\dfrac{8}{x^2-4} = \dfrac{8}{(x+2)(x-2)} = -\dfrac{2}{x+2} + \dfrac{2}{x-2}$

24 $\dfrac{5x^2 - 2x - 2}{x(x^2-1)} = \dfrac{5x^2 - 2x - 2}{x(x+1)(x-1)}$

$= \dfrac{2}{x} + \dfrac{5/2}{x+1} + \dfrac{1/2}{x-1}$

26 $\dfrac{2x^2 - 10x + 14}{(x-2)(x-3)(x-4)}$

$= \dfrac{1}{x-2} - \dfrac{2}{x-3} + \dfrac{3}{x-4}$

28 $\dfrac{5x^2 + 2x + 3}{x(x^2+x+1)} = \dfrac{3}{x} + \dfrac{2x-1}{x^2+x+1}$

30 $\dfrac{x^3 - 3x^2 + 3x - 3}{x^2 - 3x + 2} = x + \dfrac{x-3}{(x-1)(x-2)}$

$= x + \dfrac{2}{x-1} - \dfrac{1}{x-2}$

32 $\dfrac{x^5 + 2x^4 + 4x^3 + 2x^2 + x - 2}{x^4 - 1}$

$= x + 2 + \dfrac{4x^3 + 2x^2 + 2x}{(x^2+1)(x+1)(x-1)}$

$= x + 2 + \dfrac{x+1}{x^2+1} + \dfrac{1}{x+1} + \dfrac{2}{x-1}$

34 $\dfrac{x^2 + 6}{(x^2+1)(x^2+4)} = \dfrac{5/3}{x^2+1} - \dfrac{2/3}{x^2+4}$

36 $\dfrac{2x-3}{x^2 + 7x + 10} = \dfrac{2x-3}{(x+5)(x+2)}$

$= \dfrac{13/3}{x+5} - \dfrac{7/3}{x+2}$

38 $c_1 = -2/23$, $c_2 = -13/23$

40 $c_1 = 3/13$, $c_2 = 11/26$

42 $c_1 = -3/11$, $c_2 = 26/11$, $c_3 = -45/11$

44 $c_1 = 2$, $c_2 = 1$, $c_3 = -5$

46 $\displaystyle\int_0^1 \dfrac{x^3 + 2x + 2}{x^2+1}\,dx = \int_0^1 \left(x + \dfrac{x+2}{x^2+1}\right) dx$

$= \displaystyle\int_0^1 \left(x + \dfrac{1}{2}\dfrac{2x}{x^2+1} + \dfrac{2}{x^2+1}\right) dx$

$= \left.\left(\dfrac{x^2}{2} + \dfrac{1}{2}\ln(x^2+1) + 2\tan^{-1} x\right)\right|_0^1$

$= \dfrac{1}{2} + \dfrac{1}{2}\ln 2 + \dfrac{\pi}{2} = \dfrac{1}{2}(1 + \ln 2 + \pi)$

48 $\displaystyle\int_3^5 \pi\left(\dfrac{1}{\sqrt{(x^2-4)(x-1)}}\right)^2 dx$

$= \pi\displaystyle\int_3^5 \dfrac{dx}{(x+2)(x-2)(x-1)}$

$= \pi\displaystyle\int_3^5 \left(\dfrac{1/12}{x+2} + \dfrac{1/4}{x-2} - \dfrac{1/3}{x-1}\right) dx$

$= \left.\pi\left(\dfrac{1}{12}\ln|x+2| + \dfrac{1}{4}\ln|x-2| - \dfrac{1}{3}\ln|x-1|\right)\right|_3^5$

$= \pi\left(\dfrac{1}{3}\ln 2 + \dfrac{1}{4}\ln 3 - \dfrac{1}{3}\ln 4 - \dfrac{1}{12}\ln 5 + \dfrac{1}{12}\ln 7\right) = \pi\left(-\dfrac{1}{3}\ln 2 + \dfrac{1}{4}\ln 3 + \dfrac{1}{12}\ln\dfrac{7}{5}\right)$

50 $\dfrac{1}{x^2 + 2x + 2}$ is already in its partial fraction representation. So we use the substitution $u = x + 1$, $du = dx$. Thus, $\displaystyle\int_0^1 \dfrac{dx}{x^2 + 2x + 2}$

$= \displaystyle\int_0^1 \dfrac{dx}{(x+1)^2 + 1} = \int_1^2 \dfrac{du}{u^2 + 1} = \tan^{-1} u \Big|_1^2$

$= \tan^{-1} 2 - \tan^{-1} 1 = \tan^{-1} 2 - \dfrac{\pi}{4}$.

52 $\displaystyle\int_0^1 \dfrac{x^3 + x}{x^2 + 2} dx = \int_0^1 \left(x - \dfrac{x}{x^2 + 2}\right) dx$

$= \displaystyle\int_0^1 x\, dx - \dfrac{1}{2} \int_0^1 \dfrac{2x\, dx}{x^2 + 2}$

$= \left(\dfrac{x^2}{2} - \dfrac{1}{2} \ln|x^2 + 2|\right)\Big|_0^1 = \dfrac{1}{2}\left(1 + \ln \dfrac{2}{3}\right)$

54 (a) $x^2 + 3x - 5 = \left(x + \dfrac{3 - \sqrt{29}}{2}\right)\left(x + \dfrac{3 + \sqrt{29}}{2}\right)$

(b) $\displaystyle\int \dfrac{dx}{x^2 + 3x - 5}$

$= \displaystyle\int \left(\dfrac{1/\sqrt{29}}{x + \dfrac{3 - \sqrt{29}}{2}} - \dfrac{1/\sqrt{29}}{x + \dfrac{3 + \sqrt{29}}{2}}\right) dx$

$= \dfrac{1}{\sqrt{29}} \ln\left|x + \dfrac{3 - \sqrt{29}}{2}\right| - \dfrac{1}{\sqrt{29}} \ln\left|x + \dfrac{3 + \sqrt{29}}{2}\right| + C$

$= \dfrac{1}{\sqrt{29}} \ln\left|\dfrac{2x + 3 - \sqrt{29}}{2x + 3 + \sqrt{29}}\right| + C$

(c) $\displaystyle\int \dfrac{dx}{x^2 + 3x - 5} = \dfrac{1}{\sqrt{29}} \ln\left|\dfrac{2x + 3 - \sqrt{29}}{2x + 3 + \sqrt{29}}\right| + C$

56 (a) By Exercise 55, if $4x^3 + 4x^2 - 13x - 3 = 0$ has a rational root, it must be one of ± 1, ± 3, $\pm 1/2$, $\pm 3/2$, $\pm 1/4$, and $\pm 3/4$. Testing these, we find that $3/2$ is a root. Hence $4x^3 + 4x^2 - 13x - 3 = 2(x - 3/2)(2x^2 + 5x + 1)$. By the quadratic formula, the roots of $2x^2 + 5x + 1$ are $\dfrac{-5 \pm \sqrt{17}}{4}$, so $4x^3 + 4x^2 - 13x - 3$

$= 4\left(x - \dfrac{3}{2}\right)\left(x - \dfrac{-5 + \sqrt{17}}{4}\right)\left(x - \dfrac{-5 - \sqrt{17}}{4}\right)$

$= \dfrac{1}{8}(2x - 3)(4x + 5 - \sqrt{17})(4x + 5 + \sqrt{17})$.

(b) Let $P(x) = 2x^3 - x^2 - x - 3$. Note that $P(3/2) = 0$, so $P(x) = (x - 3/2)Q(x)$. To find $Q(x)$, use long division, obtaining $P(x) = \left(x - \dfrac{3}{2}\right)(2x^2 + 2x + 2) = (2x - 3)(x^2 + x + 1)$.

(Note that $x^2 + x + 1$ is indeed irreducible.)

(c) Let $P(x) = x^3 + x + 1$, so $P'(x) = 3x^2 + 1 > 0$ for all x and $P(x)$ has at most one root. Newton's method shows that the root is approximately -0.68233. Therefore $P(x) \approx (x + 0.68233)Q(x)$, and long division reveals that $Q(x) \approx x^2 - 0.68233x + 1.46557$, which is indeed irreducible. Finally, we have $P(x) \approx (x + 0.68233)(x^2 - 0.68233x + 1.46557)$.

(d) $x^3 - 8 = (x - 2)(x^2 + 2x + 4)$

58 (a) $\displaystyle\int \dfrac{x^3\, dx}{x^4 + 1} = \dfrac{1}{4} \int \dfrac{4x^3\, dx}{x^4 + 1}$

$= \dfrac{1}{4} \ln(x^4 + 1) + C$

(b) $\displaystyle\int \dfrac{x\, dx}{x^4 + 1} = \dfrac{1}{2} \int \dfrac{2x\, dx}{(x^2)^2 - 1}$. Let $u = x^2$.

7.5 Integration of Rational Functions by Partial Fractions

Then $\dfrac{1}{2} \int \dfrac{2x\, dx}{(x^2)^2 + 1} = \dfrac{1}{2} \int \dfrac{du}{u^2 + 1} =$

$\dfrac{1}{2} \tan^{-1} u + C = \dfrac{1}{2} \tan^{-1} x^2 + C.$

(c) The denominator $x^4 + 1$ factors as $(x^2 + \sqrt{2}x + 1)(x^2 - \sqrt{2}x + 1)$. Thus $\dfrac{1}{x^4 + 1} =$

$\dfrac{c_1 x + c_2}{x^2 + \sqrt{2}x + 1} + \dfrac{c_3 x + c_4}{x^2 - \sqrt{2}x + 1}$, so $1 =$

$(c_1 + c_3)x^3 + (c_2 + c_4 - c_1\sqrt{2} + c_3\sqrt{2})x^2 +$

$(c_1 - c_2\sqrt{2} + c_3 + c_4\sqrt{2})x + c_2 + c_4.$

Equating coefficients yields $c_1 = \dfrac{\sqrt{2}}{4}$, $c_3 =$

$-\dfrac{\sqrt{2}}{4}$, $c_2 = c_4 = \dfrac{1}{2}$. Thus $\dfrac{1}{x^4 + 1} =$

$\dfrac{\frac{\sqrt{2}}{4}x + \frac{1}{2}}{x^2 + \sqrt{2}x + 1} + \dfrac{-\frac{\sqrt{2}}{4}x + \frac{1}{2}}{x^2 - \sqrt{2}x + 1}$. Integrating the

first quotient yields $\displaystyle\int \dfrac{\frac{\sqrt{2}}{4}x + \frac{1}{2}}{x^2 + \sqrt{2}x + 1}\, dx =$

$\dfrac{\sqrt{2}}{4} \displaystyle\int \dfrac{x\, dx}{x^2 + \sqrt{2}x + 1} + \dfrac{1}{2} \displaystyle\int \dfrac{dx}{x^2 + \sqrt{2}x + 1} =$

$\dfrac{\sqrt{2}}{8} \displaystyle\int \dfrac{2x + \sqrt{2}}{x^2 + \sqrt{2}x + 1}\, dx - \dfrac{1}{4} \displaystyle\int \dfrac{dx}{x^2 + \sqrt{2}x + 1}$

$+ \dfrac{1}{2} \displaystyle\int \dfrac{dx}{x^2 + \sqrt{2}x + 1} =$

$\dfrac{\sqrt{2}}{8} \ln(x^2 + \sqrt{2}x + 1) + \dfrac{1}{4} \displaystyle\int \dfrac{dx}{x^2 + \sqrt{2}x + 1} =$

$\dfrac{\sqrt{2}}{8} \ln(x^2 + \sqrt{2}x + 1) +$

$\dfrac{\sqrt{2}}{4} \tan^{-1}(\sqrt{2}x + 1) + C_1.$ Similarly,

$\displaystyle\int \dfrac{-\frac{\sqrt{2}}{4}x + \frac{1}{2}}{x^2 - \sqrt{2}x + 1}\, dx = -\dfrac{\sqrt{2}}{8} \ln(x^2 - \sqrt{2}x + 1)$

$+ \dfrac{\sqrt{2}}{4} \tan^{-1}(\sqrt{2}x - 1) + C_2.$ Putting these

results together, we obtain $\displaystyle\int \dfrac{dx}{x^4 + 1} =$

$\dfrac{\sqrt{2}}{8} \ln\!\left(\dfrac{x^2 + \sqrt{2}x + 1}{x^2 - \sqrt{2}x + 1}\right) + \dfrac{\sqrt{2}}{4} \tan^{-1}(\sqrt{2}x + 1)$

$+ \dfrac{\sqrt{2}}{4} \tan^{-1}(\sqrt{2}x - 1) + C.$

7.6 Special Techniques

2   $\displaystyle\int \sin 5x \cos 2x\, dx = \int \left(\dfrac{1}{2} \sin 7x + \dfrac{1}{2} \sin 3x\right) dx$

$= -\dfrac{1}{14} \cos 7x - \dfrac{1}{6} \cos 3x + C$

4   $\displaystyle\int \cos 2\pi x \sin 5\pi x\, dx =$

$\displaystyle\int \left(\dfrac{1}{2} \sin 7\pi x + \dfrac{1}{2} \sin 3\pi x\right) dx = -\dfrac{1}{14\pi} \cos 7\pi x$

$- \dfrac{1}{6\pi} \cos 3\pi x + C.$

6   $\displaystyle\int \cos^2 5x\, dx = \int \left(\dfrac{1}{2} + \dfrac{1}{2} \cos 10x\right) dx$

$= \dfrac{1}{2} x + \dfrac{1}{20} \sin 10x + C$

8 $\int (5\cos 2x + \cos^2 7x)\, dx$

$= 5\int \cos 2x\, dx + \frac{1}{2}\int (1 + \cos 14x)\, dx$

$= \frac{5}{2}\sin 2x + \frac{1}{2}x + \frac{1}{28}\sin 14x + C$

10 $u = 3\theta,\ du = 3\, d\theta.\ \int \sec 3\theta\, d\theta = \frac{1}{3}\int \sec u\, du$

$= \frac{1}{3}\left(\frac{1}{2}\ln\left(\frac{1 + \sin u}{1 - \sin u}\right)\right) + C$

$= \frac{1}{6}\ln\left(\frac{1 + \sin 3\theta}{1 - \sin 3\theta}\right) + C$

12 $u = 4x,\ du = 4\, dx.\ \int \sec^2 4x\, dx =$

$\frac{1}{4}\int \sec^2 u\, du = \frac{1}{4}\tan u + C = \frac{1}{4}\tan 4x + C$

14 $u = 3x,\ du = 3\, dx.\ \int \frac{dx}{\cos^2 3x} = \frac{1}{3}\int \sec^2 u\, du$

$= \frac{1}{3}\tan u + C = \frac{1}{3}\tan 3x + C.$

16 $\frac{1}{2}\sin(A + B) + \frac{1}{2}\sin(A - B)$

$= \frac{1}{2}\sin A \cos B + \frac{1}{2}\cos A \sin B$

$+ \frac{1}{2}\sin A \cos B - \frac{1}{2}\cos A \sin B$

$= \sin A \cos B$

18 Let $d(\alpha)$ be the distance between the line representing the equator and the parallel line representing the latitude α, so $d(\alpha) = kF(\alpha)$ for some constant k. To find the vertical dimension of a mercator map that depicts all points on the earth except the north and south pole ($\pm 90°$ latitude), we must find how $d(\alpha)$ behaves for α near $90°$, and then multiply this result by two in order to include the southern hemisphere. Hence, $\lim\limits_{\alpha \to \pi/2^-} d(\alpha) =$

$k\lim\limits_{\alpha \to \pi/2^-} F(\alpha) = \frac{k}{2}\lim\limits_{\alpha \to \pi/2^-} \ln\left(\frac{1 + \sin\alpha}{1 - \sin\alpha}\right) = \infty$. So

the Mercator map meeting the requirements would be infinite in extent.

20 $u = \sqrt[3]{x + 1}.\ \int \frac{x^2\, dx}{\sqrt[3]{x + 1}} = \int \frac{(u^3 - 1)^2(3u^2\, du)}{u}$

$= 3\int (u^7 - 2u^4 + u)\, du$

$= \frac{3}{8}u^8 - \frac{6}{5}u^5 + \frac{3}{2}u^2 + C$

$= \frac{3}{8}(x + 1)^{8/3} - \frac{6}{5}(x + 1)^{5/3} + \frac{3}{2}(x + 1)^{2/3} + C.$

22 Let $u = \sqrt{2x + 1}$. Then $u^2 = 2x + 1,\ x = \frac{1}{2}(u^2 - 1),\ dx = u\, du.\ \int \frac{\sqrt{2x + 1}}{x}\, dx$

$= \int \frac{u(u\, du)}{\frac{1}{2}(u^2 - 1)} = 2\int \frac{u^2\, du}{u^2 - 1}$

$= 2\int \frac{(u^2 - 1) + 1}{u^2 - 1}\, du = 2\int \left(1 + \frac{1}{u^2 - 1}\right) du$

$= 2\int \left[1 + \frac{1}{2}\left(\frac{1}{u - 1} - \frac{1}{u + 1}\right)\right] du$

$= \int \left(2 + \frac{1}{u - 1} - \frac{1}{u + 1}\right) du$

$= 2u + \ln|u - 1| - \ln|u + 1| + C$

$= 2u + \ln\left|\frac{u - 1}{u + 1}\right| + C$

7.6 Special Techniques

$$= 2\sqrt{2x+1} + \ln\left|\frac{\sqrt{2x+1}-1}{\sqrt{2x+1}+1}\right| + C.$$

24 Let $u = \sqrt{x}$. Then $x = u^2$ and $dx = 2u\,du$, so

$$\int \frac{\sqrt{x}+3}{\sqrt{x}-2}\,dx = \int \frac{u+3}{u-2}\cdot 2u\,du$$

$$= 2\int \frac{u^2+3u}{u-2}\,du = 2\int \left(u + 5 + \frac{10}{u-2}\right)du$$

$$= 2\left(\frac{u^2}{2} + 5u + 10\ln|u-2|\right) + C$$

$$= u^2 + 10u + 20\ln|u-2| + C$$

$$= x + 10\sqrt{x} + 20\ln|\sqrt{x}-2| + C.$$

26 Let $u = \sqrt[3]{3x+2}$. Then $\int x(3x+2)^{5/3}\,dx =$

$$\int \left(\frac{u^3-2}{3}\right)u^5(u^2\,du) = \frac{1}{3}\int (u^{10} - 2u^7)\,du =$$

$$\frac{1}{3}\left(\frac{u^{11}}{11} - \frac{u^8}{4}\right) + C = \frac{1}{33}(3x+2)^{11/3} -$$

$$\frac{1}{12}(3x+2)^{8/3} + C = \frac{1}{44}(3x+2)^{8/3}(4x-1) + C.$$

28 Let $u = \sqrt[5]{x-3}$. Then $\int (x+2)\sqrt[5]{x-3}\,dx$

$$= \int [(u^5 + 3) + 2]u(5u^4\,du)$$

$$= 5\int (u^{10} + 5u^5)\,du = 5\left(\frac{u^{11}}{11} + \frac{5u^6}{6}\right) + C$$

$$= \frac{5}{11}(x-3)^{11/5} + \frac{25}{6}(x-3)^{6/5} + C$$

$$= \frac{5}{66}(x-3)^{6/5}(6x + 37) + C.$$

30 (a) $\int \sec^n \theta\,d\theta = \int \sec^{n-2}\theta \sec^2\theta\,d\theta$, by the suggestion. Now let $dv = \sec^2\theta\,d\theta$ and $u = \sec^{n-2}\theta$. Then $\int \sec^n \theta\,d\theta =$

$$\sec^{n-2}\theta \tan\theta - \int (n-2)\sec^{n-2}\theta \tan\theta (\tan\theta)\,d\theta$$

$$= \sec^{n-2}\theta \tan\theta -$$

$$\int (n-2)\sec^{n-2}\theta(\sec^2\theta - 1)\,d\theta. \text{ So}$$

$$(n-1)\int \sec^n\theta\,d\theta =$$

$$\sec^{n-2}\theta \tan\theta + (n-2)\int \sec^{n-2}\theta\,d\theta \text{ and}$$

the recursion follows.

(b) $\int \sec^3\theta\,d\theta = \dfrac{\sec\theta \tan\theta}{2} + \dfrac{1}{2}\int \sec\theta\,d\theta$

$$= \frac{\sec\theta\tan\theta}{2} + \frac{1}{2}\ln|\sec\theta + \tan\theta| + C$$

(c) $\int \dfrac{d\theta}{\cos^4\theta} = \int \sec^4\theta\,d\theta$

$$= \frac{\sec^2\theta \tan\theta}{3} + \frac{2}{3}\int \sec^2\theta\,d\theta$$

$$= \frac{\sec^2\theta \tan\theta}{3} + \frac{2}{3}\tan\theta + C$$

(d) $u = 2x$, $du = 2\,dx$. $\int \sec^2 2x\,dx =$

$$\frac{1}{2}\int \sec^2 u\,du = \frac{1}{2}\tan u + C$$

$$= \frac{1}{2}\tan 2x + C.$$

32 (a) Let $u = \sin\theta$ and $du = \cos\theta\,d\theta$. Then

$$\int \cot\theta\,d\theta = \int \frac{\cos\theta}{\sin\theta}\,d\theta = \int \frac{du}{u} =$$

$$\ln|u| + C = \ln|\sin\theta| + C.$$

(b) $\int \cot^2\theta\,d\theta = \int (\csc^2\theta - 1)\,d\theta$

$$= -\cot\theta - \theta + C$$

34 Write $\int \sin^n \theta \cos^n \theta \, d\theta$ as $\int \sin \theta \sin^{n-1} \theta \cos^m \theta \, d\theta$, then write $\sin^{n-1} \theta$ as $(1 - \cos^2 \theta)^{(n-1)/2}$ and use the substitution $u = \cos \theta$.

36 (a) Using the identity $\tan \dfrac{\theta}{2} = \dfrac{\sin \theta}{1 + \cos \theta}$, we have

$$\tan\left(\dfrac{\alpha}{2} + \dfrac{\pi}{4}\right) = \dfrac{\sin\left(\alpha + \dfrac{\pi}{2}\right)}{1 + \cos\left(\alpha + \dfrac{\pi}{2}\right)}$$

$$= \dfrac{\cos \alpha}{1 - \sin \alpha} = \dfrac{\sqrt{1 - \sin^2 \alpha}}{1 - \sin \alpha} =$$

$$\sqrt{\dfrac{1 + \sin \alpha}{1 - \sin \alpha}} \text{ (for } -\pi/2 < \alpha < \pi/2\text{). Thus}$$

$$\ln \tan\left(\dfrac{\alpha}{2} + \dfrac{\pi}{4}\right) = \dfrac{1}{2} \ln\left(\dfrac{1 + \sin \alpha}{1 - \sin \alpha}\right).$$

(b) Note that $\dfrac{1 + \sin \alpha}{1 - \sin \alpha} = \dfrac{(1 + \sin \alpha)^2}{1 - \sin^2 \alpha} =$

$$\left(\dfrac{1 + \sin \alpha}{\cos \alpha}\right)^2 = (\sec \alpha + \tan \alpha)^2, \text{ so}$$

$$\sqrt{\dfrac{1 + \sin \alpha}{1 - \sin \alpha}} = |\sec \alpha + \tan \alpha|. \text{ Thus}$$

$$\dfrac{1}{2} \ln\left(\dfrac{1 + \sin \alpha}{1 - \sin \alpha}\right) = \ln|\sec \alpha + \tan \alpha|, \text{ as claimed.}$$

38 First note that, for any integer $r \neq 0$,

$$\int_0^{2\pi} \sin rx \, dx = -\dfrac{1}{r} \cos rx \Big|_0^{2\pi} =$$

$$\dfrac{1}{r}(-\cos 2\pi r + \cos 0) = 0, \text{ and } \int_0^{2\pi} \cos rx \, dx =$$

$$\left(\dfrac{1}{r} \sin rx\right)\Big|_0^{2\pi} = \dfrac{1}{r}(\sin 2\pi r - \sin 0) = 0. \text{ For}$$

$r = 0$, $\int_0^{2\pi} \sin rx \, dx = \int_0^{2\pi} 0 \, dx = 0$ and

$\int_0^{2\pi} \cos rx \, dx = \int_0^{2\pi} 1 \, dx = 2\pi$. Thus,

$\int_0^{2\pi} \sin rx \, dx = 0$ for all r, while $\int_0^{2\pi} \cos rx \, dx$

$= \begin{cases} 2\pi, & \text{if } r = 0, \\ 0, & \text{if } r \neq 0. \end{cases}$ These facts will be used repeatedly in what follows.

(a) For $0 \leq m \leq n$, $\int_0^{2\pi} f(x) \cos mx \, dx =$

$$\dfrac{a_0}{2}\left(\int_0^{2\pi} \cos mx \, dx\right) +$$

$$\sum_{k=1}^{n} a_k\left(\int_0^{2\pi} \cos kx \cos mx \, dx\right) +$$

$$\sum_{k=1}^{n} b_k\left(\int_0^{2\pi} \sin kx \cos mx \, dx\right). \text{ The first}$$

integral, $\int_0^{2\pi} \cos mx \, dx$, equals 2π if $m = 0$ and equals 0 otherwise. For $m \geq 1$,

$$\int_0^{2\pi} \cos kx \cos mx \, dx =$$

$$\int_0^{2\pi} \left(\dfrac{1}{2} \cos(k-m)x + \dfrac{1}{2} \cos(k+m)x\right) dx,$$

which equals 0 when $k \neq m$ and π when $k = m$. Hence, $\sum_{k=1}^{n} a_k\left(\int_0^{2\pi} \cos kx \cos mx \, dx\right) =$

$\begin{cases} 0, & \text{if } m = 0, \\ \pi a_m, & \text{if } 1 \leq m \leq n. \end{cases}$ Adding this to

$\frac{a_0}{2}\left(\int_0^{2\pi} \cos mx \, dx\right)$, which equals πa_0 when $m = 0$ and 0 otherwise, we find that

$\frac{a_0}{2}\left(\int_0^{2\pi} \cos mx \, dx\right) +$

$\sum_{k=1}^{n} a_k\left(\int_0^{2\pi} \cos kx \cos mx \, dx\right) = \pi a_m$ for $0 \leq m \leq n$. Similarly, for $1 \leq k \leq n$, we have

$\int_0^{2\pi} \sin kx \cos mx \, dx =$

$\int_0^{2\pi}\left(\frac{1}{2}\sin(k+m)x + \frac{1}{2}\sin(k-m)x\right)dx$

$= 0$. Combining this with the preceding result, we have $\int_0^{2\pi} f(x) \cos mx \, dx = \pi a_m$.

Dividing by π gives the formula for a_m.

(b) Proceed as in (a), evaluating

$\int_0^{2\pi} f(x) \sin mx \, dx = \frac{a_0}{2}\left(\int_0^{2\pi} \sin mx \, dx\right) +$

$\sum_{k=1}^{n} a_k\left(\int_0^{2\pi} \cos kx \sin mx \, dx\right) +$

$\sum_{k=1}^{n} b_k\left(\int_0^{2\pi} \sin kx \sin mx \, dx\right)$, noting this time

that $\int_0^{2\pi} \sin mx \, dx = 0$ and

$\int_0^{2\pi} \cos kx \sin mx \, dx =$

$\int_0^{2\pi}\left(\frac{1}{2}\sin(m+k)x + \frac{1}{2}\sin(m-k)x\right)dx$

$= 0$, while $\int_0^{2\pi} \sin kx \sin mx \, dx =$

$\int_0^{2\pi}\left(\frac{1}{2}\cos(k-m)x - \frac{1}{2}\cos(k+m)x\right)dx =$

π for $k = m$ and 0 otherwise. Hence,

$\sum_{k=1}^{n} b_k\left(\int_0^{2\pi} \sin kx \sin mx \, dx\right) = \pi b_m$.

Combining these results, we get

$\int_0^{2\pi} f(x) \sin mx \, dx = \pi b_m$. Dividing by π gives the formula for b_m.

40 Let $x = 2 \sin \theta$ so $dx = 2 \cos \theta \, d\theta$. Then

$\int \sqrt{4 - x^2} \, dx = \int 2\sqrt{1 - \sin^2 \theta} \, (\cos \theta \, d\theta)$

$= 4 \int \cos^2 \theta \, d\theta = 4\left(\frac{\theta}{2} + \frac{\sin 2\theta}{4}\right) + C$

$= 2\theta + \sin 2\theta + C$

$= 2 \sin^{-1} \frac{x}{2} + \frac{x\sqrt{4 - x^2}}{2} + C$.

42 Let $x = 3 \sec \theta$ so $dx = 3 \sec \theta \tan \theta \, d\theta$. Then

$\int \frac{x^2 \, dx}{\sqrt{x^2 - 9}} = \int \frac{9 \sec^2 \theta \, (3 \sec \theta \tan \theta \, d\theta)}{3 \tan \theta}$

$= 9 \int \sec^3 \theta \, d\theta$

$= 9\left(\frac{\sec \theta \tan \theta}{2} + \frac{1}{2} \ln|\sec \theta + \tan \theta| + C_1\right)$

$= \frac{9}{2} \sec \theta \tan \theta + \frac{9}{2} \ln|\sec \theta + \tan \theta| + 9C_1$

$= \frac{1}{2}x\sqrt{x^2 - 9} + \frac{9}{2} \ln\left|\sqrt{x^2 - 9} + x\right| - \ln 3 + 9C_1$

$= \frac{1}{2}x\sqrt{x^2 - 9} + \frac{9}{2} \ln\left|\sqrt{x^2 - 9} + x\right| + C$.

44 Let $x = 2 \tan \theta$. Then $\int \frac{\sqrt{4 + x^2}}{x} dx$

$= \int \frac{2 \sec \theta}{2 \tan \theta} 2 \sec^2 \theta \, d\theta$

$$= 2\int \frac{\sec^2\theta}{\tan^2\theta}(\sec\theta\tan\theta\,d\theta), \text{ so } u = \sec\theta \text{ leads}$$

$$\text{to } \int \frac{2u^2}{u^2-1}\,du = \int\left(2 + \frac{1}{u-1} - \frac{1}{u+1}\right)du$$

$$= 2u + \ln\left|\frac{u-1}{u+1}\right| + C$$

$$= 2\sec\theta + \ln\left|\frac{\sec\theta-1}{\sec\theta+1}\right| + C$$

$$= \sqrt{4+x^2} + \ln\left(\frac{\sqrt{4+x^2}-2}{\sqrt{4+x^2}+2}\right) + C.$$

46 Let $x = a\sin\theta$. Then $\int \frac{dx}{\sqrt{a^2-x^2}} =$

$$\int \frac{a\cos\theta\,d\theta}{a\cos\theta} = \int d\theta = \theta + C$$

$$= \sin^{-1}\frac{x}{a} + C.$$

48 Let $x = a\sec\theta$. Then $\int \sqrt{x^2-a^2}\,dx =$

$$\int a\tan\theta \cdot a\sec\theta\tan\theta\,d\theta =$$

$$a^2\int \tan^2\theta\sec\theta\,d\theta = a^2\int(\sec^3\theta - \sec\theta)\,d\theta$$

$$=$$

$$a^2\left(\frac{1}{2}\sec\theta\tan\theta + \frac{1}{2}\int\sec\theta\,d\theta - \int\sec\theta\,d\theta\right)$$

$$= a^2\left(\frac{1}{2}\sec\theta\tan\theta - \frac{1}{2}\ln|\sec\theta+\tan\theta|\right) + K$$

$$= \frac{a^2}{2}\left(\frac{x}{a}\cdot\frac{\sqrt{x^2-a^2}}{a} - \ln\left|\frac{x+\sqrt{x^2-a^2}}{a}\right|\right) + K$$

$$= \frac{1}{2}x\sqrt{x^2-a^2} - \frac{1}{2}a^2\ln(x+\sqrt{x^2-a^2}) + C.$$

50 By Exercise 48 with $a = 1$, the integral equals

$$\left(\frac{1}{2}x\sqrt{x^2-1} - \frac{1}{2}\ln(x+\sqrt{x^2-1})\right)\bigg|_{\sqrt{2}}^{2}$$

$$= \frac{1}{2}\ln\left(\frac{1+\sqrt{2}}{2+\sqrt{3}}\right) + \sqrt{3} - \frac{1}{2}\sqrt{2}.$$

52 (a) Let $x = 3\sec\theta$. Then $\int \frac{x^2+\sqrt{x^2-9}}{x}\,dx$

$$= \int \frac{9\sec^2\theta + 3\tan\theta}{3\sec\theta}(3\sec\theta\tan\theta\,d\theta)$$

$$= 3\int(3\sec^2\theta\tan\theta + \tan^2\theta)\,d\theta$$

$$= 3\int \frac{3\sin\theta + \sin^2\theta\cos\theta}{\cos^3\theta}\,d\theta.$$

(b) Let $x = \sqrt{5}\tan\theta$. Then $\int \frac{x^3\sqrt{5+x^2}}{x+2}\,dx =$

$$\int \frac{(5\sqrt{5}\tan^3\theta)(\sqrt{5}\sec\theta)}{\sqrt{5}\tan\theta + 2}(\sqrt{5}\sec^2\theta\,d\theta)$$

$$= 25\sqrt{5}\int \frac{\tan^3\theta\sec^3\theta}{\sqrt{5}\tan\theta + 2}\,d\theta$$

$$= 25\sqrt{5}\int \frac{\sin^3\theta\,d\theta}{\sqrt{5}\sin\theta\cos^5\theta + 2\cos^6\theta}.$$

54 (a) $\int \frac{1+\sin\theta}{1+\cos^2\theta}\,d\theta = \int \frac{1+\dfrac{2u}{1+u^2}}{1+\left(\dfrac{1-u^2}{1+u^2}\right)^2}\cdot\frac{2\,du}{1+u^2}$

$$= 2\int \frac{1+\dfrac{2u}{1+u^2}}{1+u^2+\dfrac{(1-u^2)^2}{1+u^2}}\,du$$

7.6 Special Techniques

$$= 2\int \frac{1 + 2u + u^2}{(1+u^2)^2 + (1-u^2)^2} \, du$$

$$= \int \frac{u^2 + 2u + 1}{1 + u^4} \, du$$

(b) $\int \frac{5 + \cos\theta}{(\sin\theta)^2 + \cos\theta} \, d\theta$

$$= \int \frac{5 + \frac{1-u^2}{1+u^2}}{\left(\frac{2u}{1+u^2}\right)^2 + \frac{1-u^2}{1+u^2}} \cdot \frac{2\,du}{1+u^2}$$

$$= 2\int \frac{5 + \frac{1-u^2}{1+u^2}}{\frac{4u^2}{1+u^2} + 1 - u^2} \, du$$

$$= 2\int \frac{5 + 5u^2 + 1 - u^2}{4u^2 + 1 - u^4} \, du$$

$$= 4\int \frac{3 + 2u^2}{1 + 4u^2 - u^4} \, du$$

(c) $\int_0^{\pi/2} \frac{5\,d\theta}{2\cos\theta + 3\sin\theta}$

$$= \int_0^1 \frac{5}{2\left(\frac{1-u^2}{1+u^2}\right) + 3\left(\frac{2u}{1+u^2}\right)} \cdot \frac{2\,du}{1+u^2}$$

$$= 10\int_0^1 \frac{du}{2(1-u^2) + 6u} = 5\int_0^1 \frac{du}{1 + 3u - u^2}$$

56 Any rational function of $\tan\theta$ and $\sec\theta$ can be transformed into a rational function of $\cos\theta$ and $\sin\theta$ by multiplying the numerator and denominator by an appropriate power of $\cos\theta$. Any rational function of $\cos\theta$ and $\sin\theta$ can be transformed into a rational function of u through the substitution $u = \tan(\theta/2)$ and multiplication of the numerator and denominator by an appropriate power of $1 + u^2$, and we know that every rational function of u has an elementary antiderivative.

58 Let $u = \sqrt{x+a}$, so $u^2 - a = x$, $dx = 2u\,du$, and $\sqrt{x+b} = \sqrt{u^2 - (a-b)}$. (Without loss of generality, we may assume $a > b$.) Now

$$\int R(x, \sqrt{x+a}, \sqrt{x+b})\,dx =$$

$$\int R(u^2 - a, u, \sqrt{u^2 - (a-b)})\,2u\,du. \text{ Now let } u =$$

$\sqrt{a-b}\sec\theta$, so

$$\int R(u^2 - a, u, \sqrt{u^2 - (a-b)})\,2u\,du$$

$$= \int R(u^2 - a, u, \sqrt{u^2 - (a-b)})\,2u\,du$$

$$= \int R((a-b)\sec^2\theta - a, \sqrt{a-b}\sec\theta,$$

$\sqrt{a-b}\tan\theta)(2(a-b)\sec^2\theta\tan\theta\,d\theta)$, which is the integral of a rational function of $\tan\theta$ and $\sec\theta$. We know from Exercise 56 that this integral has an elementary antiderivative, so

$\int R(x, \sqrt{x+a}, \sqrt{x+b})\,dx$ is elementary as well.

60 (a) The radius of the circle containing the arc CD is $\cos\theta$. Hence $\widehat{CD} = \widehat{AB}\cos\theta$, so $\overline{C'D'} = \overline{A'B'} = \widehat{AB} = (\sec\theta)\widehat{CD}$.

(b) Locally, the map stretches vertical distance by the factor $F'(\theta)$. By (a) it stretches horizontal distance by the factor $\sec\theta$. Hence $F'(\theta) = \sec\theta$.

(c) $F(\alpha) = \int_0^\alpha F'(\theta)\,d\theta = \int_0^\alpha \sec\theta\,d\theta$

(d) For some constant c, the distance between

latitude lines at angles α and β is

$$c \int_\alpha^\beta \sec\theta\, d\theta = c \ln(\sec\theta + \tan\theta)\Big|_\alpha^\beta.$$

Since $3 = c \ln(\sec\theta + \tan\theta)\Big|_0^{\pi/6} = c \ln\sqrt{3}$,

we have $c = \dfrac{6}{\ln 3}$. Hence the desired distance

is $c \ln(\sec\theta + \tan\theta)\Big|_{\pi/6}^{\pi/3} =$

$$\frac{6}{\ln 3}(\ln(2+\sqrt{3}) - \ln\sqrt{3}) = \frac{6\ln(2+\sqrt{3})}{\ln 3} - 3$$

≈ 4.1925 inches.

(e) No. See Exercise 18.

7.7 What to Do in the Face of an Integral

2 Division, power rule, f'/f

4 Factor and cancel, power rule

6 Integration by parts

8 Integration by parts

10 Simplify the integrand to $2\csc 2\theta$. Or rewrite as
$\int \dfrac{1}{\sin^2\theta \cos\theta} \sin\theta\, d\theta$ and let $u = \cos\theta$.

12 Substitution ($u = x^3 + 2$), power rule

14 Substitution ($u = \cos\theta$), power rule

16 Integrand equals $\cos^2\theta = \dfrac{1}{2}(1 + \cos 2\theta)$

18 Substitution ($u = \sqrt{x}$), integration by parts

20 Power rule

22 Substitution ($x = e^u$), division, f'/f, power rule

24 Substitution ($u = 3 + \sin x$), power rule

26 Simplify, integration by parts

28 Partial fractions

30 Substitution ($u = 3 + \sqrt{x}$), power rule

32 Substitution ($u = e^x$), division, partial fractions

34 Make the substitution $u = \sqrt{x+2} - 1$. The integral can then be evaluated using the power rule and $\int \dfrac{du}{u} = \ln|u| + C$.

36 Factor $x^2 - 9$ as $(x-3)(x+3)$ and use partial fractions.

38 Substitution ($u = 3x + 2$), power rule

40 Substitution ($u = 1 + 2^x$)

42 Break in two, f'/f in one part, substitution ($x = \tan u$) in other

44 Substitution ($u = 1 + x^2$), power rule

46 Substitution ($u = x^2 - 1$), power rule

48 The substitution $u = \tan^{-1} x$ yields the integral
$\int u\, du$.

50 The substitution $u = \ln x$, $du = \dfrac{dx}{x}$ produces the integral $\int \sin u\, du$.

52 The substitution $u = x^2 + 4$, $du = 2x\, dx$ yields
$\dfrac{1}{2}\int \dfrac{du}{u^{1/2}}$. Use the power rule.

54 Write numerator as $\dfrac{1}{2}(2x+1) - \dfrac{1}{2}$, f'/f, complete the square, substitute.

56 Power rule

58 Substitution ($u = x^2 + 9$), power rule

60 (a) This integral can be evaluated when n is 1 or 2.

(b) When $n = 1$, the integral is $\int \sqrt{1+x}\, dx$. Let $u = 1 + x$, $du = dx$. $\int \sqrt{1+x}\, dx =$
$\int u^{1/2}\, du = \dfrac{2}{3}u^{3/2} + C = \dfrac{2}{3}(1+x)^{3/2} + C$.

When $n = 2$, the integral is $\int \sqrt{1+x^2}\, dx$.

This integral was shown in Example 6 in Sec 7.6 to be $\dfrac{x\sqrt{1+x^2}}{2} + \dfrac{1}{2}\ln\left(\sqrt{1+x^2}+x\right) + C$.

62 (a) The integral can be evaluated when $n = 1$ or 2.

(b) $n = 1$: Make the substitution $u = \sqrt{1-x}$, $x = 1 - u^2$, $dx = -2u\, du$. Then

$$\int (1+x)\sqrt{1-x}\, dx = \int (2-u^2)u(-2u\, du)$$

$$= \int (2u^4 - 4u^2)\, du = \frac{2}{5}u^5 - \frac{4}{3}u^3 + C$$

$$= \frac{2}{5}(1-x)^{5/2} - \frac{4}{3}(1-x)^{3/2} + C.$$

$n = 2$: $\int (1+x)^{1/2}\sqrt{1-x}\, dx =$

$\int \sqrt{1-x^2}\, dx$. By Exercise 45 of Sec. 7.6 (with $a = 1$), this equals

$\dfrac{1}{2}\left(\sin^{-1} x + x\sqrt{1-x^2}\right) + C.$

64 $\int \sqrt{1-\cos x}\, dx = \int \sqrt{2\sin^2 \dfrac{x}{2}}\, dx$

$= \pm\int \sqrt{2}\sin\dfrac{x}{2}\, dx = \pm\sqrt{2}\left(-2\cos\dfrac{x}{2}\right) + C$

$= \pm 2\sqrt{2}\cos\dfrac{x}{2} + C = \pm 2\sqrt{1+\cos x} + C.$ (The antiderivative's sign is chosen to be consistent with the application at hand.)

66 Let $x = 3\sin\theta$. Then $\int \dfrac{dx}{\sqrt{9-x^2}} = \int d\theta$

$= \theta + C = \sin^{-1}\dfrac{x}{3} + C.$

68 Let $x = 3\tan\theta$. Then $\int \dfrac{dx}{x\sqrt{x^2+9}} =$

$\dfrac{1}{3}\int \dfrac{\sec\theta}{\tan\theta}\, d\theta = \dfrac{1}{3}\int \csc\theta\, d\theta =$

$\dfrac{1}{3}\ln|\csc\theta - \cot\theta| + C = \dfrac{1}{3}\ln\left|\dfrac{\sqrt{x^2+9}-3}{x}\right| + C.$

70 Note that $\int \dfrac{dx}{x+\sqrt{x^2+25}}$

$= \int \dfrac{x-\sqrt{x^2+25}}{(x+\sqrt{x^2+25})(x-\sqrt{x^2+25})}\, dx$

$= \dfrac{1}{25}\int \left(\sqrt{x^2+25}-x\right)\, dx.$ Now let $x = 5\tan\theta.$

Then $\dfrac{1}{25}\int \left(\sqrt{x^2+25}-x\right)\, dx =$

$\dfrac{1}{25}\int (5\sec\theta - 5\tan\theta)(5\sec^2\theta\, d\theta)$

$= \int (\sec^3\theta - \tan\theta\sec^2\theta)\, d\theta =$

$\dfrac{1}{2}\sec\theta\tan\theta + \dfrac{1}{2}\ln|\sec\theta + \tan\theta| - \dfrac{1}{2}\tan^2\theta + C_1$

$= \dfrac{1}{50}x\sqrt{x^2+25} + \dfrac{1}{2}\ln\left(\sqrt{x^2+25}+x\right) - \dfrac{x^2}{50} + C.$

7.S Review Exercises

2 (a) $u = \ln x$, $dv = dx$. $\int \ln x\, dx =$

$(\ln x)x - \int x\dfrac{dx}{x} = x\ln x - x + C.$

(b) $\int \dfrac{\ln x\, dx}{x} = \int (\ln x)\dfrac{dx}{x} = \dfrac{1}{2}(\ln x)^2 + C$

(c) Not elementary.

4 (a) $\int \frac{e^x \, dx}{5e^{2x} - 3} = \int \frac{du}{5u^2 - 3}$

$= \frac{1}{2\sqrt{-5(-3)}} \ln \left| \frac{u\sqrt{5} - \sqrt{-(-3)}}{u\sqrt{5} + \sqrt{-(-3)}} \right| + C =$

$\frac{1}{2\sqrt{15}} \ln \left| \frac{e^x \sqrt{5} - \sqrt{3}}{e^x \sqrt{5} + \sqrt{3}} \right| + C$, using Formula 27

with $a = 5$ and $c = -3$.

(b) By Formula 33 on the inside front cover of the text with $p = \sqrt{3}$, $\int \frac{dx}{\sqrt{x^2 - 3}} =$

$\ln \left| x + \sqrt{x^2 - 3} \right| + C.$

6 $\int \frac{5x^4 - 5x^3 + 10x^2 - 8x + 4}{(x^2 + 1)(x - 1)} \, dx$

$= \int \left(5x + \frac{5x^2 - 3x + 4}{(x^2 + 1)(x - 1)} \right) dx$

$= \int \left(5x + \frac{2x - 1}{x^2 + 1} + \frac{3}{x - 1} \right) dx = \frac{5}{2}x^2 +$

$\ln(x^2 + 1) - \tan^{-1} x + 3 \ln |x - 1| + C$

8 Let $u = x + 3$. Then $\int_{-1}^{4} \frac{x + 2}{\sqrt{x + 3}} \, dx$

$= \int_{2}^{7} \frac{u - 1}{\sqrt{u}} \, du = \int_{2}^{7} (u^{1/2} - u^{-1/2}) \, du$

$= \left(\frac{2}{3} u^{3/2} - 2u^{1/2} \right) \Big|_{2}^{7} = \frac{1}{3}(8\sqrt{7} + 2\sqrt{2})$

12 $\frac{x^4 + 2x^2 - 2x + 2}{x^3 - 1} = x + \frac{2x^2 - x + 2}{x^3 - 1}$

$= x + \frac{x - 1}{x^2 + x + 1} + \frac{1}{x - 1}$

14 $\frac{x^4 + 3x^3 - 2x^2 + 3x - 1}{x^4 - 1} = 1 + \frac{3x^3 - 2x^2 + 3x}{x^4 - 1}$

$= 1 - \frac{1}{x^2 + 1} + \frac{2}{x + 1} + \frac{1}{x - 1}$

16 The degree of the numerator is greater than the degree of the denominator, so division is necessary. This yields $\frac{5x^3 + 11x^2 + 6x + 1}{x^2 + x} =$

$5x + 6 + \frac{1}{x^2 + x}$. Since $x^2 + x = x(x + 1)$, the

partial-fraction representation of $\frac{1}{x^2 + x}$ has the

form $\frac{c_1}{x} + \frac{c_2}{x + 1}$. Successively letting $x = 0$ and

-1 in $1 = c_1(x + 1) + c_2 x$, we obtain $c_1 = 1$ and

$c_2 = -1$. Thus, $\frac{5x^3 + 11x^2 + 6x - 1}{x^2 + x} =$

$5x + 6 + \frac{1}{x} - \frac{1}{x + 1}$.

18 (a) $\int_0^1 \sqrt[3]{x} \sqrt{x} \, dx = \int_0^1 x^{1/3} x^{1/2} \, dx = \int_0^1 x^{5/6} \, dx$

$= \frac{x^{11/6}}{11/6} \Big|_0^1 = \frac{6}{11}$

(b) Not elementary

20 (a) Let $x = u^3$. Then $\int \frac{x^{2/3}}{x + 1} \, dx =$

$\int \frac{u^2}{u^3 + 1} \cdot 3u^2 \, du$

$= \int \left(3u + \frac{1}{u + 1} - \frac{u + 1}{u^2 - u + 1} \right) du = \frac{3}{2} u^2$

7.5 Summary: Review Exercises

$$+ \ln|u+1| - \frac{1}{2}\int \frac{2u-1}{u^2-u+1}\,du -$$

$$\frac{3}{2}\int \frac{du}{\left(u-\frac{1}{2}\right)^2 + \frac{3}{4}} = \frac{3}{2}u^2 + \ln|u+1| -$$

$$\frac{1}{2}\ln(u^2-u+1) - \sqrt{3}\tan^{-1}\frac{2u-1}{\sqrt{3}} + C$$

$$= \frac{3}{2}x^{2/3} + \ln|x^{1/3}+1| -$$

$$\frac{1}{2}\ln(x^{2/3} - x^{1/3} + 1) - \sqrt{3}\tan^{-1}\left(\frac{2x^{1/3}-1}{\sqrt{3}}\right)$$

$$+ C.$$

22. $\int_0^1 (e^x+1)^3 e^x\,dx = \frac{1}{4}(e^x+1)^4 \Big|_0^1$

$$= \frac{1}{4}(e+1)^4 - 4$$

24. $\int_1^e \frac{\sqrt{\ln x}}{x}\,dx = \frac{2}{3}(\ln x)^{3/2}\Big|_1^e = \frac{2}{3}$

26. (a) Let $u = \cos\theta$; $\int \sin^5\theta\,d\theta =$

$$\int (1-\cos^2\theta)^2 \sin\theta\,d\theta$$

$$= \int (1-u^2)^2(-du)$$

$$= \int (-1 + 2u^2 - u^4)\,du$$

$$= -u + \frac{2}{3}u^3 - \frac{1}{5}u^5 + C$$

$$= -\cos\theta + \frac{2}{3}\cos^3\theta - \frac{1}{5}\cos^5\theta + C.$$ By

the recursion in Exercise 29 of Sec. 7.6,

$$\int \tan^6\theta\,d\theta = \frac{1}{5}\tan^5\theta - \int \tan^4\theta\,d\theta$$

$$= \frac{1}{5}\tan^5\theta - \frac{1}{3}\tan^3\theta + \int \tan^2\theta\,d\theta$$

$$= \frac{1}{5}\tan^5\theta - \frac{1}{3}\tan^3\theta + \tan\theta - \theta + C.$$

28. $\int \cot 3\theta\,d\theta = \frac{1}{3}\ln|\sin 3\theta| + C$

30. (a) $\int \sec^5 x \tan x\,dx = \int \sec^4 x\,(\sec x \tan x)\,dx;$

with $u = \sec x$, $du = \sec x \tan x\,dx$, this

becomes $\int u^4\,du = \frac{u^5}{5} + C =$

$$\frac{1}{5}\sec^5 x + C.$$

(b) Let $u = \cos x$, $du = -\sin x\,dx$; then

$$\int \frac{\sin x}{(\cos x)^3}\,dx = -\int \frac{du}{u^3} = \frac{1}{2}u^{-2} + C$$

$$= \frac{1}{2}\sec^2 x + C.$$

32. (a) Use $u = \frac{3}{4}x^2$ and $u = \tan\theta$ in turn:

$$\int \frac{x\,dx}{\sqrt{9x^4+16}} = \int \frac{\frac{2}{3}\,du}{4\sqrt{u^2+1}}$$

$$= \frac{1}{6}\int \frac{du}{\sqrt{u^2+1}} = \frac{1}{6}\int \frac{\sec^2\theta\,d\theta}{\sec\theta}$$

$$= \frac{1}{6}\int \sec\theta\,d\theta = \frac{1}{6}\ln|\sec\theta + \tan\theta| + K$$

$$= \frac{1}{6}\ln|u + \sqrt{u^2+1}| + K$$

$$= \frac{1}{6}\ln\left(\frac{3}{4}x^2 + \sqrt{\frac{9}{16}x^4+1}\right) + K$$

$$= \frac{1}{6} \ln\left(3x^2 + \sqrt{9x^4 + 16}\right) + C.$$

34. (a) If $u = \sqrt{1+x}$, then $x = u^2 - 1$ and $dx = 2u\, du$. Thus $\int x\sqrt{1+x}\, dx =$

$$\int (u^2 - 1)u(2u\, du) = 2\int (u^4 - u^2)\, du$$

$$= \frac{2}{5}u^5 - \frac{2}{3}u^3 + C$$

$$= \frac{2}{5}(1+x)^{5/2} - \frac{2}{3}(1+x)^{3/2} + C.$$

(b) If $x = \tan^2 \theta$, then $dx = 2\tan\theta \sec^2\theta\, d\theta$ and $\sqrt{1+x} = \sec\theta$, so $\int x\sqrt{1+x}\, dx =$

$$2\int \tan^3\theta \sec^3\theta\, d\theta$$

$$= 2\int (\sec^2\theta - 1)\sec^3\theta \tan\theta\, d\theta$$

$$= 2\int (\sec^4\theta - \sec^2\theta)\sec\theta \tan\theta\, d\theta. \text{ Let}$$

$u = \sec\theta$ and $du = \sec\theta\tan\theta\, d\theta$. The integral then becomes $2\int (u^4 - u^2)\, du =$

$$\frac{2}{5}u^5 - \frac{2}{3}u^3 + C$$

$$= \frac{2}{5}(1+x)^{5/2} - \frac{2}{3}(1+x)^{3/2} + C.$$

(c) If $u = x$ and $dv = \sqrt{1+x}\, dx$, then $du = dx$, $v = \frac{2}{3}(1+x)^{3/2}$, and $\int x\sqrt{1+x}\, dx =$

$$\frac{2}{3}x(1+x)^{3/2} - \frac{2}{3}\int (1+x)^{3/2}\, dx$$

$$= \frac{2}{3}x(1+x)^{3/2} - \frac{4}{15}(1+x)^{5/2} + C$$

$$= \frac{2}{3}(1+x)^{3/2}\left[x - \frac{2}{5}(1+x)\right] + C$$

$$= \frac{2}{3}(1+x)^{3/2}\left[\frac{3}{5}x - \frac{2}{5}\right] + C$$

$$= \frac{2}{3}(1+x)^{3/2}\left[\frac{3}{5}(1+x) - 1\right] + C$$

$$= \frac{2}{5}(1+x)^{5/2} - \frac{2}{3}(1+x)^{3/2} + C.$$

36. Make the substitution $x = 5\tan\theta$, then $dx = 5\sec^2\theta\, d\theta$ and $\int \frac{dx}{x^2\sqrt{x^2+25}} =$

$$\int \frac{5\sec^2\theta\, d\theta}{(25\tan^2\theta)(5\sec\theta)} = \frac{1}{25}\int \cot\theta \csc\theta\, d\theta$$

$$= -\frac{1}{25}\csc\theta + C = -\frac{\sqrt{x^2+25}}{25x} + C.$$

38. Let $x = 2\tan\theta$. Then $\int \frac{dx}{x\sqrt{4+x^2}} =$

$$\int \frac{2\sec^2\theta\, d\theta}{2\tan\theta\,(2\sec\theta)} = \frac{1}{2}\int \csc\theta\, d\theta$$

$$= \frac{1}{2}\ln|\csc\theta - \cot\theta| + C_1$$

$$= \frac{1}{2}\ln\left|\frac{\sqrt{x^2+4}}{x} - \frac{2}{x}\right| + C_1$$

$$= \frac{1}{2}\ln\left|\frac{\sqrt{x^2+4} - 2}{x}\right| + C.$$

40. (a) We are given that $dv = \frac{x\, dx}{\sqrt{1+x^2}}$, so let $u = x^2$, $du = 2x\, dx$, $v = \sqrt{1+x^2}$; then

$$\int \frac{x^3}{\sqrt{1+x^2}}\, dx$$

$$= x^2\sqrt{1+x^2} - \int 2x\sqrt{1+x^2}\, dx.$$

7.S Summary: Review Exercises

(b) The substitution $x = \tan \theta$, $dx = \sec^2 \theta \, d\theta$

results in $\sqrt{1 + x^2} = \sec \theta$ and $\int \dfrac{x^3}{1 + x^2} \, dx$

$= \int \dfrac{\tan^3 \theta}{\sec \theta} \sec^2 \theta \, d\theta = \int \tan^3 \theta \sec \theta \, d\theta.$

(c) If $u = \sqrt{1 + x^2}$, then $u^2 = 1 + x^2$, $2x \, dx = 2u \, du$, $x \, dx = u \, du$, and $\int \dfrac{x^3}{\sqrt{1 + x^2}} \, dx =$

$\int \dfrac{x^2 (x \, dx)}{\sqrt{1 + x^2}} = \int \dfrac{(u^2 - 1)(u \, du)}{u}$

$= \int (u^2 - 1) \, du.$

42. (a) $\dfrac{d}{dx}\left(\dfrac{1}{a} \sin^{-1} ax + C\right) = \dfrac{1}{a} \cdot \dfrac{a}{\sqrt{1 - (ax)^2}}$

$= \dfrac{1}{\sqrt{1 - a^2 x^2}}$

(b) $a = 5:\ \dfrac{1}{5} \sin^{-1} 5x + C$

(c) $a = \sqrt{3}:\ \dfrac{1}{\sqrt{3}} \sin^{-1}(\sqrt{3} x) + C$

44. $\int \dfrac{x^2 \, dx}{x^4 - 1} = \int \left(\dfrac{1/2}{x^2 + 1} - \dfrac{1/4}{x + 1} + \dfrac{1/4}{x - 1}\right) dx$

$= \dfrac{1}{2} \tan^{-1} x - \dfrac{1}{4} \ln|x + 1| + \dfrac{1}{4} \ln|x - 1| + C$

$= \dfrac{1}{2} \tan^{-1} x + \dfrac{1}{4} \ln\left|\dfrac{x - 1}{x + 1}\right| + C$

46. $\int (x^{1/4} + x^{9/20}) \, dx = \dfrac{4}{5} x^{5/4} + \dfrac{20}{29} x^{29/20} + C$

48. $\int \dfrac{dx}{x^3 + 4x} = \dfrac{1}{4} \int \left(\dfrac{1}{x} - \dfrac{x}{x^2 + 4}\right) dx$

$= \dfrac{1}{4} \ln|x| - \dfrac{1}{8} \ln(x^2 + 4) + C$

$= \dfrac{1}{8} \ln\left(\dfrac{x^2}{x^2 + 4}\right) + C$

50. $\int \tan x \, dx = -\ln|\cos x| + C$

52. $\int \dfrac{3 \, dx}{x^2 + 4x - 5} = \int \left(\dfrac{1/2}{x - 1} - \dfrac{1/2}{x + 5}\right) dx$

$= \dfrac{1}{2} \ln\left|\dfrac{x - 1}{x + 5}\right| + C$

54. $u = \ln(2x - 1)$, $dv = dx$; $\int \ln\sqrt{2x - 1} \, dx =$

$\dfrac{1}{2} \int \ln(2x - 1) \, dx$

$= \dfrac{1}{2} \ln(2x - 1) x - \dfrac{1}{2} \int x \cdot \dfrac{2}{2x - 1} \, dx$

$= \dfrac{x}{2} \ln(2x - 1) - \dfrac{1}{2} \int \left(1 + \dfrac{1}{2x - 1}\right) dx$

$= \dfrac{x}{2} \ln(2x - 1) - \dfrac{x}{2} - \dfrac{1}{4} \ln(2x - 1) + C$

$= \dfrac{1}{4}(2x - 1) \ln(2x - 1) - \dfrac{x}{2} + C$

56. $u = \tan^{-1} x$, $dv = x^3 \, dx$; $\int (\tan^{-1} x) x^3 \, dx =$

$(\tan^{-1} x) \dfrac{x^4}{4} - \int \dfrac{x^4}{4} \cdot \dfrac{dx}{1 + x^2}$

$= \dfrac{x^4}{4} \tan^{-1} x - \dfrac{1}{4} \int \left(x^2 - 1 + \dfrac{1}{1 + x^2}\right) dx$

$= \dfrac{x^4}{4} \tan^{-1} x - \dfrac{1}{12} x^3 + \dfrac{1}{4} x - \dfrac{1}{4} \tan^{-1} x + C$

$= \dfrac{1}{4}(x^4 - 1) \tan^{-1} x - \dfrac{1}{12} x^3 + \dfrac{1}{4} x + C$

58 $\int \frac{x+2}{x^2+1} dx = \frac{1}{2} \int \frac{2x\, dx}{x^2+1} + 2 \int \frac{dx}{x^2+1}$

$= \frac{1}{2} \ln(x^2+1) + 2 \tan^{-1} x + C$

60 $\int \sqrt[3]{4x+7}\, dx = \frac{1}{4} \cdot \frac{(4x+7)^{4/3}}{4/3} + C$

$= \frac{3}{16}(4x+7)^{4/3} + C$

62 $\int \frac{\ln x^4}{x} dx = 4 \int (\ln x) \frac{dx}{x} = 2(\ln x)^2 + C$

64 $\int \frac{\tan^{-1} 3x}{1+9x^2} dx = \frac{1}{6}(\tan^{-1} 3x)^2 + C$

66 Let $x = e^t$; then $\int \sin(\ln x)\, dx = \int e^t \sin t\, dt$.

Now let $u = e^t$ and $dv = \sin t\, dt$; then

$\int e^t \sin t\, dt = -e^t \cos t + \int e^t \cos t\, dt$. Next let

$U = e^t$ and $dV = \cos t\, dt$; then $\int e^t \sin t\, dt =$

$-e^t \cos t + e^t \sin t - \int e^t \sin t\, dt$, so

$\int e^t \sin t\, dt = \frac{1}{2} e^t (\sin t - \cos t) + C$

$= \frac{x}{2}(\sin(\ln x) - \cos(\ln x)) + C$.

68 $\int \frac{2+\sqrt[3]{x}}{x} dx = \int \left(\frac{2}{x} + x^{-2/3}\right) dx$

$= 2 \ln|x| + 3x^{1/3} + C$

70 $\ln(x^2 + 3x + 5) + C$

72 $\int \frac{3\, dx}{\sqrt{1-5x^2}} = \frac{3}{\sqrt{5}} \sin^{-1}(\sqrt{5} x) + C$

74 $u = \ln(4+x^2)$, $dv = dx$; $\int \ln(4+x^2)\, dx$

$= \ln(4+x^2)x - \int x \cdot \frac{2x}{4+x^2} dx$

$= x \ln(4+x^2) - \int \left(2 - \frac{8}{4+x^2}\right) dx$

$= x \ln(4+x^2) - 2x + 4 \tan^{-1} \frac{x}{2} + C$

76 $\int \sqrt{\tan \theta}\, \sec^2 \theta\, d\theta = \frac{2}{3}(\tan \theta)^{3/2} + C$

78 $\int \tan^6 \theta\, d\theta =$

$\frac{1}{5} \tan^5 \theta - \frac{1}{3} \tan^3 \theta + \tan \theta - \theta + C$, by three

applications of the recursion in Exercise 29 of Sec. 7.6.

80 $\frac{1}{2} e^{2x} + 2e^x + x + C$

82 $\int (x + x^{3/2})\, dx = \frac{1}{2} x^2 + \frac{2}{5} x^{5/2} + C$

84 $\int e^{7x}\, dx = \frac{1}{7} e^{7x} + C$

86 Let $u = \sqrt{x} + 1$. Then $\int \frac{dx}{(\sqrt{x}+1)\sqrt{x}}$

$= \int \frac{1}{u} \cdot 2\, du = 2 \ln|u| + C$

$= 2 \ln(\sqrt{x} + 1) + C$.

88 Let $x = \sin \theta$; then $\int x \sin^{-1} x\, dx =$

$\int \theta \sin \theta \cos \theta\, d\theta + C = \frac{1}{2} \int \theta (\sin 2\theta\, d\theta)$

$= \frac{1}{2} \theta\left(-\frac{1}{2} \cos 2\theta\right) - \frac{1}{2} \int \left(-\frac{1}{2} \cos 2\theta\right) d\theta$

$= -\frac{1}{4} \theta \cos 2\theta + \frac{1}{8} \sin 2\theta + C$

$$= -\frac{1}{4}\theta(1 - 2\sin^2\theta) + \frac{1}{4}\sin\theta\cos\theta + C$$

$$= \frac{1}{4}(2x^2 - 1)\sin^{-1}x + \frac{1}{4}x\sqrt{1-x^2} + C. \text{ (We}$$

used parts with $u = \theta$ and $dv = \sin 2\theta\, d\theta$.)

90 Let $u = x^2 + 1$, $du = 2x\, dx$. Then $\int \dfrac{x^7\, dx}{\sqrt{x^2 + 1}}$

$$= \int \frac{(x^2)^3\, x\, dx}{\sqrt{x^2 + 1}} = \frac{1}{2} \int \frac{(u - 1)^3\, du}{u^{1/2}}$$

$$= \frac{1}{2} \int \frac{u^3 - 3u^2 + 3u - 1}{u^{1/2}}\, du$$

$$= \frac{1}{2} \int (u^{5/2} - 3u^{3/2} + 3u^{1/2} - u^{-1/2})\, du$$

$$= \frac{1}{2}\left(\frac{2}{7}u^{7/2} - 3\cdot\frac{2}{5}u^{5/2} + 3\cdot\frac{2}{3}u^{3/2} - 2u^{1/2}\right) + C$$

$$= \frac{1}{7}(x^2 + 1)^{7/2} - \frac{3}{5}(x^2 + 1)^{5/2} + (x^2 + 1)^{3/2} -$$

$(x^2 + 1)^{1/2} + C.$

92 Let $u = x^2$. Then $\int x\sqrt{x^4 - 1}\, dx =$

$$\frac{1}{2}\int \sqrt{u^2 - 1}\, du$$

$$= \frac{1}{2}\left[\frac{1}{2}u\sqrt{u^2 - 1} - \frac{1}{2}\ln\left(u + \sqrt{u^2 - 1}\right)\right] + C$$

$$= \frac{1}{4}x^2\sqrt{x^4 - 1} - \frac{1}{4}\ln\left(x^2 + \sqrt{x^4 - 1}\right) + C, \text{ by}$$

Exercise 48 of Sec. 7.6.

94 Let $x = 3\tan\theta$. Then $\int \dfrac{dx}{\sqrt{9 + x^2}} =$

$$\int \frac{3\sec^2\theta\, d\theta}{3\sec\theta} = \int \sec\theta\, d\theta =$$

$$\ln|\sec\theta + \tan\theta| + K = \ln\left|\frac{x}{3} + \frac{\sqrt{9 + x^2}}{3}\right| + K$$

$$= \ln\left(x + \sqrt{9 + x^2}\right) + C.$$

96 $\dfrac{1}{2}\tan^{-1} 2x + C$

98 $\displaystyle\int \frac{x\, dx}{x^3 + 1} = \frac{1}{3}\int\left(\frac{x + 1}{x^2 - x + 1} - \frac{1}{x + 1}\right) dx =$

$$\frac{1}{6}\int \frac{2x - 1}{x^2 - x + 1}\, dx + \frac{1}{2}\int \frac{dx}{(x - 1/2)^2 + 3/4}$$

$$- \frac{1}{3}\int \frac{dx}{x + 1} = \frac{1}{6}\ln(x^2 - x + 1) +$$

$$\frac{1}{\sqrt{3}}\tan^{-1}\frac{2x - 1}{\sqrt{3}} - \frac{1}{3}\ln|x + 1| + C$$

100 Let $u = 2x + 1$; then $\int \dfrac{x^2\, dx}{\sqrt{2x + 1}}$

$$= \frac{1}{8}\int (u^{3/2} - 2u^{1/2} + u^{-1/2})\, du$$

$$= \frac{1}{8}\left(\frac{2}{5}u^{5/2} - \frac{4}{3}u^{3/2} + 2u^{1/2}\right) + C$$

$$= \frac{1}{120}(6u^2 - 20u + 30)\sqrt{u} + C =$$

$$\frac{1}{120}[6(2x + 1)^2 - 20(2x + 1) + 30]\sqrt{2x + 1} + C$$

$$= \frac{1}{15}(3x^2 - 2x + 2)\sqrt{2x + 1} + C.$$

102 $\displaystyle\int (x + \sin x)^2\, dx = \int (x^2 + 2x\sin x + \sin^2 x)\, dx$

$$= \frac{1}{3}x^3 + 2(-x\cos x + \sin x) +$$

$$\frac{x}{2} - \frac{\sin x \cos x}{2} + C = \frac{1}{3}x^3 + \frac{1}{2}x + 2\sin x$$

$-2x\cos x - \frac{1}{4}\sin 2x + C$. (The results of Example 4 in Sec. 7.3 and Example 2 in Sec. 7.6 were used.)

104 Let $u = \cos 3x$; then $\int \sin^3 3x \cos^2 3x \, dx$

$= \int (1 - \cos^2 3x) \cos^2 3x \sin 3x \, dx$

$= \int (1 - u^2) u^2 \left(-\frac{1}{3} du\right) = \int \left(\frac{1}{3}u^4 - \frac{1}{3}u^2\right) du$

$= \frac{1}{15}u^5 - \frac{1}{9}u^3 + C$

$= \frac{1}{15}\cos^5 3x - \frac{1}{9}\cos^3 3x + C$.

106 $\int \cos^3 x \sin^2 x \, dx$

$= \int (1 - \sin^2 x) \sin^2 x \cdot \cos x \, dx$

$= \int (\sin^2 x - \sin^4 x) \cos x \, dx$

$= \frac{1}{3}\sin^3 x - \frac{1}{5}\sin^5 x + C$

108 Make the substitution $x = 2\tan\theta$. Then $4 + x^2 = 4\sec^2\theta$ and $dx = 2\sec^2\theta \, d\theta$.

Thus, $\int \frac{dx}{(4 + x^2)^2} =$

$\int \frac{2\sec^2\theta \, d\theta}{(4\sec^2\theta)^2} = \frac{1}{8}\int \frac{d\theta}{\sec^2\theta} = \frac{1}{8}\int \cos^2\theta \, d\theta$

$= \frac{1}{8}\int \frac{1}{2}(1 + \cos 2\theta) \, d\theta$

$= \frac{1}{16}\left(\theta + \frac{\sin 2\theta}{2}\right) + C$

$= \frac{1}{16}(\theta + \sin\theta\cos\theta) + C$. From the triangle, we see that we have

$\frac{1}{16}\left(\tan^{-1}\frac{x}{2} + \frac{x}{\sqrt{4 + x^2}} \cdot \frac{2}{\sqrt{4 + x^2}}\right) + C$

$= \frac{1}{16}\left(\tan^{-1}\frac{x}{2} + \frac{2x}{4 + x^2}\right) + C$

110 $\int \frac{x^3 \, dx}{x^4 - 1} = \frac{1}{4}\ln|x^4 - 1| + C$

112 $\int \frac{dx}{x^2 + 5x - 6} = \frac{1}{7}\int \left(\frac{1}{x - 1} - \frac{1}{x + 6}\right) dx$

$= \frac{1}{7}\ln\left|\frac{x - 1}{x + 6}\right| + C$

114 Make the trigonometric substitution $x = \tan\theta$, $dx = \sec^2\theta \, d\theta$. Then $\sqrt{x^2 + 1} = \sec\theta$ and

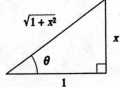

$\int \frac{\sqrt{x^2 + 1}}{x^4} dx = \int \frac{\sec\theta}{\tan^4\theta}\sec^2\theta \, d\theta =$

$\int \frac{\cos\theta}{\sin^4\theta} d\theta$. Now let $u = \sin\theta$, $du = \cos\theta \, d\theta$.

The integral becomes $\int \frac{du}{u^4} = \frac{u^{-3}}{-3} + C =$

$-\frac{1}{3}\csc^3\theta + C$. Since $\csc\theta = \frac{\sqrt{x^2 + 1}}{x}$, our

answer is $-\frac{(x^2 + 1)^{3/2}}{3x^3} + C$.

116 Let $x = \frac{1}{2}\tan\theta$; then $\int \sqrt{4x^2 + 1} \, dx =$

$$\frac{1}{2}\int \sec^3\theta\, d\theta$$

$$= \frac{1}{4}\sec\theta\tan\theta + \frac{1}{4}\ln|\sec\theta + \tan\theta| + C$$

$$= \frac{1}{2}x\sqrt{4x^2+1} + \frac{1}{4}\ln\left(2x + \sqrt{4x^2+1}\right) + C.$$

118 Let $x = \frac{1}{2}\sin\theta$; then $\int \sqrt{-4x^2+1}\, dx =$

$$\frac{1}{2}\int \cos^2\theta\, d\theta = \frac{\theta}{4} + \frac{1}{4}\sin\theta\cos\theta + C$$

$$= \frac{1}{4}\sin^{-1} 2x + \frac{1}{2}x\sqrt{1-4x^2} + C.$$

120 $\displaystyle\int \frac{dx}{2 + 3\sin x} = \int \frac{\frac{2\,du}{1+u^2}}{2 + 3\frac{2u}{1+u^2}}$

$$= \int \frac{du}{1 + 3u + u^2}$$

$$= \frac{1}{\sqrt{5}}\int \left(\frac{2}{2u + 3 - \sqrt{5}} - \frac{2}{2u + 3 + \sqrt{5}}\right) du$$

$$= \frac{1}{\sqrt{5}}\ln\left|\frac{2u + 3 - \sqrt{5}}{2u + 3 + \sqrt{5}}\right| + C$$

$$= \frac{1}{\sqrt{5}}\ln\left|\frac{2\tan(x/2) + 3 - \sqrt{5}}{2\tan(x/2) + 3 + \sqrt{5}}\right| + C$$

122 $\displaystyle\int \frac{dx}{3 + 2\sin x} = \int \frac{\frac{2\,du}{1+u^2}}{3 + 2\frac{2u}{1+u^2}}$

$$= \int \frac{2\,du}{3 + 4u + 3u^2} = \frac{1}{3}\int \frac{2\,du}{\left(u + \frac{2}{3}\right)^2 + \frac{5}{9}}$$

$$= \frac{2}{3}\frac{3}{\sqrt{5}}\tan^{-1}\frac{u + 2/3}{\sqrt{5}/3} + C$$

$$= \frac{2}{\sqrt{5}}\tan^{-1}\left(\frac{3\tan(x/2) + 2}{\sqrt{5}}\right) + C$$

124 With the aid of the solution to Exercise 28 of Sec. 7.3, we have $\int x^3 e^{-5x}\, dx$

$$= x^3 \frac{e^{-5x}}{-5} - 3x^2 \frac{e^{-5x}}{25} + 6x\frac{e^{-5x}}{-125} - 6\frac{e^{-5x}}{625} + C$$

$$= -e^{-5x}\left(\frac{1}{5}x^3 + \frac{3}{25}x^2 + \frac{6}{125}x + \frac{6}{625}\right) + C.$$

126 $u = x$, $dv = \sin 3x\, dx$; $\int x\sin 3x\, dx$

$$= x\left(-\frac{1}{3}\cos 3x\right) - \int \left(-\frac{1}{3}\cos 3x\right) dx$$

$$= -\frac{1}{3}x\cos 3x + \frac{1}{9}\sin 3x + C.$$

128 Let $x = \tan\theta$; then $\displaystyle\int \frac{2\,dx}{\sqrt{x^2+1}} = \int \frac{2\sec^2\theta\, d\theta}{\sec\theta}$

$$= 2\int \sec\theta\, d\theta = 2\ln|\sec\theta + \tan\theta| + C$$

$$= 2\ln\left(x + \sqrt{x^2+1}\right) + C.$$

130 $\displaystyle\int \frac{x^2 - 3x}{(x+1)(x-1)^2}\, dx = \int \left(\frac{1}{x+1} - \frac{1}{(x-1)^2}\right) dx$

$$= \ln|x+1| + \frac{1}{x-1} + C$$

132 $\displaystyle\int \frac{6x^3 + 2x + \sqrt{3}}{1 + 3x^2}\, dx = \int \left(2x + \frac{\sqrt{3}}{1 + 3x^2}\right) dx$

$$= x^2 + \tan^{-1}\sqrt{3}x + C$$

134 $\displaystyle\int \frac{x\,dx}{\sqrt{1-9x^2}} = -\frac{1}{9}\sqrt{1-9x^2} + C$

136 Let $u = 3x^2 - 5$ and $u = v^2$. Then $\displaystyle\int \frac{dx}{x\sqrt{3x^2-5}}$

$$= \int \frac{x\,dx}{x^2\sqrt{3x^2-5}} = \int \frac{du/6}{\frac{u+5}{3}\sqrt{u}}$$

$$= \frac{1}{2}\int \frac{du}{(u+5)\sqrt{u}} = \frac{1}{2}\int \frac{2v\,dv}{(v^2+5)v} = \int \frac{dv}{v^2+5}$$

$$= \frac{1}{\sqrt{5}} \tan^{-1}\frac{v}{\sqrt{5}} + C = \frac{1}{\sqrt{5}}\tan^{-1}\sqrt{\frac{u}{5}} + C$$

$$= \frac{1}{\sqrt{5}}\tan^{-1}\sqrt{\frac{3x^2-5}{5}} + C.$$

138 $\displaystyle\int \frac{dx}{\sin 5x} = \int \csc 5x\,dx$

$$= \frac{1}{5}\ln|\csc 5x - \cot 5x| + C$$

140 Let $u = x^3$; then $\displaystyle\int \frac{x^2\,dx}{1+3x^3+2x^6} =$

$$\int \frac{du/3}{1+3u+2u^2} = \frac{1}{3}\int \left(\frac{2}{1+2u} - \frac{1}{1+u}\right)du$$

$$= \frac{1}{3}\ln\left|\frac{1+2u}{1+u}\right| + C = \frac{1}{3}\ln\left|\frac{1+2x^3}{1+x^3}\right| + C.$$

142 Let $u = 1 - 3x^2$. Then $\displaystyle\int x^3\sqrt{1-3x^2}\,dx =$

$$\int x^2\sqrt{1-3x^2}\,x\,dx = \int \frac{1-u}{3}\sqrt{u}\left(-\frac{1}{6}\,du\right)$$

$$= \frac{1}{18}\int (u^{3/2} - u^{1/2})\,du$$

$$= \frac{1}{18}\left(\frac{2}{5}u^{5/2} - \frac{2}{3}u^{3/2}\right) + C = \frac{1}{135}(3u-5)u^{3/2} + C$$

$$= -\frac{1}{135}(9x^2+2)(1-3x^2)^{3/2} + C.$$

144 Let $u = 1 - 4x^2$. Then $\displaystyle\int \frac{x^3\,dx}{1-4x^2} =$

$$\int \frac{x^2}{1-4x^2}\,x\,dx = \int \frac{\frac{1-u}{4}}{u}\left(-\frac{1}{8}\,du\right)$$

$$= \frac{1}{32}\int \left(1 - \frac{1}{u}\right)du = \frac{1}{32}(u - \ln|u|) + K$$

$$= \frac{1}{32}(1 - 4x^2 - \ln|1-4x^2|) + K$$

$$= -\frac{1}{32}(4x^2 + \ln|1-4x^2|) + C.$$

146 $\displaystyle\int \frac{dx}{\cos^3 x} = \int \sec^3 x\,dx$, so we use the recursion

formula of Exercise 30 of Sec. 7.6 with $n = 3$:

$$\int \sec^3 x\,dx = \frac{1}{2}(\sec x\,\tan x) + \frac{1}{2}\int \sec x\,dx =$$

$$\frac{1}{2}(\sec x\,\tan x + \ln|\sec x + \tan x|) + C.$$

148 Let $w = x^3 + 1$, $dw = 3x^2\,dx$; then

$\displaystyle\int x^2 \ln(x^3+1)\,dx = \frac{1}{3}\int \ln w\,dw$. Let $u = \ln w$

and $dv = dw$; then $du = \dfrac{dw}{w}$, $v = w$, and

$$\int \ln w\,dw = w\ln w - \int w\,\frac{dw}{w}$$

$$= w\ln w - w + C_1. \text{ Hence } \int x^2 \ln(x^3+1)\,dx$$

$$= \frac{1}{3}(w\ln w - w) + C_2$$

$= \frac{1}{3}[(x^3+1)\ln(x^3+1) - (x^3+1)] + C_2$

$= \frac{1}{3}((x^3+1)\ln(x^3+1) - x^3) + C.$

150 $\int \frac{(3+x^2)^2}{x} dx = \int \frac{9+6x^2+x^4}{x} dx$

$= \int \left(\frac{9}{x} + 6x + x^3\right) dx$

$= 9\ln|x| + 3x^2 + \frac{x^4}{4} + C$

152 $\int \frac{(1+3\cos x)^2}{\sin x} dx$

$= \int \frac{1 + 6\cos x + 9\cos^2 x}{\sin x} dx$

$= \int \left(\csc x + 6\cot x + 9\frac{1-\sin^2 x}{\sin x}\right) dx$

$= \int (10\csc x + 6\cot x - 9\sin x) dx =$

$10\ln|\csc x - \cot x| + 6\ln|\sin x| + 9\cos x + C$

154 Let $x = \frac{2}{3}\sec\theta$. Then $\int \sqrt{9x^2-4}\, dx =$

$\frac{4}{3}\int \sec\theta \tan^2\theta\, d\theta$

$= \frac{4}{3}\int \sec^3\theta\, d\theta - \frac{4}{3}\int \sec\theta\, d\theta$

$= \frac{4}{3}\left(\frac{1}{2}\sec\theta\tan\theta + \frac{1}{2}\ln|\sec\theta + \tan\theta|\right) -$

$\frac{4}{3}\ln|\sec\theta + \tan\theta| + K$

$= \frac{2}{3}\sec\theta\tan\theta - \frac{2}{3}\ln|\sec\theta + \tan\theta| + K$

$= x\sqrt{\left(\frac{3x}{2}\right)^2 - 1} - \frac{2}{3}\ln\left|\frac{3x}{2} + \sqrt{\left(\frac{3x}{2}\right)^2 - 1}\right| + K$

$= \frac{1}{2}x\sqrt{9x^2-4} - \frac{2}{3}\ln\left|3x + \sqrt{9x^2-4}\right| + C.$

156 First factor $y^2 + 3y + 1$ and then replace y by x^2.

Thus $\int \frac{dx}{x^4 + 3x^2 + 1}$

$= \frac{1}{\sqrt{5}} \int \left(\frac{1}{x^2 + \frac{3-\sqrt{5}}{2}} - \frac{1}{x^2 + \frac{3+\sqrt{5}}{2}}\right) dx$

$= \frac{1}{\sqrt{5}}\left[\frac{1+\sqrt{5}}{2}\tan^{-1}\left(\frac{1+\sqrt{5}}{2}x\right) - \right.$

$\left. \frac{1-\sqrt{5}}{2}\tan^{-1}\left(\frac{1-\sqrt{5}}{2}x\right)\right] + C.$

158 $\int \ln(2x + x^2)\, dx = \int \ln|x|\, dx + \int \ln|x+2|\, dx$

$= (x\ln|x| - x) + [(x+2)\ln|x+2| - (x+2)]$

$+ K = x\ln|x| + (x+2)\ln|x+2| - 2x + C$

160 Let $u = e^{3x}$. Then $\int \frac{dx}{1 + 2e^{3x}} = \int \frac{e^{3x} dx}{e^{3x} + 2e^{6x}}$

$= \int \frac{du/3}{u + 2u^2} = \frac{1}{3}\int \left(\frac{1}{u} - \frac{2}{2u+1}\right) du$

$= \frac{1}{3}\ln|u| - \frac{1}{3}\ln|2u+1| + C$

$= x - \frac{1}{3}\ln(2e^{3x} + 1) + C.$

162 $\int \sqrt{1 + \cos 3\theta}\, d\theta = \int \sqrt{2\cos^2\frac{3\theta}{2}}\, d\theta =$

$\sqrt{2}\int \left|\cos\frac{3\theta}{2}\right| d\theta = \pm\frac{2\sqrt{2}}{3}\sin\frac{3\theta}{2} + C.$ (The sign is the same as that of $\cos \frac{3\theta}{2}$.)

164 (a) $\int x^p (1-x)^q \, dx$

$= \int (1-v^t)^p (v^t)^q (-tv^{t-1} \, dv)$

$= -t \int v^{s+t-1}(1-v^t)^p \, dv$; the final integrand is a rational function of v.

(b) Let $x = 1 - u$ and use part (a).

(c) Let $q = s/t$. Let $x = u^{-1}$ and $u = 1 + v^t$.

Then $\int x^p (1-x)^q \, dx$

$= \int u^{-p}(1-u^{-1})^q (-u^{-2} \, du)$

$= -\int u^{-(p+q)-2}(u-1)^q \, du$

$= -\int (1+v^t)^{-(p+q)-2} v^{qt} \cdot t \cdot v^{t-1} \, dv$

$= -t \int v^{s+t-1}(1+v^t)^{-(p+q)-2} \, dv$; the final integrand is a rational function of v.

166 Let $x = u^{1/n}$. Then $\int (1-x^n)^{1/m} \, dx =$

$\int (1-u)^{1/m} \frac{1}{n} u^{(1/n)-1} \, du$. By Chebyshev's theorem, this is elementary if and only if one of $1/m$, $1/n$, and $1/m + 1/n$ is an integer. If $m \geq 2$ and $n \geq 2$, then $0 \leq 1/m + 1/n \leq 1$; hence $1/m + 1/n$ is an integer only for $m = n = 2$.

168 Let $u = \sin^2 x$. Then $du = 2 \sin x \cos x \, dx = 2u^{1/2}\sqrt{1-u} \, du$, so $\int \sin^a x \, dx =$

$\int u^{a/2} \frac{du}{2u^{1/2}\sqrt{1-u}} = \frac{1}{2} \int u^{(a-1)/2}(1-u)^{-1/2} \, du$.

By Chebyshev's theorem, this is elementary if and only if either $\frac{a-1}{2}$, $\frac{1}{2}$, or $\frac{a-1}{2} + \frac{1}{2}$ is an integer; hence if and only if a is an integer.

170 $\int \sec^p x \tan^q x \, dx = \int \left(\frac{1}{\cos x}\right)^p \left(\frac{\sin x}{\cos x}\right)^q \, dx =$

$\int \sin^q x \cos^{-p-q} x \, dx$. By Exercise 169, this is elementary if and only if q is odd, $-p - q$ is odd, or $q + (-p - q) = -p$ is even; hence if and only if $p + q$ or q is odd or p is even.

172 (a) Let $x = u^{1/n}$. Then $\int \frac{x^2 \, dx}{\sqrt{1 + x^n}} =$

$\int \frac{u^{2/n}}{\sqrt{1+u}} \frac{1}{n} u^{(1/n)-1} \, du$

$= \frac{1}{n} \int u^{(3/n)-1}(1+u)^{-1/2} \, du$. For $n > 6$, $-1 < 3/n - 1 < -1/2$ and $-3/2 < (3/n - 1) + (-1/2) < -1$, so neither $3/n - 1$ nor $(3/n - 1) + (-1/2)$ is an integer. Checking the cases $1 \leq n \leq 6$ shows that $3/n - 1$ is an integer for $n = 1$ or 3 and $\left(\frac{3}{n} - 1\right) + \left(-\frac{1}{2}\right)$ is an integer for $n = 2$ or 6.

(b) $n = 1$: Let $v = 1 + x$; then $\int \frac{x^2 \, dx}{\sqrt{1+x}} =$

$\int \frac{(v-1)^2}{\sqrt{v}} \, dv = \int (v^{3/2} - 2v^{1/2} + v^{-1/2}) \, dv$

$= \frac{2}{5} v^{5/2} - \frac{4}{3} v^{3/2} + 2v^{1/2} + C$

$= \frac{2}{5}(1+x)^{5/2} - \frac{4}{3}(1+x)^{3/2} + 2(1+x)^{1/2} + C$

$= \frac{2}{15}(8 - 4x + 3x^2)\sqrt{1+x} + C$.

$n = 2$: Let $x = \sqrt{u^2 - 1}$; then $\int \dfrac{x^2\, dx}{\sqrt{1+x^2}} =$

$\int \dfrac{u^2 - 1}{u} \cdot \dfrac{u}{\sqrt{u^2-1}}\, du = \int \sqrt{u^2 - 1}\, du$

$= \dfrac{1}{2} u\sqrt{u^2-1} - \dfrac{1}{2} \ln\left(u + \sqrt{u^2 - 1}\right) + C$

$= \dfrac{1}{2} x\sqrt{1+x^2} - \dfrac{1}{2} \ln\left(x + \sqrt{1+x^2}\right)$, by

Exercise 48 in Sec. 7.6.

$n = 3$: Let $u = 1 + x^3$; then $\int \dfrac{x^2\, dx}{\sqrt{1+x^3}} =$

$\int \dfrac{du/3}{\sqrt{u}} = \dfrac{2}{3} u^{1/2} + C = \dfrac{2}{3}\sqrt{1+x^3} + C.$

$n = 6$: Let $u = x^3$ and $u = \tan \theta$; then

$\int \dfrac{x^2\, dx}{\sqrt{1+x^6}} = \int \dfrac{du/3}{\sqrt{1+u^2}} = \dfrac{1}{3} \int \dfrac{\sec^2\theta\, d\theta}{\sec\theta}$

$= \dfrac{1}{3} \int \sec\theta\, d\theta = \dfrac{1}{3} \ln|\sec\theta + \tan\theta| + C$

$= \dfrac{1}{3} \ln\left|u + \sqrt{1+u^2}\right| + C$

$= \dfrac{1}{3} \ln\left|x^3 + \sqrt{1+x^6}\right| + C.$

174 (a) Since $\int \dfrac{dr}{r\sqrt{r^6 - c^2}} = \int \dfrac{r^2\, dr}{r^3\sqrt{r^6-c^2}}$, the

substitution $u = r^3$ would help.

(b) Let $r = c^{1/3} v^{1/6}$. Then $\int \dfrac{c\, dr}{r^n\sqrt{r^6 - c^2}} =$

$\int \dfrac{c}{c^{n/3} v^{n/6} |c|\sqrt{v-1}} \cdot c^{1/3} \cdot \dfrac{1}{6} v^{-5/6}\, dv =$

$\dfrac{c^{4/3}}{6 c^{n/3}|c|} \int v^{(-n-5)/6}(v-1)^{-1/2}\, dv$. Letting $v =$

$1 + x$, Chebyshev's theorem implies that this

is elementary if and only if either $\dfrac{-n-5}{6}$ or

$\dfrac{-n-5}{6} + \left(-\dfrac{1}{2}\right)$ is an integer; that is, if and

only if $n = 3k + 1$ for some integer $k \geq 0$.

176 Let $x = \ln u$. Then $\int \dfrac{e^x}{x}\, dx = \int \dfrac{u}{\ln u} \dfrac{du}{u} =$

$\int \dfrac{1}{\ln u}\, du$. If $\int \dfrac{1}{\ln x}\, dx$ were elementary,

$\int \dfrac{e^x}{x}\, dx$ would be too.

178 (a) $(x-1)(x^2 + x + 1)$
 (b) $(x+1)(x^2 - x + 1)$
 (c) $(x - \sqrt[3]{5})(x^2 + \sqrt[3]{5}x + \sqrt[3]{25})$
 (d) $(x+2)(x^2 - 2x + 4)$
 (e) $(x-1)(x+1)(x^2 + 1)$
 (f) $(x-1)^2(x+2)$

180 (a) Follow $x = t^{-1}$ with $t = u^3 - 1$ and obtain a

rational function. $\dfrac{1}{162}\left[\dfrac{30u^8 + 39u^5 - 15u^2}{(u^3 - 1)^3}\right.$

$+ 5 \ln \dfrac{(u-1)^2}{u^2 + u + 1} + 10\sqrt{3} \tan^{-1} \dfrac{2u+1}{\sqrt{3}}\Bigg] +$

C, where $u = \sqrt[3]{1 + \dfrac{1}{x}}$, is the final answer, in

case you're interested.

(b) Follow $x = t^{-1}$ with $t = u^4 - 1$ and obtain a

rational function. $\dfrac{3u^7 + u^3}{8(u^4 - 1)^2} + \dfrac{3}{32} \ln\left|\dfrac{u-1}{u+1}\right|$

$+ \dfrac{3}{16} \tan^{-1} u + C$, where $u = \sqrt[4]{1 + \dfrac{1}{x}}$, is

the final answer.

182 (a) Not elementary

(b) Let $x = e^u$. Then $\int \cos(\ln x)\, dx =$

$\int e^u \cos u \, du = \dfrac{1}{2} e^u (\cos u + \sin u) + C =$

$\dfrac{1}{2} x(\cos(\ln x) + \sin(\ln x)) + C$, by Example 8

in Sec. 7.3.

184 (a) $u = x, \, dv = \cos x \, dx$; $\int x \cos x \, dx =$

$x \sin x - \int \sin x \, dx = x \sin x + \cos x + C$

(b) Not elementary

(c) Not elementary

(d) $\int \dfrac{\ln x^2}{x} dx = 2 \int (\ln x) \dfrac{dx}{x} = (\ln x)^2 + C$

(e) Not elementary

(f) $\int \sqrt{x-1}\sqrt{x+1}\, x \, dx = \dfrac{1}{2} \int \sqrt{x^2 - 1}\, (2x\, dx)$

$= \dfrac{1}{2} \cdot \dfrac{2}{3}(x^2 - 1)^{3/2} + C = \dfrac{1}{3}(x^2 - 1)^{3/2} + C$

186 $a = 0$: $\int d\theta = \theta + C$

$a = 1$: $\int \sqrt{1 - \sin^2 \theta}\, d\theta = \int |\cos \theta|\, d\theta$

$= \pm \sin \theta + C$

188 $b = 0$: $\int d\theta = \theta + C$

$b = 1$: $\int \sqrt{1 + \cos \theta}\, d\theta = \int \sqrt{2} \left|\cos \dfrac{\theta}{2}\right| d\theta$

$= \pm 2\sqrt{2} \sin \dfrac{\theta}{2} + C$

$b = -1$: $\int \sqrt{1 - \cos \theta}\, d\theta = \int \sqrt{2} \left|\sin \dfrac{\theta}{2}\right| d\theta$

$= \pm 2\sqrt{2} \cos \dfrac{\theta}{2} + C$

190 (a) Let $x = \sqrt{u}$, so $u = x^2$ and $dx = \dfrac{du}{2\sqrt{u}}$.

Then $\int e^{x^2} dx = \int e^u \dfrac{du}{2\sqrt{u}} =$

$\dfrac{1}{2} \int \dfrac{e^u}{\sqrt{u}} du$. If $\int \dfrac{e^u}{\sqrt{u}} du$ were elementary,

then $\int e^{x^2} dx$ would be too.

(b) Let $u = e^x$ and $dv = \dfrac{dx}{\sqrt{x}}$, so $\int \dfrac{e^x}{\sqrt{x}} dx =$

$2\sqrt{x} e^x - 2 \int \sqrt{x} e^x \, dx$. If $\int \sqrt{x} e^x \, dx$ were

elementary, then $\int \dfrac{e^x}{\sqrt{x}} dx$ would be

elementary, contradicting (a).

(c) Let $u = x$ and $dv = xe^{x^2} dx$. Then

$\int x^2 e^{x^2} dx = x \cdot \dfrac{1}{2} e^{x^2} - \int \dfrac{1}{2} e^{x^2} dx =$

$\dfrac{1}{2} x e^{x^2} - \dfrac{1}{2} \int e^{x^2} dx$. Since $\int e^{x^2} dx$ is not

elementary, $\int x^2 e^{x^2} dx$ cannot be.

8 Applications of the Definite Integral

8.1 Computing Area by Parallel Cross Sections

2. (a)

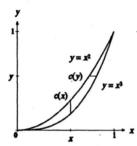

(b) $c(x) = x^2 - x^3$

(c) $c(y) = \sqrt[3]{y} - \sqrt{y}$

4. (a)

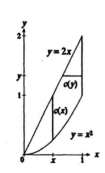

(b) $c(x) = 2x - x^2$

(c) $c(y) = \sqrt{y} - y/2$ for $0 \le y \le 1$ and $c(y) = 1 - y/2$ for $1 \le y \le 2$.

6. (a)

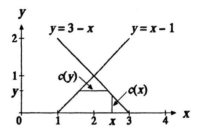

(b) The region in question is bordered by the lines $y = 0$, $y = x - 1$, and $y = -x + 3$. So $c(x) = x - 1$ for $1 \le x \le 2$ and $-x + 3$ for $2 < x \le 3$.

(c) Since $x = y + 1$ and $x = 3 - y$, $c(y) = 3 - y - (y + 1) = 2 - 2y$.

8.

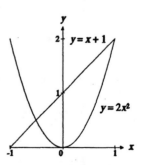

$2x^2 = x + 1$ at $x = -1/2$ and $x = 1$.

(a) $\int_{-1/2}^{1} (x + 1 - 2x^2)\, dx = \dfrac{9}{8}$

(b) $\int_0^{1/2} \left[\sqrt{\frac{y}{2}} - \left(-\sqrt{\frac{y}{2}}\right)\right] dy +$

$\int_{1/2}^2 \left(\sqrt{\frac{y}{2}} - (y-1)\right) dy = \frac{9}{8}$

10

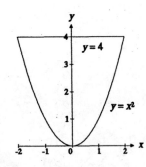

(a) $\int_{-2}^2 (4-x^2)\, dx = \frac{32}{3}$

(b) $\int_0^4 [\sqrt{y} - (-\sqrt{y})]\, dy = \frac{32}{3}$

12

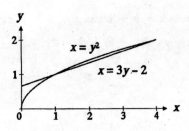

$y^2 = 3y - 2$ at $y = 1$ and $y = 2$.

(a) $\int_1^4 \left(\sqrt{x} - \frac{x+2}{3}\right) dx = \frac{1}{6}$

(b) $\int_1^2 [(3y - 2) - y^2]\, dy = \frac{1}{6}$

14 $\int_1^2 (x^3 - x^2)\, dx = \left(\frac{1}{4}x^4 - \frac{1}{3}x^3\right)\Big|_1^2 = \frac{17}{12}$

16 The region under consideration is shown in the figure. Its area is

$\int_1^2 (x^3 - (-x))\, dx =$

$\int_1^2 (x^3 + x)\, dx = \left(\frac{x^4}{4} + \frac{x^2}{2}\right)\Big|_1^2$

$= \left(\frac{16}{4} + \frac{4}{2}\right) - \left(\frac{1}{4} + \frac{1}{2}\right) = \frac{21}{4}.$

18 $\int_{\pi/2}^\pi (\sin x - \cos x)\, dx = (-\cos x - \sin x)\Big|_{\pi/2}^\pi = 2$

20 $\int_1^e [(1+x) - \ln x]\, dx$

$= \left[\left(x + \frac{1}{2}x^2\right) - (x \ln x - x)\right]\Big|_1^e = \frac{1}{2}e^2 + e - \frac{5}{2}$

22 (a) $A(b) = \int_0^b e^{-x}\cos^2 x\, dx =$

$-\frac{1}{10}e^{-x}(5 + \cos 2x - 2\sin 2x)\Big|_0^b$

$= \frac{3}{5} - \frac{1}{10}e^{-b}(5 + \cos 2b - 2\sin 2b)$

(b) $3/5$

24 Since $10^x > x > \log_{10} x$ for all $x > 0$, the area is

$\int_1^{10} (10^x - \log_{10} x)\, dx =$

$\frac{1}{\ln 10}(10^x - x(\ln x - 1))\Big|_1^{10} = \frac{10^{10} - 1}{\ln 10} - 10.$

26 The area is $\int_2^3 \frac{2x+1}{x^2+x}\, dx = \ln(x^2 + x)\Big|_2^3$

$= \ln 12 - \ln 6 = \ln 2.$

28 (a) $\int_0^{\pi/2} \sin x\, dx = 1$

(b) $\int_0^1 \left(\frac{\pi}{2} - \sin^{-1} y\right) dy = 1$

8.1 Computing Area by Parallel Cross Sections

30 (a)

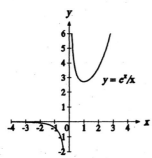

$\lim\limits_{x \to -\infty} y = 0$, $\lim\limits_{x \to 0^+} y = \infty$, and $y' =$

$\dfrac{e^x(x-1)}{x^2}$ goes from $-$ to $+$ at $x = 1$, so

$(1, e)$ is a minimum.

(b) Let $f(t) = \int_t^{t+1} \dfrac{e^x}{x}\, dx$. Then $f'(t) =$

$-\dfrac{e^t}{t} + \dfrac{e^{t+1}}{t+1}$, which equals 0 when $\dfrac{e^t}{t} =$

$\dfrac{e^{t+1}}{t+1}$. That is, when $\dfrac{t+1}{t} = \dfrac{e^{t+1}}{e^t} = e$. So

$f'(t)$ goes from $-$ to $+$ at $t = (e-1)^{-1}$.

Thus the area under $y = e^x/x$ and above $[t, t+1]$ is a minimum when $t = (e-1)^{-1}$.

32

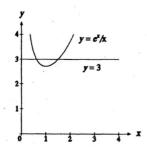

The area is $\int_{r_1}^{r_2} \left(3 - \dfrac{e^x}{x}\right) dx$, where r_1 and r_2 are

roots of $f(x) = 3 - e^x/x$. Applying Newton's method with $x_1 = 3/5$ gives $r_1 \approx 0.619$. Applying

Newton's method with $x_1 = 3/2$ gives $r_2 \approx 1.512$.

Now $\int_{r_1}^{r_2} \left(3 - \dfrac{e^x}{x}\right) dx = 3(r_2 - r_1) - \int_{r_1}^{r_2} \dfrac{e^x}{x}\, dx$

$\approx 2.679 - \int_{0.619}^{1.512} \dfrac{e^x}{x}\, dx$. Using Simpson's

method with $n = 4$ yields $\int_{0.619}^{1.512} \dfrac{e^x}{x}\, dx \approx$

$\dfrac{1.512 - 0.619}{12}\left[\dfrac{e^{0.619}}{0.619} + 4\cdot\dfrac{e^{0.842}}{0.842} + 2\cdot\dfrac{e^{1.066}}{1.066} + \right.$

$\left. 4\cdot\dfrac{e^{1.289}}{1.289} + \dfrac{e^{1.512}}{1.512}\right] \approx 2.511$. So the area is

approximately $2.679 - 2.511 = 0.168$.

34 (a) $y = 1 - \dfrac{1}{\sqrt{2}}$; $x = 1 - \dfrac{1}{\sqrt{2}}$; $x + y = \dfrac{1}{\sqrt{2}}$

(b) No

36 The equation of the line through $(-3, 9)$ and $(2, 4)$

is $y - 4 = \dfrac{9 - 4}{-3 - 2}(x - 2)$ or $y = -x + 6$. Thus

the area of the shaded region is

$\int_{-3}^{2}\left[(-x + 6) - x^2\right] dx = \left(-\dfrac{x^2}{2} + 6x - \dfrac{x^3}{3}\right)\bigg|_{-3}^{2} =$

$\dfrac{125}{6}$. The line AD must have the same slope as

BC, so its slope is -1. It is tangent to the curve $y = x^2$ at the point where the derivative $2x$ equals -1; that is, at $x = -1/2$. Thus AD contains the point $(-1/2, 1/4)$ and has as its equation $y = -x - 1/4$. Now the coordinates of the points A and D can be determined: $A = (-3, 11/4)$, $D = (2, -9/4)$. The area of the parallelogram is equal to its (vertical) base b times its (horizontal) height

h. But $b = 4 - (-9/4) = 25/4$ and $h = 2 - (-3) = 5$, so the area is $bh = \dfrac{125}{4}$. Since $\dfrac{2}{3} \cdot \dfrac{125}{4} = \dfrac{125}{6}$, the statement is proved.

37 (b) The area under the curve and above $[0, t]$ is $\int_0^t f(x)\,dx$, while the area of triangle ABC is $(1/2)tf(t)$. So $\lim\limits_{t\to 0^+} \dfrac{\text{Shaded area under curve}}{\text{Area of triangle } ABC}$

$$= \lim_{t\to 0^+} \dfrac{\int_0^t f(x)\,dx}{\frac{1}{2}tf(t)} = 2\lim_{t\to 0^+} \dfrac{\int_0^t f(x)\,dx}{tf(t)}.$$

Now $\lim\limits_{t\to 0^+}\int_0^t f(x)\,dx = \lim\limits_{t\to 0^+} tf(t) = 0$, so l'Hôpital's rule may be applied. Hence

$$2\lim_{t\to 0^+} \dfrac{\int_0^t f(x)\,dx}{tf(t)} \underset{H}{=} 2\lim_{t\to 0^+} \dfrac{f(t)}{tf'(t) + f(t)},$$

where the fundamental theorem of calculus was used as well. Here $\lim\limits_{t\to 0^+} f(t) = \lim\limits_{t\to 0^+}(tf'(t) + f(t)) = 0$, (given $f(0) = 0$) so the limit is still indeterminate. However, recall that for small t, $f'(0) \approx \dfrac{f(t) - f(0)}{t - 0} = \dfrac{f(t)}{t}$. Consequently, for t very near 0,

$$\dfrac{f(t)}{tf'(t) + f(t)} \approx \dfrac{f(t)}{t \cdot \frac{f(t)}{t} + f(t)} = \dfrac{f(t)}{2f(t)} = \dfrac{1}{2}.$$

Thus $\lim\limits_{t\to 0^+} \dfrac{f(t)}{tf'(t) + f(t)}$ appears to be $1/2$. So $2\lim\limits_{t\to 0^+} \dfrac{f(t)}{tf'(t) + f(t)}$ should be 1. Thus we expect the ratio of the two areas to be 1 as t approaches 0.

38 The desired ratio is $\dfrac{\int_0^t f(x)\,dx}{(1/2)tf(t)}$. Thus

$$\lim_{t\to 0^+} \dfrac{\int_0^t f(x)\,dx}{\frac{1}{2}tf(t)} \underset{H}{=} 2\lim_{t\to 0^+} \dfrac{f(t)}{tf'(t) + f(t)} \underset{H}{=}$$

$$2\lim_{t\to 0^+} \dfrac{f'(t)}{tf''(t) + 2f'(t)}. \text{ Now } f''(0) \approx \dfrac{f'(t) - f'(0)}{t - 0}$$

$= \dfrac{f'(t)}{t}$ when t is very near 0, so we expect

$2\lim\limits_{t\to 0^+} \dfrac{f'(t)}{tf''(t) + 2f'(t)} = \dfrac{2}{3}$, and as $t \to 0^+$, the ratio of the two areas should approach $2/3$.

40 Use partial fractions: If $\dfrac{x^2}{x^3 + x^2 + x + 1} = \dfrac{c_1}{x + 1} + \dfrac{c_2 x + c_3}{x^2 + 1}$, then $x^2 = c_1(x^2 + 1) + (x + 1)(c_2 x + c_3)$. Setting $x = -1$ yields $c_1 = 1/2$, while $x = 0$ shows that $c_3 = -1/2$ and $x = 1$ yields $c_2 = 1/2$. Hence the area is given by

$$\dfrac{1}{2}\int_1^2 \left(\dfrac{1}{x+1} + \dfrac{x-1}{x^2+1}\right)dx = \dfrac{1}{2}\int_1^2 \dfrac{dx}{x+1} +$$

$$\dfrac{1}{2}\int_1^2 \dfrac{x-1}{x^2+1}\,dx = \dfrac{1}{2}\ln(x+1)\bigg|_1^2 + \dfrac{1}{4}\int_1^2 \dfrac{2x\,dx}{x^2+1}$$

$$-\frac{1}{2}\int_1^2 \frac{dx}{x^2+1}$$

$$= \frac{1}{2}\ln\frac{3}{2} + \frac{1}{4}\ln(x^2+1)\Big|_1^2 - \frac{1}{2}\tan^{-1}x\Big|_1^2$$

$$= \frac{1}{2}\ln\frac{3}{2} + \frac{1}{4}\ln\frac{5}{2} - \frac{1}{2}\tan^{-1}2 + \frac{1}{2}\cdot\frac{\pi}{4}$$

$$= \frac{1}{4}\ln\frac{45}{8} + \frac{\pi}{8} - \frac{1}{2}\tan^{-1}2.$$

8.2 Some Pointers on Drawing

2 (a),(b)

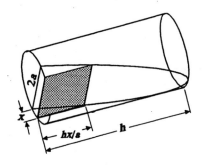

(c)

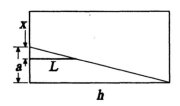

(d)

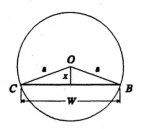

(f) From the triangle OBC in the figure in (d), we know $W = 2\sqrt{a^2 - x^2}$. From similar triangles in (c), we see that $\frac{x}{L} = \frac{a}{h}$, so $L = \frac{h}{a}x$.

Thus $A(x) = LW = \frac{2hx}{a}\sqrt{a^2 - x^2}$.

4 (a)

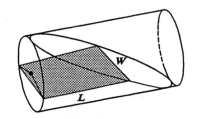

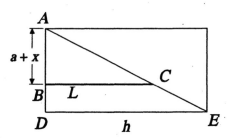

(b) From Exercise 2, $W = 2\sqrt{a^2 - x^2}$. Also, since ABC is similar to ADE, $L = \frac{h}{2a}(a + x)$. Thus $A(x) = \frac{h}{a}(a + x)\sqrt{a^2 - x^2}$.

6 (a)

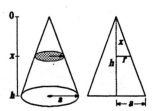

(b) From the figure, it is clear that $\frac{x}{r} = \frac{h}{a}$, so r

$= \frac{ax}{h}$ and $A(x) = \pi r^2 = \frac{\pi a^2 x^2}{h^2}$.

8 (a)

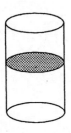

(b)

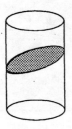

(c)

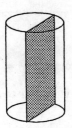

10 (a)

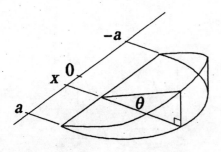

(b) From the figure, the base of the typical triangular cross section is $\sqrt{a^2 - x^2}$; its

height is $\sqrt{a^2 - x^2} \tan \theta$. Hence $A(x) = \frac{1}{2}(a^2 - x^2) \tan \theta$.

(c)

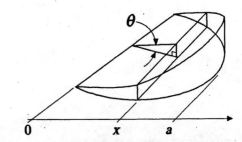

(d) From the figure, the height of the typical rectangular cross section is $x \tan \theta$, while its length is $2\sqrt{a^2 - x^2}$, so $A(x) = 2x\sqrt{a^2 - x^2} \tan \theta$.

12 (a)

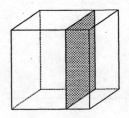

(b)

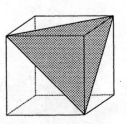

(c)

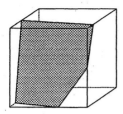

(d)

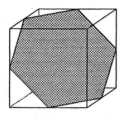

8.3 Setting Up a Definite Integral

2. The change in position of the object during the time interval dt is approximately $f(t)\,dt$. From the informal style it follows that the change in position of the object during the time interval $[a, b]$ is
$$\int_a^b f(t)\,dt.$$

4. Number of calories between r and $r + dr$ meters from center $\approx 2\pi r \cdot G(r)\,dr$, so total is
$$\int_0^a 2\pi r\,G(r)\,dr.$$

6. (a) As in Example 2, the area of the annular ring is approximately $2\pi x\,dx$. By the shorthand approach, we conclude the area of the disk is
$$\int_0^a 2\pi x\,dx.$$

(b) $\int_0^a 2\pi x\,dx = 2\pi \dfrac{x^2}{2}\Big|_0^a = \pi a^2$

8. (a)

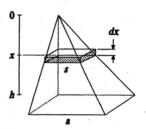

(b) The local approximation is a box with a square of side s as a base and height dx. So its volume is $s^2\,dx = \left(\dfrac{ax}{h}\right)^2 dx$. Thus the volume of the pyramid is $\int_0^h \left(\dfrac{ax}{h}\right)^2 dx$ by the informal approach.

(c) $\int_0^h \left(\dfrac{ax}{h}\right)^2 dx = \dfrac{a^2}{h^2}\int_0^h x^2\,dx = \dfrac{1}{3}a^2 h$

10. The approximating band has circumference $2\pi a \sin\theta$ and width $a\,d\theta$. An estimate of the differential surface area is thus $2\pi a^2 \sin\theta\,d\theta$. By the informal approach, the surface area of the sphere is $\int_0^\pi 2\pi a^2 \sin\theta\,d\theta = 2\pi a^2 \int_0^\pi \sin\theta\,d\theta$
$= 4\pi a^2$.

12. (a) Since everything wears out eventually, $F(t) \to 1$ as $t \to \infty$.

(b) The fraction that wear out between t and $t + dt$ equals the fraction that wear out within $t + dt$ hours minus the fraction that wear out within t hours; that is, $F(t + dt) - F(t)$. If dt

is small, this is approximately $F'(t)\,dt$.

(c) $F(1000) = 1$

(d) Let N be the number of ball bearings. Partition the interval $[0, 1000]$ into short subintervals $0 = t_0 < t_1 < \cdots < t_n = 1000$. The number of bearings that wear out during the interval $[t_i, t_{i+1}]$ is about $N \cdot f'(t_i)(t_{i+1} - t_i)$. Hence the sum of the life spans of all of the bearings is about $\sum_{i=0}^{n-1} N \cdot t_i \cdot F'(t_i)(t_{i+1} - t_i)$.

The average life span of a bearing is this quantity divided by N, about

$\sum_{i=0}^{n-1} t_i \cdot F'(t_i)(t_{i+1} - t_i)$. As the mesh of the partition approaches zero, this sum approaches $\int_0^{1000} tF'(t)\,dt$.

14 The present value is equal to the amount of money today that will equal \$1 in t years.

(a) $g(0) = 1$ dollar, since 1 dollar today will be worth 1 dollar in 0 years (that is, today).

(b) A dollar in the bank can earn interest, while the promise of a dollar cannot. The longer it takes to receive it, the less the dollar will be worth, so $g(t)$ is a decreasing function. In particular, $g(t) < g(0) = 1$.

(c) $q \cdot g(t)$, the number of dollars multiplied by the present value of \$1, is the present value of q dollars to be received in t years in the future.

(d) In the period between t and $t + dt$ years, approximately $f(t)\,dt$ dollars will be earned. Their present value is $g(t)f(t)\,dt$ dollars, according to (c).

(e) The present value of the total income earned from now until b years in the future is $\int_0^b f(t)g(t)\,dt$ dollars.

16 (a) All points in the narrow rectangle illustrated in the figure move at approximately the same speed. The rectangle is spun in a circular path of approximate radius $|x|$ at a rate of 5 times per second. So in each second each point in it covers a distance of about $5 \cdot 2\pi|x| = 10\pi|x|$ meters. Since the mass of the rectangle is $4 \cdot 10\,dx = 40\,dx$ kilograms (as in Example 3), the local estimate for the kinetic energy is $\frac{1}{2}mv^2$

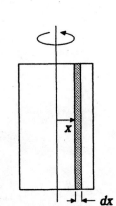

$= \frac{1}{2}(40\,dx)(10\pi|x|)^2 = 2000\pi^2 x^2\,dx$ joules.

(b) The total kinetic energy is $\int_{-3}^{3} 2000\pi^2 x^2\,dx$ joules.

(c) $\int_{-3}^{3} 2000\pi^2 x^2\,dx = 4000\pi^2 \int_0^3 x^2\,dx$

$= 4000\pi^2 \cdot \left.\frac{x^3}{3}\right|_0^3 = (4000\pi^2) \cdot 9$

$= 36{,}000\pi^2$ joules.

18 (a) $g(x)\,dx$ grams

(b) Velocity $= 7 \cdot 2\pi x$ cm/sec, so Energy $\approx \frac{1}{2} \cdot g(x)\,dx \cdot (7 \cdot 2\pi x)^2$ ergs $= 98\pi^2 x^2 g(x)\,dx$ ergs.

(c) $\int_0^b 98\pi^2 x^2 g(x)\,dx$ ergs

8.3 Setting Up a Definite Integral

20 (a)

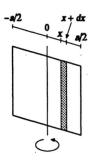

(b) The area of the rectangle is $a\,dx$ and the density of the square is $\dfrac{M}{a^2}$, so the rectangle's mass is $\dfrac{M\,dx}{a}$ and its velocity is about $5 \cdot 2\pi|x| = 10\pi|x|$. Thus its kinetic energy is

$$\frac{1}{2}\left(\frac{M\,dx}{a}\right)(10\pi|x|)^2 = \frac{50M\pi^2 x^2}{a}\,dx \text{ joules}.$$

(c) $\displaystyle\int_{-a/2}^{a/2} \frac{50M\pi^2 x^2}{a}\,dx$ joules

(d) $\displaystyle\int_{-a/2}^{a/2} \frac{50M\pi^2 x^2}{a}\,dx = \frac{100M\pi^2}{a}\int_0^{a/2} x^2\,dx$

$= \dfrac{100M\pi^2}{a} \cdot \dfrac{a^3}{24} = \dfrac{25}{6}M\pi^2 a^2$ joules

22 For convenience, let $A = \displaystyle\int_0^1 e^{x^2}\,dx$ be the area of the region, so that M/A is its density. The velocity of the vertical band in the figure is about ωx, while its mass is about $e^{x^2}\,dx \cdot \dfrac{M}{A}$, so its kinetic energy is estimated by $\dfrac{1}{2}\dfrac{Me^{x^2}\,dx}{A}\cdot(\omega x)^2 = \dfrac{M\omega^2 x^2 e^{x^2}\,dx}{2A}$.

Thus the total kinetic energy of the object is

$$\frac{M\omega^2}{2\int_0^1 e^{x^2}\,dx}\int_0^1 x^2 e^{x^2}\,dx.$$

(a) By the trapezoidal method, $\displaystyle\int_0^1 e^{x^2}\,dx \approx$

$\dfrac{1}{12}\bigl[e^{0^2} + 2e^{(1/6)^2} + 2e^{(2/6)^2} + 2e^{(3/6)^2} +$

$2e^{(4/6)^2} + 2e^{(5/6)^2} + e^{1^2}\bigr] \approx 1.4751787$ and

$\displaystyle\int_0^1 x^2 e^{x^2}\,dx \approx \dfrac{1}{12}\Bigl[0^2 e^{0^2} + 2\cdot\left(\dfrac{1}{6}\right)^2 e^{(1/6)^2} +$

$2\cdot\left(\dfrac{2}{6}\right)^2 e^{(2/6)^2} + 2\cdot\left(\dfrac{3}{6}\right)^2 e^{(3/6)^2} +$

$2\cdot\left(\dfrac{4}{6}\right)^2 e^{(4/6)^2} + 2\cdot\left(\dfrac{5}{6}\right)^2 e^{(5/6)^2} + 1^2 e^{1^2}\Bigr] \approx$

0.6527890. So the total kinetic energy is about $0.2212576 M\omega^2$.

(b) By Simpson's method, $\displaystyle\int_0^1 e^{x^2}\,dx$

$\approx \dfrac{1}{18}\bigl[e^{0^2} + 4e^{(1/6)^2} + 2e^{(2/6)^2} + 4e^{(3/6)^2} +$

$2e^{(4/6)^2} + 4e^{(5/6)^2} + e^{1^2}\bigr] \approx 1.4628735$ and

$\displaystyle\int_0^1 x^2 e^{x^2}\,dx \approx \dfrac{1}{18}\Bigl[0^2 e^{0^2} + 4\cdot\left(\dfrac{1}{6}\right)^2 e^{(1/6)^2} +$

$2\cdot\left(\dfrac{2}{6}\right)^2 e^{(2/6)^2} + 4\cdot\left(\dfrac{3}{6}\right)^2 e^{(3/6)^2} + 2\cdot\left(\dfrac{4}{6}\right)^2 e^{(4/6)^2}$

$+ 4\cdot\left(\dfrac{5}{6}\right)^2 e^{(5/6)^2} + 1^2 e^{1^2}\Bigr] \approx 0.6285547$. So

the total kinetic energy is about $0.2148356 M\omega^2$.

24 The region has density $\dfrac{M}{\displaystyle\int_1^2 e^x\,dx} = \dfrac{M}{e^2 - e}$, so the mass of the narrow rectangle is about $\dfrac{Me^x\,dx}{e^2 - e}$. Since its velocity is ωx, its kinetic energy is

$$\dfrac{1}{2}\dfrac{Me^x\,dx}{e^2 - e}(\omega x)^2$$

$$= \dfrac{M\omega^2}{2(e^2 - e)} x^2 e^x\,dx \text{ and the total}$$

kinetic energy is $\displaystyle\int_1^2 \dfrac{M\omega^2}{2(e^2 - e)} x^2 e^x\,dx$

$$= \dfrac{M\omega^2}{2(e^2 - e)}\int_1^2 x^2 e^x\,dx.\text{ From Formula 59, we}$$

have $\displaystyle\int_1^2 x^2 e^x\,dx = \left[x^2 e^x - 2xe^x + 2e^x\right]\Big|_1^2 =$

$2e^2 - e$. Finally, the kinetic energy is

$$\dfrac{M\omega^2}{2(e^2 - e)}\cdot(2e^2 - e) = \dfrac{(2e-1)M\omega^2}{2(e-1)}.$$

26   The region has density $\dfrac{M}{\displaystyle\int_2^4 \dfrac{dx}{1+x}} = \dfrac{M}{\ln 5/3}$, so

the mass of the narrow rectangle is about

$\dfrac{M}{\ln 5/3}\cdot\dfrac{dx}{1+x} = \dfrac{M\,dx}{\ln 5/3\,(1+x)}$. Since its velocity

is ωx, its kinetic energy is $\dfrac{1}{2}\cdot\dfrac{M\,dx}{\ln 5/3\,(1+x)}\cdot(\omega x)^2$

$$= \dfrac{M\omega^2}{\ln\dfrac{25}{9}}\cdot\dfrac{x^2\,dx}{1+x}\text{ and the total kinetic energy is}$$

$$\dfrac{M\omega^2}{\ln\dfrac{25}{9}}\int_2^4 \dfrac{x^2\,dx}{1+x}$$

$$= \dfrac{M\omega^2}{\ln\dfrac{25}{9}}\int_2^4 \left(x - 1 + \dfrac{1}{1+x}\right)dx$$

$$= \dfrac{M\omega^2}{\ln\dfrac{25}{9}}\left[\dfrac{x^2}{2} - x + \ln(1+x)\right]\Bigg|_2^4$$

$$= \dfrac{M\omega^2}{\ln\dfrac{25}{9}}\left(4 + \ln\dfrac{5}{3}\right) = \dfrac{\left(4 + \ln\dfrac{5}{3}\right)M\omega^2}{\ln\dfrac{25}{9}}.$$

28   The density of the solid is $\dfrac{M}{\pi a^2 h}$,

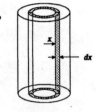

and the volume of the tubular local estimate is $2\pi rh\,dr$, so its

mass is $\dfrac{M}{\pi a^2 h}\cdot 2\pi rh\,dr =$

$\dfrac{2M}{a^2} r\,dr$. Since the velocity of the tube is ωr, its

kinetic energy is $\dfrac{1}{2}\dfrac{2M}{a^2}r\,dr\,(\omega r)^2 = \dfrac{M\omega^2}{a^2}r^3\,dr$.

Thus the total kinetic energy is $\dfrac{M\omega^2}{a^2}\displaystyle\int_0^a r^3\,dr =$

$$\dfrac{M\omega^2}{a^2}\cdot\dfrac{r^4}{4}\Bigg|_0^a = \dfrac{1}{4}M\omega^2 a^2.$$

30   The local approximation isn't good enough: The ratio between the actual area and the estimated area approaches a limit other than 1 as $dx \to 0^+$. Note

that $\displaystyle\lim_{dx \to 0^+}\dfrac{2\pi a\,dx}{2\pi r\,dx} = \dfrac{a}{r} = \dfrac{a}{\sqrt{a^2 - x^2}}$, which does

not equal 1 for $x \neq 0$.

8.3 Setting Up a Definite Integral

32 For the local approximation, select a spherical shell of radius r and thickness dr. So its volume is approximately the surface area of the shell multiplied by its "height" dr; that is, $4\pi r^2\, dr$. Now the mass of the local estimate is $4\pi r^2 g(r)\, dr$ pounds; hence the total mass of the earth would be $\int_0^{4000} 4\pi r^2 g(r)\, dr$.

8.4 Computing Volumes

2 (c)

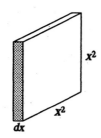

(d) $x^4\, dx$

(e) $\int_0^1 x^4\, dx$

(f) $\int_0^1 x^4\, dx = \dfrac{1}{5}$

4 (c)

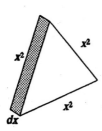

(d) The area of an equilateral triangle with side s is $\dfrac{s^2\sqrt{3}}{4}$, so the volume of the local approximation is $\dfrac{x^4\sqrt{3}}{4}\, dx$.

(e) $\int_0^1 \dfrac{x^4\sqrt{3}}{4}\, dx$

(f) $\int_0^1 \dfrac{x^4\sqrt{3}}{4}\, dx = \dfrac{\sqrt{3}}{4}\cdot\dfrac{1}{5} = \dfrac{\sqrt{3}}{20}$

6 (a)

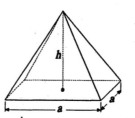

(b)

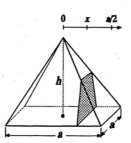

(c)

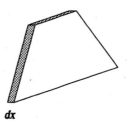

(d) From Exercise 1 of Sec. 8.2, the area of the trapezoidal base in (c) is $\dfrac{h}{2}\left(a - \dfrac{4x^2}{a}\right)$, so the volume of the local approximation is

$\dfrac{h}{2}\left(a - \dfrac{4x^2}{a}\right) dx.$

(e) $\int_{-a/2}^{a/2} \frac{h}{2}\left(a - \frac{4x^2}{a}\right) dx$

(f) $\int_{-a/2}^{a/2} \frac{h}{2}\left(a - \frac{4x^2}{a}\right) dx$

$= h \int_0^{a/2} \left(a - \frac{4x^2}{a}\right) dx$

$= h\left(a \cdot \frac{a}{2} - \frac{4}{a} \cdot \frac{(a/2)^3}{3}\right) = \frac{1}{3}a^2 h$

8 (a)

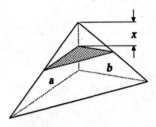

$\int_0^c \frac{1}{2} lw\, dx = \int_0^c \frac{1}{2} ab\left(\frac{x}{c}\right)^2 dx$

$= \frac{ab}{2c^2} \int_0^c x^2\, dx = \frac{abc}{6}$

(b)

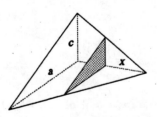

$\int_0^b \frac{1}{2} lw\, dx = \int_0^b \frac{1}{2} ac\left(\frac{x}{b}\right)^2 dx$

$= \frac{ac}{2b^2} \int_0^b x^2\, dx = \frac{abc}{6}$

10 Volume $= \int_0^1 \pi\left(\frac{1}{\sqrt{1+x^2}}\right)^2 dx = \pi \int_0^1 \frac{dx}{1+x^2}$

$= \frac{\pi^2}{4}$

12 Volume $= \int_0^1 \pi\left[(\sqrt[3]{y})^2 - (\sqrt{y})^2\right] dy$

$= \pi \int_0^1 (y^{2/3} - y)\, dy = \frac{\pi}{10}$

14 Volume $= \int_{\pi/6}^{\pi/3} \pi\left[(\sec x)^2 - (\cos x)^2\right] dx$

$= \pi \int_{\pi/6}^{\pi/3} (\sec^2 x - \cos^2 x)\, dx$

$= \pi\left[\tan x - \frac{x}{2} - \frac{\sin 2x}{4}\right]_{\pi/6}^{\pi/3} = \frac{\pi}{12}(8\sqrt{3} - \pi)$

16 (a),(b)

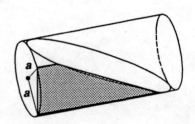

(c)

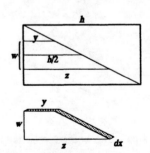

Note from the figure that $y + z = h$.

(d) Volume $= \frac{1}{2} w(y+z)\, dx$

$$= \frac{1}{2} \cdot 2\sqrt{a^2 - x^2} \cdot h \, dx = h\sqrt{a^2 - x^2} \, dx$$

(e) Total volume $= \int_{-a}^{a} h\sqrt{a^2 - x^2} \, dx$

(f) $\int_{-a}^{a} h\sqrt{a^2 - x^2} \, dx = h \int_{-a}^{a} \sqrt{a^2 - x^2} \, dx =$

$h \cdot \frac{\pi}{2} a^2 = \frac{\pi}{2} a^2 h$ since the last integral is the area of a semicircle of radius a.

18 From Example 3 of Sec. 8.2, $A(x) = \frac{h}{2a}(a^2 - x^2)$

so $V(x) = \frac{h}{2a}(a^2 - x^2) \, dx$ and the total volume is

$\int_{-a}^{a} \frac{h}{2a}(a^2 - x^2) \, dx = \frac{h}{2a} \int_{-a}^{a} (a^2 - x^2) \, dx =$

$\frac{h}{2a} \cdot \frac{4}{3} a^3 = \frac{2}{3} a^2 h$.

20 The area of the cross section parallel to the plane at distance $h - x$ is $\left(\frac{x}{h}\right)^2 A$, so $V = \int_0^h \left(\frac{x}{h}\right)^2 A \, dx$

$= \frac{Ah}{3}$.

22 If the radius is r, then a cross section at distance x from the center has radius $\sqrt{r^2 - x^2}$, so its area is $\pi(r^2 - x^2)$. The volume between the planes is

$\int_{-r/3}^{r/3} \pi(r^2 - x^2) \, dx = \pi\left(r^2 x - \frac{1}{3}x^3\right)\Big|_{-r/3}^{r/3} =$

$\frac{52\pi}{81} r^3$. The desired fraction is $\dfrac{\frac{52\pi}{81} r^3}{\frac{4\pi}{3} r^3} = \frac{13}{27}$.

24 The horizontal cross section at height x has side $\sqrt{1 - x^2}$, so $V = \int_0^1 \left(\sqrt{1 - x^2}\right)^2 dx = \frac{2}{3}$.

8.5 The Shell Technique

2 The local estimate for the volume is $2\pi R(x) c(x) \, dx$
$= 2\pi[x - (-3)]x \, dx = 2\pi x(x + 3) \, dx$. Hence the

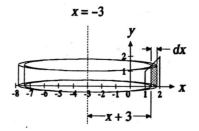

total volume is $\int_1^2 2\pi x(x + 3) \, dx =$

$2\pi\left(\dfrac{x^3}{3} + \dfrac{3x^2}{2}\right)\Big|_1^2 = \dfrac{41\pi}{3}$.

4 The local estimate for the volume is $2\pi R(y) c(y) \, dy$
$= 2\pi y\left(-\dfrac{1}{2}y + 1\right) dy = 2\pi\left(-\dfrac{1}{2}y^2 + y\right) dy$. Hence

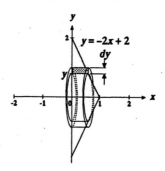

the total volume is $\int_0^2 2\pi\left(-\dfrac{1}{2}y^2 + y\right) dy =$

$2\pi\left(-\dfrac{1}{6}y^3 + \dfrac{1}{2}y^2\right)\Big|_0^2 = \dfrac{4\pi}{3}$.

6 Here $c(x) = x^2 - x^3$ and $R(x) = x - (-2)$ $= x + 2$, so the local estimate of the volume is

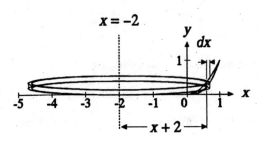

$2\pi(x+2)(x^2 - x^3) \, dx = 2\pi(2x^2 + x^3 - 2x^3 - x^4) \, dx$.

Thus the total volume is $\int_0^1 2\pi(2x^2 - x^3 - x^4) \, dx$

$= 2\pi\left(\dfrac{2x^3}{3} - \dfrac{x^4}{4} - \dfrac{x^5}{5}\right)\Big|_0^1 = \dfrac{13\pi}{30}$.

8 Here $c(x) = \sqrt[3]{x} - \sqrt{x}$ and $R(x) = x$, so the local estimate of the volume is $2\pi x(x^{1/3} - x^{1/2}) \, dx$

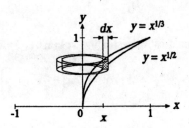

$= 2\pi(x^{4/3} - x^{3/2}) \, dx$. Thus the total volume is

$\int_0^1 2\pi(x^{4/3} - x^{3/2}) \, dx = 2\pi\left(\dfrac{3}{7}x^{7/3} - \dfrac{2}{5}x^{5/2}\right)\Big|_0^1$

$= \dfrac{2\pi}{35}$.

10 Volume $= \int_{-a}^{a} 2\pi(x+b)\left[2\sqrt{a^2 - x^2}\right] dx$

$= 4\pi \int_{-a}^{a} x\sqrt{a^2 - x^2} \, dx + 4\pi b \int_{-a}^{a} \sqrt{a^2 - x^2} \, dx$

$= 4\pi b \cdot \dfrac{\pi a^2}{2} = 2\pi^2 a^2 b$

12 (a) Volume $= \int_1^2 \pi(x + x^3)^2 \, dx$

$= \pi \int_1^2 (x^6 + 2x^4 + x^2) \, dx = \dfrac{3452\pi}{105}$

(b) Volume $= \int_1^2 \pi\left[(x + x^3)^2 - 2^2\right] dx$

$= \dfrac{3452\pi}{105} - \int_1^2 4\pi \, dx = \dfrac{3032\pi}{105}$

14 (a) A typical cross section is a circular disk with radius $R(x) = \cos x$. Hence, the volume is

$\int_0^{\pi/2} \pi(\cos x)^2 \, dx = \dfrac{\pi}{2}\int_0^{\pi/2}(1 + \cos 2x) \, dx$

$= \dfrac{\pi}{2}\left(x + \dfrac{\sin 2x}{2}\right)\Big|_0^{\pi/2} = \dfrac{\pi^2}{4}$.

(b) A typical shell is shown. On the curve $y = \cos x$, we have $x = \cos^{-1} y$. Therefore, $R(y)$

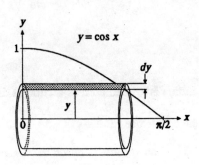

$= y$, $c(y) = \cos^{-1} y$, and the volume is

$\int_0^1 2\pi R(y) c(y) \, dy = \int_0^1 2\pi y \cos^{-1} y \, dy$. Let $u = \cos^{-1} y$ and $dv = y \, dy$. Then $du =$

$\dfrac{-dy}{\sqrt{1 - y^2}}$ and $v = \dfrac{y^2}{2}$. Thus $\int y \cos^{-1} y \, dy$

8.5 The Shell Technique

$= \frac{y^2}{2} \cos^{-1} y + \frac{1}{2} \int \frac{y^2 \, dy}{\sqrt{1-y^2}}$. Now let $y = \sin \theta$, $dy = \cos \theta \, d\theta$. Then $\int \frac{y^2 \, dy}{\sqrt{1-y^2}} =$

$\int \frac{\sin^2 \theta \cos \theta \, d\theta}{\cos \theta} = \int \sin^2 \theta \, d\theta$

$= \frac{1}{2}(\theta - \sin \theta \cos \theta) + K$

$= \frac{1}{2}\left(\sin^{-1} y - y\sqrt{1-y^2}\right) + K$. Therefore,

$\int y \cos^{-1} y \, dy = \frac{y^2}{2} \cos^{-1} y + \frac{1}{4} \sin^{-1} y - \frac{1}{4} y\sqrt{1-y^2} + C$. Finally, $2\pi \int_0^1 y \cos^{-1} y \, dy$

$= 2\pi \left(\frac{y^2}{2} \cos^{-1} y + \frac{1}{4} \sin^{-1} y - \frac{1}{4} y\sqrt{1-y^2} \right) \Big|_0^1 = \frac{\pi^2}{4}$, as in (a).

16 $V = \int_0^1 2\pi x e^{x^2} \, dx = \pi(e-1)$

18 (a) Volume $= \int_1^e \pi (\ln x)^2 \, dx = \pi \int_0^1 y^2 e^y \, dy$

$= \pi [y^2 e^y - 2(ye^y - e^y)] \Big|_0^1 = \pi(e-2)$

(b) Volume $= \int_0^1 2\pi(e - e^y) e^y \, dy$

$= 2\pi \left[\int_0^1 e^{y+1} \, dy - \int_0^1 e^{2y} \, dy \right]$

$= 2\pi \left(\frac{1}{2} e^2 - e + \frac{1}{2} \right) = \pi(e-1)^2$

20 (a) Volume $= \int_{\sqrt{3}}^{\sqrt{8}} \pi \left(\frac{1}{\sqrt{2+x^2}} \right)^2 dx$

$= \pi \int_{\sqrt{3}}^{\sqrt{8}} \frac{dx}{2+x^2} = \frac{\pi}{\sqrt{2}} \left(\tan^{-1} \sqrt{\frac{1}{2}} x \right) \Big|_{\sqrt{3}}^{\sqrt{8}}$

$= \frac{\pi}{\sqrt{2}} \left(\tan^{-1} 2 - \tan^{-1} \sqrt{\frac{3}{2}} \right)$

(b) Volume $= \int_{\sqrt{3}}^{\sqrt{8}} 2\pi x \cdot \frac{dx}{\sqrt{2+x^2}}$

$= \pi \int_{\sqrt{3}}^{\sqrt{8}} 2x(2+x^2)^{-1/2} \, dx = 2\pi(2+x^2)^{1/2} \Big|_{\sqrt{3}}^{\sqrt{8}}$

$= 2\pi(\sqrt{10} - \sqrt{5})$

22 If $y = f(x)$ then $x = f^{-1}(y)$, so the volume of the deleted part (using slabs) is $\int_1^3 \pi [f^{-1}(y)]^2 \, dy$.

Now $y = 1 + x + x^5$, so $dy = (1 + 5x^4) \, dx$; when $y = 1$, $x = 0$, and when $y = 3$, $x = 1$, so

$\int_1^3 \pi [f^{-1}(y)]^2 \, dy = \int_0^1 \pi [f^{-1}(f(x))]^2 (1 + 5x^4) \, dx$

$= \pi \int_0^1 (x^2 + 5x^6) \, dx = \frac{22\pi}{21}$. The volume of the whole cylinder is 3π, so our volume is $3\pi - \frac{22\pi}{21}$

$= \frac{41\pi}{21}$, which agrees with Example 1.

24 (a) Revolve about the y axis the region bounded by the curve $y = f(x)$, the positive x axis, and

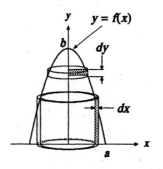

the positive y axis. We will compute the volume two ways; first by parallel cross sections and then by cylindrical shells.

(i) The cross sections are circular disks with radius x and "thickness" dy. Hence the volume is $\int_0^b \pi x^2 \, dy = \pi \int_0^b x^2 \, dy$.

(ii) A typical cylindrical shell has radius x, height y, and "thickness" dx. Hence the volume is $\int_0^a 2\pi xy \, dx = 2\pi \int_0^a xy \, dx$.

These two integrals for the volume must be equal, so $\int_0^b x^2 \, dy = \int_0^a 2xy \, dx$.

(b) Consider the integral $\int_0^a 2xy \, dx$. Let $u = y$ and $dv = 2x \, dx$. Then $du = dy$ and $v = x^2$. Note that when $x = 0$, $y = b$, and when $x = a$, $y = 0$. $\int_0^a 2xy \, dx = yx^2 \big|_{x=0}^{x=a} - \int_{x=0}^{x=a} x^2 \, dy = 0 - \int_b^0 x^2 \, dy = \int_0^b x^2 \, dy$.

8.6 The Centroid of a Plane Region

2   The area of R is $\int_{-1}^1 (1 - x^4) \, dx =$

$2 \int_0^1 (1 - x^4) \, dx = 2\left(x - \frac{x^5}{5}\right)\bigg|_0^1 = \frac{8}{5}$. Note, by symmetry, that $\bar{x} = 0$. Now $\bar{y} = \dfrac{\int_c^d y c(y) \, dy}{\text{Area of } R}$

$= \frac{5}{8} \int_0^1 y(2y^{1/4}) \, dy = \frac{5}{4} \int_0^1 y^{5/4} \, dy =$

$\frac{5}{4} \cdot \frac{4}{9} y^{9/4} \bigg|_0^1 = \frac{5}{9}$. So $(\bar{x}, \bar{y}) = (0, 5/9)$.

4   R is a triangle with base 1 and height $1/2$, so its area is $\frac{1}{2} \cdot 1 \cdot \frac{1}{2} = \frac{1}{4}$. Now by symmetry, $\bar{x} = \frac{1}{2}$. $\bar{y}$

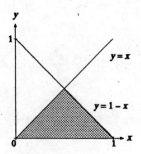

$= \dfrac{\int_c^d y c(y) \, dy}{\text{Area of } R} = 4 \int_0^{1/2} y(1 - y - y) \, dy =$

$4 \int_0^{1/2} (y - 2y^2) \, dy = 4\left(\frac{y^2}{2} - \frac{2y^3}{3}\right)\bigg|_0^{1/2} = \frac{1}{6}$. So $(\bar{x}, \bar{y}) = (1/2, 1/6)$.

6   The area of R is $\int_0^{\pi/2} \sin 2x \, dx = \dfrac{-\cos 2x}{2}\bigg|_0^{\pi/2} =$

8.6 The Centroid of a Plane Region

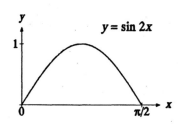

1. By symmetry, we see that $\bar{x} = \dfrac{\pi}{4}$. Now $\bar{y} =$

$$\dfrac{\dfrac{1}{2}\int_0^{\pi/2} (\sin 2x)^2\, dx}{\text{Area of } R} = \dfrac{1}{2}\int_0^{\pi/2} \dfrac{1}{2}(1 - \cos 4x)\, dx$$

$$= \dfrac{1}{4}\left(x - \dfrac{\sin 4x}{4}\right)\Big|_0^{\pi/2} = \dfrac{1}{4}\left(\dfrac{\pi}{2} - 0\right) = \dfrac{\pi}{8}, \text{ so}$$

$(\bar{x}, \bar{y}) = \left(\dfrac{\pi}{4}, \dfrac{\pi}{8}\right)$.

8 The area of the region is $\int_1^e \ln x\, dx =$

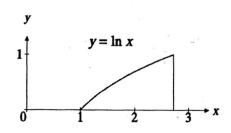

$(x \ln x - x)\Big|_1^e = [e\cdot 1 - e] - [1\cdot 0 - 1] = 1$.

Thus, $\bar{x} = \dfrac{\int_1^e x \ln x\, dx}{1} = \left(\dfrac{x^2}{2}\ln x - \dfrac{x^2}{4}\right)\Big|_1^e =$

$\dfrac{e^2}{4} + \dfrac{1}{4} = \dfrac{e^2 + 1}{4}$. $\bar{y} = \dfrac{\int_1^e \dfrac{1}{2}(\ln x)^2\, dx}{1} =$

$\dfrac{1}{2}\int_1^e (\ln x)^2\, dx$. To evaluate this integral, let $u = (\ln x)^2$ and $dv = dx$. Then $du = \dfrac{2\ln x}{x}\, dx$, $v = x$,

and $\dfrac{1}{2}\int_1^e (\ln x)^2\, dx = \dfrac{1}{2}\Bigg[x(\ln x)^2\Big|_1^e -$

$\int_1^e x\left(\dfrac{2\ln x}{x}\right)dx\Bigg] = \dfrac{e}{2} - (x\ln x - x)\Big|_1^e = \dfrac{e}{2} - 1$

$= \dfrac{e-2}{2}$. Thus $(\bar{x}, \bar{y}) = \left(\dfrac{e^2+1}{4}, \dfrac{e-2}{2}\right)$.

10 Set up coordinate axes as in the figure. When the half disk is revolved around the y axis, a sphere of

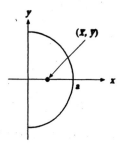

radius a results. Now $\bar{y} = 0$ by symmetry, and

$\dfrac{4}{3}\pi a^3 = 2\pi \bar{x} \cdot \dfrac{1}{2}\pi a^2$ by Pappus's theorem. Thus

$2\pi \bar{x} = \frac{8a}{3}$ and $\bar{x} = \frac{4a}{3\pi}$. Hence for the half disk, $(\bar{x}, \bar{y})$

$= (4a/3\pi, 0)$.

12

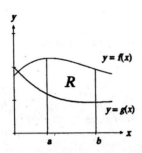

(a) The moment of R about the y axis is

$$\int_a^b xc(x)\, dx = \int_a^b x[f(x) - g(x)]\, dx.$$

(b) The following procedure is similar to that used in establishing equation (10). Consider the region R between x and $x + dx$. This region is approximately a rectangle with area $(f(x) - g(x))\, dx$. The y coordinate of the centroid of this region is approximately $\frac{f(x) + g(x)}{2}$. But the y coordinate of the centroid is equal to the moment of the rectangular region about the x axis divided by its area, so the moment of the rectangle about the x axis is $\left(\frac{f(x) + g(x)}{2}\right)(f(x) - g(x))\, dx$.

Integrating over $[a, b]$ the total moment of R about the x axis is

$$\int_a^b \frac{[f(x) + g(x)][f(x) - g(x)]}{2}\, dx =$$

$$\int_a^b \frac{(f(x))^2 - (g(x))^2}{2}\, dx.$$

14 (a) $\int_1^2 x(2x - x)\, dx = \frac{7}{3}$

(b) $\int_1^2 \frac{1}{2}[(2x)^2 - (x)^2]\, dx = \frac{7}{2}$

(c) $\int_1^2 (2x - x)\, dx = \frac{3}{2}$

(d) $\bar{x} = \frac{7/3}{3/2} = \frac{14}{9}$

(e) $\bar{y} = \frac{7/2}{3/2} = \frac{7}{3}$

16 (a) $\int_1^e x(x - 1 - \ln x)\, dx = \frac{1}{12}(4e^3 - 9e^2 - 1)$

(b) $\int_1^e \frac{1}{2}[(x - 1)^2 - (\ln x)^2]\, dx$

$= \frac{1}{6}(e^3 - 3e^2 + 5)$

(c) $\int_1^e (x - 1 - \ln x)\, dx = \frac{1}{2}(e^2 - 2e - 1)$

(d) $\bar{x} = \frac{4e^3 - 9e^2 - 1}{6(e^2 - 2e - 1)}$

8.6 The Centroid of a Plane Region

(e) $\bar{y} = \dfrac{e^3 - 3e^2 + 5}{3(e^2 - 2e - 1)}$

18 (a)

$$y = x^a$$ graph on $[0,1]$, region R.

(b) $A = \int_0^1 x^a \, dx = \dfrac{1}{a+1}$; $\bar{x}$

$= \dfrac{1}{A} \int_0^1 x \cdot x^a \, dx = \dfrac{a+1}{a+2}$; $\bar{y} =$

$\dfrac{1}{A} \int_0^1 \dfrac{1}{2}(x^a)^2 \, dx = \dfrac{a+1}{2(2a+1)}$; the centroid

is $\left(\dfrac{a+1}{a+2}, \dfrac{a+1}{2(2a+1)}\right)$.

(c) $\lim_{a \to \infty} \bar{x} = 1$, $\lim_{a \to \infty} \bar{y} = \dfrac{1}{4}$

(d) Since $\lim_{a \to \infty} (\bar{x}^a) = \lim_{a \to \infty} \left(1 - \dfrac{1}{a+2}\right)^a = \dfrac{1}{e}$,

$\lim_{a \to \infty} (\bar{x}^a - \bar{y}) = \dfrac{1}{e} - \dfrac{1}{4} > 0$. Hence, for

large a, $\bar{y} < \bar{x}^a$, so $(\bar{x}, \bar{y})$ is in R.

20 (a) $\bar{x} = \dfrac{\int_0^b x\left(\dfrac{c}{b}x\right) dx + \int_b^a x\left[\dfrac{c}{b-a}(x-a)\right] dx}{ac/2}$

$= \dfrac{b+a}{3}$ and $\bar{y} =$

$\dfrac{\dfrac{1}{2}\int_0^b \left(\dfrac{c}{b}x\right)^2 dx + \dfrac{1}{2}\int_b^a \left[\dfrac{c}{b-a}(x-a)\right]^2 dx}{ac/2}$

$= \dfrac{c}{3}$ so $(\bar{x}, \bar{y}) = \left(\dfrac{b+a}{3}, \dfrac{c}{3}\right)$.

(b) $y = \dfrac{c}{b - a/2}\left(x - \dfrac{a}{2}\right)$

(c) $\dfrac{c}{b - a/2}\left(\dfrac{b+a}{3} - \dfrac{a}{2}\right) = \dfrac{c}{b - a/2}\left(\dfrac{b}{3} - \dfrac{a}{6}\right)$

$= \dfrac{2c}{2b - a}\left(\dfrac{2b - a}{6}\right) = \dfrac{c}{3}$

8.7 Work

2 $k = F/x$, so $k = 24/3 = 8$.

(a) Work $= \int_0^3 8x \, dx = 36$ joules

(b) Work $= \int_0^4 8x \, dx = 64$ joules

4 Work $= \int_0^{0.5} 2\sqrt{x} \, dx = \dfrac{\sqrt{2}}{3}$ joules

6   Work = $\int_{20000}^{240,000} \left[\dfrac{4000}{r}\right]^2 dr$

$= -4000^2 \left[\dfrac{1}{240,000} - \dfrac{1}{20000}\right] = \dfrac{2200}{3}$

≈ 733 mi-lb

8   Let the smaller mountain have height h and radius r. Then the work W is given by

$\int_0^h (h-x)k\pi\left(\dfrac{r}{h}x\right)^2 dx = \dfrac{k\pi r^2}{h^2}\cdot\dfrac{h^4}{12} = \dfrac{1}{12}k\pi r^2 h^2$

joules. The work required to create the larger mountain is $\int_0^{2h} (2h-x)k\pi\left(\dfrac{r}{h}x\right)^2 dx =$

$\dfrac{k\pi r^2}{h^2}\cdot\dfrac{(2h)^4}{12} = 16\left(\dfrac{1}{12}k\pi r^2 h^2\right) = 16W.$

10 Let the smaller hill have height h and radius r, let the larger hill have height a and radius b, and let the density of the rubbish be k. From Exercise 8, $W = \dfrac{1}{2}k\pi r^2 h^2$ joules. For the larger hill to be similar in shape to the smaller hill, it is required that $a/h = b/r = k$ for some $k > 1$. Since the larger hill has twice the volume of the smaller hill, we have $\dfrac{1}{3}\pi b^2 a = \dfrac{2}{3}\pi r^2 h$ or $k^3\left(\dfrac{1}{3}\pi r^2 h\right) =$

$2\left(\dfrac{1}{3}\pi r^2 h\right)$, so $k = 2^{1/3}$. Now the work required to raise the larger hill is $\int_0^{2^{1/3}h} (2^{1/3}h - x)k\pi\left(\dfrac{r}{h}x\right)^2 dx$

$= \dfrac{k\pi r^2}{h^2}\cdot\dfrac{(2^{1/3}h)^4}{12} = 2^{4/3}W.$

12 Work $= \int_a^b \sigma(b-x)hc(x)\, dx$

$= \sigma h \int_a^b (b-x)c(x)\, dx$, where σ is the density of the water in pounds (mass) per cubic foot.

8.8 Improper Integrals

2   Divergent: $\int_1^\infty x^{-1/3}\, dx = \lim_{b\to\infty} \dfrac{3}{2}x^{2/3}\Big|_1^b =$

$\lim_{b\to\infty} \dfrac{3}{2}(b^{2/3} - 1) = \infty$

4   Divergent: $\int_0^\infty \dfrac{dx}{x+100} = \lim_{b\to\infty} \ln(x+100)\Big|_0^b =$

$\lim_{b\to\infty} \ln(b+100) - \ln 100 = \infty$

6   $\int_1^\infty x^{-1.01}\, dx = \lim_{b\to\infty} \int_1^b x^{-1.01}\, dx =$

$\lim_{b\to\infty} \dfrac{x^{-0.01}}{-0.01}\Big|_1^b = \lim_{b\to\infty}\left(100 - \dfrac{100}{b^{0.01}}\right) = 100.$ The improper integral is convergent.

8   $\int_0^\infty \sin 2x\, dx = \lim_{b\to\infty} \int_0^b \sin 2x\, dx =$

$\lim_{b\to\infty}\left(-\dfrac{1}{2}\cos 2x\right)\Big|_0^b = \lim_{b\to\infty}\left(\dfrac{1}{2} - \dfrac{1}{2}\cos 2b\right).$ This limit does not exist, so the improper integral is divergent.

10 For $x \geq 0$, $\dfrac{1}{\sqrt{1+x^3}} \leq \dfrac{1}{\sqrt{x^3}} = x^{3/2}$. Since

$\int_1^\infty x^{-3/2}\, dx = \lim_{b\to\infty}\int_1^b x^{-3/2}\, dx = \lim_{b\to\infty}(-2x^{-1/2})\Big|_1^b$

$= \lim_{b\to\infty}(2 - 2b^{-1/2}) = 2$, Theorem 1 implies that

$\int_1^\infty \dfrac{dx}{\sqrt{1+x^3}}$ is convergent. $\int_0^1 \dfrac{dx}{\sqrt{1+x^3}}$ is proper, so $\int_0^\infty \dfrac{dx}{\sqrt{1+x^3}}$ also converges.

12 Note that $\left|\dfrac{e^{-x}\sin x^2}{x+1}\right| \le \dfrac{e^{-x}}{x+1} \le e^{-x}$ for $x \ge 0$.

Since $\int_0^\infty e^{-x}\,dx$ converges, $\int_0^\infty \left|\dfrac{e^{-x}\sin x^2}{x+1}\right| dx$

converges. By Theorem 3, $\int_0^\infty \dfrac{e^{-x}\sin x^2}{x+1}\,dx$

converges as well.

14 $\int_0^\infty \dfrac{dx}{x^2+4} = \lim\limits_{b\to\infty} \dfrac{1}{2}\tan^{-1}\dfrac{x}{2}\Big|_1^b$

$= \lim\limits_{b\to\infty}\left[\dfrac{1}{2}\tan^{-1}\dfrac{b}{2} - \dfrac{1}{2}\tan^{-1}\dfrac{1}{2}\right]$

$= \dfrac{\pi}{4} - \dfrac{1}{2}\tan^{-1}\dfrac{1}{2}$, so $\int_0^\infty \dfrac{dx}{x^2+4}$ converges.

16 $\int_0^\infty e^{-2x}\sin 3x\,dx = \lim\limits_{b\to\infty}\int_0^b e^{-2x}\sin 3x\,dx$

$= \lim\limits_{b\to\infty} \dfrac{e^{-2x}}{13}(-2\sin 3x - 3\cos 3x)\Big|_0^b$

$= \lim\limits_{b\to\infty}\left[\dfrac{e^{-2b}}{13}(-2\sin 3b - 3\cos 3b) + \dfrac{3}{13}\right] = \dfrac{3}{13}$

18 $\int_0^\infty \dfrac{dx}{(x+1)(x+2)(x+3)}$

$= \lim\limits_{b\to\infty}\int_0^b \dfrac{dx}{(x+1)(x+2)(x+3)}$

$= \lim\limits_{b\to\infty}\int_0^b \left[\dfrac{1/2}{x+1} - \dfrac{1}{x+2} + \dfrac{1/2}{x+3}\right]dx$

$= \lim\limits_{b\to\infty}\left[\dfrac{1}{2}\ln(x+1) - \ln(x+2) + \dfrac{1}{2}\ln(x+3)\right]\Big|_0^b =$

$\lim\limits_{b\to\infty}\left[\ln\dfrac{\sqrt{(x+1)(x+3)}}{(x+2)}\right]\Big|_0^b =$

$\lim\limits_{b\to\infty}\left[\ln\sqrt{\dfrac{(b+1)(b+3)}{(b+2)^2}} - \ln\sqrt{\dfrac{3}{4}}\right] = \dfrac{1}{2}\ln\dfrac{4}{3}$, so

$\int_0^\infty \dfrac{dx}{(x+1)(x+2)(x+3)}$ converges.

20 Letting $u = x^2$ and using Formula 33 from the table of antiderivatives, we have $\int_0^\infty \dfrac{x\,dx}{\sqrt{1+x^4}} =$

$\dfrac{1}{2}\int_0^\infty \dfrac{du}{\sqrt{1+u^2}} = \lim\limits_{b\to\infty}\dfrac{1}{2}\ln\left(u + \sqrt{1+u^2}\right)\Big|_0^b = \infty$.

22 $\int_0^1 \dfrac{dx}{(x-1)^2} = \lim\limits_{b\to 1^-}\int_0^b \dfrac{dx}{(x-1)^2}$

$= \lim\limits_{b\to 1^-}\left[-(x-1)^{-1}\right]\Big|_0^b = \lim\limits_{b\to 1^-}\left[-\dfrac{1}{b-1} - 1\right] = \infty$,

so $\int_0^1 \dfrac{dx}{(x-1)^2}$ diverges.

24 The area of R is $\int_1^\infty \left(\dfrac{1}{x} - \dfrac{1}{x^2}\right)dx =$

$\lim\limits_{b\to\infty}\int_1^b \left(\dfrac{1}{x} - \dfrac{1}{x^2}\right)dx = \lim\limits_{b\to\infty}\left(\ln x + \dfrac{1}{x}\right)\Big|_1^b =$

$\lim\limits_{b\to\infty}\left(\ln b + \dfrac{1}{b} - \ln 1 - 1\right) = \infty$. The area is infinite.

26 One method involves writing $\int_0^\infty \dfrac{dx}{\sqrt{1+x^4}} =$

$\int_0^a \dfrac{dx}{\sqrt{1+x^4}} + \int_a^\infty \dfrac{dx}{\sqrt{1+x^4}}$, where $a > 0$ will be chosen in a moment. For all x, $\dfrac{1}{\sqrt{1+x^4}} < \dfrac{1}{x^2}$

and $0 \leq \displaystyle\int_a^\infty \dfrac{dx}{\sqrt{1+x^4}} \leq \int_a^\infty \dfrac{dx}{x^2} = \lim_{b\to\infty}\left(-\dfrac{1}{x}\right)\Big|_a^b$

$= \dfrac{1}{a}$. Pick a large enough so that $1/a < 0.005$

(for example). Then we can use use the error bound on Simpson's method to compute

$\displaystyle\int_0^a \dfrac{dx}{\sqrt{1+x^4}}$ with an error less than 0.005. The

combined error will then be less than 0.01, as desired.

28 $\sqrt{(a^2+x)(b^2+x)(c^2+x)} < \sqrt{x^3}$, so

$\displaystyle\int_1^\infty \dfrac{dx}{\sqrt{(a^2+x)(b^2+x)(c^2+x)}}$ converges by

comparison with $\displaystyle\int_1^\infty x^{-3/2}\,dx$. Since

$\displaystyle\int_0^1 \dfrac{dx}{\sqrt{(a^2+x)(b^2+x)(c^2+x)}}$ is finite, the given

integral converges.

30 Assume $p \neq 1$, so $\displaystyle\int_0^1 \dfrac{dx}{x^p} = \lim_{t\to 0^+}\int_t^1 \dfrac{dx}{x^p} =$

$\displaystyle\lim_{t\to 0^+}\dfrac{x^{1-p}}{1-p}\Big|_t^1 = \lim_{t\to 0^+}\left(\dfrac{1}{1-p} - \dfrac{t^{1-p}}{1-p}\right)$, which is

finite only for $p < 1$. If $p = 1$, then $\displaystyle\int_0^1 \dfrac{dx}{x} =$

$\displaystyle\lim_{t\to 0^+}\ln x\Big|_t^1 = \infty$. Thus $\displaystyle\int_0^1 \dfrac{dx}{x^p}$ converges only for

$p < 1$.

32 No. Let $f(x) = 1/x$.

34 (a) Since $\displaystyle\int_1^\infty \dfrac{1}{x^2}\,dx = \lim_{b\to\infty}\left(-\dfrac{1}{x}\right)\Big|_1^b = 1$ is

convergent and $0 \leq \left|\dfrac{\cos x}{x^2}\right| \leq \dfrac{1}{x^2}$ for

$x \neq 0$, Theorem 1 shows that $\displaystyle\int_1^\infty \left|\dfrac{\cos x}{x^2}\right| dx$

is convergent. By Theorem 3, $\displaystyle\int_1^\infty \dfrac{\cos x}{x^2}\,dx$ is

convergent.

(b) Let $u = \dfrac{1}{x}$ and $dv = \sin x\,dx$. Then $du =$

$-\dfrac{1}{x^2}\,dx$, $v = -\cos x$, and $\displaystyle\int \dfrac{\sin x}{x}\,dx$

$= \dfrac{1}{x}(-\cos x) - \displaystyle\int (-\cos x)\left(-\dfrac{1}{x^2}\,dx\right)$

$= -\dfrac{\cos x}{x} - \displaystyle\int \dfrac{\cos x}{x^2}\,dx$. Hence $\displaystyle\int_1^\infty \dfrac{\sin x}{x}\,dx$

$= \displaystyle\lim_{b\to\infty}\left(-\dfrac{\cos x}{x}\Big|_1^b - \int_1^b \dfrac{\cos x}{x^2}\,dx\right)$

$= \displaystyle\lim_{b\to\infty}\left(\cos 1 - \dfrac{\cos b}{b} - \int_1^b \dfrac{\cos x}{x^2}\,dx\right)$

$= \cos 1 - \displaystyle\lim_{b\to\infty}\int_1^b \dfrac{\cos x}{x^2}\,dx$. Since

$\displaystyle\int_1^\infty \dfrac{\cos x}{x^2}\,dx$ is convergent by part (a), so is

$\displaystyle\int_1^\infty \dfrac{\sin x}{x}\,dx$.

(c) $\displaystyle\int_0^\infty \dfrac{\sin x}{x}\,dx$ is improper for two reasons:

The integrand is undefined at $x = 0$ and the

upper limit of integration is infinite. So we write it as $\int_0^1 \frac{\sin x}{x}\,dx + \int_1^\infty \frac{\sin x}{x}\,dx$. By (b), the second integral is convergent, so we need show only that $\int_0^1 \frac{\sin x}{x}\,dx$ is convergent. Let $f(x) = \frac{\sin x}{x}$ for $x \neq 0$ and define $f(0) = 1$. Since $\lim_{x \to 0} \frac{\sin x}{x} = 1$, f is continuous on $[0, 1]$, so $\int_0^1 f(x)\,dx$ exists.

But then $\int_0^1 \frac{\sin x}{x}\,dx = \lim_{t \to 0^+} \int_t^1 \frac{\sin x}{x}\,dx = \lim_{t \to 0^+} \int_t^1 f(x)\,dx = \int_0^1 f(x)\,dx$, so $\int_0^1 \frac{\sin x}{x}\,dx$ converges. Hence $\int_0^\infty \frac{\sin x}{x}\,dx$ converges.

(d) Letting $u = e^x$, we have $du = e^x\,dx = u\,dx$, so $dx = du/u$ and $\int \sin(e^x)\,dx = \int \frac{\sin u}{u}\,du$; thus $\int_0^\infty \sin(e^x)\,dx = \lim_{b \to \infty} \int_0^b \sin(e^x)\,dx = \lim_{b \to \infty} \int_1^{e^b} \frac{\sin u}{u}\,du = \int_1^\infty \frac{\sin u}{u}\,du$, which converges by (b).

36 $P(r) = \int_0^\infty e^{-rt} t\,dt = \lim_{b \to \infty} \int_0^b t e^{-rt}\,dt$

$= \lim_{b \to \infty} \left[\frac{e^{-rt}}{r^2}(-rt - 1) \right]\Big|_0^b$

$= \lim_{b \to \infty} \left[\frac{e^{-rb}}{r^2}(-rb - 1) - \frac{e^{-r \cdot 0}}{r^2}(-r \cdot 0 - 1) \right]$

$= \lim_{b \to \infty} \left[\frac{1}{r^2} - \frac{rb + 1}{r^2 e^{rb}} \right] = \frac{1}{r^2}$. (The integral can be done with integration by parts or with Equation 63 from the integral table on the inside front cover of the text.)

38 $P(r) = \int_0^\infty e^{-rt} e^t\,dt = \int_0^\infty e^{-(r-1)t}\,dt =$

$\lim_{b \to \infty} \int_0^b e^{-(r-1)t}\,dt = \lim_{b \to \infty} \frac{-e^{-(r-1)t}}{r - 1}\Big|_0^b =$

$\lim_{b \to \infty} \left(\frac{-e^{-(r-1)b}}{r - 1} + \frac{1}{r - 1} \right) = \frac{1}{r - 1}$. (The assumption $r > 1$ was needed to ensure that $\lim_{b \to \infty} e^{-(r-1)b} = 0$.)

40 $P(r) = \int_0^\infty e^{-rt} \cos t\,dt$. Using a table of integrals or integration by parts (twice), we have

$\int e^{-rt} \cos t\,dt = \frac{e^{-rt}}{r^2 + 1}(\sin t - r \cos t) + C$.

Thus, $P(r) = \lim_{b \to \infty} \int_0^b e^{-rt} \cos t\,dt =$

$\lim_{b \to \infty} \left[\frac{e^{-rt}}{r^2 + 1}(\sin t - r \cos t) \right]\Big|_0^b =$

$\lim_{b \to \infty} \left[\frac{e^{-rb}}{r^2 + 1}(\sin b - r \cos b) - \frac{1}{r^2 + 1}(0 - r) \right]$

$= \frac{r}{r^2 + 1}$. (Notice that since $|\sin b - r \cos b| \leq |r| + 1$, $\lim_{b \to \infty} \frac{e^{-rb}}{r^2 + 1}(\sin b - r \cos b) = 0$.)

42 $Q(r) = \int_0^\infty e^{-rt} g(t)\, dt = \int_0^\infty e^{-rt} f(at)\, dt$. Let $x = at$. Then $dx = a\, dt$ and x varies from 0 to ∞ (recall that $a > 0$). Therefore $Q(r) =$

$\int_0^\infty e^{-r(x/a)} f(x)\, \dfrac{dx}{a} = \dfrac{1}{a} \int_0^\infty e^{-(r/a)x} f(x)\, dx = \dfrac{1}{a} P(r/a)$.

8.S Review Exercises

2 The cross section perpendicular to the diameter of the base and at a distance x inches from the axis of the cylinder is a right triangle with base $\sqrt{9 - x^2}$ inches and height $\dfrac{5}{3}\sqrt{9 - x^2}$ inches. Hence its area is $\dfrac{5}{6}(9 - x^2)$ and the volume is $\int_{-3}^{3} \dfrac{5}{6}(9 - x^2)\, dx$.

4 The region is described by $0 \le y \le 1$, $y^{1/2} \le x \le y^{1/3}$. Hence the moment is

$\int_0^1 (y + 2)(y^{1/3} - y^{1/2})\, dy$.

6 (a) $\int_0^1 \dfrac{dx}{(1 + x)^2} = \dfrac{1}{2}$

   (b) $\int_0^1 \pi \left(\dfrac{1}{(1 + x)^2}\right)^2 dx = \dfrac{7\pi}{24}$

   (c) $\int_0^1 2\pi x \cdot \dfrac{1}{(1 + x)^2}\, dx = \pi(2 \ln 2 - 1)$

   (d) $\int_0^1 \pi \left[\left(1 + \dfrac{1}{(1 + x)^2}\right)^2 - 1^2\right] dx = \dfrac{31\pi}{24}$

8 (a) $\int_3^4 \sqrt{x^2 - 9}\, dx = 2\sqrt{7} - \dfrac{9}{2} \ln\left(\dfrac{4 + \sqrt{7}}{3}\right)$

   (b) $\int_3^4 \pi \left(\sqrt{x^2 - 9}\right)^2 dx = \dfrac{10\pi}{3}$

   (c) $\int_3^4 2\pi x\sqrt{x^2 - 9}\, dx = \dfrac{14\sqrt{7}\pi}{3}$

   (d) $\int_3^4 \pi\left[\left(1 + \sqrt{x^2 - 9}\right)^2 - 1^2\right] dx$

   $= \pi\left[\dfrac{10}{3} + 4\sqrt{7} - 9 \ln\left(\dfrac{4 + \sqrt{7}}{3}\right)\right]$

10 $\int_{\pi/6}^{\pi/4} \dfrac{1}{2}(\sec x)^2\, dx = \dfrac{3 - \sqrt{3}}{6}$

12 $\int_0^1 \dfrac{1}{2}\left(\dfrac{1}{\sqrt{x^2 + 1}}\right)^2 dx = \dfrac{1}{2} \int_0^1 \dfrac{dx}{x^2 + 1} = \dfrac{\pi}{8}$

14 $\int_0^1 \left(\dfrac{1}{x^2 + 3x + 2} - \dfrac{1}{x^2 + 3x + 4}\right) dx$

   $= \int_0^1 \left(\dfrac{1}{x + 1} - \dfrac{1}{x + 2} - \dfrac{4}{(2x + 3)^2 + 7}\right) dx$

   $= \ln \dfrac{4}{3} - \dfrac{2}{\sqrt{7}} \tan^{-1} \dfrac{\sqrt{7}}{11}$

16 $\int_0^4 x\sqrt{2x + 1}\, dx = \dfrac{298}{15}$

18 $\int_0^1 \dfrac{1 - x}{\sqrt{4 - x^2}}\, dx = \dfrac{\pi}{6} + \sqrt{3} - 2$

20 (a) $\int_0^{\pi/4} \tan x\, dx = \dfrac{1}{2} \ln 2$

   (b) $\int_0^{\pi/4} \dfrac{1}{2}(\tan x)^2\, dx = \dfrac{4 - \pi}{8}$

   (c) $\bar{y} = \dfrac{\dfrac{4 - \pi}{8}}{\dfrac{1}{2} \ln 2} = \dfrac{4 - \pi}{4 \ln 2}$

(d) $\int_0^{\pi/4} x \tan x \, dx$

(e) $\int_0^{\pi/4} \pi[(1 + \tan x)^2 - 1^2] \, dx$

$= \pi\left(\ln 2 + 1 - \dfrac{\pi}{4}\right)$

22 Convergent. Since $\dfrac{1}{\sqrt{x}\sqrt{x+1}\sqrt{x+2}} < \dfrac{1}{\sqrt{x}\sqrt{x}\sqrt{x}} = x^{-3/2}$, $\int_1^\infty \dfrac{dx}{\sqrt{x}\sqrt{x+1}\sqrt{x+2}}$ converges by comparison with $\int_1^\infty x^{-3/2} \, dx$. Since

$\dfrac{1}{\sqrt{x}\sqrt{x+1}\sqrt{x+2}} < \dfrac{1}{\sqrt{x}\cdot 1 \cdot 1} = x^{-1/2}$,

$\int_0^1 \dfrac{dx}{\sqrt{x}\sqrt{x+1}\sqrt{x+2}}$ converges by comparison with $\int_0^1 x^{-1/2} \, dx$.

24 Let the ellipse be $\left(\dfrac{x}{a}\right)^2 + \left(\dfrac{y}{b}\right)^2 = 1$. As shown, the

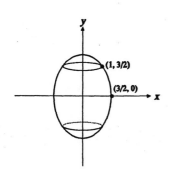

conditions imply that $\left(\dfrac{3}{2}, 0\right)$ and $\left(1, \dfrac{3}{2}\right)$ are on the ellipse, so $a = \dfrac{3}{2}$ and $b = \dfrac{9}{2\sqrt{5}}$. Hence, on the right half, $x = \sqrt{\dfrac{9}{4} - \dfrac{5}{9}y^2}$. The volume is

$\int_{-3/2}^{3/2} \pi x^2 \, dy = \pi \int_{-3/2}^{3/2} \left(\dfrac{9}{4} - \dfrac{5}{9}y^2\right) dy = \dfrac{11\pi}{2}.$

26 Introduce coordinates so that l is the y axis and A is in the left half-plane. Then $A = (-2a, d)$, $B = (a, b)$, and $C = (a, c)$ for some $a, b, c,$ and d with $a > 0$. The vertical cross section with coordinate x has length $\dfrac{b-c}{3a}(x + 2a)$. Hence $\bar{x} =$

$\int_{-2a}^{a} x \cdot \dfrac{b-c}{3a}(x + 2a) \, dx$

$= \dfrac{b-c}{3a} \int_{-2a}^{a} (x^2 + 2ax) \, dx = 0.$

28 $\int_0^\infty e^{-rx} \sin ax \, dx$

$= -\dfrac{e^{-rx}}{a^2 + r^2}(r \sin ax + a \cos ax)\Big|_0^\infty = \dfrac{a}{a^2 + r^2}$

30 (a) $\lim\limits_{b \to 0} \dfrac{\int_0^b f(x) \, dx}{b} \underset{H}{=} \lim\limits_{b \to 0} \dfrac{f(b)}{1} = f(0) = 2$

(b) $\lim\limits_{b \to \infty} \dfrac{\int_0^b f(x) \, dx}{b} \underset{H}{=} \lim\limits_{b \to \infty} \dfrac{f(b)}{1} = 3$

32 (a) 0

(b) $G(a) = \tan^{-1} ax \Big|_0^\infty = -\pi/2$

(c) $G(a) = \tan^{-1} ax \Big|_0^\infty = \pi/2$

(d)

34 Divergent: $\int_0^\infty \frac{dx}{(x-1)^2} =$

$\int_0^1 \frac{dx}{(x-1)^2} + \int_1^2 \frac{dx}{(x-1)^2} + \int_2^\infty \frac{dx}{(x-1)^2}$

$= \lim_{b \to 1^-} \left(-\frac{1}{x-1}\right)\bigg|_0^b + \lim_{a \to 1^+} \left(-\frac{1}{x-1}\right)\bigg|_a^2 +$

$\lim_{b \to \infty} \left(-\frac{1}{x-1}\right)\bigg|_2^b$, but $\lim_{b \to 1^-} \left(-\frac{1}{x-1}\right)\bigg|_0^b = \infty$.

36 (a)

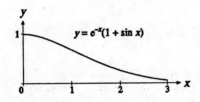

(b) $\int_0^\infty 2\pi x e^{-x}(1 + \sin x)\, dx = 2\pi \int_0^\infty x e^{-x}\, dx$

$+ 2\pi \int_0^\infty x e^{-x} \sin x\, dx = 2\pi[-(x+1)e^{-x}]\big|_0^\infty$

$+ 2\pi\left[-\frac{1}{2}e^{-x}[(x+1)\cos x + x \sin x]\right]\bigg|_0^\infty =$

3π

38 No. The integral diverges, since $\ln|2x + 1| \to \infty$

as $x \to -\frac{1}{2}$.

40 $\int_0^\infty \frac{x^{-1/2}}{1 + x}\, dx = \lim_{b \to \infty} 2\tan^{-1}\sqrt{b} - \lim_{a \to 0^+} 2\tan^{-1}\sqrt{a}$

$= \pi = \pi \csc \frac{\pi}{2}$, as claimed. (Note that the

integral is improper at both 0 and ∞.)

42 $\int_0^\infty \frac{dx}{1 + x^4} = \lim_{a \to 0^+} \int_a^1 \frac{dx}{1 + x^4} + \lim_{b \to \infty} \int_1^b \frac{dx}{1 + x^4}$

$= \lim_{a \to 0^+} \int_{1/a}^1 \frac{-dy/y^2}{1 + y^{-4}} + \lim_{b \to \infty} \int_1^{1/b} \frac{-dy/y^2}{1 + y^{-4}}$

$= \lim_{b \to \infty} \int_{1/b}^1 \frac{y^2\, dy}{1 + y^4} + \lim_{a \to 0^+} \int_1^{1/a} \frac{y^2\, dy}{1 + y^4}$

$= \int_0^\infty \frac{y^2\, dy}{1 + y^4}$

44 $\pi \int_0^{\pi/3} e^{4x} \sin^2 3x\, dx = \frac{9\pi}{104}(e^{4\pi/3} - 1)$

46 The region is described by $2 \leq x \leq 4$, $2^x \leq y \leq x^2$. Its area is $A = \int_2^4 (x^2 - 2^x)\, dx =$

$\frac{4(14 \ln 2 - 9)}{3 \ln 2}$. Hence $\bar{x} = \frac{1}{A} \int_2^4 x(x^2 - 2^x)\, dx$

$= \frac{1}{A}\left(60 - \frac{56}{\ln 2} + \frac{12}{(\ln 2)^2}\right) =$

$\frac{45(\ln 2)^2 - 42 \ln 2 + 9}{(\ln 2)(14 \ln 2 - 9)}$ and $\bar{y} =$

$\frac{1}{A} \int_2^4 \frac{1}{2}[(x^2)^2 - (2^x)^2]\, dx = \frac{1}{2A} \int_2^4 (x^4 - 4^x)\, dx$

$= \frac{1}{2A}\left(\frac{1}{5}x^5 - \frac{1}{\ln 4} \cdot 4^x\right)\bigg|_2^4 = \frac{1}{2A}\left(\frac{992}{5} - \frac{120}{\ln 2}\right) =$

$\frac{3(124 \ln 2 - 75)}{5(14 \ln 2 - 9)}$.

48 Convergent. Since $\lim_{x \to 1} \frac{\ln x}{1 - x^2} =_H \lim_{x \to 1} \frac{1/x}{-2x} =$

$-\frac{1}{2}$, the integrand is bounded on $[1/2, 1]$, so

$\int_{1/2}^1 \frac{\ln x}{1 - x^2}\, dx$ converges. For $0 < x \leq 1/2$,

$1 - x^2 \geq 3/4$, so $\left|\frac{\ln x}{1 - x^2}\right| \leq -\frac{4}{3} \ln x$. But

$\int \ln x \, dx = x \ln x - x \to 0$ as $x \to 0^+$, so $\int_0^{1/2} \left(-\frac{4}{3} \ln x\right) dx$ converges and $\int_0^{1/2} \frac{\ln x}{1 - x^2} dx$ converges by comparison.

50. (a) $v = A \cos t$ so $v^2 = A^2 \cos^2 t = A^2(1 - \sin^2 t) = A^2\left[1 - \left(\frac{x}{A}\right)^2\right] = A^2 - x^2$.

    Hence the average with respect to distance is
    $$\frac{1}{A} \int_0^A (A^2 - x^2) \, dx = \frac{2}{3} A^2.$$

    (b) The average with respect to time is
    $$\frac{1}{\pi/2} \int_0^{\pi/2} A^2 \cos^2 t \, dt = \frac{1}{\pi/2} \cdot A^2 \cdot \frac{\pi}{4} = \frac{1}{2} A^2.$$

52. Finite. $\int_1^\infty \frac{\ln x}{x^2} dx = -\frac{1 + \ln x}{x}\Big|_1^\infty = 1$, since $\lim\limits_{b \to \infty} \frac{\ln x}{x} = 0$.

54. (a) $\int_1^\infty \frac{dx}{x} = \lim\limits_{b \to \infty} \int_1^b \frac{dx}{x} = \lim\limits_{b \to \infty} \ln x \Big|_1^b = \lim\limits_{b \to \infty} \ln b = \infty$, while $\int_1^\infty \pi \left(\frac{1}{x}\right)^2 dx = \pi \lim\limits_{b \to \infty} \int_1^b \frac{dx}{x^2} = \pi \lim\limits_{b \to \infty} \left(-\frac{1}{x}\right)\Big|_1^b = \pi \lim\limits_{b \to \infty} \left(1 - \frac{1}{b}\right) = \pi$.

    (b) If we ignore such matters as surface tension and molecular size, we see that it is theoretically possible to fill the infinite "trumpet" of part (a) with a finite amount of paint. However, R fits within this volume, although its area is infinite and should therefore be impossible to paint with anything less than infinitely many buckets of paint. The apparent contradiction lies in what we mean by painting a surface. It is customary to require the coat of paint to have a uniform thickness; otherwise there is no theoretical limit to the surface that may be painted. For example, suppose you had a gallon of paint and an infinite supply of identical planks. Use half a gallon to paint one plank, one quarter of a gallon to paint the second, one-eighth of a gallon for a third, and so on. At each step you use only half of your remaining paint, so you never run out; however, the planks are getting increasingly stingy coats of paint. Suppose you had specified that R should be painted with a coat of uniform thickness 10^{-3} units. For $x > 1000$, S has cross-sectional radius $\frac{1}{x} < \frac{1}{1000}$, so such a coat of paint would not fit within the solid. This is true for any specified thickness, so the infinite region could be painted by dipping it into the finite volume only if one permitted the thickness of the coat to approach zero as $x \to \infty$.

56. (a) Improper, since integrand is undefined at $x = 0$.

    (b) Let $u = \sin x$. Then $\lim\limits_{x \to 0} \frac{\sin x}{e^{0.26 \sin x} - 1} = \lim\limits_{u \to 0} \frac{u}{e^{0.26u} - 1} \stackrel{H}{=} \lim\limits_{u \to 0} \frac{1}{0.26 e^{0.26u}} = \frac{1}{0.26}$.

(c) The integrand has a nonelementary antiderivative.

(d) $h = \frac{\pi}{12}$, so Simpson's estimate is

$$\frac{\pi}{36}\left[f(0) + 4f\left(\frac{\pi}{12}\right) + 2f\left(\frac{\pi}{6}\right) + 4f\left(\frac{\pi}{4}\right) + 2f\left(\frac{\pi}{3}\right) + 4f\left(\frac{5\pi}{12}\right) + f\left(\frac{\pi}{2}\right)\right] \approx 5.5585138,$$

where $f(x)$ is the integrand.

58 (a) $\ln f(0) = f(0) \int_0^0 f(t)\, dt = 0$, so $f(0) = 1$.

(b) Differentiating $\frac{\ln f(x)}{f(x)} = \int_0^x f(t)\, dt$ gives

$$\frac{(1 - \ln f(x))f'(x)}{[f(x)]^2} = f(x), \text{ so } f'(x) = \frac{[f(x)]^3}{1 - \ln f(x)}.$$ Hence $f'(0) = 1$.

(c) Differentiating $1 - \ln f(x) = \frac{[f(x)]^3}{f'(x)}$ gives

$$-\frac{f'(x)}{f(x)} = \frac{3[f(x)]^2[f'(x)]^2 - [f(x)]^3 f''(x)}{[f'(x)]^2}, \text{ so}$$

$$f''(x) = \frac{[f'(x)]^2}{[f(x)]^4}(3[f(x)]^3 + f'(x)).$$ Hence $f''(0) = 4$.

60 (a) $\int_{-\infty}^\infty f(x)\, dx = \int_0^\infty k e^{-kx}\, dx = -e^{-kx}\big|_0^\infty = 1$

(b) $\mu = \int_0^\infty x k e^{-kx}\, dx = -\frac{e^{-kx}}{k}(kx + 1)\bigg|_0^\infty$

$= \frac{1}{k}$

(c) $\mu_2 = \int_0^\infty \left(x - \frac{1}{k}\right)^2 k e^{-kx}\, dx$

$= -\left(x^2 + \frac{1}{k^2}\right)e^{-kx}\bigg|_0^\infty = \frac{1}{k^2}$

(d) $\sigma = \sqrt{\mu_2} = \frac{1}{k}$

62 $\int_0^\infty e^{-x^2}\, dx \approx 0.8862$, applying information from Exercise 25 of Sec. 8.8.

64 Since the force of gravity varies as $1/r$, it equals k/r for some constant k. To find k, note that when $r = 4000$ mi the force is one pound, so $k/4000 = 1$ and $k = 4000$ mi-lb. If the payload rises to a height d, the potential energy is given by $\int_{4000}^d \frac{4000}{r}\, dr$

$= 4000(\ln d - \ln 4000)$, which must equal the kinetic energy imparted to the one-pound payload. As in Sec. 8.8, the kinetic energy applied to the one-pound payload is $E = \frac{1}{2}mv^2 = \frac{1}{2}\frac{1}{0.0061}(7)^2$

≈ 4016.4 mi-lb. Solving for d, we have $d = 4000\, e^{E/4000} = 10{,}918$. Hence the payload rises a distance of approximately 6918 miles above the ground.

66 Let the force be cr^{-2} and let the initial distance be a. The work done by gravity to decrease the distance from $r + dr$ to r is approximately $cr^{-2}\, dr$. So the total work is $\int_0^a cr^{-2}\, dr = \lim_{t \to 0^+} \int_t^a cr^{-2}\, dr$

$= \lim_{t \to 0^+} (-cr^{-1})\big|_t^a = \lim_{t \to 0^+} (ct^{-1} - ca^{-1}) = \infty$.

9 Plane Curves and Polar Coordinates

9.1 Polar Coordinates

2 (a) $x = 1 \cos \frac{\pi}{6} = \frac{\sqrt{3}}{2}$, $y = 1 \sin \frac{\pi}{6} = \frac{1}{2}$; the point is $\left(\frac{\sqrt{3}}{2}, \frac{1}{2}\right)$.

(b) $(2 \cos \pi/3, 2 \sin \pi/3) = (1, \sqrt{3})$

(c) $(2 \cos(-\pi/3), 2 \sin(-\pi/3)) = (1, -\sqrt{3})$

(d) $(-2 \cos \pi/3, -2 \sin \pi/3) = (-1, -\sqrt{3})$

(e) $(2 \cos 7\pi/3, 2 \sin 7\pi/3) = (1, \sqrt{3})$

(f) $(0 \cos \pi/4, 0 \sin \pi/4) = (0, 0)$

4 (a) $\sqrt{(\sqrt{2})^2 + (\sqrt{2})^2} = 2$. $\tan \theta = \frac{\sqrt{2}}{\sqrt{2}} = 1$. Since the point is in the first quadrant, $\theta = \tan^{-1} 1 = \pi/4$; the polar coordinates are $(2, \pi/4)$.

(b) $r = \sqrt{(-1)^2 + (\sqrt{3})^2}$, $\pi + \tan^{-1}\left(\frac{\sqrt{3}}{-1}\right) = \frac{2\pi}{3}$; $\left(2, \frac{2\pi}{3}\right)$

(c) $(5, \pi)$

(d) This point is opposite that in (a), so its polar coordinates are $(2, 5\pi/4)$.

(e) $\left(3, -\frac{\pi}{2}\right)$

(f) $\left(\sqrt{2}, \frac{\pi}{4}\right)$

6 $r = \csc \theta$, so $r \sin \theta = 1$ and $y = 1$.

8 $r = 4 \cos \theta + 5 \sin \theta$ implies that $r^2 = 4r \cos \theta + 5r \sin \theta$, so $x^2 + y^2 = 4x + 5y$.

10 $y = x^2$, so $r \sin \theta = r^2 \cos^2 \theta$; then $\sin \theta = r \cos^2 \theta$, so $r = \sec \theta \tan \theta$.

12 $x^2 + y^2 = 4x$, so $r^2 = 4r \cos \theta$ and $r = 4 \cos \theta$. (Dividing by r could remove the pole from the graph, but does not in this case.)

14
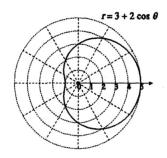

16 $r = 4^{\theta/\pi}$, $\theta > 0$
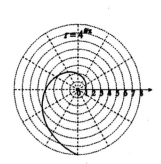

18

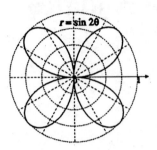

20

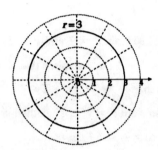

22

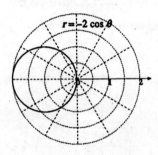

24 (a) The graph of $r = 1/\sqrt{\theta}$ is a spiral that approaches the line $y = 0$ as $\theta \to 0^+$ (since $y = r \sin \theta = (\sin \theta)/\sqrt{\theta} \to 0$ as $\theta \to 0^+$).

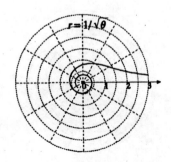

(b) $\lim_{\theta \to \infty} y = \lim_{\theta \to \infty} \dfrac{\sin \theta}{\sqrt{\theta}} = 0$

26 The curves intersect when $\sin 2\theta = \pm 1$; that is, for $\theta = \pi/4 + n\pi/2$, where n is an integer. There are four distinct points: $(1, \pi/4)$, $(1, 3\pi/4)$, $(1, 5\pi/4)$, and $(1, 7\pi/4)$ (which may be written in other ways).

28 The curves intersect when $2 \sin 2\theta = \pm 1$; that is, for $\theta = \pm \pi/12 + n\pi/2$, where n is an integer. There are eight distinct points, which may be written in many different forms: $\left(1, \pm\dfrac{\pi}{12}\right)$, $\left(1, \pm\dfrac{5\pi}{12}\right)$, $\left(1, \pm\dfrac{7\pi}{12}\right)$, and $\left(1, \pm\dfrac{11\pi}{12}\right)$.

30 $(1, 0)$, $\left(\dfrac{1}{2}, \pm\dfrac{\pi}{3}\right)$, and the pole

32

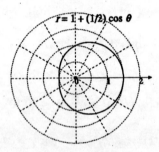

34 (a)

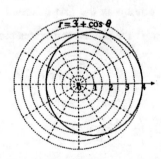

(b) $y = r \sin \theta = (3 + \cos \theta) \sin \theta$, so $\dfrac{dy}{d\theta} = 3 \cos \theta + \cos^2 \theta - \sin^2 \theta = 2(\cos \theta)^2 + 3(\cos \theta) - 1$. Setting this equal to 0 gives

9.1 Polar Coordinates

$\cos\theta = \dfrac{\sqrt{17}-3}{4}$, so $\theta = \cos^{-1}\dfrac{\sqrt{17}-3}{4}$, $r =$

$3 + \cos\theta = \dfrac{\sqrt{17}+9}{4}$, $x = r\cos\theta =$

$\dfrac{3\sqrt{17}-5}{8} \approx 0.92116$, and $y = r\sin\theta =$

$r\sqrt{1-\cos^2\theta} = \dfrac{1}{8}\sqrt{214 + 102\sqrt{17}} \approx$

3.14880.

36

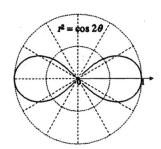

38 (a)

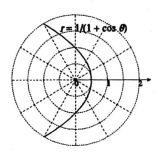

(b) $r = \dfrac{1}{1+\cos\theta}$ implies that $r + r\cos\theta = 1$,

so $\sqrt{x^2+y^2} + x = 1$ and $x^2+y^2 = (1-x)^2$; hence $y^2 = 1 - 2x$, which is the equation of a parabola.

40 We want $r = \theta$ to cross $r = 2\sin\theta$. Hence we need to find roots of $\theta - 2\sin\theta = 0$. By

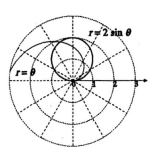

Newton's method, $\theta_{n+1} = \theta_n - \dfrac{\theta_n - 2\sin\theta_n}{1 - 2\cos\theta_n}$,

from which we obtain the solution $\theta \approx 1.8955$. The intersection point is $(1.8955, 1.8955)$ in polar and $(-0.6047, 1.7964)$ in rectangular.

9.2 Area in Polar Coordinates

2 $\displaystyle\int_0^\pi \dfrac{1}{2}(\sqrt{\theta})^2\, d\theta = \dfrac{\pi^2}{4}$

4 $\displaystyle\int_0^{\pi/2} \dfrac{1}{2}(\sqrt{\sin\theta})^2\, d\theta = \dfrac{1}{2}$

6 $\displaystyle\int_{\pi/6}^{\pi/4} \dfrac{1}{2}(\sec\theta)^2\, d\theta = \dfrac{3-\sqrt{3}}{6}$

8

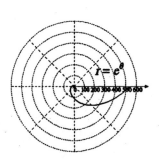

$\displaystyle\int_0^{2\pi} \dfrac{1}{2}(e^\theta)^2\, d\theta = \dfrac{1}{4}(e^{4\pi} - 1)$

10

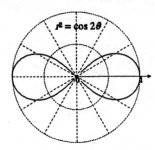

$\int_{-\pi/4}^{\pi/4} \frac{1}{2}(\sqrt{\cos 2\theta})^2 \, d\theta = \frac{1}{2}$

12 $\int_{-\pi/4}^{\pi/4} \frac{1}{2}(\cos 2\theta)^2 \, d\theta = \frac{\pi}{8}$ (Example 5, Sec. 9.1)

14 The region is obtained by deleting half of a disk of radius 1/2 from the region described by $-\pi \leq \theta$

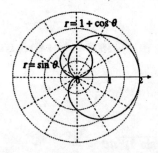

$\leq \pi/2$, $0 \leq r \leq 1 + \cos\theta$. The area is

$\int_{-\pi}^{\pi/2} \frac{1}{2}(1 + \cos\theta)^2 \, d\theta - \frac{1}{2}\pi\left(\frac{1}{2}\right)^2 = \pi + 1$.

16

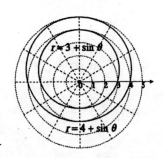

$\int_0^{2\pi} \frac{1}{2}(4 + \sin\theta)^2 \, d\theta - \int_0^{2\pi} \frac{1}{2}(3 + \sin\theta)^2 \, d\theta = 7\pi$

18 (a) The vertical line $x = 1$ has as its equation in polar coordinates, $r\cos\theta = 1$ or $r = \sec\theta$. Hence the area of the triangle is $\int_0^\beta \frac{1}{2}r^2 \, d\theta$

$= \int_0^\beta \frac{1}{2}\sec^2\theta \, d\theta$.

(b) The height of the triangle is $h = \tan\beta$. Hence the area of the triangle is $\frac{1}{2}(1)(\tan\beta) = \frac{1}{2}\tan\beta$. So, using the expression for the area found in (a), $\tan\beta = \int_0^\beta \sec^2\theta \, d\theta$.

(c) Since β was arbitrary, we may take the derivative of this equation with respect to β, using the first fundamental theorem of calculus. Then $(\tan\beta)' = \sec^2\beta$, so $(\tan x)' = \sec^2 x$, as claimed.

20 (a) Notice that as θ approaches 0, r becomes infinite. The graph is part of a spiral.

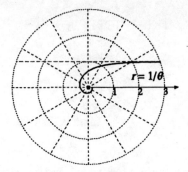

(b) For small $\alpha > 0$, the area between $\theta = \alpha$ and $\theta = \pi/2$ is $\int_\alpha^{\pi/2} \frac{1}{2}(1/\theta)^2 \, d\theta = \frac{1}{2}\int_\alpha^{\pi/2} \frac{d\theta}{\theta^2}$

$= -\frac{1}{2}\left(\frac{1}{\theta}\right)\bigg|_\alpha^{\pi/2} = \frac{1}{2\alpha} - \frac{1}{\pi}$. As α approaches 0, this becomes infinite. Hence the area

9.2 Area in Polar Coordinates

bounded by the curve $r = 1/\theta$ between $\theta = 0$ and $\theta = \pi/2$ is infinite.

22 The area is $\int_0^\alpha \frac{1}{2}(e^\theta)^2\, d\theta + \int_\alpha^{\pi/2} \frac{1}{2}(2\cos\theta)^2\, d\theta$

$= \frac{1}{2}\left[\int_0^\alpha e^{2\theta}\, d\theta + \int_\alpha^{\pi/2} 4\cos^2\theta\, d\theta\right]$, where $f(\alpha)$

$= e^\alpha - 2\cos\alpha = 0$. Using Newton's method, with $\alpha_1 = \pi/6$, gives $\alpha \approx 0.54$. Thus the area is

approximately $\frac{1}{2}\left[\int_0^{0.54} e^{2\theta}\, d\theta + \int_{0.54}^{\pi/2} 4\cos^2\theta\, d\theta\right]$

$= \frac{1}{2}\left[\frac{1}{2}e^{2\theta}\Big|_0^{0.54} + (2\theta + \sin 2\theta)\Big|_{0.54}^{\pi/2}\right] \approx 1.08.$

24 (a) No. The graph of $r = 3 + 2\cos\theta$ is a counterexample (as one can see from Example 1 and the discussion following). Here

$\int_{-\pi/2}^{\pi/2} \frac{1}{2}(3 + 2\cos\theta)^2\, d\theta = \frac{11\pi}{2} + \frac{1}{2}$, but

$\int_{\pi/2}^{3\pi/2} \frac{1}{2}(3 + 2\cos\theta)^2\, d\theta = \frac{11\pi}{2} - \frac{1}{2}.$

(b) Yes. Let P be the pole and suppose the boundary of the region is given by $r = f(\theta)$, with $f(\theta) > 0$. Choose a value for θ. Then the range from θ to $\theta + \pi$ gives, by assumption, half the region; that is, if the area is A, then $\frac{A}{2} = \int_\theta^{\theta+\pi} \frac{1}{2}[f(\phi)]^2\, d\phi$. Differentiating both sides (first fundamental theorem of calculus) yields $0 = \frac{1}{2}[f(\theta + \pi)]^2 - \frac{1}{2}[f(\theta)]^2$, so

$f(\theta + \pi) = f(\theta)$ and the chord determined by θ is bisected by P.

25 (b) Area $\geq \dfrac{\pi}{4}$

(c) Area $= \int_0^{2\pi} \frac{1}{2}[f(\theta)]^2\, d\theta$

$= \int_0^\pi \frac{1}{2}[f(\theta)]^2\, d\theta + \int_\pi^{2\pi} \frac{1}{2}[f(\theta)]^2\, d\theta$

$= \int_0^\pi \frac{1}{2}[f(\theta)]^2\, d\theta + \int_0^\pi \frac{1}{2}[f(\theta + \pi)]^2\, d\theta =$

$\frac{1}{2}\int_0^\pi ([f(\theta)]^2 + [f(\theta + \pi)]^2)\, d\theta$. Let $a =$

$f(\theta)$ and $b = f(\theta + \pi)$; then $a + b \geq 1$, so

$a^2 + b^2 \geq a^2 + (1 - a)^2 = 2\left(a - \frac{1}{2}\right)^2 + \frac{1}{2}$

$\geq \frac{1}{2}$. Hence the area is at least $\frac{1}{2}\int_0^\pi \frac{1}{2}\, d\theta$

$= \frac{\pi}{4}.$

26 (b) Area $\leq \pi/2$

(c) The statement about the chord length implies that $f(\theta) + f(\theta + \pi) \leq 1$ for all θ. Hence $[f(\theta)]^2 + [f(\theta + \pi)]^2 \leq [f(\theta) + f(\theta + \pi)]^2$

≤ 1, so Area $= \frac{1}{2}\int_0^{2\pi} [f(\theta)]^2\, d\theta$

$= \frac{1}{2}\int_0^\pi [f(\theta)]^2\, d\theta + \frac{1}{2}\int_\pi^{2\pi} [f(\theta)]^2\, d\theta$

$= \frac{1}{2}\int_0^\pi [f(\theta)]^2\, d\theta + \frac{1}{2}\int_0^\pi [f(\theta + \pi)]^2\, d\theta$

$= \frac{1}{2}\int_0^\pi ([f(\theta)]^2 + [f(\theta + \pi)]^2)\, d\theta$

$\leq \frac{1}{2}\int_0^\pi d\theta = \frac{\pi}{2}.$

9.3 Parametric Equations

2 (a)

t	-2	-1	0	1	2
x	-1	0	1	2	3
y	4	1	0	1	4

(b),(c)

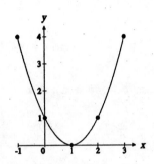

(d) $y = (x-1)^2$

4 (a)

t	0	$\pi/4$	$\pi/2$	$3\pi/4$	π
x	2	1.41	0	-1.41	-2
y	0	2.12	3	2.12	0

t	$5\pi/4$	$3\pi/2$	$7\pi/4$	2π
x	-1.41	0	1.41	2
y	-2.12	-3	-2.12	0

(b),(c)

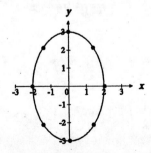

(d) $(x/2)^2 + (y/3)^2 = 1$

6 $x = t$, $y = \tan^{-1} 3t$

8 $\theta = t$, $r = 3 + \cos t$

10 $\dfrac{dy}{dx} = \dfrac{dy/dt}{dx/dt} = -\dfrac{4\sin 4t}{3\cos 3t}$, $\dfrac{d^2y}{dx^2} = \dfrac{\dfrac{d}{dt}\left(\dfrac{dy}{dx}\right)}{dx/dt}$

$= -\dfrac{16\cos 3t \cos 4t + 12\sin 3t \sin 4t}{9\cos^3 3t}$

12 $\dfrac{dy}{dx} = \dfrac{dy/dt}{dx/dt} = \dfrac{\sec^2 t}{2te^{t^2}}$, $\dfrac{d^2y}{dx^2} = \dfrac{\dfrac{d}{dt}\left(\dfrac{dy}{dx}\right)}{dx/dt}$

$= \dfrac{2te^{t^2}(2\sec^2 t \tan t) - \sec^2 t\left[e^{t^2}(4t^2 + 2)\right]}{8t^3 e^{3t^2}}$

$= \dfrac{\sec^2 t\left[2t\tan t - 2t^2 - 1\right]}{4t^3 e^{t^2}}$

14 $x = r\cos\theta = 2\cos\theta + 3\sin\theta\cos\theta$

$= \dfrac{1}{2}(4\cos\theta + 3\sin 2\theta)$,

$y = r\sin\theta = 2\sin\theta + 3\sin^2\theta$

$= \dfrac{1}{2}(3 + 4\sin\theta - 3\cos 2\theta)$,

$\dfrac{dy}{dx} = \dfrac{dy/d\theta}{dx/d\theta} = \dfrac{3\sin 2\theta + 2\cos\theta}{3\cos 2\theta - 2\sin\theta}$, $\dfrac{d^2y}{dx^2} =$

$\dfrac{\dfrac{d}{d\theta}\left(\dfrac{dy}{dx}\right)}{dx/d\theta} = \dfrac{22 + 18\sin\theta}{(3\cos 2\theta - 2\sin\theta)^3}$

16 $\dfrac{dy}{dx}\bigg|_{(1,1)} = \dfrac{dy}{dx}\bigg|_{t=0} = \dfrac{dy/dt}{dx/dt}\bigg|_{t=0}$

$= \dfrac{3\sec 3t \tan 3t}{\dfrac{(t^3 + t^2 + 1)(2t) - (t^2 + 1)(3t^2 + 2t)}{(t^3 + t^2 + 1)^2}}\bigg|_{t=0}$

$= 3\lim_{t\to 0}\dfrac{\tan 3t}{(t^3 + t^2 + 1)(2t) - (t^2 + 1)(3t^2 + 2t)}$

9.3 Parametric Equations

$= 3 \lim_{\substack{H \\ t \to 0}} \dfrac{3 \sec^2 3t}{-4t^3 - 6t} = -\infty$; the tangent at $(1, 1)$ is vertical.

18 $\dfrac{dy}{dx} = \dfrac{dy/dt}{dx/dt} = \dfrac{3e^{3t} - 2t \sin t^2}{3e^{3t} + 2 \cos 2t}$; then $\dfrac{d^2y}{dx^2} =$

$\dfrac{\frac{d}{dt}\left(\frac{dy}{dx}\right)}{dx/dt}$, where $\dfrac{d}{dt}\left(\dfrac{dy}{dx}\right) =$

$\dfrac{1}{(3e^{3t} + 2\cos 2t)^2}[(3e^{3t} + 2\cos 2t)(9e^{3t} - 2\sin t^2 - 4t^2 \cos t^2)$

$- (3e^{3t} - 2t \sin t^2)(9e^{3t} - 4 \sin 2t)]$

$= \dfrac{1}{(3e^{3t} + 2\cos 2t)^3}(6e^{3t}[2 \sin 2t + (3t-1) \sin t^2$

$- 3 \cos 2t - 2t^2 \cos t^2] -$

$4[(2t \sin 2t + \cos 2t) \sin t^2 + 2t^2 \cos 2t \cos t^2])$.

20 $\dfrac{dy}{dx} = \dfrac{dy/dt}{dx/dt} = \dfrac{2t+1}{3t^2}$, so $\dfrac{d^2y}{dx^2} = \dfrac{\frac{d}{dt}\left(\frac{dy}{dx}\right)}{dx/dt}$

$= \dfrac{\frac{3t^2(2) - (2t+1)(6t)}{9t^4}}{3t^2} = -\dfrac{2(t+1)}{9t^5}$. Now

$\dfrac{d^2y}{dx^2}$ goes from $-$ to $+$ at $t = -1$ and from $+$ to $-$ at $t = 0$, so the curve given by $x = t^3 + 1$ and $y = t^2 + t + 1$ is concave down for $t < -1$, concave up for $-1 < t < 0$, and concave down for $t > 0$.

22 (a) Here, $x = r \cos \theta = (1 + \cos \theta) \cos \theta$
$= \cos \theta + \cos^2 \theta$ and $y = r \sin \theta$
$= (1 + \cos \theta) \sin \theta = \sin \theta + \sin \theta \cos \theta$.

The slope is $\dfrac{dy}{dx} = \dfrac{dy/d\theta}{dx/d\theta}$

$= \dfrac{\cos \theta + \cos^2 \theta - \sin^2 \theta}{-\sin \theta - 2 \cos \theta \sin \theta}$

$= -\dfrac{\cos \theta + \cos 2\theta}{\sin \theta + \sin 2\theta}$.

(b) The limit as $\theta \to \pi^-$ is of the form 0/0, so l'Hôpital's rule applies:

$\lim_{\theta \to \pi^-} \left(-\dfrac{\cos \theta + \cos 2\theta}{\sin \theta + \sin 2\theta}\right)$

$= \lim_{\substack{H \\ \theta \to \pi^-}} \left(-\dfrac{-\sin \theta - 2 \sin 2\theta}{\cos \theta + 2 \cos 2\theta}\right)$

$= \lim_{\theta \to \pi^-} \dfrac{\sin \theta + 2 \sin 2\theta}{\cos \theta + 2 \cos 2\theta} = \dfrac{0}{-1 + 2} = 0$.

(c)

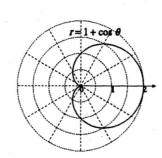

The graph forms a cusp at $\theta = \pi$.

24 (a) $(0, 3)$

(b) The horizontal speed is $\dfrac{dx}{dt}\bigg|_{t=0} = 24\big|_{t=0}$
$= 24$.

(c) The vertical speed is $\dfrac{dy}{dt}\bigg|_{t=0}$
$= (-32t + 5)\big|_{t=0} = 5$.

26 (a) $\dfrac{dy}{dx} = \dfrac{dy/d\theta}{dx/d\theta} = \dfrac{\dfrac{d}{d\theta}(e^{2\theta}\sin\theta)}{\dfrac{d}{d\theta}(e^{2\theta}\cos\theta)}$

$= \dfrac{2e^{2\theta}\sin\theta + e^{2\theta}\cos\theta}{2e^{2\theta}\cos\theta - e^{2\theta}\sin\theta}$

$= \dfrac{2\sin\theta + \cos\theta}{2\cos\theta - \sin\theta}$, and $r = e^{2\theta}$ meets $\theta = \alpha$

whenever $\theta = \alpha + 2n\pi$, where n is an integer. Since $\sin(\alpha + 2n\pi) = \sin\alpha$ and $\cos(\alpha + 2n\pi) = \cos\alpha$, the curve has the same slope at all points of intersection, namely $\dfrac{2\sin\alpha + \cos\alpha}{2\cos\alpha - \sin\alpha}$.

(b) Here $\dfrac{dy}{dx} = \dfrac{dy/d\theta}{dx/d\theta} = \dfrac{\dfrac{d}{d\theta}(\theta\sin\theta)}{\dfrac{d}{d\theta}(\theta\cos\theta)} =$

$\dfrac{\sin\theta + \theta\cos\theta}{\cos\theta - \theta\sin\theta}$, and $r = \theta$ meets $\theta = \alpha$

whenever $r = \alpha + 2n\pi$. The slope of $r = \theta$ when $r = \alpha$, $\dfrac{\sin\alpha + \alpha\cos\alpha}{\cos\alpha - \alpha\sin\alpha}$, clearly is not the same as the slope when $r = \alpha + 2\pi$, $\dfrac{\sin\alpha + (\alpha+2\pi)\cos\alpha}{\cos\alpha - (\alpha+2\pi)\sin\alpha}$.

28 (a) We have $\dfrac{x^2}{a^2} + \dfrac{y^2}{b^2} = \dfrac{(a\cos t)^2}{a^2} + \dfrac{(b\sin t)^2}{b^2}$

$= \cos^2 t + \sin^2 t = 1$, so the curve is part of the ellipse $\dfrac{x^2}{a^2} + \dfrac{y^2}{b^2} = 1$. To see that it is the entire curve, note that for $0 \le t \le \pi$, $y \ge 0$,

while for $\pi \le t \le 2\pi$, $y \le 0$. Thus, as t varies from 0 to 2π, the point (x, y) travels from $(a, 0)$ to $(-a, 0)$ along the top half of the ellipse, and then back to $(a, 0)$ along the bottom half.

(b) The area of the shaded region is $\int_0^a y\, dx$,

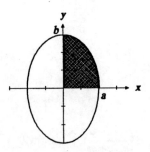

where y is the ordinate of the point with abscissa x on the upper half of the ellipse. By symmetry, the ellipse has area $4\int_0^a y\, dx$.

Let $x = a\cos t$, where t ranges from $\pi/2$ down to 0. Then $y = b\sin t$ and $dx = -a\sin t\, dt$, so the area is

$4\int_{\pi/2}^0 (b\sin t)(-a\sin t)\, dt =$

$4ab\int_0^{\pi/2}\sin^2 t\, dt = 4ab\cdot\dfrac{\pi}{4} = \pi ab$. (The shortcut in Sec. 7.6 was used to compute the last integral.)

30 From Example 2, we know $\dfrac{dy}{dx} = \dfrac{\sin\theta}{1 - \cos\theta}$. The y coordinate of the cycloid equals a (for the first time) at the smallest value of θ in $[0, 2\pi]$ such that $\cos\theta = 0$; that is, at $\theta = \pi/2$. So the slope at that point is $\left.\dfrac{dy}{dx}\right|_{\theta = \pi/2} = \dfrac{\sin\pi/2}{1 - \cos\pi/2} = 1$.

9.3 Parametric Equations

32 $\int_0^{2\pi a} 2\pi xy\, dx =$

$\int_0^{2\pi} 2\pi(a\theta - a\sin\theta)(a - a\cos\theta)(a - a\cos\theta)\, d\theta$

$= 2\pi a^3 \int_0^{2\pi} (\theta - \sin\theta)(1 - \cos\theta)^2\, d\theta = 6\pi^3 a^3$

34 (a)

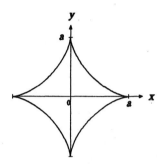

(b) $\dfrac{dy}{dx} = \dfrac{dy/dt}{dx/dt} = \dfrac{3a\sin^2 t \cos t}{-3a\cos^2 t \sin t} = -\tan t$

36 Place the circle with its center above the origin of the xy plane and tilt it about the diameter parallel to

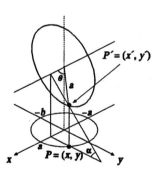

the x axis. Then the shadow of the circle will have x intercepts $(a, 0)$ and $(-a, 0)$, as shown in the figure. Let P' be a point on the circle and $P = (x, y)$ its projection (shadow) in the xy plane. From the figure we see that $x = x' = a\cos\theta$. Letting the tilt of the circle be given by α, as shown, it follows that $y/y' = \cos\alpha$, so $y = y'\cos\alpha = a\sin\theta\cos\alpha$. Thus $1 = \cos^2\theta + \sin^2\theta =$

$\left(\dfrac{x}{a}\right)^2 + \left(\dfrac{y}{a\cos\alpha}\right)^2 = \dfrac{x^2}{a^4} + \dfrac{y^2}{a^2\cos^2\alpha}$, which is

the equation of an ellipse, as claimed. Note that $b = a\cos\alpha$.

9.4 Arc Length and Speed on a Curve

2 $\dfrac{dy}{dx} = \dfrac{2}{3}x^{-1/3}$, so Arc length =

$\int_0^1 \sqrt{1 + \left(\dfrac{2}{3}x^{-1/3}\right)^2}\, dx = \dfrac{1}{27}(13\sqrt{13} - 8).$

4 $\dfrac{dy}{dx} = x - \dfrac{1}{4x}$, so $1 + \left(\dfrac{dy}{dx}\right)^2 = \left(x + \dfrac{1}{4x}\right)^2$ and

Arc length $= \int_2^3 \left(x + \dfrac{1}{4x}\right) dx = \dfrac{5}{2} + \dfrac{1}{4}\ln\dfrac{3}{2}.$

6 Arc length $= \int_0^{2\pi} \sqrt{r^2 - (r')^2}\, d\theta =$

$\int_0^{2\pi} \sqrt{e^{2\theta} + e^{2\theta}}\, d\theta = \int_0^{2\pi} \sqrt{2e^{2\theta}}\, d\theta =$

$\int_0^{2\pi} \sqrt{2}e^{\theta}\, d\theta = \sqrt{2}e^{\theta}\Big|_0^{2\pi} = \sqrt{2}(e^{2\pi} - 1)$

8 Arc length $= \int_0^{\pi} \sqrt{r^2 + (r')^2}\, d\theta$

$= \int_0^{\pi} \sqrt{\cos^4\dfrac{\theta}{2} + \left(-\cos\dfrac{\theta}{2}\sin\dfrac{\theta}{2}\right)^2}\, d\theta$

$= \int_0^{\pi} \sqrt{\cos^2\dfrac{\theta}{2}\left(\cos^2\dfrac{\theta}{2} + \sin^2\dfrac{\theta}{2}\right)}\, d\theta$

$= \int_0^{\pi} \sqrt{\cos^2\dfrac{\theta}{2}}\, d\theta = \int_0^{\pi} \cos\dfrac{\theta}{2}\, d\theta = 2\sin\dfrac{\theta}{2}\Big|_0^{\pi}$

$= 2$. Since $\cos\theta/2 \geq 0$ for $0 \leq \theta \leq \pi$, we were able to use $\sqrt{\cos^2\frac{\theta}{2}} = \cos\theta/2$.

10 Speed $= \sqrt{(3\sec 3t \tan 3t)^2 + \left(\dfrac{4}{\sqrt{1-(4t)^2}}\right)^2}$

$= \sqrt{9\sec^2 3t \tan^2 3t + \dfrac{16}{1-16t^2}}$

12 Speed $= \sqrt{(-3\csc 3t \cot 3t)^2 + \left(\dfrac{1}{1+t}\cdot\dfrac{1}{2\sqrt{t}}\right)^2}$

$= \sqrt{9\csc^2 3t \cot^2 3t + \dfrac{1}{4t(1+t)^2}}$

14 (a)

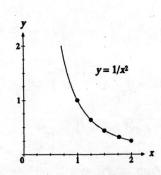

(b) Arc length $\approx \sqrt{[(4/5)^2 - 1]^2 + (5/4 - 1)^2} +$

$\sqrt{[(2/3)^2 - (4/5)^2]^2 + (3/2 - 5/4)^2} +$

$\sqrt{[1/4 - (2/3)^2]^2 + (2 - 3/2)^2} \approx 1.29$

(c) Since $\dfrac{dy}{dx} = -\dfrac{2}{x^3}$, Arc length $=$

$\int_1^2 \sqrt{1 + \left(-\dfrac{2}{x^3}\right)^2}\, dx = \int_1^2 \sqrt{1 + \dfrac{4}{x^6}}\, dx$

(d) Arc length $\approx \dfrac{1}{8}\left[\sqrt{1 + \dfrac{4}{1^6}} + \right.$

$2\sqrt{1 + \dfrac{4}{(5/4)^6}} + 2\sqrt{1 + \dfrac{4}{(3/2)^6}} +$

$\left. 2\sqrt{1 + \dfrac{4}{(7/4)^6}} + \sqrt{1 + \dfrac{4}{2^6}}\right] \approx 1.32$

(e) Arc length $\approx \dfrac{1}{12}\left[\sqrt{1 + \dfrac{4}{1^6}} + 4\sqrt{1 + \dfrac{4}{(5/4)^6}} \right.$

$+ 2\sqrt{1 + \dfrac{4}{(3/2)^6}} + 4\sqrt{1 + \dfrac{4}{(7/4)^6}} +$

$\left. \sqrt{1 + \dfrac{4}{2^6}}\right] \approx 1.30$

(f) $\int_1^2 \sqrt{1 + \dfrac{4}{x^6}}\, dx \approx 1.2968$

16 Arc length $= \int_{2\pi}^{\infty} \sqrt{(1/\theta)^2 + (-1/\theta^2)^2}\, d\theta =$

$\int_{2\pi}^{\infty} \dfrac{\sqrt{\theta^2 + 1}}{\theta^2}\, d\theta = \infty$

18 (a) $\dfrac{dy}{dx} = \dfrac{2}{3}x^{-1/3}$, so Arc length $=$

$\int_1^8 \sqrt{1 + \left(\dfrac{2}{3}x^{-1/3}\right)^2}\, dx$

$= \dfrac{1}{3}\int_1^8 \sqrt{9 + 4x^{-2/3}}\, dx$.

(b) $x = y^{3/2}$, so $\dfrac{dy}{dx} = \dfrac{3}{2}y^{1/2}$ and Arc length $=$

$\int_1^4 \sqrt{1 + \left(\dfrac{3}{2}y^{1/2}\right)^2}\, dy = \dfrac{1}{2}\int_1^4 \sqrt{4 + 9y}\, dt$.

(c) (b) is easier: $\dfrac{1}{27}(4 + 9y)^{3/2}\Big|_1^4 =$

$\frac{1}{27}(40^{3/2} - 13^{3/2})$

20 (a) Since $x = r\cos\theta = g(t)\cos(h(t))$ and $y = r\sin\theta = g(t)\sin(h(t))$ we parameterize by t.

Then $\frac{ds}{dt} = \sqrt{\left(\frac{dx}{dt}\right)^2 + \left(\frac{dy}{dt}\right)^2} =$

$\sqrt{[g'\cos h - gh'\sin h]^2 + [g'\sin h + gh'\cos h]^2}$

$= \sqrt{(g')^2 + (gh')^2}$.

(b) Here $g(t) = e^t$ and $h(t) = 5t$. Therefore the speed is $\sqrt{e^{2t} + (e^t \cdot 5)^2} = \sqrt{26 e^{2t}} = \sqrt{26}\, e^t$.

22 Since $r = a\theta$, $r' = a$, so Arc length $=$

$\int_0^{2\pi} \sqrt{r^2 + (r')^2}\, d\theta = \int_0^{2\pi} \sqrt{a^2\theta^2 + a^2}\, d\theta$

$= a\int_0^{2\pi} \sqrt{\theta^2 + 1}\, d\theta$

$= a \cdot \frac{1}{2}\left[\theta\sqrt{\theta^2 + 1} + \ln\left(\theta + \sqrt{\theta^2 + 1}\right)\right]\Big|_0^{2\pi}$

$= \frac{a}{2}\left[2\pi\sqrt{4\pi^2 + 1} + \ln\left(2\pi + \sqrt{4\pi^2 + 1}\right)\right]$.

24 $ds = \sqrt{[f(\theta + d\theta) - f(\theta)]^2 + [f(\theta + d\theta)\, d\theta]^2}$

$= \sqrt{\left(\frac{f(\theta + d\theta) - f(\theta)}{d\theta}\right)^2 + f(\theta + d\theta)^2}\, d\theta \approx$

$\sqrt{[f'(\theta)]^2 + [f(\theta)]^2}\, d\theta$, so, informally, Arc length

$= \int_\alpha^\beta \sqrt{[f(\theta)]^2 + [f'(\theta)]^2}\, d\theta$

$= \int_\alpha^\beta \sqrt{r^2 + (r')^2}\, d\theta$.

26 Average $r = \frac{1}{s}\int_0^s r\, ds =$

$\frac{1}{s}\int_0^{2\pi} r\sqrt{r^2 + (r')^2}\, d\theta \geq \frac{1}{s}\int_0^{2\pi} r \cdot r\, d\theta =$

$\frac{2}{s}\int_0^{2\pi} \frac{r^2}{2}\, d\theta = \frac{2A}{s}$. Equality occurs if and only if $r' = 0$ for all θ; that is, the curve is a circle centered at the origin.

28 (a) The length is at least $\sqrt{2}$ and can be arbitrarily large.

(b) The length is at least $\sqrt{2}$ and is less than 2.

For $f'(x) \geq 0$, $\sqrt{1 + [f'(x)]^2} \leq$

$\sqrt{1 + 2f'(x) + [f'(x)]^2} = 1 + f'(x)$, so $s =$

$\int_0^1 \sqrt{1 + [f'(x)]^2}\, dx \leq \int_0^1 [1 + f'(x)]\, dx =$

$[x + f(x)]\big|_0^1 = 2$. If equality held here, we would have $f'(x) = 0$ for all x, contradicting the fact that $f(0) \neq f(1)$.

30 Yes. To show this, extend one of the sides of P_1 until it intersects P_2 (in two points) and call the new

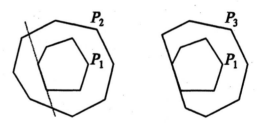

polygon P_3; it contains P_1. Clearly the perimeter of P_3 is less than the perimeter of P_2. Continuing this procedure for each side of P_1, we see that the outer polygon (with shrinking perimeter) becomes P_1. Thus the perimeter of P_1 is necessarily less than the perimeter of P_2.

9.5 The Area of a Surface of Revolution

2 $\dfrac{dy}{dx} = 3x^2$, so $\dfrac{ds}{dx} = \sqrt{1 + 9x^4}$; $R = x^3 + 1$, so

Area $= \displaystyle\int_1^2 2\pi(x^3 + 1)\sqrt{1 + 9x^4}\, dx$.

4 Area $= \displaystyle\int_1^2 2\pi x\sqrt{1 + 9x^4}\, dx$

6 Area $= \displaystyle\int_0^\pi 2\pi \sin x\sqrt{1 + \cos^2 x}\, dx$

$= 2\pi[\sqrt{2} + \ln(1 + \sqrt{2})]$

8 $\dfrac{ds}{dt} = \sqrt{\left(\dfrac{dx}{dt}\right)^2 + \left(\dfrac{dy}{dt}\right)^2} =$

$\sqrt{[e^t(\cos t - \sin t)]^2 + [e^t(\sin t + \cos t)]^2} = \sqrt{2}\,e^t$,

so Area $= \displaystyle\int_0^{\pi/2} 2\pi e^t \sin t \cdot \sqrt{2}\,e^t\, dt$

$= 2\pi\sqrt{2}\displaystyle\int_0^{\pi/2} e^{2t} \sin t\, dt = \dfrac{2}{5}\sqrt{2}\pi(2e^\pi + 1)$.

10 Here $R = y = 1/x$ and $ds = \sqrt{1 + \left(\dfrac{dy}{dx}\right)^2}\, dx =$

$\sqrt{1 + (-1/x^2)^2}\, dx = \sqrt{1 + x^{-4}}\, dx$. The surface area is $\displaystyle\int_1^2 2\pi \cdot \dfrac{1}{x}\sqrt{1 + x^{-4}}\, dx =$

$2\pi \displaystyle\int_1^2 \dfrac{\sqrt{x^4 + 1}}{x^3}\, dx$. Rewriting this as

$2\pi \displaystyle\int_1^2 \dfrac{\sqrt{1 + (x^2)^2}\, x\, dx}{(x^2)^2}$, we choose the

substitution $u = x^2$ and then the trigonometric substitution $u = \tan \theta$ (or let $x^2 = \tan \theta$ immediately).

12 $R = x$, $ds = \sqrt{1 + \left(\dfrac{dy}{dx}\right)^2}\, dx =$

$\sqrt{1 + \left(\dfrac{4}{3}x^{1/3}\right)^2}\, dx = \dfrac{1}{3}\sqrt{9 + 16x^{2/3}}\, dx$. The

surface area is $\displaystyle\int_1^8 \dfrac{2\pi}{3} x\sqrt{9 + 16x^{2/3}}\, dx$. This can be

evaluated by the trigonometric substitution $16x^{2/3} = 9\tan^2\theta$; that is, $x = \left(\dfrac{3}{4}\tan\theta\right)^3$.

14 Here $R = x$ and $ds = \sqrt{1 + \left(\dfrac{dy}{dx}\right)^2}\, dx =$

$\sqrt{1 + \left(\dfrac{1}{2}x^2 - \dfrac{1}{2}x^{-2}\right)^2}\, dx = \sqrt{\left(\dfrac{1}{2}x^2 + \dfrac{1}{2}x^{-2}\right)^2}\, dx$

$= \dfrac{1}{2}(x^2 + x^{-2})\, dx$. The surface area is

$\displaystyle\int_1^3 2\pi x \cdot \dfrac{1}{2}(x^2 + x^{-2})\, dx = \pi\displaystyle\int_1^3 \left(x^3 + \dfrac{1}{x}\right)\, dx$.

Now use the power rule and $\displaystyle\int \dfrac{1}{x}\, dx = \ln|x| + C$.

16 Here $R = y + 1 = \sqrt{1 - x^2} + 1$ and $ds =$

$\sqrt{1 + \left(\dfrac{dy}{dx}\right)^2}\, dx = \sqrt{1 + \left(\dfrac{-x}{\sqrt{1 - x^2}}\right)^2}\, dx =$

$\sqrt{1 + \dfrac{x^2}{1 - x^2}}\, dx = \dfrac{dx}{\sqrt{1 - x^2}}$. The surface area is

$\displaystyle\int_{-1}^1 2\pi(\sqrt{1 - x^2} + 1)\dfrac{dx}{\sqrt{1 - x^2}} =$

9.5 The Area of a Surface of Revolution

$2\pi \int_{-1}^{1} \left(1 + \dfrac{1}{\sqrt{1-x^2}}\right) dx$. Now let $x = \cos\theta$.

18 (a) The curve $y = \sqrt{a^2 - x^2}$ for $c \le x \le c + h$ is revolved about the x axis. Since $\dfrac{dy}{dx} = -\dfrac{x}{\sqrt{a^2-x^2}}$, $\dfrac{ds}{dx} = \dfrac{a}{\sqrt{a^2-x^2}}$, and the area is $\int_c^{c+h} 2\pi\sqrt{a^2-x^2} \cdot \dfrac{a}{\sqrt{a^2-x^2}}\, dx = 2\pi ah$.

(b) The area of the portion of a sphere between two parallel planes depends only on the distance between them, provided that both planes intersect the sphere.

20 $\int_1^3 2\pi(x^{1/5} + 1)\left[1 + \left(\dfrac{1}{5}x^{-4/5}\right)^2\right]^{1/2} dx$

$= \int_1^3 2\pi(x^{1/5} + 1) \cdot \dfrac{1}{5}(25 + x^{-8/5})^{1/2}\, dx$

$= \dfrac{2\pi}{5}\int_1^3 (x^{1/5} + 1)(25 + x^{-8/5})^{1/2}\, dx \approx 27.1100$

22 See Exercise 24 of Sec. 9.4. We note that the local approximation for area is $2\pi R\, ds$, where $R = y = r\sin\theta$, so we have $2\pi r\sin\theta\sqrt{r^2 + (r')^2}\, d\theta$ and the surface area is $\int_\alpha^\beta 2\pi r\sin\theta\sqrt{r^2 + (r')^2}\, d\theta$.

24 (a) Using the result of Exercise 21, the surface area is

$\int_0^{\pi/2} 2\pi(1+\cos\theta)\sin\theta\sqrt{(1+\cos\theta)^2 + (-\sin\theta)^2}\, d\theta$

$= 2\pi \int_0^{\pi/2} (1+\cos\theta)\sqrt{2 + 2\cos\theta}\,\sin\theta\, d\theta$

$= 2\sqrt{2}\pi \int_0^{\pi/2} (1+\cos\theta)^{3/2}\sin\theta\, d\theta$

$= 2\sqrt{2}\pi\left(-\dfrac{2}{5}(1+\cos\theta)^{5/2}\right)\Big|_0^{\pi/2}$

$= \dfrac{(32 - 4\sqrt{2})\pi}{5}$.

(b) Here $R = x = r\cos\theta = (1+\cos\theta)\cos\theta$. Also $ds = \sqrt{2 + 2\cos\theta}\, d\theta = \sqrt{2}\sqrt{1+\cos\theta}\, d\theta$, as in (a). The surface area is $\int_0^{\pi/2} 2\pi(1+\cos\theta)\cos\theta\sqrt{2}\sqrt{1+\cos\theta}\, d\theta$

$= 2\sqrt{2}\pi\int_0^{\pi/2} (1+\cos\theta)^{3/2}\cos\theta\, d\theta$

$= 2\sqrt{2}\pi\int_0^{\pi/2}\left(2\cos^2\dfrac{\theta}{2}\right)^{3/2}\left(2\cos^2\dfrac{\theta}{2} - 1\right) d\theta$

$= 8\pi\int_0^{\pi/2}\left(2\cos^5\dfrac{\theta}{2} - \cos^3\dfrac{\theta}{2}\right) d\theta$

$= \dfrac{24\sqrt{2}\pi}{5}$. (Use Formulas 49 and 50 from the table of antiderivatives.)

26 First solve the given equation for y, finding $y = (1 - x^{2/3})^{3/2}$. Thus $\dfrac{dy}{dx} = -x^{-1/3}(1 - x^{2/3})^{1/2}$. Now $R = y$ since the curve is revolved about the x axis; the surface area is

$\int_0^1 2\pi(1 - x^{2/3})^{3/2}\sqrt{1 + x^{-2/3}(1 - x^{2/3})}\, dx$

$= 2\pi\int_0^1 (1 - x^{2/3})^{3/2} x^{-1/3}\, dx$

$= \dfrac{6\pi}{5}(1 - x^{2/3})^{5/2}\Big|_0^1 = \dfrac{6\pi}{5}$.

28 (a) The volume is π, as shown in Exercise 54(a) of the Chapter 8 Review Exercises. $R = y = 1/x$, $ds = \sqrt{1 + (-1/x^2)^2}\, dx =$

$\frac{1}{x^2}\sqrt{x^4 + 1}\, dx$. The surface area is

$$\int_1^\infty 2\pi \frac{1}{x} \cdot \frac{1}{x^2}\sqrt{x^4 + 1}\, dx. \text{ Now } x^4 + 1 \geq x^4$$

so $2\pi \lim_{b \to \infty} \int_1^b \frac{1}{x^3}\sqrt{x^4 + 1}\, dx \geq$

$$2\pi \lim_{b \to \infty} \int_1^b \frac{1}{x^3} x^2\, dx = 2\pi \lim_{b \to \infty} \int_1^b \frac{1}{x}\, dx =$$

$$2\pi \lim_{b \to \infty} \ln x \Big|_1^b = 2\pi \lim_{b \to \infty} \ln b = \infty.$$

(b) No. See the discussion of "painting" in the solution to Exercise 54(b) of the Chapter 8 Review Exercises.

30 (a) $\dfrac{r_1}{L_1} = \dfrac{r_2}{L_2}$ by similar triangles.

(b) The surface of the smaller cone is, from Exercise 29, $\pi r_2 L_2$, while the surface area of the larger cone is $\pi r_1 L_1$. The area of the frustum is the difference in their areas, $\pi r_1 L_1 - \pi r_2 L_2$.

(c) $\pi r_1 L_1 - \pi r_2 L_2 = \pi r_1(L_2 + L) - \pi r_2 L_2$

$= \pi r_1 L_2 - \pi r_2 L_2 + \pi r_1 L$

$= \pi r_2 L_1 - \pi r_2 L_2 + \pi r_1 L$

$= \pi r_2(L_1 - L_2) + \pi r_1 L = \pi r_2 L + \pi r_1 L$

(d) $\pi r_2 L + \pi r_1 L = \pi L(r_1 + r_2)$

$= 2\pi L \left(\dfrac{r_1 + r_2}{2}\right) = 2\pi L r$

32 By symmetry, the centroid lies on the y axis; that is, $\bar{x} = 0$. To find $\bar{y}$, parameterize the curve by $x = a \cos \theta$ and $y = a \sin \theta$ for $0 \leq \theta \leq \pi$. Then

$$ds = \sqrt{\left(\frac{dx}{d\theta}\right)^2 + \left(\frac{dy}{d\theta}\right)^2}\, d\theta =$$

$\sqrt{(-a \sin \theta)^2 + (a \cos \theta)^2}\, d\theta = a\, d\theta.$ The length of the curve is πa, so $\bar{y} = \dfrac{1}{\pi a} \int_0^\pi a \sin \theta \cdot a\, d\theta$

$= \dfrac{a}{\pi} \int_0^\pi \sin \theta\, d\theta = \dfrac{a}{\pi}(-\cos \theta)\Big|_0^\pi = \dfrac{2a}{\pi}.$ The

centroid is $\left(0, \dfrac{2a}{\pi}\right)$.

34 Introduce an xy coordinate system so that the axis of revolution is the x axis and the center of the circle is at $(0, b)$. By symmetry, the centroid of the circle is at its center, $(0, b)$. The length of the curve is $2\pi a$ and the distance its centroid travels is $2\pi b$. By Exercise 33, the area of the doughnut is $2\pi a \cdot 2\pi b = 4\pi^2 ab$.

9.6 Curvature

2 $\dfrac{dy}{dx} = -\sin x = 0$ at $x = 0$; $\dfrac{d^2y}{dx^2} = -\cos x =$

-1 at $x = 0$. $\kappa = \dfrac{|-1|}{(1 + 0^2)^{3/2}} = 1$. Radius of

curvature = 1.

4 $\dfrac{dy}{dx} = \dfrac{1}{x} = \dfrac{1}{e}$ at $x = e$; $\dfrac{d^2y}{dx^2} = -\dfrac{1}{x^2} = -\dfrac{1}{e^2}$ at

$x = e$. Then $\kappa = \dfrac{|-1/e^2|}{[1 + (1/e)^2]^{3/2}} = \dfrac{e}{(1 + e^2)^{3/2}}$ and

radius of curvature = $\dfrac{(1 + e^2)^{3/2}}{e}$.

9.6 Curvature

6 $\dfrac{dy}{dx} = 2\sec 2x \tan 2x = 4\sqrt{3}$ at $x = \pi/6$; $\dfrac{d^2y}{dx^2} =$

$4(\sec^3 2x + \sec 2x \tan^2 2x) = 56$ at $x = \pi/6$. Then

$\kappa = \dfrac{|56|}{[1 + (4\sqrt{3})^2]^{3/2}} = \dfrac{8}{49}$ and radius of curvature

$= \dfrac{49}{8}$.

8 At $t = 2$, $\dot{x} = 2t = 4$, $\ddot{x} = 2$, $\dot{y} = 3t^2 + 4t^3 = 44$, $\ddot{y} = 6t + 12t^2 = 60$. Then $\kappa =$

$\dfrac{|4 \cdot 60 - 44 \cdot 2|}{(4^2 + 44^2)^{3/2}} = \dfrac{19}{976\sqrt{122}}$.

10 Since $\sin\theta = \dfrac{\sqrt{3}}{2}$ and $\cos\theta = \dfrac{1}{2}$, we have $\dot{x} =$

$3(\cos^2\theta)(-\sin\theta) = \dfrac{-3\sqrt{3}}{8}$, $\ddot{x} = 3(\cos\theta)(2\sin^2\theta - \cos^2\theta) = 15/8$, $\dot{y} = 3\sin^2\theta\cos\theta = 9/8$, $\ddot{y} =$

$3(\sin\theta)(2\cos^2\theta - \sin^2\theta) = -\dfrac{3\sqrt{3}}{8}$, and $\kappa =$

$\dfrac{4}{3\sqrt{3}}$.

12 $\dfrac{dy}{dx} = -\dfrac{x}{\sqrt{a^2 - x^2}}$, $\dfrac{d^2y}{dx^2} = -\dfrac{a^2}{(a^2 - x^2)^{3/2}}$, $\kappa =$

$\dfrac{1}{a}$, and radius of curvature $= a$.

14 $\dfrac{dy}{dx} = 2x$, $\dfrac{d^2y}{dx^2} = 2$, $\kappa = \dfrac{2}{(1 + 4x^2)^{3/2}}$, and

radius of curvature $= \dfrac{1}{2}(1 + 4x^2)^{3/2}$. The radius of

curvature is minimal when $x = 0$.

16 In elementary physics it's shown that centrifugal "force" is proportional to curvature. Since the curvature of the track is not continuous, the passengers will be suddenly thrown to one side as the train traverses the curve.

18 By Exercise 14, $R = \dfrac{1}{2}(1 + 4x^2)^{3/2}$. At $x = 0$, R

$= 1/2$. At $x = 2$, $R = \dfrac{1}{2} \cdot 17^{3/2}$. Hence the moment

at $(0, 0)$ is $17^{3/2}$ times as large as that at $(2, 4)$.

20 (a) $\dfrac{x^2}{a^2} + \dfrac{y^2}{b^2} = \cos^2\theta + \sin^2\theta = 1$

(b) By Exercise 19, $R = \dfrac{(a^4y^2 + b^4x^2)^{3/2}}{a^4b^4}$. At

$(a, 0)$, $R = b^2/a$. At $(0, b)$, $R = a^2/b$.

22 Here $x = (a\cos\theta)\cos\theta$ and $y = (a\cos\theta)\sin\theta$,

so $\dfrac{dx}{d\theta} = -2a\cos\theta\sin\theta = -a\sin 2\theta$ and $\dfrac{dy}{d\theta} =$

$a(\cos^2\theta - \sin^2\theta) = a\cos 2\theta$. Then $\dfrac{d^2x}{d\theta^2} = -2a\cos 2\theta$ and $\dfrac{d^2y}{d\theta^2} = -2a\sin 2\theta$. Thus $\kappa =$

$\dfrac{|\dot{x}\ddot{y} - \dot{y}\ddot{x}|}{[(\dot{x})^2 + (\dot{y})^2]^{3/2}} =$

$\dfrac{|(-a\sin 2\theta)(-2a\sin 2\theta) - (a\cos 2\theta)(-2a\cos 2\theta)|}{[(-a\sin 2\theta)^2 + (a\cos 2\theta)^2]^{3/2}}$

$= \dfrac{2a^2[\sin^2 2\theta + \cos^2 2\theta]}{a^3} = \dfrac{2}{a}$. (Not surprising

since the graph of $r = a\cos\theta$ is a circle of radius $a/2$.)

24 $x = \cos 2\theta \cos\theta$,
$y = \cos 2\theta \sin\theta$,

$\dfrac{dx}{d\theta} = -\cos 2\theta \sin \theta - 2 \sin 2\theta \cos \theta,$

$\dfrac{dy}{d\theta} = \cos 2\theta \cos \theta - 2 \sin 2\theta \sin \theta,$

$\dfrac{d^2x}{d\theta^2} = -(\cos 2\theta \cos \theta - 2 \sin 2\theta \sin \theta)$

$\qquad - 2(2 \cos 2\theta \cos \theta - \sin 2\theta \sin \theta)$

$\qquad = 4 \sin 2\theta \sin \theta - 5 \cos 2\theta \cos \theta,$

$\dfrac{d^2y}{d\theta^2} = (-\cos 2\theta \sin \theta - 2 \sin 2\theta \cos \theta)$

$\qquad - 2(2 \cos 2\theta \sin \theta + \sin 2\theta \cos \theta)$

$\qquad = -4 \sin 2\theta \cos \theta - 5 \cos 2\theta \sin \theta,$

$\kappa = \dfrac{|\dot{x}\ddot{y} - \dot{y}\ddot{x}|}{[(\dot{x})^2 + (\dot{y})^2]^{3/2}}$, where the numerator is

$(-\cos 2\theta \sin\theta - 2\sin 2\theta \cos\theta)(-4\sin 2\theta \cos\theta - 5\cos 2\theta \sin\theta)$

$- (\cos 2\theta \cos\theta - 2\sin 2\theta \sin\theta)(4\sin 2\theta \sin\theta - 5\cos 2\theta \cos\theta)$

$= 5 \cos^2 2\theta + 8 \sin^2 2\theta$ and the denominator is

$[(-\cos 2\theta \sin\theta - 2\sin 2\theta \cos\theta)^2 +$

$(\cos 2\theta \cos\theta - 2\sin 2\theta \sin\theta)^2]^{3/2} =$

$[\cos^2 2\theta + 4 \sin^2 2\theta]^{3/2}$. Hence $\kappa =$

$\dfrac{5 \cos^2 2\theta + 8 \sin^2 2\theta}{[\cos^2 2\theta + 4 \sin^2 2\theta]^{3/2}} = \dfrac{5 + 3 \sin^2 2\theta}{(1 + 3 \sin^2 2\theta)^{3/2}}.$

26 For $y = x^3$, we have $y' = 3x^2$ and $y'' = 6x$, so κ

$= \dfrac{|6x|}{[1 + (3x^2)^2]^{3/2}} = \dfrac{6|x|}{(1 + 9x^4)^{3/2}}$. At $x = 1$, $\kappa =$

$\dfrac{6 \cdot 1}{(1 + 9 \cdot 1^4)^{3/2}} = \dfrac{6}{10^{3/2}}$ and the radius of curvature

is $\dfrac{10^{3/2}}{6}$. At $x = 2$, $\kappa = \dfrac{6 \cdot 2}{(1 + 9 \cdot 2^4)^{3/2}} = \dfrac{12}{145^{3/2}}$

and the radius of curvature is $\dfrac{145^{3/2}}{12}$. Let the

maximum safe speed be $v = C\sqrt{1/\kappa}$. At $x = 1$, we

have $30 = C\sqrt{\dfrac{10^{3/2}}{6}}$, so $C = \dfrac{30\sqrt{6}}{10^{3/4}}$. Hence, at x

$= 2$, $v = \dfrac{30\sqrt{6}}{10^{3/4}} \cdot \sqrt{\dfrac{145^{3/2}}{12}} = \dfrac{30\sqrt{6} \cdot 145^{3/4}}{10^{3/4}\sqrt{12}} =$

$\dfrac{15 \cdot 29^{3/4}}{2^{1/4}} \approx 157.6$ mi/hr.

9.7 The Reflection Properties of the Conic Sections

2   $\dfrac{5\pi}{6} - \dfrac{\pi}{6} = \dfrac{2\pi}{3}$

4   $\tan \theta = \dfrac{-1/2 - 2}{1 + (-1/2)2} = \dfrac{-5/2}{0}$, so $|\tan \theta| = \infty$.

6   $\tan \theta = \dfrac{-\sqrt{3} - 1}{1 + (-\sqrt{3}) \cdot 1} = \dfrac{\sqrt{3} + 1}{\sqrt{3} - 1} = \dfrac{1}{2}(4 + 2\sqrt{3})$

$= 2 + \sqrt{3}$

8   The slope of $y = x^2$ is $2x$, which equals 2 at $x = 1$. The slope of $y = x^3$ is $3x^2$, which equals 3 at $x = 1$. The tangent of the angle is $\dfrac{3 - 2}{1 + 3 \cdot 2} = \dfrac{1}{7}$.

10 The slope of $y = \sec x$ is $\sec x \tan x = \sqrt{2}$ at $x = \pi/4$. The slope of $y = \sqrt{2} \tan x$ is $\sqrt{2} \sec^2 x = 2\sqrt{2}$ at $x = \pi/4$. The tangent of the angle is

$\dfrac{2\sqrt{2} - \sqrt{2}}{1 + 2\sqrt{2} \cdot \sqrt{2}} = \dfrac{\sqrt{2}}{5}.$

12 Position the axes so that the ellipse has the equation

$\dfrac{x^2}{a^2} + \dfrac{y^2}{b^2} = 1$ with $a \geq b$. Its foci are $F = (c, 0)$

9.7 The Reflection Properties of the Conic Sections

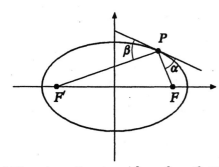

and $F' = (-c, 0)$, where $c^2 = a^2 - b^2$. If $P = (x, y)$ is on the ellipse, then the slopes of FP and $F'P$ are $\dfrac{y}{x-c}$ and $\dfrac{y}{x+c}$, respectively. To find the slope of the tangent line at P, differentiate implicitly: $\dfrac{2x}{a^2} + \dfrac{2y(dy/dx)}{b^2} = 0$; so the slope is $\dfrac{dy}{dx} = -\dfrac{b^2 x}{a^2 y}$. By Eq. (1), $\tan \alpha =$

$$\dfrac{-\dfrac{b^2 x}{a^2 y} - \dfrac{y}{x-c}}{1 + \left(-\dfrac{b^2 x}{a^2 y}\right)\dfrac{y}{x-c}} = \dfrac{-b^2 x(x-c) - y a^2 y}{a^2 y(x-c) - b^2 xy}$$

$$= \dfrac{b^2 cx - (b^2 x^2 + a^2 y^2)}{(a^2 - b^2)xy - a^2 cy} = \dfrac{b^2 cx - a^2 b^2}{c^2 xy - a^2 cy}$$

$$= \dfrac{b^2(cx - a^2)}{cy(cx - a^2)} = \dfrac{b^2}{cy}.\text{ Similarly, } \tan \beta =$$

$$\dfrac{\dfrac{y}{x+c} - \left(-\dfrac{b^2 x}{a^2 y}\right)}{1 + \dfrac{y}{x+c}\left(-\dfrac{b^2 x}{a^2 y}\right)} = \dfrac{a^2 y^2 + b^2 x^2 + b^2 cx}{a^2 xy + a^2 cy - b^2 xy}$$

$$= \dfrac{a^2 b^2 + b^2 cx}{c^2 xy + a^2 cy} = \dfrac{b^2(a^2 + cx)}{cy(a^2 + cx)} = \dfrac{b^2}{cy}.\text{ Hence}$$

$\tan \alpha = \tan \beta$, so $\alpha = \beta$.

14 (a) $\tan \theta$

(b) Slope $= \dfrac{dy}{dx} = \dfrac{dy/d\theta}{dx/d\theta}$

$$= \dfrac{f(\theta) \cos \theta + f'(\theta) \sin \theta}{-f(\theta) \sin \theta + f'(\theta) \cos \theta}$$

(c) Let $f = f(\theta)$, $f' = f'(\theta)$, $c = \cos \theta$, and $s = \sin \theta$. Then $\tan \gamma = \dfrac{\dfrac{dy}{dx} - \tan \theta}{1 + \dfrac{dy}{dx} \tan \theta} =$

$$\dfrac{\dfrac{fc + f's}{-fs + f'c} - \dfrac{s}{c}}{1 + \dfrac{fc + f's}{-fs + f'c} \cdot \dfrac{s}{c}} = \dfrac{c(fc + f's) - s(-fs + f'c)}{c(-fs + f'c) + s(fc + f's)}$$

$$= \dfrac{f}{f'} = \dfrac{f(\theta)}{f'(\theta)}.$$

(d)

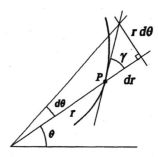

$$\tan \gamma \approx r\dfrac{d\theta}{dr} = \dfrac{r}{dr/d\theta} = \dfrac{f(\theta)}{f'(\theta)}$$

16 $r = 1 - \cos \theta$, $r' = \sin \theta$, $\tan \gamma = r/r' =$

$$\dfrac{1 - \cos \theta}{\sin \theta} = \tan \dfrac{\theta}{2}$$

18 (a) $r = 1 + \cos \theta$, $r' = -\sin \theta$, $\tan \gamma = r/r' =$

$$-\dfrac{1 + \cos \theta}{\sin \theta} = -\cot \dfrac{\theta}{2}.\text{ As } \theta \to \pi^-,$$

$\cot \theta/2 \to 0$, so $\gamma \to 0$.

(b)

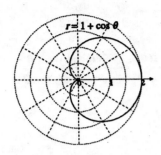

20 (a) Introduce coordinates so that the origin is at the center and one dog is at $\left(\dfrac{a}{\sqrt{2}}, 0\right)$. Since $\gamma = 3\pi/4$ for all θ, Exercise 19 gives $r = Ae^{k\theta}$ for constants A and k, which we find to be $k = -1$ and $A = \dfrac{a}{\sqrt{2}}$, so $r = \dfrac{a}{\sqrt{2}}e^{-\theta}$. Hence s

$$= \int_0^\infty \sqrt{r^2 + (r')^2}\, d\theta = \int_0^\infty ae^{-\theta}\, d\theta = a.$$

(b) Each dog initially is at distance a from the dog it is chasing. Since it runs directly toward its quarry, and the quarry's motion is perpendicular to that of the chaser, each unit of distance that the chaser runs actually brings it one unit closer to the quarry. So the total distance it runs is a.

9.S Review Exercises

2 (a) See Sec. 9.4.

(b) See Figure 1 in Sec. 9.4.

4 See the solution of Exercise 14 in Sec. 9.7.

6 (a) $ds = \sqrt{1 + e^{2x}}\, dx$

Arc length $= \int_0^1 \sqrt{1 + e^{2x}}\, dx$

Area of surface $= \int_0^1 2\pi e^x \sqrt{1 + e^{2x}}\, dx$

(revolved about x axis)

Area of surface $= \int_0^1 2\pi x\sqrt{1 + e^{2x}}\, dx$

(revolved about y axis)

(b) Let $u = e^{2x}$ and $u = v^2 - 1$. Then Arc length

$$= \int_1^{e^2} \sqrt{1 + u}\, \dfrac{du}{2u}$$

$$= \ln\left(\dfrac{\sqrt{e^2 + 1} - 1}{\sqrt{2} - 1}\right) + \sqrt{e^2 + 1} - \sqrt{2} - 1.$$

Area of surface $= 2\pi \int_1^e \sqrt{1 + w^2}\, dw$

$$= \pi\left[\ln\left(\dfrac{e + \sqrt{1 + e^2}}{1 + \sqrt{2}}\right) + e\sqrt{1 + e^2} - \sqrt{2}\right].$$

8 Multiplying by $\cos\theta + 2\sin\theta$ gives $x + 2y = 3$.

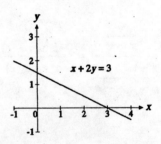

10 $x = r\cos\theta = \cos\theta + \cos^2\theta =$
$-\dfrac{1}{4} + \left(\dfrac{1}{2} + \cos\theta\right)^2$ is minimal when $\cos\theta = -\dfrac{1}{2}$.

The minimum is $-\dfrac{1}{4}$.

12 $r = \dfrac{1}{1 + \theta^2}$ so $r' = \dfrac{2\theta}{(1 + \theta^2)^2}$ and Arc length

$$= \int_0^\infty \sqrt{r^2 + (r')^2}\, d\theta = \int_0^\infty \dfrac{\sqrt{1 + 6\theta^2 + \theta^4}}{(1 + \theta^2)^2}\, d\theta.$$

For $\theta \geq 1$, $1 + 6\theta^2 + \theta^4 \leq 8\theta^4$, so the integrand

9.S Summary: Review Exercises

is less than $\dfrac{\sqrt{8\theta^4}}{(1+\theta^2)^2} < \dfrac{\sqrt{8}\theta^2}{(\theta^2)^2} = \dfrac{\sqrt{8}}{\theta^2}$. Since $\int_1^\infty \dfrac{d\theta}{\theta^2}$ converges, the arc length is finite.

14 $\dfrac{d^2y}{dx^2} = \dfrac{d}{dx}\left(\dfrac{dy}{dx}\right) = \dfrac{d}{dx}\left(\dfrac{\dot{y}}{\dot{x}}\right) = \dfrac{\dfrac{d}{dt}(\dot{y}/\dot{x})}{dx/dt}$

$= \dfrac{\dfrac{\dot{x}\ddot{y} - \dot{y}\ddot{x}}{(\dot{x})^2}}{\dot{x}} = \dfrac{\dot{x}\ddot{y} - \dot{y}\ddot{x}}{(\dot{x})^3}$

16 $\dfrac{dy}{dx} = \dfrac{1}{x} = \dfrac{1}{e}$ and $\dfrac{d^2y}{dx^2} = -\dfrac{1}{x^2} = -\dfrac{1}{e^2}$, so

$\kappa = \dfrac{e}{(e^2+1)^{3/2}}$ and $R = \dfrac{(e^2+1)^{3/2}}{e}$.

18 $\dfrac{dy}{dx} = \dfrac{1}{x}$, so $ds = \sqrt{1 + 1/x^2}\, dx$ and Arc length

$= \int_1^{\sqrt{3}} \sqrt{1 + 1/x^2}\, dx = 2 - \sqrt{2} + \ln\dfrac{1+\sqrt{2}}{\sqrt{3}}$.

20 (a) Since the curve is convex, $\dfrac{d\phi}{ds}$ does not change sign. Hence the average radius of curvature with respect to ϕ is $\dfrac{1}{2\pi}\int_0^{2\pi} R\, d\phi$

$= \dfrac{1}{2\pi}\int_0^{2\pi} \dfrac{1}{|d\phi/ds|}\, d\phi = \dfrac{1}{2\pi}\left|\int_0^{2\pi} \dfrac{ds}{d\phi}\, d\phi\right|$

$= \dfrac{1}{2\pi}(\text{Length of curve})$.

(b) The radius of curvature is a continuous function of ϕ. Let r and R be the minimum and maximum, respectively, radii of curvature on the curve. The average radius of curvature, $\bar{r}$, must lie between r and R: $r \le \bar{r} \le R$. By the intermediate-value theorem, there is a point on the curve at which the radius of curvature equals $\bar{r} = L/(2\pi)$.

22 As suggested by the given parametric equations, let G be the origin of an xy coordinate system with A lying on the positive x axis. Then the ball has coordinates $(v(\cos\theta)t, -16t^2 + v(\sin\theta)t) =$

$(x, (\tan\beta)x)$. So $t = \dfrac{x}{v\cos\theta}$, and $(\tan\beta)x =$

$-16\left(\dfrac{x}{v\cos\theta}\right)^2 + v\sin\theta\left(\dfrac{x}{v\cos\theta}\right)$. Then $(\tan\beta)x$

$= -\dfrac{16x^2}{v^2\cos^2\theta} + (\tan\theta)x$, and $x =$

$\dfrac{v^2\cos^2\theta}{16}[\tan\theta - \tan\beta]$. Hence $\dfrac{dx}{d\theta} =$

$\dfrac{v^2}{16}\dfrac{d}{d\theta}[\cos\theta\sin\theta - \tan\beta\cos^2\theta] =$

$\dfrac{v^2}{16}[\cos 2\theta + \tan\beta\sin 2\theta] = 0$ when $\tan 2\theta =$

$-\tan\beta$, or $\tan 2\theta = \tan(\beta + \pi/2)$. Thus, GB is

maximized when $\theta = \dfrac{\pi}{4} + \dfrac{\beta}{2}$.

24 (a)

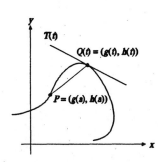

(b) $\dfrac{h(t) - h(a)}{g(t) - g(a)}, \dfrac{h'(t)}{g'(t)}$

(c) A sufficient condition is that g and h be differentiable and g' be nonzero in some interval (a, b), and that $\lim\limits_{t \to a^+} \dfrac{h(t)}{g(t)}$ exist. The result then follows by l'Hôpital's rule.

26 (a) The line containing (a, b) and the line segment whose length is the distance from (a, b) to the line $y = -x + 1$ is $y - b = x - a$, or $y = x - a + b$. These lines intersect at $\left(\dfrac{1+a-b}{2}, \dfrac{1+b-a}{2}\right)$, and the distance between (a, b) and $\left(\dfrac{1+a-b}{2}, \dfrac{1+b-a}{2}\right)$ is

$$\sqrt{\left(a - \dfrac{1+a-b}{2}\right)^2 + \left(b - \dfrac{1+b-a}{2}\right)^2}$$

$$= \sqrt{\left(\dfrac{a+b-1}{2}\right)^2 + \left(\dfrac{a+b-1}{2}\right)^2}$$

$$= \sqrt{2}\left|\dfrac{a+b-1}{2}\right|.$$

(b)

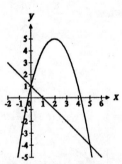

Surface area $= \int_c^d 2\pi R \, ds =$

$$\int_0^5 2\pi\sqrt{2}\left[\dfrac{x+5-(x-2)^2-1}{2}\right]\left[1 + \left(\dfrac{dy}{dx}\right)^2\right]^{1/2} dx$$

$$= \sqrt{2}\pi \int_0^5 (5x - x^2)(4x^2 - 16x + 17)^{1/2} \, dx$$

28 (a) $\dfrac{dy}{dx} = \dfrac{dy/d\theta}{dx/d\theta} = \dfrac{\dfrac{d}{d\theta}[\sin\theta + \sin\theta \cos\theta]}{\dfrac{d}{d\theta}[\cos\theta + \cos^2\theta]} =$

$\dfrac{\cos\theta + \cos 2\theta}{-\sin\theta - \sin 2\theta}$; at $\theta = \pi/2$, $\dfrac{dy}{dx} = 1$. Now $\dfrac{dy}{dx} = \dfrac{dy/dt}{dx/dt}$. At $\theta = \pi/2$, we have $1 = \dfrac{dy/dt}{5}$; so $\dfrac{dy}{dt} = 5$.

(b) The distance from the origin is, by definition, $r = 1 + \cos\theta$. Thus $\dfrac{dr}{d\theta} = -\sin\theta$.

30 (a) $\sin 3\theta = \sin(\theta + 2\theta)$

$= \sin\theta \cos 2\theta + \sin 2\theta \cos\theta$

$= \sin\theta (1 - 2\sin^2\theta) + 2\sin\theta \cos\theta \cos\theta$

$= \sin\theta - 2\sin^3\theta + 2\sin\theta (1 - \sin^2\theta)$

$= 3\sin\theta - 4\sin^3\theta$

$\cos 3\theta = \cos(\theta + 2\theta)$

$= \cos\theta \cos 2\theta - \sin\theta \sin 2\theta$

$= \cos\theta (2\cos^2\theta - 1) - 2\sin^2\theta \cos\theta$

$= 2\cos^3\theta - \cos\theta - 2(1 - \cos^2\theta)\cos\theta$

$= 4\cos^3\theta - 3\cos\theta$

(b) If $b = a/4$, then Exercise 29 says that $x = (a - b)\cos\theta + b\cos(\phi - \theta)$

$= \dfrac{3}{4}a\cos\theta + \dfrac{a}{4}\cos 3\theta$

$= \dfrac{a}{4}(3\cos\theta + \cos 3\theta) = \dfrac{a}{4}(4\cos^3\theta)$

$= a \cos^3 \theta$, where we used the identity from (a). Similarly, $y = a \sin^3 \theta$.

(c) Using the result of (b), we have $x^{2/3} + y^{2/3}$
$= (a \cos^3 \theta)^{2/3} + (a \sin^3 \theta)^{2/3}$
$= a^{2/3}(\cos^2 \theta + \sin^2 \theta) = a^{2/3}$.

(d) See graph for Exercise 34(a) of Sec. 9.3.

10 Series

10.1 An Informal Introduction to Series

2. (a) $e^{-1} \approx 1 + (-1) + \frac{(-1)^2}{2!} + \frac{(-1)^3}{3!} + \frac{(-1)^4}{4!}$

   $= \frac{3}{8} = 0.375$

   (b) $e^{-1} \approx 1 + (-1) + \cdots + \frac{(-1)^5}{5!} = \frac{11}{30}$

   ≈ 0.36667

4. (a) $\sin 1 \approx 0.8414710$

   (b) $\sin 1 \approx 1 - \frac{1^3}{3!} + \frac{1^5}{5!} \approx 0.8416667$

6. (a) $\int_{1/4}^{1/2} \frac{\sin x}{x}\, dx \approx \int_{1/4}^{1/2} \frac{x}{x}\, dx = x\big|_{1/4}^{1/2} = \frac{1}{4}$

   $= 0.25$

   (b) $\int_{1/4}^{1/2} \frac{\sin x}{x}\, dx \approx \int_{1/4}^{1/2} \frac{x - x^3/3!}{x}\, dx$

   $= \int_{1/4}^{1/2} \left(1 - \frac{1}{6}x^2\right) dx = \left(x - \frac{1}{18}x^3\right)\Big|_{1/4}^{1/2}$

   $= \frac{281}{1152} \approx 0.2439236$

   (c) $\int_{1/4}^{1/2} \frac{\sin x}{x}\, dx \approx \int_{1/4}^{1/2} \frac{x - x^3/3! + x^5/5!}{x}\, dx$

   $= \int_{1/4}^{1/2} \left(1 - \frac{1}{6}x^2 + \frac{1}{120}x^4\right) dx$

   $= \left(x - \frac{1}{18}x^3 + \frac{1}{600}x^5\right)\Big|_{1/4}^{1/2} \approx 0.2439741$

8. (a)

n	Rational Form	Decimal Form
1	$-1/2$	-0.5
2	$-5/8$	-0.625
3	$-2/3$	-0.6666667
4	$-131/192$	-0.6822917
5	$-661/960$	-0.6885417

   (b) Calculator: $\ln(1/2) = -0.6931472$

   Error: $|\ln(1/2) - (-661/960)| \approx 0.0046055$

10. Since $\sqrt{n} \geq \sqrt[3]{n}$ for $n \geq 1$ we have that $\frac{1}{\sqrt{n}} \leq$

    $\frac{1}{\sqrt[3]{n}}$ for $n \geq 1$. Thus, $\frac{1}{\sqrt{1}} + \frac{1}{\sqrt{2}} + \frac{1}{\sqrt{3}} + \cdots +$

    $\frac{1}{\sqrt{n}} \leq \frac{1}{\sqrt[3]{1}} + \frac{1}{\sqrt[3]{2}} + \frac{1}{\sqrt[3]{3}} + \cdots + \frac{1}{\sqrt[3]{n}}$. It follows

    that the expression on the right gets arbitrarily large since it was shown in this section that the expression on the left gets arbitrarily large.

12. (b) $1 - x^n = (1 - x)(1 + x + \cdots + x^{n-1})$, so

    $1 + x + \cdots + x^{n-1} = \frac{1 - x^n}{1 - x}$

    $= \frac{1}{1 - x} - \frac{x^n}{1 - x}.$

    (c) $1 + (0.3) + (0.3)^2 + (0.3)^3 + (0.3)^4 +$

    $(0.3)^5 = 1.42753$ and $\frac{1}{1 - (0.3)} \approx$

1.4285714 so the error is about
$|1.42753 - 1.4285714| = 0.0010414$.

(d) $1 + (-0.9) + (-0.9)^2 + (-0.9)^3 + (-0.9)^4 + (-0.9)^5 = 0.24661$ and $\dfrac{1}{1-(-0.9)} \approx 0.5263158$ so the error is about $|0.24661 - 0.5263158| = 0.2797058$.

14 (a) $\dfrac{1}{1+t^2} = 1 - t^2 + t^4 - t^6 + \cdots + (-1)^{n-1}t^{2n-2} + (-1)^n \dfrac{t^{2n}}{1+t^2}$, so $\int_0^x \dfrac{1}{1+t^2}\,dt =$

$\int_0^x [1 - t^2 + t^4 - t^6 + \cdots + (-1)^{n-1}t^{2n-2}]\,dt$
$+ (-1)^n \int_0^x \dfrac{t^{2n}}{1+t^2}\,dt$, resulting in $(\tan^{-1} t)\big|_0^x$

$= \left[t - \dfrac{t^3}{3} + \dfrac{t^5}{5} - \dfrac{t^7}{7} + \cdots + (-1)^{n-1}\dfrac{t^{2n-1}}{2n-1}\right]\Big|_0^x +$

$(-1)^n \int_0^x \dfrac{t^{2n}}{1+t^2}\,dt$. Hence $\tan^{-1} x =$

$x - \dfrac{x^3}{3} + \dfrac{x^5}{5} - \dfrac{x^7}{7} + \cdots + (-1)^{n-1}\dfrac{x^{2n-1}}{2n-1}$

$+ (-1)^n \int_0^x \dfrac{t^{2n}}{1+t^2}\,dt.$

(b) For $0 < x \leq 1$ we have $0 \leq \int_0^x \dfrac{t^{2n}}{1+t^2}\,dt$

$\leq \int_0^x \dfrac{t^{2n}}{1+0^2}\,dt = \dfrac{t^{2n+1}}{2n+1}\Big|_0^x = \dfrac{x^{2n+1}}{2n+1}$

$\leq \dfrac{1}{2n+1}$. But $\dfrac{1}{2n+1}$ approaches 0 as

$n \to \infty$. By comparison it follows that

$\int_0^x \dfrac{t^{2n}}{1+t^2}\,dt \to 0$ as $n \to \infty$.

(c) $1 - 1/3 + 1/5 - 1/7 + 1/9 \approx 0.835$

(d) From (c) it follows that $\pi/4 \approx 0.835$, so $\pi \approx 3.34$.

15 (a) Let $S_n = \dfrac{1}{1 \cdot 2} + \dfrac{1}{2 \cdot 3} + \dfrac{1}{3 \cdot 4} + \cdots + \dfrac{1}{n(n+1)}$.

n	S_n	Fraction	Decimal
1	1/2	1/2	0.5
2	1/2 + 1/6	2/3	0.6666667
3	1/2 + 1/6 + 1/12	3/4	0.75
4	1/2 + 1/6 + 1/12 + 1/20	4/5	0.8
5	1/2 + 1/6 + 1/12 + 1/20 + 1/30	5/6	0.8333333
6	1/2 + 1/6 + 1/12 + 1/20 + 1/30 + 1/42	6/7	0.8571429

(c) $\dfrac{1}{1 \cdot 2} + \dfrac{1}{2 \cdot 3} + \dfrac{1}{3 \cdot 4} + \cdots + \dfrac{1}{n(n+1)} =$

$\dfrac{n}{n+1}$ for $n = 1, 2, 3, \cdots$, and $\dfrac{n}{n+1} \to 1$

as $n \to \infty$.

16 (a) $\dfrac{1}{2} - \dfrac{(1/2)^3}{3} + \dfrac{(1/2)^5}{5} - \dfrac{(1/2)^7}{7} + \dfrac{(1/2)^9}{9}$

≈ 0.4636843

(b) It's easy to show that $\tan^{-1} x$
$= \pi/2 - \tan^{-1}(1/x)$ for $x > 0$; thus $\tan^{-1} 2$
$= \pi/2 - \tan^{-1}(1/2)$
$\approx 1.5707963 - 0.4636843 = 1.1071120$.

(d) Calculator: $\tan^{-1} 2 = 1.1071487$

10.2 Sequences

2 $\{1.01^n\}$: 1.01, 1.0201, 1.030301, $\cdots$; the sequence diverges since $1.01 > 1$.

4 $\{(-0.8)^n\}$: -0.8, 0.64, -0.512, $\cdots$; the sequence converges with $\lim_{n \to \infty} (-0.8)^n = 0$ since $-1 < -0.8 < 1$.

6 $\left\{\dfrac{10^n}{n!}\right\}$: 10, 50, $\dfrac{500}{3}$, $\cdots$; the sequence converges with $\lim_{n \to \infty} \dfrac{10^n}{n!} = 0$, by Example 3.

8 $\left\{\dfrac{(-1)^n}{n}\right\}$: -1, $\dfrac{1}{2}$, $\dfrac{-1}{3}$, $\cdots$; the sequence converges with $\lim_{n \to \infty} \dfrac{(-1)^n}{n} = 0$, since $-\dfrac{1}{n} \leq \dfrac{(-1)^n}{n} \leq \dfrac{1}{n}$ and both $\lim_{n \to \infty} \dfrac{-1}{n} = 0$ and $\lim_{n \to \infty} \dfrac{1}{n} = 0$.

10 $\left\{\dfrac{n}{2^n} + \dfrac{3n+1}{4n+2}\right\}$: $\dfrac{7}{6}$, $\dfrac{6}{5}$, $\dfrac{61}{56}$, $\cdots$; the sequence converges. By l'Hôpital's rule, $\lim_{x \to \infty} \dfrac{x}{2^x} \stackrel{H}{=} \lim_{x \to \infty} \dfrac{1}{2^x \ln 2} = 0$, so $\lim_{n \to \infty} \dfrac{n}{2^n} = 0$ and

$\lim_{n \to \infty} \left(\dfrac{n}{2^n} + \dfrac{3n+1}{4n+2}\right) = \lim_{n \to \infty} \dfrac{n}{2^n} + \lim_{n \to \infty} \dfrac{3 + 1/n}{4 + 2/n}$

$= 0 + \dfrac{3}{4} = \dfrac{3}{4}$.

12 $\left\{\left(\dfrac{n-1}{n}\right)^n\right\}$: 0, $\dfrac{1}{4}$, $\dfrac{8}{27}$, $\cdots$; the sequence converges with $\lim_{n \to \infty} \left(\dfrac{n-1}{n}\right)^n =$

$\lim_{n \to \infty} \left[\left(1 + \dfrac{1}{-n}\right)^{-n}\right]^{-1} = e^{-1}$.

14 Notice that the ratio of consecutive terms is given by $\dfrac{(11.8)^{n+1}}{(n+1)!} \cdot \dfrac{n!}{(11.8)^n} = \dfrac{11.8}{n+1}$, which is greater than 1 for $n = 1, 2, \cdots, 10$ and less than 1 for $n = 11, 12, \cdots$. Thus the largest term is $\dfrac{(11.8)^{11}}{11!} \approx 15471.997$.

16 (a) $(1.0006)^{1200} \approx 2.0539897 > 2$

(b) $(1.0006)^x = 2$, so $x = \dfrac{\ln 2}{\ln 1.0006} \approx 1155.592$, so choose $n = 1156$.

18 Assume that $r = 1 + x$, where x is a positive constant. Then for $n \geq 2$, $r^n = (1+x)^n = 1 + nx + a_n$, where $a_n = \dfrac{n(n-1)}{2} x^2 + \dfrac{n(n-1)(n-2)}{3 \cdot 2} x^3 + \cdots + x^n \geq 0$. Since $\lim_{n \to \infty} (1 + nx) = \infty$, we have $\lim_{n \to \infty} r^n = \lim_{n \to \infty} (1 + nx + a_n) = \infty$.

20

n	$a_{n+2} = \dfrac{1 + a_{n+1}}{a_n}$
1	1
2	3
3	$\dfrac{1+3}{1} = 4$
4	$\dfrac{1+4}{3} = \dfrac{5}{3}$
5	$\dfrac{1+5/3}{4} = \dfrac{2}{3}$
6	$\dfrac{1+2/3}{5/3} = 1$
7	$\dfrac{1+1}{2/3} = 3$

For succeeding values of n, the preceding values are repeated.

21

n	$a_n = \sum\limits_{i=n}^{2n} \dfrac{1}{i}$
1	$1 + \dfrac{1}{2} = 1.5$
2	$\dfrac{1}{2} + \dfrac{1}{3} + \dfrac{1}{4} \approx 1.0833333$
3	$\dfrac{1}{3} + \dfrac{1}{4} + \dfrac{1}{5} + \dfrac{1}{6} = 0.95$
4	$\dfrac{1}{4} + \dfrac{1}{5} + \dfrac{1}{6} + \dfrac{1}{7} + \dfrac{1}{8} \approx 0.8845238$
5	$\dfrac{1}{5} + \dfrac{1}{6} + \dfrac{1}{7} + \dfrac{1}{8} + \dfrac{1}{9} + \dfrac{1}{10} \approx 0.8456349$
6	$\dfrac{1}{6} + \dfrac{1}{7} + \dfrac{1}{8} + \dfrac{1}{9} + \dfrac{1}{10} + \dfrac{1}{11} + \dfrac{1}{12} \approx 0.8198773$

Since $\dfrac{1}{n}$ and $\dfrac{1}{2n}$ are the largest and smallest terms, respectively, in the sum

$\dfrac{1}{n} + \dfrac{1}{n+1} + \dfrac{1}{n+2} + \cdots + \dfrac{1}{2n}$, it follows that

$(n+1) \cdot \dfrac{1}{2n} < \dfrac{1}{n} + \dfrac{1}{n+1} + \cdots + \dfrac{1}{2n} <$

$(n+1) \cdot \dfrac{1}{n}$. Since $a_n - a_{n+1}$

$= \dfrac{1}{n} - \left(\dfrac{1}{2n+1} + \dfrac{1}{2n+2}\right)$

$> \dfrac{1}{n} - \left(\dfrac{1}{2n+2} + \dfrac{1}{2n+2}\right) = \dfrac{1}{n(n+1)} > 0$, we

see that a_n is a nonincreasing sequence of positive terms. It follows from Theorem 2 that this sequence has a limit. Since $\lim\limits_{n \to \infty} (n+1) \cdot \dfrac{1}{n} = 1$ and

$\lim\limits_{n \to \infty} (n+1)\dfrac{1}{2n} = \dfrac{1}{2}$, we know this limit lies

somewhere between 1/2 and 1. **Note:** It can be shown that the limit of this sequence is $\ln 2$.

22 Since $\lim\limits_{n \to \infty} x_{n+1} = \lim\limits_{n \to \infty} x_n = L$, $\lim\limits_{n \to \infty} (x_n - x_{n+1}) =$

$L - L = 0$. But $\lim\limits_{n \to \infty} (x_n - x_{n+1}) = \lim\limits_{n \to \infty} \dfrac{f(x_n)}{f'(x_n)} =$

$\dfrac{f(L)}{f'(L)}$. Thus $0 = \dfrac{f(L)}{f'(L)}$, so $f(L) = 0$.

24 Divide the interval $[0, 1]$ into n equal parts, each of length $1/n$. Let $x_i = i/n$ and $f(x) = \dfrac{1}{1+x^2}$. Then

$\lim\limits_{n \to \infty} \sum\limits_{i=1}^{\infty} \dfrac{n}{n^2 + i^2} = \lim\limits_{n \to \infty} \sum\limits_{i=1}^{n} \dfrac{1}{1+(i/n)^2} \cdot \dfrac{1}{n}$

$= \lim\limits_{n \to \infty} \sum\limits_{i=1}^{n} f(x_i) \cdot \dfrac{1}{n} = \int_0^1 f(x)\, dx$

$= \int_0^1 \dfrac{1}{1+x^2}\, dx = \tan^{-1} x \Big|_0^1 = \dfrac{\pi}{4}$.

10.2 Sequences

26. If $r = 0$, then $\lim_{n \to \infty} r^n = \lim_{n \to \infty} 0 = 0$. If $0 < r < 1$, then the sequence $r, r^2, r^3, \ldots, r^n$ is decreasing with $r^n \geq 0$ for all n. Thus, $\lim_{n \to \infty} r^n$ exists and is greater than or equal to zero. Call this limit L. Then $L = \lim_{n \to \infty} r^{n+1} = r \lim_{n \to \infty} r^n = rL$; that is, $L = rL$. Then $L(1-r) = 0$ and $L = 0$ since $1 - r > 0$. If $-1 < r < 0$, then the sequence $r, r^2, r^3, \ldots$ alternates in sign, with $r^2, r^4, r^6, \ldots, r^{2k}, \ldots$ being a decreasing sequence of positive terms, and $r, r^3, r^5, \ldots, r^{2k-1}, \ldots$ an increasing sequence of negative terms. Using the same technique as for $0 < r < 1$, it can be shown that each sequence converges to zero, so that $\lim_{n \to \infty} r^n = 0$.

28. Let $\epsilon > 0$ be given. Choose $N > 1/\epsilon$. Then for $n \geq N$ we have $\left|\dfrac{\sin n}{n} - 0\right| = \dfrac{|\sin n|}{n} \leq \dfrac{1}{n} < \dfrac{1}{N} < \epsilon$.

30. If $\lim_{n \to \infty} (-1)^n = 0$, then for each $\epsilon > 0$ there is an integer N so that $n > N$ implies $|(-1)^n - 0| = 1 < \epsilon$. However, if we choose $\epsilon = 1/2$, then for *no* value of N does $n \geq N$ imply that $1 < 1/2$. Thus, the original statement is false.

10.3 Series

2. (a) False. Apply the nth-term test.
   (b) False. Apply the nth-term test.
   (c) True. Apply the nth-term test.
   (d) False. See part (c).

4. (a) $a_4 = 5\left(-\dfrac{1}{2}\right)^4 = 0.3125$

   (b) $S_4 = 5(-1/2) + 5(-1/2)^2 + 5(-1/2)^3 + 5(-1/2)^4 = -1.5625$

   (c) $\lim_{n \to \infty} a_n = \lim_{n \to \infty} 5\left(-\dfrac{1}{2}\right)^n = 0$ by Theorem 1 of Sec. 10.2.

   (d) Applying Theorem 1 with $a = 5(-1/2)$ and $r = -1/2$, we have $\lim_{n \to \infty} S_n = \sum_{k=1}^{\infty} 5\left(-\dfrac{1}{2}\right)^k = \dfrac{5(-1/2)}{1 - (-1/2)} = -\dfrac{5}{3}$.

   (e) By (d), the series converges with sum $-5/3$.

6. $1 - \dfrac{1}{3} + \dfrac{1}{9} - \dfrac{1}{27} + \cdots = \dfrac{1}{1 - \left(-\dfrac{1}{3}\right)} = \dfrac{3}{4}$

8. $\sum_{n=1}^{\infty} 10^n$ diverges since $\lim_{n \to \infty} 10^n \neq 0$.

10. $\sum_{n=1}^{\infty} 7(-1.01)^n$ diverges since $\lim_{n \to \infty} 7(-1.01)^n \neq 0$.

12. $-\dfrac{3}{2} + \dfrac{3}{4} - \dfrac{3}{8} + \cdots = \dfrac{-3/2}{1 - (-1/2)} = -1$

14. $\sum_{n=1}^{\infty} \dfrac{1}{\left(1 + \dfrac{1}{n}\right)^n}$ diverges since $\lim_{n \to \infty} \dfrac{1}{\left(1 + \dfrac{1}{n}\right)^n} = \dfrac{1}{e} \neq 0$.

16. $\sum_{n=1}^{\infty} \dfrac{n}{2n+1}$ diverges since $\lim_{n \to \infty} a_n = \lim_{n \to \infty} \dfrac{n}{2n+1} = \dfrac{1}{2} \neq 0$.

18 $\sum_{n=1}^{\infty} 100\left(-\frac{8}{9}\right)^n = \frac{100(-8/9)}{1-(-8/9)} = -\frac{800}{17}$

20 $\sum_{n=1}^{\infty} (4^{-n} + n^{-1}) = \sum_{n=1}^{\infty} \left[\left(\frac{1}{4}\right)^n + \frac{1}{n}\right]$ diverges since

$\sum_{n=1}^{\infty} \frac{1}{n}$ diverges.

22 The total distance traveled is $6 + 2(0.9)(6) +$
$2(0.9)^2(6) + 2(0.9)^3(6) + \cdots = 6 + \frac{10.8}{1-(0.9)}$
$= 6 + 108 = 114$ feet.

24 $0.3333\cdots = \frac{3}{10} + \frac{3}{100} + \frac{3}{1000} + \cdots$

$= \frac{3/10}{1 - 1/10} = \frac{1}{3}$, as expected.

26 By Theorem 1, $\sum_{n=1}^{\infty} ar^{n-1} = a + ar + ar^2 + \cdots$

$+ ar^{n-1} + \cdots = \frac{a}{1-r}$. By the formula in

Appendix C for the sum of a finite geometric

series, $S_n = \frac{a(1-r^n)}{1-r}$, so $\sum_{k=1}^{\infty} ar^{k-1} - S_n$

$= \frac{a}{1-r} - \frac{a(1-r^n)}{1-r} = \frac{ar^n}{1-r}$.

28 (a)

Dose	Amount Active
1	A
2	$A + Ae^{-6k}$
3	$A + Ae^{-6k} + Ae^{-12k}$
4	$A + Ae^{-6k} + Ae^{-12k} + Ae^{-18k}$
$\vdots$	$\vdots$
n	$A + Ae^{-6k} + Ae^{-12k} + \cdots + Ae^{-6(n-1)k}$

(b) $\lim_{n\to\infty} S_n = \sum_{n=1}^{\infty} A(e^{-6k})^{n-1} = \frac{A}{1-e^{-6k}} < \infty$

30

Transactions	Billions of $
1	1
2	$1 + (0.8)$
3	$1 + (0.8) + (0.8)^2$
4	$1 + (0.8) + (0.8)^2 + (0.8)^3$
$\vdots$	$\vdots$
n	$1 + (0.8) + (0.8)^2 + \cdots + (0.8)^{n-1}$

(a) $S_n = 1 + (0.8) + (0.8)^2 + \cdots + (0.8)^{n-1}$

(b) $\lim_{n\to\infty} S_n = \sum_{n=1}^{\infty} (0.8)^{n-1} = \frac{1}{1-(0.8)}$

$= 5$ billion dollars

(c) If 80% is replaced by 90%, then $\lim_{n\to\infty} S_n =$

$\sum_{n=1}^{\infty} (0.9)^{n-1} = \frac{1}{1-(0.9)} = 10$ billion

dollars.

10.3 Series

32. (b) $\sum_{k=1}^{8} \frac{n}{2^n}$

    $= \frac{1}{2} + \frac{2}{4} + \frac{3}{8} + \frac{4}{16} + \frac{5}{32} + \frac{6}{64} + \frac{7}{128} + \frac{8}{256}$

    $= 1.9609375$

34. (a)

n	S_n
1	$\frac{1}{3} \approx 0.3333333$
2	$\frac{1}{3} + \frac{2}{9} \approx 0.5555556$
3	$\frac{1}{3} + \frac{2}{9} + \frac{3}{27} \approx 0.6666667$
4	$\frac{1}{3} + \frac{2}{9} + \frac{3}{27} + \frac{4}{81} \approx 0.7160494$
5	$\frac{1}{3} + \frac{2}{9} + \frac{3}{27} + \frac{4}{81} + \frac{5}{243} \approx 0.7366255$
6	$\frac{1}{3} + \frac{2}{9} + \frac{3}{27} + \frac{4}{81} + \frac{5}{243} + \frac{6}{729} \approx 0.7448560$

    $\sum_{n=1}^{\infty} n3^{-n}$ apparently converges.

    (b) By Exercise 33, the sum is 3/4.

    (c) $\frac{1}{3} + \frac{2}{3^2} + \frac{3}{3^3} + \frac{4}{3^4} + \cdots + \frac{n}{3^n}$

    $< \frac{1}{3} + \frac{2}{3^2} + \frac{2^2}{3^3} + \cdots + \frac{2^n}{3^n} < \frac{1/3}{1 - 2/3} = 1$,

    so $\sum_{n=1}^{\infty} n3^{-n} \leq 1$.

10.4 The Integral Test

2. Let $f(x) = \frac{1}{x^{0.9}}$, which is continuous, positive, and decreasing for $x \geq 1$; since $\int_1^{\infty} \frac{1}{x^{0.9}} \, dx =$

   $\lim_{b \to \infty} \frac{x^{0.1}}{0.1} \Big|_1^b = \infty$, $\sum_{n=1}^{\infty} \frac{1}{n^{0.9}}$ diverges.

4. Let $f(x) = \frac{1}{x^2 + 1}$, which is continuous, positive, and decreasing for $x \geq 1$; since $\int_1^{\infty} \frac{1}{x^2 + 1} \, dx =$

   $\lim_{b \to \infty} \tan^{-1} x \Big|_1^b = \frac{\pi}{2} - \frac{\pi}{4} = \frac{\pi}{4}$, $\sum_{n=1}^{\infty} \frac{1}{n^2 + 1}$ converges.

6. Let $f(x) = \frac{1}{x + 1000}$, which is continuous, positive, and decreasing for $x \geq 1$; since

   $\int_1^{\infty} \frac{1}{x + 1000} \, dx = \lim_{b \to \infty} \ln(x + 1000) \Big|_1^b = \infty$,

   $\sum_{n=1}^{\infty} \frac{1}{n + 1000}$ diverges.

8. Let $f(x) = \frac{x^3}{e^x}$, which is continuous, positive, and decreasing for $x \geq 3$; since $\int_3^{\infty} x^3 e^{-x} \, dx =$

   $\lim_{b \to \infty} [-(x^3 + 3x^2 + 6x + 6)e^{-x}] \Big|_3^{\infty} = \frac{78}{e^3}$, $\sum_{n=3}^{\infty} \frac{n^3}{e^n}$

   converges, hence $\sum_{n=1}^{\infty} \frac{n^3}{e^n}$ converges.

10 $\sum_{n=1}^{\infty} \frac{1}{n^3}$ converges since $p = 3 > 1$.

12 $\sum_{n=1}^{\infty} \frac{1}{n^{0.999}}$ diverges since $p = 0.999 < 1$.

14 If $p \le 1$ then $\frac{1}{n^p} \ge \frac{1}{n}$, so the p-series diverges by comparison with the harmonic series.

16 (a) $S_4 = 1 + \frac{1}{16} + \frac{1}{81} + \frac{1}{256} \approx 1.0788$

(b) $\frac{1}{375} = \int_5^{\infty} \frac{1}{x^4}\,dx < R_4 < \int_4^{\infty} \frac{1}{x^4}\,dx$

$= \frac{1}{192}$

(c) From (a), $1.07875 < S_4 < 1.07885$, so

$1.0814 < 1.07875 + \frac{1}{375} < \sum_{n=1}^{\infty} \frac{1}{n^4} <$

$1.07885 + \frac{1}{192} < 1.0841$

18 (a) $S_4 = \frac{1}{2} + \frac{1}{6} + \frac{1}{12} + \frac{1}{20} = 0.8$

(b) $-\ln \frac{5}{6} = \int_5^{\infty} \frac{1}{x^2 + x}\,dx < R_4 <$

$\int_4^{\infty} \frac{1}{x^2 + x}\,dx = -\ln \frac{4}{5}$

(c) $0.9823 < 0.8 - \ln \frac{5}{6} < \sum_{n=1}^{\infty} \frac{1}{n^2 + n} <$

$0.8 - \ln \frac{4}{5} < 1.0232$

20 (a) $\int_{1001}^{\infty} \frac{1}{x^3}\,dx < R_{1000} < \int_{1000}^{\infty} \frac{1}{x^3}\,dx$, that is,

$\frac{1}{2(1001)^2} < R_{1000} < \frac{1}{2(1000)^2}$.

(b) $R_n < \int_n^{\infty} \frac{1}{x^3}\,dx = \frac{1}{2n^2}$. To make this less than 0.0001, choose $n > \sqrt{5000} \approx 70.7$. Any integer from 71 on up will do.

22 (a) $R_n < \int_n^{\infty} \frac{1}{x^5}\,dx = \frac{1}{4n^4}$. To make this less than 0.0001, choose $n > 2500^{1/4} \approx 7.07$. Any integer from 8 on up will do.

(b) By part (a), S_8 will give the prescribed accuracy: $S_8 = \frac{1}{1^5} + \frac{1}{2^5} + \frac{1}{3^5} + \cdots + \frac{1}{8^5} \approx 1.03688$.

24 Let the real number a fall between integers N and $N + 1$ and let n be any integer larger than $N + 1$:

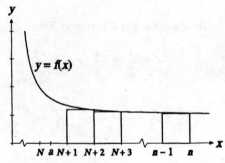

From the figure $f(N + 2) + f(N + 3) + \cdots + f(n)$ $< \int_a^n f(x)\,dx$, so that $S_n = f(1) + f(2) + \cdots + f(n) < f(1) + \cdots + f(N) + f(N + 1) + \int_a^n f(x)\,dx$. If $\int_a^{\infty} f(x)\,dx = I < \infty$ then $S_n < f(1) + \cdots + f(N) + f(N + 1) + I$ for $n > N + 1$ so that S_n converges, that is, $\sum_{n=1}^{\infty} a_n$ converges.

From this diagram $\int_{N+1}^{n} f(x)\, dx < f(N+1) + \cdots$

$+ f(n-1) < \sum_{i=N+1}^{n} f(i)$ for $n > N+1$ so that

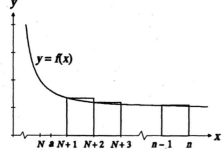

$\int_{a}^{n} f(x)\, dx = \int_{a}^{N+1} f(x)\, dx + \int_{N+1}^{n} f(x)\, dx <$

$\int_{a}^{N+1} f(x)\, dx + \sum_{i=N+1}^{\infty} f(i)$. Thus, if $\int_{a}^{\infty} f(x)\, dx$

diverges, then $\sum_{n=N}^{\infty} a_n$ diverges, so $a_1 + a_2 + \cdots$

$+ a_{N-1} + \sum_{n=N}^{\infty} a_n = \sum_{n=1}^{\infty} a_n$ diverges.

26. Let $f(x) = \dfrac{1}{\sqrt{x}}$ so that $\int_{1}^{n+1} \dfrac{1}{\sqrt{x}}\, dx <$

$f(1) + \cdots + f(n) < a_1 + \int_{1}^{n} \dfrac{1}{\sqrt{x}}\, dx$, that is,

$2\sqrt{n+1} - 2 = 2\sqrt{x}\big|_{1}^{n+1} < \sum_{i=1}^{n} \dfrac{1}{\sqrt{i}} < 1 + 2\sqrt{x}\big|_{1}^{n}$

$= 2\sqrt{n} - 1$.

28. (a) From inequalities (1) and (2) we know that

$\int_{1}^{n+1} f(x)\, dx < \sum_{k=1}^{n} f(k) <$

$f(1) + \int_{1}^{n} f(x)\, dx$. Letting $n \to \infty$ gives

$\int_{1}^{\infty} f(x)\, dx < \sum_{k=1}^{\infty} f(k) < f(1) + \int_{1}^{\infty} f(x)\, dx$.

(Ordinarily, *equality* may also occur after we take the limit; but in this case there is always the discrepancy between the sum of the rectangles and the area under the curve. See Figs. 3 and 4 of the text.)

(b) Let $f(x) = \dfrac{1}{x^2}$; then from part (a) $\int_{1}^{\infty} \dfrac{1}{x^2}\, dx$

$< \sum_{n=1}^{\infty} \dfrac{1}{n^2} < 1 + \int_{1}^{\infty} \dfrac{1}{x^2}\, dx$, that is, $1 <$

$\sum_{n=1}^{\infty} \dfrac{1}{n^2} < 2$.

30. (a) We wish to show $\sum_{i=1}^{n} a_i \leq \prod_{i=1}^{n} (1 + a_i)$: If n

$= 1$ this is true since $a_1 \leq 1 + a_1$. Assume

the inequality is true for $n = k$, that is, $\sum_{i=1}^{k} a_i$

$\leq \prod_{i=1}^{k} (1 + a_i)$. If $n = k+1$ then $\sum_{i=1}^{k+1} a_i =$

$\sum_{i=1}^{k} a_i + a_{k+1} \leq \prod_{i=1}^{k} (1 + a_i) + a_{k+1}$

$= (1 + a_1)(1 + a_2)\cdots(1 + a_k) + a_{k+1}$

$\leq (1 + a_1)(1 + a_2)\cdots(1 + a_k)(1 + a_{k+1})$

$= \prod_{i=1}^{k+1} (1 + a_i)$.

This completes the proof by mathematical induction.

(b) If $\lim_{n \to \infty} \prod_{i=1}^{n} (1 + a_i) = L$, then by part (a) S_n

$$= \sum_{i=1}^{n} a_i \le \prod_{i=1}^{n}(1+a_i) \le L,$$ where S_n is an increasing sequence. Thus $\lim_{n\to\infty} S_n =$

$$\lim_{n\to\infty}\sum_{i=1}^{n} a_i = \sum_{n=1}^{\infty} a_n$$ exists and is less than or equal to L.

32 (a) For $0 < x < 1$, $\dfrac{1}{1-x} = 1 + x + x^2 + x^3 + \cdots$ so that $\dfrac{1}{1-(1/p_i)} =$

$$1 + \left(\frac{1}{p_i}\right) + \left(\frac{1}{p_i}\right)^2 + \left(\frac{1}{p_i}\right)^3 + \cdots.$$

(b) $\dfrac{1}{1-(1/p_1)} \cdot \dfrac{1}{1-(1/p_2)} \cdots \dfrac{1}{1-(1/p_m)} =$

$$\left(1 + \frac{1}{p_1} + \frac{1}{p_1^2} + \frac{1}{p_1^3} + \cdots\right)\left(1 + \frac{1}{p_2} + \frac{1}{p_2^2} + \frac{1}{p_2^3} + \cdots\right)\cdots$$

$$\left(1 + \frac{1}{p_m} + \frac{1}{p_m^2} + \frac{1}{p_m^3} + \cdots\right) = \sum_{n=1}^{\infty} \frac{1}{n},$$ since the denominators on the left-hand side run through all possible prime factorizations, that is, all possible integers $n > 0$.

(c) Assume that there are m primes $p_1, p_2, \cdots,$ and p_m and that the equality in (b) is true. The product on the left-hand side is a finite number while the infinite series on the right-hand side is infinitely large. This is a clear contradiction.

10.5 Comparison Tests

2 $\sum_{n=1}^{\infty} \dfrac{n+2}{(n+1)\sqrt{n}}$ diverges since $\dfrac{n+2}{(n+1)\sqrt{n}} >$

$\dfrac{n+1}{(n+1)\sqrt{n}} = \dfrac{1}{\sqrt{n}}$ and $\sum_{n=1}^{\infty} \dfrac{1}{\sqrt{n}}$ diverges (p-series with $p = 1/2$).

4 $\sum_{n=1}^{\infty} \dfrac{1}{n 2^n}$ converges since $\dfrac{1}{n 2^n} \le \dfrac{1}{2^n}$ and $\sum_{n=1}^{\infty} \dfrac{1}{2^n}$

$$= \sum_{n=1}^{\infty} \left(\frac{1}{2}\right)^n$$ converges (geometric series).

6 $\sum_{n=1}^{\infty} \dfrac{2^n + n}{3^n}$ converges since $\lim_{n\to\infty} \dfrac{(2^n + n)/3^n}{2^n/3^n} =$

$\lim_{n\to\infty} \dfrac{2^n + n}{2^n} = 1 + 0 = 1$ and $\sum_{n=1}^{\infty} \dfrac{2^n}{3^n} =$

$\sum_{n=1}^{\infty} \left(\dfrac{2}{3}\right)^n$ converges (geometric series).

8 $\sum_{n=1}^{\infty} \dfrac{(1 + 1/n)^n}{n^2}$ converges since $\lim_{n\to\infty} \dfrac{(1+1/n)^n/n^2}{1/n^2}$

$= \lim_{n\to\infty} \left(1 + \dfrac{1}{n}\right)^n = e$ and $\sum_{n=1}^{\infty} \dfrac{1}{n^2}$ converges (p-series with $p = 2$).

10 $\sum_{n=1}^{\infty} \dfrac{2^n}{n^2}$ diverges by the nth-term test for divergence, since $\lim_{n\to\infty} \dfrac{2^n}{n^2} = \infty$.

12 $\sum_{n=1}^{\infty} \dfrac{1}{n!}$ converges since $\dfrac{1}{n!} \le \dfrac{1}{2^n}$ for $n = 4, 5,$

10.5 Comparison Tests

$6, \cdots$ and $\sum_{n=4}^{\infty} \dfrac{1}{2^n} = \sum_{n=4}^{\infty} \left(\dfrac{1}{2}\right)^n$ converges (geometric series).

14 $\sum_{n=1}^{\infty} \dfrac{n^2(2^n + 1)}{3^n + 1}$ converges since

$$\lim_{n \to \infty} \dfrac{n^2(2^n + 1)/(3^n + 1)}{(3/4)^n} = \lim_{n \to \infty} \dfrac{n^2}{(9/8)^n} = 0 \text{ and}$$

$\sum_{n=1}^{\infty} \left(\dfrac{3}{4}\right)^n$ converges (geometric series).

16 $\sum_{n=1}^{\infty} \dfrac{\ln n}{n}$ diverges by comparison with the harmonic series.

18 $\sum_{n=1}^{\infty} \dfrac{5^n}{n^n}$ converges since $5^n/n^n \leq 5^n/6^n = (5/6)^n$ for $n \geq 6$ and $\sum_{n=6}^{\infty} \left(\dfrac{5}{6}\right)^n$ converges (geometric series).

20 $\sum_{n=2}^{\infty} \dfrac{1}{\sqrt{n} \ln n}$ diverges since $\lim_{n \to \infty} \dfrac{\frac{1}{\sqrt{n} \ln n}}{1/n} =$

$\lim_{n \to \infty} \dfrac{\sqrt{n}}{\ln n} \underset{H}{=} \lim_{n \to \infty} \dfrac{\frac{1}{2\sqrt{n}}}{1/n} = \lim_{n \to \infty} \dfrac{\sqrt{n}}{2} = \infty$ and

$\sum_{n=2}^{\infty} \dfrac{1}{n}$ diverges (harmonic series).

22 $\sum_{n=1}^{\infty} \dfrac{n^2 e^n}{\pi^n}$ converges by limit-comparison with

$\sum_{n=1}^{\infty} \left(\dfrac{9}{10}\right)^n$. Since $\dfrac{9\pi}{10e} \approx 1.04 > 1$, $\lim_{n \to \infty} \dfrac{n^2 e^n/\pi^n}{(9/10)^n}$

$= \lim_{n \to \infty} \dfrac{n^2}{[9\pi/(10e)]^n} = 0.$

24 $\sum_{k=1}^{\infty} \dfrac{4}{2k^2 - k}$ converges since $\lim_{k \to \infty} \dfrac{\frac{4}{2k^2 - k}}{1/k^2} =$

$\lim_{k \to \infty} \dfrac{4k^2}{2k^2 - k} = 2$ and $\sum_{k=1}^{\infty} \dfrac{1}{k^2}$ converges (p-series with $p = 2$).

26 $\sum_{n=1}^{\infty} \csc \dfrac{1}{n}$ diverges since $\lim_{n \to \infty} \csc \dfrac{1}{n} = \infty$, running afoul of the nth-term test for divergence.

28 $\sum_{n=1}^{\infty} \left(\dfrac{n}{2n - 1}\right)^n$ converges since $\lim_{n \to \infty} \dfrac{\left(\dfrac{n}{2n - 1}\right)^n}{(1/2)^n} =$

$\lim_{n \to \infty} \left(\dfrac{2n}{2n - 1}\right)^n = \lim_{n \to \infty} \left(\dfrac{1}{1 - \dfrac{1}{2n}}\right)^n = \dfrac{1}{e^{-1/2}} = e^{1/2}$

and $\sum_{n=1}^{\infty} \left(\dfrac{1}{2}\right)^n$ converges (geometric series).

30 No conclusion can be drawn. The series $\sum_{n=1}^{\infty} \dfrac{1}{n^2}$ is convergent and both $\lim_{n \to \infty} \dfrac{1/n^{3/2}}{1/n^2} = \lim_{n \to \infty} \sqrt{n} = \infty$

and $\lim_{n \to \infty} \dfrac{1/\sqrt{n}}{1/n^2} = \lim_{n \to \infty} n^{3/2} = \infty$, but $\sum_{n=1}^{\infty} \dfrac{1}{n^{3/2}}$

converges and $\sum_{n=1}^{\infty} \dfrac{1}{\sqrt{n}}$ diverges.

32 If $\sum_{n=1}^{\infty} b_n$ is divergent, then $\sum_{n=1}^{\infty} 3b_n$ is divergent. If

$3b_n \le a_n$, then $\sum\limits_{n=1}^{\infty} a_n$ is divergent.

34 No conclusion can be drawn. The series $\sum\limits_{n=1}^{\infty} \dfrac{1}{\sqrt{n}}$ diverges with $\dfrac{1}{\sqrt{n}} \to 0$ as $n \to \infty$. Also, $\dfrac{1}{2n} \le \left(\dfrac{1}{\sqrt{n}}\right)^2 = \dfrac{1}{n}$, with $\sum\limits_{n=1}^{\infty} \dfrac{1}{2n}$ diverging. On the other hand, $\dfrac{1}{2n^2} < \left(\dfrac{1}{\sqrt{n}}\right)^2 = \dfrac{1}{n}$ with $\sum\limits_{n=1}^{\infty} \dfrac{1}{2n^2}$ converging.

36 If $k = 1$ then $\sum\limits_{n=2}^{\infty} \dfrac{1}{n \ln n}$ diverges since $\int_2^{\infty} \dfrac{1}{x \ln x} dx = \lim\limits_{b \to \infty} \ln(\ln x)\big|_2^b = \infty$. If $0 < k < 1$, then $\sum\limits_{n=2}^{\infty} \dfrac{1}{n^k \ln n}$ diverges by comparison with the series for $k = 1$. If $k > 1$, then $\sum\limits_{n=2}^{\infty} \dfrac{1}{n^k \ln n}$ converges by comparison with $\sum\limits_{n=2}^{\infty} \dfrac{1}{n^k}$.

38 For any value of p, $\sum\limits_{n=1}^{\infty} n^p e^{-n}$ converges since $\lim\limits_{n \to \infty} \dfrac{n^p e^{-n}}{1/e^{n/2}} = \lim\limits_{n \to \infty} \dfrac{n^p}{e^{n/2}} = 0$ (by repeated use of l'Hôpital's rule) and $\sum\limits_{n=1}^{\infty} \dfrac{1}{e^{n/2}} = \sum\limits_{n=1}^{\infty} \left(\dfrac{1}{e^{1/2}}\right)^n$ converges (geometric series).

40 (a) $\sum\limits_{k=n+1}^{\infty} \dfrac{1}{k!} = \dfrac{1}{(n+1)!}\left[1 + \dfrac{1}{n+2} + \dfrac{1}{(n+2)(n+3)} + \dfrac{1}{(n+2)(n+3)(n+4)} + \cdots\right]$

$< \dfrac{1}{(n+1)!}\left[1 + \dfrac{1}{n+2} + \dfrac{1}{(n+2)^2} + \dfrac{1}{(n+2)^3} + \cdots\right]$

(b) If $n = 4$ in part (a) then $0 < \sum\limits_{k=5}^{\infty} \dfrac{1}{k!}$

$< \dfrac{1}{5!}\left[1 + \dfrac{1}{6} + \dfrac{1}{6^2} + \dfrac{1}{6^3} + \cdots\right] = \dfrac{1}{5!} \cdot \dfrac{1}{1 - \dfrac{1}{6}}$ so

that $1 + 1 + \dfrac{1}{2!} + \dfrac{1}{3!} + \dfrac{1}{4!} < \sum\limits_{k=0}^{\infty} \dfrac{1}{k!} <$

$1 + 1 + \dfrac{1}{2!} + \dfrac{1}{3!} + \dfrac{1}{4!} + \dfrac{1}{5!} \cdot \dfrac{1}{1 - \dfrac{1}{6}}$.

(c) From part (b), $2.708 < \sum\limits_{k=0}^{\infty} \dfrac{1}{k!} < 2.72$.

(d) From (a), if $n = 6$ then $\sum\limits_{k=n+1}^{\infty} \dfrac{1}{k!} <$

$\dfrac{1}{7!} \dfrac{1}{1 - 1/8} = \dfrac{8}{7 \cdot 7!} = \dfrac{1}{4410} < 0.0003$.

(e) Since $\sum\limits_{k=0}^{6} \dfrac{1}{k!} = \dfrac{1957}{720} = 2.71805555\cdots$, it follows from (d) that $2.718 < \sum\limits_{k=0}^{\infty} \dfrac{1}{k!} <$

2.7184. So to three places, the sum is 2.718.

10.5 Comparison Tests

42 $\sum_{n=1}^{\infty} \frac{\ln n}{n^2} = \frac{\ln 2}{4} + \sum_{n=3}^{\infty} \frac{\ln n}{n^2} < \frac{1}{4} + \int_2^{\infty} \frac{\ln x}{x^2} dx$

$= \frac{1}{4} + \lim_{b \to \infty} \left(-\frac{1 + \ln x}{x} \right)\Big|_2^b = \frac{1}{4} + \frac{1 + \ln 2}{2} <$

$\frac{1}{4} + \frac{2}{2} = 1.25.$ (Smaller bounds can be obtained with more work; the sum is about 0.937548.)

44 $\sum_{n=k+1}^{\infty} \frac{1}{n 2^n} < \frac{1}{(k+1)2^{k+1}} \left(1 + \frac{1}{2} + \frac{1}{2^2} + \cdots \right) =$

$\frac{1}{(k+1)2^k} < 0.0005$ for $k = 8$. Since $\sum_{n=1}^{8} \frac{1}{n 2^n} \approx$

0.692750, we have $0.6927 < \sum_{n=1}^{\infty} \frac{1}{n 2^n} <$

$0.6928 + 0.0005 = 0.6933$. Hence, to three places, the sum is 0.693. (In fact, it equals $\ln 2$.)

10.6 Ratio Tests

2 $\lim_{n \to \infty} \frac{a_{n+1}}{a_n} = \lim_{n \to \infty} \frac{(n+2)^2}{(n+1)2^{n+1}} \cdot \frac{n 2^n}{(n+1)^2} =$

$\lim_{n \to \infty} \left(\frac{n+2}{n+1} \right)^2 \cdot \frac{n}{n+1} \cdot \frac{1}{2} = 1^2 \cdot 1 \cdot \frac{1}{2} = \frac{1}{2} < 1,$ so

the series converges.

4 $\lim_{n \to \infty} \frac{a_{n+1}}{a_n} = \lim_{n \to \infty} \frac{(n+1)!}{3^{n+1}} \cdot \frac{3^n}{n!} = \lim_{n \to \infty} \frac{n+1}{3} = \infty,$

so $\sum_{i=1}^{\infty} \frac{n!}{3^n}$ diverges.

6 $\lim_{n \to \infty} \frac{a_{n+1}}{a_n} = \lim_{n \to \infty} \frac{(n+1)!}{(n+1)^{n+1}} \cdot \frac{n^n}{n!} = \lim_{n \to \infty} \frac{n^n}{(n+1)^n}$

$= \lim_{n \to \infty} \frac{1}{\left(1 + \frac{1}{n}\right)^n} = \frac{1}{e} < 1,$ so $\sum_{i=1}^{\infty} \frac{n!}{n^n}$

converges.

8 (a) $\lim_{n \to \infty} \frac{a_{n+1}}{a_n} = \lim_{n \to \infty} \frac{2^{n+1}}{(n+1)^5} \cdot \frac{n^5}{2^n} =$

$\lim_{n \to \infty} 2 \left(\frac{n}{n+1} \right)^5 = 2 > 1,$ so $\sum_{n=1}^{\infty} \frac{2^n}{n^5}$ diverges.

(b) $\lim_{n \to \infty} a_n = \lim_{n \to \infty} \frac{2^n}{n^5} = \infty,$ so $\sum_{n=1}^{\infty} \frac{2^n}{n^5}$ diverges by the nth-term test.

10 $\lim_{n \to \infty} \frac{a_{n+1}}{a_n} = \lim_{n \to \infty} \frac{x^{n+1}}{n+1} \cdot \frac{n}{x^n} = \lim_{n \to \infty} \frac{n}{n+1} \cdot x = 1 \cdot x$

$= x$ so that $\sum_{n=1}^{\infty} \frac{x^n}{n}$ converges (a) for $0 < x < 1$

the series and diverges (b) for $x \geq 1$. (For $x = 1$ it is the harmonic series.)

12 $\lim_{n \to \infty} \frac{a_{n+1}}{a_n} = \lim_{n \to \infty} \frac{2^{n+1} x^{n+1}}{(n+1)!} \cdot \frac{n!}{2^n x^n} = \lim_{n \to \infty} \frac{2x}{n+1}$

$= 0,$ so the series converges for all positive values of x.

14 $\lim_{n \to \infty} (a_n)^{1/n} = \lim_{n \to \infty} \left[\frac{(1 + 1/n)^n (2n+1)^n}{(3n+1)^n} \right]^{1/n} =$

$\lim_{n \to \infty} \left(1 + \frac{1}{n} \right) \cdot \left(\frac{2n+1}{3n+1} \right) = 1 \cdot \frac{2}{3} = \frac{2}{3} < 1,$ so

$\sum_{n=1}^{\infty} \frac{\left(1 + \frac{1}{n}\right)^n (2n+1)^n}{(3n+1)^n}$ converges.

16 The function $f(x) = \frac{\ln x}{x^2}$ is continuous, positive,

and decreasing for $x > \sqrt{e}$ and $R_n < \int_n^\infty \frac{\ln x}{x^2}\, dx$

$= \lim_{b \to \infty} \left. \frac{-\ln x}{x} \right|_n^b + \int_n^\infty \frac{1}{x^2}\, dx = \frac{\ln n}{n} - \lim_{b \to \infty} \left. \frac{1}{x} \right|_n^b =$

$\frac{\ln n}{n} + \frac{1}{n}$. If $n = 4$ then $R_4 < 0.597$ and $S_4 =$

$\frac{\ln 1}{1^2} + \frac{\ln 2}{2^2} + \frac{\ln 3}{3^2} + \frac{\ln 4}{4^2} \approx 0.382$, so that

$\sum_{n=1}^\infty \frac{\ln n}{n^2} < 0.382 + 0.597 = 0.979.$

18 For $n \geq 1$, $\frac{a_{n+1}}{a_n} = \frac{(n+3)/2^{n+1}}{(n+2)/2^n} = \frac{1}{2} + \frac{1}{2(n+2)}$

$\leq \frac{2}{3}$. Hence $a_n \leq \left(\frac{2}{3}\right)^{n-1} a_1 = \left(\frac{2}{3}\right)^{n-2}$, so

$\sum_{n=1}^\infty \frac{n+2}{2^n} \leq \sum_{n=1}^\infty \left(\frac{2}{3}\right)^{n-2} = \frac{9}{2} = 4.5.$ (The actual sum is 4.)

20 Note that $\frac{n}{n^3 + 1} = \frac{1}{n^2 + 1/n} < \frac{1}{n^2}$. Thus

$\sum_{n=1}^\infty \frac{n}{n^3 + 1} = \frac{1}{2} + \sum_{n=2}^\infty \frac{n}{n^3 + 1} < \frac{1}{2} + \sum_{n=2}^\infty \frac{1}{n^2}$

$< \frac{1}{2} + \int_1^\infty \frac{dx}{x^2} = \frac{3}{2} = 1.5.$ (Smaller bounds can be obtained with more work. The actual sum is about 1.1116439.)

22 The function $f(x) = \frac{x}{x^2 + 1}$ is continuous, positive, and decreasing for $x > 1$ so that the mth partial sum $S_m = \sum_{k=1}^m f(k) > \int_1^{m+1} f(x)\, dx$, that

is, $S_m > \int_1^{m+1} \frac{x}{x^2 + 1}\, dx = \left. \frac{1}{2} \ln(1 + x^2) \right|_1^{m+1} =$

$\frac{1}{2} \ln[(m+1)^2 + 1] - \frac{1}{2} \ln 2.$ To make $S_m > 1000$

choose $\frac{1}{2} \ln[(m+1)^2 + 1] - \frac{1}{2} \ln 2 > 1000$; that

is, $\ln[(m+1)^2 + 1] > 2000 + \ln 2$, so

$[(m+1)^2 + 1] > 2e^{2000}$, and therefore $(m+1)^2$

$> 2e^{2000} - 1.$ Thus $m + 1 > \sqrt{2e^{2000} - 1}$ so

choose $m > \sqrt{2e^{2000} - 1} - 1.$

24 The mth partial sum is $S_m = \sum_{n=1}^m \frac{(n+2)^2}{(n+1)\sqrt{n}} >$

$\sum_{n=1}^m \sqrt{n} > \int_0^m \sqrt{x}\, dx = \frac{2}{3} m^{3/2},$ so pick $m \geq$

$1500^{2/3} \approx 131.04$; that is, choose $m \geq 132.$ (In fact, any $m \geq 125$ works.)

26 (a) $\sum_{n=1}^\infty \frac{1}{n}$ diverges (harmonic series); if $\gamma = n^{1/n}$

then $\ln \gamma = \frac{\ln n}{n}$, so $\lim_{n \to \infty} (\ln \gamma) = \lim_{n \to \infty} \frac{\ln n}{n}$

$\underset{H}{=} \lim_{n \to \infty} \frac{1}{n} = 0,$ so that $\lim_{n \to \infty} n^{1/n} = 1.$ Thus,

$\lim_{n \to \infty} \left(\frac{1}{n}\right)^{1/n} = \lim_{n \to \infty} \frac{1}{n^{1/n}} = \frac{1}{1} = 1.$

(b) $\sum_{n=1}^\infty \frac{1}{n^2}$ converges (p-series with $p = 2$);

$\lim_{n \to \infty} \left(\frac{1}{n^2}\right)^{1/n} = \left(\lim_{n \to \infty} \left(\frac{1}{n}\right)^{1/n}\right)^2 = 1^2 = 1,$ by

part (a).

28 (a) $\sum_{n=1}^{\infty} \frac{1}{n}$ diverges (harmonic series) and

$$\lim_{n\to\infty} \frac{a_{n+1}}{a_n} = \lim_{n\to\infty} \frac{1}{n+1} \cdot \frac{n}{1} = 1.$$

(b) $\sum_{n=1}^{\infty} \frac{1}{n^2}$ converges (p-series with $p = 2$) and

$$\lim_{n\to\infty} \frac{a_{n+1}}{a_n} = \lim_{n\to\infty} \frac{1}{(n+1)^2} \cdot \frac{n^2}{1} = \lim_{n\to\infty} \left(\frac{n}{n+1}\right)^2 = 1^2 = 1.$$

10.7 Tests for Series with Both Positive and Negative Terms

2 $\sum_{n=1}^{\infty} (-1)^n \cdot \frac{1}{1+2^{-n}}$ diverges since

$$\lim_{n\to\infty} (-1)^n \frac{1}{1+2^{-n}} = \lim_{n\to\infty} (-1)^n \frac{1}{1+1/2^n} \neq 0$$

(nth-term test).

4 $\sum_{n=1}^{\infty} (-1)^{n+1} \frac{5^n}{n!}$ converges since $a_n = \frac{5^n}{n!}$ is a positive, decreasing sequence for $n \geq 5$ with $a_n \to 0$ as $n \to \infty$.

6 $\sum_{n=1}^{\infty} (-1)^{n+1} \sqrt{n}$ diverges since $\lim_{n\to\infty} (-1)^{n+1} \sqrt{n} \neq 0$

(nth-term test).

8 $\sum_{n=1}^{\infty} (-1)^{n+1} \frac{1}{n^2}$ converges since $a_n = 1/n^2$ is a positive, decreasing sequence with $a_n \to 0$ as $n \to \infty$.

10 (a) $\sum_{n=1}^{\infty} (-1)^{n+1} \cdot \frac{1}{n 2^n} \approx S_6$

$$= \frac{1}{2} - \frac{1}{8} + \frac{1}{24} - \frac{1}{64} + \frac{1}{160} - \frac{1}{384}$$

$$= 0.4046875$$

(b) $|R_6| < |a_7| = \frac{1}{7 \cdot 2^7} = \frac{1}{896} < 0.00112$

12 $\sum_{n=1}^{\infty} (-1)^n \ln \frac{1}{n} = \sum_{n=1}^{\infty} (-1)^{n+1} \ln n$ diverges since

$\lim_{n\to\infty} \ln n \neq 0$ (nth-term test).

14 $\sum_{n=1}^{\infty} \frac{\sin n}{n^{1.01}}$ converges absolutely since $\left|\frac{\sin n}{n^{1.01}}\right| \leq \frac{1}{n^{1.01}}$ and $\sum_{n=1}^{\infty} \frac{1}{n^{1.01}}$ converges (p-series with $p = 1.01$).

16 $\sum_{n=1}^{\infty} (-1)^n \cos \frac{\pi}{n^2}$ diverges since $\lim_{n\to\infty} \cos \frac{\pi}{n^2} \neq 0$

(nth-term test).

18 $\sum_{n=1}^{\infty} (-1)^n \frac{2^n}{n!}$ converges absolutely since

$$\lim_{n\to\infty} \left| \frac{(-1)^{n+1} \frac{2^{n+1}}{(n+1)!}}{(-1)^n \frac{2^n}{n!}} \right| = \lim_{n\to\infty} \frac{2}{n+1} = 0 < 1.$$

20 $\sum_{n=1}^{\infty} \frac{(-3)^n (1+n^2)}{n!}$ converges absolutely since

$$\lim_{n\to\infty} \left| \frac{\frac{(-3)^{n+1}[1+(n+1)^2]}{(n+1)!}}{\frac{(-3)^n (1+n^2)}{n!}} \right|$$

$$= \lim_{n \to \infty} \frac{3(n^2 + 2n + 2)}{(n+1)(n^2+1)} = 0 < 1.$$

22 $\sum_{n=1}^{\infty} (-1)^n \frac{n+5}{n^2}$ converges conditionally. Since a_n

$= \frac{n+5}{n^2}$ is a positive decreasing sequence with

$a_n \to 0$ as $n \to \infty$, $\sum_{n=1}^{\infty} (-1)^n \frac{n+5}{n^2}$ converges; but

$\frac{n+5}{n^2} > \frac{n+0}{n^2} = \frac{1}{n}$ and $\sum_{n=1}^{\infty} \frac{1}{n}$ diverges

(harmonic series) so that $\sum_{n=1}^{\infty} \frac{n+5}{n^2}$ diverges.

24 $\sum_{n=1}^{\infty} \frac{(-1)^n}{n^{1/3}}$ converges conditionally: $a_n = \frac{1}{n^{1/3}}$ is a

positive decreasing sequence with $a_n \to 0$ as $n \to \infty$

so $\lim_{n \to \infty} \sum_{n=1}^{\infty} \frac{(-1)^n}{n^{1/3}}$ converges; but $\sum_{n=1}^{\infty} \frac{1}{n^{1/3}}$ diverges

(p-series with $p = 1/3$).

26 $\sum_{n=1}^{\infty} \frac{(-\pi)^{2n+1}}{(2n+1)!}$ converges absolutely since

$$\lim_{n \to \infty} \left| \frac{\frac{(-\pi)^{2(n+1)+1}}{(2(n+1)+1)!}}{\frac{(-\pi)^{2n+1}}{(2n+1)!}} \right| = \lim_{n \to \infty} \frac{\pi^{2n+3}}{(2n+3)!} \cdot \frac{(2n+1)!}{\pi^{2n+1}}$$

$$= \lim_{n \to \infty} \frac{\pi^2}{(2n+3)(2n+2)} = 0 < 1.$$

28 (a) $S_6 = \frac{1}{2} - \frac{1}{4} + \frac{1}{8} - \frac{1}{16} + \frac{1}{32} - \frac{1}{64}$

$= 0.328125.$

(b) $|R_6| < |a_7| = \frac{1}{128} = 0.0078125.$

(c) $R_6 = \sum_{n=7}^{\infty} \frac{(-1)^{n+1}}{2^n} = \frac{1/2^7}{1-(-1/2)} = \frac{1}{192}$

$= 0.005208333\cdots.$

30 The series $\sum_{n=0}^{\infty} \frac{(-1)^n}{n!}$ is a strictly decreasing

alternating series so that $|R_n| < |a_{n+1}| =$

$\frac{1}{(n+1)!}$. To ensure that $|R_n| < 0.005$ choose

$\frac{1}{(n+1)!} < 0.005$, so $200 < (n+1)!$ and $n = 5$

suffices. So we compute $S_5 =$

$1 - 1 + \frac{1}{2!} - \frac{1}{3!} + \frac{1}{4!} - \frac{1}{5!} = 0.36666\cdots.$ To

two decimal places, the sum is 0.37.

32 (a) If $P(n)/Q(n) \to 0$ as $n \to \infty$, then the degree of $P(x)$ must be less than the degree of $Q(x)$. (If $P(x)$ and $Q(x)$ are equal in degree, then $P(n)/Q(n) \to a/b$, where a and b are the leading coefficients of $P(x)$ and $Q(x)$, respectively. If the degree of $P(x)$ is greater than that of $Q(x)$, then $|P(n)/Q(n)| \to \infty$ as $n \to \infty$.) Conversely, if the degree of $P(x)$ is less than that of $Q(x)$, then $P(n)/Q(n) \to 0$.

(b) Let $P(x) = ax^r +$ (lower order terms) and $Q(x) = bx^s +$ (lower order terms), where a and b are nonzero. By the limit-comparison test, $\sum_{n=1}^{\infty} \frac{P(n)}{Q(n)}$ converges if and only if the p-series $\sum_{n=1}^{\infty} \frac{1}{n^{s-r}}$ converges. This happens if

10.7 Tests for Series with Both Positive and Negative Terms

and only if $s - r > 1$. That is, $\sum_{n=1}^{\infty} \frac{P(n)}{Q(n)}$ converges if and only if the degree of $Q(x)$ exceeds the degree of $P(x)$ by at least 2. The convergence is absolute since the terms all have the same sign for large n.

(c) If the degree of $P(x)$ is greater than or equal to the degree of $Q(x)$, the terms do not approach 0, so the series diverges. If the degree of $Q(x)$ exceeds the degree of $P(x)$ by at least 2, then the series converges absolutely, by (b). If the degree of $Q(x)$ is one greater than that of $P(x)$, then the series converges by the alternating-series test, while the corresonding series of positive terms diverges by part (b). Hence $\sum_{n=1}^{\infty} (-1)^n \frac{P(n)}{Q(n)}$ converges conditionally if and only if the degree of $Q(x)$ is one more than that of $P(x)$.

34 (a) Assume $\lim_{n \to \infty} \left| \frac{a_{n+1}}{a_n} \right| = L > 1$, and pick any number r satisfying $1 < r < L$. Then there exists an integer N so that $\left| \frac{a_{n+1}}{a_n} \right| > r$ for $n \geq N$, so that $|a_{n+1}| > r|a_n|$ for $n \geq N$. It follows that $|a_{N+1}| > r|a_N|$, $|a_{N+2}| > r|a_{N+1}| > r^2|a_N|$, $|a_{N+3}| > r|a_{N+2}| > r^3|a_N|$, ..., and in general, $|a_{N+k}| > r^k|a_N|$ for $k = 1, 2, 3, 4, \ldots$. Thus, $\lim_{n \to \infty} |a_n| = \lim_{k \to \infty} |a_{N+k}| \geq \lim_{k \to \infty} r^k|a_N| = \infty$ since $r > 1$.

(b) If $\lim_{n \to \infty} a_n = 0$, then $\lim_{n \to \infty} |a_n| = 0$. But this contradicts part (a). Thus, $\lim_{n \to \infty} a_n \neq 0$.

36 (a) $\lim_{n \to \infty} \left| \frac{\frac{x^{n+1}}{(n+1)!}}{x^n/n!} \right| = \lim_{n \to \infty} \frac{|x|}{n+1} = 0 < 1$ for all values of x, so that $\sum_{n=1}^{\infty} \frac{x^n}{n!}$ converges absolutely for all values of x.

(b) $\lim_{n \to \infty} \left| \frac{\frac{x^{n+1}}{(n+1)^2}}{x^n/n^2} \right| = \lim_{n \to \infty} \left(\frac{n}{n+1} \right)^2 |x| = |x|$.

If $|x| < 1$, that is, if $-1 < x < 1$, then $\sum_{n=1}^{\infty} \frac{x^n}{n^2}$ converges absolutely. And if $|x| > 1$, that is, if $x < -1$ or $x > 1$, then the series diverges. If $x = \pm 1$, then $\sum_{n=1}^{\infty} \left| \frac{x^n}{n^2} \right| = \sum_{n=1}^{\infty} \frac{1}{n^2}$ converges absolutely (p-series with $p = 2$). Hence $\sum_{n=1}^{\infty} \left| \frac{x^n}{n^2} \right|$ converges absolutely for $|x| \leq 1$ and diverges for $|x| > 1$.

38 Yes.

10.S Review Exercises

2. $\sum_{n=1}^{\infty} \frac{(-1)^n}{3^n}$ converges.

Geometric series: $\sum_{n=1}^{\infty} \left(\frac{-1}{3}\right)^n = \frac{-1/3}{1-(-1/3)}$

$= -\frac{1}{4}.$

Alternating-series test: $\frac{1}{3^n}$ is a positive, decreasing sequence with $\frac{1}{3^n} \to 0$ as $n \to \infty$.

Absolute-ratio test: $\lim_{n\to\infty} \left|\frac{a_{n+1}}{a_n}\right| = \lim_{n\to\infty} \frac{1}{3^{n+1}} \cdot \frac{3^n}{1}$

$= \frac{1}{3} < 1.$

4. $\sum_{k=1}^{\infty} \frac{\sqrt{k}}{k^2-2}$ converges.

Limit-comparison test: $\lim_{k\to\infty} \frac{\frac{\sqrt{k}}{k^2-2}}{1/k^{3/2}} =$

$\lim_{k\to\infty} \frac{k^2}{k^2-2} = 1$ and $\sum_{k=1}^{\infty} \frac{1}{k^{3/2}}$ converges (p-series with $p = 3/2$).

6. $\sum_{n=1}^{\infty} \left[\frac{2}{3+1/n}\right]^n$ converges.

Comparison test: $\left[\frac{2}{3+1/n}\right]^n < \left(\frac{2}{3}\right)^n$ and

$\sum_{n=1}^{\infty} \left(\frac{2}{3}\right)^n$ converges (geometric series).

Limit-comparison test: $\lim_{n\to\infty} \frac{\left[\frac{2}{3+1/n}\right]^n}{(2/3)^n} =$

$\lim_{n\to\infty} \frac{3^n}{(3+1/n)^n} = \lim_{n\to\infty} \frac{1}{[(1+1/(3n))^{3n}]^{1/3}} = \frac{1}{e^{1/3}}$

and $\sum_{n=1}^{\infty} \left(\frac{2}{3}\right)^n$ converges (geometric series).

Root test: $\lim_{n\to\infty} (a_n)^{1/n} = \lim_{n\to\infty} \left(\left[\frac{2}{3+1/n}\right]^n\right)^{1/n}$

$= \lim_{n\to\infty} \frac{2}{3+1/n} = \frac{2}{3} < 1.$

Ratio test: $\lim_{n\to\infty} \frac{a_{n+1}}{a_n} = \lim_{n\to\infty} \left(\frac{2}{3+\frac{1}{n+1}}\right)^{n+1}\left(\frac{3+1/n}{2}\right)^n$

$= \lim_{n\to\infty} \frac{2}{3+\frac{1}{n+1}} \cdot \frac{\left(1+\frac{1}{3n}\right)^n}{\left(1+\frac{1}{3(n+1)}\right)^n} = \frac{2}{3} \cdot \frac{e^{1/3}}{e^{1/3}} = \frac{2}{3}$

$< 1.$

8. $\sum_{n=1}^{\infty} \frac{10^n}{n!}$ converges.

Ratio test: $\lim_{n\to\infty} \frac{a_{n+1}}{a_n} = \lim_{n\to\infty} \frac{10^{n+1}}{(n+1)!} \cdot \frac{n!}{10^n} =$

$\lim_{n\to\infty} \frac{10}{n+1} = 0 < 1.$

Comparison test: $\frac{10^n}{n!} \leq \frac{10^n}{11! \, 11^{n-11}} =$

$\frac{11^{11}}{11!}\left(\frac{10}{11}\right)^n$ for $n \geq 28$ and $\sum_{n=28}^{\infty} \frac{10^n}{11^n} =$

$\sum_{n=28}^{\infty} \left(\frac{10}{11}\right)^n$ converges (geometric series).

Limit-comparison test: $\lim_{n \to \infty} \frac{10^n/n!}{(10/11)^n} = \lim_{n \to \infty} \frac{11^n}{n!} =$

$0 < 1$ and $\sum_{n=1}^{\infty} \left(\frac{10}{11}\right)^n$ converges.

10 $\sum_{k=1}^{\infty} \frac{\sin^2 k}{k^2}$ converges.

Comparison test: $\frac{\sin^2 k}{k^2} \leq \frac{1}{k^2}$ and $\sum_{k=1}^{\infty} \frac{1}{k^2}$

converges (p-series with $p = 2$).

12 $\sum_{n=1}^{\infty} (-1)^n \frac{(n+1)^2}{n!}$ converges.

Alternating-series test: $p_n = \frac{(n+1)^2}{n!}$ is a positive,

decreasing sequence for $n \geq 2$ and $p_n \to 0$ as $n \to \infty$.

Absolute-ratio test: $\lim_{n \to \infty} \left| \frac{a_{n+1}}{a_n} \right| =$

$\lim_{n \to \infty} \frac{(n+2)^2}{(n+1)!} \cdot \frac{n!}{(n+1)^2} = \lim_{n \to \infty} \frac{1}{n+1} \left(\frac{n+2}{n+1}\right)^2 =$

$0 \cdot 1^2 = 0 < 1.$

14 $\sum_{n=0}^{\infty} (-1)^n \frac{\pi^{2n+1}}{2^{2n+1}(2n+1)!}$ converges.

Alternating-series test: $p_n = \frac{\pi^{2n+1}}{2^{2n+1}(2n+1)!}$ is a

positive, decreasing sequence with $p_n \to 0$ as $n \to \infty$.

Absolute-ratio test: $\lim_{n \to \infty} \left| \frac{a_{n+1}}{a_n} \right| =$

$\lim_{n \to \infty} \frac{\pi^{2n+3}}{2^{2n+3}(2n+3)!} \cdot \frac{2^{2n+1}(2n+1)!}{\pi^{2n+1}} =$

$\lim_{n \to \infty} \frac{\pi^2}{2^2} \cdot \frac{1}{(2n+3)(2n+2)} = 0 < 1.$

16 $\sum_{n=1}^{\infty} \cos \frac{1}{n}$ diverges.

nth-term test: $\lim_{n \to \infty} a_n = \lim_{n \to \infty} \cos \frac{1}{n} = \cos 0$

$= 1 \neq 0.$

18 $\sum_{n=1}^{\infty} \frac{(-2)^n}{n}$ diverges.

Absolute-ratio test: $\lim_{n \to \infty} \left| \frac{a_{n+1}}{a_n} \right| = \lim_{n \to \infty} \frac{2^{n+1}}{n+1} \cdot \frac{n}{2^n}$

$= \lim_{n \to \infty} \frac{n}{n+1} \cdot 2 = 2 > 1.$

nth-term test: $\lim_{n \to \infty} a_n = \lim_{n \to \infty} \frac{(-2)^n}{n} \neq 0.$

20 $\sum_{n=0}^{\infty} \frac{5n^2 - 3n + 1}{2n^3 + n^2 - 1}$ diverges.

Limit-comparison test: $\lim_{n \to \infty} \frac{\frac{5n^2 - 3n + 1}{2n^3 + n^2 - 1}}{1/n} =$

$\lim_{n \to \infty} \frac{5n^3 - 3n^2 + n}{2n^3 + n^2 - 1} = \frac{5}{2}$ and $\sum_{n=1}^{\infty} \frac{1}{n}$ diverges

(harmonic series).

Comparison test: $\frac{5n^2 - 3n + 1}{2n^3 + n^2 - 1} >$

$$\frac{5n^2 - 3n^2 + 1}{2n^3 + n^2 + 1} > \frac{2n^2}{2n^3 + n^3 + n^3} = \frac{1}{2n} \text{ for}$$

$n \geq 1$ and $\sum_{n=1}^{\infty} \frac{1}{2n}$ diverges (harmonic series).

22 $\sum_{n=1}^{\infty} \ln\left(\frac{n+1}{n}\right)$ diverges.

Partial sums: $\ln\left(\frac{n+1}{n}\right) = \ln(n+1) - \ln n$ so

that $S_n = (\ln(n+1) - \ln n) + (\ln n - \ln(n-1))$
$+ \cdots + (\ln 3 - \ln 2) + (\ln 2 - \ln 1)$
$= \ln(n+1)$. Then $\lim_{n \to \infty} S_n = \lim_{n \to \infty} \ln(n+1) = \infty$.

Integral test: $\int_1^{\infty} \ln \frac{x+1}{x} dx$

$= \int_1^{\infty} [\ln(x+1) - \ln x] dx$

$= \lim_{b \to \infty} [(x+1)\ln(x+1) - (x+1) - (x \ln x - x)]\Big|_1^b$

$= \lim_{b \to \infty} [\ln(x+1)^{x+1} - \ln x^x - 1]\Big|_1^b$

$= \lim_{b \to \infty} \left[\ln \frac{(x+1)^{x+1}}{x^x} - 1\right]\Big|_1^b$

$= \lim_{b \to \infty} \left[\ln\left((x+1)\cdot\left(1 + \frac{1}{x}\right)^x\right) - 1\right]\Big|_1^b = \infty$.

Limit-comparison test: $\lim_{n \to \infty} \frac{\ln\left(\frac{n+1}{n}\right)}{1/n} \underset{H}{=}$

$\lim_{n \to \infty} \frac{\frac{n}{n+1}\cdot\frac{-1}{n^2}}{-1/n^2} = \lim_{n \to \infty} \frac{n}{n+1} = 1$ and $\sum_{n=1}^{\infty} \frac{1}{n}$

diverges (harmonic series).

24 $\sum_{n=0}^{\infty} (-1)^n \frac{\pi^{2n}}{(2n)!}$ converges.

Alternating-series test: $p_n = \frac{\pi^{2n}}{(2n)!}$ is a positive,

decreasing sequence for $n \geq 1$ with $p_n \to 0$, as $n \to \infty$.

Absolute-ratio test: $\lim_{n \to \infty} \left|\frac{a_{n+1}}{a_n}\right| =$

$\lim_{n \to \infty} \frac{\pi^{2(n+1)}}{(2(n+1))!} \cdot \frac{(2n)!}{\pi^{2n}} = \lim_{n \to \infty} \frac{\pi^2}{(2n+2)(2n+1)} = 0$

< 1.

Absolute-convergence test: $|a_n| = \frac{\pi^{2n}}{(2n)!} < \frac{\pi^{2n}}{4^{2n}}$

for $n \geq 5$ and $\sum_{n=5}^{\infty} \frac{\pi^{2n}}{4^{2n}} = \sum_{n=5}^{\infty} \left(\frac{\pi^2}{16}\right)^n$ converges

(geometric series).

26 $\sum_{n=1}^{\infty} \frac{\ln n}{n}$ diverges.

Comparison test: $\frac{\ln n}{n} > \frac{1}{n}$ for $n \geq 3$ and $\sum_{n=3}^{\infty} \frac{1}{n}$

diverges (harmonic series).

Limit comparison test: $\lim_{n \to \infty} \frac{(\ln n)/n}{1/n} = \lim_{n \to \infty} \ln n = \infty$ and $\sum_{n=1}^{\infty} \frac{1}{n}$ diverges.

Integral test: $f(x) = \frac{\ln x}{x}$ is positive, continuous,

and decreasing for $x \geq e$ and $\int_e^{\infty} \frac{\ln x}{x} dx =$

10.S Summary: Review Exercises

$\lim\limits_{b \to \infty} \frac{1}{2}(\ln x)^2 \Big|_e^b = \infty.$

28 $\sum\limits_{n=1}^{\infty} \frac{\sqrt{n+1} - \sqrt{n}}{n}$

$= \sum\limits_{n=1}^{\infty} \frac{\sqrt{n+1} - \sqrt{n}}{n} \cdot \frac{\sqrt{n+1} + \sqrt{n}}{\sqrt{n+1} + \sqrt{n}}$

$= \sum\limits_{n=1}^{\infty} \frac{1}{n(\sqrt{n+1} + \sqrt{n})}$ converges.

Comparison test: $\frac{1}{n(\sqrt{n+1} + \sqrt{n})} < \frac{1}{n(0 + \sqrt{n})}$

$= \frac{1}{n^{3/2}}$ and $\sum\limits_{n=1}^{\infty} \frac{1}{n^{3/2}}$ converges (p-series with

$p = 3/2$).

Limit-comparison test: $\lim\limits_{n \to \infty} \frac{\frac{1}{n(\sqrt{n+1} + \sqrt{n})}}{1/n^{3/2}} =$

$\lim\limits_{n \to \infty} \frac{1}{\sqrt{1 + 1/n} + 1} = \frac{1}{2}$ and $\sum\limits_{n=1}^{\infty} \frac{1}{n^{3/2}}$ converges

(p-series with $p = 3/2$).

30 $\sum\limits_{n=1}^{\infty} \sin \frac{1}{n}$ diverges.

Limit-comparison test: $\lim\limits_{n \to \infty} \frac{\sin \frac{1}{n}}{1/n} = 1$ and $\sum\limits_{n=1}^{\infty} \frac{1}{n}$

diverges (harmonic series).

32 $\sum\limits_{n=1}^{\infty} \frac{(-1)^n n^2}{(2n)!}$ converges.

Absolute-ratio test: $\lim\limits_{n \to \infty} \left| \frac{a_{n+1}}{a_n} \right|$

$= \lim\limits_{n \to \infty} \frac{(n+1)^2}{(2(n+1))!} \cdot \frac{(2n)!}{n^2}$

$= \lim\limits_{n \to \infty} \frac{(n+1)^2}{n^2(2n+2)(2n+1)} = 0 < 1.$

Alternating-series test: $p_n = \frac{n^2}{(2n)!}$ is a positive,

decreasing sequence with $p_n \to 0$ as $n \to \infty$.

Absolute-convergence test: $|a_n| = \frac{n^2}{(2n)!} < \frac{n^2}{n^4}$

$= \frac{1}{n^2}$ and $\sum\limits_{n=1}^{\infty} \frac{1}{n^2}$ converges (p-series with $p = 2$)

34 For $n \geq 1$, $R_n < \int_n^{\infty} \frac{x}{3^x} dx = \int_n^{\infty} x \cdot 3^{-x} dx$

$= \lim\limits_{b \to \infty} \left(\frac{-x}{3^x \ln 3} - \frac{1}{3^x (\ln 3)^2} \right) \Big|_n^b$

$= \frac{n}{3^n \ln 3} + \frac{1}{3^n (\ln 3)^2} = \frac{1 + n \ln 3}{3^n (\ln 3)^2}.$ We need

$\frac{1 + n \ln 3}{3^n (\ln 3)^2} < 0.005$, so choose $n = 7$ to ensure

that $S_7 = \frac{1}{3} + \frac{2}{9} + \frac{3}{27} + \frac{4}{81} + \frac{5}{243} + \frac{6}{729} + \frac{7}{2187}$

≈ 0.75 estimates $\sum\limits_{n=1}^{\infty} \frac{n}{3^n}$ to two places. (In fact,

the sum equals $3/4$ by Exercise 33 in Sec. 10.3.)

36 $R_n = \sum\limits_{k=n+1}^{\infty} \frac{(1/2)^k}{k!} < \frac{1}{(n+1)!} \sum\limits_{k=n+1}^{\infty} \left(\frac{1}{2} \right)^k =$

$\frac{1}{2^n(n+1)!}$. For $n = 4$, $R_n < \frac{1}{1920} < 0.00053.$

Since $S_4 = \frac{83}{128} = 0.6484375$, we conclude that

0.65 approximates the sum to two places. (In fact, the sum equals $\sqrt{e} - 1 \approx 0.6487$.)

38 (a) $|a_n| = \left|\dfrac{\cos[(2n+1)t]}{(2n+1)^2}\right| \le \dfrac{1}{(2n+1)^2} < \dfrac{1}{4n^2}$; since $\sum_{n=1}^{\infty} \dfrac{1}{4n^2}$ converges (p-series with $p = 2$), $\sum_{n=0}^{\infty} \dfrac{\cos[(2n+1)t]}{(2n+1)^2}$ converges by the absolute-convergence test for all t.

(b) $\sum_{n=0}^{\infty} \dfrac{\cos(2n+1)t}{(2n+1)^2} = \dfrac{\pi^2 - 2\pi t}{8}$, so $t = 0$ implies $\dfrac{1}{1^2} + \dfrac{1}{3^2} + \dfrac{1}{5^2} + \cdots = \dfrac{\pi^2}{8}$.

40 (a) Experiments with a calculator suggest that the limit is large. See (b).

(b) From Sec. 10.4, we know that $\int_1^\infty f(x)\,dx <$
$\sum_{n=1}^{\infty} f(n) < f(1) + \int_1^\infty f(x)\,dx$, where $f(x)$ is a positive, decreasing function. Using $f(x) = x^{-p}$, we have $\int_1^\infty \dfrac{1}{x^p}\,dx < \sum_{n=1}^{\infty} \dfrac{1}{n^p} <$
$1 + \int_1^\infty \dfrac{1}{x^p}\,dx$. Hence $\dfrac{1}{p-1} < \zeta(p) < \dfrac{p}{p-1}$, as claimed. Since $\lim_{p \to 1^+} \dfrac{1}{p-1} = \infty$ and $\lim_{p \to 1^+} \dfrac{p}{p-1} = \infty$, it now follows that $\lim_{p \to 1^+} \zeta(p) = \infty$.

(c) Since $\dfrac{1}{p-1} < \zeta(p) < \dfrac{p}{p-1}$, $1 < \zeta(p)(p-1) < p$. But $\lim_{p \to 1^+} p = 1$, so
$\lim_{p \to 1^+} \zeta(p)(p-1) = 1$.

42 (b) The region of area a_1 lies within the triangle whose vertices are $(1, f(1))$, $(2, f(2))$, and

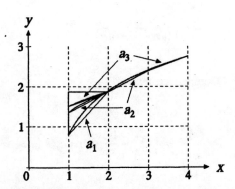

$(1, f(2))$. The region of area a_2 is enclosed by the curve $y = f(x)$ and a chord; note that it can be translated along its chord into the indicated triangle without overlapping the region of area a_1. Subsequent regions can also be translated into the triangle. Since they lie within the triangle without overlapping, the triangle's area is larger.

44 From part (b) of Exercise 43, $C =$
$\lim_{n \to \infty} \left[n \ln n - n + 1 - \ln n! + \ln\sqrt{n}\right]$
$= \lim_{n \to \infty} \left[\ln n^n - \ln e^n + \ln e - \ln n! + \ln \sqrt{n}\right]$
$= \lim_{n \to \infty} \ln \dfrac{n^n \cdot e \cdot \sqrt{n}}{e^n n!} = \ln\left(\lim_{n \to \infty} \dfrac{(n/e)^n e \sqrt{n}}{n!}\right)$ so that
$\lim_{n \to \infty} \dfrac{(n/e)^n e \sqrt{n}}{n!} = e^C$. Thus, $\lim_{n \to \infty} \dfrac{n!}{(n/e)^n e \sqrt{n}} =$

10.S Summary: Review Exercises

$\dfrac{1}{e^C}$, so $\lim\limits_{n\to\infty} \dfrac{n!}{(e/e^C)(n/e)^n \sqrt{n}} = 1$ implies that

$\lim\limits_{n\to\infty} \dfrac{n!}{k\left(\dfrac{n}{e}\right)^n \sqrt{n}} = 1$, where $k = \dfrac{e}{e^C} = e^{1-C}$.

46 (a) $2\cdot 4\cdot 6\cdot 8\cdots 2n = (2\cdot 1)(2\cdot 2)(2\cdot 3)(2\cdot 4)\cdots(2n)$
$= 2^n\cdot n!$

(b) $1\cdot 3\cdot 5\cdot 7\cdots(2n-1)$

$= \dfrac{1\cdot 2\cdot 3\cdot 4\cdot 5\cdot 6\cdot 7 \cdots (2n-2)(2n-1)\cdot 2n}{2\cdot 4\cdot 6\cdot 8 \cdots 2n}$

$= \dfrac{(2n)!}{2^n n!}$ by part (a).

(c) Taking the square root of both sides of the equation in Exercise 45(f) gives $\sqrt{\dfrac{\pi}{2}}$

$= \lim\limits_{n\to\infty} \dfrac{2\cdot 4\cdot 6\cdot 8 \cdots 2n}{3\cdot 5\cdot 7\cdot 9 \cdots (2n-1)} \cdot \dfrac{1}{\sqrt{2n+1}}$

$= \lim\limits_{n\to\infty} \dfrac{2^n n!}{\dfrac{(2n)!}{2^n n!} \sqrt{2n+1}} = \lim\limits_{n\to\infty} \dfrac{(n!)^2 4^n}{(2n)! \sqrt{2n+1}}$.

48 (b) If $a_1 = a_2 = 2$, then all $a_n = 2$, so $\lim\limits_{n\to\infty} a_n = 2$. If $a_1 = -1$ or $a_2 = -1$, or $a_1 + a_2 = -2$, then $\lim\limits_{n\to\infty} a_n = -1$. Otherwise, $\lim\limits_{n\to\infty} a_n$ does not exist. See the note at the end of this solutiong for more information.

(c) If $a_n = \dfrac{2 + a_{n-1}}{a_{n-2}}$ and $\lim\limits_{n\to\infty} a_n = L$, then $L = \dfrac{2+L}{L}$, so $L^2 - L - 2 = 0$, resulting in

$(L-2)(L+1) = 0$, so $L = 2$ or $L = -1$ are the only possible limits.

Note: This exercise can be used as an interesting application after level curves have been discussed in Sec. 14.6. The points (a_n, a_{n+1}) lie on a level curve of the function f whose formula is $f(x, y) = x + y + \dfrac{(x+y+1)(x+y+2)}{xy}$. Examination of the limits of this function gives an analytic basis for the exploratory results of (b).

50 The area of the shaded region, γ, is less than the sum of the areas of the rectangles with vertices

$\left(n, \dfrac{1}{n+1}\right)$, $\left(n, \dfrac{1}{n}\right)$, $\left(n+1, \dfrac{1}{n+1}\right)$, and $\left(n+1, \dfrac{1}{n}\right)$

and greater than the sum of the areas of the triangles lying inside the shaded region: $\gamma <$

$\left(1 - \dfrac{1}{2}\right) + \left(\dfrac{1}{2} - \dfrac{1}{3}\right) + \left(\dfrac{1}{3} - \dfrac{1}{4}\right) + \left(\dfrac{1}{4} - \dfrac{1}{5}\right) + \cdots$

$= \sum\limits_{k=1}^{\infty} \left(\dfrac{1}{k} - \dfrac{1}{k+1}\right) = \lim\limits_{n\to\infty} \sum\limits_{k=1}^{n} \left(\dfrac{1}{k} - \dfrac{1}{k+1}\right) =$

$\lim\limits_{n\to\infty} \left(1 - \dfrac{1}{n+1}\right) = 1$ and $\gamma > \dfrac{1}{2}\left(1 - \dfrac{1}{2}\right) +$

$\dfrac{1}{2}\left(\dfrac{1}{2} - \dfrac{1}{3}\right) + \dfrac{1}{2}\left(\dfrac{1}{3} - \dfrac{1}{4}\right) + \dfrac{1}{2}\left(\dfrac{1}{4} - \dfrac{1}{5}\right) + \cdots =$

$\dfrac{1}{2}$; that is, $1/2 < \gamma < 1$.

52 $\sum\limits_{n=1}^{\infty} (e^{1/n} - 1)$ diverges since $\lim\limits_{n\to\infty} \dfrac{e^{1/n} - 1}{1/n} \overset{H}{=}$

$\lim\limits_{n\to\infty} \dfrac{e^{1/n}\cdot(-1/n^2)}{-1/n^2} = \lim\limits_{n\to\infty} e^{1/n} = e^0 = 1$ and

$\sum\limits_{n=1}^{\infty} \dfrac{1}{n}$ diverges.

11 Power Series and Complex Numbers

11.1 Taylor Series

2 $f(x) = \dfrac{1}{1+x}$, $f'(x) = -\dfrac{1}{(1+x)^2}$, $f''(x) =$

$\dfrac{2}{(1+x)^3}$, so $a_0 = f(1) = \dfrac{1}{2}$, $a_1 = f'(1) = -\dfrac{1}{4}$, a_2

$= \dfrac{f''(1)}{2} = \dfrac{1}{8}$ and $P_1(x; 1) = \dfrac{1}{2} - \dfrac{1}{4}(x-1) =$

$\dfrac{3}{4} - \dfrac{1}{4}x$, $P_2(x; 1) = \dfrac{1}{2} - \dfrac{1}{4}(x-1) + \dfrac{1}{8}(x-1)^2$.

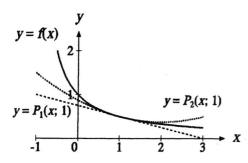

4 $f(x) = \ln(1+x)$, $f'(x) = \dfrac{1}{1+x}$, $f''(x) =$

$-\dfrac{1}{(1+x)^2}$, $f'''(x) = \dfrac{2}{(1+x)^3}$ so $a_0 = f(1) =$

$\ln 2$, $a_1 = f'(1) = \dfrac{1}{2}$, $a_2 = \dfrac{f''(1)}{2} = -\dfrac{1}{8}$, $a_3 = \dfrac{f'''(1)}{3!}$

$= \dfrac{1}{24}$ and $P_1(x; 1) = \ln 2 + \dfrac{1}{2}(x-1)$, $P_2(x; 1) =$

$\ln 2 + \dfrac{1}{2}(x-1) - \dfrac{1}{8}(x-1)^2$, $P_3(x; 1) =$

$\ln 2 + \dfrac{1}{2}(x-1) - \dfrac{1}{8}(x-1)^2 + \dfrac{1}{24}(x-1)^3$.

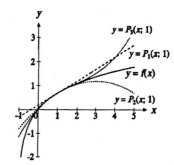

6 $f(x) = f'(x) = f''(x) = f'''(x) = f^{(4)}(x) = e^x$, so

$a_0 = f(2) = e^2$, $a_1 = f'(2) = e^2$, $a_2 = \dfrac{f''(2)}{2}$

$= \dfrac{e^2}{2}$, $a_3 = \dfrac{f'''(2)}{3!} = \dfrac{e^2}{6}$, $a_4 = \dfrac{f^{(4)}(2)}{4!} = \dfrac{e^2}{24}$

so $P_1(x; 2) = e^2 + e^2(x-2)$, $P_2(x; 2) =$

$e^2 + e^2(x-2) + \dfrac{e^2}{2}(x-2)^2$, $P_3(x; 2) = e^2 +$

$e^2(x-2) + \dfrac{e^2}{2}(x-2)^2 + \dfrac{e^2}{6}(x-2)^3$, $P_4(x; 2) =$

$e^2 + e^2(x-2) + \dfrac{e^2}{2}(x-2)^2 + \dfrac{e^2}{6}(x-2)^3 +$

$\dfrac{e^2}{24}(x-2)^4$.

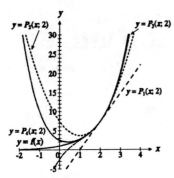

8 $f(x) = \tan^{-1} x$, $f'(x) = \dfrac{1}{1+x^2}$, $f''(x) = \dfrac{-2x}{(1+x^2)^2}$,

$f'''(x) = \dfrac{2(3x^2 - 1)}{(1+x^2)^3}$ so $a_0 = f(-1) = -\dfrac{\pi}{4}$, a_1

$= f'(-1) = \dfrac{1}{2}$, $a_2 = \dfrac{f''(-1)}{2} = \dfrac{1}{4}$, $a_3 =$

$\dfrac{f'''(-1)}{3!} = \dfrac{1}{12}$ and $P_1(x; -1) = -\dfrac{\pi}{4} + \dfrac{1}{2}(x+1)$,

$P_2(x; -1) = -\dfrac{\pi}{4} + \dfrac{1}{2}(x+1) + \dfrac{1}{4}(x+1)^2$,

$P_3(x; -1) =$

$-\dfrac{\pi}{4} + \dfrac{1}{2}(x+1) + \dfrac{1}{4}(x+1)^2 + \dfrac{1}{12}(x+1)^3$.

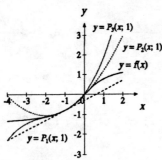

10 $f(x) = \sin x$, $f'(x) = \cos x$, $f''(x) = -\sin x$, $f'''(x) = -\cos x$, $f^{(4)}(x) = \sin x$, $f^{(5)}(x) = \cos x$, $f^{(6)}(x) = -\sin x$, $f^{(7)}(x) = -\cos x$, so $a_0 = f(0) = 0$, $a_1 = f'(0) = 1$, $a_2 = \dfrac{f''(0)}{2} = 0$, $a_3 = \dfrac{f'''(0)}{3!} = \dfrac{-1}{3!}$,

$a_4 = \dfrac{f^{(4)}(0)}{4!} = 0$, $a_5 = \dfrac{f^{(5)}(0)}{5!} = \dfrac{1}{5!}$, $a_6 = \dfrac{f^{(6)}(0)}{6!} = 0$, $a_7 = \dfrac{f^{(7)}(0)}{7!} = -\dfrac{1}{7!}$ and $P_7(x; 0) =$

$x - \dfrac{x^3}{3!} + \dfrac{x^5}{5!} - \dfrac{x^7}{7!}$.

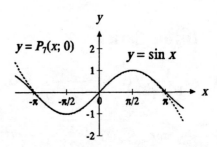

12 $f(x) = \sin x$, $f'(x) = \cos x$, $f''(x) = -\sin x$, $f'''(x) = -\cos x$ so $a_0 = f\left(\dfrac{\pi}{4}\right) = \dfrac{\sqrt{2}}{2}$, $a_1 = f'\left(\dfrac{\pi}{4}\right) = \dfrac{\sqrt{2}}{2}$, $a_2 = \dfrac{1}{2}f''\left(\dfrac{\pi}{4}\right) = -\dfrac{\sqrt{2}}{4}$, $a_3 = \dfrac{1}{3!}f'''\left(\dfrac{\pi}{4}\right)$

$= -\dfrac{\sqrt{2}}{12}$ and $P_3\left(x; \dfrac{\pi}{4}\right) =$

$\dfrac{\sqrt{2}}{2} + \dfrac{\sqrt{2}}{2}\left(x - \dfrac{\pi}{4}\right) - \dfrac{\sqrt{2}}{4}\left(x - \dfrac{\pi}{4}\right)^2 - \dfrac{\sqrt{2}}{12}\left(x - \dfrac{\pi}{4}\right)^3$.

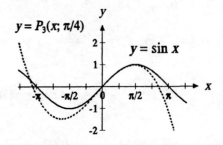

14 $f(x) = \ln(1-x)$, $f'(x) = \dfrac{1}{x-1}$, $f''(x) = -\dfrac{1}{(x-1)^2}$, $f'''(x) = \dfrac{2}{(x-1)^3}$, $f^{(4)}(x) = -\dfrac{3 \cdot 2}{(x-1)^4}$,

$f^{(5)}(x) = \frac{4 \cdot 3 \cdot 2}{(x-1)^5}, \cdots, f^{(n)}(x) = (-1)^{n+1}\frac{(n-1)!}{(x-1)^n}$

for $n \geq 1$ so that $a_0 = f(0) = 0$, $a_1 = f'(0) = -1$, $a_2 = \frac{f''(0)}{2} = -\frac{1}{2}$, $a_3 = \frac{f'''(0)}{3!} = -\frac{1}{3}$, $a_4 = \frac{f^{(4)}(0)}{4!} = -\frac{1}{4}, \cdots, a_n = \frac{f^{(n)}(0)}{n!} = -\frac{1}{n}$ for $n \geq 1$

and the Maclaurin series for $\ln(1-x)$ is $-x - \frac{1}{2}x^2 - \frac{1}{3}x^3 - \frac{1}{4}x^4 - \cdots - \frac{1}{n}x^n - \cdots$.

16 $f(x) = \sin x$, $f'(x) = \cos x$, $f''(x) = -\sin x$, $f'''(x) = -\cos x$, $f^{(4)}(x) = \sin x, \cdots$ so $a_0 = f(0) = 0$, $a_1 = f'(0) = 1$, $a_2 = \frac{f''(0)}{2!} = 0$, $a_3 = \frac{f'''(0)}{3!} = -\frac{1}{3!}$, $a_4 = \frac{f^{(4)}(0)}{4!} = 0$, $a_5 = \frac{f^{(5)}(0)}{5!} = \frac{1}{5!}, \cdots$

and the Maclaurin series for $\sin x$ is $x - \frac{x^3}{3!} + \frac{x^5}{5!} - \frac{x^7}{7!} + \cdots + (-1)^n\frac{x^{2n+1}}{(2n+1)!} + \cdots$.

18 $f(x) = \frac{1}{1-x}$, $f'(x) = \frac{1}{(1-x)^2}$, $f''(x) = \frac{2}{(1-x)^3}$, $f'''(x) = \frac{3 \cdot 2}{(1-x)^4}$, $f^{(4)}(x) = \frac{4 \cdot 3 \cdot 2}{(1-x)^5}, \cdots, f^{(n)}(x) = \frac{n!}{(1-x)^{n+1}}$ for $n \geq 0$ so $a_0 = f(0) = 1$, $a_1 = f'(0) = 1$, $a_2 = \frac{f''(0)}{2} = 1$, $a_3 = \frac{f'''(0)}{3!} = 1, \cdots, a_n = \frac{f^{(n)}(0)}{n!} = 1$ for $n \geq 0$ and

the Maclaurin series for $\frac{1}{1-x}$ is $1 + x + x^2 + x^3 + \cdots + x^n + \cdots$.

20 (a) $f(x) = \ln x$, $f'(x) = \frac{1}{x}$, $f''(x) = -\frac{1}{x^2}$, $f'''(x) = \frac{2}{x^3}$, $f^{(4)}(x) = -\frac{3 \cdot 2}{x^4}$, $f^{(5)}(x) = \frac{4 \cdot 3 \cdot 2}{x^5}, \cdots, f^{(n)}(x) = (-1)^{n+1}\frac{(n-1)!}{x^n}$ for $n \geq 1$, so $a_0 = f(1) = 0$, $a_1 = f'(1) = 1$, $a_2 = \frac{f''(1)}{2} = -\frac{1}{2}$, $a_3 = \frac{f'''(1)}{3!} = \frac{1}{3}, \cdots$, $a_n = \frac{f^{(n)}(1)}{n!} = \frac{(-1)^{n+1}}{n}$ for $n \geq 1$ and

$P_{10}(x; 1) = (x-1) - \frac{1}{2}(x-1)^2 + \frac{1}{3}(x-1)^3 - \frac{1}{4}(x-1)^4 + \frac{1}{5}(x-1)^5 - \frac{1}{6}(x-1)^6 + \frac{1}{7}(x-1)^7 - \frac{1}{8}(x-1)^8 + \frac{1}{9}(x-1)^9 - \frac{1}{10}(x-1)^{10}$.

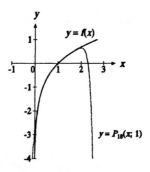

(b)

x	$\ln x$	$P_{10}(x; 1)$
1/2	−0.69314718	−.0693064856
2	0.69314718	0.64563492
4	1.386294361	−4311.246429

22 The degree of $P_n(x; a)$ is at most n.

24 (a) $f(x) = (1 + x)^4, f'(x) = 4(1 + x)^3, f''(x) = 12(1 + x)^2, f'''(x) = 24(1 + x), f^{(4)}(x) = 24$ so $a_0 = f(0) = 1, a_1 = f'(0) = 4, a_2 = \frac{f''(0)}{2} = 6, a_3 = \frac{f'''(0)}{3!} = 4, a_4 = \frac{f^{(4)}(0)}{4!} = 1$ and $P_4(x; 0) = 1 + 4x + 6x^2 + 4x^3 + x^4$.

(b) $(1 + x)^4 = (1 + 2x + x^2)(1 + 2x + x^2) = 1 + 4x + 6x^2 + 4x^3 + x^4$.

26 $f(x) = \cos x, f'(x) = -\sin x, f''(x) = -\cos x, f'''(x) = \sin x, f^{(4)}(x) = \cos x, f^{(5)}(x) = -\sin x, f^{(6)}(x) = -\cos x$ so $a_0 = f\left(\frac{\pi}{4}\right) = \frac{\sqrt{2}}{2}, a_1 = f'\left(\frac{\pi}{4}\right) = -\frac{\sqrt{2}}{2}, a_2 = \frac{1}{2}f''\left(\frac{\pi}{4}\right) = -\frac{\sqrt{2}}{4}, a_3 = \frac{1}{3!}f'''\left(\frac{\pi}{4}\right) = \frac{\sqrt{2}}{12}, a_4 = \frac{1}{4!}f^{(4)}\left(\frac{\pi}{4}\right) = \frac{\sqrt{2}}{48}, a_5 = \frac{1}{5!}f^{(5)}\left(\frac{\pi}{4}\right) = -\frac{\sqrt{2}}{240}, a_6 = \frac{1}{6!}f^{(6)}\left(\frac{\pi}{4}\right) = -\frac{\sqrt{2}}{1440}$

and $P_6\left(x; \frac{\pi}{4}\right) = \frac{\sqrt{2}}{2} - \frac{\sqrt{2}}{2}\left(x - \frac{\pi}{4}\right) - \frac{\sqrt{2}}{4}\left(x - \frac{\pi}{4}\right)^2 + \frac{\sqrt{2}}{12}\left(x - \frac{\pi}{4}\right)^3 + \frac{\sqrt{2}}{48}\left(x - \frac{\pi}{4}\right)^4 - \frac{\sqrt{2}}{240}\left(x - \frac{\pi}{4}\right)^5 - \frac{\sqrt{2}}{1440}\left(x - \frac{\pi}{4}\right)^6$.

28 (a) $f(x) = \frac{1}{x}, f'(x) = -\frac{1}{x^2}, f''(x) = \frac{2}{x^3}, f'''(x) = -\frac{6}{x^4}, f^{(4)}(x) = \frac{24}{x^5}$ so $a_0 = f(1) = 1, a_1 = f'(1) = -1, a_2 = \frac{f''(1)}{2} = 1, a_3 = \frac{f'''(1)}{3!} = -1, a_4 = \frac{f^{(4)}(1)}{4!} = 1$ and $P_4(x; 1) = 1 - (x - 1) + (x - 1)^2 - (x - 1)^3 + (x - 1)^4$.

(b)

(c)

x	$1/x$	$P_4(x; 1)$
0.1	10.0000	4.0951
0.5	2.0000	1.9375
0.9	1.1111	1.1111
1	1.0000	1.0000
1.1	0.9091	0.9091
2	0.5000	1.0000
3	0.3333	11.0000

30 No. If there were such a polynomial $p(x) = a_0 + a_1x + a_2x^2 + \cdots + a_nx^n$, then $0 = \frac{d^{n+1}}{dx^{n+1}}(p(x)) = \frac{d^{n+1}}{dx^{n+1}}(\ln x) = (-1)^n \frac{n!}{x^{n+1}}$. This is not possible for any n, no matter what the interval for x.

32 (a) If m is an odd number, then $x^m = -(-x)^m$. If $p(x)$ is a polynomial containing only odd powers of x, then clearly $p(x) = -p(-x)$, that is, $p(x)$ is an odd function.

(b) If f is an odd function, then $P_n(x; 0)$ is an odd function: It is easily shown that if g is an odd function, then g' is an even function; and if g is an even function, then g' is an odd function. In addition, if g is an even function, then $g'(0) = 0$ (if $g'(0)$ exists). Since f is an odd function, it follows that $f(0) = 0, f''(0) = 0, f^{(4)}(0) = 0, \cdots$ so that $P_n(x; 0) = f(0) + f'(0)x + \dfrac{f''(0)}{2!}x^2 + \dfrac{f'''(0)}{3!}x^3 + \cdots + \dfrac{f^{(n)}(0)}{n!}x^n$ is composed of only odd powers of x. By Exercise 32(a), $P_n(x; 0)$ is odd.

34 $f(x) = (1 + x)^n, f'(x) = n(1 + x)^{n-1}, f''(x) = n(n-1)(1+x)^{n-2}, f'''(x) = n(n-1)(n-2)(1+x)^{n-3}, \cdots, f^{(n)}(x) = n(n-1)(n-2)\cdots(2)(1+x)^{n-n} = n(n-1)(n-2)\cdots(2) = n!$. If we use the notation $\binom{m}{k} = \dfrac{m!}{k!(m-k)!}$ then we have $a_0 = f(0) = 1 = \binom{n}{0}, a_1 = f'(0) = n = \binom{n}{1}, a_2 = \dfrac{f''(0)}{2} = \dfrac{n(n-1)}{2} = \binom{n}{2}, \cdots, a_n = \dfrac{f^{(n)}(0)}{n!} = \dfrac{n!}{n!} = 1 = \binom{n}{n}$ and the Maclaurin series for $(1 + x)^n$ is

$\binom{n}{0} + \binom{n}{1}x + \binom{n}{2}x^2 + \cdots + \binom{n}{n-1}x^{n-1} + \binom{n}{n}x^n.$

11.2 The Error in Taylor Series

2 $R_n(x; 0) = \dfrac{f^{(n+1)}(c_n)}{(n+1)!}x^{n+1}$ for some c_n between 0 and x. Since $f^{(n)}(x) = (-1)^n e^{-x}$, $|R_n(x; 0)| = \dfrac{e^{-c_n}}{(n+1)!}|x|^{n+1}$. For $x \geq 0$, $c_n \geq 0$ so $e^{-c_n} \leq 1$ and $|R_n(x; 0)| \leq \dfrac{x^{n+1}}{(n+1)!}$ which approaches 0 as $n \to \infty$. For $x < 0$, $c_n \geq x$ so $e^{-c_n} \leq e^{-x}$ and $|R_n(x; 0)| \leq e^{-x}\dfrac{(-x)^{n+1}}{(n+1)!}$ which also approaches 0 as $n \to \infty$.

4 For $f(x) = \dfrac{1}{1+x}$, $f^{(n)}(x) = \dfrac{(-1)^n n!}{(1+x)^{n+1}}$ so that

$|R_n(x; 0)| = \left|\dfrac{f^{(n+1)}(c_n)}{(n+1)!}x^{n+1}\right| = \dfrac{|x|^{n+1}}{|1+c_n|^{n+2}}$. If $0 \leq c_n \leq x < 1$, then $\dfrac{|x|^{n+1}}{|1+c_n|^{n+2}} = \dfrac{1}{|1+c_n|}\left|\dfrac{x}{1+c_n}\right|^{n+1} < \left|\dfrac{x}{1+c_n}\right|^{n+1} \to 0$ as $n \to \infty$, since $\left|\dfrac{x}{1+c_n}\right| < 1$; if $-\dfrac{1}{2} < x \leq c_n \leq 0$, then

$\dfrac{|x|^{n+1}}{|1+c_n|^{n+2}} = \dfrac{1}{|1+c_n|}\left|\dfrac{x}{1+c_n}\right|^{n+1} < 2\left|\dfrac{x}{1+c_n}\right|^{n+1}$

$\to 0$ as $n \to \infty$, since $\left|\dfrac{x}{1+c_n}\right| < 1$. In both cases, $\lim_{n \to \infty} R_n(x; 0) = 0$, so $f(x) = \dfrac{1}{1+x}$ is represented by its Maclaurin series.

6 $f(x) = \cos x$, so $|f^{(n)}(a)| \leq 1$ for all values of n and a; thus, $|R_n(x; a)| = \left|\dfrac{f^{(n+1)}(c_n)(x-a)^{n+1}}{(n+1)!}\right| \leq$

$\dfrac{|x-a|^{n+1}}{(n+1)!} \to 0$ as $n \to \infty$.

8 $e^2 = 1 + 2 + \dfrac{2^2}{2!} + \dfrac{2^3}{3!} + \dfrac{2^4}{4!} + \cdots$, and if

$R_n(2; 0) = \dfrac{f^{(n+1)}(c_n)}{(n+1)!} \cdot 2^{n+1} = e^{c_n} \cdot \dfrac{2^{n+1}}{(n+1)!} <$

$e^2 \cdot \dfrac{2^{n+1}}{(n+1)!} < \dfrac{3^2 \cdot 2^{n+1}}{(n+1)!} < 0.0005$, then n must

be at least 10. Thus $1 + 2 + \dfrac{2^2}{2!} + \dfrac{2^3}{3!} + \cdots + \dfrac{2^{10}}{10!}$

≈ 7.389 estimates e^2 to three decimal places.

10 $\sin x = \dfrac{\sqrt{2}}{2} + \dfrac{\sqrt{2}}{2}\left(x - \dfrac{\pi}{4}\right) - \dfrac{\sqrt{2}/2}{2!}\left(x - \dfrac{\pi}{4}\right)^2 -$

$\dfrac{\sqrt{2}/2}{3!}\left(x - \dfrac{\pi}{4}\right)^3 + \cdots$ so $\sin 40° = \sin\dfrac{2\pi}{9} = \dfrac{\sqrt{2}}{2}$

$- \dfrac{\sqrt{2}}{2}\left(\dfrac{\pi}{36}\right) - \dfrac{\sqrt{2}}{4}\left(\dfrac{\pi}{36}\right)^2 + \dfrac{\sqrt{2}}{12}\left(\dfrac{\pi}{36}\right)^3 + \cdots$. For

$f(x) = \sin x$, $|f^{(n+1)}(c_n)| \leq 1$, so that if

$\left|R_n\left(\dfrac{2\pi}{9}; \dfrac{\pi}{4}\right)\right| = \dfrac{|f^{(n+1)}(c_n)|}{(n+1)!} \cdot \left|\dfrac{2\pi}{9} - \dfrac{\pi}{4}\right|^{n+1} \leq$

$\dfrac{(\pi/36)^{n+1}}{(n+1)!} < 0.0005$, then n must be at least 2.

Thus, $\dfrac{\sqrt{2}}{2} - \dfrac{\sqrt{2}}{2}\left(\dfrac{\pi}{36}\right) - \dfrac{\sqrt{2}}{4}\left(\dfrac{\pi}{36}\right)^2 \approx 0.643$

estimates $\sin\dfrac{2\pi}{9}$ to three decimal places.

12 (a) Let $x_1 = \dfrac{\pi}{4}$ and $x_2 = \dfrac{47\pi}{180}$ so that $dx =$

$x_2 - x_1 = \dfrac{\pi}{90}$. The differential $df =$

$f'\left(\dfrac{\pi}{4}\right)dx = \left(\cos\dfrac{\pi}{4}\right)\dfrac{\pi}{90} = \dfrac{\sqrt{2}}{2} \cdot \dfrac{\pi}{90} \approx$

0.02468 and $\Delta f = f\left(\dfrac{47\pi}{180}\right) - f\left(\dfrac{\pi}{4}\right)$ with df

$\approx \Delta f$, that is, $\sin\dfrac{47\pi}{180} - \sin\dfrac{\pi}{4} \approx 0.02468$;

thus $\sin\dfrac{47\pi}{180} \approx 0.73179$.

(b) $f(x) = \sin x$, $f'(x) = \cos x$, $f''(x) = -\sin x$,

so $a_0 = f\left(\dfrac{\pi}{4}\right) = \dfrac{\sqrt{2}}{2}$, $a_1 = f'\left(\dfrac{\pi}{4}\right) = \dfrac{\sqrt{2}}{2}$,

$a_2 = \dfrac{1}{2!}f''\left(\dfrac{\pi}{4}\right) = -\dfrac{\sqrt{2}}{4}$ and $P_1\left(x; \dfrac{\pi}{4}\right) =$

$\dfrac{\sqrt{2}}{2} + \dfrac{\sqrt{2}}{2}\left(x - \dfrac{\pi}{4}\right)$. Thus, $P_1\left(\dfrac{47\pi}{180}; \dfrac{\pi}{4}\right) \approx$

0.73179.

(c) The error in (b) is $\left|R_1\left(\dfrac{47\pi}{180}; \dfrac{\pi}{4}\right)\right| =$

$\left|\dfrac{f''(c_n)}{2!}\left(\dfrac{\pi}{90}\right)^2\right| = \dfrac{\pi^2}{16{,}200}|-\sin c_n| \leq \dfrac{\pi^2}{16{,}200}$

≈ 0.00061.

(d) (See part (b).) $P_2\left(\dfrac{47\pi}{180}; \dfrac{\pi}{4}\right) =$

$\dfrac{\sqrt{2}}{2} + \dfrac{\sqrt{2}}{2}\left(x - \dfrac{\pi}{4}\right) - \dfrac{\sqrt{2}}{4}\left(x - \dfrac{\pi}{4}\right)^2$ so

$P_2\left(\dfrac{47\pi}{180}; \dfrac{\pi}{4}\right) \approx 0.7313587$.

11.2 The Error in Taylor Series

(e) The error in (d) is $\left|R_2\left(\frac{47\pi}{180}; \frac{\pi}{4}\right)\right| =$

$\left|\frac{f'''(c_n)}{3!}\left(\frac{\pi}{90}\right)^3\right| = \frac{\pi^3}{4,374,000}|-\cos c_n| \leq$

$\frac{\pi^3}{4,374,000} \approx 0.000007.$

(f) Calculator: $\sin 47° = \sin\frac{47\pi}{180} \approx 0.7313537$

and $P_2\left(\frac{47\pi}{180}; \frac{\pi}{4}\right) \approx 0.7313587$ so exact error

is 0.0000050 and estimated error is 0.000007.

14 $\cos x = 1 - \frac{x^2}{2!} + \frac{x^4}{4!} - \cdots \pm \frac{x^{2n}}{(2n)!} + \cdots$ so that

$\frac{1 - \cos x}{x} = \frac{x}{2!} - \frac{x^3}{4!} + \frac{x^5}{6!} - \cdots \pm \frac{x^{2n+1}}{(2(n+1))!}$

$+ \cdots$ is a decreasing, alternating series for x in

$[0, 1]$ with $|R_n| < |a_{n+1}| = \frac{x^{2(n+1)+1}}{(2(n+2))!} \leq$

$\frac{1}{(2(n+2))!}$. If $\int_0^1 \frac{1}{(2(n+2))!} dx = \frac{1}{(2(n+2))!} <$

0.0005, then n must be at least 2, so that

$\int_0^1 \left(\frac{x}{2!} - \frac{x^3}{4!} + \frac{x^5}{6!}\right) dx = \left(\frac{x^2}{4} - \frac{x^4}{96} + \frac{x^6}{4320}\right)\Big|_0^1$

≈ 0.240 estimates $\int_0^1 \frac{1 - \cos x}{x} dx$ to three

decimal places.

16 (a) If $\left|\int_b^\infty \frac{\sin(x^6/100)}{x^6} dx\right| \leq$

$\int_b^\infty \frac{|\sin(x^6/100)|}{x^6} dx \leq \int_b^\infty \frac{1}{x^6} dx = \frac{1}{5b^5}$

< 0.001, then $b = 2.9$ works.

(b) $\int_0^{2.9} \frac{\sin(x^6/100)}{x^6} dx =$

$\int_0^{2.9} \frac{0.01x^6 - (0.01)^3 \frac{x^{18}}{3!} + (0.01)^5 \frac{x^{30}}{5!} - \cdots}{x^6} dx$

$= \int_0^{2.9} \left(0.01 - \frac{0.01^3}{3!}x^{12} + \frac{0.01^5}{5!}x^{24} - \frac{0.01^7}{7!}x^{36} + \cdots\right) dx$

$= (0.01)(2.9) - \frac{(0.01)^3}{3!} \cdot \frac{(2.9)^{13}}{13} +$

$\frac{(0.01)^5}{5!} \cdot \frac{(2.9)^{25}}{25} - \frac{(0.01)^7}{7!} \cdot \frac{(2.9)^{37}}{37} + \cdots$ is a

decreasing, alternating series. Thus, if

$\frac{(0.01)^{2n+1}}{(2n+1)!} \cdot \frac{(2.9)^{12n+1}}{12n+1} < 0.001$, then n must

be at least 5. Hence, $(0.01)(2.9) -$

$\frac{(0.01)^3}{3!} \cdot \frac{(2.9)^{13}}{13} + \frac{(0.01)^5}{5!} \cdot \frac{(2.9)^{25}}{25} -$

$\frac{(0.01)^7}{7!} \cdot \frac{(2.9)^{37}}{37} + \frac{(0.01)^9}{9!} \cdot \frac{(2.9)^{49}}{49} \approx 0.0236$

estimates $\int_0^{2.9} \frac{\sin(x^6/100)}{x^6} dx$ with error at

most 0.001.

(c) From parts (a) and (b), $\int_0^\infty \frac{\sin(x^6/100)}{x^6} dx =$

$\int_0^{2.9} \frac{\sin(x^6/100)}{x^6} dx + \int_{2.9}^\infty \frac{\sin(x^6/100)}{x^6} dx$

≈ 0.0236 with error at most 0.002.

18 $|R_n(x; 0)| = \left|\dfrac{f^{(n+1)}(c_n)x^{n+1}}{(n+1)!}\right| \leq \dfrac{2^{n+1}|x|^{n+1}}{(n+1)!} =$

$\dfrac{(2|x|)^{n+1}}{(n+1)!} \to 0$ as $n \to \infty$ for all x.

20 (a) $\displaystyle\int_0^1 \dfrac{1}{1+x^3}\,dx = \int_0^1 \dfrac{1}{(x+1)(x^2-x+1)}\,dx$

$= \displaystyle\int_0^1 \left(\dfrac{1/3}{x+1} + \dfrac{-x/3 + 2/3}{x^2-x+1}\right) dx$

$= \dfrac{1}{3}\ln(x+1)\Big|_0^1 - \dfrac{1}{3}\displaystyle\int_0^1 \dfrac{x-2}{x^2-x+1}\,dx$

$= \dfrac{1}{3}\ln 2 - \left[\dfrac{1}{6}\ln(x^2-x+1) - \dfrac{1}{\sqrt{3}}\arctan\dfrac{2x-1}{\sqrt{3}}\right]_0^1$

$= \dfrac{1}{3}\ln 2 + \dfrac{1}{\sqrt{3}}\cdot\dfrac{\pi}{6} - \dfrac{1}{\sqrt{3}}\left(-\dfrac{\pi}{6}\right) =$

$\dfrac{1}{3}\ln 2 + \dfrac{\pi}{3\sqrt{3}} \approx 0.8356.$

(b) $h = 1/6$, so Simpson's estimate for

$\displaystyle\int_0^1 \dfrac{1}{1+x^3}\,dx$ is $\dfrac{1/6}{3}\Big[f(0) + 4f\left(\dfrac{1}{6}\right) +$

$2f\left(\dfrac{1}{3}\right) + 4f\left(\dfrac{1}{2}\right) + 2f\left(\dfrac{2}{3}\right) + 4f\left(\dfrac{5}{6}\right) + f(1)\Big]$

$= \dfrac{1}{18}[15.0423] \approx 0.8357.$

(c) $h = 1/6$ so the trapezoidal estimate for

$\displaystyle\int_0^1 \dfrac{1}{1+x^3}\,dx$ is $\dfrac{1/6}{2}\Big[f(0) + 2f\left(\dfrac{1}{6}\right) +$

$2f\left(\dfrac{1}{3}\right) + 2f\left(\dfrac{1}{2}\right) + 2f\left(\dfrac{2}{3}\right) + 2f\left(\dfrac{5}{6}\right) + f(1)\Big]$

$\approx \dfrac{1}{12}(10.0069) \approx 0.8339.$

(d) $\dfrac{1}{1+x^3} = 1 - x^3 + x^6 - x^9 + x^{12} -$

$x^{15} + \cdots$ so $\displaystyle\int_0^1 \dfrac{1}{1+x^3}\,dx \approx$

$\displaystyle\int_0^1 [1 - x^3 + x^6 - x^9 + x^{12} - x^{15}]\,dx$

$= \left[x - \dfrac{x^4}{4} + \dfrac{x^7}{7} - \dfrac{x^{10}}{10} + \dfrac{x^{13}}{13} - \dfrac{x^{16}}{16}\right]_0^1$

$= 0.8073.$

22 If $t = 1/x^2$, then $x = 1/\sqrt{t}$, $e^{-1/x^2} = e^{-t}$, and $x^n = t^{-n/2}$. Hence $\displaystyle\lim_{x\to 0}\dfrac{e^{-1/x^2}}{x^n} = \lim_{t\to\infty}\dfrac{e^{-t}}{t^{-n/2}} = \lim_{t\to\infty}\dfrac{t^{n/2}}{e^t}$

$= 0$ (which can be demonstrated by Exercise 49 of Sec. 6.5 or by repeated applications of l'Hôpital's rule.)

24 (a) $f'(0) = \displaystyle\lim_{h\to 0}\dfrac{f(0+h) - f(0)}{h} =$

$\displaystyle\lim_{h\to 0}\dfrac{e^{-1/h^2} - 0}{h} = 0$, by Exercise 22.

(b) $f''(0) = \displaystyle\lim_{h\to 0}\dfrac{f'(0+h) - f'(0)}{h} =$

$\displaystyle\lim_{h\to 0}\dfrac{e^{-1/h^2}\cdot\dfrac{2}{h^3} - 0}{h} = \lim_{h\to 0}\dfrac{2e^{-1/h^2}}{h^4} = 0$ (by Exercise 22).

(c) Since all derivatives of $f(x)$ are of the form $\dfrac{P(x)e^{-1/x^2}}{Q(x)}$, where $P(x)$ and $Q(x)$ are polynomials, it follows by induction (from Exercise 23) that $f^{(n)}(0) = 0$ for $n = 3, 4, 5,$ ….

26 By Theorem 2, $f(x) - P_3(x; 0) = \dfrac{f^{(4)}(c(x))}{4!} x^4$,

where $c(x)$ is between 0 and x. That is, $f(x) = f(0) + f'(0)x + \dfrac{f''(0)}{2} x^2 + \dfrac{f'''(0)}{6} x^3 +$

$\dfrac{f^{(4)}(c(x))}{4!} x^4$. Hence $\int_{-1}^{1} f(x)\, dx$

$= 2f(0) + \dfrac{1}{3} f''(0) + \int_{-1}^{1} \dfrac{1}{24} f^{(4)}(c(x)) x^4\, dx.$

Since $|f^{(4)}(c(x))| \leq M_4$, $\left| \int_{-1}^{1} \dfrac{1}{24} f^{(4)}(c(x)) x^4\, dx \right|$

$\leq \dfrac{M_4}{24} \int_{-1}^{1} x^4\, dx = \dfrac{M_4}{60}$. Also, $f\!\left(\dfrac{1}{\sqrt{3}}\right) = f(0) +$

$f'(0) \cdot \dfrac{1}{\sqrt{3}} + \dfrac{1}{2} f''(0) \cdot \dfrac{1}{3} + \dfrac{1}{6} f'''(0) \cdot \dfrac{1}{3\sqrt{3}} +$

$\dfrac{1}{24} f^{(4)}(c(1/\sqrt{3})) \cdot \dfrac{1}{9}$, $f(-1/\sqrt{3}) = f(0) - f'(0) \cdot \dfrac{1}{\sqrt{3}}$

$+ \dfrac{1}{2} f''(0) \cdot \dfrac{1}{3} - \dfrac{1}{6} f'''(0) \cdot \dfrac{1}{3\sqrt{3}} +$

$\dfrac{1}{24} f^{(4)}(c(-1/\sqrt{3})) \cdot \dfrac{1}{9}$, so $f\!\left(\dfrac{1}{\sqrt{3}}\right) + f\!\left(-\dfrac{1}{\sqrt{3}}\right) = 2f(0)$

$+ \dfrac{1}{3} f''(0) + \dfrac{1}{216} [f^{(4)}(c(1/\sqrt{3})) + f^{(4)}(c(-1/\sqrt{3}))].$

Note that $\left| \dfrac{1}{216} [f^{(4)}(c(1/\sqrt{3})) + f^{(4)}(c(-1/\sqrt{3}))] \right|$

$\leq \dfrac{1}{216}(M_4 + M_4) = \dfrac{M_4}{108}$. Hence,

$\left| \int_{-1}^{1} f(x)\, dx - f\!\left(\dfrac{1}{\sqrt{3}}\right) - f\!\left(-\dfrac{1}{\sqrt{3}}\right) \right|$

$\leq \left| \int_{-1}^{1} f(x)\, dx - \left(2f(0) + \dfrac{1}{3} f''(0) \right) \right| +$

$\left| 2f(0) + \dfrac{1}{3} f''(0) - [f(1/\sqrt{3}) + f(-1/\sqrt{3})] \right|$

$\leq \dfrac{M_4}{60} + \dfrac{M_4}{108} = \dfrac{7 M_4}{270}.$

11.3 Why the Error in Taylor Series is Controlled by a Derivative

2 Assume that $R(a) = 0$, $R'(a) = 0$, and $m \leq R''(x) \leq M$ for x in $[a, b]$, where m and M are constants. Show that $\dfrac{m}{2!}(b - a)^2 \leq R(b) \leq \dfrac{M}{2!}(b - a)^2$.

Since $m \leq R''(x) \leq M$, $\int_a^t m\, dx \leq \int_a^t R''(x)\, dx$

$\leq \int_a^t M\, dx$ so that $mx \big|_a^t \leq R'(x) \big|_a^t \leq Mx \big|_a^t$,

that is, $m(t - a) \leq R'(t) - R'(a) = R'(t) \leq M(t - a)$, or $m(x - a) \leq R'(x) \leq M(x - a)$.

Similarly, $\int_a^t m(x - a)\, dx \leq \int_a^t R'(x)\, dx \leq$

$\int_a^t M(x - a)\, dx$ so that $\dfrac{m}{2!}(x - a)^2 \Big|_a^t \leq R(x) \big|_a^t$

$\leq \dfrac{M}{2!}(x - a)^2 \Big|_a^t$, that is, $\dfrac{m}{2!}(t - a)^2 \leq$

$R(t) - R(a) = R(t) \leq \dfrac{M}{2!}(t - a)^2$, or $\dfrac{m}{2!}(x - a)^2$

$\leq R(x) \leq \dfrac{M}{2!}(x - a)^2$. In particular, for $x = b$,

$\dfrac{m}{2!}(b - a)^2 \leq R(b) \leq \dfrac{M}{2!}(b - a)^2.$

4 Assume that $R(a) = 0$, $R'(a) = 0$, $R''(a) = 0$, $R'''(a) = 0$, $R^{(4)}(a) = 0$, and $m \le R^{(5)}(x) \le M$ for x in $[a, b]$, where m and M are constants. Show that $\frac{m}{5!}(b - a)^5 \le R(b) \le \frac{M}{5!}(b - a)^5$. Since $m \le R^{(5)}(x) \le M$, $\int_a^t m \, dx \le \int_a^t R^{(5)}(x) \, dx \le \int_a^t M \, dx$ so that $mx \big|_a^t \le R^{(4)}(x) \big|_a^t \le Mx \big|_a^t$, that is, $m(t - a) \le R^{(4)}(t) - R^{(4)}(a) = R^{(4)}(t) \le M(t - a)$, or $m(x - a) \le R^{(4)}(x) \le M(x - a)$. Similarly, $\int_a^t m(x - a) \, dx \le \int_a^t R^{(4)}(x) \, dx \le \int_a^t M(x - a) \, dx$ so that $\frac{m}{2}(x - a)^2 \big|_a^t \le R'''(x) \big|_a^t \le \frac{M}{2}(x - a)^2 \big|_a^t$, that is, $\frac{m}{2}(t - a)^2 \le R'''(t) - R'''(a) = R'''(t) \le \frac{M}{2}(t - a)^2$, or $\frac{m}{2}(x - a)^2 \le R'''(x) \le \frac{M}{2}(x - a)^2$. Similarly, $\int_a^t \frac{m}{2}(x - a)^2 \, dx \le \int_a^t R'''(x) \, dx \le \int_a^t \frac{M}{2}(x - a)^2 \, dx$ so that $\frac{m}{3!}(x - a)^3 \big|_a^t \le R''(x) \big|_a^t \le \frac{M}{3!}(x - a)^3 \big|_a^t$, that is, $\frac{m}{3!}(t - a)^3 \le R''(t) - R''(a) = R''(t) \le \frac{M}{3!}(t - a)^3$, or $\frac{m}{3!}(x - a)^3 \le R''(x) \le \frac{M}{3!}(x - a)^3$. Similarly, $\int_a^t \frac{m}{3!}(x - a)^3 \, dx \le \int_a^t R''(x) \, dx \le \int_a^t \frac{M}{3!}(x - a)^3 \, dx$ so that $\frac{m}{4!}(x - a)^4 \big|_a^t \le R'(x) \big|_a^t \le \frac{M}{4!}(x - a)^4 \big|_a^t$, that is, $\frac{m}{4!}(t - a)^4 \le R'(t) - R'(a) = R'(t) \le \frac{M}{4!}(t - a)^4$, or $\frac{m}{4!}(x - a)^4 \le R'(x) \le \frac{M}{4!}(x - a)^4$. Similarly, $\int_a^t \frac{m}{4!}(x - a)^4 \, dx \le \int_a^t R'(x) \, dx \le \int_a^t \frac{M}{4!}(x - a)^4 \, dx$ so that $\frac{m}{5!}(x - a)^5 \big|_a^t \le R(x) \big|_a^t \le \frac{M}{5!}(x - a)^5 \big|_a^t$, that is, $\frac{m}{5!}(t - a)^5 \le R(t) - R(a) = R(t) \le \frac{M}{5!}(t - a)^5$, or $\frac{m}{5!}(x - a)^5 \le R(x) \le \frac{M}{5!}(x - a)^5$. In particular, for $x = b$, $\frac{m}{5!}(b - a)^5 \le R(b) \le \frac{M}{5!}(b - a)^5$.

6 (a) $m \le f''(x) \le M$, so $\int_a^t m \, dx \le \int_a^t f''(x) \, dx \le \int_a^t M \, dx$, $m(x - a) \le f'(x) - f'(a) \le M(x - a)$, $\int_a^t m(x - a) \, dx \le \int_a^t [f'(x) - f'(a)] \, dx \le \int_a^t M(x - a) \, dx$, $\frac{m}{2}(x - a)^2 \le f(x) - f(a) - f'(a)(x - a) \le \frac{M}{2}(x - a)^2$.

11.3 Why the Error in Taylor Series is Controlled by a Derivative

(b) $m \le f'''(x) \le M$, so $\int_a^t m\,dx \le$

$\int_a^t f'''(x)\,dx \le \int_a^t M\,dx$, $m(x-a) \le$

$f''(x) - f''(a) \le M(x-a)$, $\int_a^t m(x-a)\,dx$

$\le \int_a^t [f''(x) - f''(a)]\,dx \le \int_a^t M(x-a)\,dx$,

$\frac{m}{2}(x-a)^2 \le f'(x) - f'(a) - f''(a)(x-a) \le$

$\frac{M}{2}(x-a)^2$, $\int_a^t \frac{m}{2}(x-a)^2\,dx \le$

$\int_a^t [f'(x) - f'(a) - f''(a)(x-a)]\,dx \le$

$\int_a^t \frac{M}{2}(x-a)^2\,dx$, $\frac{m}{3!}(x-a)^3 \le$

$f(x) - f(a) - f'(a)(x-a) - \frac{f''(a)}{2!}(x-a)^2 \le$

$\frac{M}{3!}(x-a)^3$.

(c) $m \le f^{(4)}(x) \le M$, so $\int_a^t m\,dx \le$

$\int_a^t f^{(4)}(x)\,dx \le \int_a^t M\,dx$, $m(x-a) \le$

$f'''(x) - f'''(a) \le M(x-a)$, $\int_a^t m(x-a)\,dx$

$\le \int_a^t [f'''(x) - f'''(a)]\,dx \le$

$\int_a^t M(x-a)\,dx$, $\frac{m}{2!}(x-a)^2 \le$

$f''(x) - f''(a) - f'''(a)(x-a) \le \frac{M}{2!}(x-a)^2$,

$\int_a^t \frac{m}{2!}(x-a)^2\,dx \le$

$\int_a^t [f''(x) - f''(a) - f'''(a)(x-a)]\,dx \le$

$\int_a^t \frac{M}{2!}(x-a)^2\,dx$, $\frac{m}{3!}(x-a)^3 \le$

$f'(x) - f'(a) - f''(a)(x-a) - \frac{f'''(a)}{2!}(x-a)^2$

$\le \frac{M}{3!}(x-a)^3$, $\int_a^t \frac{m}{3!}(x-a)^3\,dx \le$

$\int_a^t \left[f'(x) - f'(a) - f''(a)(x-a) - \frac{f'''(a)}{2!}(x-a)^2\right]dx$

$\le \int_a^t \frac{M}{3!}(x-a)^3\,dx$, $\frac{m}{4!}(x-a)^4 \le f(x) -$

$f(a) - f'(a)(x-a) - \frac{f''(a)}{2!}(x-a)^2 -$

$\frac{f'''(a)}{3!}(x-a)^3 \le \frac{M}{4!}(x-a)^4$.

(d) In general, $m \le f^{(n+1)}(x) \le M$ implies

$\frac{m}{(n+1)!}(x-a)^{n+1} \le f(x) - f(a) -$

$f'(a)(x-a) - \frac{f''(a)}{2!}(x-a)^2 - \cdots -$

$\frac{f^{(n)}(a)}{n!}(x-a)^n \le \frac{M}{(n+1)!}(x-a)^{n+1}$.

8 (a) The total error is $E =$

$\int_a^b f(x)\,dx - \sum_{k=0}^{n-1} f\left(a + kh + \frac{h}{2}\right)h =$

$\sum_{k=0}^{n-1} \left[\int_{a+kh}^{a+(k+1)h} f(x)\,dx - f\left(a + kh + \frac{h}{2}\right)h\right]$.

Letting c denote one of the midpoints

$a + kh + \frac{h}{2}$, we wish to estimate

$\int_{c-h/2}^{c+h/2} f(x)\,dx - f(c)h$. Adding these estimates for all n values of c will give an estimate of E.

(b) $E(0) = \int_c^c f(x)\,dx - f(c) \cdot 0 = 0$.

$E'(t) = \dfrac{d}{dt}\left(\int_c^{c+t/2} f(x)\,dx + \int_{c-t/2}^c f(x)\,dx\right) - f(c) = \dfrac{1}{2}f\!\left(c + \dfrac{t}{2}\right) - \dfrac{1}{2}\!\left[f\!\left(c - \dfrac{t}{2}\right)\right] - f(c) =$

$\dfrac{1}{2}f\!\left(c + \dfrac{t}{2}\right) + \dfrac{1}{2}f\!\left(c - \dfrac{t}{2}\right) - f(c)$, so $E'(0) = 0$.

(c) By the mean-value theorem, there is a number T in $(c - t/2, c + t/2)$ such that $f''(T) = \dfrac{1}{t}\!\left[f'\!\left(c + \dfrac{t}{2}\right) - f'\!\left(c - \dfrac{t}{2}\right)\right]$. For this T, $E''(t) = \dfrac{t}{4}f''(T)$.

(d) From (c), we have $\dfrac{1}{4}m_2 t \le E''(t) \le \dfrac{1}{4}M_2 t$ for $0 \le t \le h$. Integrating and using $E'(0) = 0$ gives $\dfrac{1}{8}m_2 t^2 \le E'(t) \le \dfrac{1}{8}M_2 t^2$.

Integrating again and using $E(0) = 0$ gives $\dfrac{1}{24}m_2 h^3 \le E(h) \le \dfrac{1}{24}M_2 h^3$.

(e) E is the sum of n quantities, each of which is between $\dfrac{1}{24}m_2 h^3$ and $\dfrac{1}{24}M_2 h^3$, so E is between $\dfrac{n}{24}m_2 h^3 = \dfrac{1}{24}m_2(b-a)h^2$ and $\dfrac{n}{24}M_2 h^3 = \dfrac{1}{24}M_2(b-a)^2$.

10 (a) By Equation (4) of Sec. 5.4, Simpson's estimate is $\dfrac{h}{3}[f(a) + 4f(a+h) + 2f(a+2h) + 4f(a+3h) + \cdots + f(b)] =$

$\dfrac{h}{3}\sum_{k=0}^{n/2-1}[f(a+2kh) + 4f(a+(2k+1)h) + f(a+(2k+2)h)]$. The integral is $\int_a^b f(x)\,dx = \sum_{k=0}^{n/2-1}\int_{a+2kh}^{a+(2k+2)h} f(x)\,dx$, so the error is $E =$

$\sum_{k=0}^{n/2-1}\Bigg(\int_{a+2kh}^{a+(2k+2)h} f(x)\,dx - \dfrac{h}{3}[f(a+2kh) + 4f(a+(2k+1)h) + f(a+(2k+2)h)]\Bigg)$.

Letting c denote one of the points $a + (2k+1)h$, we wish to estimate

$\int_{c-h}^{c+h} f(x)\,dx - \dfrac{h}{3}[f(c-h) + 4f(c) + f(c+h)]$. Adding these estimates for all $n/2$ values of c will give an estimate of E.

(b) $E'(t) = f(c+t) - (-f(c-t)) - \dfrac{t}{3}[-f'(c-t) + f'(c+t)] - \dfrac{1}{3}[f(c-t) + 4f(c) + f(c+t)] =$

$\dfrac{2}{3}[f(c+t) + f(c-t)] - \dfrac{4}{3}f(c) - \dfrac{t}{3}[f'(c+t) - f'(c-t)]$

(c) $E''(t) = \frac{2}{3}[f'(c+t) - f'(c-t)] -$

$\frac{t}{3}[f''(c+t) - (-f''(c-t))] -$

$\frac{1}{3}[f'(c+t) - f'(c-t)] = \frac{1}{3}[f'(c+t) - f'(c-t)]$

$- \frac{t}{3}[f''(c+t) + f''(c-t)]$

(d) $E'''(t) = \frac{1}{3}[f''(c+t) - (-f''(c-t))] -$

$\frac{t}{3}[f'''(c+t) - f'''(c-t)] -$

$\frac{1}{3}[f''(c+t) - f''(c-t)] =$

$-\frac{t}{3}[f'''(c+t) - f'''(c-t)]$

(e) By the mean-value theorem, there is a number T in $(c-t, c+t)$ such that $f^{(4)}(T) = \frac{f'''(c+t) - f'''(c-t)}{2t}$, so $E'''(t) =$

$-\frac{2t^2}{3} f^{(4)}(T)$.

(f) Substitute $t = 0$ in the equations in (a), (b), and (c).

(g) From (e), we have $-\frac{2}{3}t^2 m_4 \geq E'''(t) \geq$

$-\frac{2}{3}t^2 M_4$. Integrating and using $E''(0) = 0$

gives $-\frac{2}{9}t^3 m_4 \geq E''(t) \geq -\frac{2}{9}t^3 M_4$.

Integrating again and using $E'(0) = 0$ gives

$-\frac{1}{18}t^4 m_4 \geq E'(t) \geq -\frac{1}{18}t^4 M_4$. Integrating

a third time and using $E(0) = 0$ gives

$-\frac{1}{90}h^5 m_4 \geq E(h) \geq -\frac{1}{90}h^5 M_4$.

(h) E is the sum of $n/2$ quantities, each of which

is between $-\frac{1}{90}h^5 m_4$ and $-\frac{1}{90}h^5 M_4$, so E is

between $\frac{n}{2}\left(-\frac{1}{90}h^5 m_4\right) = -\frac{m_4(b-a)h^4}{180}$ and

$\frac{n}{2}\left(-\frac{1}{90}h^5 M_4\right) = -\frac{M_4(b-a)h^4}{180}$.

11.4 Power Series and Radius of Convergence

2 The absolute value of the ratio of successive terms

is $\sqrt{\frac{n}{n+1}}|x|$, which approaches $|x|$ as $n \to \infty$.

Hence, the series converges absolutely for $|x| <$
1. For $x = 1$ it diverges by the test for p series.
For $x = -1$ it converges by the alternating-series test.

4 The absolute value of the ratio of successive terms

is $\left(1 + \frac{1}{n}\right)^2 \cdot \frac{1}{e}|x|$, which approaches $\frac{|x|}{e}$ as $n \to$

∞. Hence, the series converges absolutely for $\frac{|x|}{e}$

< 1, or $|x| < e$, and diverges for $\frac{|x|}{e} > 1$, or

$|x| > e$. For $x = e$ and $x = -e$ the series
diverges by the nth-term test.

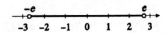

6 By the absolute-ratio test, the radius of convergence is 1. The series diverges for $x = 1$ and converges for $x = -1$.

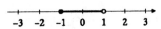

8 The radius of convergence is ∞.

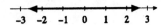

10 The series converges only for $x = 0$.

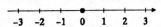

12 By substituting $2x$ for x in the series in Exercise 6, we obtain convergence for $-1/2 \le x < 1/2$.

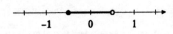

14 (a) True, since the radius of convergence is at least 5.
   (b) True, since the radius of convergence is at least 5.
   (c) Cannot be determined.
   (d) Cannot be determined.
   (e) Diverges, since the radius of convergence is no greater than 8.
   (f) Cannot be determined.

16 (a) Cannot be determined; it could be the endpoint of an interval of convergence.
   (b) Converges, since the radius of convergence is at least 6.
   (c) Converges, since the radius of convergence is at least 6.

18 The radius of convergence is 3 by the absolute-ratio test. The series diverges for $x = 4$ and converges for $x = -2$.

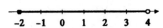

20 The radius of convergence is 1. The series diverges for $x = 5$ and converges for $x = 3$.

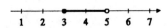

22 The radius of convergence is 1. For $x = 6$, the series is $\sum_{n=1}^{\infty} \dfrac{1}{n \ln n}$, which diverges since

$$\int_2^{\infty} \frac{dx}{x \ln x} = \lim_{b \to \infty} \ln(\ln x)\Big|_2^b = \infty.$$ For $x = 4$, the

series converges by the alternating-series test.

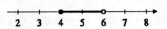

24 The radius of convergence is 1. For $x = 0$ or -2, it diverges by the nth-term test.

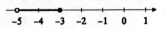

26 The radius of convergence is 1. If $x = -5$, then

$$\sum_{n=0}^{\infty} \frac{1}{n+2}$$ diverges, and if $x = -3$, then

$$\sum_{n=0}^{\infty} \frac{(-1)^n}{n+2}$$ converges.

28 The radius of convergence is 1. If $x = -3$, then

11.4 Power Series and Radius of Convergence

$\sum_{n=0}^{\infty} (-1)^n \frac{n^2+1}{n^3+1}$ converges. If $x = -1$, then

$\sum_{n=0}^{\infty} \frac{n^2+1}{n^3+1}$ diverges.

$\begin{array}{c}\leftarrow\!\!+\!\!-\!\!+\!\!-\!\!+\!\!-\!\!\bullet\!\!-\!\!+\!\!-\!\!\circ\!\!-\!\!+\!\!-\!\!+\!\!\rightarrow \\ -5\ -4\ -3\ -2\ -1\ \ 0\ \ 1 \end{array}$

30 From Equation (4) with $r = 1/3$, the first five terms of the Maclaurin series are

$1 + \frac{1}{3}x - \frac{1}{9}x^2 + \frac{5}{81}x^3 - \frac{10}{243}x^4 + \cdots$.

32 From Equation (4) with $r = -4$, the first five terms of the Maclaurin series are

$1 - 4x + 10x^2 - 20x^3 + 35x^4 - \cdots$.

34 The radius of convergence is $R \geq 4$, so the series also converges for $|x - 3| < 4$; that is, for $-1 < x < 7$.

36 Since both $\sum a_n x^n$ and $\sum b_n x^n$ converge for $|x| < 3$, so does their sum; hence the radius of convergence R of $\sum (a_n + b_n)x^n$ is at least 3. If $R > 3$, then both $\sum b_n x^n$ and $\sum (a_n + b_n)x^n$ converge whenever $|x|$ is less than the lesser of 5 and R, so the same is true of their difference, $\sum a_n x^n$. But the lesser of 5 and R is greater than 3, contradicting the fact that $\sum a_n x^n$ has radius of convergence 3. Hence $R = 3$.

38 (a) The first four terms in the Maclaurin series for $(1 + x)^{-2}$ are $1 - 2x + 3x^2 - 4x^3 + \cdots$.

   (b) The nth-term in the Maclaurin series is $b_n = \frac{(-2)(-3)(-4)\cdots(-1-n)}{1 \cdot 2 \cdot 3 \cdots n} = (-1)^n(n+1)$ for $n = 0, 1, 2, \cdots$.

   (c) The Maclaurin series for $(1 - x)^{-2}$ is $1 + 2x + 3x^2 + 4x^3 + \cdots$.

40 From Equation (4) with $r = -1/2$, $(1 - x)^{-1/2} = 1 + \frac{1}{2}x + \frac{3}{8}x^2 + \cdots$ so that $m_0\left(1 - \frac{v^2}{c^2}\right)^{-1/2} =$

$m_0\left(1 + \frac{1}{2}\left(\frac{v^2}{c^2}\right) + \frac{3}{8}\left(\frac{v^2}{c^2}\right)^2 + \cdots\right) =$

$m_0\left(1 + \frac{1}{2}\cdot\frac{v^2}{c^2} + \frac{3}{8}\cdot\frac{v^4}{c^4} + \cdots\right)$, as Feynman said.

42 (a) Arc length $= \int_0^{\pi/2} \sqrt{\left(\frac{dx}{dt}\right)^2 + \left(\frac{dy}{dt}\right)^2}\, dt$

   $= \int_0^{\pi/2} \sqrt{(-a\sin t)^2 + (b\cos t)^2}\, dt$

   $= \int_0^{\pi/2} \sqrt{a^2 \sin^2 t + b^2(1 - \sin^2 t)}\, dt$

   $= \int_0^{\pi/2} b\sqrt{1 - [1 - (a/b)^2]\sin^2 t}\, dt$

   (b) $(1 - x)^{1/2} =$

   $1 - \frac{1}{2}x - \frac{1}{8}x^2 - \frac{1}{16}x^3 - \frac{5}{128}x^4 - \frac{7}{256}x^5 - \cdots$,

   so $E = \int_0^{\pi/2} \sqrt{1 - (k^2 \sin^2 \theta)}\, d\theta \approx$

   $\int_0^{\pi/2} \left[1 - \frac{1}{2}k^2 \sin^2 \theta - \frac{1}{8}k^4 \sin^4 \theta - \right.$

   $\left. \frac{1}{16}k^6 \sin^6 \theta - \frac{5}{128}k^8 \sin^8 \theta - \right.$

   $\left. \frac{7}{256}k^{10} \sin^{10} \theta \right] d\theta = \frac{\pi}{2} - \frac{1}{2}k^2 \frac{\pi}{4} -$

   $\frac{1}{8}k^4 \frac{3\pi}{16} - \frac{1}{16}k^6 \frac{5\pi}{32} - \frac{5}{128}k^8 \frac{35\pi}{256} -$

$$\frac{7}{256}k^{10} \frac{63\pi}{512} = \frac{\pi}{2} - \frac{\pi}{8}k^2 - \frac{3\pi}{128}k^4 -$$

$$\frac{5\pi}{512}k^6 - \frac{175\pi}{32,768}k^8 - \frac{441\pi}{131,072}k^{10}.$$

11.5 Manipulating Power Series

2 $\quad \dfrac{d}{dx}(e^x) = \dfrac{d}{dx}\left(1 + x + \dfrac{x^2}{2!} + \dfrac{x^3}{3!} + \dfrac{x^4}{4!} + \cdots\right)$

$$= 0 + 1 + \frac{2x}{2!} + \frac{3x^2}{3!} + \frac{4x^3}{4!} + \cdots$$

$$= 1 + x + \frac{x^2}{2!} + \frac{x^3}{3!} + \cdots = e^x$$

4 (a) Since $\dfrac{1}{1+x^3} = 1 - x^3 + x^6 - x^9 + \cdots$ for

$|x| < 1$, Theorem 3 implies that $\displaystyle\int_0^x \frac{dt}{1+t^3}$

$$= x - \frac{x^4}{4} + \frac{x^7}{7} - \frac{x^{10}}{10} + \cdots.$$

(b) $\displaystyle\int_0^{0.7} \frac{dt}{1+t^3}$

$$= (0.7) - \frac{(0.7)^4}{4} + \frac{(0.7)^7}{7} - \frac{(0.7)^{10}}{10} + \cdots$$

(c) The series in (b) is decreasing and alternating. If $|R_n| < |a_{n+1}| = \dfrac{(0.7)^{3n+4}}{3n+4} < 0.0005$,

then n must be at least 4. Thus, $(0.7) -$

$$\frac{(0.7)^4}{4} + \frac{(0.7)^7}{7} - \frac{(0.7)^{10}}{10} + \frac{(0.7)^{13}}{13} \approx$$

0.6496 estimates $\displaystyle\int_0^{0.7} \frac{dt}{1+t^3}$ to three places.

(d) Factor $1 + t^3$ into $(1+t)(1-t+t^2)$ and use partial fractions.

(e) $\dfrac{1}{2}\ln 17 - \dfrac{1}{4}\ln 79 + \dfrac{\pi}{12\sqrt{3}}$

$\qquad - \dfrac{1}{2\sqrt{3}}\tan^{-1}\dfrac{2}{5\sqrt{3}}$

6 $\quad 1 - \cos x = \dfrac{x^2}{2!} - \dfrac{x^4}{4!} + \dfrac{x^6}{6!} - \cdots$ and $\dfrac{1}{1-x^2}$

$= 1 + x^2 + x^4 + x^6 + \cdots$. By long division,

$$\frac{1 - \cos x}{1 - x^2} = \frac{\dfrac{x^2}{2} - \dfrac{x^4}{24} + \dfrac{x^6}{720} - \dfrac{x^8}{40320} + \cdots}{1 - x^2}$$

$$= \frac{1}{2}x^2 + \frac{11}{24}x^4 + \frac{331}{720}x^6 + \frac{2707}{8064}x^8 + \cdots.$$

8 By long division, $\dfrac{x}{\cos x} = \dfrac{x}{1 - \dfrac{x^2}{2} + \dfrac{x^4}{24} + \cdots}$

$$= x + \frac{1}{2}x^3 + \frac{5}{24}x^5 + \cdots.$$

10 By long division, $\dfrac{\sin 3x}{\sin 2x} =$

$$\frac{3x - \dfrac{9}{2}x^3 + \dfrac{81}{40}x^5 - \cdots}{2x - \dfrac{4}{3}x^3 + \dfrac{4}{15}x^5 - \cdots}$$

$= \dfrac{3}{2} - \dfrac{5}{4}x^2 - \dfrac{1}{48}x^4 - \cdots$, so that $\displaystyle\lim_{x \to 0}\dfrac{\sin 3x}{\sin 2x}$

$= \displaystyle\lim_{x \to 0}\left(\dfrac{3}{2} - \dfrac{5}{4}x^2 - \dfrac{1}{48}x^4 - \cdots\right) = \dfrac{3}{2}.$

11.5 Manipulating Power Series

12 Since $\sin x = x - \dfrac{x^3}{6} + \cdots$, $\dfrac{x}{\sin x} = 1 + \dfrac{x^2}{6} + \cdots$. Since $\ln(1+x) = x - \dfrac{x^2}{2} + \cdots$,

$\dfrac{x}{\ln(1+x)} = 1 + \dfrac{x}{2} + \cdots$. Hence,

$\displaystyle\lim_{x \to 0} \left(\dfrac{1}{\sin x} - \dfrac{1}{\ln(1+x)} \right)$

$= \displaystyle\lim_{x \to 0} \dfrac{1}{x} \left[\dfrac{x}{\sin x} - \dfrac{x}{\ln(1+x)} \right]$

$= \displaystyle\lim_{x \to 0} \dfrac{1}{x} \left[\left(1 + \dfrac{x^2}{6} + \cdots\right) - \left(1 + \dfrac{x}{2} + \cdots\right) \right]$

$= \displaystyle\lim_{x \to 0} \dfrac{1}{x} \left(-\dfrac{x}{2} + \cdots \right) = -\dfrac{1}{2}$.

14 Since $\sin x = x + \cdots$, $1 - \cos x =$

$1 - \left(1 - \dfrac{1}{2}x^2 + \cdots\right) = \dfrac{1}{2}x^2 + \cdots$, and $e^{x^3} - 1$

$= (1 + x^3 + \cdots) - 1 = x^3 + \cdots$,

$\displaystyle\lim_{x \to 0} \dfrac{\sin x\, (1 - \cos x)}{e^{x^3} - 1} = \lim_{x \to 0} \dfrac{(x + \cdots)\left(\dfrac{1}{2}x^2 + \cdots\right)}{x^3 + \cdots}$

$= \dfrac{1}{2}$.

16 $\displaystyle\int_0^{1/2} \sqrt{x}\, e^{-x}\, dx =$

$\displaystyle\int_0^{1/2} \left(x^{1/2} - x^{3/2} + \dfrac{x^{5/2}}{2} - \dfrac{x^{7/2}}{3!} + \cdots \right) dx =$

$\left. \left(\dfrac{2}{3}x^{3/2} - \dfrac{2}{5}x^{5/2} + \dfrac{2}{7}\dfrac{x^{7/2}}{2} - \dfrac{2}{9}\dfrac{x^{9/2}}{3!} + \cdots \right) \right|_0^{1/2}$ is a

decreasing, alternating series. The error in a partial sum is bounded by the first omitted term. We find

$\dfrac{2}{3}\left(\dfrac{1}{2}\right)^{3/2} - \dfrac{2}{5}\left(\dfrac{1}{2}\right)^{5/2} + \dfrac{1}{7}\left(\dfrac{1}{2}\right)^{7/2} - \dfrac{1}{27}\left(\dfrac{1}{2}\right)^{9/2} +$

$\dfrac{1}{132}\left(\dfrac{1}{2}\right)^{11/2} \approx 0.17615$ estimates $\displaystyle\int_0^{1/2} \sqrt{x}\, e^{-x}\, dx$ to

four decimal places.

18 $\tan^{-1} x = x - \dfrac{x^3}{3} + \dfrac{x^5}{5} - \dfrac{x^7}{7} + \cdots$

(a) The coefficient of x^{100} is 0 so that $0 = \dfrac{f^{(100)}(0)}{100!}$, that is, $f^{(100)}(0) = 0$.

(b) The coefficient of x^{101} is $\dfrac{1}{101}$ so that $\dfrac{1}{101} = \dfrac{f^{(101)}(0)}{101!}$, that is, $f^{(101)}(0) = 100!$.

20 This assertion is a generalization of the fact that

$\displaystyle\lim_{x \to a} [f(x) + g(x)] = \lim_{x \to a} f(x) + \lim_{x \to a} g(x)$.

22 (a) $\ln 3 = 2\left(\dfrac{1}{2} + \dfrac{(1/2)^3}{3} + \dfrac{(1/2)^5}{5} + \dfrac{(1/2)^7}{7} + \cdots \right) =$

$1 + \dfrac{(1/2)^2}{3} + \dfrac{(1/2)^4}{5} + \dfrac{(1/2)^6}{7} + \cdots$; to have

$\displaystyle\sum_{n=k}^{\infty} \dfrac{(1/2)^{2n}}{2n+1} < \sum_{n=k}^{\infty} \left(\dfrac{1}{2}\right)^{2n} = \sum_{n=k}^{\infty} \left(\dfrac{1}{4}\right)^n =$

$\dfrac{1}{3}\left(\dfrac{1}{4}\right)^{k-1} < 0.005$, then k must be at least 5.

Thus, $1 + \dfrac{(1/2)^2}{3} + \dfrac{(1/2)^4}{5} + \dfrac{(1/2)^6}{7} + \dfrac{(1/2)^8}{9}$

≈ 1.098 gives 1.10 as a two-place estimate for $\ln 3$.

(b) The first omitted term in part (a) is $\frac{(1/2)^{10}}{11} \approx$ 0.000088778. The error in part (a) is larger than this term.

24 Refer to the statement of Exercise 22.

(a) $\ln 2 = \ln(1 + 1)$
$$= 1 - \frac{1}{2} + \frac{1}{3} - \frac{1}{4} + \frac{1}{5} - \cdots$$

(b) $\ln 2 = -\ln\left(1 - \frac{1}{2}\right)$
$$= \frac{1}{2} + \frac{1}{2 \cdot 2^2} + \frac{1}{3 \cdot 2^3} + \frac{1}{4 \cdot 2^4} + \frac{1}{5 \cdot 2^5} + \cdots$$

(c) $\ln 2 = \ln\left(\frac{1 + 1/3}{1 - 1/3}\right) =$
$$2\left(\frac{1}{3} + \frac{1}{3 \cdot 3^3} + \frac{1}{5 \cdot 3^5} + \frac{1}{7 \cdot 3^7} + \cdots\right)$$

(d) Let $x_i = 1 + \frac{i}{n}$. Then $\ln 2 = \int_1^2 \frac{1}{x} dx$
$$= \lim_{n \to \infty} \sum_{i=1}^n f(x_i) \frac{i}{n} = \lim_{n \to \infty} \sum_{i=1}^n \frac{i}{n + i}.$$

(e) $x_{n+1} = x_n - 1 + 2e^{-x_n}$ and $\ln 2 = \lim_{n \to \infty} x_n$.

26 If $f(x) = a_0 + a_1 x + a_2 x^2 + a_3 x^3 + \cdots$, then $f(0) = a_0 = 0$. If $g(x) = b_0 + b_1 x + b_2 x^2 + b_3 x^3 + \cdots$, then $g(0) = b_0 = 0$, and $g'(x) = b_1 + 2b_2 x + 3b_3 x^2 + \cdots$, so $g'(0) = b_1 \neq 0$. Thus, $\lim_{x \to 0} \frac{f(x)}{g(x)}$

$$= \lim_{x \to 0} \frac{x(a_1 + a_2 x + a_3 x^2 + \cdots)}{x(b_1 + b_2 x + b_3 x^2 + \cdots)} = \frac{a_1}{b_1}, \text{ and}$$

$$\lim_{x \to 0} \frac{f'(x)}{g'(x)} = \lim_{x \to 0} \frac{a_1 + 2a_2 x + 3a_3 x^2 + \cdots}{b_1 + 2b_2 x + 3b_3 x^2 + \cdots} = \frac{a_1}{b_1}.$$

28 Since $\ln(x + 1) = x - \frac{x^2}{2} + \frac{x^3}{3} - \cdots$, $\ln(x + p)$

$$= \ln\left[p\left(\frac{x}{p} + 1\right)\right] = \ln p + \ln\left(\frac{x}{p} + 1\right)$$

$$= \ln p + \frac{x}{p} - \frac{x^2}{2p^2} + \frac{x^3}{3p^3} - \cdots \text{ and } \ln(y + q)$$

$$= \ln q + \frac{y}{q} - \frac{y^2}{2q^2} + \frac{y^3}{3q^3} - \cdots.$$

11.6 Complex Numbers

2 (a) $(2 + 3i)^2 = 4 + 12i + 9i^2 = -5 + 12i$

(b) $\frac{4}{3 - i} = \frac{4(3 + i)}{(3 - i)(3 + i)} = \frac{6}{5} + \frac{2}{5}i$

(c) $(1 + i)(3 - i) = 4 + 2i$

(d) $\frac{1 + 5i}{2 - 3i} = \frac{(1 + 5i)(2 + 3i)}{(2 - 3i)(2 + 3i)} = -1 + i$

4 (a)

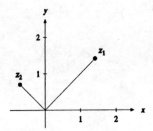

(b) $z_1 z_2 =$
$$2\left(\cos \frac{\pi}{4} + i \sin \frac{\pi}{4}\right) \cdot 1\left(\cos \frac{3\pi}{4} + i \sin \frac{3\pi}{4}\right)$$
$$= 2(\cos \pi + i \sin \pi) = -2$$

(c) $z_1 = 2\left(\frac{1}{\sqrt{2}} + \frac{1}{\sqrt{2}}i\right) = \sqrt{2} + \sqrt{2}i$, z_2
$$= 1\left(-\frac{1}{\sqrt{2}} + \frac{1}{\sqrt{2}}i\right) = -\frac{1}{\sqrt{2}} + \frac{1}{\sqrt{2}}i$$

(d) $z_1 z_2 = (\sqrt{2} + \sqrt{2}i)\left(-\frac{1}{\sqrt{2}} + \frac{1}{\sqrt{2}}i\right) =$

$-1 + 0i + i^2 = -2$

6 (a) $i^3 = i^2 i = -i$

(b) $i^4 = (i^2)^2 = 1$

(c) $i^5 = i^4 i = i$

(d) $i^{73} = (i^4)^{18} i = i$

8 (a)

Number	z^2	z^3	z^4	z^5	z^6
Magnitude	0.81	0.729	0.6561	0.59049	0.531441
Argument	$\pi/2$	$3\pi/4$	π	$5\pi/4$	$3\pi/2$

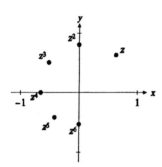

(b) Since the magnitude, $(0.9)^n$, approaches 0 as $n \to \infty$, so does z^n.

10 Note that $8 + 8\sqrt{3}i = 16\left(\cos\frac{\pi}{3} + i \sin\frac{\pi}{3}\right)$ if $z = r(\cos\theta + i \sin\theta)$ with $r > 0$, then $z^4 = 8 + 8\sqrt{3}i$ if and only if $r = 2$ and $4\theta = \frac{\pi}{3} + 2\pi n$, hence if and only if $r = 2$ and $\theta = \frac{\pi}{12} + \frac{n\pi}{2}$. The roots are therefore

$2(\cos\theta + i \sin\theta)$ with $\theta = \frac{\pi}{12}, \frac{7\pi}{12}, \frac{13\pi}{12}$, or $\frac{19\pi}{12}$.

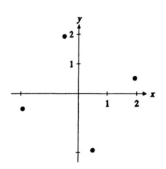

12 $P = 4\sqrt{2}\left(\cos\frac{\pi}{4} + i \sin\frac{\pi}{4}\right)$, so $P^{-1} = \frac{1}{4\sqrt{2}}\left(\cos\left(-\frac{\pi}{4}\right) + i \sin\left(-\frac{\pi}{4}\right)\right) = \frac{1}{8} - \frac{1}{8}i$.

14 (a) $x = \frac{-1 \pm \sqrt{1^2 - 4 \cdot 1 \cdot 1}}{2} = -\frac{1}{2} \pm \frac{\sqrt{3}}{2}i$

(b)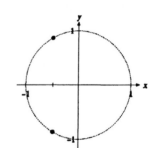

16 (a) $3\left(-\frac{1}{\sqrt{2}} + i\frac{1}{\sqrt{2}}\right) = -\frac{3}{\sqrt{2}} + \frac{3}{\sqrt{2}}i$

(b) $2\left(\frac{\sqrt{3}}{2} + i\frac{1}{2}\right) = \sqrt{3} + i$

(c) $10(-1 + i \cdot 0) = -10$

(d) $0.185 + 0.075i$

18 (a) $\cos\left(\frac{5\pi}{4} - \frac{3\pi}{4}\right) + i \sin\left(\frac{5\pi}{4} - \frac{3\pi}{4}\right) = i$

(b) $\dfrac{-\frac{1}{\sqrt{2}} - \frac{1}{\sqrt{2}}i}{-\frac{1}{\sqrt{2}} + \frac{1}{\sqrt{2}}i} = i$

20 (a) $3^2 - 4^2 i^2 = 25$

(b) $\dfrac{(3 + 5i)(-2 - i)}{(-2 + i)(-2 - i)} = -\dfrac{1}{5} - \dfrac{13}{5}i$

(c) $\dfrac{2 - i}{(2 + i)(2 - i)} = \dfrac{2}{5} - \dfrac{1}{5}i$

(d) $\cos\left(20 \cdot \dfrac{\pi}{12}\right) + i \sin\left(20 \cdot \dfrac{\pi}{12}\right) =$

$\cos\left(\dfrac{5\pi}{3}\right) + i \sin\left(\dfrac{5\pi}{3}\right) = \dfrac{1}{2} - \dfrac{\sqrt{3}}{2}i$

(e) $r^{-1}(\cos\theta - i\sin\theta)$

(f) $\text{Re}(r^{10}(\cos 10\theta + i\sin 10\theta)) = r^{10}\cos 10\theta$

(g) $\dfrac{3\left(\dfrac{\sqrt{3}}{2} + \dfrac{1}{2}i\right)(5 + 12i)}{(5 - 12i)(5 + 12i)}$

$= \dfrac{3}{338}[(5\sqrt{3} - 12) + (5 + 12\sqrt{3})i]$

22 (a)
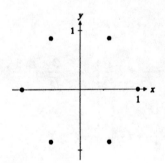

(b)
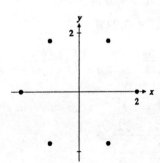

24 (a) $0 \le |z_1 + z_2| \le 2$

(b) $|z_1 z_2| = 1$

26 (a) $\arg \bar{z} = -\theta$

(b) $\arg(1/z) = -\theta$

28 (a) $z^2 = \dfrac{1}{2} + i + \dfrac{1}{2}i^2 = i$

(b) $z^2 = \left(\cos\dfrac{\pi}{4} + i\sin\dfrac{\pi}{4}\right)^2$

$= \cos\dfrac{\pi}{2} + i\sin\dfrac{\pi}{2} = i$

30 (a)
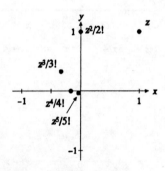

(b) Since $\left|\dfrac{z^n}{n!}\right| = \dfrac{|z|^n}{n!} = \dfrac{(\sqrt{2})^n}{n!} \to 0$ as $n \to \infty$,

$\dfrac{z^n}{n!} \to 0$ also.

32 (a) The chosen points should lie on the curve shown in the figure.

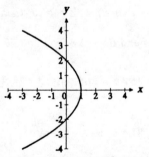

(b) $z^2 = (1 + it)^2 = (1 - t^2) + (2t)i$, so $x = 1 - t^2$, $y = 2t$.

(c) From part (b), the curve is the parabola $x = 1 - \left(\dfrac{y}{2}\right)^2 = 1 - \dfrac{y^2}{4}$.

34 (a)

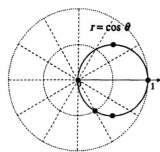

(b) In rectangular coordinates, five points on the graph are $(0, 0)$, $\left(\frac{1}{2}, \frac{1}{2}\right)$, $(1, 0)$, $\left(\frac{1}{2}, -\frac{1}{2}\right)$, $\left(\frac{1}{4}, -\frac{\sqrt{3}}{4}\right)$; and viewing them as complex numbers z: $0 + 0i$, $\frac{1}{2} + \frac{1}{2}i$, $1 + 0i$, $\frac{1}{2} - \frac{1}{2}i$, and $\frac{1}{4} - \frac{\sqrt{3}}{4}i$, the squares are z^2: $0 + 0i$, $0 + \frac{1}{2}i$, $1 + 0i$, $0 - \frac{1}{2}i$, and $-\frac{1}{8} - \frac{\sqrt{3}}{8}i$.

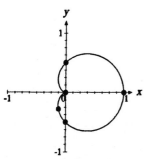

(c) Let (ρ, ϕ) be a point in polar coordinates that equals the square (as a complex number) of the point (r, θ), where $r = \cos \theta$. By DeMoivre's law, $\rho = r^2$ and $\phi = 2\theta$. Now (r, θ) can be written as $z = x + iy = r \cos \theta + ir \sin \theta = \cos^2 \theta + i \sin \theta \cos \theta$, so that z^2

$= \cos^4 \theta - \sin^2 \theta \cos^2 \theta + i \cdot 2 \sin \theta \cos^3 \theta$

$= \cos^2 \theta \cos 2\theta + i \cos^2 \theta \sin 2\theta$. Hence the rectangular coordinates of (ρ, ϕ) are $x = \cos^2 \theta \cos 2\theta$ and $y = \cos^2 \theta \sin 2\theta$, so $\rho^2 = x^2 + y^2 = \cos^4 \theta$, that is, $\rho = \cos^2 \theta = \frac{1}{2}(1 + \cos 2\theta)$. But $2\theta = \phi$, so $\rho = \frac{1}{2}(1 + \cos \phi)$. The graph is a cardioid.

36 The roots are $\dfrac{-i \pm \sqrt{i^2 - 4 \cdot 1 \cdot (3 - i)}}{2} =$

$\dfrac{-i \pm \sqrt{4i - 13}}{2}$. Numerically, they are about $0.274 + 1.324i$ and $-0.274 - 2.324i$.

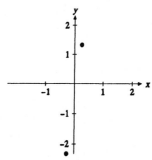

38 (a) By the quadratic formula, the roots of $ax^2 + bx + c = 0$ are $z_1 = \frac{1}{2a}\left(-b + i\sqrt{4ac - b^2}\right)$ and $z_2 = \frac{1}{2a}\left(-b - i\sqrt{4ac - b^2}\right)$. Clearly, $z_1 + z_2 = -\frac{b}{a}$ and $z_1 z_2 = \frac{1}{4a^2}[b^2 + (4ac - b^2)] = \frac{c}{a}$.

(b) $a(x - z_1)(x - z_2) = a(x^2 - (z_1 + z_2)x + z_1 z_2)$
$= a\left(x^2 + \frac{b}{a}x + \frac{c}{a}\right) = ax^2 + bx + c$.

(c) $\dfrac{1}{a(z_1-z_2)}\left[\dfrac{1}{x-z_1}-\dfrac{1}{x-z_2}\right]$

$=\dfrac{1}{a(z_1-z_2)}\left[\dfrac{x-z_2-x+z_1}{(x-z_1)(x-z_2)}\right]$

$=\dfrac{1}{a(z_1-z_2)}\left[\dfrac{z_1-z_2}{(x-z_1)(x-z_2)}\right]$

$=\dfrac{1}{a(x-z_1)(x-z_2)}=\dfrac{1}{ax^2+bx+c}.$

11.7 The Relation Between the Exponential and the Trigonometric Functions

2   $5e^{\pi i/4} = 5\left(\cos\dfrac{\pi}{4}+i\sin\dfrac{\pi}{4}\right)=\dfrac{5}{\sqrt{2}}+\dfrac{5}{\sqrt{2}}i$, so

$\text{Re}(5e^{\pi i/4})=\text{Im}(5e^{\pi i/4})=5/\sqrt{2}.$

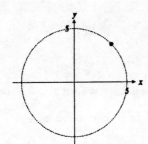

4   $e^{2+3i}=e^2(\cos 3+i\sin 3)$, so $\text{Re}(e^{2+3i})=e^2\cos 3$ and $\text{Im}(e^{2+3i})=e^2\sin 3.$

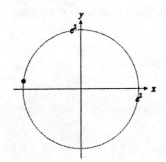

6   $2e^{\pi i}\cdot 3e^{-\pi i/3}=6e^{2\pi i/3}=6\left(\cos\dfrac{2\pi}{3}+i\sin\dfrac{2\pi}{3}\right)$

$=6\left(-\dfrac{1}{2}+\dfrac{\sqrt{3}}{2}i\right)=-3+3i\sqrt{3}.$ The real part is

-3; the imaginary part is $3\sqrt{3}.$

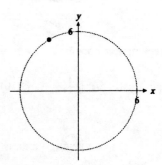

8   $3e^{i\pi/4}$

10   $7\left(\cos\dfrac{\pi}{3}+i\sin\dfrac{\pi}{3}\right)=7e^{i\pi/3}$

12   $e^{1+(9\pi/4)i}=e\cdot e^{i\pi/4}=e\left(\cos\dfrac{\pi}{4}+i\sin\dfrac{\pi}{4}\right)$

$=\dfrac{e}{\sqrt{2}}+\dfrac{e}{\sqrt{2}}i$

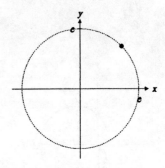

14 $e^{-1+(17\pi/6)i} = e^{-1}e^{5\pi i/6} = e^{-1}\left(\cos\dfrac{5\pi}{6} + i\sin\dfrac{5\pi}{6}\right)$

$= -\dfrac{\sqrt{3}}{2e} + \dfrac{i}{2e}$

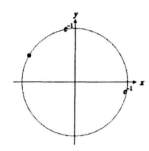

16 $\left|[e^{(a+ib)}]^n\right| = \left|e^{an}e^{ibn}\right| = e^{an} = (e^a)^n$. This approaches 0 if and only if $e^a < 1$, that is, if and only if $a < 0$.

18 Let $z = x + iy$. Then $e^z = e^x(\cos y + i\sin y) = -1$ if and only if $e^x = 1$ and $\cos y + i\sin y = -1$; that is, if and only if $x = 0$ and $y = \pi + 2\pi n$. Hence $e^z = -1$ if and only if $z = (2n+1)\pi i$ for some integer n.

20 (a)

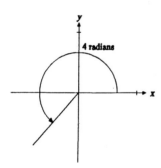

(b)

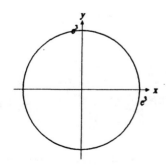

22 $1 + \cos\theta + \cos 2\theta + \cdots + \cos(n-1)\theta =$

$\operatorname{Re}(1 + e^{\theta i} + e^{2\theta i} + \cdots + e^{(n-1)\theta i}) = \operatorname{Re}\left(\dfrac{1 - e^{n\theta i}}{1 - e^{\theta i}}\right)$

$= \operatorname{Re}\left(\dfrac{(1-\cos n\theta) - i\sin n\theta}{(1-\cos\theta) - i\sin\theta}\right) =$

$\operatorname{Re}\left(\dfrac{1 - \cos\theta - \cos n\theta + \cos\theta\cos n\theta + \sin\theta\sin n\theta + i(\cdots)}{1 - 2\cos\theta + \cos^2\theta + \sin^2\theta}\right)$

$= \dfrac{1 - \cos\theta - \cos n\theta + \cos(n-1)\theta}{2 - 2\cos\theta}$

24 Let $z_1 = Ai$ and $z_2 = Bi$. Then $\cos(A+B) + i\sin(A+B) = e^{(A+B)i} = e^{Ai}e^{Bi} = (\cos A + i\sin A)(\cos B + i\sin B) = (\cos A\cos B - \sin A\sin B) + i(\sin A\cos B + \cos A\sin B)$. Hence $\cos(A+B) = \cos A\cos B - \sin A\sin B$ and $\sin(A+B) = \sin A\cos B + \cos A\sin B$.

26 Since $\displaystyle\sum_{n=0}^{\infty}\left|\dfrac{\sin n\theta}{n!}\right| < \sum_{n=0}^{\infty}\dfrac{1}{n!} = e$, $\displaystyle\sum_{n=0}^{\infty}\dfrac{\sin n\theta}{n!}$ converges by the absolute convergence test. To evaluate its sum, observe that $\displaystyle\sum_{n=0}^{\infty}\dfrac{\sin n\theta}{n!} =$

$\displaystyle\sum_{n=1}^{\infty}\dfrac{1}{n!}\operatorname{Im}(e^{in\theta}) = \operatorname{Im}\left(\sum_{n=0}^{\infty}\dfrac{1}{n!}e^{in\theta}\right) = \operatorname{Im}(e^{e^{i\theta}}) =$

$\operatorname{Im}(e^{\cos\theta + i\sin\theta}) =$

$\operatorname{Im}(e^{\cos\theta}[\cos(\sin\theta) + i\sin(\sin\theta)]) =$

$e^{\cos\theta}\sin(\sin\theta)$.

28 If $e^z = e^{z'}$, then $e^{z'-z} = 1$ so, by Exercise 17, $z' - z = 2\pi ni$ for some integer n. Hence, if L is one value of $\ln b$, all others are of the form $L + 2\pi ni$.

(a) Since $e^{\pi i} = -1$, $\ln(-1) = \pi i + 2\pi ni = (2n+1)\pi i$ for integers n. Hence $(-1)^i =$

$e^{i \ln(-1)} = e^{i(2n+1)\pi i} = e^{-(2n+1)\pi}$. That is, the values of $(-1)^i$ are $\cdots, e^{-3\pi}, e^{-\pi}, e^{\pi}, e^{3\pi}, e^{5\pi}, \cdots$.

(b) Let L be the real natural log of 10. Then $10^{1/2}$
$= e^{(1/2) \ln 10} = e^{(L + 2n\pi i)/2} = e^{L/2} e^{n\pi i} = \sqrt{10}(-1)^n$. The values of $10^{1/2}$ are $\sqrt{10}$ and $-\sqrt{10}$.

(c) Let L be the real natural logarithm of 10. Then $10^3 = e^{3 \ln 10} = e^{3(L+2n\pi i)} = e^{3L} e^{6n\pi i} = e^{3L} = 1000$. The only value of 10^3 is 1000.

30 Define $\sin z = \dfrac{e^{iz} - e^{-iz}}{2i}$ and $\cos z = \dfrac{e^{iz} + e^{-iz}}{2}$. Then $|\sin(-i)| = \left|\dfrac{e - e^{-1}}{2i}\right| = \dfrac{1}{2}(e - e^{-1}) \approx 1.1752 > 1$ and $|\cos(-i)| = \left|\dfrac{e + e^{-1}}{2}\right| = \dfrac{1}{2}(e + e^{-1}) \approx 1.5431 > 1$.

32 $\int x e^{ax} \cos bx \, dx = \int \text{Re}(x e^{ax} e^{ibx}) \, dx$

$= \text{Re}\left(\int x e^{(a+ib)x} \, dx\right) = \text{Re}\left(\dfrac{e^{(a+ib)x}[(a+ib)x - 1]}{(a+ib)^2}\right)$

$= \text{Re}\left(\dfrac{e^{ax}(\cos bx + i \sin bx)[(ax-1) + ibx](a-ib)^2}{(a+ib)^2(a-ib)^2}\right)$

$= \dfrac{e^{ax}}{(a^2+b^2)^2}[(a(a^2+b^2)x - a^2 + b^2) \cos bx + (b(a^2+b^2)x - 2ab) \sin bx]$

34 z can be written in the polar form $re^{i\phi}$, where $r = |z|$ and $\phi = \arg z$. Then $e^{i\theta} z = e^{i\theta} r e^{i\phi} = r e^{i(\theta + \phi)}$, which is a complex number whose magnitude is the same as that of z, but whose argument has increased by θ; that is, the new number is obtained by rotating z about the pole by θ radians.

11.S Review Exercises

2 (a) $f(x) = \int_0^x \dfrac{\sin t}{t} \, dt$

$= \int_0^x \dfrac{0 + t + 0t^2 - t^3/3! + 0t^4 + \cdots}{t} \, dt$

$= \int_0^x \left(1 - \dfrac{t^2}{3!} + \cdots\right) dt = \left(t - \dfrac{t^3}{3 \cdot 3!} + \cdots\right)\bigg|_0^x$

$= x - \dfrac{x^3}{3 \cdot 3!} + \cdots$

(b) $a_n = \dfrac{x^{2n+1} \cdot (-1)^n}{(2n+1)(2n+1)!}$ for $n = 0, 1, 2, \cdots$.

(c) $\int_0^1 \dfrac{\sin t}{t} \, dt = 1 - \dfrac{1}{3 \cdot 3!} + \dfrac{1}{5 \cdot 5!} - \dfrac{1}{7 \cdot 7!} + \cdots$ is a decreasing, alternating series. Since $\dfrac{1}{7 \cdot 7!} < 0.0005$, $1 - \dfrac{1}{3 \cdot 3!} + \dfrac{1}{5 \cdot 5!} \approx 0.946$ estimates $\int_0^1 \dfrac{\sin t}{t} \, dt$ to three decimal places.

4 Consider $\sum_{n=1}^{\infty} e^{-(n-1)(N+K)\alpha}$ as a geometric series with initial term $e^{-(1-1)(N+K)\alpha} = 1$ and ratio

$\dfrac{\exp(-n(N+K)\alpha)}{\exp(-(n-1)(N+K)\alpha)} = \exp[-(N+K)\alpha]$. Its sum is therefore $\dfrac{1}{1 - e^{-(N+K)\alpha}}$ (provided that

$e^{-(N+K)\alpha} < 1$). This establishes the result in the economics text.

6 (a) $f(x) = \sin^2 x, f^{(0)}(0) = 0; f^{(1)}(x) = 2 \sin x \cos x = \sin 2x$ and $f^{(1)}(0) = 0; f^{(2)}(x) = 2 \cos 2x$ and $f^{(2)}(0) = 2; f^{(3)}(x) = -4 \sin 2x$ and $f^{(3)}(0) = 0; f^{(4)}(x) = -8 \cos 2x$ and $f^{(4)}(0) = -8; f^{(5)}(x) = 16 \sin 2x$ and $f^{(5)}(0) = 0; f^{(6)}(x) = 32 \cos 2x$ and $f^{(6)}(0) = 32$. We finally have three nonzero terms:

$$\sin^2 x = \frac{2x^2}{2!} - \frac{8x^4}{4!} + \frac{32x^6}{6!} - \cdots$$

$$= x^2 - \frac{x^4}{3} + \frac{2x^6}{45} - \cdots.$$

(b) $\sin^2 x = \frac{1}{2}(1 - \cos 2x)$

$$= \frac{1}{2}\left(1 - 1 + \frac{4x^2}{2!} - \frac{16x^4}{4!} + \frac{64x^6}{6!} - \cdots\right)$$

$$= \frac{1}{2}\left(2x^2 - \frac{2x^4}{3} + \frac{4x^6}{45} - \cdots\right)$$

$$= x^2 - \frac{x^4}{3} + \frac{2x^6}{45} - \cdots$$

8 $P_3(x; 0) = 1 - x + x^2 - x^3$

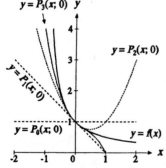

10 $P_3(x; 0) = x - x^3/6$

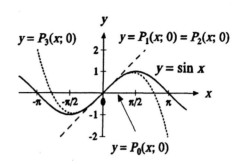

12 $|R_n(x; 0)| = \left|\frac{f^{(n+1)}(c_n)}{(n+1)!}x^{n+1}\right| = \frac{e^{c_n}}{(n+1)!}|x|^{n+1}$,

where c_n is between 0 and x. This is at most

$\frac{e^x x^{n+1}}{(n+1)!}$ for $x > 0$ and at most $\frac{e^0 |x|^{n+1}}{(n+1)!}$ for $x \leq 0$; $\lim_{n \to \infty} \frac{e^x x^{n+1}}{(n+1)!} = e^x \cdot 0 = 0$ and $\lim_{n \to \infty} \frac{e^0 |x|^{n+1}}{(n+1)!} = 1 \cdot 0 = 0$, so that $\lim_{n \to \infty} R_n(x; 0) = 0$.

14 For $f(x) = \sin 2x$, $f^{(n)}(x)$ is $\pm 2^n$ times $\sin 2x$ or $\cos 2x$. Thus, $|R_n(x; 0)| = \left|\frac{f^{(n+1)}(c_n)}{(n+1)!}x^{n+1}\right| <$

$\frac{2^{n+1}|x|^{n+1}}{(n+1)!} = \frac{(2|x|)^{n+1}}{(n+1)!} \to 0$ as $n \to \infty$.

16 See Exercise 4 in Sec. 11.2.

18 $e^{1/10} \approx 1 + \frac{1}{10} + \frac{1}{2!}\left(\frac{1}{10}\right)^2 = 1.105$. The error is

$$\sum_{n=3}^{\infty} \frac{1}{n! \, 10^n} < \sum_{n=3}^{\infty} \frac{1}{6 \cdot 10^n} = \frac{1/6000}{1 - 1/10} \approx$$

0.000185. (This bound can improved considerably.)

20 $\sin 28° = \sin \frac{7\pi}{45} \approx \frac{7\pi}{45} - \frac{1}{3!}\left(\frac{7\pi}{45}\right)^3 + \frac{1}{5!}\left(\frac{7\pi}{45}\right)^5$

≈ 0.469472879. The series is decreasing and alternating, so the error is less than the next term,

$\dfrac{1}{7!}\left(\dfrac{7\pi}{45}\right)^7 \approx 0.00000132.$

22. Since $\ln(1-x) = -x - \dfrac{x^2}{2} - \dfrac{x^3}{3} - \cdots$, the series converges to $-\ln(1-2x)$ for $-\dfrac{1}{2} < x < \dfrac{1}{2}$. $R = \dfrac{1}{2}$. Convergence is absolute for $|x| < \dfrac{1}{2}$ and conditional for $x = -\dfrac{1}{2}$.

24. The series converges absolutely to e^{x-3} for all x. $R = \infty$.

26. For $x \neq 0$, $\lim\limits_{n\to\infty} |(-n)^n x^n| = \infty$, so the series diverges for such x. It converges absolutely to 0 for $x = 0$. Hence $R = 0$.

28. This geometric series converges absolutely to
$$\dfrac{1}{1 - \dfrac{3x-2}{4}} = \dfrac{4}{6 - 3x} \text{ for } \left|\dfrac{3x-2}{4}\right| < 1; \text{ that is,}$$
for $\left|x - \dfrac{2}{3}\right| < \dfrac{4}{3}$. It diverges elsewhere. $R = \dfrac{4}{3}$.

30. Note that $\sum\limits_{n=1}^{\infty}(-2)^{-n} - \sum\limits_{n=1}^{10}(-2)^{-n} = \sum\limits_{n=11}^{\infty}(-2)^{-n}$.

(a) $\sum\limits_{n=11}^{\infty}(-2)^{-n}$ is a decreasing, alternating series, so it converges and its sum lies between every consecutive pair of partial sums. The first partial sum is the first term $(-2)^{-11} = -\dfrac{1}{2048}$ and the second partial sum is

$(-2)^{-11} + (-2)^{-12} = -\dfrac{1}{4096}$. The sum therefore lies between these two numbers.

(b) $\left|\sum\limits_{n=11}^{\infty}(-2)^{-n}\right| \leq \sum\limits_{n=11}^{\infty} 2^{-n} = \dfrac{2^{-11}}{1 - (1/2)} = 2^{-10}$
$= \dfrac{1}{1024}$

(c) $\sum\limits_{n=11}^{\infty}(-2)^{-n}$ is geometric with initial term $(-2)^{-11}$ and ratio $-\dfrac{1}{2}$; its sum is $\dfrac{(-2)^{-11}}{1 - (-1/2)}$
$= -\dfrac{1}{3072}.$

(d) For $|x| < 1$, $\sum\limits_{n=0}^{\infty}(-1)^n x^n$ converges to $f(x)$
$= (1+x)^{-1}$. Also, $P_{10}(x;0) = \sum\limits_{n=0}^{10}(-1)^n x^n$,
so $R_{10}(x;0) = f(x) - P_{10}(x;0)$
$= f(x) - \sum\limits_{n=0}^{10}(-1)^n x^n$
$= \sum\limits_{n=0}^{\infty}(-1)^n x^n - \sum\limits_{n=0}^{10}(-1)^n x^n$. In particular, the quantity given in the exercise equals $R_{10}\left(\dfrac{1}{2};0\right)$. By Theorem 1 in Sec. 11.2,
$R_{10}\left(\dfrac{1}{2};0\right) = \dfrac{f^{(11)}(c)}{11!}\left(\dfrac{1}{2} - 0\right)^{11} = \dfrac{f^{(11)}(c)}{2^{11} \cdot 11!}$ for some c between 0 and $\dfrac{1}{2}$. But, by an easy induction, $f^{(k)}(x) = (-1)^k k!(1+x)^{-(k+1)}$, so

$R_{10}\left(\frac{1}{2}; 0\right) = \frac{(-1)^{11} 11!(1 + c)^{-12}}{2^{11} \cdot 11!} =$

$-2^{-11}(1 + c)^{-12}$. Since c is between 0 and $\frac{1}{2}$,

$(1 + c)^{-12}$ is between 1 and $\frac{2^{12}}{3^{12}}$, so $R_{10}\left(\frac{1}{2}; 0\right)$

is between $-\frac{1}{2^{11}}$ and $-\frac{2}{3^{12}}$.

32 $f(x) = x^2 + x + 2, f^{(1)}(x) = 2x + 1, f^{(2)}(x) = 2$,
and $f^{(3)}(x) = f^{(4)}(x) = f^{(5)}(x) = \cdots = 0$; thus $f(5) =$
$32, f^{(1)}(5) = 11, f^{(2)}(5) = 2$, and $x^2 + x + 2 =$
$32 + 11(x - 5) + \frac{2}{2!}(x - 5)^2 = 32 + 11(x - 5)$
$+ (x - 5)^2$.

34 A Taylor series is a power series which can be used in place of its associated function. Since it "acts" like a polynomial, it's useful in estimating values of the function and integrals of the function.

36 By the absolute-ratio test,

$\lim_{n \to \infty} \left| \frac{2^{n+1}(x - 1)^{n+1}}{[1 + 1/(n + 1)]^{n+1}} \cdot \frac{(1 + 1/n)^n}{2^n(x - 1)^n} \right| =$

$\lim_{n \to \infty} \frac{(1 + 1/n)^n}{[1 + 1/(n + 1)]^{n+1}} \left| \frac{x - 1}{1/2} \right| = \frac{e}{e}\left| \frac{x - 1}{1/2} \right| < 1$,

so $-1 < \frac{x - 1}{1/2} < 1$ or $\frac{1}{2} < x < \frac{3}{2}$. The

radius of convergence is $\frac{1}{2}$. At $x = \frac{1}{2}$ or $x = \frac{3}{2}$,

the absolute value of the nth term approaches $\frac{1}{e}$,

so the series diverges by the nth-term test for divergence.

38 By the ratio test, $\lim_{n \to \infty} \left| \frac{(2n + 3)x^{n+1}}{(n + 1)!} \cdot \frac{n!}{(2n + 1)x^n} \right| =$

$\lim_{n \to \infty} \frac{2n + 3}{2n + 1} \cdot \frac{|x|}{n + 1} = 0$, so the radius of

convergence is infinite.

40 By the ratio test, $\lim_{n \to \infty} \left| \frac{(x - 1)^{n+1}}{\ln(n + 1)} \cdot \frac{\ln n}{(x - 1)^n} \right| =$

$\lim_{n \to \infty} \frac{\ln n}{\ln(n + 1)} |x - 1| = |x - 1| < 1$, so $-1 <$

$x - 1 < 1$ or $0 < x < 2$. The radius of
convergence is 1. At $x = 0$, the series is a
decreasing, alternating series and thus converges
conditionally. At $x = 2$, however, it diverges by
comparison with the harmonic series.

42 By the ratio test, $\lim_{n \to \infty} \left| \frac{(x + 3)^{n+1}}{(n + 1)^{1/3}} \cdot \frac{n^{1/3}}{(x + 3)^n} \right| =$

$\lim_{n \to \infty} \left(\frac{n}{n + 1}\right)^{1/3} |x + 3| = |x + 3| < 1$, so $-1 <$

$x + 3 < 1$ or $-4 < x < -2$. The radius of
convergence is 1. At $x = -2$, we get the p series
with $p = \frac{1}{3}$, which diverges, while at $x = -4$ it is
a decreasing, alternating series that converges
conditionally.

44 The general term is $a_n = \left(\frac{n + 1}{n + 3}\right)^{n^2} x^n$; by the root

test, we have $\lim_{n \to \infty} |a_n|^{1/n} = \lim_{n \to \infty} \left(\frac{n + 1}{n + 3}\right)^n |x| =$

$|x| \lim_{n \to \infty} \left(\frac{n + 1}{n + 3}\right)^n$. Now let $y = \left(\frac{n + 1}{n + 3}\right)^n$ and

consider $\lim_{n \to \infty} \ln y = \lim_{n \to \infty} n \ln\left(\frac{n + 1}{n + 3}\right) =$

$$\lim_{n\to\infty} \frac{\ln(n+1) - \ln(n+3)}{1/n} \underset{H}{=} \lim_{n\to\infty} \frac{\frac{1}{n+1} - \frac{1}{n+3}}{-1/n^2}$$

$$= \lim_{n\to\infty} \frac{2}{(n+1)(n+3)} \cdot \frac{n^2}{-1} = -2. \text{ Therefore}$$

$y \to e^{-2}$ and the result of the root test is $|x|e^{-2} < 1$, so $|x| < e^2$; the radius of convergence is e^2. At $x = \pm e^2$, the absolute value of the general term of the series is $\left(\frac{n+1}{n+3}\right)^{n^2} e^{2n}$; call this w and consider

$\ln w = n^2 \ln\left(\frac{n+1}{n+3}\right) + 2n$. Then $\lim_{n\to\infty} \ln w =$

$$\lim_{n\to\infty} \frac{\ln\left(\frac{n+1}{n+3}\right) + \frac{2}{n}}{1/n^2} \underset{H}{=} \lim_{n\to\infty} \frac{\frac{2}{(n+1)(n+3)} - \frac{2}{n^2}}{-2/n^3}$$

$$= \lim_{n\to\infty} \frac{4n^2 + 3n}{(n+1)(n+3)} = 4. \text{ Thus the } n\text{th term}$$

approaches e^4 in absolute value and the series diverges.

46 Since $|7 - 2| = |(-3) - 2| = 5$, the radius of convergence is 5.

48 $\dfrac{(e^{x^2} - 1)^2}{1 - x^2/2 - \cos x}$

$$= \frac{[(1 + x^2 + \cdots)^2 - 1]^2}{1 - x^2/2 - (1 - x^2/2 + x^4/24 - \cdots)}$$

$$= \frac{x^4 + \cdots}{-\frac{1}{24}x^4 + \cdots} \to -24 \text{ as } x \to 0.$$

50 (a) $\dfrac{\pi}{4} \approx 4\left(\dfrac{1}{5} - \dfrac{1}{3 \cdot 5^3} + \dfrac{1}{5 \cdot 5^5}\right) -$

$\left(\dfrac{1}{70} - \dfrac{1}{3 \cdot 70^3} + \dfrac{1}{5 \cdot 70^5}\right) +$

$\left(\dfrac{1}{99} - \dfrac{1}{3 \cdot 99^3} + \dfrac{1}{5 \cdot 99^5}\right) \approx 0.78540526$

(b) The error in using the first three terms of the alternating, decreasing series for $\tan^{-1} x$ is less than $\dfrac{1}{7}x^7$. Hence the error in (a) is at

most $4 \cdot \dfrac{1}{7}\left(\dfrac{1}{5}\right)^7 + \dfrac{1}{7}\left(\dfrac{1}{70}\right)^7 + \dfrac{1}{7}\left(\dfrac{1}{99}\right)^7 <$

0.0000074.

(c) Compute the tangent of the right-hand side using the identity $\tan(\tan^{-1} a + \tan^{-1} b) = \dfrac{a + b}{1 - ab}$ repeatedly.

52 (a) $\displaystyle\sum_{n=1}^{\infty} \frac{x^n}{n^2}$

(b) $\displaystyle\sum_{n=0}^{\infty} x^n$

(c) $\displaystyle\sum_{n=1}^{\infty} \frac{(-1)^n}{n} x^n$

54 (a) $\tan^{-1} x = x - \dfrac{x^3}{3} + \dfrac{x^5}{5} - \dfrac{x^7}{7} + \cdots$ so that

$\dfrac{f^{(99)}(0)}{99!} = -\dfrac{1}{99}$ or $f^{(99)}(0) = -98!$; $f^{(100)}(0)$

$= 0$; and $\dfrac{f^{(101)}(0)}{101!} = \dfrac{1}{101}$ or $f^{(101)}(0) = 100!$.

(b) $e^{x^2} = 1 + x^2 + \dfrac{x^4}{2!} + \dfrac{x^6}{3!} + \dfrac{x^8}{4!} + \cdots$ so that

11.S Summary: Review Exercises

$f^{(99)}(0) = 0$; $f^{(100)}(0) = \dfrac{100!}{50!}$; and $f^{(101)}(0) = 0$.

56. (a) $e^{-x^2} = 1 - x^2 + \dfrac{x^4}{2!} - \dfrac{x^6}{3!} + \dfrac{x^8}{4!} - \cdots$

    (b) $\displaystyle\int_0^1 e^{-x^2}\, dx$

    $= \displaystyle\int_0^1 \left(1 - x^2 + \dfrac{x^4}{2} - \dfrac{x^6}{6} + \cdots\right) dx$

    $= \left(x - \dfrac{x^3}{3} + \dfrac{x^5}{10} - \dfrac{x^7}{42} + \cdots\right)\Big|_0^1$

    $= 1 - \dfrac{1}{3} + \dfrac{1}{10} - \dfrac{1}{42} + \cdots$

    (c) Part (b) is a decreasing, alternating series with nth term $\dfrac{(-1)^n}{(2n+1)n!}$ for $n = 0, 1, 2, \cdots$. If $\dfrac{1}{(2n+1)n!} < 0.0005$, then n must be at least 6. Thus,

    $1 - \dfrac{1}{3} + \dfrac{1}{5\cdot 2!} - \dfrac{1}{7\cdot 3!} + \dfrac{1}{9\cdot 4!} - \dfrac{1}{11\cdot 5!} \approx$

    0.747 estimates $\displaystyle\int_0^1 e^{-x^2}\, dx$ to three decimal places.

    (d) For $f(x) = e^{-x^2}$, $M_4 \le 12$, so that if the error $\dfrac{(b-a)M_4 h^4}{180} = \dfrac{1}{15n^4} < 0.0005$, then n must be at least 4. The Simpson estimate with $n = 4$ is $\dfrac{1}{12}\left[f(0) + 4f\left(\dfrac{1}{4}\right) + 2f\left(\dfrac{1}{2}\right) + 4f\left(\dfrac{3}{4}\right) + f(1)\right] \approx 0.747$. It estimates $\displaystyle\int_0^1 e^{-x^2}\, dx$ to three decimal places.

58. (a) Since $f^{(2)}(a) \ne 0$ and $f^{(2)}$ is continuous, we can choose h sufficiently small so that $f^{(2)}(x)$ is either always positive or negative for x in the interval $[a - h, a + h]$. In either case, $f^{(1)}$ will be one-to-one on the interval because its derivative does not change sign. Now suppose that $f(a + h) = f(a) + hf'(a + \theta h)$ and $f(a + h) = f(a) + hf'(a + \phi h)$, where θ and ϕ are both in $[0, 1]$. By subtracting these equations we obtain $f'(a + \theta h) = f'(a + \phi h)$. Since f' is one-to-one, it follows that $\theta = \phi$. θ is unique.

    (b) By (a), $hf'(a + \theta h) = f(a + h) - f(a)$, where $0 \le \theta \le 1$. However, we also have $f(a + h)$
    $= f(a) + f'(a)h + \dfrac{1}{2}f''(c)h^2$ for some value of c between a and $a + h$, so $hf'(a + \theta h) = f'(a)h + \dfrac{1}{2}f''(c)h^2$. Dividing by h, we have $f'(a + \theta h) = f'(a) + \dfrac{h}{2}f''(c)$. But we may also write $f'(a + \theta h) = f'(a) + f''(d)\theta h$ for some value of d between a and $a + \theta h$. Observe that $c \to a$ and $d \to a$ as $h \to 0$. Hence $f'(a) + \dfrac{h}{2}f''(c) = f'(a) + f''(d)\theta h$, so $\dfrac{h}{2}f''(c) = f''(d)\theta h$ and $\theta = \dfrac{(h/2)f''(c)}{f''(d)h} = \dfrac{f''(c)}{2f''(d)}$. As $h \to 0$, we have $\theta \to \dfrac{f''(a)}{2f''(a)} = \dfrac{1}{2}$, as claimed.

60 $b^2/2$ is correct. In the second argument, the approximation used for e^{-b} is not good enough; if we take one more term of its Maclaurin series, we find $\int_0^b xe^{-x}\,dx = 1 - e^{-b}(1 + b)$

$\approx 1 - \left(1 - b + \frac{1}{2}b^2\right)(1 + b) = \frac{1}{2}b^2 - \frac{1}{2}b^3 \approx \frac{1}{2}b^2.$

62 Since $f^{(4)}(x) = 0$, f must be the polynomial with Maclaurin series $3 + 2x + \frac{5}{2}x^2 + \frac{1}{12}x^3$.

64 (a) $f''(0) = -f(0) = -A$
(b) $f'''(0) = -f'(0) = -B$
(c) $f^{(n)}(0) = (-1)^{n/2} f(0) = (-1)^{n/2} A$ for n even and $f^{(n)}(0) = (-1)^{(n-1)/2} f'(0) = (-1)^{(n-1)/2} B$ for n odd.
(d) The Maclaurin series for $f(x)$ is $A + Bx - \frac{A}{2!}x^2 - \frac{B}{3!}x^3 + \frac{A}{4!}x^4 + \frac{B}{5!}x^5 - \frac{A}{6!}x^6 - \frac{B}{7!}x^7 + \cdots$

(e) Since f and f' are differentiable, they are continuous. For $0 \leq t \leq x$, let N_1 be the maximum value of $|f(t)|$ and N_2 the maximum value of $|f'(t)|$; let M be the larger of N_1 and N_2. Then $|R_n(x; 0)| = \left|\frac{f^{(n+1)}(c)}{(n+1)!}x^{n+1}\right| \leq$

$M \cdot \frac{|x|^{n+1}}{(n+1)!} \to M \cdot 0 = 0$ as $n \to \infty$.

(f) $f(x) = A\left(1 - \frac{x^2}{2!} + \frac{x^4}{4!} - \cdots\right) +$

$B\left(x - \frac{x^3}{3!} + \frac{x^5}{5!} - \cdots\right) = A\cos x + B\sin x.$

(g) $f(x) = A\cos x + B\sin x$, so $f'(x) = -A\sin x + B\cos x$, and $f''(x) = -A\cos x - B\sin x = -f(x)$.

66 $\int_0^1 \frac{\sin^{-1} x}{\sqrt{1-x^2}}\,dx = \frac{1}{2}(\sin^{-1} x)^2\Big|_0^1 = \frac{\pi^2}{8}$

68 $\sin^{-1} x = \int_0^x \frac{1}{\sqrt{1-t^2}}\,dt$

$= \int_0^x \left[1 + \frac{1}{2}t^2 + \frac{1\cdot 3}{2\cdot 4}t^4 + \frac{1\cdot 3\cdot 5}{2\cdot 4\cdot 6}t^6 + \cdots\right]dt$

$= x + \frac{1}{2\cdot 3}x^3 + \frac{1\cdot 3}{2\cdot 4\cdot 5}x^5 + \frac{1\cdot 3\cdot 5}{2\cdot 4\cdot 6\cdot 7}x^7 + \cdots$ for $|x| < 1$.

70 Using the results of Exercises 68, 69, and 45 (of the Chapter 10 Review Exercises), we have

$\int_0^1 \frac{\sin^{-1} x}{\sqrt{1-x^2}}\,dx$

$= \int_0^1 \frac{x + \frac{1}{6}x^3 + \frac{3}{40}x^5 + \frac{5}{112}x^7 + \cdots}{\sqrt{1-x^2}}\,dx =$

$\int_0^{\pi/2}\left[\sin\theta + \frac{1}{6}\sin^3\theta + \frac{3}{40}\sin^5\theta + \frac{5}{112}\sin^7\theta + \cdots\right]d\theta$

$= \frac{1}{1^2} + \frac{1}{3^2} + \frac{1}{5^2} + \frac{1}{7^2} + \cdots.$

12 Vectors

12.1 The Algebra of Vectors

2

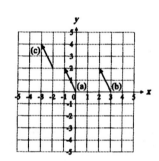

4 (a) The head is located at (1, 1, 1).
(b) The head is located at (3, 4, 5).

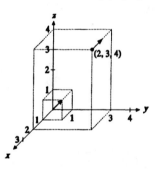

6 (a) The head is located at (0, 1, 1).
(b) The head is located at (−1, 0, 0).

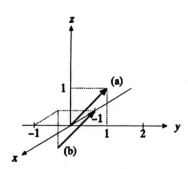

8

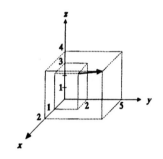

$\overrightarrow{PQ} = i + 3j + k$

$|\overrightarrow{PQ}| = \sqrt{1^2 + 3^2 + 1^2} = \sqrt{11}$

10

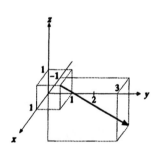

$\overrightarrow{PQ} = -2i + 2j - 3k$

$|\overrightarrow{PQ}| = \sqrt{(-2)^2 + 2^2 + (-3)^2} = \sqrt{17}$

12 (a) $A = -\frac{\sqrt{2}}{2}i - \frac{\sqrt{2}}{2}j$

(b) $A = -2i$

(c) $A = 2i + 2j$

(d) $A = -\frac{1}{2}j$

14 (a)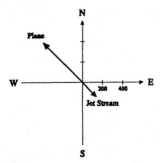

(c) 350 mph

16 (a) $A + B = \langle 5/2, 10/3, -1/6 \rangle$
$A - B = \langle -3/2, -8/3, 1/2 \rangle$

(b) $A + B = i + 8j + 10k$
$A - B = 3i - 2j - 2k$

18 (a) $c = 2$
(b) $c = -1$
(c) $c = 0$
(d) $c = 1/11$

20 (a) 12
(b) 18
(c) 1
(d) 60
(e) 0

22 (a) $u = \frac{1}{3}(2i - 2j + k)$

24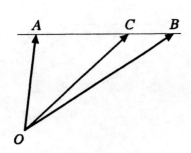

To show $C = \frac{1}{3}A + \frac{2}{3}B$, we first recognize that

$\overrightarrow{AC} = 2\overrightarrow{CB}$, $\overrightarrow{AC} = C - A$ and $\overrightarrow{CB} = B - C$.

Since $\overrightarrow{AC} = 2\overrightarrow{CB}$, $C - A = 2(B - C)$ and

therefore $C = \frac{1}{3}A + \frac{2}{3}B$.

26 $-2i + 6j - 12k = -2(i - 3j + 6k)$

28 (a) 8
(b) 2

30 (a)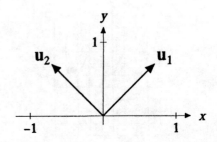

Note that $\|u_1\| = 1$ and $\|u_2\| = 1$, so u_1

and u_2 are unit vectors. u_1 has slope $\frac{\sqrt{2}/2}{\sqrt{2}/2} =$

1 and u_2 has slope $\frac{-\sqrt{2}/2}{\sqrt{2}/2} = -1$, so the

vectors are perpendicular.

(b) $i = \frac{1}{\sqrt{2}}u_1 - \frac{1}{\sqrt{2}}u_2$

(c) $j = \frac{1}{\sqrt{2}}u_1 + \frac{1}{\sqrt{2}}u_2$

(d) $-2i + 3j = \frac{1}{\sqrt{2}}u_1 + \frac{5}{\sqrt{2}}u_2$

32 (a)

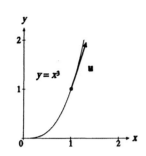

(b) $\mathbf{u} = \dfrac{1}{\sqrt{10}}\mathbf{i} + \dfrac{3}{\sqrt{10}}\mathbf{j}$

34

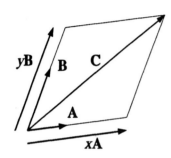

The parallelogram in the figure shows how $x\mathbf{A}$ and $y\mathbf{B}$ could be found so $\mathbf{C} = x\mathbf{A} + y\mathbf{B}$.

36 (a) Let Q_A be the midpoint of BC; from Exercise 24, $\overrightarrow{OP} = \dfrac{1}{3}\mathbf{A} + \dfrac{2}{3}\overrightarrow{OQ_A}$. By Exercise 23,

$\overrightarrow{OQ_A} = \dfrac{1}{2}(\mathbf{B} + \mathbf{C})$, so $\overrightarrow{OP} =$

$\dfrac{1}{3}\mathbf{A} + \dfrac{2}{3} \cdot \dfrac{1}{2}(\mathbf{B} + \mathbf{C}) = \dfrac{1}{3}(\mathbf{A} + \mathbf{B} + \mathbf{C})$.

(b) Let Q_B be the midpoint of AC and Q_C the midpoint of AB. Then by the argument of (a), if P_B is the point on BQ_B twice as from B as from Q_B, then $\overrightarrow{OP_B} = \dfrac{1}{3}(\mathbf{A} + \mathbf{B} + \mathbf{C})$, hence $P = P_B$. By the same argument, if P_C is the point on CQ_C twice as far from C as from Q_C,

then $\overrightarrow{OP_C} = \dfrac{1}{3}(\mathbf{A} + \mathbf{B} + \mathbf{C})$ and so $P = P_B$ $= P_C$. That is, the three median lines all intersect at P.

38 In the quadrilateral $ABCD$, let P, Q, R, and S be the midpoints of AB, BC, CD, and DA,

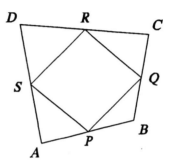

respectively. Considering triangle ABC, we see, by Example 8, that PQ is parallel to AC (and half as long). Considering triangle ACD, RS is parallel to AC (and half its length). Thus PQ and RS are parallel to one another (and of equal length). Similarly considering triangles ABD and BCD, PS and QR are each parallel to BD (and half its length), so they are parallel to one another (and of equal length). Therefore $PQRS$ is a parallelogram.

40 The equality holds when $x_1 = kx_2$ and $y_1 = ky_2$ for some nonnegative real number k or when $x_2 = y_2 = 0$. Let $\mathbf{A} = \langle x_1, y_1 \rangle$ and $\mathbf{B} = \langle x_2, y_2 \rangle$. By Exercise 39, we have $\|\mathbf{A} + \mathbf{B}\| \leq \|\mathbf{A}\| + \|\mathbf{B}\|$; that is, $\sqrt{(x_1 + x_2)^2 + (y_1 + y_2)^2} \leq \sqrt{x_1^2 + y_1^2} + \sqrt{x_2^2 + y_2^2}$. Squaring both sides, subtracting $x_1^2 + x_2^2 + y_1^2 + y_2^2$, and dividing by 2 gives the result.

12.2 Projections

2 (a)

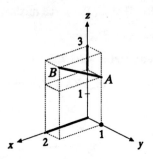

(b) $\|\text{proj}_i(2i + k)\| = 2$
$\|\text{proj}_j(2i + k)\| = 0$
$\|\text{proj}_k(2i + k)\| = 1$

(c) $|\overrightarrow{AB}| = \sqrt{5}$

4

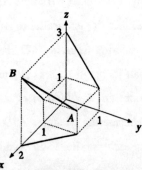

$\|\text{proj of } i - j + 2k \text{ on } xy \text{ plane}\| = \sqrt{2}$
$\|\text{proj of } i - j + 2k \text{ on } xz \text{ plane}\| = \sqrt{5}$
$\|\text{proj of } i - j + 2k \text{ on } yz \text{ plane}\| = \sqrt{5}$

6 (b) $|\overrightarrow{AB}| = \sqrt{5}$

8 (b) $5 = \cos\dfrac{\pi}{3}(\text{Area of } U) = \dfrac{1}{2}(\text{Area of } U)$, so

Area of $U = 10$.

10 (a)

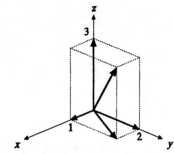

(b) $\text{proj}_i(i + 2j + 3k) = i$
$\text{proj}_j(i + 2j + 3k) = 2j$
proj of $i + 2j + 3k$ on xy plane $= i + 2j$

12 (a),(b)

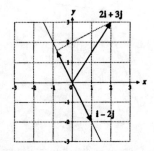

(c) $x = -4/5, y = 8/5$

14 (a) $-j$
(b) $-k$
(c) $-k$
(d) $-j - k$
(e) $2i - k$

16 $\text{proj}_B A = 0$ if B is orthogonal to A.

18 In order for $\|\text{proj}_B A\|$ to equal $\|A\|$, A must be parallel to B.

20 $\text{proj}_{-i}(i + 3j + k) = i$
$\text{orth}_{-i}(i + 3j + k) = 3j + k$

12.2 Projections

22 (a)

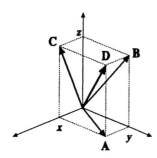

(b) $A = x\mathbf{i} + y\mathbf{j}$
$B = y\mathbf{j} + z\mathbf{k}$
$C = x\mathbf{i} + z\mathbf{k}$

(c) $D = \frac{1}{2}(A + B + C)$

24 With notation as in Exercise 22, we have $d = \sqrt{x^2 + y^2}$, $e = \sqrt{y^2 + z^2}$, $f = \sqrt{x^2 + z^2}$, and the length of the line segment is $\|D\| = \sqrt{\frac{1}{2}(d^2 + e^2 + f^2)}$.

12.3 The Dot Product of Two Vectors

2 $2 \cdot 3 \cos \frac{3\pi}{4} = -3\sqrt{2}$

4 $A \cdot B = 0$

6 $A \cdot B = (0.3)(2) + (0.5)(-1.5) = 0.6 - 0.75 = -0.15$

8 $A \cdot B = (1)(2) + (1)(3) + (1)(-5) = 2 + 3 - 5 = 0$

10 (a)

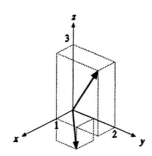

(c) The dot product is 0. Therefore, the vectors are perpendicular.

12 (a)

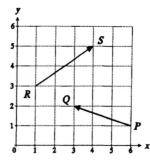

(c) $\cos \theta = \frac{(-3)(3) + (1)(2)}{\sqrt{(-3)^2 + 1^2}\sqrt{3^2 + 2^2}} = \frac{-7}{\sqrt{130}}$

(c) $\theta \approx 2.2318$ radians $\approx 127.9°$

14 $\theta = \cos^{-1} \frac{0}{\sqrt{11}\sqrt{54}} = \cos^{-1} 0 = \pi/2$.

16 $\overrightarrow{AB} = \langle 0, -2, 6 \rangle$ and $\overrightarrow{CD} = \langle 2, 2, 1 \rangle$.

$\theta = \cos^{-1} \frac{2}{\sqrt{40}\sqrt{9}} \approx 1.4652$ radians $\approx 83.9°$

18 The component of $2\mathbf{i} - 4\mathbf{j} + 5\mathbf{k}$ on $3\mathbf{i} - \mathbf{j} - \sqrt{2}\mathbf{k}$
is $(2\mathbf{i} - 4\mathbf{j} + 5\mathbf{k}) \cdot \frac{3\mathbf{i} - \mathbf{j} - \sqrt{2}\mathbf{k}}{\|3\mathbf{i} - \mathbf{j} - \sqrt{2}\mathbf{k}\|} =$

$(2\mathbf{i} - 4\mathbf{j} + 5\mathbf{k}) \cdot \frac{3\mathbf{i} - \mathbf{j} - \sqrt{2}\mathbf{k}}{2\sqrt{3}} = \frac{10 - 5\sqrt{2}}{2\sqrt{3}}$ and the projection is $\dfrac{10-5\sqrt{2}}{2\sqrt{3}}\left(\dfrac{3\mathbf{i}-\mathbf{j}-\sqrt{2}\mathbf{k}}{2\sqrt{3}}\right)$

$= \dfrac{10-5\sqrt{2}}{12}(3\mathbf{i}-\mathbf{j}-\sqrt{2}\mathbf{k})$.

20 (b) Use $(2, -1)$ and $(6, 1)$ to create the vector $4\mathbf{i} + 2\mathbf{j}$. The absolute value of the component of $-4\mathbf{i} + 5\mathbf{j}$ on this vector is the desired length. The component is $(-4\mathbf{i} + 5\mathbf{j}) \cdot \dfrac{4\mathbf{i} + 2\mathbf{j}}{2\sqrt{5}}$

$= \dfrac{-6}{2\sqrt{5}} = -\dfrac{3}{\sqrt{5}}$, so the length is $\dfrac{3}{\sqrt{5}}$.

22 (a) The vector $\mathbf{C}$ is just the projection of $3\mathbf{i} + 4\mathbf{j}$ on $2\mathbf{i} - \mathbf{j}$, which is $(3\mathbf{i} + 4\mathbf{j}) \cdot \dfrac{2\mathbf{i}-\mathbf{j}}{\sqrt{5}}\left(\dfrac{2\mathbf{i}-\mathbf{j}}{\sqrt{5}}\right)$

$= \dfrac{2}{5}(2\mathbf{i} - \mathbf{j}) = \dfrac{4}{5}\mathbf{i} - \dfrac{2}{5}\mathbf{j}$. Then $\mathbf{D}$ is just the orthogonal vector component, which equals

$3\mathbf{i} + 4\mathbf{j} - \left(\dfrac{4}{5}\mathbf{i} - \dfrac{2}{5}\mathbf{j}\right) = \dfrac{11}{5}\mathbf{i} + \dfrac{22}{5}\mathbf{j}$.

24 $2\mathbf{B}$ and $-\mathbf{B}$ are parallel to $\mathbf{B}$, so the projection of $\mathbf{A}$ on those two vectors is the same as its projection on $\mathbf{B}$: $\mathbf{i} - \mathbf{j} + 4\mathbf{k}$.

26 For $\text{comp}_\mathbf{B} \mathbf{A}$ to equal $\text{comp}_\mathbf{A} \mathbf{B}$, we must have $\mathbf{A} \cdot \dfrac{\mathbf{B}}{|\mathbf{B}|} = \mathbf{B} \cdot \dfrac{\mathbf{A}}{|\mathbf{A}|}$. Since $\mathbf{A} \cdot \mathbf{B} = \mathbf{B} \cdot \mathbf{A}$, this happens if and only if $\|\mathbf{A}\| = \|\mathbf{B}\|$.

28 (a) $(3\mathbf{i} + 4\mathbf{j} + 2\mathbf{k}) \cdot \dfrac{\mathbf{i} + \mathbf{j} + 3\mathbf{k}}{\sqrt{11}} = \dfrac{13}{\sqrt{11}}$

(b) $(3\mathbf{i} + 4\mathbf{j} + 2\mathbf{k}) \cdot \dfrac{-\mathbf{i} - \mathbf{j} - 3\mathbf{k}}{\sqrt{11}} = -\dfrac{13}{\sqrt{11}}$

(c) $(3\mathbf{i} + 4\mathbf{j} + 2\mathbf{k}) \cdot \dfrac{2\mathbf{i} + 2\mathbf{j} + 6\mathbf{k}}{2\sqrt{11}} = \dfrac{13}{\sqrt{11}}$

30 $\text{proj}_\mathbf{B} \mathbf{A} = (\mathbf{i} + 2\mathbf{j} + 3\mathbf{k}) \cdot \dfrac{\mathbf{i} + \mathbf{j}}{\sqrt{2}}\left(\dfrac{\mathbf{i} + \mathbf{j}}{\sqrt{2}}\right) =$

$\dfrac{3}{2}(\mathbf{i} + \mathbf{j}) = \dfrac{3}{2}\mathbf{i} + \dfrac{3}{2}\mathbf{j}$, so $\text{orth}_\mathbf{B} \mathbf{A} = \mathbf{i} + 2\mathbf{j} + 3\mathbf{k}$

$- \left(\dfrac{3}{2}\mathbf{i} + \dfrac{3}{2}\mathbf{j}\right) = -\dfrac{1}{2}\mathbf{i} + \dfrac{1}{2}\mathbf{j} + 3\mathbf{k}$.

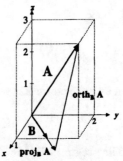

32 To find the distance of $P = (1, 2, -3)$ from the line through $Q = (2, 1, 4)$ and $R = (1, 5, -2)$, let $\mathbf{A} = \overrightarrow{RP} = -3\mathbf{j} - \mathbf{k}$ and $\mathbf{B} = \overrightarrow{RQ} = \mathbf{i} - 4\mathbf{j} + 6\mathbf{k}$. Then the distance we want is $\|\text{orth}_\mathbf{B} \mathbf{A}\|$. Now $\text{orth}_\mathbf{B} \mathbf{A} = \mathbf{A} - \text{proj}_\mathbf{B} \mathbf{A} = -3\mathbf{j} - \mathbf{k} -$

$(-3\mathbf{j} - \mathbf{k}) \cdot \dfrac{\mathbf{i} - 4\mathbf{j} + 6\mathbf{k}}{\sqrt{53}}\left(\dfrac{\mathbf{i} - 4\mathbf{j} + 6\mathbf{k}}{\sqrt{53}}\right)$

$= -3\mathbf{j} - \mathbf{k} - \dfrac{6}{53}(\mathbf{i} - 4\mathbf{j} + 6\mathbf{k})$

$= -\dfrac{6\mathbf{i} + 135\mathbf{j} + 89\mathbf{k}}{53}$, so the distance is $\dfrac{\sqrt{26{,}182}}{53}$

≈ 3.053.

34 $\mathbf{i} - \mathbf{j} - 2\mathbf{k}$ is only one of many possibilities.

36 $\overrightarrow{AF} \cdot \overrightarrow{BD} = \langle 0, 2, 2 \rangle \cdot \langle -2, -2, 2 \rangle = 0$, the vectors are perpendicular, and the cosine of the angle between them is $\cos\dfrac{\pi}{2} = 0$.

12.3 The Dot Product of Two Vectors

38 $\overrightarrow{MD} \cdot \overrightarrow{MF} = \langle -1, -2, 2 \rangle \cdot \langle -1, 0, 2 \rangle = 5$, $\|\overrightarrow{MD}\| = 3$, and $\|\overrightarrow{MF}\| = \sqrt{5}$, so $\cos\theta = \dfrac{5}{3\sqrt{5}} = \dfrac{\sqrt{5}}{3}$.

40 Let $P = (1, 2, 3)$, $Q = (1, 4, 2)$, and $R = (2, 1, -4)$. Since $\overrightarrow{PQ} = \langle 0, 2, -1 \rangle$ is perpendicular to $\overrightarrow{QR} = \langle 1, -3, -6 \rangle$, the distance from P to the line through Q and R is just the length of $\overrightarrow{PQ}$, $\sqrt{5}$.

42 $\mathbf{u}_1 = \cos\theta_1\,\mathbf{i} + \sin\theta_1\,\mathbf{j}$ and $\mathbf{u}_2 = \cos\theta_2\,\mathbf{i} + \sin\theta_2\,\mathbf{j}$ are the unit vectors pointing in the θ_1 and θ_2 directions, respectively. The angle between $\mathbf{u}_1$ and $\mathbf{u}_2$ is therefore $\theta_1 - \theta_2$ (where we assume $\theta_1 > \theta_2$, without any loss of generality). But the cosine of the angle between $\mathbf{u}_1$ and $\mathbf{u}_2$ is also given by $\dfrac{\mathbf{u}_1 \cdot \mathbf{u}_2}{|\mathbf{u}_1||\mathbf{u}_2|} = \mathbf{u}_1 \cdot \mathbf{u}_2$

$= \langle \cos\theta_1, \sin\theta_1 \rangle \cdot \langle \cos\theta_2, \sin\theta_2 \rangle$

$= \cos\theta_1 \cos\theta_2 + \sin\theta_1 \sin\theta_2$.

44 Let the vertices of the tetrahedron be O, A, B, and C, and let $\mathbf{A} = \overrightarrow{OA}$, $\mathbf{B} = \overrightarrow{OB}$, and $\mathbf{C} = \overrightarrow{OC}$.

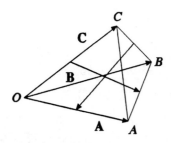

Then the midpoints of two opposite sides are given by $\dfrac{1}{2}\mathbf{A}$ and $\dfrac{1}{2}(\mathbf{B} + \mathbf{C})$, so that the vector $\dfrac{1}{2}(\mathbf{A} - \mathbf{B} - \mathbf{C})$ represents the (directed) line segment connecting the two midpoints. Similarly, the vectors $\dfrac{1}{2}(\mathbf{A} + \mathbf{B})$ and $\dfrac{1}{2}\mathbf{C}$ give the midpoints of two more opposite sides and the vector $\dfrac{1}{2}(\mathbf{A} + \mathbf{B} - \mathbf{C})$ represents the segment connecting the midpoints. The line segments are perpendicular if and only if $(\mathbf{A} - \mathbf{B} - \mathbf{C}) \cdot (\mathbf{A} + \mathbf{B} - \mathbf{C}) = 0$. Rewriting this as $(\mathbf{A} - \mathbf{C} - \mathbf{B}) \cdot (\mathbf{A} - \mathbf{C} + \mathbf{B}) = \|\mathbf{A} - \mathbf{C}\|^2 - \|\mathbf{B}\|^2 = 0$, we see that the line segments are perpendicular if and only if $\mathbf{A} - \mathbf{C}$ and $\mathbf{B}$ have the same length; but these two vectors represent the two remaining sides of the tetrahedron.

46 (a) Yes. By Exercise 45 (restricting it to two dimensions), we know that $\mathbf{A} = (\mathbf{A} \cdot \mathbf{u}_1)\mathbf{u}_1 + (\mathbf{A} \cdot \mathbf{u}_2)\mathbf{u}_2$, so if $\mathbf{A} \cdot \mathbf{u}_1 = 0$ and $\mathbf{A} \cdot \mathbf{u}_2 = 0$, then $\mathbf{A} = 0\mathbf{u}_1 + 0\mathbf{u}_2 = \mathbf{0}$.

(b) Yes. Since $\mathbf{A} \cdot \mathbf{v}_1 = 0$ and $\mathbf{A} \cdot \mathbf{v}_2 = 0$, $\mathbf{A}$ is perpendicular to both $\mathbf{v}_1$ and $\mathbf{v}_2$. But in the plane, two nonzero vectors perpendicular to the same nonzero vector must be parallel. However, $\mathbf{v}_1$ and $\mathbf{v}_2$ are nonparallel unit vectors. Hence $\mathbf{A}$ must be $\mathbf{0}$.

48 (a) $\langle 20, 0, 7, 9, 15 \rangle \cdot \langle 50, 70, 30, 20, 10 \rangle = 1540$, so it costs $1540 to produce the appliances.

(b) $2\langle 20, 0, 7, 9, 15 \rangle = \langle 40, 0, 14, 18, 30 \rangle$

50 Let $\mathbf{A} = \langle a, b, c \rangle$ and $\mathbf{B} = \langle d, e, f \rangle$; then $\mathbf{A} \cdot \mathbf{B} = ad + be + cf$. Now $x\mathbf{A} = \langle xa, xb, xc \rangle$ and $y\mathbf{B}$

$= \langle yd, ye, yf \rangle$, so $x\mathbf{A} \cdot y\mathbf{B} = xa \cdot yd + xb \cdot ye + xc \cdot yf = xy(ad + be + cf) = xy(\mathbf{A} \cdot \mathbf{B})$, as claimed.

52 $\mathbf{F} = 10\left(\dfrac{2\mathbf{i} + 3\mathbf{j} + \mathbf{k}}{\|2\mathbf{i} + 3\mathbf{j} + \mathbf{k}\|}\right) = \dfrac{10}{\sqrt{14}}(2\mathbf{i} + 3\mathbf{j} + \mathbf{k})$

while $\mathbf{R} = \mathbf{i} + 2\mathbf{j} + 2\mathbf{k}$, so the work is $\mathbf{F} \cdot \mathbf{R} = \dfrac{10}{\sqrt{14}}(2 + 6 + 2) = \dfrac{100}{\sqrt{14}}$ joules.

12.4 Lines and Planes

2 $2(x - 1) - (y - 0) = 0$ simplifies to $2x - y = 2$.

4 $(x - 2) + 3(y + 1) = 0$ simplifies to $x + 3y + 1 = 0$.

6 $\pi \mathbf{i} - \sqrt{2}\mathbf{j}$

8 $2\mathbf{i} + 5\mathbf{j}$

10 $\dfrac{|2 \cdot \frac{3}{2} - \frac{2}{3} + 5|}{\sqrt{4+1}} = \dfrac{|3 - \frac{2}{3} + 5|}{\sqrt{5}} = \dfrac{22}{3\sqrt{5}}$

12 A normal is $2\mathbf{i} - 3\mathbf{j} - \mathbf{k}$.

The unit normal is $\pm \dfrac{1}{\sqrt{14}}(2\mathbf{i} - 3\mathbf{j} - \mathbf{k})$.

14 $\dfrac{|1 + 2 \cdot 2 - 3 \cdot 3 + 5|}{\sqrt{1 + 4 + 9}} = \dfrac{1}{\sqrt{14}}$

16 The equation of the plane is $x + y + z - 5 = 0$.

The distance from the origin is $\dfrac{|-5|}{\sqrt{1+1+1}} = \dfrac{5}{\sqrt{3}}$.

18 A unit vector parallel to the line through $(1, 3, 2)$ and $(4, -1, 5)$ is $\dfrac{(4-1)\mathbf{i} + (-1-3)\mathbf{j} + (5-2)\mathbf{k}}{\sqrt{9 + 16 + 9}} =$

$\dfrac{3\mathbf{i} - 4\mathbf{j} + 3\mathbf{k}}{\sqrt{34}}$. The direction cosines are therefore

$\cos \alpha = \dfrac{3}{\sqrt{34}}$, $\cos \beta = -\dfrac{4}{\sqrt{34}}$, and $\cos \gamma = \dfrac{3}{\sqrt{34}}$. (Alternatively, one can use the opposite unit vector and obtain direction cosines with the opposite signs.)

20 (a) $\overrightarrow{P_0 P_1} = \langle 3 - 2, 0 - 1, 4 - 5 \rangle$

$= \langle 1, -1, -1 \rangle$, so $\dfrac{\overrightarrow{P_0 P_1}}{|\overrightarrow{P_0 P_1}|} =$

$\dfrac{1}{\sqrt{3}}\langle 1, -1, -1 \rangle$. The direction cosines of $\overrightarrow{P_0 P_1}$

are $\cos \alpha = \dfrac{1}{\sqrt{3}}$, $\cos \beta = -\dfrac{1}{\sqrt{3}}$, $\cos \gamma = -\dfrac{1}{\sqrt{3}}$. The direction angles (in radians) are $\alpha \approx 0.96$, $\beta \approx 2.19$, and $\gamma \approx 2.19$.

(b) $\overrightarrow{P_1 P_0} = \langle -1, 1, 1 \rangle$, so $\dfrac{\overrightarrow{P_1 P_0}}{|\overrightarrow{P_1 P_0}|} =$

$\dfrac{1}{\sqrt{3}}\langle -1, 1, 1 \rangle$. The direction cosines of $\overrightarrow{P_1 P_0}$

are $\cos \alpha = -\dfrac{1}{\sqrt{3}}$, $\cos \beta = \dfrac{1}{\sqrt{3}}$, and $\cos \gamma = \dfrac{1}{\sqrt{3}}$. The direction angles are $\alpha \approx 2.19$, $\beta \approx 0.96$, $\gamma \approx 0.96$.

22 A vector parallel to L is $3\mathbf{i} + 3\mathbf{j} + 4\mathbf{k}$.

(a) $x = 1 + 3t$, $y = 2 + 3t$, $z = 3 + 4t$

(b) $\mathbf{P} = \mathbf{i} + 2\mathbf{j} + 3\mathbf{k} + t(3\mathbf{i} + 3\mathbf{j} + 4\mathbf{k})$

12.4 Lines and Planes

24 A vector parallel to the line is $-3\mathbf{i} + 4\mathbf{j} - 3\mathbf{k}$, so $\dfrac{x-7}{-3} = \dfrac{y+1}{4} = \dfrac{z-5}{-3}$ must be an equation of the line.

26 $1 = \cos^2\alpha + \cos^2\beta + \cos^2\gamma = 3\cos^2\alpha$, so $\cos\alpha = \pm\dfrac{1}{\sqrt{3}}$. Therefore $\alpha = \cos^{-1}\dfrac{\sqrt{3}}{3}$ or $\alpha = \pi - \cos^{-1}\dfrac{\sqrt{3}}{3}$.

28 The vectors $\langle 2, 3, 4 \rangle$ and $\langle 3, -1, 2 \rangle$ are normals to the two planes; since the angle between two planes is the same as the angle between their normals, $\cos\theta = \dfrac{\mathbf{A}\cdot\mathbf{B}}{|\mathbf{A}||\mathbf{B}|} = \dfrac{11}{\sqrt{29}\sqrt{14}}$, $\theta \approx 0.993 \approx 56.9°$.

30 Parametric equations for the line are $x = 1 + t$, $y = 2 - t$, $z = 1 + 2t$, while an equation of the plane is $2(x - 1) + 5(y + 2) + 7(z + 3) = 0$, which simplifies to $2x + 5y + 7z = -29$. By substitution, $2(1 + t) + 5(2 - t) + 7(1 + 2t) = -29$, so $11t = -48$ and thus $t = -\dfrac{48}{11}$, which gives $x = 1 + \left(-\dfrac{48}{11}\right) = -\dfrac{37}{11}$, $y = 2 - \left(-\dfrac{48}{11}\right) = \dfrac{70}{11}$, $z = 1 + 2\left(-\dfrac{48}{11}\right) = -\dfrac{85}{11}$, so the point is $\left(-\dfrac{37}{11}, \dfrac{70}{11}, -\dfrac{85}{11}\right)$.

32 Parametric equations for the line are $x = 1 + t$, $y = 2 - t$, and $z = 4 - 5t$. Substituting into the plane equation $x + 2y + 5z = 0$ yields $(1 + t) + 2(2 - t) + 5(4 - 5t) = 0$, so $26t = 25$ and $t = \dfrac{25}{26}$. Hence the solution is $\left(\dfrac{51}{26}, \dfrac{27}{26}, -\dfrac{21}{26}\right)$.

34 If a line is parallel to a given line, then it is parallel to a vector in the direction of the given line. One such vector is $\langle 3, -1, -8 \rangle$. Therefore parametric equations for the desired line are $x = 1 + 3t$, $y = 3 - t$, and $z = 4 - 8t$.

36 Select a point $P_0 = (x_0, y_0, z_0)$ on the plane $Ax + By + Cz + D = 0$. Let $\mathbf{u}$ be a unit vector perpendicular to the plane; $\mathbf{u} = \dfrac{A\mathbf{i} + B\mathbf{j} + C\mathbf{k}}{\sqrt{A^2 + B^2 + C^2}}$ is a good choice. The vector $\overrightarrow{P_0P_1}$ from P_0 to P_1 has a component normal to the plane whose length is the desired distance; that is, the distance is

$$|\text{comp}_\mathbf{u} \overrightarrow{P_0P_1}| = |\overrightarrow{P_0P_1}\cdot\mathbf{u}|$$

$$= \left|\langle x_1 - x_0, y_1 - y_0, z_1 - z_0\rangle\cdot\dfrac{A\mathbf{i} + B\mathbf{j} + C\mathbf{k}}{\sqrt{A^2 + B^2 + C^2}}\right|$$

$$= \left|\dfrac{A(x_1 - x_0) + B(y_1 - y_0) + C(z_1 - z_0)}{\sqrt{A^2 + B^2 + C^2}}\right|$$

$$= \left|\dfrac{Ax_1 + By_1 + Cz_1 - (Ax_0 + By_0 + Cz_0)}{\sqrt{A^2 + B^2 + C^2}}\right|$$

$$= \left|\dfrac{Ax_1 + By_1 + Cz_1 + D}{\sqrt{A^2 + B^2 + C^2}}\right|.$$

38 The vector $\mathbf{A} = 3\mathbf{i} - 2\mathbf{j} + 3\mathbf{k}$ is parallel to the line through $(1, 3, 2)$ and $(4, 1, 5)$, while the vector $\mathbf{N} = \mathbf{i} - \mathbf{j} - 2\mathbf{k}$ is perpendicular to the plane $x - y - 2z + 15 = 0$. The angle between $\mathbf{A}$ and $\mathbf{N}$ is $\theta = \cos^{-1}\left(-\dfrac{1}{\sqrt{22}\sqrt{6}}\right) \approx 1.6579$ radians $\approx$

40 The point $P = (x, y, z)$ on the line through $R = (1, 2, 5)$ and $S = (3, 1, 1)$ that is closest to $Q = (2, -1, 5)$ is the point such that $\overrightarrow{QP} = \langle x - 2, y + 1, z - 5 \rangle$ is perpendicular to $\overrightarrow{RS} = \langle 2, -1, -4 \rangle$; that is, $\overrightarrow{QP} \cdot \overrightarrow{RS} = 2(x - 2) - (y + 1) - 4(z - 5) = 2x - y - 4z + 15 = 0$. Parametric equations for the line through R and S are $x = 1 + 2t$, $y = 2 - t$, and $z = 5 - 4t$; therefore $2(1 + 2t) - (2 - t) - 4(5 - 4t) + 15 = 2 + 4t - 2 + t - 20 + 16t + 15 = 21t - 5 = 0$, so $t = 5/21$. Hence $P = \left(\dfrac{31}{21}, \dfrac{37}{21}, \dfrac{85}{21}\right)$.

42 $\overrightarrow{P_0 P_1} = (x_1 - x_0)\mathbf{i} + (y_1 - y_0)\mathbf{j} + (z_1 - z_0)\mathbf{k}$ is a vector from the line to the given point and $\mathbf{A} = a_1\mathbf{i} + a_2\mathbf{j} + a_3\mathbf{k}$ is a vector parallel to the line.

The distance from P_1 to the line is $\|\text{orth}_\mathbf{A} \overrightarrow{P_0 P_1}\|$

$= \|\overrightarrow{P_0 P_1} - \text{proj}_\mathbf{A} \overrightarrow{P_0 P_1}\|$

$= \left\|\overrightarrow{P_0 P_1} - \left(\overrightarrow{P_0 P_1} \cdot \dfrac{\mathbf{A}}{|\mathbf{A}|}\right) \dfrac{\mathbf{A}}{|\mathbf{A}|}\right\|$.

44 Note that $\gamma = 90° - 75° = 15°$, so $\cos \gamma = \cos 15°$. To find $\cos \alpha$ and $\cos \beta$, let $\mathbf{A} = \overrightarrow{OS} = x\mathbf{i} + y\mathbf{j} + z\mathbf{k}$. Then $\cos \alpha = \dfrac{x}{|\mathbf{A}|}$ and $\cos \beta = \dfrac{y}{|\mathbf{A}|}$. Since the angle between $\mathbf{A}$ and its projection on the xy plane, $x\mathbf{i} + y\mathbf{j}$, is $75°$, we have

$(x\mathbf{i} + y\mathbf{j} + z\mathbf{k})\cdot(x\mathbf{i} + y\mathbf{j}) = x^2 + y^2 = \|\mathbf{A}\|\|x\mathbf{i} + y\mathbf{j}\| \cos 75° = \|\mathbf{A}\| \sqrt{x^2 + y^2} \cos 75°$,

so $\|\mathbf{A}\| = \dfrac{\sqrt{x^2 + y^2}}{\cos 75°}$. The angle between $x\mathbf{i} + y\mathbf{j}$ and $\mathbf{i}$ is $60°$, so $\dfrac{(x\mathbf{i} + y\mathbf{j}) \cdot \mathbf{i}}{\sqrt{x^2 + y^2}} = \dfrac{x}{\sqrt{x^2 + y^2}} = \cos 60° = \dfrac{1}{2}$. Hence $2x = \sqrt{x^2 + y^2}$, which yields $y = \sqrt{3} x$. Therefore $\cos \alpha = \dfrac{x}{\sqrt{x^2 + y^2}} \cos 75°$

$= \dfrac{1}{2} \cos 75°$ and $\cos \beta = \dfrac{y}{\sqrt{x^2 + y^2}} \cos 75° = \dfrac{\sqrt{3}}{2} \cos 75°$. (With various trigonometric identities, one could compute these in terms of radicals: $\cos \alpha = \dfrac{\sqrt{6} - \sqrt{2}}{8}$, $\cos \beta = \dfrac{3\sqrt{2} - \sqrt{6}}{8}$, and $\cos \gamma = \dfrac{\sqrt{6} + \sqrt{2}}{4}$, but that is not necessary.)

12.5 Determinants

2 By Theorem 1, $\begin{vmatrix} 3 & 3 \\ 7 & 7 \end{vmatrix} = 0$.

4 By Theorem 1, $\begin{vmatrix} 1 & 3 & 1 \\ 2 & 1 & 2 \\ 4 & 5 & 4 \end{vmatrix} = 0$.

6 $\begin{vmatrix} 0 & 0 & 0 \\ 1 & 5 & 9 \\ 3 & -1 & 2 \end{vmatrix} = 0 \begin{vmatrix} 5 & 9 \\ -1 & 2 \end{vmatrix} - 0 \begin{vmatrix} 1 & 9 \\ 3 & 2 \end{vmatrix} + 0 \begin{vmatrix} 1 & 5 \\ 3 & -1 \end{vmatrix} = 0$

8 0, by Theorem 1.

10 $\begin{vmatrix} 1 & 2 & 3 \\ -1 & 4 & -2 \\ 5 & 1 & 0 \end{vmatrix} =$

$1\begin{vmatrix} 4 & -2 \\ 1 & 0 \end{vmatrix} - 2\begin{vmatrix} -1 & -2 \\ 5 & 0 \end{vmatrix} + 3\begin{vmatrix} -1 & 4 \\ 5 & 1 \end{vmatrix}$

$= 1 \cdot 2 - 2 \cdot 10 + 3(-21) = -81$

12 $\begin{vmatrix} 2 & 3 \\ 4 & 6 \end{vmatrix} = 0$

14 abs $\begin{vmatrix} -1 & -1 \\ 2 & 3 \end{vmatrix} = |-3 + 2| = 1$

16 $\begin{vmatrix} a_1 & b_1 & a_1 \\ a_2 & b_2 & a_2 \\ a_3 & b_3 & a_3 \end{vmatrix} =$

$a_1\begin{vmatrix} b_2 & a_2 \\ b_3 & a_3 \end{vmatrix} - b_1\begin{vmatrix} a_2 & a_2 \\ a_3 & a_3 \end{vmatrix} + a_1\begin{vmatrix} a_2 & b_2 \\ a_3 & b_3 \end{vmatrix}$

$= a_1(b_2 a_3 - a_2 b_3) - b_1(0) + a_1(a_2 b_3 - b_2 a_3)$

$= a_1(b_2 a_3 - a_2 b_3) + a_1(-1)(b_2 a_3 - a_2 b_3)$

$= (a_1 - a_1)(b_2 a_3 - a_2 b_3) = 0$

18 $\begin{vmatrix} a_1 & a_2 & a_3 \\ b_1 & b_2 & b_3 \\ c_1 & c_2 & c_3 \end{vmatrix}$

$= a_1\begin{vmatrix} b_2 & b_3 \\ c_2 & c_3 \end{vmatrix} - a_2\begin{vmatrix} b_1 & b_3 \\ c_1 & c_3 \end{vmatrix} + a_3\begin{vmatrix} b_1 & b_2 \\ c_1 & c_2 \end{vmatrix}$

$= a_1(b_2 c_3 - b_3 c_2) - a_2(b_1 c_3 - b_3 c_1) + a_3(b_1 c_2 - b_2 c_1)$

$= a_1 b_2 c_3 - a_1 b_3 c_2 - a_2 b_1 c_3 + a_2 b_3 c_1 + a_3 b_1 c_2 - a_3 b_2 c_1$

$\begin{vmatrix} c_1 & c_2 & c_3 \\ b_1 & b_2 & b_3 \\ a_1 & a_2 & a_3 \end{vmatrix}$

$= c_1\begin{vmatrix} b_2 & b_3 \\ a_2 & a_3 \end{vmatrix} - c_2\begin{vmatrix} b_1 & b_3 \\ a_1 & a_3 \end{vmatrix} + c_3\begin{vmatrix} b_1 & b_2 \\ a_1 & a_2 \end{vmatrix}$

$= c_1(b_2 a_3 - b_3 a_2) - c_2(b_1 a_3 - b_3 a_1) + c_3(b_1 a_2 - b_2 a_1)$

$= c_1 b_2 a_3 - c_1 b_3 a_2 - c_2 b_1 a_3 + c_2 b_3 a_1 + c_3 b_1 a_2 - c_3 b_2 a_1$

$= a_3 b_2 c_1 - a_2 b_3 c_1 - a_3 b_1 c_2 + a_1 b_3 c_2 + a_2 b_1 c_3 - a_1 b_2 c_3$

$= -(a_1 b_2 c_3 - a_1 b_3 c_2 - a_2 b_1 c_3 + a_2 b_3 c_1 + a_3 b_1 c_2 - a_3 b_2 c_1).$

Hence, $\begin{vmatrix} c_1 & c_2 & c_3 \\ b_1 & b_2 & b_3 \\ a_1 & a_2 & a_3 \end{vmatrix} = -\begin{vmatrix} a_1 & a_2 & a_3 \\ b_1 & b_2 & b_3 \\ c_1 & c_2 & c_3 \end{vmatrix}.$

20 As in Exercise 18, $\begin{vmatrix} a_1 & a_2 & a_3 \\ b_1 & b_2 & b_3 \\ c_1 & c_2 & c_3 \end{vmatrix} =$

$a_1 b_2 c_3 - a_1 b_3 c_2 - a_2 b_1 c_3 + a_2 b_3 c_1 + a_3 b_1 c_2 - a_3 b_2 c_1.$

Similarly, $\begin{vmatrix} a_1 & b_1 & c_1 \\ a_2 & b_2 & c_2 \\ a_3 & b_3 & c_3 \end{vmatrix} =$

$a_1\begin{vmatrix} b_2 & c_2 \\ b_3 & c_3 \end{vmatrix} - b_1\begin{vmatrix} a_2 & c_2 \\ a_3 & c_3 \end{vmatrix} + c_1\begin{vmatrix} a_2 & b_2 \\ a_3 & b_3 \end{vmatrix}$

$= a_1(b_2 c_3 - c_2 b_3) - b_1(a_2 c_3 - c_2 a_3) + c_1(a_2 b_3 - b_2 a_3)$

$= a_1 b_2 c_3 - a_1 b_3 c_2 - a_2 b_1 c_3 + a_3 b_1 c_2 + a_2 b_3 c_1 - a_3 b_2 c_1.$

Comparison of these quantities shows that the determinants are equal.

22. As the figure shows, adding $k\mathbf{B}$ to $\mathbf{A}$ creates a new parallelogram whose area is the same as the parallelogram defined by $\mathbf{A}$ and $\mathbf{B}$, because the change in $\mathbf{A}$ causes the loss of a triangular region that exactly matches the new triangular region that is included.

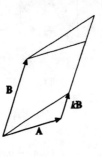

24. As in the proof of Theorem 3, the area is

$\|\mathbf{A}\| \|\mathbf{B}\| \sin \theta$. We know that $\cos \theta = \dfrac{\mathbf{A} \cdot \mathbf{B}}{|\mathbf{A}||\mathbf{B}|}$

$= \dfrac{a_1 b_1 + a_2 b_2 + a_3 b_3}{\sqrt{a_1^2 + a_2^2 + a_3^2} \sqrt{b_1^2 + b_2^2 + b_3^2}}$, so $\sin^2 \theta =$

$1 - \cos^2 \theta = 1 - \dfrac{(a_1 b_1 + a_2 b_2 + a_3 b_3)^2}{(a_1^2 + a_2^2 + a_3^2)(b_1^2 + b_2^2 + b_3^2)} =$

$\dfrac{(a_1^2 + a_2^2 + a_3^2)(b_1^2 + b_2^2 + b_3^2) - (a_1 b_1 + a_2 b_2 + a_3 b_3)^2}{(a_1^2 + a_2^2 + a_3^2)(b_1^2 + b_2^2 + b_3^2)}.$

Upon expanding the products and collecting like terms, the numerator becomes

$a_1^2 b_2^2 - 2a_1 b_1 a_2 b_2 + a_2^2 b_1^2 +$

$a_1^2 b_3^2 - 2a_1 b_1 a_3 b_3 + a_3^2 b_1^2 +$

$a_2^2 b_3^2 - 2a_2 b_2 a_3 b_3 + a_3^2 b_2^2 = (a_1 b_2 - a_2 b_1)^2 +$

$(a_1 b_3 - a_3 b_1)^2 + (a_2 b_3 - a_3 b_2)^2$. If we now take the square root and multiply by $\|\mathbf{A}\|$ and $\|\mathbf{B}\|$, we obtain the area formula

$|\mathbf{A}||\mathbf{B}|\sqrt{(a_1 b_1 - a_2 b_2)^2 + (a_1 b_3 - a_3 b_1)^2 + (a_2 b_3 - a_3 b_2)^2}.$

12.6 The Cross Product of Two Vectors

2.

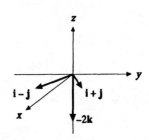

$\begin{vmatrix} \mathbf{i} & \mathbf{j} & \mathbf{k} \\ 1 & 1 & 0 \\ 1 & -1 & 0 \end{vmatrix} = \mathbf{k} \begin{vmatrix} 1 & 1 \\ 1 & -1 \end{vmatrix} = -2\mathbf{k}$

4.

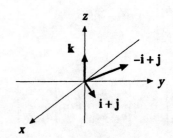

$\begin{vmatrix} \mathbf{i} & \mathbf{j} & \mathbf{k} \\ 0 & 0 & 1 \\ 1 & 1 & 0 \end{vmatrix} = -\mathbf{i} + \mathbf{j}$

6. $\begin{vmatrix} \mathbf{i} & \mathbf{j} & \mathbf{k} \\ 1 & -1 & 0 \\ 0 & 1 & 4 \end{vmatrix} = -4\mathbf{i} - 4\mathbf{j} + \mathbf{k}$. To see that this is perpendicular to $\mathbf{A}$ and $\mathbf{B}$, compute the dot products: $(\mathbf{i} - \mathbf{j}) \cdot (-4\mathbf{i} - 4\mathbf{j} + \mathbf{k}) = 0$ and $(\mathbf{j} + 4\mathbf{k}) \cdot (-4\mathbf{i} - 4\mathbf{j} + \mathbf{k}) = 0$.

8. $\begin{vmatrix} \mathbf{i} & \mathbf{j} & \mathbf{k} \\ 1 & -1 & 5 \\ 2 & 3 & 3 \end{vmatrix} = -18\mathbf{i} + 7\mathbf{j} + 5\mathbf{k};$

12.6 The Cross Product of Two Vectors

$\|-18\mathbf{i} + 7\mathbf{j} + 5\mathbf{k}\| = \sqrt{398}$

10 Area $= \dfrac{1}{2}$ mag $\begin{vmatrix} \mathbf{i} & \mathbf{j} & \mathbf{k} \\ 1 & 2 & 3 \\ 2 & -1 & 2 \end{vmatrix} = \dfrac{1}{2} \| \langle 7, 4, -5 \rangle \|$

$= \dfrac{3\sqrt{10}}{2}$

12 Volume $=$ abs $\begin{vmatrix} 3 & 4 & 2 \\ 2 & 3 & 4 \\ 1 & -1 & -1 \end{vmatrix} = 17$

14 Volume $=$ abs $\begin{vmatrix} 3 & 3 & 2 \\ 1 & 4 & -1 \\ 1 & 2 & 3 \end{vmatrix} = 26$

16 Let $\mathbf{A} = \langle 2-1, 1-3, 1+1 \rangle = \langle 1, -2, 2 \rangle$ and $\mathbf{B} = \langle 2-1, 1-3, 1-4 \rangle = \langle 1, -2, -3 \rangle$; then a normal to the plane is given by $\mathbf{A} \times \mathbf{B} = \begin{vmatrix} \mathbf{i} & \mathbf{j} & \mathbf{k} \\ 1 & -2 & 2 \\ 1 & -2 & -3 \end{vmatrix} = \langle 10, 5, 0 \rangle$. Any multiple of $\langle 2, 1, 0 \rangle$ is perpendicular to the plane.

18 $\mathbf{A} = \langle 4-1, 1-2, 0-1 \rangle = \langle 3, -1, -1 \rangle$ is parallel to the first line and $\mathbf{B} = \langle 2-3, 6-5, -3-2 \rangle = \langle -1, 1, -5 \rangle$ is parallel to the second line, so $\mathbf{A} \times \mathbf{B} = \begin{vmatrix} \mathbf{i} & \mathbf{j} & \mathbf{k} \\ 3 & -1 & -1 \\ -1 & 1 & -5 \end{vmatrix} =$

$6\mathbf{i} + 16\mathbf{j} + 2\mathbf{k}$ is perpendicular to both lines, as is any other multiple of $3\mathbf{i} + 8\mathbf{j} + \mathbf{k}$.

20 (a) $\mathbf{A}$ is perpendicular to $\mathbf{A} \times \mathbf{B}$, so $\text{proj}_{\mathbf{A} \times \mathbf{B}} \mathbf{A} = \mathbf{0}$.

   (b) $\text{orth}_{\mathbf{A} \times \mathbf{B}} \mathbf{A} = \mathbf{A} - \text{proj}_{\mathbf{A} \times \mathbf{B}} \mathbf{A} = \mathbf{A} - \mathbf{0} = \mathbf{A}$

22 (a) If $\mathbf{A} = x\mathbf{i} + y\mathbf{j} + z\mathbf{k}$, then $\mathbf{A} \times \mathbf{B} = \begin{vmatrix} \mathbf{i} & \mathbf{j} & \mathbf{k} \\ x & y & z \\ cx & cy & cz \end{vmatrix} = \mathbf{0}$.

   (b) $\|\mathbf{A} \times \mathbf{B}\|$ gives the area of the parallelogram defined by $\mathbf{A}$ and $\mathbf{B}$. Since $\mathbf{A}$ and $\mathbf{B}$ are parallel, the resulting parallelogram has area 0, so $\mathbf{A} \times \mathbf{B} = \mathbf{0}$.

24 What are you looking in here for? It's algebra! You can do it.

26 In the proof of Theorem 4, we see that $\mathbf{A} \cdot (\mathbf{B} \times \mathbf{C})$
$= \begin{vmatrix} a_1 & a_2 & a_3 \\ b_1 & b_2 & b_3 \\ c_1 & c_2 & c_3 \end{vmatrix}$. By reversing the rows for $\mathbf{B}$ and $\mathbf{C}$, we get $\mathbf{A} \cdot (\mathbf{C} \times \mathbf{B})$, which, by Theorem 2(a), must be the negative of $\mathbf{A} \cdot (\mathbf{B} \times \mathbf{C})$. Also, $(\mathbf{A} \times \mathbf{B}) \cdot \mathbf{C} = \mathbf{C} \cdot (\mathbf{A} \times \mathbf{B}) = \begin{vmatrix} c_1 & c_2 & c_3 \\ a_1 & a_2 & a_3 \\ b_1 & b_2 & b_3 \end{vmatrix}$, which involves two row swaps (switch the second and third, then the first and second) in the determinant for $\mathbf{A} \cdot (\mathbf{B} \times \mathbf{C})$, so it has the same sign as $\mathbf{A} \cdot (\mathbf{B} \times \mathbf{C})$.

28 The points $P_0 = (0, 0, 0)$, $P_1 = (x_1, y_1, z_1)$, $P_2 = (x_2, y_2, z_2)$, and $P_3 = (x_3, y_3, z_3)$ lie on the same plane if and only if the parallelepiped spanned by $\overrightarrow{P_0 P_1} = \langle x_1, y_1, z_1 \rangle$, $\overrightarrow{P_0 P_2} = \langle x_2, y_2, z_2 \rangle$, and $\overrightarrow{P_0 P_3} = \langle x_3, y_3, z_3 \rangle$ has volume 0. By Theorem

4, this occurs if and only if $\begin{vmatrix} x_1 & y_1 & z_1 \\ x_2 & y_2 & z_2 \\ x_3 & y_3 & z_3 \end{vmatrix} = 0.$

30 Yes. Since $\|A \times B\| = \|A\|\|B\|\sin\theta = 0$ and $\|A\| \neq 0$, we have either $\|B\| = 0$ or $\sin\theta = 0$. Since $A \cdot B = \|A\|\|B\|\cos\theta = 0$, we have either $\|B\| = 0$ or $\cos\theta = 0$. Since $\cos^2\theta + \sin^2\theta = 1$, $\sin\theta$ and $\cos\theta$ cannot both be 0, so we must have $B = 0$.

32 By the identity for the vector triple product, $E \times (C \times D) = (E \cdot D)C - (E \cdot C)D$. Substituting $E = A \times B$, $(A \times B) \times (C \times D) = [(A \times B) \cdot D]C - [(A \times B) \cdot C]D$, as claimed.

34 (a) $j + k$ is one of many possibilities. (Crossing $3i - j + k$ with any nonzero vector will produce others.)

(b) Divide your vector in (a) by its magnitude.

36 (a) k_1 is perpendicular to the face determined by v_2 and v_3, so it is a multiple of $v_2 \times v_3$. The magnitude of k_1 is just the reciprocal of the height of the parallelepiped from the face determined by v_2 and v_3 to the opposite face. Since $\|v_2 \times v_3\|$ is the area of the given face and the volume of the parallelepiped is $|v_1 \cdot (v_2 \times v_3)|$, the reciprocal of the height is $\dfrac{\|v_2 \times v_3\|}{|v_1 \cdot (v_2 \times v_3)|}$. Hence $k_1 =$

$\dfrac{\|v_2 \times v_3\|}{v_1 \cdot (v_2 \times v_3)} \left(\dfrac{v_2 \times v_3}{|v_2 \times v_3|} \right) = \dfrac{v_2 \times v_3}{v_1 \cdot (v_2 \times v_3)}$ is a

perfectly good choice. The equations for k_2 and k_3 are obtained similarly.

(b) We have $k_2 \times k_3$

$= \dfrac{v_3 \times v_1}{v_1 \cdot (v_2 \times v_3)} \times \dfrac{v_1 \times v_2}{v_1 \cdot (v_2 \times v_3)}$

$= \dfrac{(v_3 \times v_1) \times (v_1 \times v_2)}{[v_1 \cdot (v_2 \times v_3)]^2}$. By Exercises 32 and

26, this equals

$\dfrac{((v_3 \times v_1) \cdot v_2)v_1 - ((v_3 \times v_1) \cdot v_1)v_2}{[v_1 \cdot (v_2 \times v_3)]^2}$

$= \dfrac{(v_2 \cdot (v_3 \times v_1))v_1 - 0v_2}{[v_1 \cdot (v_2 \times v_3)]^2} = \dfrac{(v_1 \cdot (v_2 \times v_3))v_1}{[v_1 \cdot (v_2 \times v_3)]^2}$

$= \dfrac{v_1}{v_1 \cdot (v_2 \times v_3)}$. Hence $k_1 \cdot (k_2 \times k_3)$

$= \dfrac{v_2 \times v_3}{v_1 \cdot (v_2 \times v_3)} \cdot \dfrac{v_1}{v_1 \cdot (v_2 \times v_3)} = \dfrac{v_1 \cdot (v_2 \times v_3)}{[v_1 \cdot (v_2 \times v_3)]^2}$

$= \dfrac{1}{v_1 \cdot (v_2 \times v_3)}.$

(c) Yes. From (b), $\dfrac{k_2 \times k_3}{k_1 \cdot (k_2 \times k_3)} =$

$\dfrac{v_1}{v_1 \cdot (v_2 \times v_3)} \div \dfrac{1}{v_1 \cdot (v_2 \times v_3)} = v_1.$

38 (a) Note that $A \times (B \times C)$ is normal to $B \times C$, which is perpendicular to both B and C. Hence $A \times (B \times C)$ must be in the same plane as B and C, so $A \times (B \times C) = xB + yC$ for some scalars x and y.

(b) Observe that $A \cdot [A \times (B \times C)] = 0$, so $0 = A \cdot [xB + yC] = x(A \cdot B) + y(A \cdot C)$.

(c) Since A is not perpendicular to C, $A \cdot C \neq 0$. Hence (b) implies that $y = -\dfrac{(A \cdot B)x}{A \cdot C}$. Thus

$$A \times (B \times C) = xB - \frac{(A \cdot B)x}{A \cdot C}C =$$

$$\frac{x}{A \cdot C}[(A \cdot C)B - (A \cdot B)C]. \text{ Let } z = \frac{x}{A \cdot C}.$$

40 If $\mathcal{P}$ is the plane perpendicular to A, then $\text{proj}_\mathcal{P}(B + C) = \text{proj}_\mathcal{P} B + \text{proj}_\mathcal{P} C = B_2 + C_2$, so $A \times (B + C) = \|A\|(B_2 + C_2) = \|A\|B_2 + \|A\|C_2 = A \times B + A \times C$.

42 (a) Let $P = xi + yj$, $P_1 = x_1i + x_1j$, $P_{12} = \overrightarrow{P_1P_2}$
$= (x_2 - x_1)i + (y_2 - y_1)j$, and $P_{13} = \overrightarrow{P_1P_3}$
$= (x_3 - x_1)i + (y_3 - y_1)j$. For each point $P = (x, y)$, there are unique numbers a and b such that $P = P_1 + aP_{12} + bP_{13}$. To compute b, use the cross product: $P \times P_{12} =$
$P_1 \times P_{12} + aP_{12} \times P_{12} + bP_{13} \times P_{12} =$
$P_1 \times P_{12} + bP_{13} \times P_{12}$. Hence
$(P - P_1) \times P_{12} = bP_{13} \times P_{12}$. In coordinates,
$[(x - x_1)(y_2 - y_1) - (x_2 - x_1)(y - y_1)]k =$
$b[(x_3 - x_1)(y_2 - y_1) - (x_2 - x_1)(y_3 - y_1)]k$,
so $b = \dfrac{(x - x_1)(y_2 - y_1) - (x_2 - x_1)(y - y_1)}{(x_3 - x_1)(y_2 - y_1) - (x_2 - x_1)(y_3 - y_1)}$.

P and P_3 are on the same side of P_1P_2 if and only if $(x - x_1)(y_2 - y_1) - (x_2 - x_1)(y - y_1)$ and $(x_3 - x_1)(y_2 - y_1) - (x_2 - x_1)(y_3 - y_1)$ have the same sign. Proceeding similarly, one can devise test for whether P and P_1 are on the same side of P_2P_3 and whether P and P_2 are on the same side of P_1P_3. If all three conditions hold at the same time, P is inside the triangle.

12.7 More on Lines and Planes

2 A normal to the plane is $\begin{vmatrix} i & j & k \\ 3 & 5 & -2 \\ 2 & -1 & 3 \end{vmatrix} =$

$13i - 13j - 13k$ or $N = \langle 1, -1, -1 \rangle$. A vector in the direction from the plane to the point is $A = \langle 1, 2, 2 \rangle$. The distance is $|\text{comp}_N A|$

$= \left| \langle 1, 2, 2 \rangle \cdot \dfrac{\langle 1, -1, -1 \rangle}{\sqrt{3}} \right| = \sqrt{3}.$

4 Let $P = (1, 2, 3)$, $Q = (-2, -1, 3)$, and $R = (4, 1, 2)$. Then $A = \overrightarrow{QP} = \langle 1 + 2, 2 + 1, 3 - 3 \rangle = \langle 3, 3, 0 \rangle$ is a vector from the line to the given point and $B = \overrightarrow{QR} = \langle 4 + 2, 1 + 1, 2 - 3 \rangle = \langle 6, 2, -1 \rangle$ is a vector parallel to the line. The distance is $\|\text{orth}_B A\| = \|A - \text{proj}_B A\|$

$= \left\| \langle 3, 3, 0 \rangle - \langle 3, 3, 0 \rangle \cdot \dfrac{\langle 6, 2, -1 \rangle}{\sqrt{41}} \left(\dfrac{\langle 6, 2, -1 \rangle}{\sqrt{41}} \right) \right\|$

$= \left\| \langle 3, 3, 0 \rangle - \left\langle \dfrac{144}{41}, \dfrac{48}{41}, -\dfrac{24}{41} \right\rangle \right\|$

$= \left\| \left\langle -\dfrac{21}{41}, \dfrac{75}{41}, \dfrac{24}{41} \right\rangle \right\| = \dfrac{\sqrt{6642}}{41}.$

6 A point on the first line is $P_1 = (2, 4, 1)$ and a vector parallel to the line is $A = i + j + k$, while $P_2 = (1, 3, 2)$ is on the second line and $B = 2i - j - k$ is parallel to it. Then $N = A \times B = 3j - 3k$ is perpendicular to both lines. The distance between the two lines is thus

$\|\text{proj}_N \overrightarrow{P_1P_2}\|$

$$= \left\| \langle -1, -1, 1 \rangle \cdot \frac{\langle 0, 3, -3 \rangle}{3\sqrt{2}} \left(\frac{\langle 0, 3, -3 \rangle}{3\sqrt{2}} \right) \right\|$$

$$= \| \langle 0, -1, 1 \rangle \| = \sqrt{2}.$$

8. (a) Use $\mathbf{N} = \overrightarrow{P_1P_2} \times \overrightarrow{P_1P_3}$ as a normal vector and choose one of the points to plug into the formula of Theorem 3 of Sec. 12.4.

    (b) $P_1 = (2, 2, 1)$, $P_2 = (0, 1, 5)$, and $P_3 = (2, -1, 0)$, so $\overrightarrow{P_1P_2} = \langle -2, -1, 4 \rangle$, $\overrightarrow{P_1P_3} = \langle 0, -3, -1 \rangle$, $\mathbf{N} = \overrightarrow{P_1P_2} \times \overrightarrow{P_1P_3} = \langle 13, -2, 6 \rangle$. Then $13(x - 2) - 2(y - 2) + 6(z - 1) = 0$ or $13x - 2y + 6z = 28$ is an equation of the plane.

10. (a) If the line through P_1 and P_2 is parallel to $Ax + By + Cz + D = 0$, then $\overrightarrow{P_1P_2}$ and $\mathbf{N} = \langle A, B, C \rangle$ should be perpendicular. Check that $\overrightarrow{P_1P_2} \cdot \mathbf{N} = 0$.

    (b) $P_1 = (1, -2, 3)$ and $P_2 = (5, 3, 0)$, so $\overrightarrow{P_1P_2} = \langle 4, 5, -3 \rangle$. We have $\mathbf{N} = \langle 2, -1, 1 \rangle$, so $\overrightarrow{P_1P_2} \cdot \mathbf{N} = 8 - 5 - 3 = 0$. The line through P_1 and P_2 is parallel to the plane $2x - y + z + 3 = 0$.

12. (a) If the plane through P_1, P_2, and P_3 is parallel to the plane through Q_1, Q_2, and Q_3, then the normals $\overrightarrow{P_1P_2} \times \overrightarrow{P_1P_3}$ and $\overrightarrow{Q_1Q_2} \times \overrightarrow{Q_1Q_3}$ must be parallel. Check that one is a scalar multiple of the other.

    (b) $P_1 = (1, 2, 3)$, $P_2 = (4, 1, -1)$, $P_3 = (2, 0, 1)$, $Q_1 = (2, 3, 4)$, $Q_2 = (5, 2, 0)$, and $Q_3 = (3, 1, 2)$, so $\mathbf{N}_1 = \overrightarrow{P_1P_2} \times \overrightarrow{P_1P_3} = \langle 3, -1, -4 \rangle \times \langle 1, -2, -2 \rangle = \langle -6, 2, -5 \rangle$ and $\mathbf{N}_2 = \overrightarrow{Q_1Q_2} \times \overrightarrow{Q_1Q_3} = \langle 3, -1, -4 \rangle \times \langle 1, -2, -2 \rangle = \langle -6, 2, -5 \rangle = \mathbf{N}_1$. The planes are parallel.

14. (a) Cross $\mathbf{N}_1 = \langle A_1, B_1, C_1 \rangle$ and $\mathbf{N}_2 = \langle A_2, B_2, C_2 \rangle$.

    (b) Choose the value of one variable arbitrarily and solve the resulting system of two equations for the remaining two variables. (If the equations have no solution, try a different variable. At least one of the three will work.)

    (c) The planes are $2x - y + 3z + 4 = 0$ and $3x + 2y + 5z = 0$, so $\mathbf{N}_1 = \langle 2, -1, 3 \rangle$, $\mathbf{N}_2 = \langle 3, 2, 5 \rangle$, and $\mathbf{N}_1 \times \mathbf{N}_2 = \langle -11, -1, 7 \rangle$. Choose $z = 0$. Then $2x - y = -4$ and $3x + 2y = 0$ has the solution $x = -8/7$ and $y = 12/7$. Thus $x = -\frac{8}{7} - 11t$, $y = \frac{12}{7} - t$, and $z = 7t$ are parametric equations for L.

16. The vector $\mathbf{A} = \langle 2, -1, 1 \rangle$ is parallel to the line through $(1, 2, 1)$ and $(-1, 3, 0)$, while $\mathbf{B} = \langle 1, 1, -2 \rangle$ is normal to the plane $x + y - 2z = 0$. The angle θ between $\mathbf{A}$ and $\mathbf{B}$ is given by

$$\cos^{-1} \frac{\mathbf{A} \cdot \mathbf{B}}{|\mathbf{A}||\mathbf{B}|} = \cos^{-1}\left(-\frac{1}{6}\right) \approx 1.7382 \text{ radians} \approx 99.59°.$$ The line makes an angle of about $9.59°$ with the plane.

18. (a) Two

(b) Choose either $\dfrac{A\mathbf{i} + B\mathbf{j} + C\mathbf{k}}{\sqrt{A^2 + B^2 + C^2}}$ or

$-\dfrac{A\mathbf{i} + B\mathbf{j} + C\mathbf{k}}{\sqrt{A^2 + B^2 + C^2}}$.

(c) $\dfrac{3\mathbf{i} - 2\mathbf{j} + 4\mathbf{k}}{\sqrt{9 + 4 + 16}} = \dfrac{1}{\sqrt{29}}(3\mathbf{i} - 2\mathbf{j} + 4\mathbf{k})$

20 (a) Choose arbitrary values for two variables and solve for the third. (If a variable has a coefficient of 0 it cannot be the one you solve for.)

(b) Let $x = y = 0$ in $3x - y + z + 10 = 0$. Then $z = -10$, so $(0, 0, -10)$ is a point on the plane.

22 The cross product of $\mathbf{N}_1 = \langle A_1, B_1, C_1 \rangle$ and $\mathbf{N}_2 = \langle A_2, B_2, C_2 \rangle$ is parallel to L. The components of $\dfrac{\mathbf{N}_1 \times \mathbf{N}_2}{|\mathbf{N}_1 \times \mathbf{N}_2|}$ are the direction cosines.

24 (a) $\|\mathbf{A} \times \mathbf{B}\|$

(b) $\|\langle 2, 3, 1 \rangle \times \langle 4, -1, 5 \rangle\|$
$= \|\langle 16, -6, -14 \rangle\| = \sqrt{488} = 2\sqrt{122}$

26 $\mathbf{B} = \mathbf{A} - 2(\mathbf{A} \cdot \mathbf{n})\mathbf{n}$ lies in the plane defined by $\mathbf{A}$ and $\mathbf{n}$. The vectors $-\mathbf{A}$ and $\mathbf{B}$ are supposed to

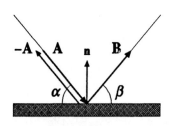

make equal angles with the unit vector $\mathbf{n}$. Note that $\|\mathbf{B}\|^2 = \mathbf{B} \cdot \mathbf{B} = [\mathbf{A} - 2(\mathbf{A} \cdot \mathbf{n})\mathbf{n}] \cdot [\mathbf{A} - 2(\mathbf{A} \cdot \mathbf{n})\mathbf{n}] = \mathbf{A} \cdot \mathbf{A} - 4(\mathbf{A} \cdot \mathbf{n})(\mathbf{A} \cdot \mathbf{n}) + 4(\mathbf{A} \cdot \mathbf{n})^2(\mathbf{n} \cdot \mathbf{n}) = \|\mathbf{A}\|^2$, so

$\|\mathbf{B}\| = \|\mathbf{A}\|$. Also $\mathbf{B} \cdot \mathbf{n} = [\mathbf{A} - 2(\mathbf{A} \cdot \mathbf{n})\mathbf{n}] \cdot \mathbf{n} = \mathbf{A} \cdot \mathbf{n} - 2(\mathbf{A} \cdot \mathbf{n})(\mathbf{n} \cdot \mathbf{n}) = -\mathbf{A} \cdot \mathbf{n}$, so the angle which $\mathbf{B}$ makes with $\mathbf{n}$ is $\cos^{-1}\left(\dfrac{-\mathbf{A} \cdot \mathbf{n}}{\|\mathbf{B}\|}\right) = \cos^{-1}\left(\dfrac{-\mathbf{A} \cdot \mathbf{n}}{\|\mathbf{A}\|}\right)$.

But this is the same angle that $-\mathbf{A}$ makes with $\mathbf{n}$, showing that $-\mathbf{A}$ and $\mathbf{B}$ satisfy the condition $\alpha = \beta$, as claimed.

28 (a) In Exercise 44 of Sec. 12.4, we found the direction cosines of the vector $\overrightarrow{OS}$ to be $\cos \alpha = \dfrac{1}{2} \cos 75°$, $\cos \beta = \dfrac{\sqrt{3}}{2} \cos 75°$, and $\cos \gamma = \cos 15°$. The direction cosines of $\mathbf{N}$ are $\cos \alpha' = \cos 90°$, $\cos \beta' = \cos 68°$, and $\cos \gamma' = \cos 22°$. Thus the cosine of the angle between $\overrightarrow{OS}$ and $\mathbf{N}$ is $\cos \alpha \cos \alpha' + \cos \beta \cos \beta' + \cos \gamma \cos \gamma'$

$= \dfrac{\sqrt{3}}{2} \cos 75° \cos 68° + \cos 15° \cos 22°$

≈ 0.9796.

(b) The angle is $\cos^{-1} 0.9796 \approx 11.6°$.

30 (a) Using O to represent the origin, we see that $\mathbf{A} = \overrightarrow{OP_1} + t\overrightarrow{P_1Q_1}$ traces out the line L_1 that passes through P_1 and Q_1. Similarly, $\mathbf{B} = \overrightarrow{OP_2} + s\overrightarrow{P_2Q_2}$ traces out the line L_2 that passes through P_2 and Q_2. We need to find t and s such that the corresponding points R_1 and R_2 on L_1 and L_2 give a vector $\overrightarrow{R_1R_2}$ that is normal to both lines. (It will also give the minimum distance between L_1 and L_2.) For the

desired values of t and s, we have $\mathbf{A} = \overrightarrow{OR_1}$
$= \overrightarrow{OP_1} + t\overrightarrow{P_1Q_1}$ and $\mathbf{B} = \overrightarrow{OR_2} = \overrightarrow{OP_2} + s\overrightarrow{P_2Q_2}$
so $\overrightarrow{R_1R_2} = \mathbf{B} - \mathbf{A} = \overrightarrow{OP_2} - \overrightarrow{OP_1} + s\overrightarrow{P_2Q_2} - t\overrightarrow{P_1Q_1} = \overrightarrow{P_1P_2} + s\overrightarrow{P_2Q_2} - t\overrightarrow{P_1Q_1}$. Note that $\overrightarrow{R_1R_2}$ is perpendicular to both $\overrightarrow{P_1Q_1}$ and $\overrightarrow{P_2Q_2}$, so we use the dot product to obtain two equations: $0 = \overrightarrow{P_1Q_1} \cdot \overrightarrow{P_1P_2} + s\overrightarrow{P_1Q_1} \cdot \overrightarrow{P_2Q_2} - t|\overrightarrow{P_1Q_1}|^2$
and $0 = \overrightarrow{P_2Q_2} \cdot \overrightarrow{P_1P_2} + s|\overrightarrow{P_2Q_2}|^2 - t\overrightarrow{P_1Q_1} \cdot \overrightarrow{P_2Q_2}$. With two equations in two unknowns, we can solve for t and s to find R_1 and R_2.

(b) We have $P_1 = (3, 2, 1)$, $Q_1 = (1, 1, 1)$, $P_2 = (0, 2, 0)$, and $Q_2 = (2, 1, -1)$, so $\overrightarrow{P_1Q_1} = \langle -2, -1, 0 \rangle$, $\overrightarrow{P_2Q_2} = \langle 2, -1, -1 \rangle$, and $\overrightarrow{P_1P_2} = \langle -3, 0, -1 \rangle$. By (a), $0 = \langle -2, -1, 0 \rangle \cdot \langle -3, 0, -1 \rangle + s\langle -2, -1, 0 \rangle \cdot \langle 2, -1, -1 \rangle - t\|\langle -2, -1, 0 \rangle\|^2 = 6 - 3s - 5t$, so $3s + 5t = 6$. Similarly, $0 = \langle 2, -1, -1 \rangle \cdot \langle -3, 0, -1 \rangle + s\|\langle 2, -1, -1 \rangle\|^2 - t\langle -2, -1, 0 \rangle \cdot \langle 2, -1, -1 \rangle = -5 + 6s + 3t$, so $6s + 3t = 5$. Solving for t and s, we have $s = 1/3$ and $t = 1$. Then $\mathbf{A}$

$= \overrightarrow{OR_1} = \langle 3, 2, 1 \rangle + 1\langle -2, -1, 0 \rangle = \langle 1, 1, 1 \rangle$ and $\mathbf{B} = \overrightarrow{OR_2} = \langle 0, 2, 0 \rangle + \frac{1}{3}\langle 2, -1, -1 \rangle = \left\langle \frac{2}{3}, \frac{5}{3}, -\frac{1}{3} \right\rangle$, so $R_1 = (1, 1, 1)$, and $R_2 = \left(\frac{2}{3}, \frac{5}{3}, -\frac{1}{3} \right)$.

12.S Review Exercises

2 (a) -2

(b) 3

(c) 1.2

(d) $\dfrac{(-2\mathbf{i} + 3\mathbf{j}) \cdot (4\mathbf{i} - 5\mathbf{j})}{\sqrt{16 + 25}} = -\dfrac{23}{\sqrt{41}}$

6 (a) Let $\mathbf{B} = \mathbf{A} \times \mathbf{C}$, where $\mathbf{C}$ is not parallel to $\mathbf{A}$.

(b) Take the vector $\mathbf{A} \times \mathbf{B}$.

8 The vector $2\mathbf{i} - \mathbf{j} + 3\mathbf{k}$ is perpendicular to the plane, so the line $\mathbf{P} = t(2\mathbf{i} - \mathbf{j} + 3\mathbf{k})$ is perpendicular to the plane and passes through the origin. Substituting into the equation of the plane yields $2(2t) + t + 3(3t) + 12 = 0$ and $t = -6/7$, so the required point is $\left(-\dfrac{12}{7}, \dfrac{6}{7}, -\dfrac{18}{7} \right)$.

10 $2x + y = 4$ and $x + 3y = -2$, so $x = \dfrac{14}{5}$, $y = -8/5$.

12 $\begin{vmatrix} a_1 & 0 & a_3 \\ b_1 & 0 & b_3 \\ c_1 & 0 & c_3 \end{vmatrix} = a_1 \begin{vmatrix} 0 & b_3 \\ 0 & c_3 \end{vmatrix} - 0 \begin{vmatrix} b_1 & b_3 \\ c_1 & c_3 \end{vmatrix} + a_3 \begin{vmatrix} b_1 & 0 \\ c_1 & 0 \end{vmatrix}$

$= 0$

12.5 Summary: Review Exercises

14. If **A** is the vector in question and the distance from the point to the plane is D, then $D = \|\mathbf{A}\| \cos 45°$. Now $D = \dfrac{|1 - 12 + 15 + 4|}{\sqrt{1 + 16 + 25}} = \dfrac{8}{\sqrt{42}}$. Hence $\|\mathbf{A}\| = \dfrac{8}{\sqrt{42} \cos 45°} = \dfrac{8}{\sqrt{21}}$.

16. A vector parallel to the line of intersection is
$$\begin{vmatrix} \mathbf{i} & \mathbf{j} & \mathbf{k} \\ 2 & 5 & 1 \\ 3 & -1 & 4 \end{vmatrix} = 21\mathbf{i} - 5\mathbf{j} - 17\mathbf{k}.$$

  (a) $21, -5, -17$

  (b) $\dfrac{21}{\sqrt{755}}, -\dfrac{5}{\sqrt{755}}, -\dfrac{17}{\sqrt{755}}$ or their negatives.

  (c) $\alpha \approx 40.158°, \beta \approx 100.484°, \gamma \approx 128.221°$, or $\alpha \approx 139.842°, \beta \approx 79.516°, \gamma \approx 51.779°$

  (d) $(10, -1, -5)$ is one such point.

18. $\mathbf{N}_1 = \langle 1, -3, 4\rangle$, $\mathbf{N}_2 = \langle 2, 1, 1\rangle$, $\cos\theta = \dfrac{\mathbf{N}_1 \cdot \mathbf{N}_2}{|\mathbf{N}_1||\mathbf{N}_2|} = \dfrac{2 - 3 + 4}{\sqrt{26}\sqrt{6}} = \dfrac{3}{\sqrt{156}}$, $\theta = \cos^{-1}\dfrac{3}{\sqrt{156}} \approx 1.33$ radians or $76.1°$

20. An equation for the plane is $2x + 4y + 5z - 1 = 0$; $2\cdot 4 + 4\cdot 5 + 5(-7) - 1 = -8 \neq 0$, so the point does not lie on the plane.

22. (a) $\text{comp}_\mathbf{i}\,\mathbf{A} = \|\text{proj}_\mathbf{i}\,\mathbf{A}\| = \|\langle 2, 0, 0\rangle\| = 2$

  (b) $\text{comp}_{-\mathbf{i}}\,\mathbf{A} = -\|\text{proj}_\mathbf{i}\,\mathbf{A}\| = -\|\langle 2, 0, 0\rangle\| = -2$

  (c) $\text{proj}_\mathbf{i}\,\mathbf{A} = \langle 2, 0, 0\rangle$

  (d) $\text{proj}_{-\mathbf{i}}\,\mathbf{A} = \langle 2, 0, 0\rangle$

  (e) $\text{proj}_{\mathbf{i}+\mathbf{j}}\,\mathbf{A} = \left(\mathbf{A} \cdot \dfrac{\mathbf{i} + \mathbf{j}}{\|\mathbf{i} + \mathbf{j}\|}\right)\dfrac{\mathbf{i} + \mathbf{j}}{\|\mathbf{i} + \mathbf{j}\|}$
  $= \dfrac{5}{\sqrt{2}}\dfrac{\mathbf{i} + \mathbf{j}}{\sqrt{2}} = \dfrac{5}{2}\mathbf{i} + \dfrac{5}{2}\mathbf{j}$

24. (a)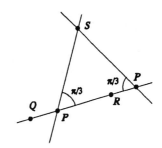

  (b) We know that $\overrightarrow{PS} \cdot \overrightarrow{QR} = \cos\dfrac{\pi}{3}\,|\overrightarrow{PS}||\overrightarrow{QR}|$
  $= \dfrac{1}{2}|\overrightarrow{PS}||\overrightarrow{QR}|$. Also, P lies on the line L through Q and R, so $\overrightarrow{OP} = \overrightarrow{OQ} + t\overrightarrow{QR}$ for some t. The dot product gives us an equation which we can solve for t.

  (c) We have $Q = (1, 0, 1)$, $R = (2, -1, 2)$, and $S = (3, 2, -4)$, so $\overrightarrow{QR} = \mathbf{i} - \mathbf{j} + \mathbf{k}$ and the line L is given by $x = 1 + t$, $y = -t$, and $z = 1 + t$. Hence $\overrightarrow{PS} = [3 - (1 + t)]\mathbf{i} + [2 - (-t)]\mathbf{j} + [-4 - (1 + t)]\mathbf{k} = (2 - t)\mathbf{i} + (2 + t)\mathbf{j} - (5 + t)\mathbf{k}$. Thus the equation $\overrightarrow{PS} \cdot \overrightarrow{QR} = \dfrac{1}{2}|\overrightarrow{PS}||\overrightarrow{QR}|$ becomes $-5 - 3t = \dfrac{1}{2}\sqrt{(2 - t)^2 + (2 + t)^2 + (5 + t)^2}\sqrt{3}$, so $25 + 30t + 9t^2 = \dfrac{3}{4}(33 + 10t + 3t^2)$, which reduces to $27t^2 + 90t + 1 = 0$, so $t =$

$\dfrac{-15 \pm \sqrt{222}}{9}$. The two possibilities for P are thus $\left(\dfrac{-6 \pm \sqrt{222}}{9}, \dfrac{-15 \pm \sqrt{222}}{9}, \dfrac{-6 \pm \sqrt{222}}{9}\right)$.

26 They lie on the same side of the plane if and only if $Ax_1 + By_1 + Cz_1 + D$ and $Ax_2 + By_2 + Cz_2 + D$ have the same sign.

28 (a) The volume is $\text{abs}\begin{vmatrix} 1 & 2 & 3 \\ 2 & 1 & 1 \\ 3 & 3 & 1 \end{vmatrix} = 9$.

(b) $\text{mag}\begin{vmatrix} \mathbf{i} & \mathbf{j} & \mathbf{k} \\ 1 & 2 & 3 \\ 3 & 3 & 1 \end{vmatrix} = \|-7\mathbf{i} + 8\mathbf{j} - 3\mathbf{k}\| = \sqrt{122}$

(c) Call the angle θ. Since $\mathbf{B} \times \mathbf{C} = -2\mathbf{i} + \mathbf{j} + 3\mathbf{k}$ is normal to the face spanned by $\mathbf{B}$ and $\mathbf{C}$, we have $\sin\theta = \dfrac{\text{comp}_{\mathbf{B} \times \mathbf{C}} \mathbf{A}}{\|\mathbf{A}\|} = \dfrac{\mathbf{A} \cdot (\mathbf{B} \times \mathbf{C})}{\|\mathbf{A}\|\|\mathbf{B} \times \mathbf{C}\|} = \dfrac{9}{\sqrt{14}\sqrt{14}} = \dfrac{9}{14}$. Therefore, $\theta = \sin^{-1}(9/14) \approx 40.005°$.

30 $\mathbf{u}_2, -\mathbf{u}_2$

32 $\overrightarrow{P_1P_4} \cdot \left(\overrightarrow{P_1P_2} \times \overrightarrow{P_1P_3}\right)$ and $\overrightarrow{P_1P_5} \cdot \left(\overrightarrow{P_1P_2} \times \overrightarrow{P_1P_3}\right)$ have the same sign.

34 From $\mathbf{A}_1 + \mathbf{A}_2 = \mathbf{A}'_1 + \mathbf{A}'_2$, it follows that $\mathbf{A}_1 - \mathbf{A}'_1 = \mathbf{A}'_2 - \mathbf{A}_2$. But $\mathbf{A}_1 - \mathbf{A}'_1$ is parallel to $\mathbf{B}$, while $\mathbf{A}'_2 - \mathbf{A}_2$ is perpendicular to $\mathbf{B}$. Hence this vector must be $\mathbf{0}$, so $\mathbf{A}_1 - \mathbf{A}'_1 = \mathbf{A}'_2 - \mathbf{A}_2 = \mathbf{0}$. Therefore $\mathbf{A}_1 = \mathbf{A}'_1$ and $\mathbf{A}_2 = \mathbf{A}'_2$.

36 Assuming that $\mathbf{N}_1 = (A_1\mathbf{i} + B_1\mathbf{j} + C_1\mathbf{k}) \times (A_2\mathbf{i} + B_2\mathbf{j} + C_2\mathbf{k}) \neq \mathbf{0}$ and $\mathbf{N}_2 = (A_3\mathbf{i} + B_3\mathbf{j} + C_3\mathbf{k}) \times (A_4\mathbf{i} + B_4\mathbf{j} + C_4\mathbf{k}) \neq \mathbf{0}$ then the intersections are perpendicular if $\mathbf{N}_1 \cdot \mathbf{N}_2 = 0$.

40 $\overrightarrow{P_0P_1} = \langle -1, 2, 1 \rangle$. The equations for L are $x = 2 - t$, $y = 1 + 2t$, $z = 1 + t$. Also $\overrightarrow{P_2P} = \langle (2 - t) - 3, (1 + 2t) - 1, (1 + t) - 5 \rangle = \langle -1 - t, 2t, -4 + t \rangle$. $\overrightarrow{P_2P}$ will be perpendicular to L if and only if $\overrightarrow{P_2P} \cdot \langle -1, 2, 1 \rangle = \langle -1 - t, 2t, -4 + t \rangle \cdot \langle -1, 2, 1 \rangle = 0$. Hence $1 + t + 4t - 4 + t = 0$, so $6t - 3 = 0$, and $t = 1/2$. Therefore, $P = (3/2, 2, 3/2)$.

42 We have $\mathbf{A} = 2\mathbf{i} + 3\mathbf{j} + 4\mathbf{k}$ and $\mathbf{n} = \dfrac{\mathbf{i} - \mathbf{j} + 2\mathbf{k}}{\sqrt{6}}$. By Exercise 26 of Sec. 12.7, the vector we want is $\mathbf{B} = \mathbf{A} - 2(\mathbf{A} \cdot \mathbf{n})\mathbf{n} = 2\mathbf{i} + 3\mathbf{j} + 4\mathbf{k} - 2\left[(2\mathbf{i} + 3\mathbf{j} + 4\mathbf{k}) \cdot \dfrac{\mathbf{i} - \mathbf{j} + 2\mathbf{k}}{\sqrt{6}}\right]\dfrac{\mathbf{i} - \mathbf{j} + 2\mathbf{k}}{\sqrt{6}}$
$= 2\mathbf{i} + 3\mathbf{j} + 4\mathbf{k} - \dfrac{7}{3}(\mathbf{i} - \mathbf{j} + 2\mathbf{k})$
$= -\dfrac{1}{3}\mathbf{i} + \dfrac{16}{3}\mathbf{j} - \dfrac{2}{3}\mathbf{k}$.

44 Suppose that the bottom square has vertices in the xy plane with coordinates $(0, 0, 0)$, $(3, 0, 0)$, $(3, 3, 0)$, and $(0, 3, 0)$. The coordinates of the vertices of the top square would be $(-1, -1, 10)$, $(4, -1, 10)$, $(4, 4, 10)$, and $(-1, 4, 10)$. The plane containing $(0, 0, 0)$, $(3, 0, 0)$ and $(-1, -1, 10)$ has a normal $3\mathbf{i} \times (-\mathbf{i} - \mathbf{j} + 10\mathbf{k}) = -30\mathbf{j} - 3\mathbf{k}$, so we may use $\mathbf{N}_1 = -10\mathbf{j} - \mathbf{k}$. The plane passing through $(0, 0, 0)$, $(0, 3, 0)$, and $(-1, -1, 10)$ has a normal $3\mathbf{j} \times (-\mathbf{i} - \mathbf{j} + 10\mathbf{k}) = 30\mathbf{i} + 3\mathbf{k}$, so we may use $\mathbf{N}_2 = 10\mathbf{i} + \mathbf{k}$. The angle between N_1 and N_2 is $\theta = \cos^{-1}\left(-\dfrac{1}{101}\right)$

$\approx 1.5807 \approx 90.57°$.

46 (a) Given **A**, **B**, and **C**, we know that $\text{proj}_N \mathbf{C} =$

$\text{proj}_{\mathbf{A}\times\mathbf{B}} \mathbf{C} = \left(\mathbf{C} \cdot \dfrac{\mathbf{A}\times\mathbf{B}}{\|\mathbf{A}\times\mathbf{B}\|}\right) \dfrac{\mathbf{A}\times\mathbf{B}}{\|\mathbf{A}\times\mathbf{B}\|}$

$= \dfrac{\mathbf{C}\cdot(\mathbf{A}\times\mathbf{B})}{\|\mathbf{A}\times\mathbf{B}\|^2}(\mathbf{A}\times\mathbf{B})$

(b) $\mathbf{C} - \dfrac{\mathbf{C}\cdot(\mathbf{A}\times\mathbf{B})}{\|\mathbf{A}\times\mathbf{B}\|^2}(\mathbf{A}\times\mathbf{B})$

48 Taking 0° longitude to be the negative x direction and 90° to be the positive y direction (see the

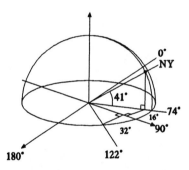

figure), we can compute coordinates for the locations of New York and San Francisco. Since the earth's radius is 4000, the z coordinate for New York is $4000 \sin 41°$. As shown in the figures, the y coordinate is $4000 \cos 41° \cos 16°$ and the x coordinate is $-4000 \cos 41° \sin 16°$.

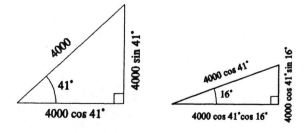

Hence the position vector of New York is $\mathbf{P} \approx$

$4000(-0.2080\mathbf{i} + 0.7255\mathbf{j} + 0.6561\mathbf{k})$. Similarly, the position vector of San Francisco is $\mathbf{Q} \approx 4000(0.4176\mathbf{i} + 0.6683\mathbf{j} + 0.6157\mathbf{k})$.

(a) The angle between **P** and **Q** is approximately 36.6924°, or 0.6404 radians, so the great circle distance between New York and San Francisco is about $4000(0.6404) = 2{,}561$ miles.

(b) The straight-line distance between New York and San Francisco is $\|\mathbf{P} - \mathbf{Q}\| \approx 2518$ miles.

13 The Derivative of a Vector Function

13.1 The Derivative of a Vector Function

2. (a) $G(0) = i$
   $G(1) = 3i + 4j$
   $G(2) = 5i + 8j$

   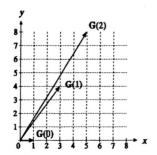

   (b) $y = 4\left(\dfrac{x-1}{2}\right)$ is a line.

4. (a) $\Delta G = (6.03i + 8.0802j) - (6i + 8j)$
   $= 0.03i + 0.0802j$

   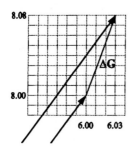

   (b) $\dfrac{\Delta G}{0.01} = 3.00i + 8.02j$

   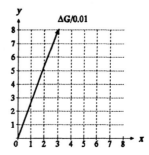

   (c) $G'(2) = 3i + 8j$

6. (a) $r(t) = 2ti + 4t^2 j$
   (b) See (d).
   (c) $\|r'(1)\| = \|2i + 8j\| = \sqrt{2^2 + 8^2} = 2\sqrt{17}$
   (d)

   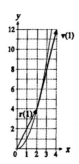

8. (a) Distance traveled $\approx \|G(2.01) - G(2)\|$
   $= \|0.03i - 0.01j + 0.02k\| = \dfrac{\sqrt{14}}{100}$

   (b) $G'(2) \approx \dfrac{\Delta G}{0.01} = \dfrac{G(2.01) - G(2)}{0.01}$
   $= \dfrac{0.03i - 0.01j + 0.02k}{0.01} = 3i - j + 2k$

(c) Speed $= \|G'(2)\| \approx \|3i - j + 2k\|$
$= \sqrt{14}$

10 $v(t) = r'(t) = (3 \cos 5t)'i + (2 \sin 5t)'j + (t^2)'k$
$= (-15 \sin 5t)i + (10 \cos 5t)j + (2t)k$, Speed $=$
$\|v(t)\| = \sqrt{225 \sin^2 5t + 100 \cos^2 5t + 4t^2}$
$= \sqrt{100 + 125 \sin^2 5t + 4t^2}$

12 $v(t) = r'(t) = (\sec^2 3t)'i + (\sqrt{1 + t^2})'j$
$= [2 \sec 3t (\sec 3t \tan 3t)(3)]i + \left(\dfrac{2t}{2\sqrt{1 + t^2}}\right)j$
$= (6 \sec^2 3t \tan 3t)i + \dfrac{t}{\sqrt{1 + t^2}}j$, Speed $=$
$\|v(t)\| = \sqrt{36 \sec^4 3t \tan^2 3t + \dfrac{t^2}{1 + t^2}}$

14 (a)

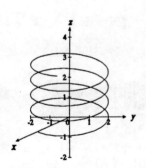

(b) $\|G'(t)\| = |-8\pi \sin 4\pi t\, i + 8\pi \cos 4\pi t\, j + k\|$
$= \sqrt{64\pi^2 + 1}$

(c) $\dfrac{G'(t)}{|G'(t)|}$

$= \dfrac{-8\pi \sin 4\pi t\, i + 8\pi \cos 4\pi t\, j + k}{\sqrt{64\pi^2 + 1}}$

16 (a) $x = 4t$ and $y = 16t^2$, so $y = x^2$, as claimed.
(b)

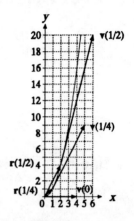

(c) $v(t) = 4i + 32tj$, $\lim\limits_{t \to \infty} |v(t)| = \infty$,

$\lim\limits_{t \to \infty} \dfrac{v(t)}{|v(t)|} = j$

18 (a) $\left(\dfrac{x}{2}\right)^2 + y^2 = (\cos t^2)^2 + (\sin t^2)^2 = 1$

(b) $r(t) = 2 \cos t^2\, i + \sin t^2\, j$
$v(t) = r'(t) = -4t \sin t^2\, i + 2t \cos t^2\, j$

(c) $\|v(t)\| = \sqrt{16t^2 \sin^2 t^2 + 4t^2 \cos^2 t^2} =$
$2t\sqrt{3 \sin^2 t^2 + 1}$; $\lim\limits_{t \to \infty} |v(t)| = \infty$.

20 $r(t) = 100ti + (100\sqrt{3}t - 16t^2)j$, $v(t)$
$= 100i + (100\sqrt{3} - 32t)j$

(a) $r(0) = 0$, $v(0) = 100i + 100\sqrt{3}j$

(b) The ball reaches its maximum height at $t =$
$\dfrac{100\sqrt{3}}{32} = \dfrac{25\sqrt{3}}{8}$. $r\left(\dfrac{25\sqrt{3}}{8}\right) =$
$\dfrac{625\sqrt{3}}{2}i + \dfrac{1875}{4}j$; $v\left(\dfrac{25\sqrt{3}}{8}\right) = 100i$.

(c) The ball strikes the ground at $t = \dfrac{25\sqrt{3}}{4}$.

$$r\left(\frac{25\sqrt{3}}{4}\right) = 625\sqrt{3}\,i, \quad v\left(\frac{25\sqrt{3}}{4}\right)$$

$$= 100i - 100\sqrt{3}\,j.$$

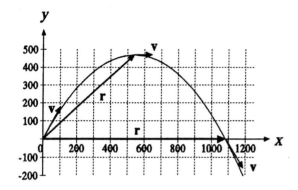

22. (a) $r(t) = (x_0 + 3t\cos\alpha)i + (y_0 + 3t\cos\beta)j + (z_0 + 3t\cos\gamma)k$

    (b) $v(t) = 3\cos\alpha\,i + 3\cos\beta\,j + 3\cos\gamma\,k$

24. (b) $G'(t) = i + 2tj + 3t^2k$ and $H'(t) = 2ti + 3t^2j + 4t^3k$, so $G'(1) = i + 2j + 3k$ and $H'(1) = 2i + 3j + 4k$; hence $\cos\theta =$

    $$\frac{G'(1)\cdot H'(1)}{|G'(1)||H'(1)|} = \frac{20}{\sqrt{406}}, \text{ so } \theta =$$

    $$\cos^{-1}\frac{20}{\sqrt{406}} \approx 6.98°.$$

26. (a) The tangent line to the curve follows the path traced out by $G'(1) = 2i + 3j + 12k$ (with its tail at $(1, 3, 4)$, the coordinates of the head of $G(1)$). So the line is given parametrically by $x = 1 + 2(t-1)$, $y = 3 + 3(t-1)$, and $z = 4 + 12(t-1)$. The spaceship is located at $(2t - 1, 3t, 12t - 8)$ for $t > 1$.

    (b) No, since $2t - 1 = 9$ and $3t = 15$ for $t = 5$, but $12t - 8 = 52 \ne 50$.

    (c) We must minimize the distance (or its square) from $(2t - 1, 3t, 12t - 8)$ to $(9, 15, 50)$. Now the square of the distance is given by

$$(2t - 1 - 9)^2 + (3t - 15)^2 + (12t - 8 - 50)^2$$
$$= 13(t - 5)^2 + 4(6t - 29)^2, \text{ and}$$

$$\frac{d}{dt}[13(t-5)^2 + 4(6t-29)^2] =$$

$26(t - 5) + 48(6t - 29) = 0$ when $t =$

$\frac{761}{157}$. At this value of t, the distance is

$$\sqrt{13\left(\frac{761}{157} - 5\right)^2 + 4\left(6\cdot\frac{761}{157} - 29\right)^2} \approx 0.5755.$$

28. $\Delta G = G(t + \Delta t) - G(t)$ is an estimate of the distance the particle travels during the time Δt. So $\frac{\Delta G}{\Delta t}$ is approximately the velocity of the particle at time t (distance/time). Since $\frac{\Delta G}{\Delta t}$ represents the approximate velocity of the particle at time t, $\left\|\frac{\Delta G}{\Delta t}\right\|$ represents an estimate of its speed at time t.

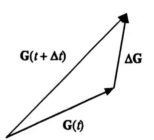

30. $\int \sin^2 3t\, dt = \frac{t}{2} - \frac{\sin 6t}{12} + C_1$ and

    $\int \frac{t}{3t^2 + 1}\, dt = \frac{1}{6}\ln(3t^2 + 1) + C_2$. Hence, $r(t)$

    $= \left(\frac{t}{2} - \frac{\sin 6t}{12}\right)i + \left[\frac{1}{6}\ln(3t^2 + 1) + 1\right]j.$

32. $\dfrac{1}{4} \displaystyle\int \dfrac{4t^3}{t^4 + 1}\, dt = \dfrac{1}{4} \ln(t^4 + 1) + C_1$ and

$\displaystyle\int \ln(t + 1)\, dt = (t + 1)\ln(t + 1) - t + C_2$, so

$\mathbf{r}(t) = \dfrac{1}{4}\ln(t^4 + 1)\mathbf{i} + ((t+1)\ln(t+1) - t)\mathbf{j}$.

34. $\displaystyle\int \dfrac{t}{(t+1)(t+2)(t+3)}\, dt = -\dfrac{1}{2}\ln|t+1| +$

$2\ln|t+2| - \dfrac{3}{2}\ln|t+3| + C_1$ and

$\displaystyle\int \dfrac{t^2}{(t+2)^3}\, dt = \ln|t+2| + \dfrac{4}{t+2} -$

$\dfrac{2}{(t+2)^2} + C_2$. Hence $\mathbf{r}(t) = \left[-\dfrac{1}{2}\ln|t+1| +\right.$

$2\ln|t+2| - \dfrac{3}{2}\ln|t+3| - 2\ln 2 +$

$\left.\dfrac{3}{2}\ln 3 + 1\right]\mathbf{i} +$

$\left[\ln|t+2| + \dfrac{4}{t+2} - \dfrac{2}{(t+2)^2} - \ln 2 - \dfrac{5}{2}\right]\mathbf{j}$.

36. $\displaystyle\int t^3 e^{-t}\, dt = -e^{-t}(t^3 + 3t^2 + 6t + 6) + C_1$

$\displaystyle\int (1+t)(2+t)\, dt = 2t + \dfrac{3}{2}t^2 + \dfrac{1}{3}t^3 + C_2$

Hence $\mathbf{r}(t) = [-e^{-t}(t^3 + 3t^2 + 6t + 6) + 8]\mathbf{i} +$

$\left[2t + \dfrac{3}{2}t^2 + \dfrac{1}{3}t^3 - 1\right]\mathbf{j}$.

13.2 Properties of the Derivative of a Vector Function

2. $(\mathbf{G}\cdot\mathbf{H})' = \mathbf{G}\cdot\mathbf{H}' + \mathbf{H}\cdot\mathbf{G}' =$
$[3t\mathbf{i} + \cos 3t\,\mathbf{j}]\cdot[2t\mathbf{i} + 3\cos 3t\,\mathbf{j}] +$
$[t^2\mathbf{i} + \sin 3t\,\mathbf{j}]\cdot[3\mathbf{i} - 3\sin 3t\,\mathbf{j}] = 9t^2 + 3\cos 6t$
$(\mathbf{G}\cdot\mathbf{H})' = ([3t\mathbf{i} + (\cos 3t)\mathbf{j}]\cdot[t^2\mathbf{i} + (\sin 3t)\mathbf{j}])'$
$= [3t^3 + \dfrac{1}{2}\sin 6t]' = 9t^2 + 3\cos 6t$

4. $(\mathbf{G}\cdot\mathbf{H})' = \mathbf{G}\cdot\mathbf{H}' + \mathbf{H}\cdot\mathbf{G}'$

$= \left[\tan^{-1} 3t\,\mathbf{i} + \dfrac{1}{t}\mathbf{j}\right]\cdot\left[2\sec^2 2t\,\mathbf{i} + \dfrac{1}{1+t}\mathbf{j}\right] +$

$[(\tan 2t)\mathbf{i} + \ln(1+t)\mathbf{j}]\cdot\left[\dfrac{3}{1+9t^2}\mathbf{i} - \dfrac{1}{t^2}\mathbf{j}\right]$

$= 2\sec^2 2t\,\tan^{-1} 3t +$

$\dfrac{1}{t(t+1)} + \dfrac{3\tan 2t}{1+9t^2} - \dfrac{\ln(1+t)}{t^2}$

$(\mathbf{G}\cdot\mathbf{H})' = \left(\left[\tan^{-1} 3t\,\mathbf{i} + \dfrac{1}{t}\mathbf{j}\right]\cdot[\tan 2t\,\mathbf{i} + \ln(1+t)\mathbf{j}]\right)'$

$= \left(\tan^{-1} 3t\,\tan 2t + \dfrac{\ln(1+t)}{t}\right)' = \dfrac{3\tan 2t}{1+9t^2} +$

$2\sec^2 2t\,\tan^{-1} 3t + \dfrac{1}{t(t+1)} - \dfrac{\ln(1+t)}{t^2}$

6. $[t^3\mathbf{G}(t)]' = t^3\mathbf{G}'(t) + 3t^2\mathbf{G}(t)$
$= (2t^3 + 9t^2)\mathbf{i} - 3t^3\mathbf{j} + 4t^3\mathbf{k} = \mathbf{0}$ for $t = 0$.

8. (a) $\Delta(\mathbf{G}\cdot\mathbf{H}) = \mathbf{G}(t + \Delta t)\cdot\mathbf{H}(t + \Delta t) - \mathbf{G}\cdot\mathbf{H}$
$= (\mathbf{G} + \Delta\mathbf{G})\cdot(\mathbf{H} + \Delta\mathbf{H}) - \mathbf{G}\cdot\mathbf{H}$
$= \mathbf{G}\cdot\mathbf{H} + \mathbf{G}\cdot\Delta\mathbf{H} + \Delta\mathbf{G}\cdot\mathbf{H} + \Delta\mathbf{G}\cdot\Delta\mathbf{H} - \mathbf{G}\cdot\mathbf{H}$
$= \mathbf{G}\cdot\Delta\mathbf{H} + \Delta\mathbf{G}\cdot\mathbf{H} + \Delta\mathbf{G}\cdot\Delta\mathbf{H}$

(b) $\displaystyle\lim_{\Delta t \to 0} \dfrac{\Delta(\mathbf{G}\cdot\mathbf{H})}{\Delta t}$

$= \displaystyle\lim_{\Delta t \to 0} \dfrac{\mathbf{G}\cdot\Delta\mathbf{H} + \Delta\mathbf{G}\cdot\mathbf{H} + \Delta\mathbf{G}\cdot\Delta\mathbf{H}}{\Delta t}$

$$= \lim_{\Delta t \to 0} \left(\mathbf{G} \cdot \frac{\Delta \mathbf{H}}{\Delta t} + \frac{\Delta \mathbf{G}}{\Delta t} \cdot \mathbf{H} + \Delta \mathbf{G} \cdot \frac{\Delta \mathbf{H}}{\Delta t} \right)$$

$$= \mathbf{G} \cdot \mathbf{H}' + \mathbf{G}' \cdot \mathbf{H}$$

12 $\Delta(\mathbf{G} + \mathbf{H}) = (\mathbf{G} + \mathbf{H})(t + \Delta t) - (\mathbf{G} + \mathbf{H})(t)$
$= \mathbf{G}(t + \Delta t) + \mathbf{H}(t + \Delta t) - \mathbf{G}(t) - \mathbf{H}(t)$
$= [\mathbf{G}(t + \Delta t) - \mathbf{G}(t)] + [\mathbf{H}(t + \Delta t) - \mathbf{H}(t)]$
$= \Delta \mathbf{G} + \Delta \mathbf{H}$, so $\lim_{\Delta t \to 0} \frac{\Delta(\mathbf{G} + \mathbf{H})}{\Delta t} =$

$\lim_{\Delta t \to 0} \left(\frac{\Delta \mathbf{G}}{\Delta t} + \frac{\Delta \mathbf{H}}{\Delta t} \right) = \mathbf{G}'(t) + \mathbf{H}'(t).$

14 (a) $\mathbf{r} = t\mathbf{i} + t^2\mathbf{j} + t\mathbf{k}$, so $\mathbf{v}(t) = \mathbf{r}'(t) = \mathbf{i} + 2t\mathbf{j} + \mathbf{k}$. Hence the speed is $\|\mathbf{v}(t)\|$
$= \sqrt{1^2 + (2t)^2 + 1^2} = \sqrt{4t^2 + 2}.$

(b) The point (t, t^2, t) is above the point $(t, t^2, 0)$ in the xy plane, but $(t, t^2, 0)$ is on the parabola $y = x^2$, so the particle's path lies above a parabola in the xy plane.

(c) Observe that the x and z coordinates of any point on the curve are equal. Therefore the curve lies in the plane $x = z$.

(d)

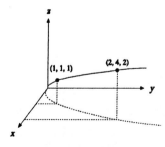

16 Since $\|\mathbf{r}\|^2 = \|\mathbf{G}(t)\|^2 = \mathbf{G}(t) \cdot \mathbf{G}(t)$, by Theorem 2 we have $\frac{d}{dt}(\|\mathbf{r}\|^2) = [\mathbf{G}(t) \cdot \mathbf{G}(t)]' =$

$\mathbf{G}(t) \cdot \mathbf{G}'(t) + \mathbf{G}'(t) \cdot \mathbf{G}(t) = 2\mathbf{G}(t) \cdot \mathbf{G}'(t) = 0$. The derivative of $\|\mathbf{r}\|^2$ is zero, so $\|\mathbf{r}\|^2$ does not vary with time. It follows that $\|\mathbf{r}\|$ is constant, which means that the path $\mathbf{r} = \mathbf{G}(t)$ lies on a sphere of radius $\|\mathbf{r}\|$.

18 $\mathbf{v}(t)$ is constant, so it is equal to $a\mathbf{i} + b\mathbf{j} + c\mathbf{k}$ for suitable constants a, b, and c. Now $\mathbf{r}(t) = x(t)\mathbf{i} + y(t)\mathbf{j} + z(t)\mathbf{k}$ and $\mathbf{v}(t) = x'(t)\mathbf{i} + y'(t)\mathbf{j} + z'(t)\mathbf{k}$, so $x'(t) = a$, $y'(t) = b$, and $z'(t) = c$. Thus $x(t) = at + a_0$, $y(t) = bt + b_0$, and $z(t) = ct + c_0$, for some constants a_0, b_0, and c_0, and $\mathbf{r}(t) = (at + a_0)\mathbf{i} + (bt + b_0)\mathbf{j} + (ct + c_0)\mathbf{k} = (a\mathbf{i} + b\mathbf{j} + c\mathbf{k})t + (a_0\mathbf{i} + b_0\mathbf{j} + c_0\mathbf{k}) = \mathbf{v}(t)t + \mathbf{r}(0)$. Recall from Sec. 12.4 that this is the equation of the line through the point $\mathbf{r}(0)$ that is parallel to the constant vector $\mathbf{v}(t)$; that is, $\mathbf{r}(t)$ lies on a straight line.

20 (a) $\mathbf{G}'(t) = \frac{1}{2}(1 + \sqrt{3t})^{-1/2}(\sqrt{3})\left(\frac{1}{2}t^{-1/2}\right)\mathbf{i} +$

$\frac{[(1 + t^3)(2)\sin 4t\, (4)\cos 4t - \sin^2 4t\, (3t^2)]\mathbf{j}}{(1 + t^3)^2}$

$= \frac{\sqrt{3}}{4}[t(1 + \sqrt{3t})]^{-1/2}\mathbf{i} +$

$\frac{[8(1 + t^3)\cos 4t \sin 4t - 3t^2 \sin^2 4t]}{(1 + t^3)^2}\mathbf{j}$

(b) $\mathbf{G}'(t) = \left[\frac{3t^2}{2\sqrt{t} + 2t} + 2t \ln 3 \ln(1 + \sqrt{t})3^{t^2}\right]\mathbf{i}$

$+ \frac{3}{\sqrt{1 - 9t^2}}\mathbf{j}$

(c) $\mathbf{G}'(t) = \frac{5}{|t|\sqrt{25t^2 - 1}}\mathbf{i} + 6\sec^3 2t \tan 2t\, \mathbf{j}$

13.3 The Acceleration Vector

2 $r(t) = \cos\dfrac{\pi e^t}{2} \mathbf{i} + \sin\dfrac{\pi e^t}{2} \mathbf{j}$

$\mathbf{v}(t) = -\dfrac{\pi}{2} e^t \sin\dfrac{\pi e^t}{2} \mathbf{i} + \dfrac{\pi}{2} e^t \cos\dfrac{\pi e^t}{2} \mathbf{j}$

$\mathbf{a}(t) = \left[-\dfrac{\pi^2}{4} e^{2t} \cos\dfrac{\pi e^t}{2} - \dfrac{\pi}{2} e^t \sin\dfrac{\pi e^t}{2} \right] \mathbf{i} +$

$\left[-\dfrac{\pi^2}{4} e^{2t} \sin\dfrac{\pi e^t}{2} + \dfrac{\pi}{2} e^t \cos\dfrac{\pi e^t}{2} \right] \mathbf{j}$

$\mathbf{r}(0) = \mathbf{j}$, $\mathbf{v}(0) = -\dfrac{\pi}{2}\mathbf{i}$, and $\mathbf{a}(0) = -\dfrac{\pi}{2}\mathbf{i} - \dfrac{\pi^2}{4}\mathbf{j}$.

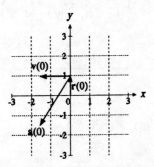

4 (a) A circle of radius 10

(b) $\mathbf{v} = -20\pi \sin 2\pi t\, \mathbf{i} + 20\pi \cos 2\pi t\, \mathbf{j}$

$\mathbf{a} = -40\pi^2 \cos 2\pi t\, \mathbf{i} - 40\pi^2 \sin 2\pi t\, \mathbf{j}$

(c) (**v** and **a** are not plotted to scale in the figure because their magnitudes are too great.)

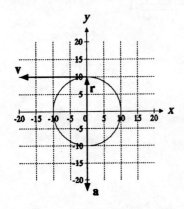

(d) $\mathbf{a} = -4\pi^2(10\cos 2\pi t\, \mathbf{i} + 10\sin 2\pi t\, \mathbf{j})$
$= -4\pi \mathbf{r}$

6 (a) $\mathbf{a} = -\cos t\, \mathbf{i} + \sin t\, \mathbf{j}$, which is a unit vector.

(b)
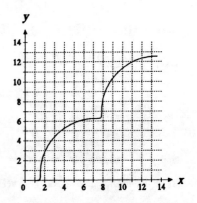

(c)
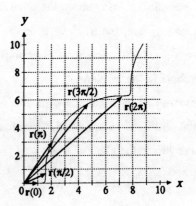

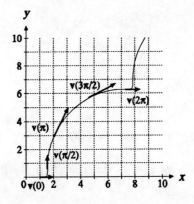

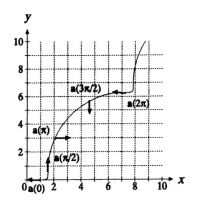

8 (a) $\mathbf{v}(t) = 2\mathbf{i} + 3\mathbf{j} + 26t\mathbf{k}$

   (b) $\mathbf{a}(t) = 26\mathbf{k}$

10 (a) $(m\mathbf{v})' = m(\mathbf{v})' + m'\mathbf{v} = m\mathbf{a} + m'\mathbf{v}$

   (b) If m is constant then $m' = 0$, so $\mathbf{F} = m\mathbf{a}$.

12 (a) If $\|\mathbf{v}\| = k$, a constant, then $\mathbf{v}\cdot\mathbf{v} = k^2$ and $0 = (\mathbf{v}\cdot\mathbf{v})' = \mathbf{v}\cdot\mathbf{v}' + \mathbf{v}'\cdot\mathbf{v} = 2\mathbf{v}'\cdot\mathbf{v} = 2\mathbf{a}\cdot\mathbf{v}$, which implies that $\mathbf{a}$ and $\mathbf{v}$ are perpendicular.

   (b) If $\mathbf{a}$ and $\mathbf{v}$ were not perpendicular, then $\mathbf{a}$ would have a component in the $\mathbf{v}$ direction. So $\|\mathbf{v}\|$ would not be constant.

14 (a) $\mathbf{r}(t) = \cos t^2\, \mathbf{i} + \sin t^2\, \mathbf{j}$, so $\mathbf{v}(t) = -2t\sin t^2\, \mathbf{i} + 2t\cos t^2\, \mathbf{j}$ and $\mathbf{a}(t) = (-2\sin t^2 - 4t^2\cos t^2)\mathbf{i} + (2\cos t^2 - 4t^2\sin t^2)\mathbf{j}$.

   (b) $\mathbf{r}\cdot\mathbf{a} = (\cos t^2)(-2\sin t^2 - 4t^2\cos t^2) + (\sin t^2)(2\cos t^2 - 4t^2\sin t^2) = -2\sin t^2 \cos t^2 - 4t^2\cos^2 t^2 + 2\sin t^2 \cos t^2 - 4t^2\sin^2 t^2 = -4t^2(\cos^2 t^2 + \sin^2 t^2) = -4t^2 \le 0$.

16 (a) 4. ($\|\mathbf{a}\| = v^2/r$. See also Exercise 28.)

   (b) Half as much

18 The distance traveled in one orbit or radius r is $2\pi r$. If the orbital speed is v, then the orbital time (or period) is given by $t = \dfrac{2\pi r}{v}$. From part (b) of Exercise 17, we know that $v = \sqrt{k/r}$, so $t = \dfrac{2\pi r}{\sqrt{k/r}}$

$= \dfrac{2\pi r^{3/2}}{\sqrt{k}} = \dfrac{2\pi r^{3/2}}{4000\sqrt{0.006}} \approx 0.02028 r^{3/2}$ seconds.

At the surface of the earth $r = 4000$, so $t \approx 0.02028(4000)^{3/2} \approx 5130$ seconds $= 85.5$ minutes. One hundred miles above the earth's surface $r = 4100$, so $t \approx 0.02028(4100)^{3/2} \approx 5324$ seconds ≈ 88.7 minutes. One thousand miles above the earth's surface $r = 5000$, so $t \approx 0.02028(5000)^{3/2} \approx 7170$ seconds ≈ 119.5 minutes.

20 By Exercise 18, $r = (t/0.02028)^{2/3}$, where r is the orbital radius in miles and t is the orbital period in seconds. We have $t = 92$ minutes $= 5520$ seconds, so $r = (5520/0.02028)^{2/3} \approx 4200.0$ miles. The satellite is 200 miles above the surface of the earth.

22 (a) $\dfrac{dy}{dx} = \dfrac{\dot{y}}{\dot{x}}$, $\dfrac{d^2y}{dx^2} = \dfrac{\dot{x}\ddot{y} - \dot{y}\ddot{x}}{\dot{x}^3}$

   (b) $\dfrac{dy}{dx} = \dfrac{3}{2}$, $\dfrac{d^2y}{dx^2} = \dfrac{5}{8}$, $\dfrac{1}{\kappa} = \dfrac{[1 + (dy/dx)^2]^{3/2}}{|d^2y/dx^2|}$

   $= \dfrac{(13/4)^{3/2}}{5/8} = \dfrac{13\sqrt{13}}{5}$

24 $\int \dfrac{t\, dt}{(1-t^2)^{3/2}} = \dfrac{1}{(1-t^2)^{1/2}} + C_1$; $\mathbf{v}(0) = 4\mathbf{i} + 2\mathbf{j}$, so $C_1 = 3$.

$\int \left(\dfrac{1}{(1-t^2)^{1/2}} + 3\right) dt = \sin^{-1} t + 3t + C_2$;

$\mathbf{r}(0) = 2\mathbf{i}$, so $C_2 = 2$.

$\int \dfrac{-4\, dt}{(1+2t)^2} = \dfrac{2}{1+2t} + C_3$; $\mathbf{v}(0) = 4\mathbf{i} + 2\mathbf{j}$, so

$C_3 = 0$. $\int \dfrac{2\, dt}{1+2t} = \ln(1+2t) + C_4$; $\mathbf{r}(0) = 2\mathbf{i}$, so $C_4 = 0$.

Hence $r(t) = [\sin^{-1} t + 3t + 2]i + [\ln(1 + 2t)]j$.

26 With notation as in Exercise 25, we have $x''(t) = 2^t(\ln 2)^2$, so $x'(t) = (\ln 2)^2 \int 2^t \, dt = (\ln 2)2^t + C_0$. Since $x'(0) = 3 + \ln 2$, $C_0 = 3$. Hence $x(t) = \int [(\ln 2)2^t + 3] \, dt = 2^t + 3t + C_1$. Since $x(0) = 2$, $C_1 = 1$ and $x(t) = 2^t + 3t + 1$. Next, $y''(t) = -\dfrac{16t}{(1 + 4t^2)^2}$, so $y'(t) = -\int \dfrac{16t \, dt}{(1 + 4t^2)^2} = -2 \int \dfrac{8t \, dt}{(1 + 4t^2)^2} = 2(1 + 4t^2)^{-1} + C_2$. Since $y'(0) = 2$, $C_2 = 0$. Hence $y(t) = \int \dfrac{2 \, dt}{1 + 4t^2} = \tan^{-1} 2t + C_3$. Since $y(0) = 0$, $C_3 = 0$ and $y(t) = \tan^{-1} 2t$. Thus, $r(t) = (2^t + 3t + 1)i + \tan^{-1} 2t \, j$.

28 (a) From PRO, we see that $(v\Delta t)^2 + r^2 = (r + s)^2$. So $(v\Delta t)^2 = 2rs + s^2$. But s is

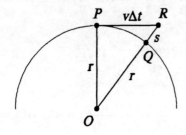

small, so s^2 is negligible. Then $s \approx \dfrac{(v\Delta t)^2}{2r}$.

(b) From (a), $s \cdot \dfrac{2}{(\Delta t)^2} = s \cdot \dfrac{1}{(\Delta t/\sqrt{2})^2} \approx \dfrac{v^2}{r}$.

Since $s \cdot \dfrac{1}{(\Delta t/\sqrt{2})^2} \approx \|a\|$, $\|a\| \approx \dfrac{v^2}{r}$.

30 The velocity is 1000 km/hr = $\dfrac{10^7 \text{ m}}{3600 \text{ s}} \approx$ 2777.78 m/s, and the desired acceleration is $a = 7 \cdot (9.8)$ m/s^2 = 68.6 m/s^2. Hence the corresponding radius is $\dfrac{v^2}{a} = \dfrac{(2777.78 \text{ m/s})^2}{68.6 \text{ m/s}^2} \approx 1.12 \times 10^5$ m.

13.4 The Components of Acceleration

2 $T = \dfrac{i - 16tj}{\sqrt{1 + 256t^2}}$, so

$T(1) = \dfrac{i - 16j}{\sqrt{257}}$,

$N(1) = \dfrac{-16i - j}{\sqrt{257}}$. See the figure at the right.

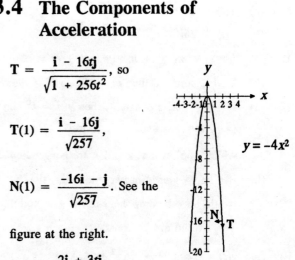

4 $T(t) = \dfrac{2i + 3tj}{\sqrt{4 + 9t^2}}$, so

$T(1) = \dfrac{2i + 3j}{\sqrt{13}}$; $N(1) = \dfrac{-3i + 2j}{\sqrt{13}}$

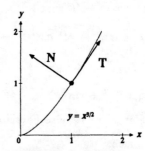

6 $T(t) = \dfrac{(1 - \cos t)i + (\sin t)j}{\sqrt{2 - 2\cos t}}$

$T(1) = \dfrac{(1 - \cos 1)i + \sin 1 \, j}{\sqrt{2 - 2\cos 1}}$

13.4 The Components of Acceleration

$$N(1) = \frac{\sin 1\, i - (1 - \cos 1)j}{\sqrt{2 - 2\cos 1}}$$

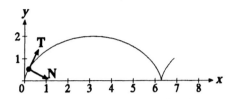

8 False

10 (a)

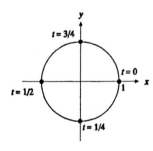

(b) $v(t) = 2\pi(-\sin 2\pi t\, i - \cos 2\pi t\, j)$
$T(t) = -\sin 2\pi t\, i - \cos 2\pi t\, j$
$N(t) = -\cos 2\pi t\, i + \sin 2\pi t\, j$
$r(0) = i,\ v(0) = -2\pi j,\ T(0) = -j,$
$N(0) = -i$

(c)

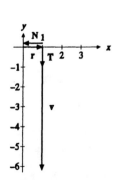

12 $v \cdot N = 0,\ v \cdot T > 0$

14 $a_T = -2,\ a_N = 9/2$

16 $a_T = 3,\ a_N = \sqrt{25 - 9} = 4$

18 $v(t) = 10\pi t(-\sin \pi t^2\, i + \cos \pi t^2\, j)$
$a(t) = [-20\pi^2 t^2 \cos \pi t^2 - 10\pi \sin \pi t^2]i +$
$[-20\pi^2 t^2 \sin \pi t^2 + 10\pi \cos \pi t^2]j$
$v(1) = -10\pi j,\ T(1) = -j,\ a(1) = 20\pi^2 i - 10\pi j,$
$a_T = a(1) \cdot T = 10\pi,\ a_N = 20\pi^2,\ \kappa = \dfrac{1}{r} = \dfrac{1}{5}$

20 (a) $a \cdot v \le 0$
    (b) $a \cdot v \ge 0$

22 $v(t) = -3\sin t\, i + 4\cos t\, j,$
$a(t) = -3\cos t\, i - 4\sin t\, j,$
$T(t) = \dfrac{-3\sin t\, i + 4\cos t\, j}{\sqrt{9\sin^2 t + 16\cos^2 t}},\ a_T =$

$\dfrac{-7\sin t \cos t}{\sqrt{9\sin^2 t + 16\cos^2 t}},\ a_N = \dfrac{9\sin^2 t + 16\cos^2 t}{\sqrt{9\cos^2 t + 16\sin^2 t}},$

$r = \sqrt{9\cos^2 t + 16\sin^2 t}$

24 The figure shows one possibility. Since the particle is slowing down, **a** must have a southward (downward) component.

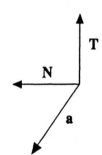

26 (a)

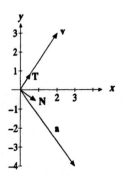

(b) $v = \|\mathbf{v}\| = \|2\mathbf{i} + 3\mathbf{j}\| = \sqrt{4 + 9} = \sqrt{13}$.

Now $\dfrac{d^2s}{dt^2} = a_T = \mathbf{a} \cdot \mathbf{T} = \mathbf{a} \cdot \dfrac{\mathbf{v}}{v} =$

$(3\mathbf{i} - 4\mathbf{j}) \cdot \left(\dfrac{2\mathbf{i} + 3\mathbf{j}}{\sqrt{13}}\right) = -\dfrac{6}{\sqrt{13}}$.

(c) $\mathbf{T} = \dfrac{1}{\sqrt{13}}(2\mathbf{i} + 3\mathbf{j})$ and $\mathbf{N}$ points into the fourth quadrant, so $\mathbf{N} = \dfrac{1}{\sqrt{13}}(3\mathbf{i} - 2\mathbf{j})$ and

$\mathbf{a} \cdot \mathbf{N} = \dfrac{3 \cdot 3 + (-4)(-2)}{\sqrt{13}} = \dfrac{17}{\sqrt{13}}$. But $\mathbf{a} \cdot \mathbf{N} =$

$a_N = \dfrac{v^2}{r} = \dfrac{13}{r}$, so $\dfrac{17}{\sqrt{13}} = \dfrac{13}{r}$ or

$r = \dfrac{13\sqrt{13}}{17}$.

28 (a) $\|\mathbf{v}\| = \sqrt{(t \cos t)^2 + (t \sin t)^2} = t$, $\|\mathbf{a}\|$
$= \sqrt{(\cos t - t \sin t)^2 + (\sin t + t \cos t)^2}$
$= \sqrt{1 + t^2}$.

(b) $a_T = \mathbf{a} \cdot \mathbf{T} = (\cos t - t \sin t)\cos t +$
$(\sin t + t \cos t) \sin t = 1$

$a_N = \mathbf{a} \cdot \mathbf{N} = (\cos t - t \sin t)(-\sin t) +$
$(\sin t + t \cos t) \cos t = t$

(c) $r = \dfrac{v^2}{a_N} = \dfrac{t^2}{t} = t$

30 (a)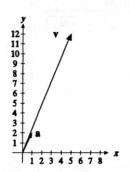

(c) $\mathbf{T} = \dfrac{5}{13}\mathbf{i} + \dfrac{12}{13}\mathbf{j}$, $\|\mathbf{a}\| = \sqrt{5}$, $a_T =$

$(\mathbf{i} + 2\mathbf{j}) \cdot \left(\dfrac{5}{13}\mathbf{i} + \dfrac{12}{13}\mathbf{j}\right) = \dfrac{29}{13}$, $a_N = \dfrac{2}{13}$

32 (a) $\mathbf{v}(t) = \mathbf{i} + 3t^2\mathbf{j}$; $\mathbf{T}(t) = \dfrac{\mathbf{i} + 3t^2\mathbf{j}}{\sqrt{1 + 9t^4}}$

(b) $\mathbf{T}'(t) = -\dfrac{18t^3}{(1 + 9t^4)^{3/2}}\mathbf{i} + \dfrac{6t}{(1 + 9t^4)^{3/2}}\mathbf{j}$

$= \dfrac{6t}{(1 + 9t^4)^{3/2}}(-3t^2\mathbf{i} + \mathbf{j})$. Since $\mathbf{T}'(0) = \mathbf{0}$,

$\mathbf{N}(0)$ is undefined.

(c) $y = x^3$ has an inflection point at $x = 0$.

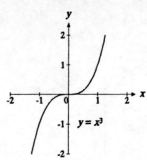

34 Since $\mathbf{T}$ is a unit vector, if we place $\mathbf{T}$ at the origin, its tip describes a path lying on the unit circle. For a unit circle, the angle ϕ equals the arc length s (if we agree to measure ϕ and s from a common starting point; otherwise they differ by a constant). Hence $d\mathbf{T}/d\phi$ is a unit vector.

13.4 The Components of Acceleration

36 (a) Here $x = t$ and $y = f(x) = f(t)$. So $\dfrac{dx}{dt} = 1$,

and $\dfrac{d^2x}{dt^2} = 0$. From Exercise 35, $r^2 =$

$$\dfrac{v^4}{\left(\dfrac{d^2x}{dt^2}\right)^2 + \left(\dfrac{d^2y}{dt^2}\right)^2 - \left(\dfrac{d^2s}{dt^2}\right)^2}$$

$= \dfrac{v^4}{\left(\dfrac{d^2y}{dx^2}\right)^2 - \left(\dfrac{dv}{dt}\right)^2} = \dfrac{v^4}{\left(\dfrac{d^2y}{dx^2}\right)^2 - \left(\dfrac{dv}{dx}\right)^2}$ since

$v = \dfrac{ds}{dt}$. Also, $v = \sqrt{\left(\dfrac{dx}{dt}\right)^2 + \left(\dfrac{dy}{dt}\right)^2}$

$= \sqrt{1 + \left(\dfrac{dy}{dx}\right)^2}$ and $\dfrac{dv}{dx} =$

$\left(1 + \left(\dfrac{dy}{dx}\right)^2\right)^{-1/2} \cdot \dfrac{dy}{dx} \cdot \dfrac{d^2y}{dx^2}$. Thus, $r^2 =$

$$\dfrac{\left(1 + \left(\dfrac{dy}{dx}\right)^2\right)^2}{\left(\dfrac{d^2y}{dx^2}\right)^2 - \left(1 + \left(\dfrac{dy}{dx}\right)^2\right)^{-1}\left(\dfrac{dy}{dx}\right)^2\left(\dfrac{d^2y}{dx^2}\right)^2}$$

$$= \dfrac{\left(1 + \left(\dfrac{dy}{dx}\right)^2\right)^2}{\left(\dfrac{d^2y}{dx^2}\right)^2\left[1 - \left(1 + \left(\dfrac{dy}{dx}\right)^2\right)^{-1}\left(\dfrac{dy}{dx}\right)^2\right]}$$

$= \dfrac{\left(1 + \left(\dfrac{dy}{dx}\right)^2\right)^3}{\left(\dfrac{d^2y}{dx^2}\right)^2}$ and $r = \dfrac{\left(1 + \left(\dfrac{dy}{dx}\right)^2\right)^{3/2}}{\left|\dfrac{d^2y}{dx^2}\right|}$.

(b) Equation (1) says that $\kappa = \dfrac{|\mathbf{v} \times \mathbf{a}|}{|\mathbf{v}|^3}$, so $r =$

$\dfrac{|\mathbf{v}|^3}{|\mathbf{v} \times \mathbf{a}|}$. We have $\mathbf{r} = x\mathbf{i} + y\mathbf{j}$, so $\mathbf{v} =$

$\mathbf{i} + y'\mathbf{j}$ and $\mathbf{a} = y''\mathbf{j}$. Now $\|\mathbf{v}\| =$

$\sqrt{1 + (y')^2}$ and $\mathbf{v} \times \mathbf{a} = y''\mathbf{k}$, yielding $r =$

$\dfrac{(1 + (y')^2)^{3/2}}{|y''|}$, the same result as (a).

38 If $\kappa = \left\|\dfrac{d\mathbf{T}}{ds}\right\|$, then $\dfrac{d\mathbf{T}}{dt} = \left\|\dfrac{d\mathbf{T}}{dt}\right\|\mathbf{N} = \left\|\dfrac{d\mathbf{T}}{ds}\right\|\left|\dfrac{ds}{dt}\right|\mathbf{N}$

$= \kappa v \mathbf{N} = \dfrac{v}{r}\mathbf{N}$, preserving the result of Theorem

2 (which was originally derived for planar curves only). Thus this definition of κ is consistent with the planar case, but extends an important property of curvature to space curves (without depending of ϕ, which was defined only for planar curves).

40 (a) **B** has constant magnitude, so its derivative is perpendicular to it.

(b) Since **B** and **B'** are perpendicular, **B'** lies in the plane containing **T** and **N**. Hence **B'** = $x\mathbf{T} + y\mathbf{N}$ for some scalars x and y.

(c) $\mathbf{B} \cdot \mathbf{T} = 0$, so $0 = (\mathbf{B} \cdot \mathbf{T})' = \mathbf{B} \cdot \mathbf{T}' + \mathbf{B}' \cdot \mathbf{T}$
$= \mathbf{B} \cdot (\|\mathbf{T}'\|\mathbf{N}) + \mathbf{B}' \cdot \mathbf{T} = \|\mathbf{T}'\|\mathbf{B} \cdot \mathbf{N} + \mathbf{B}' \cdot \mathbf{T}$
$= 0 + \mathbf{B}' \cdot \mathbf{T} = \mathbf{B}' \cdot \mathbf{T}$; hence **B'** and **T** are perpendicular, so **B'** is parallel to **N**.

42 (a) Since **N** has constant magnitude, $d\mathbf{N}/ds$ is perpendicular to **N**. Therefore $d\mathbf{N}/ds$ must lie in the plane determined by **T** and **B**; that is, $d\mathbf{N}/ds = p\mathbf{T} + q\mathbf{B}$ for some scalars p and q.

(b) $\mathbf{B} \cdot \mathbf{N} = 0$, so $0 = \dfrac{d}{ds}(\mathbf{B} \cdot \mathbf{N})$

$$= \mathbf{B} \cdot \frac{d\mathbf{N}}{ds} + \frac{d\mathbf{B}}{ds} \cdot \mathbf{N}$$

$$= \mathbf{B} \cdot (p\mathbf{T} + q\mathbf{B}) + (-\tau \mathbf{N}) \cdot \mathbf{N} = q - \tau;$$

hence $q = \tau$.

(c) $\mathbf{N} \cdot \mathbf{T} = 0$, so $0 = \frac{d}{ds}(\mathbf{N} \cdot \mathbf{T})$

$$= \mathbf{N} \cdot \frac{d\mathbf{T}}{ds} + \frac{d\mathbf{N}}{ds} \cdot \mathbf{T}$$

$$= \mathbf{N} \cdot (\kappa \mathbf{N}) + (p\mathbf{T} + q\mathbf{B}) \cdot \mathbf{T} = \kappa + p; \text{ hence}$$

$p = -\kappa$.

13.5 Newton's Law Implies Kepler's Three Laws

Note: All exercises from this section are solved in the *Student's Solutions Manual*.

13.S Review Exercises

2    $\mathbf{G}'(t) = \mathbf{i} + 2t\mathbf{j} + 3t^2\mathbf{k}$; $\mathbf{G}'(1) = \mathbf{i} + 2\mathbf{j} + 3\mathbf{k}$, so the astronaut travels along the line $(\mathbf{i} + \mathbf{j} + \mathbf{k}) + t(\mathbf{i} + 2\mathbf{j} + 3\mathbf{k})$

(a) $(\mathbf{i} + \mathbf{j} + \mathbf{k}) + 2(\mathbf{i} + 2\mathbf{j} + 3\mathbf{k}) = 3\mathbf{i} + 5\mathbf{j} + 7\mathbf{k}$

(b) $(5-1)\mathbf{i} + (8-1)\mathbf{j} + (9-1)\mathbf{k} = 4\mathbf{i} + 7\mathbf{j} + 8\mathbf{k}$ is a vector from $(1, 1, 1)$ to $(5, 8, 9)$.

$[\langle 4, 7, 8 \rangle \times \langle 1, 2, 3 \rangle] \times \langle 1, 2, 3 \rangle = \langle -14, -14, 14 \rangle$ is a vector perpendicular to the line and lying in the plane containing the line and the point. The distance between the point and plane is

$$\left| \frac{(4\mathbf{i} + 7\mathbf{j} + 8\mathbf{k}) \cdot (-14\mathbf{i} - 14\mathbf{j} + 14\mathbf{k})}{\sqrt{3} \cdot 14^2} \right| = \sqrt{3}.$$

4   (a) $v = \|\mathbf{v}\| = \|2\mathbf{i} + 3\mathbf{j}\| = \sqrt{2^2 + 3^2} = \sqrt{13}$

(b) $\mathbf{T} = \frac{\mathbf{v}}{v} = \frac{2\mathbf{i} + 3\mathbf{j}}{\sqrt{13}}$; since $\mathbf{N}$ is a unit vector perpendicular to $\mathbf{T}$, we know that $\mathbf{N} = \frac{\pm(3\mathbf{i} - 2\mathbf{j})}{\sqrt{13}}$. The particle is turning to the left (as can be seen from the figure below) and $\mathbf{T}$ points into the first quadrant, so $\mathbf{N}$ must point into the second quadrant: $\mathbf{N} = \frac{-3\mathbf{i} + 2\mathbf{j}}{\sqrt{13}}$.

(c)

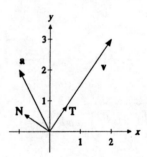

(d) $\mathbf{a} \cdot \mathbf{N} = (-\mathbf{i} + 2\mathbf{j}) \cdot (-3\mathbf{i} + 2\mathbf{j}) \cdot \frac{1}{\sqrt{13}} =$

$$\frac{(-1)(-3) + 2 \cdot 2}{\sqrt{13}} = \frac{7}{\sqrt{13}}$$

(e) $r = \frac{v^2}{a_N} = \frac{v^2}{\mathbf{a} \cdot \mathbf{N}} = \frac{13}{7/\sqrt{13}} = \frac{13\sqrt{13}}{7}$

(f) $\mathbf{a} \cdot \mathbf{T} = (-\mathbf{i} + 2\mathbf{j}) \cdot (2\mathbf{i} + 3\mathbf{j}) \cdot \frac{1}{\sqrt{13}}$

$$= \frac{(-1)(2) + 2 \cdot 3}{\sqrt{13}} = \frac{4}{\sqrt{13}}$$

(g) Since $\mathbf{a} \cdot \mathbf{T} > 0$, the particle is speeding up;

$$\frac{d^2s}{dt^2} = a_T = \mathbf{a} \cdot \mathbf{T} = \frac{4}{\sqrt{13}}.$$

(h) From $\mathbf{v} = 2\mathbf{i} + 3\mathbf{j} = \frac{dx}{dt}\mathbf{i} + \frac{dy}{dt}\mathbf{j}$, we know

$dy/dt = 3$.

(i) From (h), we know $\dfrac{dx}{dt} = 2$ and $\dfrac{dy}{dt} = 3$.

Hence $\dfrac{dy}{dx} = \dfrac{dy/dt}{dx/dt} = \dfrac{3}{2}$.

(j) From $\mathbf{a} = -\mathbf{i} + 2\mathbf{j} = \dfrac{d^2x}{dt^2}\mathbf{i} + \dfrac{d^2y}{dt^2}\mathbf{j}$, we know $\dfrac{d^2y}{dt^2} = 2$.

(k) $\kappa = \dfrac{|\mathbf{v} \times \mathbf{a}|}{\|\mathbf{v}\|^3} = \dfrac{|(2\mathbf{i} + 3\mathbf{j}) \times (-\mathbf{i} + 2\mathbf{j})|}{(\sqrt{13})^3}$

$= \dfrac{|7\mathbf{k}|}{13\sqrt{13}} = \dfrac{7}{13\sqrt{13}}$

6 (b) $\mathbf{G}'(t) = 2t\mathbf{i} + 6\mathbf{j}$, $\mathbf{H}'(t) = 3t^2\mathbf{i} + \mathbf{j}$ and
$\mathbf{G}'(1)\cdot\mathbf{H}'(1) = (2\mathbf{i} + 6\mathbf{j})\cdot(3\mathbf{i} + \mathbf{j}) = 6 + 6$
$= 12 = \sqrt{4 + 36}\sqrt{9 + 1}\cos\theta = \sqrt{400}\cos\theta$
$= 20\cos\theta$. Hence $\cos\theta = \dfrac{12}{20} = \dfrac{3}{5}$ and θ
$= \cos^{-1} 0.6 \approx 53.13°$.

8 $2\mathbf{i} + 5\mathbf{j} + 7\mathbf{j} + 4t\left(\dfrac{1}{3}\mathbf{i} + \dfrac{2}{3}\mathbf{j} + \dfrac{2}{3}\mathbf{k}\right)$

$= \left(2 + \dfrac{4}{3}t\right)\mathbf{i} + \left(5 + \dfrac{8}{3}t\right)\mathbf{j} + \left(7 + \dfrac{8}{3}t\right)\mathbf{k}$

10 (a) By symmetry, the circular arc connecting $(-1, 0)$ and $(1, 0)$ must lie on the circle

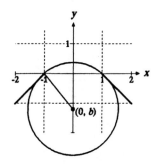

$x^2 + (y - b)^2 = 1 + b^2$. By implicit differentiation, $\dfrac{dy}{dx} = \dfrac{x}{b - y}$. At $(-1, 0)$, we must have $\dfrac{x}{b - y} = \dfrac{-1}{b - 0} = 1$, so $b = -1$.

At $(1, 0)$, we must have $\dfrac{x}{b - y} = \dfrac{1}{b - 0} = -1$, so $b = -1$. The results are consistent, showing that the line segments can be connected smoothly by an arc of radius $\sqrt{2}$.

(b) At $(-1, 0)$, the track suddenly changes from having 0 curvature to having a curvature of $1/\sqrt{2}$. Everyone in the car will feel a sudden rightward acceleration of $v^2/\sqrt{2}$, where v is the velocity of the train. A passenger standing up facing the engine could fall over to the left, while a passenger leaning against the left wall of the train would feel it suddenly press against him.

(c) The point $(1, 0)$ lies on $y = A + Bx^2 + Cx^4$, so $A = -B - C$. Now $y' = 2Bx + 4Cx^3$ and

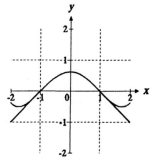

$y'' = 2B + 12Cx^2$. In order to make the curvature equal to 0 at $(-1, 0)$ and $(1, 0)$, we need $y'' = 0$ at those points; that is, $2B + 12C = 0$, so $B = -6C$. Hence the curve must be of the form $y = 5C - 6Cx^2 + Cx^4$. To make

the slopes match at $(-1, 0)$ and $(1, 0)$, we need $y' = -12Cx + 4Cx^3$ to equal 1 at the first point and -1 at the second. In either case, we find that $C = 1/8$, so the desired curve is $y = \frac{5}{8} - \frac{3}{4}x^2 + \frac{1}{8}x^4$.

12 The first astronaut's trajectory after she shuts off her rockets is given by $\mathbf{G}(1) + \mathbf{G}'(1)(t - 1) = \mathbf{j} + \mathbf{k} + (2\mathbf{i} + \mathbf{j} + \mathbf{k})(t - 1) = (2t - 2)\mathbf{i} + t\mathbf{j} + t\mathbf{k}$, while the second astronaut's trajectory after he shuts off his rockets is $\mathbf{F}(2) + \mathbf{F}'(2)(t - 2) = \mathbf{i} + 2\mathbf{k} + (\pi\mathbf{j} + \mathbf{k})(t - 2) = \mathbf{i} + \pi(t - 2)\mathbf{j} + t\mathbf{k}$. The square of the distance between them at time t is $(2t - 3)^2 + (\pi t - 2\pi - t)^2 + 0^2$. Differentiating with respect to t and setting the result equal to 0 shows that the astronauts are at their point of closest approach when $t = \frac{2\pi^2 - 2\pi + 6}{\pi^2 - 2\pi + 5} \approx 2.2659$, at which time their distance is approximately 2.0959.

14 (a) Since $\mathbf{a}$ is always pointing at $(0, 0, 0)$, this says that $\mathbf{a}$ and $\mathbf{r}$ point in opposite directions; that is, $\mathbf{a} = -q\mathbf{r}$ (where q is a positive scalar function). Thus $\frac{d}{dt}(\mathbf{r} \times \mathbf{v}) = \mathbf{r}' \times \mathbf{v} + \mathbf{r} \times \mathbf{v}'$
$= \mathbf{v} \times \mathbf{v} + \mathbf{r} \times \mathbf{a} = \mathbf{0} + \mathbf{r} \times (-q\mathbf{r}) = -q(\mathbf{r} \times \mathbf{r}) = \mathbf{0}$. Hence $\mathbf{r} \times \mathbf{v}$ is constant, as claimed.

(b) By (a), $\mathbf{r} \times \mathbf{v} = \mathbf{C}$ for some nonzero constant vector $\mathbf{C}$. Note that $\mathbf{r} \cdot \mathbf{C} = \mathbf{r} \cdot (\mathbf{r} \times \mathbf{v}) = (\mathbf{r} \times \mathbf{r}) \cdot \mathbf{C} = \mathbf{0} \cdot \mathbf{C} = 0$, so $\mathbf{r}$ is perpendicular to $\mathbf{C}$. Hence $\mathbf{r}$ describes a curve lying in a plane with normal $\mathbf{C}$.

16 The tangential vector component of $\mathbf{a}$ is $(\mathbf{a} \cdot \mathbf{T})\mathbf{T} = \left(\mathbf{a} \cdot \frac{\mathbf{v}}{|\mathbf{v}|}\right)\frac{\mathbf{v}}{|\mathbf{v}|}$, so the normal vector component of $\mathbf{a}$ is $\mathbf{a} - \left(\mathbf{a} \cdot \frac{\mathbf{v}}{|\mathbf{v}|}\right)\frac{\mathbf{v}}{|\mathbf{v}|}$, which points in the same direction as $\mathbf{N}$. By making this into a unit vector, we obtain $\mathbf{N}$ itself: $\mathbf{N} = \dfrac{\mathbf{a} - \left(\mathbf{a} \cdot \frac{\mathbf{v}}{|\mathbf{v}|}\right)\frac{\mathbf{v}}{|\mathbf{v}|}}{\left\|\mathbf{a} - \left(\mathbf{a} \cdot \frac{\mathbf{v}}{|\mathbf{v}|}\right)\frac{\mathbf{v}}{|\mathbf{v}|}\right\|}$.

14 Partial Derivatives

14.1 Graphs

2

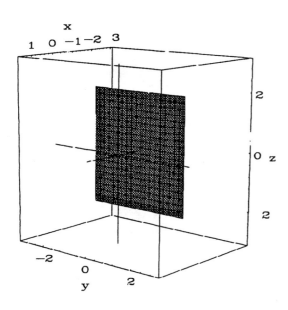

6

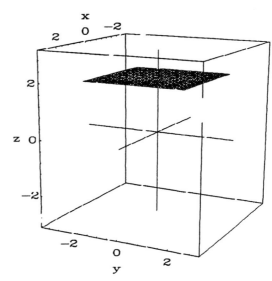

4

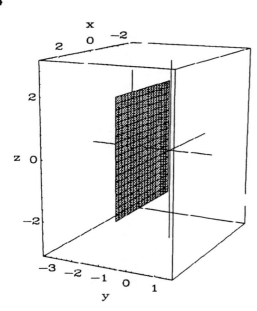

8

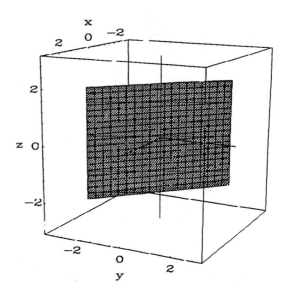

10

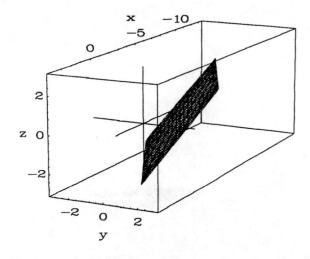

12

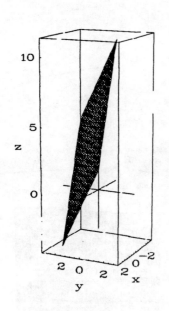

14 Equation: $(x-1)^2 + (y-1)^2 + (z-0)^2$
$= (\sqrt{2})^2$ or $x^2 - 2x + y^2 - 2y + z^2 = 0$

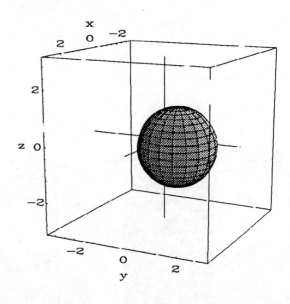

16 Equation: $(x-1)^2 + (y-2)^2 + (z-(-1))^2$
$= 3^2$ or $x^2 + y^2 + z^2 - 2x - 4y + 2z = 3$

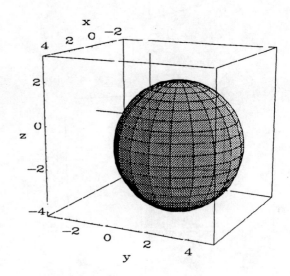

14.1 Graphs

18

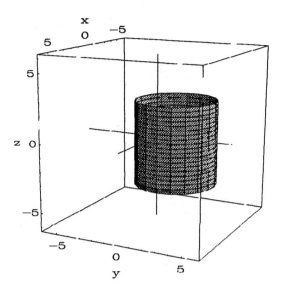

20

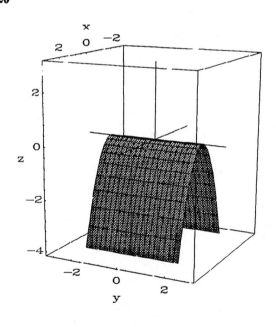

22

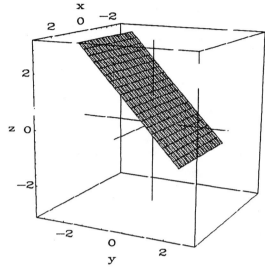

24

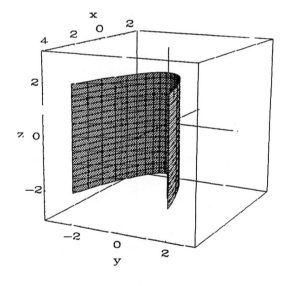

26

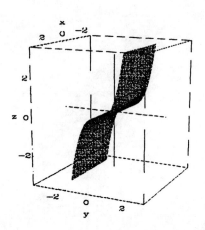

28

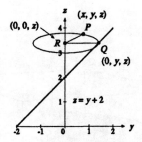

Since $\overline{PR} = \sqrt{x^2 + y^2} = \overline{QR}$, we have $\sqrt{x^2 + y^2} = z - 2$ or $z = 2 + \sqrt{x^2 + y^2}$ as the equation of the surface.

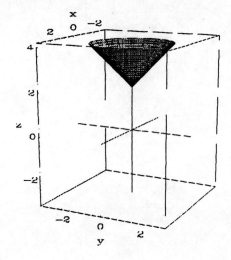

30 (a)

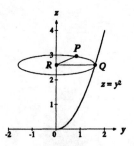

Since $\overline{PR} = \sqrt{x^2 + y^2} = \overline{QR}$, we have $\sqrt{x^2 + y^2} = \sqrt{z}$, or $x^2 + y^2 = z$ as the equation of the surface.

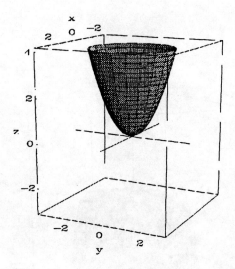

(b)

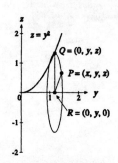

Since $\overline{PR} = \sqrt{x^2 + z^2} = \overline{QR}$, we have $\sqrt{x^2 + z^2} = y^2$ or $x^2 + z^2 = y^4$ as the equation of the surface.

32

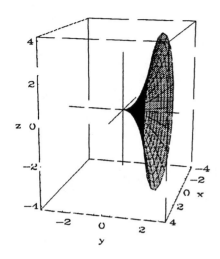

(b) The lines $z = 2y + 1$ and $y + z = 1$ in the yz plane cross at the point $P = (0, 0, 1)$ on the z axis. Using the point $Q = (0, 1, 3)$ on the first line and $R = (0, -1, 2)$ on the second line, we see that the angle θ between the two lines is given by $\cos \theta =$

$$\frac{\vec{PQ} \cdot \vec{PR}}{|\vec{PQ}||\vec{PR}|} = \frac{\langle 0, 1, 2 \rangle \cdot \langle 0, -1, 1 \rangle}{|\langle 0, 1, 2 \rangle||\langle 0, -1, 1 \rangle|} =$$

$\frac{1}{\sqrt{10}}$. For any point $S = (x, y, z)$ that lies on the surface of the double cone, we must have

$$\frac{1}{\sqrt{10}} = \frac{\vec{PS} \cdot \vec{PR}}{|\vec{PS}||\vec{PR}|}$$

$$= \frac{\langle x, y, z-1 \rangle \cdot \langle 0, -1, 1 \rangle}{\|\langle x, y, z-1 \rangle\|\|\langle 0, -1, 1 \rangle\|}$$

$$= \frac{-y + (z-1)}{\sqrt{x^2 + y^2 + (z-1)^2}\sqrt{2}}. \text{ Hence}$$

$$\sqrt{x^2 + y^2 + (z-1)^2} = \sqrt{5}[-y + (z-1)].$$

Squaring and collecting like terms yields
$x^2 - 4y^2 - 4(z-1)^2 + 10y(z-1) = 0$.

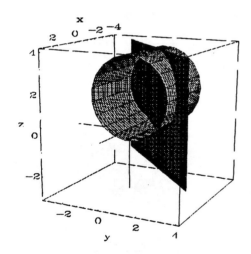

34 (a)

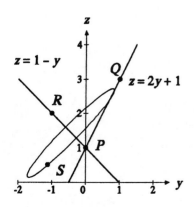

14.2 Quadric Surfaces

2

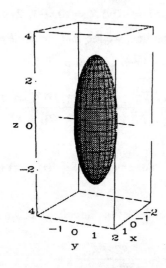

The trace in the plane $z = 0$ is a circle of radius 1, while the trace in the plane $z = 2$ is a circle of radius $\sqrt{5}/3$.

4

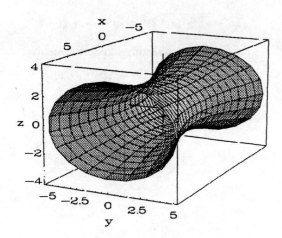

The trace in the plane $x = 8$ is the ellipse $4y^2 + 9z^2 = 100$.

6

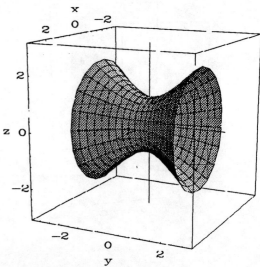

The trace in the plane $y = 0$ is a unit circle. In the planes $y = \pm\sqrt{3}$ they are circles of radius 2.

8

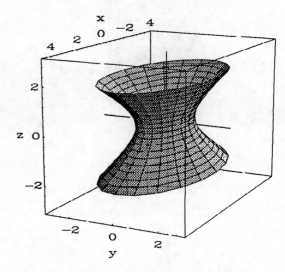

The trace in the plane $z = 0$ is the ellipse $x^2 + 4y^2 = 4$. In the planes $z = \pm\sqrt{3}$ they are the ellipse $x^2 + 4y^2 = 16$.

10 $3z^2 = x^2 + y^2$

12 $\dfrac{1}{3}y^2 = x^2 + z^2$

14 The half-vertex angle is $\alpha = \tan^{-1}\sqrt{\frac{3}{4}} \approx 40.9°$.

The trace in the plane $z = 1$ is a circle of radius $\sqrt{3}/2$ while the trace in the plane $x = 2$ is a hyperbola (as shown in the figure).

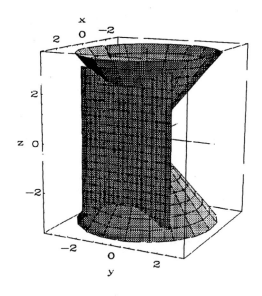

16 The half-vertex angle is $\alpha = \tan^{-1}\sqrt{3} = \pi/3 = 60°$, while the trace in the plane $y = 2$ is a circle of radius $2\sqrt{3}$ and the trace in the plane $z = 1$ is a hyperbola.

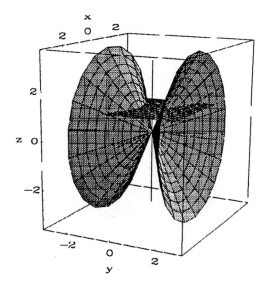

18 Let $P = (x, y, z)$ be a point on the surface of revolution. Let $Q = (0, y, z')$ be a point on the

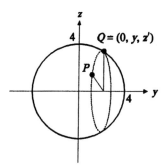

curve $y^2 + (z')^2 = 16$ from which P is derived. From the figure, we see $Q = (0, y, z') = (0, y, \sqrt{x^2 + z^2})$. Hence the surface is

$y^2 + \left(\sqrt{x^2 + z^2}\right)^2 = 16$ or $x^2 + y^2 + z^2 = 16$.

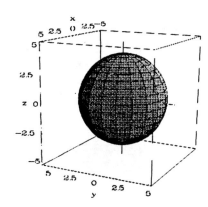

20 Let $P = (x, y, z)$ be a point on the surface of revolution. Let $Q = (x', 0, z)$ be a point on the

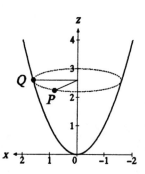

curve $z = (x')^2$, where $x' \geq 0$, from which P is derived. Now $x' = \sqrt{x^2 + y^2}$; so $Q =$

($\sqrt{x^2 + y^2}$, 0, z). Hence the equation for the surface is $z = \left(\sqrt{x^2 + y^2}\right)^2$ or $z = x^2 + y^2$. The surface is a paraboloid.

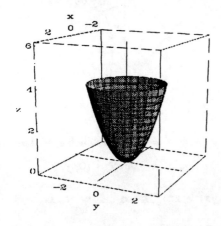

24

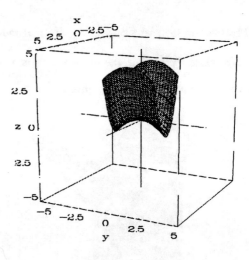

22

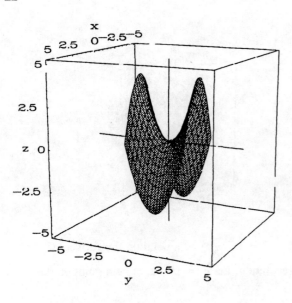

26

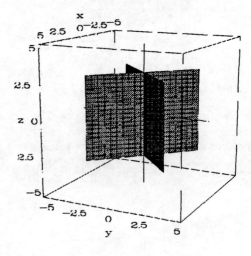

28

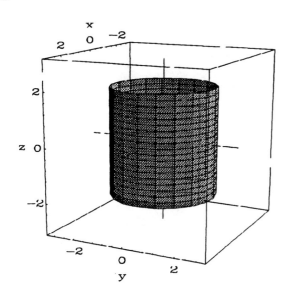

30 The plane $y = k$ is always parallel to the xz plane. The formula for the trace of $x^2 + y^2 - z^2 = 1$ in the $y = k$ plane is given by $x^2 + k^2 - z^2 = 1$ or $x^2 - z^2 = 1 - k^2$, which is a hyperbola (provided that $k^2 \neq 1$).

32 (a)

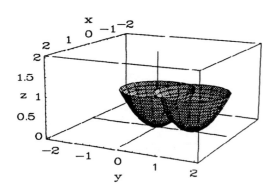

(b) The surfaces intersect in a curve that lies in the plane given by $x^2 + y^2 = x^2 + (y - 1)^2$, or $y = 1/2$, which is parallel to the xz plane.

(c) The curve is the trace of either $z = x^2 + y^2$ or $z = x^2 + (y - 1)^2$ in the plane $y = 1/2$; that is, the curve is given by $z = x^2 + (1/2)^2$ or $z = x^2 + 1/4$, which is a parabola.

34 Traces of $\dfrac{x^2}{a^2} + \dfrac{y^2}{b^2} - \dfrac{z^2}{c^2} = 1$ in planes parallel to the xz plane and yz plane are hyperbolas; so this surface cannot be a surface of revolution about the y or x axis. However, the trace of the hyperboloid in the plane $z = k$ is given by $\dfrac{x^2}{a^2} + \dfrac{y^2}{b^2} - \dfrac{k^2}{c^2} = 1$ or $\dfrac{x^2}{a^2} + \dfrac{y^2}{b^2} = 1 + \dfrac{k^2}{c^2}$. Since $1 + \dfrac{k^2}{c^2} \geq 1$, this trace is a circle if and only if $|a| = |b|$. There is no restriction on c. Under these constraints, this surface is a surface of revolution about the z axis.

36 The graph of $x^2 + y^2 - z^2 = 0$ is a double cone with half-angle $45°$ that contains the z axis. The graph of $x^2 + y^2 - z^2 = 1$ is a hyperboloid of one sheet that contains the z axis and approaches the double cone above asymptotically. The graph of $x^2 + y^2 - z^2 = -1$ is a hyperboloid of two sheets that does not intersect the x and y axes, and it approaches the double cone above asymptotically.

38 (a) The trace is given by $z = kx$, which is a line for $k \neq 0$.

40 (a)

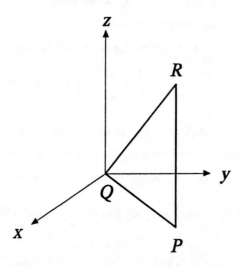

(b) $P = (3, 3, 0)$, $Q = (0, 0, 0)$, $R = (3, 3, 3)$

(c) Area $= \frac{1}{2} |\overrightarrow{QR} \times \overrightarrow{QP}|$

$= \frac{1}{2} |\langle 3, 3, 3 \rangle \times \langle 3, 3, 0 \rangle| = \frac{9\sqrt{2}}{2}$.

42

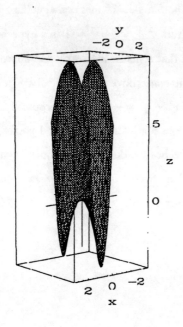

44

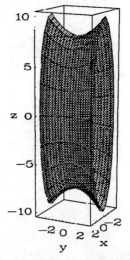

46 (a) To remove the z dependence, we substitute $z = x + y$ into $\frac{x^2}{1} + \frac{y^2}{4} + \frac{z^2}{9} = 1$. Then we have $\frac{x^2}{1} + \frac{y^2}{4} + \frac{(x+y)^2}{9} = 1$. This is the equation of C^*.

(b) The curve C^* is an ellipse.

14.3 Functions and Their Level Curves

2 Let $z = f(x, y) = x + 1$. Then the graph of f consists of all points (x, y, z) such that $z = x + 1$. Given k, the trace of f in the plane $y = k$ is the line $z = x + 1$. Rewriting $z = x + 1$ as $-x + z - 1 = 0$, we see that the graph of f is a plane.

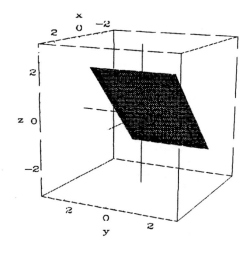

4 Let $z = f(x, y) = -2$. The graph is that of the constant function $f(x, y) = -2$. Thus the graph of f is a plane parallel to the xy plane.

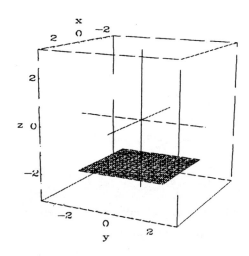

6 Let $z = f(x, y) = y^2$. The graph is the set of all points (x, y, z) under the constraint that $z = y^2$. For all k, the trace of f in the plane $x = k$ is the parabola $z = y^2$. Therefore, the graph of f is a parabolic cylinder.

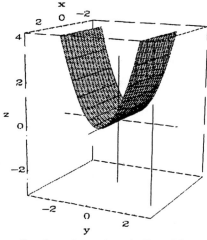

8 Let $z = f(x, y) = 2x - y + 1$. Rewriting $z = 2x - y + 1$ as $-2x + y + z - 1 = 0$, it is clear that the graph of f is a plane. When $x = y = 0$, $z = 1$; when $x = z = 0$, $y = 1$; when $y = z = 0$, $x = -1/2$. Hence the intercepts of the plane are $(-1/2, 0, 0)$, $(0, 1, 0)$, and $(0, 0, 1)$. These three points completely determine the plane.

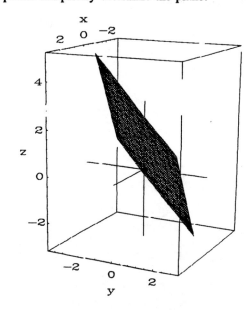

10 Let $z = f(x, y) = \sqrt{x^2 + y^2}$. Note that $z \geq 0$ for all x and y. When $z = k \geq 0$, we have $\sqrt{x^2 + y^2} = k$ or $x^2 + y^2 = k^2$, a circle of radius k. When $y = 0$, $z = \sqrt{x^2} = |x|$. Similarly, when $x = 0$, $z = \sqrt{y^2} = |y|$. The graph of f is a right circular cone with half-vertex angle $\pi/4$, vertex at the origin, and axis coinciding with the z axis.

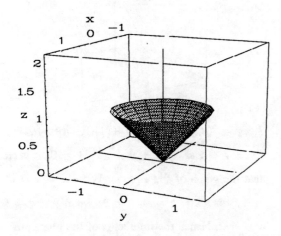

12 Since $f(x, y) = x + 2y$, the level curves of $f(x, y) = -1, 0, 1$ and 2 are, respectively, the lines $x + 2y = -1$, $x + 2y = 0$, $x + 2y = 1$, and $x + 2y = 2$.

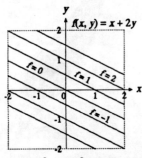

14 Since $f(x, y) = x^2 - 2y^2$, the level curves of $f(x, y) = -1, 1,$ and 2 are, respectively, the hyperbolas $x^2 - 2y^2 = -1$, $x^2 - 2y^2 = 1$, and $x^2 - 2y^2 = 2$. The asymptotes for the hyperbolas are the level curves for $f(x, y) = 0$, the lines $x = \pm\sqrt{2}y$.

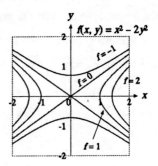

16 Note that $f(1, 2) = 1^2 + 3 \cdot 2^2 = 13$, so the corresponding level curve is $f(x, y) = 13$. The level curve is the ellipse $x^2 + 3y^2 = 13$.

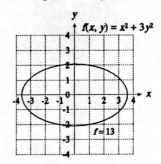

18 Observe that $f(2, 3) = 2^2 - 3^2 = -5$, so the corresponding level curve is $f(x, y) = -5$. The level curve is the hyperbola $y^2 - x^2 = 5$.

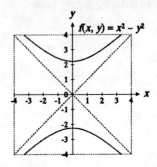

20 (a) From the figure, it appears that the ratio is
$$\frac{500 \text{ langleys}}{350 \text{ langleys}} = \frac{10}{7}.$$

(b) The intensity of solar radiation measured is proportional to the distance from the sun that the measurement is made. This is illustrated by the increasing trend of measured solar radiation in the U.S. as one travels from north to south. The two sharp bends may be attributed to the mountain ranges in those two areas, specifically the Sierra Nevada and Rocky Mountains.

22 (a) The level surface $x + y + z = 1$ is a plane.
   (b) The level surface $x^2 + y^2 + z^2 = 1$ is a sphere.
   (c) The level surface $x^2 + y^2 - z^2 = 1$ is a hyperboloid in one sheet.
   (d) The level surface $x^2 - y^2 - z^2 = 1$ is a hyperboloid in two sheets.

14.4 Limits and Continuity

2 $\lim\limits_{(x,y)\to(1,1)} \dfrac{x^2}{x^2 + y^2} = \dfrac{1^2}{1^2 + 1^2} = \dfrac{1}{2}$

4 Along the line $y = x$, we see that
$$\lim_{(x,y)\to(0,0)} \frac{xy}{x^2 + y^2} = \lim_{x\to 0} \frac{x\cdot x}{x^2 + x^2} = \lim_{x\to 0} \frac{x^2}{2x^2}$$
$= \dfrac{1}{2}$. Along the line $y = 0$, $\lim\limits_{(x,y)\to(0,0)} \dfrac{xy}{x^2 + y^2} =$
$\lim\limits_{x\to 0} \dfrac{x\cdot 0}{x^2 + 0^2} = 0$. Since $\dfrac{1}{2} \neq 0$,
$\lim\limits_{(x,y)\to(0,0)} \dfrac{xy}{x^2 + y^2}$ does not exist.

6 Along the line $y = 0$, $\lim\limits_{(x,y)\to(0,0)} (x^2)^y = \lim\limits_{x\to 0} (x^2)^0$
$= 1$. Along the line $x = 0$, $\lim\limits_{(x,y)\to(0,0)} (x^2)^y =$
$\lim\limits_{y\to 0} (0^2)^y = 0$. Since $1 \neq 0$, $\lim\limits_{(x,y)\to(0,0)} (x^2)^y$ does not exist.

8 Along the line $y = x$, $\lim\limits_{(x,y)\to(0,0)} (1 + x)^{1/y} =$
$\lim\limits_{x\to 0} (1 + x)^{1/x} = e$, by definition. Along the line x
$= 0$, $\lim\limits_{(x,y)\to(0,0)} (1 + x)^{1/y} = \lim\limits_{y\to 0} (1 + 0)^{1/y} = 1$.
Since $e \neq 1$, $\lim\limits_{(x,y)\to(0,0)} (1 + x)^{1/y}$ does not exist.

10 (a) The domain of f is the set of all points $P = (x, y)$ excluding the origin.
   (b) For every point $P_0 = (x_0, y_0)$ in the domain of f, there exists a disk with center P_0 and positive radius that is contained in the domain of f. Also, $\lim\limits_{P\to P_0} f(P) = \lim\limits_{P\to P_0} \dfrac{1}{x^2 + 2y^2} =$
   $\dfrac{1}{x_0^2 + 2y_0^2} = f(P_0)$, since $x_0 \neq 0$ and $y_0 \neq 0$.
   Thus f is continuous.

12 (a) The domain of f consists of all points $P = (x, y)$ such that $x^2 + y^2 \geq 25$.
   (b) f is continuous, although we have to treat the boundary of the domain—the circle $x^2 + y^2 = 25$—as a special case, since we cannot center disks on such points which will be in the domain.

14 (a) The domain of f consists of all points $P = (x, y)$ such that $x^2 + y^2 < 49$.
   (b) For every point $P_0 = (x_0, y_0)$ in the domain of f, there exists a disk with center P_0 and positive radius that is contained in the domain of f. Also, $\lim\limits_{P\to P_0} f(P) = \lim\limits_{P\to P_0} \dfrac{1}{\sqrt{49 - x^2 - y^2}}$
   $= \dfrac{1}{\sqrt{49 - x_0^2 - y_0^2}} = f(P_0)$ since $x_0^2 + y_0^2 <$
   49, so f is continuous throughout its domain.

16 R is bordered by the circle $x^2 + y^2 = 1$. Hence the circle $x^2 + y^2 = 1$ is the boundary of R.

18 The function $\dfrac{1}{x+y}$ is defined everywhere except along the line $x + y = 0$. Therefore, the line $x + y$ is the boundary of R.

20 R is bordered by the line $y = x$. Thus the line $y = x$ is the boundary of R.

22 (a) Let the disk have radius at most 0.002. For any $P = (x, y)$ in this disk,

$\sqrt{(x-1)^2 + (y-1)^2} < 0.002$ (by the distance formula). So $|x - 1| = \sqrt{(x-1)^2}$

$\leq \sqrt{(x-1)^2 + (y-1)^2} < 0.002$, and similarly, $|y - 1| < 0.002$. Now $|f(P) - 5|$
$= |2x + 3y - 5| = |2x - 2 + 3y - 3| \leq$
$|2x - 2| + |3y - 3| = 2|x - 1| + 3|y - 1| <$
$2 \cdot 0.002 + 3 \cdot 0.002 = 0.01$, where the triangle inequality was used.

(b) Let $\epsilon > 0$ be given. Choose $\delta = \epsilon/5 > 0$. For any $P = (x, y)$ in the disk with radius δ and center $(1, 1)$, $|x - 1| < \delta$ and $|y - 1| < \delta$. Hence $|f(P) - 5| = |2x + 3y - 5| =$
$|2x - 2 + 3y - 3| \leq |2x - 2| + |3y - 3| =$
$2|x - 1| + 3|y - 1| < 2 \cdot \delta + 3 \cdot \delta = 5\delta =$
$5 \cdot \dfrac{\epsilon}{5} = \epsilon$, and such a disk has thus been found.

(c) $\lim\limits_{(x,y) \to (1,1)} f(x, y) = 5$

24 (a) The domain of f is all points $P = (x, y)$ excluding the origin.

(b) Along the line $y = 2x$, $f(x, 2x) = \dfrac{10x^3}{2x^4 + 12x^2}$

$= \dfrac{10x}{2x^2 + 12} \to 0$ as $P \to (0, 0)$.

(c) Along the line $y = 3x$, $f(x, 3x) =$
$\dfrac{15x^3}{2x^4 + 27x^2} = \dfrac{15x}{2x^2 + 27} \to 0$ as $P \to (0, 0)$.

(d) Along the parabola $y = x^2$, $f(x, x^2) =$
$\dfrac{5x^2 \cdot x^2}{2x^4 + 3(x^2)^2} = \dfrac{5x^4}{5x^4} \to 1$ as $P \to (0, 0)$.

(e) No. Since $0 \neq 1$, the limit cannot exist.

14.5 Partial Derivatives

2 $f(x, y) = x^2 y + 4$, so $f_x = 2xy$ and $f_y = x^2$.

4 $f(x, y) = 6x - 7y$, so $f_x = 6$ and $f_y = -7$.

6 $f(x, y) = \ln(x + 2y)$; $f_x = \dfrac{1}{x + 2y}$; $f_y =$
$\dfrac{1}{x + 2y}(2) = \dfrac{2}{x + 2y}$

8 $f(x, y) = \tan^{-1} \dfrac{y}{x}$; $f_x = \dfrac{1}{1 + (y/x)^2}\left(-\dfrac{y}{x^2}\right)$
$= -\dfrac{y}{x^2 + y^2}$; $f_y = \dfrac{1}{1 + (y/x)^2}\left(\dfrac{1}{x}\right) = \dfrac{1}{x + y^2/x}$
$= \dfrac{x}{x^2 + y^2}$

10 $f(x, y) = \dfrac{x^2 + \cos 3y}{1 + x}$;

$f_x = \dfrac{(1+x)(2x) - (x^2 + \cos 3y)(1)}{(1+x)^2}$

$= \dfrac{x^2 + 2x - \cos 3y}{(1+x)^2}$; $f_y = \dfrac{(-\sin 3y)(3)}{1 + x}$

$= -\dfrac{3 \sin 3y}{1 + x}$

14.5 Partial Derivatives

12 $f(x, y) = \sin^{-1}(x + 3y); f_x = \dfrac{1}{\sqrt{1 - (x + 3y)^2}};$

$f_y = \dfrac{3}{\sqrt{1 - (x + 3y)^2}}$

14 $f(x, y) = \dfrac{1}{\sqrt{x^2 + y^2}} = (x^2 + y^2)^{-1/2};$

$f_x = -\dfrac{1}{2}(x^2 + y^2)^{-3/2}(2x) = -x(x^2 + y^2)^{-3/2};$

$f_y = -\dfrac{1}{2}(x^2 + y^2)^{-3/2}(2y) = -y(x^2 + y^2)^{-3/2}$

16 $f(x, y) = x^4 y^7; f_x = 4x^3 y^7; f_{xx} = 12x^2 y^7; f_{xy} = 28x^3 y^6; f_y = 7x^4 y^6; f_{yy} = 42x^4 y^5; f_{yx} = 28x^3 y^6 = f_{xy}$

18 $f(x, y) = \sin(x^2 y); f_x = 2xy \cos(x^2 y);$

$f_{xx} = 2xy(-\sin(x^2 y))(2xy) + 2y \cos(x^2 y)$
$= -4x^2 y^2 \sin(x^2 y) + 2y \cos(x^2 y);$

$f_{xy} = 2xy(-\sin(x^2 y))(x^2) + 2x \cos(x^2 y)$
$= -2x^3 y \sin(x^2 y) + 2x \cos(x^2 y)$

$f_y = x^2 \cos(x^2 y);$

$f_{yy} = x^2(-\sin(x^2 y))(x^2) = -x^4 \sin(x^2 y);$

$f_{yx} = x^2(-\sin(x^2 y))(2xy) + 2x \cos(x^2 y)$
$= -2x^3 y \sin(x^2 y) + 2x \cos(x^2 y) = f_{xy}$

20 $f(x, y) = \dfrac{x}{y}; f_x = \dfrac{1}{y}; f_{xx} = 0; f_{xy} = -\dfrac{1}{y^2};$

$f_y = -\dfrac{x}{y^2}; f_{yy} = \dfrac{2x}{y^3}; f_{yx} = -\dfrac{1}{y^2} = f_{xy}$

22 The desired slope is $\dfrac{\partial z}{\partial x}$ evaluated at $(\pi/4, \pi/2);$

that is, $\dfrac{\partial z}{\partial x}\bigg|_{(\pi/4, \pi/2)} = -\sin(x + 2y)|_{(\pi/4, \pi/2)} =$

$-\sin \dfrac{5\pi}{4} = \dfrac{\sqrt{2}}{2}.$

24 The requested slope is $\dfrac{\partial z}{\partial x}\bigg|_{(1,0)} =$

$(2xe^{xy} + x^2 y e^{xy})|_{(1,0)} = 2 + 0 = 2.$

26 The slope is $\dfrac{\partial z}{\partial y}\bigg|_{(0,1)} = -\dfrac{x}{y^2} e^{x/y} \bigg|_{(0,1)} = 0.$

28 (a)

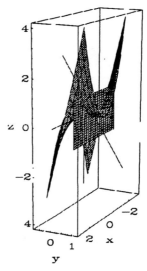

(b) The equation of the trace is $z = \dfrac{x^2}{2}$. The

slope equals $\dfrac{\partial z}{\partial x}\bigg|_{(1, \frac{1}{2}, \frac{1}{2})} = x|_{(1, \frac{1}{2}, \frac{1}{2})} = 1.$

30 From the definition of the partial derivative,

$\dfrac{\partial f}{\partial x}\bigg|_{(x_0, y_0)} \approx \dfrac{f(x_0 + \Delta x, y_0) - f(x_0, y_0)}{\Delta x}$ and

$\dfrac{\partial f}{\partial y}\bigg|_{(x_0, y_0)} \approx \dfrac{f(x_0, y_0 + \Delta y) - f(x_0, y_0)}{\Delta y}$ for Δx and

Δy small. Hence $\dfrac{\partial f}{\partial x}\bigg|_{(2,3)} \approx \dfrac{f(1.98, 3) - f(2, 3)}{1.98 - 2}$

$= \dfrac{1.03 - 1}{-0.02} = \dfrac{0.03}{-0.02} = -\dfrac{3}{2}$ and $\dfrac{\partial f}{\partial y}\bigg|_{(2,3)} \approx$

$$\frac{f(2, 3.04) - f(2, 3)}{3.04 - 3} = \frac{0.98 - 1}{0.04} = \frac{-0.02}{0.04}$$

$$= -\frac{1}{2}.$$

32 From the definition of the partial derivative,

$$\frac{\partial}{\partial x}(x^2y) = \lim_{\Delta x \to 0} \frac{(x + \Delta x)^2 y - x^2 y}{\Delta x}$$

$$= \lim_{\Delta x \to 0} \frac{x^2 y + 2xy\Delta x + y(\Delta x)^2 - x^2 y}{\Delta x}$$

$$= \lim_{\Delta x \to 0} \frac{2xy\Delta x + y(\Delta x)^2}{\Delta x} = \lim_{\Delta x \to 0} (2xy + y\Delta x)$$

$$= 2xy, \text{ which equals } 2ab \text{ at the point } (a, b).$$

34 $T_x = -\frac{1}{2}(x^2 + y^2)^{-3/2}(2x) = -x(x^2 + y^2)^{-3/2}$, and

$$T_{xx} = -x\left[-\frac{3}{2}(x^2 + y^2)^{-5/2}(2x)\right] - (x^2 + y^2)^{-3/2}$$

$= 3x^2(x^2 + y^2)^{-5/2} - (x^2 + y^2)^{-3/2}$. By symmetry,
$T_{yy} = 3y^2(y^2 + x^2)^{-5/2} - (y^2 + x^2)^{-3/2}$. Then
$T_{xx} + T_{yy} = 3x^2(x^2 + y^2)^{-5/2} - (x^2 + y^2)^{-3/2} + 3y^2(y^2 + x^2)^{-5/2} - (y^2 + x^2)^{-3/2}$
$= (3x^2 + 3y^2)(x^2 + y^2)^{-5/2} - 2(x^2 + y^2)^{-3/2}$
$= 3(x^2 + y^2)^{-3/2} - 2(x^2 + y^2)^{-3/2} = (x^2 + y^2)^{-3/2}$
$\neq 0$. So $T(x, y)$ does not satisfy the given differential equation.

36 (a) By definition, $g'(y) = \lim_{\Delta y \to 0} \frac{g(y + \Delta y) - g(y)}{\Delta y}$

$$= \lim_{\Delta y \to 0} \frac{\int_a^b f(x, y + \Delta y)\, dx - \int_a^b f(x, y)\, dx}{\Delta y}$$

$$= \lim_{\Delta y \to 0} \int_a^b \frac{f(x, y + \Delta y) - f(x, y)}{\Delta y}\, dx$$

$$= \int_a^b \left[\lim_{\Delta y \to 0} \frac{f(x, y + \Delta y) - f(x, y)}{\Delta y}\right] dx$$

$$= \int_a^b \frac{\partial f}{\partial y}\, dx, \text{ assuming that the limit and}$$

integral sign can be interchanged.

(b) Here $g(y) = \int_a^b x^3 y^2 \, dx = \frac{1}{4}y^2 x^4 \Big|_a^b =$

$\frac{1}{4}y^2(b^4 - a^4)$, and $\frac{d}{dy}g(y) = \frac{1}{2}y(b^4 - a^4)$.

Now $\frac{\partial f}{\partial y} = 2x^3 y$, and $\int_a^b \frac{\partial f}{\partial y}\, dx =$

$$\int_a^b 2x^3 y \, dx = \frac{1}{2}yx^4 \Big|_a^b = \frac{1}{2}y(b^4 - a^4) =$$

$\frac{d}{dy}g(y)$, as claimed in (a).

38 $f(x, y) = a + bx + cy + dx^2 + exy + ky^2$

(a) $f(0, 0) = a$

(b) $\frac{\partial f}{\partial x}(0, 0) = (b + 2dx + ey)|_{(0,0)} = b$

(c) $\frac{\partial f}{\partial y}(0, 0) = (c + ex + 2ky)|_{(0,0)} = c$

(d) $\frac{\partial^2 f}{\partial x^2} = 2d$, so $\frac{1}{2}\frac{\partial^2 f}{\partial x^2} = d$, in particular, d

$$= \frac{1}{2}\frac{\partial^2 f}{\partial x^2}(0, 0).$$

(e) $\frac{\partial^2 f}{\partial x \partial y} = \frac{\partial}{\partial x}\left(\frac{\partial f}{\partial y}\right) = \frac{\partial}{\partial x}(c + ex + 2ky) = e$;

in particular, $e = \frac{\partial^2 f}{\partial x \partial y}(0, 0)$.

(f) $\frac{\partial^2 f}{\partial y^2} = 2k$ and $k = \frac{1}{2}\frac{\partial^2 f}{\partial y^2}$, in particular,

$k = \frac{1}{2}\frac{\partial^2 f}{\partial y^2}(0, 0)$.

14.5 Partial Derivatives

40 (a) Since for small Δy, $\dfrac{\partial u}{\partial y}\bigg|_{(x_0, y_0)} \approx$

$\dfrac{u(x_0, y_0 + \Delta y) - u(x_0, y_0)}{\Delta y}$, it follows that

$u(x_0, y_0 + \Delta y) \approx \Delta y \dfrac{\partial u}{\partial y}\bigg|_{(x_0, y_0)} + u(x_0, y_0)$.

Thus $u(1, 2.01) \approx \Delta y \dfrac{\partial u}{\partial y}\bigg|_{(1,2)} + u(1, 2)$

$= 0.01 \cdot 1.2 + 3 = 3.012$.

(b) Similar to (a), we have $u(x_0 + \Delta x, y_0) \approx$

$\Delta x \dfrac{\partial u}{\partial x}\bigg|_{(x_0, y_0)} + u(x_0, y_0)$. Then $u(0.98, 2) \approx$

$\Delta x \dfrac{\partial u}{\partial x}\bigg|_{(1,2)} + u(1, 2) = (-0.02) \cdot 2 + 3 =$

2.96.

(c) By (b), $u(1.02, 2) \approx \Delta x \dfrac{\partial u}{\partial x}\bigg|_{(1,2)} + u(1, 2)$

$= 0.02 \cdot 2 + 3 = 3.04$. Assuming $\dfrac{\partial u}{\partial x}\bigg|_{(1,2)} \approx$

$\dfrac{\partial u}{\partial x}\bigg|_{(1.02, 2)}$, we know from (a) that

$u(1.02, 2.03) \approx \Delta y \dfrac{\partial u}{\partial x}\bigg|_{(1.02, 2)} + u(1.02, 2)$

$\approx \Delta y \dfrac{\partial u}{\partial x}\bigg|_{(1,2)} + u(1.02, 2)$

$\approx 0.03 \cdot 2 + 3.04 = 3.10$.

14.6 The Chain Rule

2 By Theorem 2, $\dfrac{dz}{dt} = z_x \dfrac{dx}{dt} + z_y \dfrac{dy}{dt} =$

$(e^y)(1) + (xe^y)(3) = e^{1+3t} + 3te^{1+3t} =$

$(1 + 3t)e^{1+3t}$. Writing z as a function of t gives z

$= xe^y = te^{1+3t}$. Therefore, $\dfrac{dz}{dt} = t \cdot 3e^{1+3t} + e^{1+3t}$

$= (1 + 3t)e^{1+3t}$, which verifies the chain rule in Theorem 2.

4 Applying Theorem 2, we have $\dfrac{dz}{dt}$

$= z_x \dfrac{dx}{dt} + z_y \dfrac{dy}{dt} = \dfrac{1}{x + 3y} \cdot 2t + \dfrac{3}{x + 3y}(3 \sec^2 3t)$

$= \dfrac{2t}{t^2 + 3 \tan 3t} + \dfrac{9 \sec^2 3t}{t^2 + 3 \tan 3t}$

$= \dfrac{2t + 9 \sec^2 3t}{t^2 + 3 \tan 3t}$.

Writing z as a function of t yields $z = \ln(x + 3y)$

$= \ln(t^2 + 3 \tan 3t)$. Then $\dfrac{dz}{dt} = \dfrac{2t + 9 \sec^2 3t}{t^2 + 3 \tan 3t}$,

as before.

6 From Theorem 3 we have $\dfrac{\partial z}{\partial t} = z_x x_t + z_y y_t$

$= \cos(x + 3y) \dfrac{1}{2\sqrt{tu}} + 3\cos(x + 3y) \dfrac{1}{2\sqrt{t}}$

$= \dfrac{1 + 3\sqrt{u}}{2\sqrt{tu}} \cos(x + 3y)$

$= \dfrac{1 + 3\sqrt{u}}{2\sqrt{tu}} \cos\left(\dfrac{\sqrt{t} + 3\sqrt{tu} + 3u}{\sqrt{u}}\right)$.

Writing z as a function of t and u gives $z =$

$\sin(x + 3y) = \sin\left(\dfrac{\sqrt{t} + 3\sqrt{tu} + 3u}{\sqrt{u}}\right)$. Then $\dfrac{\partial z}{\partial t} =$

$\cos\left(\dfrac{\sqrt{t} + 3\sqrt{tu} + 3u}{\sqrt{u}}\right)\left[\dfrac{1}{\sqrt{u}}\left(\dfrac{1}{2\sqrt{t}} + \dfrac{3\sqrt{u}}{2\sqrt{t}}\right)\right]$

$= \dfrac{1 + 3\sqrt{u}}{2\sqrt{tu}} \cos\left(\dfrac{\sqrt{t} + 3\sqrt{tu} + 3u}{\sqrt{u}}\right)$, as before.

8 (a) There are two middle variables: g and h.

   (b) There are three bottom variables: t_1, t_2 and t_3.

   (c) $\dfrac{\partial z}{\partial t_3} = \dfrac{\partial z}{\partial g}\cdot\dfrac{\partial g}{\partial t_3} + \dfrac{\partial z}{\partial h}\cdot\dfrac{\partial h}{\partial t_3}$

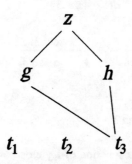

10 By Theorem 2, $\dfrac{dz}{dt} = z_x\dfrac{dx}{dt} + z_y\dfrac{dy}{dt} =$

$3\cdot 4 + 2(-3) = 6$.

12 (a) Theorem 3 applies, so $\dfrac{\partial z}{\partial v} = z_x\cdot x_v + z_y\cdot y_v =$

$z_x(-2v) + z_y(2v) = 2v(z_y - z_x)$ and $\dfrac{\partial z}{\partial u} =$

$z_x\cdot x_u + z_y\cdot y_u = z_x(2u) + z_y(-2u) =$

$2u(z_x - z_y)$. (See the diagram.) Hence

$u\dfrac{\partial z}{\partial v} + v\dfrac{\partial z}{\partial u} = 2uv(z_y - z_x) + 2uv(z_x - z_y)$

$= 0$.

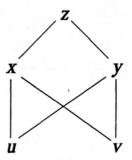

(b) By Theorem 3, $\dfrac{\partial z}{\partial v} = z_x\cdot x_v + z_y\cdot y_v =$

$\cos(x + 2y)(-2v) + 2\cos(x + 2y)(2v)$

$= 2v\cos(x + 2y) =$

$2v\cos(u^2 - v^2 + 2(v^2 - u^2)) = 2v\cos(v^2 - u^2)$

and $\dfrac{\partial z}{\partial u} = z_x\cdot x_u + z_y\cdot y_u = \cos(x + 2y)(2u) +$

$2\cos(x + 2y)(-2u) = -2u\cos(x + 2y) =$

$-2u\cos(v^2 - u^2)$. Then $u\dfrac{\partial z}{\partial v} + v\dfrac{\partial z}{\partial u} = 2uv$

$\cos(v^2 - u^2) - 2uv\cos(v^2 - u^2) = 0$, and (a) is verified.

14 (a) $w = f(x - y, y - z, z - x)$, so $\dfrac{\partial w}{\partial x}$

$= f_1\cdot\dfrac{\partial}{\partial x}(x - y) + f_2\cdot\dfrac{\partial}{\partial x}(y - z) + f_3\cdot\dfrac{\partial}{\partial x}(z - x)$

$= f_1\cdot 1 + f_2\cdot 0 + f_3\cdot(-1) = f_1 - f_3$, $\dfrac{\partial w}{\partial y} =$

$f_1\cdot(-1) + f_2\cdot 1 + f_3\cdot 0 = -f_1 + f_2$, and $\dfrac{\partial w}{\partial z}$

$= f_1\cdot 0 + f_2\cdot(-1) + f_3\cdot 1 = -f_2 + f_3$.

Therefore $\dfrac{\partial w}{\partial x} + \dfrac{\partial w}{\partial y} + \dfrac{\partial w}{\partial z} = 0$.

(b) $w = f(s, t, u) = s^2 + t^2 - u$, where $s =$

$x - y$, $t = y - z$, and $u = z - x$, so $\dfrac{\partial w}{\partial x} =$

14.6 The Chain Rule

$w_s s_x + w_t t_x + w_u u_x = 2s \cdot 1 + 2t \cdot 0 + 1 \cdot (-1)$

$= 2s - 1$, $\dfrac{\partial w}{\partial y} = 2s \cdot (-1) + 2t \cdot 1 + 1 \cdot 0 =$

$-2s + 2t$, and $\dfrac{\partial w}{\partial z} = 2s \cdot 0 + 2t \cdot (-1) + 1 \cdot 1$

$= -2t + 1$. Therefore, $\dfrac{\partial w}{\partial x} + \dfrac{\partial w}{\partial y} + \dfrac{\partial w}{\partial z}$

$= (2s - 1) + (-2s + 2t) + (-2t + 1) =$

0, and (a) is verified.

16 Let $u = mx + y$ so that $z = f(u)$. Using the results of Exercise 15 with $a = m$, $b = 1$, and $c = d = 0$, we have $\dfrac{\partial^2 z}{\partial x^2}$

$= m^2 \dfrac{\partial^2 f}{\partial u^2} + 2 \cdot m \cdot 0 \dfrac{\partial^2 f}{\partial u \, \partial v} + 0^2 \dfrac{\partial^2 f}{\partial v^2} = m^2 \dfrac{d^2 f}{du^2}$,

$\dfrac{\partial^2 z}{\partial x \, \partial y}$

$= m \cdot 1 \dfrac{\partial^2 f}{\partial u^2} + (m \cdot 0 + 1 \cdot 0) \dfrac{\partial^2 f}{\partial u \, \partial v} + 0 \cdot 0 \dfrac{\partial^2 f}{\partial v^2}$

$= m \dfrac{d^2 f}{du^2}$, and $\dfrac{\partial^2 z}{\partial y^2}$

$= 1^2 \dfrac{\partial^2 f}{\partial u^2} + 2 \cdot 1 \cdot 0 \dfrac{\partial^2 f}{\partial u \, \partial v} + 0^2 \dfrac{\partial^2 f}{\partial u^2} = \dfrac{d^2 f}{du^2}$. Now

z is a solution to the given partial differential

equation if $a \dfrac{\partial^2 z}{\partial x^2} + b \dfrac{\partial^2 z}{\partial x \, \partial y} + c \dfrac{\partial^2 z}{\partial y^2}$

$= am^2 \dfrac{d^2 f}{du^2} + bm \dfrac{d^2 f}{du^2} + c \dfrac{d^2 f}{du^2}$

$= (am^2 + bm + c) \dfrac{d^2 f}{du^2} = 0$. That is, either

$am^2 + bm + c = 0$ or $\dfrac{d^2 f}{du^2} = 0$ for z to be a

solution. If $\dfrac{d^2 f}{du^2} = 0$, it follows that $z = f(u) =$

$c_1 u + c_2 = c_1 (mx + y) + c_2$ for any constants c_1 and c_2. In summary, for z to be a solution, either $am^2 + bm + c = 0$ or $z = c_1 (mx + y) + c_2$ for any constants c_1 and c_2.

18 (a) For $u(x, t) = $ $\dfrac{\partial u}{\partial t} = ke^{kt}g(x)$, $\dfrac{\partial u}{\partial x} = e^{kt}g'(x)$

and $\dfrac{\partial^2 u}{\partial x^2} = e^{kt}g''(x) = e^{kt}g(x)$ since $g''(x) =$

$g(x)$. Then $k \dfrac{\partial^2 u}{\partial x^2} = ke^{kt}g(x) = \dfrac{\partial u}{\partial t}$, as

claimed.

(b) $g(x) = 3e^{-x} + 4e^x$, so $g'(x) = -3e^{-x} + 4e^x$, and $g''(x) = 3e^{-x} + 4e^x = g(x)$.

20 (a) The relation between rectangular and polar coordinates is given by the equations $x = r \cos \theta$ and $y = r \sin \theta$. By the chain rule,

$\dfrac{\partial z}{\partial r} = \dfrac{\partial z}{\partial x} \cdot \dfrac{\partial x}{\partial r} + \dfrac{\partial z}{\partial y} \cdot \dfrac{\partial y}{\partial r}$

$= \dfrac{\partial z}{\partial x} \cdot \dfrac{\partial}{\partial r}(r \cos \theta) + \dfrac{\partial z}{\partial y} \cdot \dfrac{\partial}{\partial r}(r \sin \theta)$

$= \cos \theta \dfrac{\partial z}{\partial x} + \sin \theta \dfrac{\partial z}{\partial y}$.

(b) Similarly, by the chain rule, $\dfrac{\partial z}{\partial \theta}$

$= \dfrac{\partial z}{\partial x} \cdot \dfrac{\partial x}{\partial \theta} + \dfrac{\partial z}{\partial y} \cdot \dfrac{\partial y}{\partial \theta}$

$= \dfrac{\partial z}{\partial x} \cdot \dfrac{\partial}{\partial \theta}(r \cos \theta) + \dfrac{\partial z}{\partial y} \cdot \dfrac{\partial}{\partial \theta}(r \sin \theta)$

$$= -r\sin\theta\,\frac{\partial z}{\partial x} + r\cos\theta\,\frac{\partial z}{\partial y}.$$

(c) Observe that $\left(\dfrac{\partial z}{\partial r}\right)^2 + \dfrac{1}{r^2}\left(\dfrac{\partial z}{\partial \theta}\right)^2 =$

$$\left(\cos\theta\,\frac{\partial z}{\partial x} + \sin\theta\,\frac{\partial z}{\partial y}\right)^2 +$$

$$\frac{1}{r^2}\left(-r\sin\theta\,\frac{\partial z}{\partial x} + r\cos\theta\,\frac{\partial z}{\partial y}\right)^2$$

$$= \cos^2\theta\left(\frac{\partial z}{\partial x}\right)^2 + 2\cos\theta\,\frac{\partial z}{\partial x}\sin\theta\,\frac{\partial z}{\partial y} +$$

$$\sin^2\theta\left(\frac{\partial z}{\partial y}\right)^2 + \frac{1}{r^2}\left(r^2\sin^2\theta\left(\frac{\partial z}{\partial x}\right)^2 +\right.$$

$$\left.2(-r\sin\theta)\frac{\partial z}{\partial x}(r\cos\theta)\frac{\partial z}{\partial y} + r^2\cos^2\theta\left(\frac{\partial z}{\partial y}\right)^2\right)$$

$$= \cos^2\theta\left(\frac{\partial z}{\partial x}\right)^2 + 2\cos\theta\sin\theta\,\frac{\partial z}{\partial x}\cdot\frac{\partial z}{\partial y} +$$

$$\sin^2\theta\left(\frac{\partial z}{\partial y}\right)^2 + \sin^2\theta\left(\frac{\partial z}{\partial x}\right)^2 -$$

$$2\sin\theta\cos\theta\,\frac{\partial z}{\partial x}\cdot\frac{\partial z}{\partial y} + \cos^2\theta\left(\frac{\partial z}{\partial y}\right)^2$$

$$= (\sin^2\theta + \cos^2\theta)\left[\left(\frac{\partial z}{\partial x}\right)^2 + \left(\frac{\partial z}{\partial y}\right)^2\right]$$

$$= \left(\frac{\partial z}{\partial x}\right)^2 + \left(\frac{\partial z}{\partial y}\right)^2, \text{ which we were to show.}$$

22 $\dfrac{dV}{dt} = yz\,\dfrac{\partial x}{\partial t} + xz\,\dfrac{\partial y}{\partial t} + xy\,\dfrac{\partial z}{\partial t}$

$= 8\cdot 4\cdot 2 + 3\cdot 4\cdot(-5) + 3\cdot 8\cdot 2 = 52$ ft³/sec.

24 By definition, $\dfrac{\partial z}{\partial t} =$

$$\lim_{\Delta t \to 0}\frac{z(x(t+\Delta t, u), y(t+\Delta t, u)) - z(x(t, u) - y(t, u))}{\Delta t}.$$

Since u is fixed, the numerator in the limit becomes

$f_x(x(t, u), y(t, u))[x(t + \Delta t, u) - x(t, u)]$

$+ f_y(x(t, u), y(t, u))[y(t + \Delta t, u) - y(t, u)]$

$+ \epsilon_1(x(t + \Delta t, u) - x(t, u))$

$+ \epsilon_2(y(t + \Delta t, u) - y(t, u))$ by Theorem 1, where

$\epsilon_1 \to 0$ and $\epsilon_2 \to 0$ as $\Delta t \to 0$. The expression in the limit therefore becomes

$f_x(x(t, u), y(t, u))\left[\dfrac{x(t + \Delta t), u) - x(t, u)}{\Delta t}\right]$

$+ f_y(x(t, u), y(t, u))\left[\dfrac{y(t + \Delta t, u) - y(t, u)}{\Delta t}\right]$

$+ \epsilon_1\dfrac{x(t + \Delta t, u) - x(t, u)}{\Delta t}$

$+ \epsilon_2\dfrac{y(t + \Delta t, u) - y(t, u)}{\Delta t}$. Hence in the limit as

$\Delta t \to 0$, $\dfrac{\partial z}{\partial t} = f_x(x(t, u), y(t, u))\dfrac{\partial x}{\partial t} +$

$f_y(x(t, u), y(t, u))\dfrac{\partial y}{\partial t} + 0\cdot\dfrac{\partial x}{\partial t} + 0\cdot\dfrac{\partial y}{\partial t}$

$= \dfrac{\partial f}{\partial x}\cdot\dfrac{\partial x}{\partial t} + \dfrac{\partial f}{\partial y}\cdot\dfrac{\partial y}{\partial t}$. In a similar fashion it may

be shown that $\dfrac{\partial z}{\partial u} = \dfrac{\partial f}{\partial x}\cdot\dfrac{\partial x}{\partial u} + \dfrac{\partial f}{\partial y}\cdot\dfrac{\partial y}{\partial u}$.

26 (a) $\dfrac{\partial r}{\partial x} = \dfrac{\partial}{\partial x}((x^2 + y^2)^{1/2}) = \dfrac{1}{2}(x^2 + y^2)^{-1/2}(2x)$

$= x(x^2 + y^2)^{-1/2} = \dfrac{x}{r} = \dfrac{r\cos\theta}{r} = \cos\theta$

14.6 The Chain Rule

(b) $\dfrac{\partial r}{\partial x} = \dfrac{\partial}{\partial x}\left(\dfrac{x}{\cos\theta}\right) = \dfrac{1}{\cos\theta}$

(c) In (a), r was treated as a function of x and y, while in (b) it was treated as a function of x and θ.

28 By the chain rule, $u_r = u_x \cdot x_r + u_y \cdot y_r = \cos\theta\, u_x + \sin\theta\, u_y$, $ru_r = r\cos\theta\, u_x + r\sin\theta\, u_y = xu_x + yu_y$, $(ru_r)_r = (xu_x + yu_y)_r = x_r \cdot u_x + x(u_{xx}\cdot x_r + u_{xy}\cdot y_r) + y_r \cdot u_y + y(u_{yx}\cdot x_r + u_{yy}\cdot y_r)$
$= \cos\theta\, u_x + r\cos^2\theta\, u_{xx} + r\cos\theta\sin\theta\, u_{xy} + \sin\theta\, u_y + r\sin\theta\cos\theta\, u_{yx} + r\sin^2\theta\, u_{yy}$
$= \cos\theta\, u_x + \sin\theta\, u_y + 2r\sin\theta\cos\theta\, u_{xy} + r\cos^2\theta\, u_{xx} + r\sin^2\theta\, u_{yy}$, and $\dfrac{1}{r}(ru_r)_r =$

$\dfrac{\cos\theta}{r}u_x + \dfrac{\sin\theta}{r}u_y + 2\sin\theta\cos\theta\, u_{xy} + \cos^2\theta\, u_{xx}$

$+ \sin^2\theta\, u_{yy}$. Also $u_\theta = u_x \cdot x_\theta + u_y \cdot y_\theta = (-r\sin\theta)u_x + (r\cos\theta)u_y = -yu_x + xu_y$, $u_{\theta\theta} = (-yu_x + xu_y)_\theta = -y_\theta \cdot u_x - yu_{x\theta} + x_\theta \cdot u_y + xu_{y\theta}$
$= -r\cos\theta\, u_x - y(u_{xx}\cdot x_\theta + u_{xy}\cdot y_\theta) + (-r\sin\theta)u_y + x(u_{yx}\cdot x_\theta + u_{yy}\cdot y_\theta) =$
$-r\cos\theta\, u_x - (-r^2\sin^2\theta\, u_{xx} + r^2\cos\theta\sin\theta\, u_{xy})$
$- r\sin\theta\, u_y + (-r^2\cos\theta\sin\theta\, u_{yx} + r^2\cos^2\theta\, u_{yy})$
$= -r\cos\theta\, u_x - r\sin\theta\, u_y - 2r^2\cos\theta\sin\theta\, u_{xy} + r^2\sin^2\theta\, u_{xx} + r^2\cos^2\theta\, u_{yy}$, and $\dfrac{1}{r^2}u_{\theta\theta} =$

$-\dfrac{\cos\theta}{r}u_x - \dfrac{\sin\theta}{r}u_y - 2\cos\theta\sin\theta\, u_{xy} + \sin^2\theta\, u_{xx} + \cos^2\theta\, u_{yy}$. So $\dfrac{1}{r}(ru_r)_r + \dfrac{1}{r^2}u_{\theta\theta} =$

$\dfrac{\cos\theta}{r}u_x + \dfrac{\sin\theta}{r}u_y + 2\sin\theta\cos\theta\, u_{xy} +$

$\cos^2\theta\, u_{xx} + \sin^2\theta\, u_{yy} + \left(-\dfrac{\cos\theta}{r}u_x - \dfrac{\sin\theta}{r}u_y\right.$

$- 2\sin\theta\cos\theta\, u_{xy} + \sin^2\theta\, u_{xx} + \cos^2\theta\, u_{yy})$
$= (\cos^2\theta + \sin^2\theta)u_{xx} + (\sin^2\theta + \cos^2\theta)u_{yy} = u_{xx} + u_{yy}$.

30 The path from (x, y) through $(x, y + \Delta y)$ to $(x + \Delta x, y + \Delta y)$ could be used. First consider the path from (x, y) to $(x, y + \Delta y)$; on this path x is constant. By the mean-value theorem, there exists a number c_1 between y and $y + \Delta y$ such that

$f(x, y + \Delta y) - f(x, y) = \dfrac{\partial f}{\partial y}(x, c_1)\,\Delta y$. Next

consider the path from $(x, y + \Delta y)$ to $(x + \Delta x, y + \Delta y)$; on this path the y coordinate is constant. Again by the mean-value theorem there exists a c_2 between x and $x + \Delta x$ such that

$f(x + \Delta x, y + \Delta y) - f(x, y + \Delta y)$

$= \dfrac{\partial f}{\partial x}(c_2, y + \Delta y)\,\Delta x$. Hence we can write Δf

$= f(x + \Delta x, y + \Delta y) - f(x, y)$
$= f(x + \Delta x, y + \Delta y) - f(x, y + \Delta y) +$
$f(x, y + \Delta y) - f(x, y)$
$= \dfrac{\partial f}{\partial x}(c_2, y + \Delta y)\,\Delta x + \dfrac{\partial f}{\partial y}(x, c_1)\,\Delta y$. Assuming

$\dfrac{\partial f}{\partial x}$ and $\dfrac{\partial f}{\partial y}$ are continuous at (x, y), we know that

$\dfrac{\partial f}{\partial y}(x, c_1)\,\Delta y = \dfrac{\partial f}{\partial y}(x, y) + \epsilon_1$ and $\dfrac{\partial f}{\partial x}(c_2, y + \Delta y)$

$= \dfrac{\partial f}{\partial x}(x, y) + \epsilon_2$ where $\epsilon_1, \epsilon_2 \to 0$ as $\Delta x, \Delta y \to 0$.

In summary, $\Delta f = \dfrac{\partial f}{\partial x}(x, y)\,\Delta x + \dfrac{\partial f}{\partial y}(x, y)\,\Delta y +$

$\epsilon_1\Delta x + \epsilon_2\Delta y$, which is exactly Equation (7).

32 (a) $r = 2$

(b) $r = -1$

(c) $r = 0$

34 (a) $f(x, y) = x^2(\ln x - \ln y)$ and $r = 2$. Now f_x
$$= 2x \ln(x/y) + x^2 \cdot \frac{y}{x} \cdot \frac{1}{y} = 2x \ln(x/y) + x,$$
$$xf_x = 2x^2 \ln(x/y) + x^2, f_y = x^2 \cdot y/x \cdot (-x/y^2)$$
$$= -x^2/y, \text{ and } yf_y = -x^2. \text{ Then } xf_x + yf_y$$
$$= 2x^2 \ln(x/y) + x^2 - x^2 = 2(x^2 \ln(x/y))$$
$$= 2(x^2(\ln x - \ln y)) = 2f(x, y). \text{ This verifies Euler's Theorem.}$$

(b) $f(x, y) = (x^2 + y^2)^{-1/2}$ and $r = -1$. Now $f_x = -x(x^2 + y^2)^{-3/2}$, $xf_x = -x^2(x^2 + y^2)^{-3/2}$, $f_y = -y(x^2 + y^2)^{-3/2}$, and $yf_y = -y^2(x^2 + y^2)^{-3/2}$, so
$$xf_x + yf_y = -x^2(x^2 + y^2)^{-3/2} - y^2(x^2 + y^2)^{-3/2}$$
$$= -(x^2 + y^2)(x^2 + y^2)^{-3/2} = -(x^2 + y^2)^{-1/2}$$
$$= (-1)f(x, y).$$

(c) $f(x, y) = \sin(y/x)$ and $r = 0$. Now $f_x = \cos(y/x)\left(-\frac{y}{x^2}\right) = -\frac{y \cos(y/x)}{x^2}, xf_x = -\frac{y}{x} \cos \frac{y}{x}, f_y = \cos(y/x)\left(\frac{1}{x}\right) = \frac{\cos(y/x)}{x},$ and
$$yf_y = \frac{y}{x} \cos \frac{y}{x}. \text{ Then } xf_x + yf_y =$$
$$-\frac{y}{x} \cos \frac{y}{x} + \frac{y}{x} \cos \frac{y}{x} = 0 = 0 \cdot f(x, y).$$

14.7 Directional Derivatives and the Gradient

2 Let $f(x, y) = x^4 y^5$.

(a) $\mathbf{u} = 0\mathbf{i} + 1\mathbf{j}$;
$$\frac{\partial f}{\partial x}(1, 1) \cos \theta + \frac{\partial f}{\partial y}(1, 1) \sin \theta = \frac{\partial f}{\partial y}(1, 1)$$
$$= 5x^4 y^4 \big|_{(1,1)} = 5$$

(b) $\mathbf{u} = 0\mathbf{i} + (-1)\mathbf{j}$;
$$\frac{\partial f}{\partial x}(1, 1) \cos \theta + \frac{\partial f}{\partial y}(1, 1) \sin \theta = -\frac{\partial f}{\partial y}(1, 1)$$
$$= -5$$

(c) $\mathbf{u} = \cos \frac{\pi}{3} \mathbf{i} + \sin \frac{\pi}{3} \mathbf{j}$;
$$\frac{\partial f}{\partial x}(1, 1) \cos \frac{\pi}{3} + \frac{\partial f}{\partial y}(1, 1) \sin \frac{\pi}{3} =$$
$$\left[\frac{1}{2}(4x^3 y^5) + \frac{\sqrt{3}}{2}(5x^4 y^4)\right]_{(1,1)} = 2 + \frac{5\sqrt{3}}{2}$$
$$= \frac{4 + 5\sqrt{3}}{2}$$

4 Let $f(x, y, z) = x^2 y z^3$.

(a) $\mathbf{u} = \frac{\mathbf{i} + \mathbf{j} + \mathbf{k}}{\sqrt{1^2 + 1^2 + 1^2}} = \frac{1}{\sqrt{3}}(\mathbf{i} + \mathbf{j} + \mathbf{k})$;
$$\frac{\partial f}{\partial x} \cos \alpha + \frac{\partial f}{\partial y} \cos \beta + \frac{\partial f}{\partial z} \cos \gamma$$
$$= \frac{1}{\sqrt{3}}\left(\frac{\partial f}{\partial x} + \frac{\partial f}{\partial y} + \frac{\partial f}{\partial z}\right)$$
$$= \frac{1}{\sqrt{3}}(2xyz^3 + x^2 z^3 + 3x^2 yz^2)$$
$$= \frac{xz^2}{\sqrt{3}}(2yz + xz + 3xy)$$

14.7 Directional Derivatives and the Gradient

(b) $\mathbf{u} = \dfrac{2\mathbf{i} - \mathbf{j} + 2\mathbf{k}}{\sqrt{2^2 + (-1)^2 + 2^2}} = \dfrac{1}{3}(2\mathbf{i} - \mathbf{j} + 2\mathbf{k});$

$\dfrac{\partial f}{\partial x}\cos\alpha + \dfrac{\partial f}{\partial y}\cos\beta + \dfrac{\partial f}{\partial z}\cos\gamma$

$= \dfrac{1}{3}\left(2\dfrac{\partial f}{\partial x} - \dfrac{\partial f}{\partial y} + 2\dfrac{\partial f}{\partial z}\right)$

$= \dfrac{1}{3}(4xyz^3 - x^2z^3 + 6x^2yz^2)$

$= \dfrac{xz^2}{3}(4yz - xz + 6xy)$

(c) $\mathbf{u} = \dfrac{\mathbf{i} + \mathbf{k}}{\sqrt{1^2 + 0^2 + 1^2}} = \dfrac{1}{\sqrt{2}}(\mathbf{i} + \mathbf{k});$

$\dfrac{\partial f}{\partial x}\cos\alpha + \dfrac{\partial f}{\partial y}\cos\beta + \dfrac{\partial f}{\partial z}\cos\gamma$

$= \dfrac{1}{\sqrt{2}}\left(\dfrac{\partial f}{\partial x} + \dfrac{\partial f}{\partial z}\right) = \dfrac{1}{\sqrt{2}}(2xyz^3 + 3x^2yz^2)$

$= \dfrac{xyz^2}{\sqrt{2}}(2z + 3x)$

6 (a) $\nabla f(1,1) = \left(\dfrac{\partial f}{\partial x}\mathbf{i} + \dfrac{\partial f}{\partial y}\mathbf{j}\right)\bigg|_{(1,1)} = 3\mathbf{i} - 3\mathbf{j}$

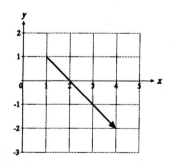

(b) $|\nabla f(1,1)| = |3\mathbf{i} - 3\mathbf{j}| = \sqrt{3^2 + 3^2} = 3\sqrt{2}$

(c) $\mathbf{u} = \dfrac{\nabla f(1,1)}{|\nabla f(1,1)|} = \dfrac{3\mathbf{i} - 3\mathbf{j}}{3\sqrt{2}} = \dfrac{1}{\sqrt{2}}\mathbf{i} - \dfrac{1}{\sqrt{2}}\mathbf{j}$

8 $f_x = \dfrac{-x}{(x^2 + y^2)^{3/2}}, f_y = \dfrac{-y}{(x^2 + y^2)^{3/2}}$

(a) $-\dfrac{1}{5^{3/2}}\mathbf{i} + \dfrac{-2}{5^{3/2}}\mathbf{j} = -\dfrac{1}{5\sqrt{5}}\mathbf{i} - \dfrac{2}{5\sqrt{5}}\mathbf{j}$

(b) $\dfrac{-3}{9^{3/2}}\mathbf{i} = -\dfrac{1}{9}\mathbf{i}$

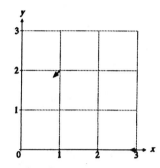

10 Let $\mathbf{u} = \dfrac{\nabla f(a,b)}{|\nabla f(a,b)|}$. Then $D_\mathbf{u} f$ at (a,b) is

$\nabla f(a,b)\cdot\mathbf{u} = |\nabla f(a,b)|\,|\mathbf{u}|\cos 0° = |\nabla f(a,b)|.$
Select a unit vector $\mathbf{v}$ such that $\mathbf{u}\cdot\mathbf{v} = 0$.
Moreover, $\nabla f(a,b)\cdot\mathbf{v} = 0$. Hence $D_\mathbf{v} f$ at (a,b) is
$\nabla f(a,b)\cdot\mathbf{v} = 0$. We conclude that there is always
a direction in which the directional derivative of f
at (a,b) is 0.

12 The largest directional derivative of f at (a,b,c) is

$|\nabla f(a,b,c)| = |f_x\mathbf{i} + f_y\mathbf{j} + f_z\mathbf{k}| =$

$|2\mathbf{i} + 3\mathbf{j} + 4\mathbf{k}| = \sqrt{2^2 + 3^2 + 4^2} = \sqrt{29}.$

14 (a) The directional derivatives can be estimated in
the direction of $\mathbf{v} = 0.1\mathbf{i} + 0.2\mathbf{j} + 0.1\mathbf{k}$ and
its opposite.

(b) $\nabla f\cdot\dfrac{\mathbf{v}}{\|\mathbf{v}\|} \approx \dfrac{3.1 - 3}{\|\mathbf{v}\|} = \dfrac{0.1}{\sqrt{0.01 + 0.04 + 0.01}}$

$= 1/\sqrt{6}$; the other directional derivative is, of
course, approximately $-1/\sqrt{6}$.

16. Let $T(x, y, z)$ be the temperature at (x, y, z). We are given that $T_x = 0.03°/\text{cm}$, $T_y = -0.02°/\text{cm}$, and $T_z = 0.05°/\text{cm}$. The unit vector in the direction of $\langle 2, 5, 1 \rangle$ is $\mathbf{u} = \dfrac{\langle 2, 5, 1 \rangle}{\sqrt{2^2 + 5^2 + 1^2}} = \dfrac{1}{\sqrt{30}} \langle 2, 5, 1 \rangle$. Then $D_\mathbf{u} T = \nabla T \cdot \mathbf{u}$

$= T_x \cdot \dfrac{2}{\sqrt{30}} + T_y \cdot \dfrac{5}{\sqrt{30}} + T_z \cdot \dfrac{1}{\sqrt{30}}$

$= \dfrac{0.06 - 0.10 + 0.05}{\sqrt{30}} \approx 0.0018°/\text{cm}$, which is positive. Hence it is getting warmer.

18. Here $\mathbf{u} = \dfrac{-0.01\mathbf{i} - 0.02\mathbf{j} + 0.01\mathbf{k}}{\sqrt{(-0.01)^2 + (-0.02)^2 + (0.01)^2}} = \dfrac{1}{\sqrt{6}}(-\mathbf{i} + 2\mathbf{j} + \mathbf{k})$ is the unit vector pointing from $(1, 1, 2)$ to $(0.99, 0.98, 2.01)$. Then $D_\mathbf{u} f = -3$. The distance from $(1, 1, 2)$ to $(0.99, 0.98, 2.01)$ is $\dfrac{\sqrt{6}}{100}$, so $f(0.99, 0.98, 2.01)$

$\approx D_\mathbf{u} f \cdot \dfrac{\sqrt{6}}{100} + f(1, 1, 2) = \dfrac{400 - 3\sqrt{6}}{100}$.

20. $\nabla f = 3x^2 yz \mathbf{i} + x^3 z \mathbf{j} + x^3 y \mathbf{k}$ and $\nabla f(2, 1, -1) = -12\mathbf{i} - 8\mathbf{j} + 8\mathbf{k}$, so $D_\mathbf{u} f =$

$\dfrac{(-12\mathbf{i} - 8\mathbf{j} + 8\mathbf{k}) \cdot (2\mathbf{i} - \mathbf{k})}{\sqrt{5}} = \dfrac{-24 - 8}{\sqrt{5}} = -\dfrac{32}{\sqrt{5}}$.

The directional derivative will have its largest value in the direction of $-12\mathbf{i} - 8\mathbf{j} + 8\mathbf{k}$; that is, $\|\nabla f\| = \sqrt{144 + 64 + 64} = 4\sqrt{17}$.

22. $\nabla f = \dfrac{2x}{1 + x^2 + y + z}\mathbf{i} + \dfrac{1}{1 + x^2 + y + z}\mathbf{j} + \dfrac{1}{1 + x^2 + y + z}\mathbf{k}$, so $\nabla f(1, 1, 1) =$

$\dfrac{1}{2}\mathbf{i} + \dfrac{1}{4}\mathbf{j} + \dfrac{1}{4}\mathbf{k}$ and $\nabla f(1, 1, 1) \cdot (-\mathbf{i}) = -\dfrac{1}{2}$. The maximal directional derivative is $\|\nabla f(1, 1, 1)\|$

$= \sqrt{\dfrac{1}{4} + \dfrac{1}{16} + \dfrac{1}{16}} = \dfrac{\sqrt{6}}{4}$.

24. $\nabla f = x^x(1 + \ln x)ye^{z^2}\mathbf{i} + x^x e^{z^2}\mathbf{j} + 2x^x yze^{z^2}\mathbf{k}$ and

$\nabla f(1, 1, 0) = \mathbf{i} + \mathbf{j}$, so $\nabla f(1, 1, 0) \cdot \dfrac{\mathbf{i} - \mathbf{j} + \mathbf{k}}{\sqrt{3}}$

$= 0$. The maximal directional derivative is $\|\nabla f(1, 1, 0)\| = \sqrt{2}$.

26. (a) $f(x, y, z) = x^2 + y^2 + z^2$, so $f_x = 2x$, $f_y = 2y$, and $f_z = 2z$. Therefore $\nabla f(x, y, z) = 2x\mathbf{i} + 2y\mathbf{j} + 2z\mathbf{k}$, $\nabla f(2, 0, 0) = 4\mathbf{i}$, $\nabla f(0, 2, 0) = 4\mathbf{j}$, and $\nabla f(0, 0, 2) = 4\mathbf{k}$.

28. Let $\nabla f(a, b, c) = A\mathbf{i} + B\mathbf{j} + C\mathbf{k}$. Then any unit vector $\mathbf{u}$ in the plane $A(x - a) + B(y - b) + C(z - c) = 0$ is perpendicular to $\nabla f(a, b, c)$. Therefore $D_\mathbf{u} f = 0$ for any such $\mathbf{u}$. Thus there are infinitely many $\mathbf{u}$ such that $D_\mathbf{u} f = 0$.

30. $\nabla T = 2\mathbf{i} - 2\mathbf{j}$, so

(a) the bug should move in the direction of $2\mathbf{i} - 2\mathbf{j}$;

(b) the bug should move in the direction of $-2\mathbf{i} + 2\mathbf{j}$;

(c) the bug should move in the direction of $\mathbf{i} + \mathbf{j}$ or $-\mathbf{i} - \mathbf{j}$, since $(2\mathbf{i} - 2\mathbf{j}) \cdot (\mathbf{i} + \mathbf{j}) = 0$.

32 $\nabla f = \dfrac{-x}{(x^2+y^2+z^2)^{3/2}}\mathbf{i} + \dfrac{-y}{(x^2+y^2+z^2)^{3/2}}\mathbf{j} +$

$\dfrac{-z}{(x^2+y^2+z^2)^{3/2}}\mathbf{k}$. Therefore, $\nabla f = \dfrac{-\mathbf{r}}{|\mathbf{r}|^3}$.

34 Let $Q = (a + \Delta x, b + \Delta y, c + \Delta z)$. Then $\overrightarrow{PQ} =$

$\Delta x \mathbf{i} + \Delta y \mathbf{j} + \Delta z \mathbf{k}$ and $\nabla f \cdot \overrightarrow{PQ} =$

$f_x(a, b, c)\Delta x + f_y(a, b, c)\Delta y + f_z(a, b, c)\Delta z =$

$df \approx \Delta f$ when Δx, Δy, and Δz are small.

36 (a)

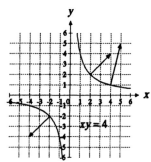

(b) $\nabla f(x, y) = y\mathbf{i} + x\mathbf{j}$

(d) The gradient ∇f has slope x/y. The level curve $xy = k$ has slope dy/dx, where $(xy)' = xy' + y = 0$, so $dy/dx = -y/x$, showing that the gradient is normal to the level curve.

38 $f(x, y) = \sin \alpha x \cos \beta y$, so $f_x = \alpha \cos \alpha x \cos \beta y$,
$f_y = -\beta \sin \alpha x \sin \beta y$, and $\mathbf{E} = -\nabla f =$
$-\alpha \cos \alpha x \cos \beta y\, \mathbf{i} + \beta \sin \alpha x \sin \beta y\, \mathbf{j}$.

40 Note that $\dfrac{\partial g}{\partial \theta} = -f_x(a, b)\sin\theta + f_y(a, b)\cos\theta =$

0 when $\theta = \tan^{-1}\dfrac{f_y(a, b)}{f_x(a, b)}$. For this θ, $g(\theta) =$

$f_x(a, b) \cdot \dfrac{f_x(a, b)}{\sqrt{[f_x(a, b)]^2 + [f_y(a, b)]^2}} +$

$f_y(a, b) \cdot \dfrac{f_y(a, b)}{\sqrt{[f_x(a, b)]^2 + [f_y(a, b)]^2}}$

$= \sqrt{[f_x(a, b)]^2 + [f_y(a, b)]^2}$. Since $\dfrac{\partial^2 g}{\partial \theta^2} =$

$-f_x(a, b)\cos\theta - f_y(a, b)\sin\theta < 0$ for θ

$= \tan^{-1}\dfrac{f_y(a, b)}{f_x(a, b)}$, the maximum value of $g(\theta)$ is

$\sqrt{[f_x(a, b)]^2 + [f_y(a, b)]^2}$.

14.8 Normals and the Tangent Plane

2 Let $f(x, y) = x^2 + y^2$. Then $\nabla f = 2x\mathbf{i} + 2y\mathbf{j}$
$= -6\mathbf{i} + 8\mathbf{j}$ at $(-3, 4)$.

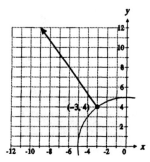

4 Let $f(x, y) = x^2 - y^2$. Then $\nabla f = 2x\mathbf{i} - 2y\mathbf{j}$
$= 6\mathbf{i} + 2\sqrt{8}\mathbf{j}$ at $(3, -\sqrt{8})$.

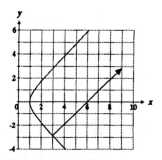

6 $\nabla f = (y + z)\mathbf{i} + (x + z)\mathbf{j} + (x + y)\mathbf{k}$
$\nabla f(1, 2, 3) = 5\mathbf{i} + 4\mathbf{j} + 3\mathbf{k}$

8 $\nabla f = 2x\mathbf{i} - \mathbf{k}$, $\nabla f(1, 3, 1) = 2\mathbf{i} - \mathbf{k}$

10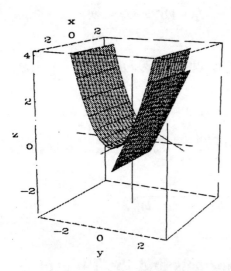

12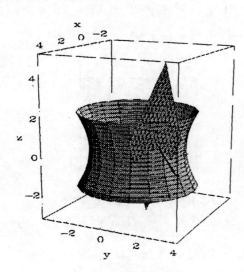

14 $\nabla f = 2x\mathbf{i} + 6y\mathbf{j} - 2z\mathbf{k}$ and $\nabla f(3, 1, 2) = 6\mathbf{i} + 6\mathbf{j} - 4\mathbf{k}$. Hence $6x + 6y - 4z = 16$ is an equation of the plane.

16 $\nabla f = \left(1 - \dfrac{x^2}{yz}\right)^{-1/2} \cdot \dfrac{1}{\sqrt{yz}}\mathbf{i} +$

$\left(1 - \dfrac{x^2}{yz}\right)^{-1/2} \cdot \dfrac{x}{\sqrt{z}} \cdot \left(-\dfrac{1}{2}y^{-3/2}\right)\mathbf{j} +$

$\left(1 - \dfrac{x^2}{yz}\right)^{-1/2} \cdot \dfrac{x}{\sqrt{y}}\left(-\dfrac{1}{2}z^{-3/2}\right)\mathbf{k}$

$= (yz - x^2)^{-1/2}\left[\mathbf{i} - \dfrac{x}{2y}\mathbf{j} - \dfrac{x}{2z}\mathbf{k}\right]$, $\nabla f(1, 2, 2) =$

$(4 - 1)^{-1/2}\left[\mathbf{i} - \dfrac{1}{4}\mathbf{j} - \dfrac{1}{4}\mathbf{k}\right] = \dfrac{1}{\sqrt{3}}\left(\mathbf{i} - \dfrac{1}{4}\mathbf{j} - \dfrac{1}{4}\mathbf{k}\right)$.

Hence $\dfrac{1}{\sqrt{3}}(x - 1) - \dfrac{1}{4\sqrt{3}}(y - 2) - \dfrac{1}{4\sqrt{3}}(z - 2) =$

0, or $\dfrac{1}{\sqrt{3}}x - \dfrac{1}{4\sqrt{3}}y - \dfrac{1}{4\sqrt{3}}z = 0$ is an equation

for the tangent plane.

18 $\Delta z = \sqrt{(3.02)^2 + (3.99)^2} - \sqrt{3^2 + 4^2} \approx 0.00405$,

$dz = \dfrac{x}{\sqrt{x^2 + y^2}}\bigg|_{(3,4)}(0.02) + \dfrac{y}{\sqrt{x^2 + y^2}}\bigg|_{(3,4)}(-0.01)$

$= 0.004$.

20 $\Delta T \approx T_L \Delta L + T_g \Delta g$

$= \dfrac{2\pi}{\sqrt{g}}\dfrac{1}{2\sqrt{L}}\Delta L + 2\pi\sqrt{L}\left(-\dfrac{1}{2}g^{-3/2}\right)\Delta g$

$= \dfrac{\pi}{\sqrt{gL}}\Delta L - \dfrac{\pi}{g}\sqrt{\dfrac{L}{g}}\Delta g$,

$\left|\dfrac{\Delta T}{T}\right| \approx \left|\dfrac{\pi}{\sqrt{gL}} \cdot \dfrac{\sqrt{g}}{2\pi\sqrt{L}}\Delta L - \dfrac{\pi}{g}\sqrt{\dfrac{L}{g}} \cdot \dfrac{\sqrt{g}}{2\pi\sqrt{L}}\Delta g\right|$

$= \left|\dfrac{1}{2}\dfrac{\Delta L}{L} - \dfrac{1}{2}\dfrac{\Delta g}{g}\right| \leq \dfrac{1}{2}\left|\dfrac{\Delta L}{L}\right| + \dfrac{1}{2}\left|\dfrac{\Delta g}{g}\right|$

$= \dfrac{1}{2}(0.03) + \dfrac{1}{2}(0.02) = 0.025$.

22 $\Delta V = (x + \Delta x)(y + \Delta y)(z + \Delta z) - xyz$
$= xy\Delta z + xz\Delta y + x\Delta y\Delta z + yz\Delta x + y\Delta x\Delta z +$
$z\Delta x\Delta y + \Delta x\Delta y\Delta z$; $dV = yz\Delta x + xz\Delta y + xy\Delta z$

24 Let $g(x, y, z) = z - xy$; then $\nabla g = -y\mathbf{i} - x\mathbf{j} + \mathbf{k}$ provides a normal vector to the surface $z = xy$. In particular, $\nabla g(2, 3, 6) = -3\mathbf{i} - 2\mathbf{j} + \mathbf{k}$ is a normal at $(2, 3, 6)$. Any vector perpendicular to $-3\mathbf{i} - 2\mathbf{j} + \mathbf{k}$ will be tangent to the surface. One example is $2\mathbf{i} - 3\mathbf{j}$, yielding $x = 2 + 2t$, $y = 3 - 3t$, and $z = 6$ as parametric equations for a tangent line. (One way to get vectors perpendicular to $-3\mathbf{i} - 2\mathbf{j} + \mathbf{k}$ is to cross it with any randomly chosen vector.)

26 (a), (b)

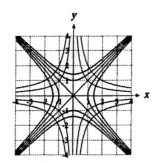

(c) $\nabla f \cdot \nabla g = (y\mathbf{i} + x\mathbf{j}) \cdot (2x\mathbf{i} - 2y\mathbf{j}) = 0$

(d) Stream lines of air flow

28 $\nabla f = 2xyz\mathbf{i} + x^2z\mathbf{j} + x^2y\mathbf{k}$, $\nabla f(1, 1, 1) = 2\mathbf{i} + \mathbf{j} + \mathbf{k}$, $\nabla g = (y + z)\mathbf{i} + (x + z)\mathbf{j} + (x + y)\mathbf{k}$, $\nabla g(1, 1, 1) = 2\mathbf{i} + 2\mathbf{j} + 2\mathbf{k}$. $\nabla f \times \nabla g$ is perpendicular to the normals of both planes, so

$$\nabla f \times \nabla g = \begin{vmatrix} \mathbf{i} & \mathbf{j} & \mathbf{k} \\ 2 & 1 & 1 \\ 2 & 2 & 2 \end{vmatrix} = -2\mathbf{j} + 2\mathbf{k}$$ is parallel to

the line of intersection. An equation for the line is $x = 1$, $y = 1 - 2t$, $z = 1 + 2t$.

30 Suppose the vector $\mathbf{v}$ is tangent at the point (a, b, c) to the curve formed by the intersection of the surfaces $f(x, y, z) = 0$ and $g(x, y, z) = 0$. Then $\mathbf{v}$ must lie in the tangent planes of both surfaces at (a, b, c) and so must be perpendicular to the normal vectors to the surfaces at the same point. Normal vectors to the surfaces are provided by ∇f and ∇g; if we let $\mathbf{v} = \nabla f \times \nabla g$, then $\mathbf{v}$ has the desired properties.

34 (a) $2 \cdot 1 + 3 \cdot 2 - 3 = 5$, so the point $(1, 2, 3)$ lies on the plane $2x + 3y - z = 5$. The point also lies on the sphere $x^2 + y^2 + z^2 = 14$ because $1^2 + 2^2 + 3^2 = 14$.

(b) We know that $2\mathbf{i} + 3\mathbf{j} - \mathbf{k}$ is a normal to the plane $2x + 3y - z = 5$. To find a normal to the sphere, let the function $f(x, y, z) = x^2 + y^2 + z^2$. Then $\nabla f = 2x\mathbf{i} + 2y\mathbf{j} + 2z\mathbf{k}$, so at the point $(1, 2, 3)$ a normal to the sphere is $2\mathbf{i} + 4\mathbf{j} + 6\mathbf{k}$. To find the angle between the surfaces at the point $(1, 2, 3)$ we compute the angle between their normals. Let θ denote the desired angle; then $\cos \theta$

$$= \frac{(2\mathbf{i} + 3\mathbf{j} - \mathbf{k}) \cdot (2\mathbf{i} + 4\mathbf{j} + 6\mathbf{k})}{\|2\mathbf{i} + 3\mathbf{j} - \mathbf{k}\| \|2\mathbf{i} + 4\mathbf{j} + 6\mathbf{k}\|}$$

$$= \frac{2 \cdot 2 + 3 \cdot 4 + (-1) \cdot 6}{\sqrt{2^2 + 3^2 + (-1)^2} \sqrt{2^2 + 4^2 + 6^2}} = \frac{10}{\sqrt{14}\sqrt{56}}$$

$$= \frac{10}{28} = \frac{5}{14}, \text{ so } \theta = \cos^{-1} \frac{5}{14} \approx 1.2056 \approx 69.08°.$$

36 (a) Two vectors are $\mathbf{i} + 2\mathbf{k}$ and $\mathbf{j} + 3\mathbf{k}$.

(b) A normal is $(\mathbf{i} + 2\mathbf{k}) \times (\mathbf{j} + 3\mathbf{k}) = -2\mathbf{i} - 3\mathbf{j} + \mathbf{k}$.

(c) $f(3.02, 4.99)$

$$\approx f(3, 5) + \frac{(2\mathbf{i} + 3\mathbf{j}) \cdot (0.02\mathbf{i} - 0.01\mathbf{j})}{\sqrt{(0.02)^2 + (-0.01)^2}}$$

$= 7.4472.$

14.9 Critical Points and Extrema

2 $f_x = 2x$ and $f_y = -2y$, so $(0, 0)$ is the only critical point. $D = 2(-2) - 0 = -4 < 0$, so f has no relative extrema; $(0, 0)$ is a saddle point.

4 $f_x = 4x^3 + 16x = x(4x^2 + 16), f_y = 2y - 4 = 2(y - 2), f_{xx} = 12x^2 + 16, f_{yy} = 2,$ and $f_{xy} = 0$. The only critical point is $(0, 2)$. $D = 16 \cdot 2 - 0 = 32 > 0$ and $f_{xx} = 16 > 0$; so f has a relative minimum of -4 at $(0, 2)$.

6 $f_x = 2x + 2y + 4, f_y = 2x + 4y, f_{xx} = 2, f_{yy} = 4,$ and $f_{xy} = 2$. The only critical point is $(-4, 2)$. $D = 8 - 4 > 0$ and $f_{xx} = 2 > 0$, so f has a relative minimum of -8 at $(-4, 2)$.

8 $f_x = -8x - y, f_y = -x - 6y, f_{xx} = -8, f_{yy} = -6,$ and $f_{xy} = -1$. The only critical point is $(0, 0)$. $D = (-8)(-6) - 1 > 0$, and $f_{xx} = -8 < 0$, so f has a relative maximum of 0 at $(0, 0)$.

10 $f_x = 3x^2 + 3y, f_y = -3y^2 + 3x, f_{xx} = 6x, f_{yy} = -6y, f_{xy} = 3$. The critical points are $(0, 0)$ and $(1, -1)$. $D(0, 0) = 0 \cdot 0 - 9 < 0$, which indicates a saddle. $D(1, -1) = 6 \cdot 6 - 9 > 0$ and $f_{xx} = 6 > 0$, so f has a relative minimum of -1 at $(1, -1)$.

12 $D = 2 \cdot 4 - 9 < 0$, so f has a saddle point at (a, b).

14 $D = 3 \cdot 4 - 4 > 0$ and $f_{xx} = 3 > 0$, so f has a relative minimum at (a, b).

16 $D = 3(-4) - 4 < 0$, so f has a saddle point at (a, b).

18 $f_x = 3y - 3x^2, f_y = 3x - 3y^2, f_{xx} = -6x, f_{yy} = -6y, f_{xy} = 3$. The critical points are $(0, 0)$ and $(1, 1)$. For $(0, 0)$, $D = 0 \cdot 0 - 9 < 0$, so f has a saddle point at $(0, 0)$. $D(1, 1) = (-6)(-6) - 9 > 0, f_{xx} = -6 < 0$, so f has a relative maximum of 1 at $(1, 1)$.

20 $f_x = 6y - 2xy - y^2, f_y = 6x - x^2 - 2xy, f_{xx} = -2y, f_{yy} = -2x,$ and $f_{xy} = 6 - 2x - 2y$. The critical points are $(0, 0), (0, 6), (6, 0)$ and $(2, 2)$. For each of the points $(0, 0), (0, 6)$, and $(6, 0)$, at least one of f_{xx} and f_{yy} is zero, and $f_{xy} \neq 0$ for all these points. Then for $(0, 0), (0, 6),$ and $(6, 0), D = -f_{xy}^2 < 0$, and $(0, 0), (0, 6),$ and $(6, 0)$ are saddle points of f. Also, $D(2, 2) = (-4)(-4) - 4 > 0, f_{xx} = -4 < 0$, so f has a relative maximum of 8 at $(2, 2)$.

22 $f_x = y \cdot 2^{xy} \ln 2, f_y = x \cdot 2^{xy} \ln 2, f_{xx} = y^2 \cdot 2^{xy} (\ln 2)^2, f_{yy} = x^2 \cdot 2^{xy} (\ln 2)^2,$ and $f_{xy} = xy \cdot 2^{xy} (\ln 2)^2 + 2^{xy} \ln 2$. The only critical point is $(0, 0)$. $D(0, 0) = 0 \cdot 0 - (\ln 2)^2$, so f has a saddle point at $(0, 0)$.

24 $f_x = 1 - \dfrac{8}{x^2 y}, f_y = 1 - \dfrac{8}{xy^2}, f_{xx} = \dfrac{16}{x^3 y}, f_{yy} = \dfrac{16}{xy^3}, f_{xy} = \dfrac{8}{x^2 y^2}$. The only critical point is $(2, 2)$.

Now $D(2, 2) = 1 \cdot 1 - \dfrac{1}{4} > 0$, and $f_{xx} = 1 > 0$, so f has a relative minimum of 6 at $(2, 2)$.

26 $V = lwh$ and the cost is $C = 4lh + 4wh + 6lw$. $V = 1$, so $l = \dfrac{1}{wh}$ and $C = 4 \cdot \dfrac{1}{w} + 4wh + 6 \cdot \dfrac{1}{h}$.

To minimize C we compute $C_w = -\dfrac{4}{w^2} + 4h, C_h = 4w - \dfrac{6}{h^2}, C_{ww} = \dfrac{8}{w^3}, C_{hh} = \dfrac{12}{h^3},$ and $C_{wh} = 4$. The only critical point that makes sense is $w = \sqrt[3]{\dfrac{2}{3}}$ and $h = \sqrt[3]{\dfrac{9}{4}}$. At this point, $D = 12 \cdot \dfrac{16}{3} - 16 > 0$ and $C_{ww} = 12 > 0$, so C is a minimum at this point. Therefore the minimum cost is $8\sqrt[3]{\frac{3}{2}} + 6\sqrt[3]{\frac{4}{9}} \approx 13.74$ cents when $l = \sqrt[3]{\frac{2}{3}} \approx 0.87$ ft and $w = \sqrt[3]{\frac{2}{3}} \approx 0.87$ ft, and $h = \sqrt[3]{\frac{9}{4}} \approx 1.31$ ft. Expect to spend 14¢.

28 Call the dimensions x, y, and z. The volume is then $V = xyz$. The surface area is $2xy + 2xz + 2yz = 12$, so $xy + (x + y)z = 6$ and $z = \frac{6 - xy}{x + y}$.

Hence $V = xy \cdot \frac{6 - xy}{x + y} = \frac{6xy - x^2y^2}{x + y}$. When V is maximal, we must have $V_x = V_y = 0$, so $V_x = \frac{(x + y)(6y - 2xy^2) - (6xy - x^2y^2)(1)}{(x + y)^2} = \frac{y^2(6 - x^2 - 2xy)}{(x + y)^2} = 0$ and $V_y = \frac{x^2(6 - y^2 - 2xy)}{(x + y)^2} = 0$. Since $x \neq 0$ and $y \neq 0$, it follows that $6 - x^2 - 2xy = 0$ and $6 - y^2 - 2xy = 0$; subtracting these equations yields $x^2 - y^2 = 0$, so $x = y$. Hence $0 = 6 - x^2 - 2x^2 = 6 - 3x^2$, so $x = y = \sqrt{2}$ and $z = \frac{6 - 2}{2\sqrt{2}} = \sqrt{2}$. The box is a cube.

30 (a) The revenue is $600r + 900t$, so the profit is $P(r, t) = 600r + 900t - 2r^2 - rt - 2t^2$ dollars.

(b) We must maximize $P(r, t)$ for $r \geq 0$, $t \geq 0$. Note that $P(r, t) < 900(r + t) - \frac{1}{4}(r + t)^2$.

Hence $P(r, t) \to -\infty$ as $r + t \to \infty$, so $P(r, t)$ must have a maximum either at a critical point with $r > 0$, $t > 0$, or on the boundary. We need $P_r = 600 - 4r - t = 0$ and $P_t = 900 - r - 4t$, so $4r + t = 600$ and $r + 4t = 900$. Thus $r = 100$ and $t = 200$. At this point, $P = 120{,}000$. The boundary has two pieces: $t = 0$, $r \geq 0$ and $r = 0$, $t \geq 0$. On the first, $P = 600r - 2r^2$ so we have $P_r = 600 - 4r$, which is positive for $r < 150$ and negative for $r > 150$; the maximum on this edge is $P(150, 0) = 45{,}000$. On the second piece, $P = 900t - 2t^2$, so $P_t = 900 - 4t$, which is positive for $t < 225$ and negative for $t > 225$; hence the maximum is $P(0, 225) = 101{,}250$. The overall maximum therefore occurs for $r = 100$, $t = 200$. The discriminant is $D = P_{rr}P_{tt} - (P_{rt})^2 = (-4)(-4) - (-1)^2 = 15$. Since $D > 0$ and $P_{rr} = -4 < 0$, the point is a local maximum, as expected.

32 Let $f(x, y) = x^2 + kxy + 3y^2$. Then $f_x = 2x + ky$, $f_y = kx + 6y$, $f_{xx} = 2$, $f_{yy} = 6$, $f_{xy} = k$, and $D = 12 - k^2$. Since, at $(0, 0)$, $f_x = f_y = 0$ and $f_{xx} > 0$, f will have a minimum at $(0, 0)$ if $k^2 < 12$, f will not if $k^2 > 12$, and may or may not if $k^2 = 12$. In the final case $k = \pm 2\sqrt{3}$, so $f(x, y) = (x \pm \sqrt{3}y)^2 \geq 0$; hence f does actually have minimum at $(0, 0)$. So f has a minimum at $(0, 0)$ if and only if $k^2 \leq 12$; that is, for $-2\sqrt{3} \leq k \leq 2\sqrt{3}$.

34 (a) $f_x = 2x(2 - 2x^2 - y^2)e^{-x^2-y^2}$ and $f_y = 2y(1 - 2x^2 - y^2)e^{-x^2-y^2}$. Setting these to 0 gives $x(2 - 2x^2 - y^2) = 0$ and $y(1 - 2x^2 - y^2)$

$= 0$. Since $2 - 2x^2 - y^2$ and $1 - 2x^2 - y^2$ cannot both be 0, there are five critical points: $(0, 0)$, $(0, \pm 1)$, and $(\pm 1, 0)$.

(b) Let $t = x^2 + y^2$. Then $0 \le f(x, y) = (2x^2 + y^2)e^{-x^2-y^2} \le (2x^2 + 2y^2)e^{-x^2-y^2} = 2te^{-t}$. Since $te^{-t} \to 0$ as $t \to \infty$, $f(x, y) \to 0$ as $x^2 + y^2 \to \infty$.

(c) Since $f(x, y) \ge 0$ for all x and y, while $f(0, 0) = 0$, the minimum is 0.

(d) Since $f(x, y) \to 0$ as $x^2 + y^2 \to \infty$, there must be a maximum at a critical point. Since $f(0, 0) = 0$, $f(0, 1) = f(0, -1) = \dfrac{1}{e}$, and $f(1, 0) = f(-1, 0) = 2/e$, the maximum is $2/e$.

36 First, check for critical points: $f_x = 6x + 2y = 0$ when $y = -3x$; $f_y = -8y + 2x = 0$ when $x = 4y$. If $f_x = f_y = 0$, then $x = 4y = 4(-3x) = -12x$ or $13x = 0$, so $x = y = 0$. There are thus no critical points except on the boundary. Now examine the edges: On E_1, $x = 0$, so $f(x, y) = f(0, y) = -4y^2$. The maximum value of $-4y^2$ for y in $[0, 1]$ is 0 and occurs for $y = 0$, that is, at $(0, 0)$. On E_2, $y = 0$, so $f(x, y) = f(x, 0) = 3x^2$. The maximum value is 3 at $x = 1$, that is, at $(1, 0)$. On E_3, $x = 1$, so $f(x, y) = f(1, y) = 3 - 4y^2 + 2y$. Since
$$\dfrac{d}{dy}(3 - 4y^2 + 2y) = -8y + 2 = 0 \text{ at } y = \dfrac{1}{4}$$
and
$$\dfrac{d^2}{dy^2}(3 - 4y^2 + 2y) = -8 < 0,$$
we see that $f\left(1, \dfrac{1}{4}\right) = \dfrac{13}{4}$ is a maximum value. Checking the endpoints reveals that $f(1, 0) = 3$ and $f(1, 1) = 1$, so the maximum on E_3 is $\dfrac{13}{4}$ at $\left(1, \dfrac{1}{4}\right)$. On E_4, $y = 1$ so $f(x, y) = f(x, 1) = 3x^2 - 4 + 2x$. Since
$$\dfrac{d}{dx}(3x^2 - 4 + 2x) = 6x + 2 = 0 \text{ for } x = -\dfrac{1}{3},$$
which does not lie in the interval, the maximum must occur at an endpoint. Comparing $f(0, 1) = -4$ and $f(1, 1) = 1$, we find that the maximum on E_4 is 1 at $(1, 1)$. Combining the results for the edges, we find that the maximum in the square region is $f\left(1, \dfrac{1}{4}\right) = \dfrac{13}{4}$.

38 Let $f(x, y) = -x + 3y + 6$. Note that $f_x = -1 \ne 0$ and $f_y = 3 \ne 0$, so there are no critical points.

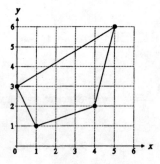

The maximum of f over the region must therefore occur on the region's boundary. Note further that the boundaries are all straight lines that may be represented by equations of the form $y = ax + b$. Upon substituting $f(x, y)$ for y in terms of x, we will obtain an expression for x that is a linear function of x (namely, $f(x, ax + b) = (3a - 1)x + (3b + 6)$). The extrema of a linear function always occur at endpoints, so we need only check the values of f at the endpoints of the edges; that is, at the vertices of the region. Now $f(1, 1) = 8$, $f(4, 2)$

= 8, $f(5, 6) = 19$, and $f(0, 3) = 15$, so the maximum value is 19 and occurs at $(5, 6)$. (The minimum is 8 and occurs everywhere along the edge joining $(1, 1)$ and $(4, 2)$.)

40 $D = z_{xx}z_{yy} - z_{xy}^2 = 36 - z_{xy}^2$. Since $z_x = z_y = 0$ and $z_{xx} > 0$, z has a relative minimum at (x_0, y_0) if $36 - z_{xy}^2 > 0$; that is, for $-6 < z_{xy} < 6$.

42 (a) $f(x, y) = x^2 + 2xy + y^2 = (x + y)^2$, so f is never less than 0, which occurs at the origin (and everywhere else on the line $y = -x$).

(b) On the line $y = -x$ the function has a maximum at the origin. On the line $x = 0$ it has a minimum.

(c) Use the negative of the function in (a).

44
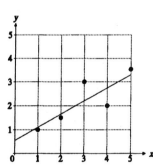
$y = 0.55x + 0.55$

46 The total area is $f(x_1, x_2) = x_1(1 - x_1) + (x_2 - x_1)(1 - x_2) = x_2 - x_1^2 - x_2^2 + x_1x_2$. To
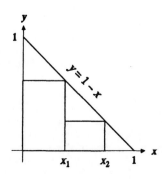
maximize f we find that $f_{x_1} = -2x_1 + x_2$, $f_{x_2} = 1 - 2x_2 + x_1$, $f_{x_1x_1} = -2$, $f_{x_2x_2} = -2$, and $f_{x_1x_2} = 1$. The only critical point is $x_1 = \frac{1}{3}$, $x_2 = \frac{2}{3}$.

$D = 4 - 1 > 0$ and $f_{x_1x_1} = -2 < 0$, so f is maximal when $x_1 = \frac{1}{3}$ and $x_2 = \frac{2}{3}$. The maximum area is $\frac{1}{3}$.

48 Let $f(x, y, z) = xyz$. Now $x + y + z = 1$, so $f = xy(1 - x - y)$ and $f_x = y - 2xy - y^2$, $f_y = x - x^2 - 2xy$, $f_{xx} = -2y$, $f_{yy} = -2x$, and $f_{xy} = 1 - 2x - 2y$. The critical points are $(0, 0)$, $(1, 0)$, $(0, 1)$, and $\left(\frac{1}{3}, \frac{1}{3}\right)$.

(a) The minimum value of f is zero, which occurs whenever any of x, y, or z is zero.

(b) For $\left(\frac{1}{3}, \frac{1}{3}\right)$, $D = \left(\frac{2}{3}\right)^2 - \left(\frac{1}{3}\right)^2 > 0$ and $f_{xx} = -\frac{2}{3} < 0$, so f has a maximum of $\frac{1}{27}$ at $\left(\frac{1}{3}, \frac{1}{3}, \frac{1}{3}\right)$.

50 Following the notation in the proof of case 1 of Theorem 3, we are given that $AC - B^2 > 0$ and $A < 0$. We must show that $Ax^2 + 2Bxy + Cy^2 \leq 0$, which is equivalent to showing that $A(Ax^2 + 2Bxy + Cy^2) \geq 0$, since $A < 0$. The remainder of the proof of case 2 is then identical to the proof of case 1.

14.10 Lagrange Multipliers

2 $L(x, y, \lambda) = x^2 + y^2 - \lambda(2x + 3y - 6)$. We need to solve the equations $L_x = 2x - 2\lambda = 0$, $L_y = 2y - 3\lambda = 0$, and $L_\lambda = -2x - 3y + 6 = 0$. $x = \lambda$ and $y = \frac{3}{2}\lambda$, so $L_\lambda = -2\lambda - \frac{9}{2}\lambda + 6 = -\frac{13}{2}\lambda + 6 = 0$. Hence $\lambda = -6\left(-\frac{2}{13}\right) = \frac{12}{13}$, $x = \frac{12}{13}$, and $y = \frac{18}{13}$. $x^2 + y^2$ has a minimum of $\frac{468}{169} = \frac{36}{13}$ at $\left(\frac{12}{13}, \frac{18}{13}\right)$.

4 $L(x, y, \lambda) = x + 2y - \lambda(x^2 + 2y^2 - 8)$. We need to solve the equations $L_x = 1 - 2\lambda x = 0$, $L_y = 2 - 4\lambda y = 0$, and $L_\lambda = -(x^2 + 2y^2 - 8) = 0$. $x = \frac{1}{2\lambda}$ and $y = \frac{1}{2\lambda}$, so $L_\lambda = -\left(\frac{1}{4\lambda^2} + \frac{2}{4\lambda^2} - 8\right) = 0$. Hence, $8\lambda^2 = \frac{3}{4}$ and $\lambda^2 = \frac{3}{32}$. Therefore, $x = \frac{2\sqrt{6}}{3}$ and $y = \frac{2\sqrt{6}}{3}$, or $x = \frac{-2\sqrt{6}}{3}$ and $y = -\frac{2\sqrt{6}}{3}$. The maximum, $2\sqrt{6}$, occurs at $\left(\frac{2\sqrt{6}}{3}, \frac{2\sqrt{6}}{3}\right)$.

6 Minimize $S = 2xy + 2yz + 2xz$, subject to $V = xyz = 1$. $L(x, y, z, \lambda) = 2xy + 2yz + 2xz - \lambda(xyz - 1)$, $L_x = 2y + 2z - \lambda yz = 0$, $L_y = 2x + 2z - \lambda xz = 0$, and $L_z = 2x + 2y - \lambda xy = 0$. Solving each equation for λ, we obtain $\lambda = \frac{2}{z} + \frac{2}{y}$, $\lambda = \frac{2}{z} + \frac{2}{x}$, and $\lambda = \frac{2}{y} + \frac{2}{x}$. Equating the first two, we find that $\frac{2}{y} = \frac{2}{x}$, so $x = y$. Similarly, $x = z$. Hence $x = y = z$, and $xyz = 1$ implies that $x = y = z = 1$; the minimum surface area is 6.

8 $L(x, y, z, \lambda) = x + y + 2z - \lambda(x^2 + y^2 + z^2 - 9)$, $L_x = 1 - 2\lambda x = 0$, $L_y = 1 - 2\lambda y = 0$, and $L_z = 2 - 2\lambda z = 0$. Then $x = \frac{1}{2\lambda}$, $y = \frac{1}{2\lambda}$, and $z = \frac{1}{\lambda}$, so $\frac{1}{4\lambda^2} + \frac{1}{4\lambda^2} + \frac{1}{\lambda^2} = 9$ and $\lambda^2 = \frac{1}{6}$. We want x, y, and z to be positive, so $x = \frac{\sqrt{6}}{2}$, $y = \frac{\sqrt{6}}{2}$, $z = \sqrt{6}$, giving a maximum of $3\sqrt{6}$.

10 Maximize $V = xyz$, with $S = 2xy + 2xz + 2yz = 6$. $L(x, y, z, \lambda) = xyz - \lambda(2xy + 2xz + 2yz - 6)$. $L_x = yz - 2\lambda(y + z) = 0$, $L_y = xz - 2\lambda(x + z) = 0$, and $L_z = xy - 2\lambda(x + y) = 0$. Solving $L_x = yz - 2\lambda y - 2\lambda z = 0$ for y yields $y = \frac{2\lambda z}{z - 2\lambda}$. Similarly, $x = \frac{2\lambda z}{z - 2\lambda}$, so $x = y$. It then follows that $x = y = z = 1$ and $V = 1$ is the maximum.

12 Minimize $x^2 + y^2 + z^2$ with the constraint $xyz = 1$. $L(x, y, z, \lambda) = x^2 + y^2 + z^2 - \lambda(xyz - 1)$. $L_x = 2x - \lambda yz = 0$, $L_y = 2y - \lambda xz = 0$, and $L_z = 2z - \lambda xy = 0$. Since $xyz = 1$, $xy = \frac{1}{z}$, $xz = \frac{1}{y}$, and $yz = \frac{1}{x}$. Hence, $2x - \frac{\lambda}{x} = 0$ and $\lambda = 2x^2$. Similarly, $\lambda = 2y^2 = 2z^2$. Therefore $x^2 = y^2$

14.10 Lagrange Multipliers

$= z^2$. The minimum distance will be $\sqrt{3}$, at any of the points $(1, 1, 1)$, $(1, -1, -1)$, $(-1, 1, -1)$, and $(-1, -1, 1)$.

14 Minimize $x^2 + y^2 + z^2$ with the constraints $2y + 4z - 5 = 0$ and $4(x^2 + y^2) - z^2 = 0$. $L(x, y, z, \lambda, \mu) = x^2 + y^2 + z^2 - \lambda(2y + 4z - 5) - \mu(4(x^2 + y^2) - z^2)$, $L_x = 2x - 8\mu x = 2x(1 - 4\mu) = 0$, $L_y = 2y - 2\lambda - 8\mu y = 0$, and $L_z = 2z - 4\lambda + 2\mu z = 0$. $L_y = 0$ implies that $y = \dfrac{2\lambda}{2 - 8\mu}$, $L_z = 0$ implies that $z = \dfrac{4\lambda}{2 + 2\mu}$, and $L_x = 0$ implies that $x(2 - 8\mu) = 0$, and either $x = 0$ or $\mu = \dfrac{1}{4}$. If $x = 0$, then $L_\mu = 0$ implies that $4y^2 = z^2$ and $z = \pm 2y$. Using L_λ, we see that $y = \dfrac{1}{2}$ or $-\dfrac{5}{6}$, and $z = 1$ or $\dfrac{5}{3}$. If $\mu = \dfrac{1}{4}$, then $2y = 8\mu y$, and $\lambda = 0$. This gives $2z = -\dfrac{1}{2}z$ and $z = 0$. But if $z = 0$, then $y = \dfrac{5}{2}$, and $L_\mu \neq 0$. Hence, there are two possible solutions: $\left(0, \dfrac{1}{2}, 1\right)$ and $\left(0, -\dfrac{5}{6}, \dfrac{5}{3}\right)$. The former gives the minimum value of $\dfrac{5}{4}$.

16 $L(l, w, h, \lambda) = 4lh + 4wh + 6lw - \lambda(lwh - 1)$, $L_l = 4h + 6w - \lambda wh = 0$, $L_w = 4h + 6l - \lambda lh = 0$, $L_h = 4l + 4w - \lambda lw = 0$, and $L_\lambda = -(lwh - 1) = 0$. Since $lwh = 1$, $l = \dfrac{1}{wh}$, and

solving for λ gives $\lambda = \dfrac{4l + 4w}{lw} = \dfrac{4 + 4w^2h}{w}$, $\lambda = \dfrac{4h + 6l}{lh} = \dfrac{6 + 4wh^2}{h}$, and $\lambda = \dfrac{4h + 6w}{wh}$.

Solving for w and h yields $w = \sqrt[3]{\dfrac{2}{3}}$ and $h = \sqrt[3]{\dfrac{9}{4}}$, and $l = \dfrac{1}{wh} = \sqrt[3]{\dfrac{2}{3}}$.

18 (a) $L(x, y, z, \lambda) = x^2 + y^2 + z^2 - \lambda(x + y + z - 1)$, $L_x = 2x - \lambda$, $L_y = 2y - \lambda$, $L_z = 2z - \lambda$, and $L_\lambda = -(x + y + z - 1)$. So $x = y = z$, and $x + y + z = 3x = 1$ implies $x = y = z = 1/3$. The minimum is $\left(\dfrac{1}{3}\right)^2 + \left(\dfrac{1}{3}\right)^2 + \left(\dfrac{1}{3}\right)^2 = \dfrac{1}{3}$.

(b) By the same reasoning as Exercise 29 of Sec. 14.9, the largest $x^2 + y^2 + z^2$ can be subject to $x + y + z = 1$ (x, y, z nonnegative) is 1 when (x, y, z) is $(1, 0, 0)$, $(0, 1, 0)$, or $(0, 0, 1)$.

20 Let $f(x, y, z) = x + 2y + 3z$ and $g(x, y, z) = x + 3y + 9z$. Then $\nabla f \times \nabla g = (\mathbf{i} + 2\mathbf{j} + 3\mathbf{k}) \times (\mathbf{i} + 3\mathbf{j} + 9\mathbf{k}) = 9\mathbf{i} - 6\mathbf{j} + \mathbf{k}$ is a vector parallel to the line of intersection of $f(x, y, z) = 6$ and $g(x, y, z) = 9$. Note that $(0, 3, 0)$ lies on this line. Then the squared distance is

$$\left| -3\mathbf{j} - \left(-3\mathbf{j} \cdot \dfrac{9\mathbf{i} - 6\mathbf{j} + \mathbf{k}}{\sqrt{9^2 + (-6)^2 + 1^2}}\right) \dfrac{9\mathbf{i} - 6\mathbf{j} + \mathbf{k}}{\sqrt{9^2 + (-6)^2 + 1^2}} \right|^2$$

$$= \left| -3\mathbf{j} - \dfrac{18}{\sqrt{118}} \dfrac{9\mathbf{i} - 6\mathbf{j} + \mathbf{k}}{\sqrt{118}} \right|^2$$

$$= \left\| \frac{81}{59}\mathbf{i} - \frac{123}{59}\mathbf{j} - \frac{9}{59}\mathbf{k} \right\|^2 = \frac{369}{59}.$$

22 (a)

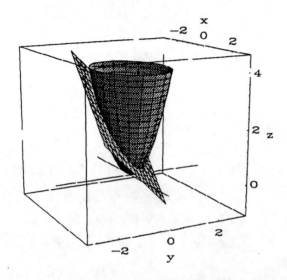

(d) $L(x, y, z, \lambda, \mu) = z - \lambda(x + y + z - 1) - \mu(z - x^2 - 2y^2)$, $L_x = -\lambda + 2\mu x$, $L_y = -\lambda + 4\mu y$, $L_z = 1 - \lambda - \mu$, $L_\lambda = -(x + y + z - 1)$, and $L_\mu = -(z - x^2 - 2y^2)$.

Setting these partial derivatives to 0, we have

$$x = \frac{\lambda}{2\mu} = \frac{1-\mu}{2\mu}, \; y = \frac{\lambda}{4\mu} = \frac{1-\mu}{4\mu}, \; z$$

$$= 1 - x - y = 1 - \frac{1-\mu}{2\mu} - \frac{1-\mu}{4\mu} =$$

$$\frac{4\mu - 2(1-\mu) - (1-\mu)}{4\mu} = \frac{7\mu - 3}{4\mu}, \text{ and } z$$

$$= x^2 + 2y^2 = \left(\frac{1-\mu}{2\mu}\right)^2 + 2\left(\frac{1-\mu}{4\mu}\right)^2 =$$

$$\frac{2(1-\mu)^2 + (1-\mu)^2}{8\mu^2} = \frac{3(1-\mu)^2}{8\mu^2}. \text{ (If } \mu =$$

0, then $\lambda = 0$, which is impossible.) Then

$$\frac{7\mu - 3}{4\mu} = \frac{3(1-\mu)^2}{8\mu^2}, \; 11\mu^2 - 3 = 0, \text{ and}$$

$\mu = \pm\sqrt{\frac{3}{11}}$. We choose $\mu = -\sqrt{\frac{3}{11}}$ so that

z is maximized. Hence the highest point of intersection is

$$\left(-\frac{\sqrt{11} + \sqrt{3}}{2\sqrt{3}}, -\frac{\sqrt{11} + \sqrt{3}}{4\sqrt{3}}, \frac{3\sqrt{11} + 7\sqrt{3}}{4\sqrt{3}}\right).$$

24 (a)

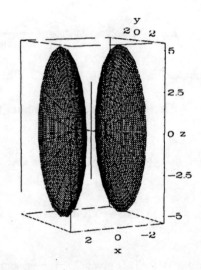

(c) $L(x, y, z, \lambda) = (x - 1)^2 + (y - 1)^2 + (z - 1)^2 - \lambda(x^2 - y^2/4 - z^2/9 - 1)$, $L_x = 2(x - 1) - 2\lambda x$, $L_y = 2(y - 1) + \frac{\lambda}{2}y$, $L_z = 2(z - 1) + \frac{2\lambda}{9}z$, $L_\lambda = -(x^2 - y^2/4 - z^2/9 - 1)$. Equating L_x, L_y, and L_z to 0 gives $x = \frac{1}{1 - \lambda}$, $y = \frac{4}{4 + \lambda}$, and $z = \frac{9}{9 + \lambda}$.

Substituting the values into $L_\lambda = 0$ yields

$$\left(\frac{1}{1-\lambda}\right)^2 - \frac{1}{4}\left(\frac{4}{4+\lambda}\right)^2 - \frac{1}{9}\left(\frac{9}{9+\lambda}\right)^2 = 1,$$

which simplifies to $\lambda^6 + 24\lambda^5 + 202\lambda^4 + 572\lambda^3 - 383\lambda^2 - 3384\lambda + 468 = 0$. The pertinent solution is $\lambda = 0.1366372$, which leads to $x = 1.1582616, y = 0.9669690, z = 0.9850451$, and a minimum distance of 0.1623620.

26 Maximize $\sin \alpha \sin \beta \sin \gamma$ subject to $\alpha + \beta + \gamma = \pi$. Let $L(\alpha, \beta, \gamma, \lambda) = \sin \alpha \sin \beta \sin \gamma - \lambda(\alpha + \beta + \gamma - \pi)$. Then $L_\alpha = \cos \alpha \sin \beta \sin \gamma - \lambda$, $L_\beta = \sin \alpha \cos \beta \sin \gamma - \lambda$, and $L_\gamma = \sin \alpha \sin \beta \cos \gamma - \lambda$. Setting these equal to 0 gives $\cos \alpha \sin \beta \sin \gamma = \sin \alpha \cos \beta \sin \gamma = \sin \alpha \sin \beta \cos \gamma$. Dividing by $\sin \alpha \sin \beta \sin \gamma$ yields $\cot \alpha = \cot \beta = \cot \gamma$. Notice that α, β, γ all lie in the open interval $(0, \pi)$. Since $\cot x$ is a decreasing function on this interval, we must have $\alpha = \beta = \gamma$. Since $\alpha + \beta + \gamma = \pi$, all three must equal $\dfrac{\pi}{3}$; the triangle is equilateral.

28 $L(x, y, z, \lambda, \mu) = x + 2y + 3z - \lambda(x^2 + y^2 + z^2 - 1) - \mu(x + y + z)$, $L_x = 1 - 2\lambda x - \mu = 0$, $L_y = 2 - 2\lambda y - \mu = 0$, $L_z = 3 - 2\lambda z - \mu = 0$, $L_\lambda = -(x^2 + y^2 + z^2 - 1) = 0$, and $L_\mu = -(x + y + z) = 0$. Hence $x = \dfrac{1 - \mu}{2\lambda}$, $y = \dfrac{2 - \mu}{2\lambda}$, and $z = \dfrac{3 - \mu}{2\lambda}$. (If $\lambda = 0$, then $\mu = 1, \mu = 2$, and $\mu = 3$, which is impossible.) Therefore

$$\dfrac{1 - \mu}{2\lambda} + \dfrac{2 - \mu}{2\lambda} + \dfrac{3 - \mu}{2\lambda} = \dfrac{6 - 3\mu}{2\lambda} = 0, \text{ so}$$

$\mu = 2$. Also, $\dfrac{(1 - \mu)^2 + (2 - \mu)^2 + (3 - \mu)^2}{4\lambda^2}$

$= \dfrac{(1 - 2)^2 + (2 - 2)^2 + (3 - 2)^2}{4\lambda^2} = \dfrac{2}{4\lambda^2} = 1$,

so $\lambda^2 = \dfrac{1}{2}$. Possible solutions are $\left(-\dfrac{1}{\sqrt{2}}, 0, \dfrac{1}{\sqrt{2}}\right)$ and $\left(\dfrac{1}{\sqrt{2}}, 0, -\dfrac{1}{\sqrt{2}}\right)$. The maximum is $\sqrt{2}$ at

$\left(-\dfrac{1}{\sqrt{2}}, 0, \dfrac{1}{\sqrt{2}}\right)$.

30 (a) Let $L(x_1, \cdots, x_n, y_1, \cdots, y_n, \lambda, \mu) =$

$$\sum_{i=1}^{n} x_i y_i - \lambda\left(\sum_{i=1}^{n} x_i^2 - 1\right) - \mu\left(\sum_{i=1}^{n} y_i^2 - 1\right).$$

Then $L_{x_i} = y_i - 2\lambda x_i$ and $L_{y_i} = x_i - 2\mu y_i$. Setting $L_{x_i} = 0$ for all i results in $y_i = 2\lambda x_i$, while setting $L_{y_i} = 0$ yields $x_i = 2\mu y_i$. By the first result, $1 = \sum_{i=1}^{n} y_i^2 = (2\lambda)^2 \sum_{i=1}^{n} x_i^2 = (2\lambda)^2$, so $2\lambda = \pm 1$. Thus either $y_i = x_i$ for all values of i, or $y_i = -x_i$ for all values of i. In the first case, we have $\sum_{i=1}^{n} x_i y_i = \sum_{i=1}^{n} x_i^2 = 1$. In the second case, $\sum_{i=1}^{n} x_i y_i = \sum_{i=1}^{n} (-x_i^2) = -1$. Hence, the maximum value is 1 and occurs whenever $y_i = x_i$ for all values of i. (Observe that this is not just a single point, but an entire class of points that satisfy a certain condition.)

(b) Let $A = \left(\sum_{i=1}^{n} a_i^2\right)^{1/2}$, $B = \left(\sum_{i=1}^{n} b_i^2\right)^{1/2}$, $x_i = a_i/A$ and $y_i = b_i/B$ for $1 \leq i \leq n$. Then we have $\sum_{i=1}^{n} x_i^2 = 1$ and $\sum_{i=1}^{n} y_i^2 = 1$, so, by (a),

$\sum_{i=1}^{n} x_i y_i \leq 1$. That is, $\sum_{i=1}^{n} a_i b_i \leq AB =$

$\left(\sum_{i=1}^{n} a_i^2\right)^{1/2} \left(\sum_{i=1}^{n} b_i^2\right)^{1/2}$.

(c) Let $\mathbf{a} = a_1\mathbf{i} + a_2\mathbf{j} + a_3\mathbf{k}$ and $\mathbf{b} = b_1\mathbf{i} + b_2\mathbf{j} + b_3\mathbf{k}$. Then $\sum_{i=1}^{3} a_i b_i = \mathbf{a} \cdot \mathbf{b}$, $\left(\sum_{i=1}^{3} a_i^2\right)^{1/2} = |\mathbf{a}|$, and $\left(\sum_{i=1}^{3} b_i^2\right)^{1/2} = |\mathbf{b}|$. If θ is the angle between $\mathbf{a}$ and $\mathbf{b}$, then $\mathbf{a} \cdot \mathbf{b} = \|\mathbf{a}\| \|\mathbf{b}\| \cos\theta \leq \|\mathbf{a}\| \|\mathbf{b}\|$, so the desired inequality follows.

32 (a) Let $L(x_1, \cdots, x_n, y_1, \cdots, y_n, \lambda, \mu) =$

$\sum_{i=1}^{n} x_i y_i - \lambda\left(\sum_{i=1}^{n} x_i^p - 1\right) - \mu\left(\sum_{i=1}^{n} y_i^q - 1\right)$.

Then $L_{x_i} = y_i - \lambda p x_i^{p-1}$ and $L_{y_i} = x_i - \mu q y_i^{q-1}$. Setting $L_{x_i} = 0$ for all i yields $y_i = \lambda p x_i^{p-1}$. Hence $x_i y_i = \lambda p x_i^p$ and

$\sum_{i=1}^{n} x_i y_i = \sum_{i=1}^{n} \lambda p x_i^p = \lambda p \sum_{i=1}^{n} x_i^p = \lambda p$.

Setting $L_{y_i} = 0$ for all i yields $x_i = \mu q y_i^{q-1}$ and similarly we have $\sum_{i=1}^{n} x_i y_i = \mu q$. Thus $\lambda p = \mu q$. Now $y_i = \lambda p x_i^{p-1} = \lambda p (\mu q)^{p-1} y_i^{(q-1)(p-1)}$. Since $\frac{1}{p} + \frac{1}{q} = 1$, it follows that $q - 1 = \frac{q}{p}$ and $p - 1 = \frac{p}{q}$.

The above equation reduces to $y_i = \lambda p(\mu q)^{p-1} y_i = (\lambda p)^p y_i$ and $y_i[1 - (\lambda p)^p] = 0$. So either $y_i = 0$ for all i or $(\lambda p)^p = 1$. If $y_i = 0$ for all i, then $x_i y_i = 0$ for all i and $\sum_{i=1}^{n} x_i y_i$ clearly is not maximized. Hence $(\lambda p)^p = 1$ and $\lambda p = 1 = \mu q$. Then $\sum_{i=1}^{n} x_i y_i$ takes on a maximum value of 1 when $y_i = x_i^{p-1}$ (or $x_i = y_i^{q-1}$) for all i.

(b) Let $A = \left(\sum_{i=1}^{n} a_i^p\right)^{1/p}$, $B = \left(\sum_{i=1}^{n} b_i^q\right)^{1/q}$, $x_i = \frac{a_i}{A}$, and $y_i = \frac{b_i}{B}$ for $1 \leq i \leq n$. Then we have $\sum_{i=1}^{n} x_i^p = \sum_{i=1}^{n} y_i^q = 1$, so by (a) $\sum_{i=1}^{n} x_i y_i \leq 1$. Therefore $\sum_{i=1}^{n} x_i y_i = \frac{1}{AB} \sum_{i=1}^{n} a_i b_i \leq 1$, and it follows that $\sum_{i=1}^{n} a_i b_i \leq AB =$

$\left(\sum_{i=1}^{n} a_i^p\right)^{1/p} \left(\sum_{i=1}^{n} b_i^q\right)^{1/q}$.

34 $L(c_1, \cdots, c_N, E^1, \cdots, E^N, \lambda) =$

$\sum_{i=1}^{N} c_i E^i - \lambda\left(\sum_{i=1}^{N} H^i - H\right)$, $\frac{\partial L}{\partial E^i} = c_i - \lambda \frac{\partial H^i}{\partial E^i}$

$= 0$. Thus for all i, $\lambda = \frac{c_i}{\partial H^i/\partial E^i}$ and $\frac{1}{\lambda} =$

$\frac{1}{c_i} \cdot \frac{\partial H^i}{\partial E^i}$. Since $\frac{1}{\lambda}$ is not a function of i, we have

$\frac{1}{c_i} \cdot \frac{\partial H^i}{\partial E^i} = \frac{1}{\lambda}$, which is constant for all i.

14.11 The Chain Rule Revisited

2 By the chain rule, $\left(\frac{\partial u}{\partial y}\right)_z =$

$\left(\frac{\partial u}{\partial x}\right)_{y,z} \frac{\partial x}{\partial y} + \left(\frac{\partial u}{\partial y}\right)_{x,z} \frac{\partial y}{\partial y} + \left(\frac{\partial u}{\partial z}\right)_{x,y} \frac{\partial z}{\partial y}$. Since on the

left-hand side of the equality $u = f(h(y, z), y, z) = k$, we have $\left(\frac{\partial u}{\partial y}\right)_z = 0$. On the right-hand side of

the equation, $\frac{\partial x}{\partial y} = \left(\frac{\partial x}{\partial y}\right)_z$, $\frac{\partial y}{\partial y} = 1$, and $\frac{\partial z}{\partial y} = 0$.

Then $0 = \left(\frac{\partial u}{\partial x}\right)_{y,z} \left(\frac{\partial x}{\partial y}\right)_z + \left(\frac{\partial u}{\partial y}\right)_{x,z}$, and solving for

$\left(\frac{\partial x}{\partial y}\right)_z$, $\left(\frac{\partial x}{\partial y}\right)_z = -\frac{(\partial u/\partial y)_{x,z}}{(\partial u/\partial x)_{y,z}}$.

4 $u = f(x, y, z) = 5x + 2y + 3z; \left(\frac{\partial u}{\partial z}\right)_{x,y} = 3$,

$\left(\frac{\partial u}{\partial x}\right)_{y,z} = 5$, and $-\frac{(\partial u/\partial z)_{x,y}}{(\partial u/\partial x)_{y,z}} = -\frac{3}{5}$. Now

consider $u = f(x, y, z) = 5x + 2y + 3z = k = 6$.

Then $x = \frac{6 - 2y - 3z}{5}$ and $\left(\frac{\partial x}{\partial z}\right)_y = -\frac{3}{5}$

$= -\frac{(\partial u/\partial z)_{x,y}}{(\partial u/\partial x)_{y,z}}$.

6 (a) We proceed as in Example 3. Let $E = f(T, P)$ and $T = g(V, P)$. We also have $P = P + 0 \cdot V$

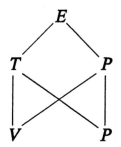

$= h(V, P)$. So E is a function of T and P and each of these is a function of V and P. By

Theorem 3 of Sec. 14.6, we get $\left(\frac{\partial E}{\partial V}\right)_P$

$= \left(\frac{\partial E}{\partial T}\right)_P \left(\frac{\partial T}{\partial V}\right)_P + \left(\frac{\partial E}{\partial P}\right)_T \left(\frac{\partial P}{\partial V}\right)_P$

$= \left(\frac{\partial E}{\partial T}\right)_P \left(\frac{\partial T}{\partial V}\right)_P + 0 = \left(\frac{\partial E}{\partial T}\right)_P \left(\frac{\partial T}{\partial V}\right)_P$.

(b) Again using Theorem 3, $\left(\frac{\partial E}{\partial P}\right)_V$

$= \left(\frac{\partial E}{\partial T}\right)_P \left(\frac{\partial T}{\partial P}\right)_V + \left(\frac{\partial E}{\partial P}\right)_T \left(\frac{\partial P}{\partial P}\right)_V$

$= \left(\frac{\partial E}{\partial T}\right)_P \left(\frac{\partial T}{\partial P}\right)_V + \left(\frac{\partial E}{\partial P}\right)_T$.

8 By Exercise 5, showing that $\left(\frac{\partial P}{\partial T}\right)_V \left(\frac{\partial T}{\partial V}\right)_P \left(\frac{\partial V}{\partial P}\right)_T =$

-1 is equivalent to showing $\left(\frac{\partial V}{\partial T}\right)_P \left(\frac{\partial T}{\partial V}\right)_P = 1$. But

if we interchange the roles of P and V in the

argument in Exercise 7, $\left(\frac{\partial V}{\partial T}\right)_P \left(\frac{\partial T}{\partial V}\right)_P = 1$ follows.

Thus $\left(\frac{\partial P}{\partial T}\right)_V \left(\frac{\partial T}{\partial V}\right)_P \left(\frac{\partial V}{\partial P}\right)_T = -1$.

10 By the chain rule, $\left(\frac{\partial u}{\partial x}\right)_y =$

$\left(\frac{\partial u}{\partial x}\right)_{y,z} \frac{\partial x}{\partial x} + \left(\frac{\partial u}{\partial y}\right)_{x,z} \frac{\partial y}{\partial x} + \left(\frac{\partial u}{\partial z}\right)_{x,y} \left(\frac{\partial z}{\partial x}\right)_y$, or, more

clearly, $\frac{\partial G}{\partial x} = \frac{\partial F}{\partial x} \cdot \frac{\partial x}{\partial x} + \frac{\partial F}{\partial y} \cdot \frac{\partial y}{\partial x} + \frac{\partial F}{\partial z} \cdot \frac{\partial f}{\partial x} =$

$\frac{\partial F}{\partial x} + \frac{\partial F}{\partial z} \frac{\partial f}{\partial x}$. Then, since $G(x, y) = x^2 y$, we have $\frac{\partial f}{\partial x}$

$= \frac{\frac{\partial G}{\partial x} - \frac{\partial F}{\partial x}}{\partial F/\partial z} = \frac{2xy - \partial F/\partial x}{\partial F/\partial z}$.

14.S Review Exercises

2 (a)

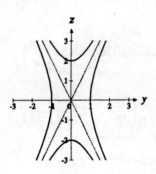

(b)

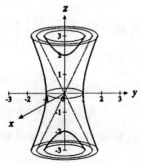

(c) Surface of revolution for $y^2 - \frac{z^2}{4} = 1$:

$x^2 + y^2 - \frac{z^2}{4} = 1$

Surface of revolution for $y^2 - \frac{z^2}{4} = -1$:

$-x^2 - y^2 + \frac{z^2}{4} = 1$

Surface of revolution for $y = 2z$: $4z^2 = x^2 + y^2$

4 (a)

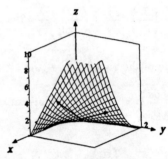

(b) $z_x = 4xy$, which equals 4 at $(1, 1, 2)$. This is the slope at $(1, 1, 2)$ in a direction parallel to the positive x axis, so any vector of slope 4, such as $\mathbf{i} + 4\mathbf{k}$ is tangent to the surface at $(1, 1, 2)$. See (a).

(c) $z_y = 2x^2$, which equals 2 at $(1, 1, 2)$, so any vector of slope 2, such as $\mathbf{j} + 2\mathbf{k}$, is tangent to the surface at $(1, 1, 2)$. See (a).

(d) If we cross the vectors found in (b) and (c), we obtain a vector normal to the surface; thus $-4\mathbf{i} - 2\mathbf{j} + \mathbf{k}$ is perpendicular at $(1, 1, 2)$.

(e) $D_{\mathbf{u}} f = \nabla f \cdot \mathbf{u} = (4\mathbf{i} + 2\mathbf{j} - \mathbf{k}) \cdot \frac{\mathbf{j} + 2\mathbf{k}}{\sqrt{5}} = 0$

(f) $D_{\mathbf{u}} f = \nabla f \cdot \mathbf{u} = \nabla f \cdot \frac{\nabla f}{|\nabla f|} = \|\nabla f\| = \sqrt{21}$

(g) $D_{\mathbf{u}} f = \nabla f \cdot \mathbf{u}$

$= (4\mathbf{i} + 2\mathbf{j} - \mathbf{k}) \cdot \frac{2\mathbf{i} + 3\mathbf{j} + 4\mathbf{k}}{\sqrt{29}} = \frac{10}{\sqrt{29}}$

6 Subtracting the two equations, $x^2 + y^2 = 1$ and $x^2 + z^2 = 1$, we get $y^2 - z^2 = 0$, so $z = \pm y$. The intersection has two parts, one in the plane $z = y$, the other in the plane $z = -y$. Each part consists of those points in its plane that are directly above or below the circle $x^2 + y^2 = 1$.

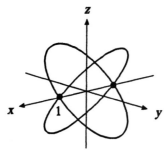

8 (a) $\dfrac{\partial z}{\partial x} = 2xy$

(b) $\dfrac{\partial z}{\partial x} = 2xue^{3x} + 3x^2ue^{3x} = xue^{3x}(2 + 3x)$

(c) $\left(\dfrac{\partial z}{\partial x}\right)_y = 2xy$, $\left(\dfrac{\partial z}{\partial x}\right)_u = xue^{3x}(2 + 3x)$

10 By the chain rule, $\dfrac{\partial u}{\partial u} = \dfrac{\partial u}{\partial x} \cdot \dfrac{\partial x}{\partial u} + \dfrac{\partial u}{\partial y} \cdot \dfrac{\partial y}{\partial u}$ and

$\dfrac{\partial u}{\partial v} = \dfrac{\partial u}{\partial x} \cdot \dfrac{\partial x}{\partial v} + \dfrac{\partial u}{\partial y} \cdot \dfrac{\partial y}{\partial v}$. Since $\dfrac{\partial u}{\partial u} = 1$ and $\dfrac{\partial u}{\partial v} = 0$, the equations immediately follow.

12 (a)

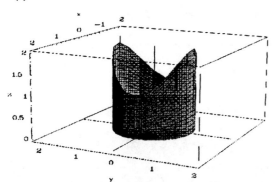

(b) The local approximation for the area of the cylindrical region can best be found by taking angle θ as our parameter as it goes from 0 to 2π. The local approximation is like a rectangle of base $d\theta$ and height $x^2 + 2y^2 = \cos^2\theta + 2\sin^2\theta = \dfrac{3}{2} - \dfrac{1}{2}\cos 2\theta$. Area =

$\int_0^{2\pi} \left(\dfrac{3}{2} - \dfrac{1}{2}\cos 2\theta\right) d\theta = \left(\dfrac{3}{2}\theta - \dfrac{1}{4}\sin 2\theta\right)\Big|_0^{2\pi}$

$= 3\pi.$

14 (a)

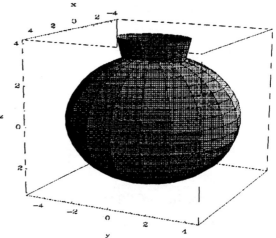

(b) P is satisfied by the simultaneous equations $y = x$, $z = 2x^2 + y^2$, and $x^2 + y^2 + 2z^2 = 20$. Thus $z = 3y^2$ and $z^2 = 10 - y^2$. Then

$3z^2 + z - 30 = 0$ and $z = \dfrac{-1 + \sqrt{361}}{6} = 3$.

So $P = (1, 1, 3)$.

(c) $f(x, y, z) = 2x^2 + y^2 - z$; $\nabla f(1, 1, 3) = 4\mathbf{i} + 2\mathbf{j} - \mathbf{k}$

(d) $g(x, y, z) = x^2 + y^2 + 2z^2$; $\nabla g(1, 1, 3) = 2\mathbf{i} + 2\mathbf{j} + 12\mathbf{k}$

(e) $\nabla f \times \nabla g = 26\mathbf{i} - 50\mathbf{j} + 4\mathbf{k}$

16 (a) By Sec. 14.8 (p. 842) with $f(x, y) = x^2 + y^2$, the vector $f_x\mathbf{i} + f_y\mathbf{j} - \mathbf{k} = 2x\mathbf{i} + 2y\mathbf{j} - \mathbf{k}$ is normal to the surface. Hence, at $(1, 1, 2)$, $2\mathbf{i} + 2\mathbf{j} - \mathbf{k}$ is a normal vector, so

$\dfrac{2\mathbf{i} + 2\mathbf{j} - \mathbf{k}}{\sqrt{4 + 4 + 1}} = \dfrac{2}{3}\mathbf{i} + \dfrac{2}{3}\mathbf{j} - \dfrac{1}{3}\mathbf{k}$ is a unit normal.

(b) Let $f(x, y, z) = x + y^2 + z^3$. Then $\nabla f = \mathbf{i} + 2y\mathbf{j} + 3z^2\mathbf{k}$ is normal to the level surfaces of f. At $(1, 1, 1)$, $\nabla f = \mathbf{i} + 2\mathbf{j} + 3\mathbf{k}$, which is a vector of magnitude $\sqrt{1 + 4 + 9} = \sqrt{14}$, so $\dfrac{\mathbf{i} + 2\mathbf{j} + 3\mathbf{k}}{\sqrt{14}}$ is a unit normal.

(c) Let $f(x, y, z) = x^2 + y^2 + z^2$; then $\nabla f = 2x\mathbf{i} + 2y\mathbf{j} + 2z\mathbf{k}$ is a normal to the level surfaces of f. Its length is $2\sqrt{x^2 + y^2 + z^2}$, so a unit normal vector is $\dfrac{x\mathbf{i} + y\mathbf{j} + z\mathbf{k}}{\sqrt{x^2 + y^2 + z^2}}$.

18 (a) The vector $\mathbf{v}(t) = \mathbf{G}'(t)$ is tangent (parallel) to the curve $\mathbf{r} = \mathbf{G}(t)$. The vector $\mathbf{T}(t) = \mathbf{v}(t)/\|\mathbf{v}(t)\|$ has constant magnitude—it's a unit vector—so its derivative is perpendicular to it. The derivative is also perpendicular to the curve.

(b) The curve $f(x, y) = 0$ is a level curve of the function $z = f(x, y)$. Hence the gradient ∇f is perpendicular to the curve.

(c) Let the surfaces be given by $f(x, y, z) = 0$ and $g(x, y, z) = 0$. The gradients ∇f and ∇g are perpendicular to the curve formed by the intersection of the surfaces, so $\nabla f \times \nabla g$ is parallel to it.

(d) The answer given in (a) for plane curves is equally valid for curves in space.

20 (a) $\dfrac{\partial^2 (xy)}{\partial x\, \partial y} = \dfrac{\partial}{\partial x}\left(\dfrac{\partial (xy)}{\partial y}\right) = \dfrac{\partial}{\partial x}(x) = 1$

(b) $f(x, y) = xy + g(x) + h(y)$

22 By Exercise 33 of Sec. 14.6, we have $xf_x + yf_y = nf$. Differentiating with respect to x, we obtain $(xf_{xx} + f_x) + yf_{yx} = nf_x$, so, by the equality of the mixed partial derivatives, $xf_{xx} + yf_{xy} = (n - 1)f_x$. Similarly, differentiating with respect to y yields $xf_{xy} + (yf_{yy} + f_y) = nf_y$, so $xf_{xy} + yf_{yy} = (n - 1)f_y$. Combining these equations, we have $x^2 f_{xx} + 2xy f_{xy} + y^2 f_{yy} = x(xf_{xx} + yf_{xy}) + y(xf_{xy} + yf_{yy}) = x(n - 1)f_x + y(n - 1)f_y = (n - 1)(xf_x + yf_y) = (n - 1)nf$.

24 (a) First we establish that $\dfrac{f''}{f} = \dfrac{g'}{a^2 g}$ for all x and t. We have $u_x = f'g$ and $u_{xx} = f''g$, while $u_t = fg' = a^2 u_{xx} = a^2 f''g$, and the equality follows. Since the equality holds for all x and t, we know $\dfrac{f''(x)}{f(x)} = \dfrac{g'(0)}{a^2 g(0)} = k$ for all x.

Since $\dfrac{g'(t)}{a^2 g(t)} = \dfrac{f''(x)}{f(x)} = k$, we conclude that

$$\dfrac{f''}{f} = \dfrac{g'}{a^2 g} = k.$$

(b) $g(t) = Ce^{ka^2 t}$

(c) Let $f(x) = c_1 \sin \sqrt{-k}\, x + c_2 \cos \sqrt{-k}\, x$. We have $f'(x) = c_1 \sqrt{-k} \cos \sqrt{-k}\, x - c_2 \sqrt{-k} \sin \sqrt{-k}\, x$ and $f''(x) = c_1 \sqrt{-k}(-\sqrt{-k} \sin \sqrt{-k}\, x) - c_2 \sqrt{-k}(\sqrt{-k} \cos \sqrt{-k}\, x) = k(c_1 \sin \sqrt{-k}\, x + c_2 \cos \sqrt{-k}\, x)$. Thus $\dfrac{f''(x)}{f(x)} = k$ and $f(x)$ satisfies the equation $f''/f = k$.

14.S Summary: Review Exercises

26 The argument supposes that $\lim_{\Delta t \to 0} \frac{\Delta z}{\Delta x} = \frac{\partial z}{\partial x}$, which is not in general true. Given Δt, we have $\Delta z = \Delta f = f(x(t + \Delta t), y(t + \Delta t)) - f(x(t), y(t))$. In the difference quotient $\frac{\partial z}{\partial x}$ it is required that y be held constant; since y varies with t, this condition is not met.

28 (a) $z = f(x, y) = g(r, \theta)$, since $x = r\cos\theta$ and $y = r\sin\theta$. By the chain rule, $z_r = z_x x_r + z_y y_r = z_x \cos\theta + z_y \sin\theta$; then $z_{rr} = (z_x)_r \cos\theta + (z_y)_r \sin\theta = z_{xx} \cos^2\theta + z_{xy}(2\sin\theta\cos\theta) + z_{yy}\sin^2\theta$; $z_\theta = z_x x_\theta + z_y y_\theta = z_x(-r\sin\theta) + z_y(r\cos\theta)$; $z_{\theta\theta} = (-r\sin\theta)(z_x)_\theta + z_x(-r\cos\theta) + (r\cos\theta)(z_y)_\theta + z_y(-r\sin\theta)$
$= [z_{xx}\sin^2\theta - z_{xy}(2\sin\theta\cos\theta) + z_{yy}\cos^2\theta]r^2 - rz_r$. Then $\frac{1}{r}z_r + z_{rr} + \frac{1}{r^2}z_{\theta\theta}$
$= \frac{1}{r}z_r + z_{xx}\cos^2\theta + z_{xy}(2\sin\theta\cos\theta) + z_{yy}\sin^2\theta + z_{xx}\sin^2\theta - z_{xy}(2\sin\theta\cos\theta) + z_{yy}\cos^2\theta - \frac{1}{r}z_r = z_{xx} + z_{yy}$, as claimed.

(b) If $z(t, r, \theta) = u(r)\sin(\alpha t + a)\sin(\beta\theta + b)$, then $z_t = u(r)[\alpha\cos(\alpha t + a)\sin(\beta\theta + b)]$,
$z_{tt} = u(r)[-\alpha^2 \sin(\alpha t + a)\sin(\beta\theta + b)]$,
$z_r = u'(r)\sin(\alpha t + a)\sin(\beta\theta + b)$,
$z_{rr} = u''(r)\sin(\alpha t + a)\sin(\beta\theta + b)$,
$z_\theta = u(r)[\beta\sin(\alpha t + a)\cos(\beta\theta + b)]$, and
$z_{\theta\theta} = u(r)[-\beta^2 \sin(\alpha t + a)\sin(\beta\theta + b)]$.
In order for Equation (2) to hold, we require
$\frac{1}{c^2}z_{tt} = \frac{1}{r}z_r + z_{rr} + \frac{1}{r^2}z_{\theta\theta}$ or $-\frac{\alpha^2}{c^2}z =$
$\left(\frac{u'}{r} + u''\right)\sin(\alpha t + a)\sin(\beta\theta + b) - \frac{\beta^2}{r^2}z$,
$\left(\frac{\alpha^2}{c^2} - \frac{\beta^2}{r^2}\right)z + \left(\frac{u'}{ru} + \frac{u''}{u}\right)z = 0$ (assuming $u(r) \neq 0$), and $\frac{\alpha^2}{c^2} - \frac{\beta^2}{r^2} + \frac{u'}{ru} + \frac{u''}{u} = 0$.

Hence u must satisfy the equation
$$u'' + \frac{u'}{r} + u\left(\frac{\alpha^2}{c^2} - \frac{\beta^2}{r^2}\right) = 0.$$

30 Let $f(x, y, z) = 2x^2 + 3y^2 + z^2$. Then $\nabla f = 4x\mathbf{i} + 6y\mathbf{j} + 2z\mathbf{k}$ is normal to the level surfaces of f. At $(1, 1, 1)$, $\mathbf{A} = \nabla f(1, 1, 1) = 4\mathbf{i} + 6\mathbf{j} + 2\mathbf{k}$ is normal to the surface $2x^2 + 3y^2 + z^2 = 6$. Similarly, letting $g(x, y, z) = x^3 + y^3 + z^3$, $\nabla g = 3x^2\mathbf{i} + 3y^2\mathbf{j} + 3z^2\mathbf{k}$ is normal to the level surfaces of g. At $(1, 1, 1)$, $\mathbf{B} = \nabla g(1, 1, 1) = 3\mathbf{i} + 3\mathbf{j} + 3\mathbf{k}$ is normal to the surface $x^3 + y^3 + z^3 = 3$. The angle between the surfaces is the angle between $\mathbf{A}$ and $\mathbf{B}$. Denote this angle by θ. Recall that $\cos\theta =$
$\frac{\mathbf{A} \cdot \mathbf{B}}{|\mathbf{A}||\mathbf{B}|} = \frac{(4\mathbf{i} + 6\mathbf{j} + 2\mathbf{k}) \cdot (3\mathbf{i} + 3\mathbf{j} + 3\mathbf{k})}{\sqrt{4^2 + 6^2 + 2^2}\sqrt{3^2 + 3^2 + 3^2}}$
$= \frac{4 \cdot 3 + 6 \cdot 3 + 2 \cdot 3}{\sqrt{56}\sqrt{27}} = \frac{36}{6\sqrt{42}}$, so $\theta = \cos^{-1}\frac{6}{\sqrt{42}}$
≈ 0.3876 radians $\approx 22.21°$.

32 $\Delta f = f(0 + 0.01, 2 - 0.01) - f(0, 2)$
$= f(0.01, 1.99) - f(0, 2) \approx 1.59217 - 1.57080$
≈ 0.021. $f_x = e^{xy}\left(y\cos^{-1}x - \frac{1}{\sqrt{1-x^2}}\right) = \pi - 1$
at $(0, 2)$, $f_y = xe^{xy}\cos^{-1}x = 0$ at $(0, 2)$, and $df = (\pi - 1)(0.01) + 0(-0.01) = \frac{\pi - 1}{100} \approx 0.021$.

34 $\Delta f = f(1.01, 1.99) - f(1, 2) \approx 16.15758 - 16 \approx 0.158$. $f(x, y) = x^3y^4$, $f_x = 3x^2y^4$, $f_y = 4x^3y^3$, $f_x(1, 2) = 3 \cdot 1 \cdot 16 = 48$, $f_y(1, 2) = 4 \cdot 1 \cdot 8 = 32$, $\Delta x = 0.1$, and $\Delta y = -0.1$, so $df = 48(0.1) + 32(-0.1) = 1.6$.

36 $\Delta f = f(1.99, 3.02) - f(2, 3) \approx 2.08318 - 2.07944 \approx 0.004$. $f(x, y) = \ln(x + 2y)$, $f_x = \dfrac{1}{x + 2y}$, $f_y = \dfrac{2}{x + 2y}$, $f_x(2, 3) = \dfrac{1}{8}$, $f_y(2, 3) = \dfrac{1}{4}$, $\Delta x = -0.1$, and $\Delta y = 0.2$, so $df = \dfrac{1}{8}(-0.1) + \dfrac{1}{4}(0.2) = 0.0375$.

38 $f(x, y) = x^3y^2$, $f_x = 3x^2y^2$, $f_y = 2x^3y$, and $\dfrac{df}{f} = \dfrac{3x^2y^2 \Delta x + 2x^3y \Delta y}{x^3y^2} = 3 \cdot \dfrac{\Delta x}{x} + 2 \cdot \dfrac{\Delta y}{y}$. The possible measurement errors in x and y are 3% and 4%, respectively, which means that $\left|\dfrac{\Delta x}{x}\right| \leq 0.03$ and $\left|\dfrac{\Delta y}{y}\right| \leq 0.04$. The measurement error in f is $\left|\dfrac{\Delta f}{f}\right|$, which is approximated by $\left|\dfrac{df}{f}\right| = \left|3 \cdot \dfrac{\Delta x}{x} + 2 \cdot \dfrac{\Delta y}{y}\right| \leq 3\left|\dfrac{\Delta x}{x}\right| + 2\left|\dfrac{\Delta y}{y}\right| \leq 3(0.03) + 2(0.04) = 0.17$. The maximum possible error in f is about 17%. (In fact, it can be almost 18.2%.)

40 $f(x, y) = x^5y$, $f_x = 5x^4y$, $f_y = x^5$, and $\dfrac{df}{f} = \dfrac{5x^4y \Delta x + x^5 \Delta y}{x^5y} = 5 \cdot \dfrac{\Delta x}{x} + \dfrac{\Delta y}{y}$. Now $\left|\dfrac{\Delta x}{x}\right| \leq 0.03$ and $\left|\dfrac{\Delta y}{y}\right| \leq 0.04$, so the relative error is approximated by $\left|\dfrac{df}{f}\right| = \left|5 \cdot \dfrac{\Delta x}{x} + \dfrac{\Delta y}{y}\right| \leq 5\left|\dfrac{\Delta x}{x}\right| + \left|\dfrac{\Delta y}{y}\right| \leq 5(0.03) + (0.04) = 0.19$. The maximum error is about 19%. (In fact, it can be almost 20.6%.)

42 Let $P_1 = (x, y, z)$ be a point on the smooth curve. We know $\overrightarrow{P_0P_1}$ is perpendicular to the curve when $(\nabla F \times \nabla G) \cdot \overrightarrow{P_0P_1} = 0$. The problem is equivalent to showing that $(x - a)(F_yG_z - F_zG_y) + (y - b)(F_zG_x - F_xG_z) + (z - c)(F_xG_y - F_yG_x) = 0$. Letting $L(x, y, z, \lambda, \mu) = (x - a)^2 + (y - b)^2 + (z - c)^2 - \lambda(F(x, y, z)) - \mu(G(x, y, z))$, $L_x = 2(x - a) - \lambda F_x - \mu G_x = 0$, $L_y = 2(y - b) - \lambda F_y - \mu G_y = 0$, and $L_z = 2(z - c) - \lambda F_z - \mu G_x = 0$, we see that $2(x - a)(F_yG_z - F_zG_y) = (\lambda F_x + \mu G_x)(F_yG_z - G_yF_z)$, $2(y - b)(F_zG_x - F_xG_z) = (\lambda F_y + \mu G_y)(F_zG_x - F_xG_z)$ and $2(z - c)(F_xG_y - F_yG_x) = (\lambda F_z + \mu G_z)(F_xG_y - F_yG_x)$. So P_1 is nearest P_0 when $2(x - a)(F_yG_z - G_yF_z) + 2(y - b)(F_xG_y - F_yG_x) + 2(z - c)(F_xG_y - F_yG_x) = 0$, and it follows from above that $\overrightarrow{P_0P_1}$ is perpendicular to the curve when P_1 is nearest P_0.

44 Let $f(x, y, z) = 2xy + 2xz + 2yz$, $x = 4$, $y = 2$, $z = 3$, $\Delta x = 0.01$, $\Delta y = -0.03$, and $\Delta z = -0.02$. Then the surface area is approximately $f(x, y, z) + df = f(4, 2, 3) + (2y + 2z)\Delta x + (2x + 2z)\Delta y + (2x + 2y)\Delta z = 52 + 10(0.01) + 14(-0.03) + 12(-0.02) = 51.44$.

46 See the discussion of Lagrange multipliers on

p. 861 in Sec. 14.10.

48 (d) Let $f(x, y, z) = 2x^2 + 3y^2 - z$. Then at (x_0, y_0, z_0) on the surface $z = 2x^2 + 3y^2$, $\nabla f = 4x_0\mathbf{i} + 6y_0\mathbf{j} - \mathbf{k}$ and the tangent plane to $z = 2x^2 + 3y^2$ at (x_0, y_0, z_0) is $4x_0(x - x_0) + 6y_0(y - y_0) - (z - z_0) = 0$. We know that the points $(1, 0, 0)$ and $(0, 1, 0)$ lie in this plane; so $4x_0(1 - x_0) + 6y_0(-y_0) - (-z_0) = 0$ and $4x_0(-x_0) + 6y_0(1 - y_0) - (-z_0) = 0$. These equations reduce to $4x_0 = z_0$ and $6y_0 = z_0$ since $z_0 = 2x_0^2 + 3y_0^2$. So $x_0 = \frac{3}{2}y_0$ and substituting this expression for x_0 into the first equation above yields two solutions: $x_0 = y_0 = z_0 = 0$ and $x_0 = \frac{6}{5}$, $y_0 = \frac{4}{5}$, $z_0 = \frac{24}{5}$.

Thus the two planes touch the paraboloid at the points $(0, 0, 0)$ and $\left(\frac{6}{5}, \frac{4}{5}, \frac{24}{5}\right)$.

50 Let $P = (a, b, c)$ and $Q = (x, y, z)$. We must show that, at the point Q, $\nabla f(x, y, z)$ is parallel to $\overrightarrow{PQ}$. Let $L(x, y, z, \lambda) = (x - a)^2 + (y - b)^2 + (z - c)^2 - \lambda(f(x, y, z))$. Then $L_x = 2(x - a) - \lambda f_x$, $L_y = 2(y - b) - \lambda f_y$, and $L_z = 2(z - c) - \lambda f_z$. Setting these equal to zero results in $x - a = \frac{\lambda f_x}{2}$, $y - b = \frac{\lambda f_y}{2}$, and $z - c = \frac{\lambda f_z}{2}$. Then $\overrightarrow{PQ} = (x - a)\mathbf{i} + (y - b)\mathbf{j} + (z - c)\mathbf{k} = \frac{\lambda}{2}(f_x\mathbf{i} + f_y\mathbf{j} + f_z\mathbf{k}) = \frac{\lambda}{2}\nabla f(x, y, z)$, and hence $\overrightarrow{PQ}$ is parallel to $\nabla f(x, y, z)$ and therefore perpendicular to the surface Q. (Actually, if there is a point R on S whose distance from P is maximal, then $\overrightarrow{PQ}$ is also perpendicular to S at R; in other words, we need not worry about whether the solution for λ above yields a maximum or a minimum.)

52 The gradient ∇f should be long because $|\nabla f|$ is an indicator of how rapidly f is changing.

54 Let $L(x, y, z, \lambda, \mu) = x + 2y + z - \lambda(x + y + 2z - 2) - \mu(x^2 + y^2 - 9)$. Setting $L_x = 1 - \lambda - 2\mu x = 0$, $L_y = 2 - \lambda - 2\mu y = 0$, and $L_z = 1 - 2\lambda = 0$, we find that $\lambda = \frac{1}{2}$, $x = \frac{1}{4\mu}$, and $y = \frac{3}{4\mu}$ so that $y = 3x$. Setting $L_\mu = 0$ we obtain $x^2 + y^2 = 10x^2 = \frac{10}{16\mu^2} = 9$. Then $\mu = \pm\frac{\sqrt{10}}{12}$. If $\mu = \frac{\sqrt{10}}{12}$, then $x = \frac{3\sqrt{10}}{10}$, $y = \frac{9\sqrt{10}}{10}$, and setting $L_\lambda = 0$ gives $z = \frac{2 - x - y}{2} = \frac{5 - 3\sqrt{10}}{5}$. If $\mu = -\frac{\sqrt{10}}{12}$, then $x = -\frac{3\sqrt{10}}{10}$, $y = -\frac{9\sqrt{10}}{10}$, and $z = \frac{5 + 3\sqrt{10}}{5}$. Calculating $x + 2y + z$ at both these points yields a maximum value of $\frac{2 + 3\sqrt{10}}{2}$ when $x = \frac{3\sqrt{10}}{10}$, $y = \frac{9\sqrt{10}}{10}$, and $z = \frac{5 - 3\sqrt{10}}{5}$.

56 (a) $\frac{\partial f}{\partial y} = \frac{\partial}{\partial y}\left(\int_x^y g(z, t)\, dt\right) = g(z, y)$

(b) $\dfrac{\partial f}{\partial x} = \dfrac{\partial}{\partial x}\left(\int_x^y g(z, t)\, dt\right)$

$= \dfrac{\partial}{\partial x}\left(-\int_y^x g(z, t)\, dt\right) = -g(z, x)$

(c) $\dfrac{\partial f}{\partial z} = \dfrac{\partial}{\partial z}\left(\int_x^y g(z, t)\, dt\right) = \int_x^y \dfrac{\partial g}{\partial z}(z, t)\, dt$

58 Here $z = f(x, y) = g(u, v)$. Hence by the chain rule, $z_u = f_x x_u + f_y y_u = x f_x$, $z_{uu} = f_x(x) + x(f_{xx} \cdot x)$ $= x f_x + x^2 f_{xx}$, $z_v = f_x x_v + f_y y_v = y f_y$, $z_{vv} = f_y(y) + y(f_{yy} \cdot y) = y f_y + y^2 f_{yy}$. Then $z_{uu} + z_{vv} = x^2 f_{xx} + y^2 f_{yy} + x f_x + y f_y = x^2 z_{xx} + y^2 z_{yy} + x z_x + y z_y$, where $z = g(u, v)$ on the left-hand side of the equation and $z = f(x, y)$ on the right-hand side.

60 (a) $\dfrac{\partial x}{\partial \theta} = \dfrac{\partial}{\partial \theta}(r \cos \theta) = -r \sin \theta$

(b) $\dfrac{\partial \theta}{\partial x} = \dfrac{\partial}{\partial x}\left(\tan^{-1} \dfrac{y}{x}\right) = \dfrac{-y/x^2}{1 + (y/x)^2}$

$= \dfrac{-y}{x^2 + y^2} = \dfrac{-r \sin \theta}{r^2} = -\dfrac{\sin \theta}{r}$

(c) $x = f(r, \theta)$, $\theta = g(x, y)$ is a function of two different variables. Hence (a) and (b) do not contradict each other.

62 $f_x = \dfrac{r x^{r-1}}{1 + x^r + y^r}$ and $f_y = \dfrac{r y^{r-1}}{1 + x^r + y^r}$. Hence

$f_{xy} = r x^{r-1} \cdot \dfrac{\partial}{\partial y}\left(\dfrac{1}{1 + x^r + y^r}\right)$

$= r x^{r-1} \dfrac{-r y^{r-1}}{(1 + x^r + y^r)^2} =$

$-\dfrac{r x^{r-1}}{1 + x^r + y^r} \cdot \dfrac{r y^{r-1}}{1 + x^r + y^r} = -f_x f_y$, so

$f_{xy} + f_x f_y = 0$.

64 Let $f = x^2 + 2y^2 + 3z^2$ and $g = 2x^2 - y^2 + z^2$, then $\nabla f = 2x\mathbf{i} + 4y\mathbf{j} + 6z\mathbf{k}$ and $\nabla g = 4x\mathbf{i} - 2y\mathbf{j} + 2z\mathbf{k}$. Then $\nabla f(1, 2, 3) = 2\mathbf{i} + 8\mathbf{j} + 18\mathbf{k}$ is normal to $x^2 + 2y^2 + 3z^2 = 36$ and $\nabla g(1, 2, 3) = 4\mathbf{i} - 4\mathbf{j} + 6\mathbf{k}$ is normal to $2x^2 - y^2 + z^2 = 7$, so

$\nabla f(1, 2, 3) \times \nabla g(1, 2, 3) = \begin{vmatrix} \mathbf{i} & \mathbf{j} & \mathbf{k} \\ 2 & 8 & 18 \\ 4 & -4 & 6 \end{vmatrix}$

$= 120\mathbf{i} + 60\mathbf{j} - 40\mathbf{k} = 20(6\mathbf{i} + 3\mathbf{j} - 2\mathbf{k})$ is tangent to both surfaces, as is the vector $6\mathbf{i} + 3\mathbf{j} - 2\mathbf{k}$.

66 (a) See Exercise 23 of Sec. 14.8.

(b) Since $D_\mathbf{T} u = \nabla F \cdot \mathbf{T}$ and $\mathbf{T} = \mathbf{G}'(s)$ by definition, it follows from (a) that $\dfrac{du}{ds} = D_\mathbf{T} u$.

68 (a) To find $\dfrac{\partial V}{\partial T}$, differentiate both sides of the given equation with respect to T, holding P fixed:

$\left(P + \dfrac{a}{V^2}\right)\dfrac{\partial}{\partial T}(V - b) + (V - b)\dfrac{\partial}{\partial T}\left(P + \dfrac{a}{V^2}\right)$

$= c,\ \left(P + \dfrac{a}{V^2}\right)\dfrac{\partial V}{\partial T} + (V - b) \cdot \dfrac{-2a}{V^3} \cdot \dfrac{\partial V}{\partial T} = c,$

$\left(P + \dfrac{a}{V^2} - \dfrac{2a}{V^2} + \dfrac{2ab}{V^3}\right)\dfrac{\partial V}{\partial T} = c,$

$\dfrac{PV^3 - aV + 2ab}{V^3} \cdot \dfrac{\partial V}{\partial T} = c,\ \dfrac{\partial V}{\partial T} =$

$\dfrac{cV^3}{PV^3 - aV + 2ab}$. Next, since $T =$

14.S Summary: Review Exercises

$$\frac{V-b}{c}\left(P + \frac{a}{V^2}\right), \frac{\partial T}{\partial P} = \frac{V-b}{c}. \text{ Finally, since}$$

$$P = \frac{cT}{V-b} - \frac{a}{V^2}, \frac{\partial P}{\partial V} = -\frac{cT}{(V-b)^2} + \frac{2a}{V^3}$$

$$= \frac{2a(V-b)^2 - cTV^3}{V^3(V-b)^2}$$

$$= \frac{2a(V-b)^2 - (P + a/V^2)(V-b)V^3}{V^3(V-b)^2}$$

$$= \frac{2a(V-b) - (P + a/V^2)V^3}{V^3(V-b)}$$

$$= \frac{aV - 2ab - PV^3}{V^3(V-b)}.$$

(b) $\frac{\partial V}{\partial T} \cdot \frac{\partial T}{\partial P} \cdot \frac{\partial P}{\partial V} =$

$$\frac{cV^3}{PV^3 - aV + 2ab} \cdot \frac{V-b}{c} \cdot \frac{aV - 2ab - PV^3}{V^3(V-b)}$$

$$= -1$$

70 Note that $f(x, y) = x^4 - x^2y^2 + y^4 = (x^2 - y^2)^2 + x^2y^2 \geq 0$ for all choices of x and y. Since $f(0, 0) = 0$, the minimum value is 0.

72 Let x be the side of the square, r the radius of the circle, and y the side of the equilateral triangle. Then $4x + 2\pi r + 3y = 1$, so $y = \frac{1}{3}(1 - 4x - 2\pi r)$ and $A = x^2 + \pi r^2 + \frac{\sqrt{3}}{4}y^2 =$

$$x^2 + \pi r^2 + \frac{\sqrt{3}}{36}(1 - 8x - 4\pi r + 16x^2 + 16\pi rx$$

$$+ 4\pi^2 r^2). \ A_x = 2x + \frac{\sqrt{3}}{36}(-8 + 32x + 16\pi r),$$

$$A_r = 2\pi r + \frac{\sqrt{3}}{36}(-4\pi + 16\pi x + 8\pi^2 r),$$

$$A_{xx} = 2 + \frac{8\sqrt{3}}{9}, A_{rr} = 2\pi + \frac{2\sqrt{3}}{9}\pi^2,$$

$$A_{xr} = \frac{4\sqrt{3}}{9}\pi. \text{ First, we find critical points on the "interior" of the intervals for } x \text{ and } r. \text{ Setting } A_x =$$

0 yields $r = \frac{1}{16\pi}\left(-\frac{72x}{\sqrt{3}} + 8 - 32x\right)$. Substituting

this value of r into $A_r = 0$ gives $0 =$

$$\frac{1}{8}\left(-\frac{72x}{\sqrt{3}} + 8 - 32x\right) +$$

$$\frac{\sqrt{3}}{36}\left(-4\pi + 16\pi x + \frac{\pi}{2}\left(-\frac{72x}{\sqrt{3}} + 8 - 32x\right)\right) =$$

$$1 + \left(-\frac{9}{\sqrt{3}} - 4 - \pi\right)x, \text{ and } x = \frac{\sqrt{3}}{9 + (4+\pi)\sqrt{3}}.$$

Then $r = \frac{1}{2\pi} - \frac{1}{16\pi}\left(\frac{72 + 32\sqrt{3}}{\sqrt{3}}x\right)$

$$= \frac{1}{2\pi} - \frac{1}{16\pi}\left(\frac{72 + 32\sqrt{3}}{9 + (4+\pi)\sqrt{3}}\right)$$

$$= \frac{1}{2\pi}\left(1 - \frac{9 + 4\sqrt{3}}{9 + (4+\pi)\sqrt{3}}\right) = \frac{\sqrt{3}/2}{9 + (4+\pi)} \text{ and}$$

$$y = \frac{1}{3}\left(1 - \frac{4\sqrt{3}}{9 + (4+\pi)\sqrt{3}} - 1 + \frac{9 + 4\sqrt{3}}{9 + (4+\pi)\sqrt{3}}\right) =$$

$$\frac{3}{9 + (4+\pi)\sqrt{3}}. \text{ For these values of } x, r, \text{ and } y,$$

$$A = x^2 + \pi r^2 + \frac{\sqrt{3}}{4}y^2 = \frac{3 + 3\pi/4 + 9\sqrt{3}/4}{[9 + (4+\pi)\sqrt{3}]^2} \approx$$

0.02026. Since $D = A_{xx}A_{rr} - (A_{xr})^2$

$= \left(2 + \frac{8\sqrt{3}}{9}\right)\left(2\pi + \frac{2\sqrt{3}\pi^2}{9}\right) - \frac{48\pi^2}{81} > 0$ and $A_{xx} >$

0, the area is minimal at this point. We now check for extrema on the "boundary": at $x = r = 0$, $A = \frac{\sqrt{3}}{4}y^2$

$= \frac{\sqrt{3}}{36} \approx 0.04811$; at $x = y = 0$, $A = \pi r^2 = \frac{1}{4\pi}$

≈ 0.07958; at $y = r = 0$, $A = x^2 = \frac{1}{16}$

$= 0.0625$.

(a) From above, the minimum occurs when $x = \frac{\sqrt{3}}{9 + (4 + \pi)\sqrt{3}}$, $r = \frac{1}{2}x$, and $y = \sqrt{3}x$.

(b) The maximum occurs when $x = y = 0$ and $r = \frac{1}{2\pi}$.

74 (a) $(\cos\theta + i\sin\theta)(x + iy) = x\cos\theta + iy\cos\theta + ix\sin\theta + i^2 y\sin\theta = (x\cos\theta - y\sin\theta) + i(x\sin\theta + y\cos\theta)$, showing that after a rotation by θ, the point (x, y) becomes the point $(x\cos\theta - y\sin\theta, x\sin\theta + y\cos\theta)$.

(b) The line is given parametrically by $x = \frac{1}{2}$, $y = t$, and $z = 2t$. When the point $\left(\frac{1}{2}, t, 2t\right)$ is rotated about the z axis, a circle centered at the origin with radius $\sqrt{\left(\frac{1}{2}\right)^2 + t^2} = \sqrt{\frac{1}{4} + t^2}$ is produced. Since $z = 2t$ is on this

circle, we have $x^2 + y^2 = \frac{1}{4} + t^2 = \frac{1}{4} + \frac{z^2}{4}$, and an equation for the surface is

$4x^2 + 4y^2 - z^2 = 1$.

(c) The surface is a hyperboloid of one sheet.

76 (a) $f(x, y) = \frac{1}{\sqrt{x^2 + y^2}}$, so $f_x =$

$-\frac{1}{2}(x^2 + y^2)^{-3/2}(2x) = \frac{-x}{(x^2 + y^2)^{3/2}}$.

Similarly, $f_y = \frac{-y}{(x^2 + y^2)^{3/2}}$; therefore $\nabla f =$

$\frac{-x\mathbf{i} - y\mathbf{j}}{(x^2 + y^2)^{3/2}}$.

(b) Note that $f = \frac{1}{r}$. By Exercise 75(d), $\nabla f =$

$\frac{\partial f}{\partial r}\mathbf{u}_r + \frac{1}{r} \cdot \frac{\partial f}{\partial \theta}\mathbf{u}_\theta$. But $\frac{\partial f}{\partial r} = -\frac{1}{r^2}$ and $\frac{\partial f}{\partial \theta}$

$= 0$, so $\nabla f = -\frac{1}{r^2}\mathbf{u}_r$.

(c) Observe that $\frac{-x\mathbf{i} - y\mathbf{j}}{(x^2 + y^2)^{3/2}} = \frac{-\mathbf{r}}{r^3} = -\frac{1}{r^2} \cdot \frac{\mathbf{r}}{r}$

$= -\frac{1}{r^2}\mathbf{u}_r$; thus the same value for f is given by the expressions in rectangular and polar coordinates.

15 Definite Integrals over Plane and Solid Regions

15.1 The Definite Integral of a Function over a Region in the Plane

2. (a) Mass $\approx ((0.5)^2 + (0.5)^2)\cdot 1 + ((1.5)^2 + (0.5)^2)\cdot 1 + ((2.5)^2 + (0.5)^2)\cdot 1 + ((3.5)^2 + (0.5)^2)\cdot 1 + ((0.5)^2 + (1.5)^2)\cdot 1 + ((1.5)^2 + (1.5)^2)\cdot 1 + ((2.5)^2 + (1.5)^2)\cdot 1 + ((3.5)^2 + (1.5)^2)\cdot 1 = 52$ grams.

   (b) Mass $\approx (1^2 + 1^2)\cdot 1 + (2^2 + 1^2)\cdot 1 + (3^2 + 1^2)\cdot 1 + (4^2 + 1^2)\cdot 1 + (1^2 + 2^2)\cdot 1 + (2^2 + 2^2)\cdot 1 + (3^2 + 2^2)\cdot 1 + (4^2 + 2^2)\cdot 1 = 80$ grams.

   (c) Mass $\approx (0^2 + 0^2)\cdot 1 + (1^2 + 0^2)\cdot 1 + (2^2 + 0^2)\cdot 1 + (3^2 + 0^2)\cdot 1 + (0^2 + 1^2)\cdot 1 + (1^2 + 1^2)\cdot 1 + (2^2 + 1^2)\cdot 1 + (3^2 + 1^2)\cdot 1 = 32$ grams.

4. (a) $\int_R f(P)\,dA \approx 2\cdot 4 + 4\cdot 4 + 2\cdot 4 + 4\cdot 4 = 48$

   (b) Since $1 \leq f(P) \leq 5$, $\int_R 1\,dA \leq \int_R f(P)\,dA \leq \int_R 5\,dA$. Thus $16 = \int_R 1\,dA \leq \int_R f(P)\,dA \leq \int_R 5\,dA = 80$.

6. $m \leq f(P) \leq M$ implies $\lim_{mesh \to 0} \sum_{i=1}^{n} mA_i \leq \lim_{mesh \to 0} \sum_{i=1}^{n} f(P_i)A_i \leq \lim_{mesh \to 0} \sum_{i=1}^{n} MA_i$. So $\int_R m\,dA \leq \int_R f(P)\,dA \leq \int_R M\,dA$. Thus, $mA = \int_R m\,dA \leq \int_R f(P)\,dA \leq \int_R M\,dA = MA$.

8. (a) $\int_R e^{xy}\,dA \approx (0.2)^2[e^{(0.1)(0.1)} + e^{(0.3)(0.1)} + e^{(0.5)(0.1)} + e^{(0.7)(0.1)} + e^{(0.1)(0.3)} + e^{(0.3)(0.3)} + e^{(0.5)(0.3)} + e^{(0.7)(0.3)} + e^{(0.1)(0.5)} + e^{(0.3)(0.5)} + e^{(0.5)(0.5)} + e^{(0.7)(0.5)} + e^{(0.1)(0.7)} + e^{(0.3)(0.7)} + e^{(0.5)(0.7)} + e^{(0.7)(0.7)}] \approx 0.7583$

   (b) Average $\approx \dfrac{0.7583}{\text{Area of } R} = \dfrac{0.7583}{0.64} \approx 1.1849$

   (c) Since $1 \leq f(P) \leq e^{0.64}$, by Exercise 6, $(0.64)\cdot 1 \leq \int_R f(P)\,dA \leq (0.64)e^{0.64}$.

10. (a) $z = \sqrt{x^2 + y^2}$

    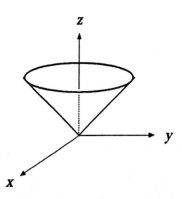

    (b) For all points P in R, $z = \sqrt{x^2 + y^2} \leq \sqrt{2}$. Therefore $V = \int_R z\,dA \leq \int_R \sqrt{2}\,dA = \sqrt{2}$.

(c) Use the upper right-hand corner of each square as the sampling point. Then the

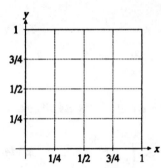

estimate for V is $\left[\sqrt{\left(\frac{1}{4}\right)^2 + \left(\frac{1}{4}\right)^2} + \right.$

$\sqrt{\left(\frac{1}{4}\right)^2 + \left(\frac{1}{2}\right)^2} + \sqrt{\left(\frac{1}{4}\right)^2 + \left(\frac{3}{4}\right)^2}$

$+ \sqrt{\left(\frac{1}{4}\right)^2 + 1^2} + \sqrt{\left(\frac{1}{2}\right)^2 + \left(\frac{1}{4}\right)^2}$

$+ \sqrt{\left(\frac{1}{2}\right)^2 + \left(\frac{1}{2}\right)^2} + \sqrt{\left(\frac{1}{2}\right)^2 + \left(\frac{3}{4}\right)^2}$

$+ \sqrt{\left(\frac{1}{2}\right)^2 + 1^2} + \sqrt{\left(\frac{3}{4}\right)^2 + \left(\frac{1}{4}\right)^2}$

$+ \sqrt{\left(\frac{3}{4}\right)^2 + \left(\frac{1}{2}\right)^2} + \sqrt{\left(\frac{3}{4}\right)^2 + \left(\frac{3}{4}\right)^2}$

$+ \sqrt{\left(\frac{3}{4}\right)^2 + 1^2} + \sqrt{1^2 + \left(\frac{1}{4}\right)^2}$

$+ \sqrt{1^2 + \left(\frac{1}{2}\right)^2}$

$\left. + \sqrt{1^2 + \left(\frac{3}{4}\right)^2} + \sqrt{1^2 + 1^2} \right]\left(\frac{1}{16}\right) \approx 0.9272.$

(d) Use the lower left-hand corner of each square as the sampling point. Then the estimate for V

is $\left[\sqrt{0^0 + 0^2} + \sqrt{0^2 + \left(\frac{1}{4}\right)^2} + \sqrt{0^2 + \left(\frac{1}{2}\right)^2} \right.$

$+ \sqrt{0^2 + \left(\frac{3}{4}\right)^2} + \sqrt{\left(\frac{1}{4}\right)^2 + 0^2}$

$+ \sqrt{\left(\frac{1}{4}\right)^2 + \left(\frac{1}{4}\right)^2} + \sqrt{\left(\frac{1}{4}\right)^2 + \left(\frac{1}{2}\right)^2}$

$+ \sqrt{\left(\frac{1}{4}\right)^2 + \left(\frac{3}{4}\right)^2} + \sqrt{\left(\frac{1}{2}\right)^2 + 0^2}$

$+ \sqrt{\left(\frac{1}{2}\right)^2 + \left(\frac{1}{4}\right)^2} + \sqrt{\left(\frac{1}{2}\right)^2 + \left(\frac{1}{2}\right)^2}$

$+ \sqrt{\left(\frac{1}{2}\right)^2 + \left(\frac{3}{4}\right)^2} + \sqrt{\left(\frac{3}{4}\right)^2 + 0^2}$

$+ \sqrt{\left(\frac{3}{4}\right)^2 + \left(\frac{1}{4}\right)^2} + \sqrt{\left(\frac{3}{4}\right)^2 + \left(\frac{1}{2}\right)^2}$

$\left. + \sqrt{\left(\frac{3}{4}\right)^2 + \left(\frac{3}{4}\right)^2} \right]\left(\frac{1}{16}\right) \approx 0.6015.$

12 By definition $f_1 = \frac{1}{A_1}\int_{R_1} f(P)\, dA$, $f_2 = \frac{1}{A_2}\int_{R_2} f(P)\, dA$, and the average value of f over R,

$f_{av} = \dfrac{\int_R f(P)\, dA}{A_1 + A_2}$. But $\int_R f(P)\, dA =$

$\int_{R_1} f(P)\, dA + \int_{R_2} f(P)\, dA$. Therefore $f_{av} =$

$\dfrac{f_1 A_1 + f_2 A_2}{A_1 + A_2}.$

14 $b - a \leq \overline{PQ} \leq b + a$. Therefore $\dfrac{1}{b + a} \leq$

$f(P) \leq \dfrac{1}{b - a}$. Therefore $\int_R \dfrac{dA}{b + a} \leq$

$\int_R f(P)\, dA \leq \int_R \dfrac{dA}{b - a}$, which implies that

$\dfrac{\pi a^2}{b + a} \leq \int_R f(P)\, dA \leq \dfrac{\pi a^2}{b - a}$.

16 (a) Estimate of the area of R is $2^2 \cdot \dfrac{73}{100} = 2.92$.

(b) $\int_R f(P)\, dA = f_{av}(\text{Area of } R) \approx (2.31)(2.92)$
     $= 6.7452$

18 (b) If a region has diameter d its area is at most
     $\dfrac{\pi d^2}{4}$.

(c) Yes

(d) As small as you want.

15.2 Computing $\int_R f(P)\, dA$ using Rectangular Coordinates

2 (a) $0 \leq x \leq 1, 0 \leq y \leq x$;
        $1 \leq x \leq 2, 0 \leq y \leq -x + 2$

(b) $0 \leq y \leq 1, y \leq x \leq -y + 2$

4

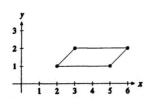

(a) The description of R by vertical cross sections requires three separate statements: $2 \leq x \leq 3, 1 \leq y \leq x - 1$; $3 \leq x \leq 5, 1 \leq y \leq 2$; $5 \leq x \leq 6, x - 4 \leq y \leq 2$.

(b) The description by horizontal cross sections is $1 \leq y \leq 2, y + 1 \leq x \leq y + 4$.

6 (a) $1 \leq x \leq 3, x - 1 \leq y \leq -\dfrac{3}{2}x + \dfrac{15}{2}$

(b) $0 \leq y \leq 2, 1 \leq x \leq y + 1$;
     $2 \leq y \leq 3, 1 \leq x \leq 3$;
     $3 \leq y \leq 6, 1 \leq x \leq -\dfrac{2}{3}y + 5$

8 (a) $-1 \leq x \leq 1$,
        $-\sqrt{4 - 4x^2} \leq y \leq \sqrt{4 - 4x^2}$

(b) $-2 \leq y \leq 2$,
     $-\sqrt{1 - \dfrac{1}{4}y^2} \leq x \leq \sqrt{1 - \dfrac{1}{4}y^2}$

10 (a) $0 \leq x \leq 1, 1 - x \leq y \leq e^x$

(b) $0 \leq y \leq 1, 1 - y \leq x \leq 1$;
     $1 \leq y \leq e, \ln y \leq x \leq 1$

12 (a) $1 \leq x \leq 2, x \leq y \leq 5 - x$

(b) $1 \leq y \leq 2, 1 \leq x \leq y$;
     $2 \leq y \leq 3, 1 \leq x \leq 2$;
     $3 \leq y \leq 4, 1 \leq x \leq 5 - y$

14

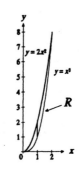

$1 \leq y \leq 2, 1 \leq x \leq \sqrt[3]{y}$;

$2 \leq y \leq 8, \sqrt{\dfrac{y}{2}} \leq x \leq \sqrt[3]{y}$

16

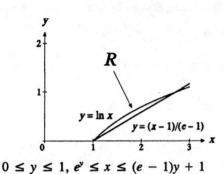

$0 \leq y \leq 1,\ e^y \leq x \leq (e-1)y + 1$

18 $\int_1^2 \int_x^{2x} dy\, dx = \int_1^2 y \Big|_x^{2x} dx = \int_1^2 x\, dx = \frac{x^2}{2}\Big|_1^2$

$= 2 - 1/2 = 3/2.$

20 $\int_1^2 \int_0^y e^{x+y}\, dx\, dy = \int_1^2 e^{x+y}\Big|_0^y dy$

$= \int_1^2 (e^{2y} - e^y)\, dy = \frac{e^{2y}}{2} - e^y \Big|_1^2$

$= \frac{e^4}{2} - e^2 - \left(\frac{e^2}{2} - e\right) = \frac{e^4}{2} - \frac{3e^2}{2} + e$

22 $\int_0^1 \int_0^x y \sin \pi x\, dy\, dx = \int_0^1 \frac{1}{2}y^2 \sin \pi x \Big|_0^x dx$

$= \int_0^1 \frac{1}{2}x^2 \sin \pi x\, dx$

$= \frac{1}{2}\left[\frac{2x}{\pi^2}\sin \pi x - \frac{\pi^2 x^2 - 2}{\pi^3}\cos \pi x\right]\Big|_0^1$

$= \frac{1}{2}\left[\frac{2}{\pi^2}\cdot 0 - \frac{\pi^2 - 2}{\pi^3}(-1)\right] - \frac{1}{2}\left[0 - \frac{-2}{\pi^3}(1)\right]$

$= \frac{1}{2}\left[\frac{\pi^2 - 2}{\pi^3} - \frac{2}{\pi^3}\right] = \frac{\pi^2 - 4}{2\pi^3}$

24 (a) The graph is shown in the figure below.

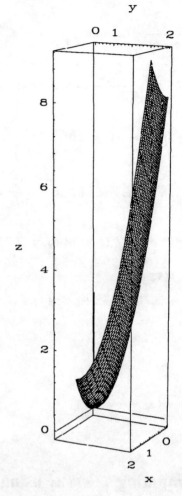

(b) R can be described by $0 \leq x \leq 1$, $0 \leq y \leq 2$.

(c) The volume is given by $V = \int_R z\, dA$

$= \int_0^1 \int_0^2 (x^2 + 2y^2)\, dy\, dx.$

(d) $\int_0^1 \int_0^2 (x^2 + 2y^2)\, dy\, dx$

$= \int_0^1 \left(x^2 y + \frac{2}{3}y^3\right)\Big|_0^2 dx = \int_0^1 \left(2x^2 + \frac{16}{3}\right)dx$

$= \frac{2}{3}x^3 + \frac{16}{3}x \Big|_0^1 = \frac{2}{3} + \frac{16}{3} = 6$

(e) $V = \int_0^2 \int_0^1 (x^2 + 2y^2)\, dx\, dy$

$= \int_0^2 \frac{1}{3}x^3 + 2y^2 x \Big|_0^1 dy = \int_0^2 \left(\frac{1}{3} + 2y^2\right) dy$

$= \frac{1}{3}y + \frac{2}{3}y^3 \Big|_0^2 = \frac{2}{3} + \frac{16}{3} = 6$

26 (a)

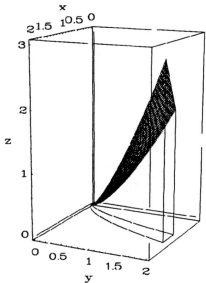

(b) Using vertical cross sections x varies from 0 to $\sqrt{2}$. For x in $[0, 1]$, y ranges from x^2 to $2x^2$. For x in $[1, \sqrt{2}]$, y ranges from x^2 to 2. Therefore the description of R by vertical cross sections is $0 \leq x \leq 1$, $x^2 \leq y \leq 2x^2$ and $1 \leq x \leq \sqrt{2}$, $x^2 \leq y \leq 2$.

Using horizontal cross sections, y varies from 0 to 2 and x varies from $\sqrt{y/2}$ to $\sqrt{y}$.

(c) See (d).

(d) $V = \int_R z\, dA$

$= \int_0^1 \int_{x^2}^{2x^2} xy\, dy\, dx + \int_1^{\sqrt{2}} \int_{x^2}^{2} xy\, dy\, dx$

$= \int_0^1 \frac{1}{2}xy^2 \Big|_{x^2}^{2x^2} dx + \int_1^{\sqrt{2}} \frac{1}{2}xy^2 \Big|_{x^2}^{2} dx$

$= \int_0^1 \frac{3}{2}x^5\, dx + \int_1^{\sqrt{2}} \left(2x - \frac{1}{2}x^5\right) dx$

$= \frac{1}{4}x^6 \Big|_0^1 + \left(x^2 - \frac{1}{12}x^6\right)\Big|_1^{\sqrt{2}} = \frac{1}{4} + \frac{5}{12} = \frac{2}{3}$

(e) $V = \int_0^2 \int_{\sqrt{y/2}}^{\sqrt{y}} xy\, dx\, dy = \int_0^2 \frac{1}{2}x^2 y \Big|_{\sqrt{y/2}}^{\sqrt{y}} dy$

$= \int_0^2 \frac{1}{2}y\left(y - \frac{y}{2}\right) dy = \int_0^2 \frac{1}{4}y^2\, dy = \frac{2}{3}$

28 Mass $= \int_0^1 \int_0^x \frac{1}{1+x^2}\, dy\, dx = \frac{1}{2}\ln 2$

30 The average temperature is $\dfrac{\int_R T(x,y)\, dA}{\text{Area of } R}$. The

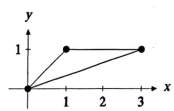

description of the region R is simpler using horizontal cross sections, so y varies from 0 to 1. x varies from the line through $(0, 0)$ and $(1, 1)$ to the line through $(0, 0)$ and $(3, 1)$. These lines have equations $x = y$ and $x = 3y$ so R can be described as follows: $0 \leq y \leq 1$, $y \leq x \leq 3y$. Thus

$\int_R T(x, y)\, dA = \int_0^1 \int_y^{3y} e^{x-y}\, dx\, dy$

$= \int_0^1 e^{x-y} \Big|_y^{3y} dy = \int_0^1 (e^{2y} - 1)\, dy$

$= \frac{1}{2}e^{2y} - y \Big|_0^1 = \frac{1}{2}e^2 - 1 - \frac{1}{2} = \frac{1}{2}(e^2 - 3)$.

The area of R is 1. Thus the average temperature is

$\frac{1}{2}(e^2 - 3) \approx 2.19$.

32

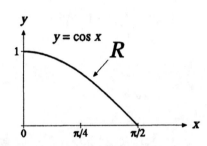

$\int_0^1 \int_0^{\cos^{-1} y} x^2 \, dx \, dy$

34

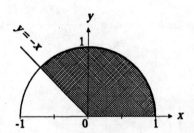

$\int_0^{1/\sqrt{2}} \int_{-y}^{\sqrt{1-y^2}} x^3 y \, dx \, dy$

$+ \int_{1/\sqrt{2}}^1 \int_{-\sqrt{1-y^2}}^{\sqrt{1-y^2}} x^3 y \, dx \, dy$

36

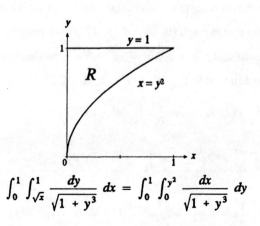

$\int_0^1 \int_{\sqrt{x}}^1 \frac{dy}{\sqrt{1+y^3}} \, dx = \int_0^1 \int_0^{y^2} \frac{dx}{\sqrt{1+y^3}} \, dy$

$= \int_0^1 \frac{y^2}{\sqrt{1+y^3}} \, dy = \frac{2}{3}\sqrt{1+y^3}\Big|_0^1 = \frac{2}{3}\sqrt{2} - \frac{2}{3}$

38

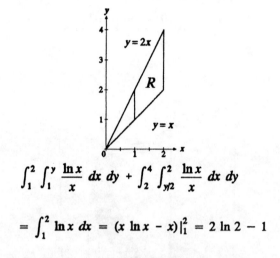

$\int_1^2 \int_1^y \frac{\ln x}{x} \, dx \, dy + \int_2^4 \int_{y/2}^2 \frac{\ln x}{x} \, dx \, dy$

$= \int_1^2 \ln x \, dx = (x \ln x - x)\Big|_1^2 = 2 \ln 2 - 1$

40 $\int_R f(P) \, dA = \lim_{N \to 0} \int_N^1 \int_{y^2}^y \frac{\sin y}{y} \, dx \, dy$

$= \lim_{N \to 0} \int_N^1 (\sin y - y \sin y) \, dy = 1 - \sin 1$

42 Using the definition of the average value of a function, $f(t) = \dfrac{\int_R T(x, y, t) \, dA}{\text{Area of } R}$

$= \frac{2}{\pi^2} \int_0^\pi \int_0^{\pi/2} e^{-tx} \sin(x + 3y) \, dy \, dx$

$= \frac{2}{\pi^2} \int_0^\pi e^{-tx} \left[-\frac{1}{3} \cos(x + 3y)\right]_0^{\pi/2} dx$

$= -\frac{2}{3\pi^2} \int_0^\pi e^{-tx} \left(\cos\left(x + \frac{3\pi}{2}\right) - \cos x\right) dx$

$= -\frac{2}{3\pi^2} \int_0^\pi e^{-tx} (\sin x - \cos x) \, dx =$

$-\frac{2}{3\pi^2} \left[\frac{e^{-tx}}{t^2+1}(-t \sin x - \cos x + t \cos x - \sin x)\right]_0^\pi$

15.2 Computing $\int_R f(P)\, dA$ using Rectangular Coordinates

$$= -\frac{2}{3\pi^2}\left[\frac{e^{-tx}}{t^2+1}((t-1)\cos x - (t+1)\sin x)\right]_0^\pi$$

$$= -\frac{2}{3\pi^2}\left[\frac{e^{-\pi t}}{t^2+1}(1-t) - \frac{1}{t^2+1}(t-1)\right] =$$

$$-\frac{2}{3\pi^2(t^2+1)}(e^{-\pi t}+1)(1-t).$$

$$\frac{df}{dt} = -\frac{2}{3\pi^2}\left(\left(\frac{1}{t^2+1}\right)'(e^{-\pi t}+1)(1-t) + \right.$$

$$\left.\frac{1}{t^2+1}(e^{-\pi t}+1)'(1-t) + \frac{1}{t^2+1}(e^{-\pi t}+1)(1-t)'\right)$$

$$= -\frac{2}{3\pi^2}\left(\frac{-2t}{(t^2+1)^2}(e^{-\pi t}+1)(1-t) + \right.$$

$$\left.\frac{1}{t^2+1}(-\pi e^{-\pi t})(1-t) + \left(\frac{1}{t^2+1}\right)(e^{-\pi t}+1)(-1)\right)$$

$$= -\frac{2}{3\pi^2}\left((e^{-\pi t}+1)\frac{2t}{(t^2+1)^2}(t-1) + \right.$$

$$\left.\frac{1}{t^2+1}(\pi e^{-\pi t})(t-1) - \frac{1}{t^2+1}(e^{-\pi t}+1)\right)$$

$$= -\frac{2}{3\pi^2(t^2+1)}\left((e^{-\pi t}+1)\frac{2t}{t^2+1}(t-1) + \right.$$

$$\left.(\pi e^{-\pi t})(t-1) - (e^{-\pi t}+1)\right).$$

44 Note that $2x^3y^2$ is an odd function with respect to x and even with respect to y. Its integral over R will be 0. That is, $\int_{-1}^1 \int_{-1}^1 (2x^3y^2 + 7)\, dy\, dx =$

$\int_{-1}^1 \int_{-1}^1 7\, dy\, dx = 7(\text{Area of } R) = 28.$

15.3 Moments and Centers of Mass

2 The description of R is simpler using horizontal cross sections: $0 \le y \le 1$ $y \le x \le 2 - y$. The total mass is $\int_0^1 \int_y^{2-y} y\, dx\, dy = \int_0^1 yx\Big|_y^{2-y} dy$

$$= \int_0^1 (y(2-y)-y^2)\, dy = \int_0^1 (2y - 2y^2)\, dy$$

$$= y^2 - \frac{2}{3}y^3\Big|_0^1 = 1 - 2/3 = 1/3. \text{ Therefore } \bar{x} =$$

$$3\int_0^1 \int_y^{2-y} xy\, dx\, dy = 3\int_0^1 \frac{1}{2}yx^2\Big|_y^{2-y} dy$$

$$= 3\int_0^1 \left(\frac{1}{2}y(2-y)^2 - \frac{1}{2}y^3\right) dy$$

$$= 3\int_0^1 (2y - 2y^2)\, dy = 3\cdot 1/3 = 1 \text{ which was}$$

obvious from symmetry arguments. $\bar{y} =$

$$3\int_0^1 \int_y^{2-y} y^2\, dx\, dy = 3\int_0^1 y^2 x\Big|_y^{2-y} dy$$

$$= 3\int_0^1 (y^2(2-y) - y^3)\, dy$$

$$= 3\int_0^1 (2y^2 - 2y^3)\, dy = 3\left(\frac{2}{3}y^3 - \frac{1}{2}y^4\right)\Big|_0^1$$

$= 3(2/3 - 1/2) = 3(1/6) = 1/2$. So the center of mass is $(1, 1/2)$.

4 Use Formula 63 (with $a = \ln 2$ and $a = 2\ln 2$) in the table of antiderivatives. $M_y =$

$\int_0^1 \int_{2^x}^{1+x} x(x+y)\, dy\, dx =$

$\frac{31}{24} - \frac{3}{\ln 2} + \frac{35}{8(\ln 2)^2} - \frac{2}{(\ln 2)^3}$. Total mass $= M$

$= \int_0^1 \int_{2^x}^{1+x} (x+y)\, dy\, dx$

$= 2 - \dfrac{11}{4\ln 2} + \dfrac{1}{(\ln 2)^2}$. Thus, $\bar{x} = \dfrac{M_y}{M} =$

$\dfrac{31(\ln 2)^3 - 72(\ln 2)^2 + 105 \ln 2 - 48}{48(\ln 2)^3 - 66(\ln 2)^2 + 24(\ln 2)} \approx 0.5617.$

Similarly, $\bar{y} = \dfrac{M_x}{M}$

$= \dfrac{141(\ln 2)^2 - 128(\ln 2) + 27}{144(\ln 2)^2 - 198(\ln 2) + 72} \approx 1.5274.$

6   We need the antiderivative of $x^n e^x$ for various values of n. See Formulas 62 and 63 in the table of antiderivatives. The mass of the lamina is

$\displaystyle\int_0^2 \int_0^{x^2} e^x \, dy \, dx = \int_0^2 y e^x \Big|_{y=0}^{y=x^2} dx$

$= \displaystyle\int_0^2 x^2 e^x \, dx = (x^2 - 2x + 2)e^x \Big|_0^2$

$= (4 - 4 + 2)e^2 - (0 - 0 + 2)e^0 = 2(e^2 - 1)$. Now

$M_x = \displaystyle\int_0^2 \int_0^{x^2} y e^x \, dy \, dx = \int_0^2 \dfrac{y^2}{2} e^x \Big|_{y=0}^{y=x^2} dx$

$= \displaystyle\int_0^2 \dfrac{1}{2} x^4 e^x \, dx$

$= \dfrac{1}{2}(x^4 - 4x^3 + 12x^2 - 24x + 24)e^x \Big|_0^2$

$= \dfrac{1}{2}(16 - 32 + 48 - 48 + 24)e^2 - \dfrac{1}{2}(24)e^0$

$= 4e^2 - 12$, so $\bar{y} = \dfrac{M_x}{\text{Mass}} = \dfrac{4e^2 - 12}{2(e^2 - 1)} =$

$\dfrac{2e^2 - 6}{e^2 - 1}$. Finally, $M_y = \displaystyle\int_0^2 \int_0^{x^2} x e^x \, dy \, dx$

$= \displaystyle\int_0^2 xy e^x \Big|_{y=0}^{y=x^2} dx = \int_0^2 x^3 e^x \, dx$

$= (x^3 - 3x^2 + 6x - 6)e^x \Big|_0^2 = 2e^2 + 6$, so $\bar{x} =$

$\dfrac{2e^2 + 6}{2(e^2 - 1)} = \dfrac{e^2 + 3}{e^2 - 1}.$

8   The description of R is simpler using horizontal cross sections. Clearly $0 \le y \le 1$. The equation

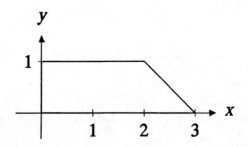

of the line through $(2, 1)$ and $(3, 0)$ is $y = -x + 3$ so $0 \le x \le 3 - y$. The total mass is

$\displaystyle\int_0^1 \int_0^{3-y} \sin x \, dx \, dy = \int_0^1 -\cos x \Big|_0^{3-y} dy$

$= \displaystyle\int_0^1 [1 - \cos(3 - y)] \, dy = y + \sin(3 - y) \Big|_0^1$

$= 1 + \sin 2 - \sin 3 \approx 1.768$. Now evaluate the integrals for the numerators.

$M_y = \displaystyle\int_0^1 \int_0^{3-y} x \sin x \, dx \, dy$

$= \displaystyle\int_0^1 (\sin x - x \cos x) \Big|_0^{3-y} dy$

$= \displaystyle\int_0^1 [\sin(3 - y) - (3 - y) \cos(3 - y)] \, dy$

$= \displaystyle\int_0^1 \sin(3 - y) \, dy - 3 \int_0^1 \cos(3 - y) \, dy -$

$\displaystyle\int_0^1 -y \cos(3 - y) \, dy = \cos(3 - y) \Big|_0^1 +$

$3 \sin(3 - y) \Big|_0^1 - [y \sin(3 - y) - \cos(3 - y)] \Big|_0^1$

$= 2 \cos 2 - 2 \cos 3 + 2 \sin 2 - 3 \sin 3$

≈ 2.543. $M_x = \int_0^1 \int_0^{3-y} y \sin x \, dx \, dy$

$= \int_0^1 -y \cos x \Big|_0^{3-y} dy$

$= \int_0^1 -y[\cos(3-y) - 1] \, dy$

$= \int_0^1 -y \cos(3-y) \, dy + \int_0^1 y \, dy$

$= [y \sin(3-y) - \cos(3-y)]\Big|_0^1 + \frac{1}{2} y^2 \Big|_0^1$

$= \sin 2 - \cos 2 + \cos 3 + \frac{1}{2} \approx 0.835$. Thus

$(\bar{x}, \bar{y}) \approx \left(\frac{2.543}{1.768}, \frac{0.835}{1.768}\right) \approx (1.438, 0.472)$.

10 Position the triangle R as in the figure. We know from Exercise 20 of Sec. 8.6 that the centroid is at

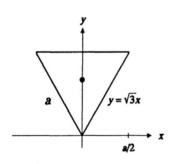

$(0, a/\sqrt{3})$. The description of R is $0 \leq y \leq \frac{\sqrt{3}a}{2}$

and $-\frac{y}{\sqrt{3}} \leq x \leq \frac{y}{\sqrt{3}}$. Note that $\sigma = \frac{M}{\sqrt{3}a^2/4} = \frac{4M}{\sqrt{3}a^2}$. By symmetry, $I =$

$2\sigma \int_0^{\sqrt{3}a/2} \int_0^{y/\sqrt{3}} [x^2 + (y - a/\sqrt{3})^2] \, dx \, dy$

$= 2\sigma \int_0^{\sqrt{3}a/2} \left[\frac{x^3}{3} + x\left(y - \frac{a}{\sqrt{3}}\right)^2\right]\Big|_0^{y/\sqrt{3}} dy$

$= 2\sigma \int_0^{\sqrt{3}a/2} \left(\frac{10y^3}{9\sqrt{3}} - \frac{2}{3}ay^2 + \frac{a^2 y}{3\sqrt{3}}\right) dy$

$= 2\sigma \left(\frac{10y^4}{36\sqrt{3}} - \frac{2ay^3}{9} + \frac{a^2 y^2}{6\sqrt{3}}\right)\Big|_0^{\sqrt{3}a/2}$

$= 2 \frac{4M}{\sqrt{3}a^2}\left(\frac{10a^4}{64\sqrt{3}} - \frac{a^4\sqrt{3}}{12} + \frac{a^4}{8\sqrt{3}}\right)$

$= \frac{8Ma^2}{3}\left(\frac{5}{32} - \frac{1}{4} + \frac{1}{8}\right) = \frac{Ma^2}{12}$.

12 Position the square as in the figure. Then $I = \int_R r^2 \sigma(P) \, dA = \frac{M}{a^2} \int_{-a/2}^{a/2} \int_{-a/2}^{a/2} (x^2 + y^2) \, dy \, dx$

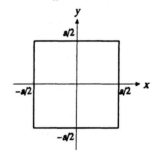

$= \frac{4M}{a^2} \int_0^{a/2} \int_0^{a/2} (x^2 + y^2) \, dy \, dx$

$= \frac{4M}{a^2} \int_0^{a/2} \left(\frac{a}{2}x^2 + \frac{1}{3}\frac{a^3}{8}\right) dx$

$= \frac{4M}{a^2}\left(\frac{a}{2}\frac{1}{3}\frac{a^3}{8} + \frac{1}{3}\frac{a^3}{8}\frac{a}{2}\right) = \frac{Ma^2}{6}$.

14 $I = \int_R r^2 \sigma(P) \, dA = \int_0^1 \int_{x^3}^{x^2} (x^2 + y^2) y e^{-x} \, dy \, dx$

$= \int_0^1 \int_{x^3}^{x^2} (x^2 y e^{-x} + y^3 e^{-x}) \, dy \, dx$

$$= \int_0^1 \left(x^2 e^{-x} \frac{y^2}{2} + e^{-x} \frac{y^4}{4}\right)\bigg|_{x^3}^{x^2} dx$$

$$= \int_0^1 \left(\frac{1}{2}x^6 e^{-x} + \frac{1}{4}x^8 e^{-x} - \frac{1}{2}x^8 e^{-x} - \frac{1}{4}x^{12} e^{-x}\right) dx =$$

$\int_0^1 e^{-x}\left(\frac{1}{2}x^6 - \frac{1}{4}x^8 - \frac{1}{4}x^{12}\right) dx$. With the aid of a symbolic math program (recommended!) or the tabular technique from Exercise 28 of Sec. 7.3 (see p. 181 of this manual), we find that

$\int_0^1 e^{-x}\left(\frac{x^6}{2} - \frac{x^8}{4} - \frac{x^{12}}{4}\right) dx = -e^{-x}(x^6/2 - x^8/4$
$- x^{12}/4 + 3x^5 - 2x^7 - 3x^{11} + 15x^4 - 14x^6 -$
$33x^{10} + 60x^3 - 84x^5 - 330x^9 + 180x^2 - 420x^4 -$
$2970x^8 + 360x - 1680x^3 - 23{,}760x^7 + 360 -$
$5040x^2 - 165{,}690x^6 - 10{,}080x - 994{,}140x^5 -$
$10080 - 4{,}970{,}700x^4 - 19{,}882{,}800x^3 -$
$59{,}648{,}400x^2 - 119{,}296{,}800x - 119{,}296{,}800)\big|_0^1$
$= 324{,}308{,}848 e^{-1} - 119{,}306{,}520 \approx 37.7690$.

16 Choose axes such that the center of mass is at the origin; that way M_x and M_y will be 0. Let L_2 intersect the plane at (m, n). Then

$$I_2 = \int_R [(x - m)^2 + (y - n)^2] \sigma(x, y) \, dA$$

$$= \int_R (x^2 - 2mx + m^2 + y^2 - 2ny + n^2) \sigma(x, y) \, dA$$

$$= \int_R (x^2 + y^2) \sigma(x, y) \, dA +$$

$$\int_R (m^2 + n^2) \sigma(x, y) \, dA - 2m \int_R x\sigma(x, y) \, dA$$

$$- 2n \int_R y\sigma(x, y) \, dA = I_1 + a^2 M.$$

18 Divide the lamina into 2 sections; draw a line at $x = 2$. Then since it is homogenous $\bar{x}_1 = 1$, $\bar{y}_1 = $

$3/2$, $\bar{x}_2 = 4$, and $\bar{y}_2 = 1$. Also σ is constant so $\bar{x}$

$$= \frac{M_1 \bar{x}_1 + M_2 \bar{x}_2}{M_1 + M_2} = \frac{(6\sigma)(1) + (8\sigma)(4)}{14\sigma} = \frac{38\sigma}{14\sigma}$$

$$= \frac{19}{7}, \bar{y} = \frac{M_1 \bar{y}_1 + M_2 \bar{y}_2}{M_1 + M_2}$$

$$= \frac{(6\sigma)(3/2) + (8\sigma)(1)}{14\sigma} = \frac{17}{14}.$$

20 (a) By placing the origin at the center of mass, we automatically have $\bar{x} = 0$ and $\bar{y} = 0$; but

$$\bar{x} = \frac{M_y}{\text{Mass}} \text{ and } \bar{y} = \frac{M_x}{\text{Mass}}, \text{ so } M_x = M_y$$

$= 0$.

(b) By Theorem 2 of Sec. 12.4, the distance of the point (x, y) from the line $ax + by = 0$ is $\frac{|ax + by|}{\sqrt{a^2 + b^2}}$. The quantity $ax + by$ is negative for (x, y) on one side of L and is positive on the other side; since we want the sign of the lever arm to vary in this manner, we drop the absolute value bars. Now the moment of a mass about a line is the integral of the product of the lever arm and the density function taken over the planar region occupied by the mass. Thus, we have Moment about $L = M_L$

$$= \int_R \frac{ax + by}{\sqrt{a^2 + b^2}} \sigma(P) \, dA.$$

(c) By (b), $M_L = \frac{1}{\sqrt{a^2 + b^2}} \int_R (ax + by) \sigma(P) \, dA$

$$= \frac{a}{\sqrt{a^2 + b^2}} \int_R x\sigma(P) \, dA +$$

15.3 Moments and Centers of Mass

$\frac{b}{\sqrt{a^2+b^2}} \int_R y\sigma(P)\, dA = 0$, since the two integrals are M_y and M_x, respectively, which were shown in (a) to be zero.

15.4 Computing $\int_R f(P)\, dA$ using Polar Coordinates

2

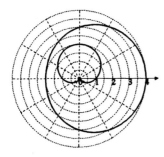

$0 \leq \theta \leq 2\pi,\ 1 + \sin\theta \leq r \leq 3 + \cos\theta$

4

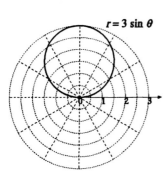

$0 \leq \theta \leq \pi,\ 0 \leq r \leq 3\sin\theta$

6

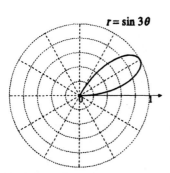

$0 \leq \theta \leq \pi/3,\ 0 \leq r \leq \sin 3\theta$

8 (a) A circle of radius 2 at the origin

(b) $-2 \leq x \leq 2,\ -\sqrt{4-x^2} \leq y \leq \sqrt{4-x^2}$

(c) $0 \leq \theta \leq 2\pi,\ 0 \leq r \leq 2$

10 (a) $\frac{\pi}{4} \leq \theta \leq \frac{\pi}{2},\ \csc\theta \leq r \leq 2\csc\theta$

(b) $1 \leq y \leq 2,\ 0 \leq x \leq y$

(c) $0 \leq x \leq 1,\ 1 \leq y \leq 2$;
$1 \leq x \leq 2,\ x \leq y \leq 2$

12

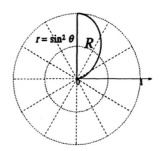

$\int_R r^2\, dA = \int_0^{\pi/2} \int_0^{\sin^2\theta} r^3\, dr\, d\theta$

$= \int_0^{\pi/2} \frac{1}{4} \sin^8\theta\, d\theta = \frac{35\pi}{1024}$

14

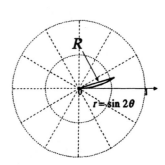

$\int_R r^2\, dA = \int_0^{0.3} \int_0^{\sin 2\theta} r^3\, dr\, d\theta$

$= \frac{1}{4} \int_0^{0.3} \sin^4 2\theta\, d\theta$

$= \frac{1}{4}\left(\frac{3\theta}{8} - \frac{\sin 4\theta}{8} + \frac{\sin 8\theta}{64}\right)\Big|_0^{0.3} \approx 0.00164$

16

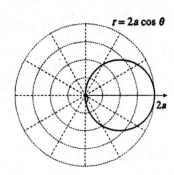

$$\int_R y^2 \, dA = \int_{-\pi/2}^{\pi/2} \int_0^{2a \cos\theta} r^3 \sin^2\theta \, dr \, d\theta = \frac{\pi a^4}{4}$$

18

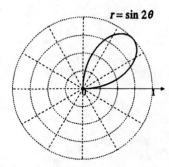

$$\int_R y^2 \, dA = \int_0^{\pi/2} \int_0^{\sin 2\theta} r^3 \sin^2\theta \, dr \, d\theta = \frac{3\pi}{128}$$

20 Let R be the region within the leaf. Since the density is constant, $\bar{x} = \dfrac{\int_R x \, dA}{\int_R dA}$. $\int_R x \, dA =$

$$\int_{-\pi/6}^{\pi/6} \int_0^{\cos 3\theta} (r \cos\theta) \, r \, dr \, d\theta$$

$$= \int_{-\pi/6}^{\pi/6} \frac{1}{3} (\cos 3\theta)^3 \cos\theta \, d\theta = \frac{27\sqrt{3}}{320}. \quad \int_R dA =$$

$$\int_{-\pi/6}^{\pi/6} \int_0^{\cos 3\theta} r \, dr \, d\theta = \frac{\pi}{12}. \text{ Thus } \bar{x} = \frac{27\sqrt{3}}{320} \cdot \frac{12}{\pi}$$

$$= \frac{81\sqrt{3}}{80\pi}. \text{ By symmetry, } \bar{y} = 0. \text{ Hence the center}$$

of mass is $\left(\dfrac{81\sqrt{3}}{80\pi}, 0\right)$.

22 $\int_R f(P) \, dA = \int_{\pi/6}^{\pi/4} \int_2^3 r^2 \sin\theta \, dr \, d\theta$

$$= \int_{\pi/6}^{\pi/4} \frac{19}{3} \sin\theta \, d\theta = \frac{19}{6}(\sqrt{3} - \sqrt{2}) \text{ and Area of}$$

$$R = \int_{\pi/6}^{\pi/4} \int_2^3 r \, dr \, d\theta = \int_{\pi/6}^{\pi/4} \frac{5}{2} \, d\theta = \frac{5\pi}{24}.$$

Thus $\dfrac{\int_R f(P) \, dA}{\text{Area of } R} = \dfrac{76}{5\pi}(\sqrt{3} - \sqrt{2}).$

24 $\int_R f(P) \, dA = \int_0^{\pi} \int_0^{1+\cos\theta} r^2 \sin\theta \, dr \, d\theta +$

$$\int_{\pi}^{2\pi} \int_0^{1+\cos\theta} (-r^2 \sin\theta) \, dr \, d\theta = \frac{8}{3} \text{ and Area of}$$

$$R = \int_0^{2\pi} \int_0^{1+\cos\theta} r \, dr \, d\theta = \frac{3\pi}{2}. \text{ Thus}$$

$$\frac{\int_R f(P) \, dA}{\text{Area of } R} = \frac{8}{3} \cdot \frac{2}{3\pi} = \frac{16}{9\pi}.$$

26 $\int_0^1 \int_0^{\sqrt{1-x^2}} x^3 \, dy \, dx = \int_0^{\pi/2} \int_0^1 r^4 \cos^3\theta \, dr \, d\theta$

$$= \frac{2}{15}$$

28 $\int_1^2 \int_{x/\sqrt{3}}^{\sqrt{3}x} (x^2 + y^2)^{3/2} \, dy \, dx$

$$= \int_{\pi/6}^{\pi/3} \int_{\sec\theta}^{2\sec\theta} (r^2)^{3/2} \, r \, dr \, d\theta$$

$$= \frac{341\sqrt{3}}{20} - \frac{527}{180} + \frac{93}{40} \ln \frac{2 + \sqrt{3}}{\sqrt{3}}$$

30 The projection of the region to the xy plane is the disk whose boundary is $x^2 + y^2 = x + y$ or, equivalently, $0 \leq \theta \leq \pi$, $0 \leq r \leq \cos\theta + \sin\theta$. Thus the volume is

$$\int_0^{\pi} \int_0^{\cos\theta + \sin\theta} (r \cos\theta + r \sin\theta - r^2) \, r \, dr \, d\theta = \frac{\pi}{8}.$$

32 $\int_R dA = \int_0^{2\pi} \int_0^a r\, dr\, d\theta = \int_0^{2\pi} \frac{1}{2} a^2\, d\theta = \pi a^2$

34 Position the disk as in the figure. Then $I_z =$

$\int_R r^2 \sigma(P)\, dA = \frac{M}{\pi a^2} \int_{-\pi/2}^{\pi/2} \int_0^{2a\cos\theta} r^3\, dr\, d\theta$

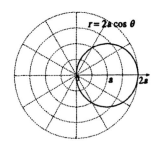

$= \frac{4M}{\pi a^2} \int_{-\pi/2}^{\pi/2} a^4 \cos^4 \theta\, d\theta = \frac{4Ma^2}{\pi} \left(\frac{3\pi}{8}\right)$

$= \frac{3Ma^2}{2}.$

36 Position the disk as in Exercise 34. By Exercise 15(b) of Sec. 15.3, $I_z = I_x + I_y$; hence $I_y = I_z - I_x$

$= Ma^2 \left(\frac{3}{2} - \frac{1}{4}\right) = \frac{5Ma^2}{4}$, where we used the

result of Exercise 35.

38 The computation in Example 5 proceeds as before until the point where $(1 - \cos^2 \theta)^{3/2}$ was replaced by $\sin^3 \theta$, since this identity is not valid for $-\pi/2 \leq \theta < 0$. We obtain instead the integral

$\frac{a^3}{3} \int_{-\pi/2}^{\pi/2} [1 - (1 - \cos^2 \theta)^{3/2}]\, d\theta$

$= \frac{a^3}{3} \int_{-\pi/2}^{\pi/2} (1 - |\sin^3 \theta|)\, d\theta$

$= \frac{a^3}{3} \int_{-\pi/2}^{0} (1 + \sin^3 \theta)\, d\theta +$

$\frac{a^3}{3} \int_0^{\pi/2} (1 - \sin^3 \theta)\, d\theta$

$= \frac{a^3}{3} \left(\theta - \cos\theta + \frac{1}{3}\cos^3\theta\right)\bigg|_{-\pi/2}^{0} + \frac{a^3}{18}(3\pi - 4)$

$= \frac{a^3}{3}\left[-1 + \frac{1}{3} - \left(-\frac{\pi}{2}\right)\right] + \frac{a^3}{18}(3\pi - 4)$

$= \frac{a^3}{18}(3\pi - 4) + \frac{a^3}{18}(3\pi - 4) = \frac{a^3}{9}(3\pi - 4),$

which agrees with the result of Example 5.

40 The area under the curve is $\int_{-\infty}^{\infty} \frac{1}{\sqrt{2\pi}} e^{-x^2/2}\, dx$. Let

$y = x/\sqrt{2}$. Then the area is $\frac{2}{\sqrt{2\pi}} \int_0^{\infty} e^{-y^2} \sqrt{2}\, dy$

$= \frac{2\sqrt{2}}{\sqrt{2\pi}} \cdot \frac{\sqrt{\pi}}{2} = 1.$

42 $V \neq \int_\alpha^\beta A(\theta)\, d\theta,\ V = \int_\alpha^\beta A(\theta)\, r\, d\theta$

44 Let the center of the disk R be the pole of a polar coordinate system. Then the function to be averaged is just r. The area of R is A, so its radius is $\sqrt{A/\pi}$. Thus the average distance is

$\frac{1}{\text{Area of } R} \int_R r\, dA = \frac{1}{A} \int_0^{2\pi} \int_0^{\sqrt{A/\pi}} r^2\, dr\, d\theta$

$= \frac{1}{A} \int_0^{2\pi} \frac{1}{3}(A/\pi)^{3/2}\, d\theta = \frac{1}{A} \cdot \frac{1}{3} \cdot \frac{A^{3/2}}{\pi^{3/2}} \cdot 2\pi = \frac{2\sqrt{A}}{3\sqrt{\pi}}$

$\approx 0.376\sqrt{A}.$

46 Place the square R with its center at the origin and its sides parallel to the x and y axes. The function

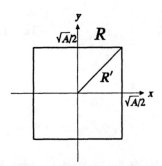

to be averaged is the distance $r = \sqrt{x^2 + y^2}$. Observe that the average distance from the center to points in the square is the same as the average distance from the center to points in the triangle R'. (The eight congruent triangles that partition R all have the same average, so R must share that average.) R' is described by $0 \leq \theta \leq \pi/4$, $0 \leq r \leq \dfrac{\sqrt{A}}{2} \sec \theta$, and the area of R' is $A/8$, so the desired average is $\dfrac{1}{\text{Area of } R'} \int_{R'} r \, dA$

$$= \frac{8}{A} \int_0^{\pi/4} \int_0^{\frac{\sqrt{A}}{2} \sec \theta} r^2 \, dr \, d\theta =$$

$$\frac{8}{A} \int_0^{\pi/4} \frac{1}{3} \left(\frac{\sqrt{A}}{2} \sec \theta \right)^3 d\theta = \frac{\sqrt{A}}{3} \int_0^{\pi/4} \sec^3 \theta \, d\theta$$

$$= \frac{\sqrt{A}}{3} \cdot \frac{1}{2} [\sec \theta \tan \theta + \ln |\sec \theta + \tan \theta|] \Big|_0^{\pi/4}$$

$$= \frac{\sqrt{A}}{6} [\sqrt{2} + \ln(\sqrt{2} + 1)] \approx 0.383\sqrt{A}.$$ (This agrees with the text because $\tan \dfrac{3\pi}{8} = 1 + \sqrt{2}$.)

48 Place the region R as in Exercise 44. Observe that the average for the quarter disk in the first quadrant is equal to the average for the entire region. The "metropolitan" distance to be averaged is $x + y = r \cos \theta + r \sin \theta = r(\cos \theta + \sin \theta)$. The area of the quarter disk is $A/4$, so the average is

$$\frac{4}{A} \int_0^{\pi/2} \int_0^{\sqrt{A/\pi}} r^2 (\cos \theta + \sin \theta) \, dr \, d\theta$$

$$= \frac{4}{A} \cdot \frac{1}{3} \cdot \frac{A^{3/2}}{\pi^{3/2}} \int_0^{\pi/2} (\cos \theta + \sin \theta) \, d\theta$$

$$= \frac{4\sqrt{A}}{3\pi^{3/2}} (\sin \theta - \cos \theta) \Big|_0^{\pi/2} = \frac{8\sqrt{A}}{3\pi^{3/2}} \approx 0.479\sqrt{A}.$$

50 (a) The definition of $\int_R f(P) \, dA$ uses the limit of approximating sums as the mesh approaches 0. The partition of R by rays and circular arcs is simply one such possibility for creating an approximation.

(b) With θ's value fixed at θ_i, the given sum approximates the integral of $f(r, \theta_i) r$ as r ranges from a to b, with $[a, b]$ partitioned by $r_0, r_1, \cdots, r_n$.

(c) As shown in (b), the inner sum of (3) is an approximation for $\int_a^b f(r, \theta_i) r \, dr$, a function of θ, which we can call $g(\theta)$. Using g in the inner sum, (3) becomes $\sum_{i=1}^n g(\theta_i)(\theta_i - \theta_{i-1})$.

(d) The outer sum of (3) takes values of $g(\theta)$ as θ ranges from α to β, producing an approximating sum for $\int_\alpha^\beta g(\theta) \, d\theta$.

(e) As already noted, (3) is an approximating sum for $\int_R f(P) \, dA$. By (b), (c), and (d), (3) is

15.4 Computing $\int_R f(P)\, dA$ using Polar Coordinates

also an approximating sum for the iterated

integral $\int_\alpha^\beta \int_a^b f(r, \theta)\, r\, dr\, d\theta$.

15.5 The Definite Integral of a Function over a Region in Space

2 Estimated mass $= 16 \cdot 6 + 16 \cdot 14 + 16 \cdot 22 + 16 \cdot 14$
$= 896$ grams

4 The average distance is $\dfrac{\int_R f(P)\, dV}{\text{Volume of } R}$, where $f(P)$

$= \overline{P_0 P}$.

6 (a) It is less than or equal to 8 times the volume of R.

(b) It is less than or equal to 8.

8 Mesh $= \sqrt{2^2 + 2^2 + 2^2} = 2\sqrt{3}$ cm

10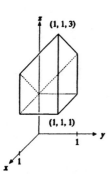

12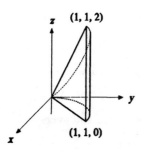

14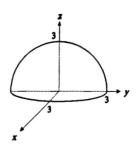

16 $\int_0^1 \int_{x^3}^{x^2} \int_0^{x+y} z\, dz\, dy\, dx$

$= \int_0^1 \int_{x^3}^{x^2} \dfrac{1}{2}(x+y)^2\, dy\, dx$

$= \int_0^1 \dfrac{1}{6}[(x+x^2)^3 - (x+x^3)^3]\, dx$

$= \int_0^1 \dfrac{1}{6}(3x^4 + x^6 - 3x^7 - x^9)\, dx = \dfrac{5}{112}$

18 $\int_0^1 \int_0^x \int_0^3 (x^2 + y^2)\, dz\, dy\, dx$

$= \int_0^1 \int_0^x 3(x^2 + y^2)\, dy\, dx = \int_0^1 4x^3\, dx = 1$

20 (a) $0 \le x \le 1, 0 \le y \le 1,$
$1 - y \le z \le y + 1$

(b) $0 \le x \le 1, 1 \le z \le 2, z - 1 \le y \le 1;$
$0 \le x \le 1, 0 \le z \le 1, 1 - z \le y \le 1$

22 (a)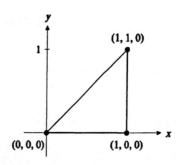

(b) Top: $z = -x + 2y + 2$
Bottom: $z = -x - y + 2$

(c) $0 \le x \le 1, 0 \le y \le x,$
$-x - y + 2 \le z \le -x + 2y + 2$

24 $\int_R z \, dV = \int_0^2 \int_0^3 \int_0^{x+2y} z \, dz \, dy \, dx = 58$

26 Average $= \dfrac{1}{a^3} \int_0^a \int_0^a \int_0^a (x^2 + y^2 + z^2) \, dz \, dy \, dx$

$= a^2$

28 Total mass $= \int_0^1 \int_0^{-2x+2} \int_0^{xy} (x+y) \, dz \, dy \, dx$

$= \dfrac{1}{5}$

30 (a) $-r \leq x \leq r$, $-\sqrt{r^2 - x^2} \leq y \leq \sqrt{r^2 - x^2}$,

$\dfrac{h}{r}\sqrt{x^2 + y^2} \leq z \leq h$

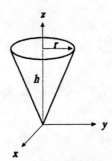

(b) $\int_R z \, dV = \int_{-r}^r \int_{-\sqrt{r^2-x^2}}^{\sqrt{r^2-x^2}} \int_{\frac{h}{r}\sqrt{x^2+y^2}}^h z \, dz \, dy \, dx$

$= \dfrac{\pi h^2 r^2}{4}$, $\bar{z} = \dfrac{\pi h^2 r^2}{4} \cdot \dfrac{3}{\pi r^2 h} = \dfrac{3h}{4}$

32 $\int_R \dfrac{e^{-x}}{\sqrt{y}} \, dV = \int_0^1 \int_{x^2}^1 \int_0^{2y} \dfrac{e^{-x}}{\sqrt{y}} \, dz \, dy \, dx$

$= \dfrac{20}{3}(3e^{-1} - 1)$. $\int_R dV = \int_0^1 \int_{x^2}^1 \int_0^{2y} dz \, dy \, dx$

$= 4/5$. So the average temperature is

$\dfrac{25}{3}(3e^{-1} - 1)$.

34 The moment of inertia about the line is

$\int_{-c/2}^{c/2} \int_{-b/2}^{b/2} \int_{-a/2}^{a/2} \dfrac{M}{abc}(y^2 + z^2) \, dx \, dy \, dz = \dfrac{M}{12}(b^2 + c^2)$.

36 Let the mutually perpendicular lines be the axes of an xyz coordinate system. The sum of the moments of inertia is then $I_x + I_y + I_z = \int_R (y^2 + z^2) \, \delta \, dV$

$+ \int_R (x^2 + z^2) \, \delta \, dV + \int_R (x^2 + y^2) \, \delta \, dV$

$= \int_R (2x^2 + 2y^2 + 2z^2) \, \delta \, dV$. Observe that the

integrand is 2δ times the square of the distance to the origin, and does not depend on the choice of axes.

38 The plane $x = 0$ divides the region R into two hemispheres: one in which x is always positive, and one in which x is always negative. By symmetry, the contributions of the hemispheres to the integral are equal in magnitude and opposite to sign; thus

$\int_R x \, dV = 0$.

40 (a) The average value of f over the plane region

$R(x)$ is $\dfrac{\int_{R(x)} f(P) \, dA}{\text{Area of } R(x)}$. But the region $R(x)$ is

described by $y_1(x) \leq y \leq y_2(x)$, $z_1(x, y) \leq z \leq z_2(x, y)$, where x is fixed. By Sec. 15.2,

$\int_R f(P) \, dA = \int_{y_1(x)}^{y_2(x)} \int_{z_1(x, y)}^{z_2(x, y)} f(x, y, z) \, dz \, dy$.

Hence the average density in $R(x)$ is as claimed.

(b) The mass of the solid region between $R(x)$ and $R(x + \Delta x)$ equals the volume of the region times the average density in the region. For $x \leq x \leq x + \Delta x$, $R(x)$ has approximately the same area and average density as $R(x + \Delta x)$. Hence the volume of the solid region is about (Area of $R(x)) \cdot \Delta x$ and its average density is, in part (a), about

$$\frac{1}{\text{Area of } R(x)} \int_{y_1(x)}^{y_2(x)} \int_{z_1(x,y)}^{z_2(x,y)} f(x, y, z)\, dz\, dy.$$

Hence the mass of the region is approximately the product of the volume and the average density; that is, $[(\text{Area of } R(x)) \cdot \Delta x] \times$

$$\frac{1}{\text{Area of } R(x)} \int_{y_1(x)}^{y_2(x)} \int_{z_1(x,y)}^{z_2(x,y)} f(x, y, z)\, dz\, dy$$

$$= \int_{y_1(x)}^{y_2(x)} \int_{z_1(x,y)}^{z_2(x,y)} f(x, y, z)\, dz\, dy\, \Delta x.$$

(c) Part (b) gave the "local approximation" to the mass between $R(x)$ and $R(x + \Delta x)$. By the informal approach, $\int_R f(P)\, dV = $ Mass of R

$$= \int_{x_1}^{x_2} \int_{y_1(x)}^{y_2(x)} \int_{z_1(x,y)}^{z_2(x,y)} f(x, y, z)\, dz\, dy\, dx.$$

15.6 Computing $\int_R f(P)\, dV$ using Cylindrical Coordinates

2

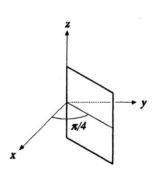

4

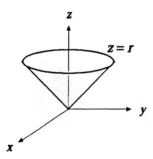

6

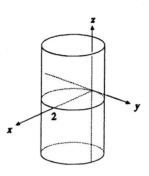

8

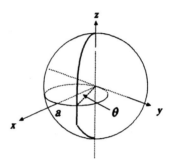

10 $-\dfrac{\pi}{2} \le \theta \le \dfrac{\pi}{2},\ 0 \le r \le a\cos\theta,\ -\sqrt{a^2 - r^2} \le z \le \sqrt{a^2 - r^2}$

12 $0 \le \theta \le 2\pi,\ 0 \le r \le 3,\ r \le z \le 3$

14 $0 \le \theta \le \pi/3,\ 0 \le r \le \sin 3\theta,$
$-\sqrt{1 - r^2\cos 2\theta} \le z \le \sqrt{1 - r^2\cos 2\theta}$

16

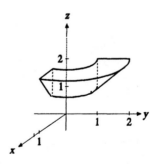

18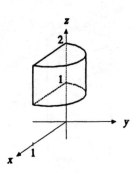

20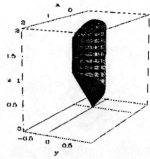

22 $-\frac{\pi}{8} \leq \theta \leq \frac{\pi}{8}, 0 \leq r \leq \cos 4\theta, 0 \leq z \leq \sqrt{4 - r^2}$

24 $0 \leq \theta \leq \frac{\pi}{4}, 0 \leq r \leq \sec \theta, 0 \leq z \leq 1 - r\cos\theta$

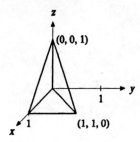

26 $\int_0^{2\pi} \int_0^1 \int_{-\sqrt{a^2 - r^2}}^{\sqrt{a^2 - r^2}} z^2 r \, dz \, dr \, d\theta$

$= \int_0^{2\pi} \int_0^1 \frac{z^3 r}{3} \Big|_{-\sqrt{a^2 - r^2}}^{\sqrt{a^2 - r^2}} dr \, d\theta$

$= \int_0^{2\pi} \int_0^1 \frac{2r}{3}(a^2 - r^2)^{3/2} dr \, d\theta$

$= -\frac{2}{3} \int_0^{2\pi} \frac{1}{5}(a^2 - r^2)^{5/2} \Big|_0^1 d\theta$

$= -\frac{2}{15} \int_0^{2\pi} [(a^2 - 1)^{5/2} - a^5] d\theta$

$= \frac{2}{15}(a^5 - (a^2 - 1)^{5/2}) 2\pi$

$= \frac{4\pi}{15}(a^5 - (a^2 - 1)^{5/2})$

28 From Exercise 27 the mass in R is $\pi/8$. Then $\bar{z} =$

$\frac{8}{\pi} \int_0^{2\pi} \int_0^{1/\sqrt{2}} \int_r^{\sqrt{1 - r^2}} z^2 r \, dz \, dr \, d\theta =$

$\frac{8}{\pi} \int_0^{2\pi} \int_0^{1/\sqrt{2}} \frac{1}{3} z^3 r \Big|_r^{\sqrt{1 - r^2}} dr \, d\theta$

$= \frac{8}{3\pi} \int_0^{2\pi} \int_0^{1/\sqrt{2}} [r(1 - r^2)^{3/2} - r^4] dr \, d\theta$

$= \frac{8}{3\pi} \int_0^{2\pi} \left(-\frac{1}{5}(1 - r^2)^{5/2} - \frac{1}{5}r^5\right)\Big|_0^{1/\sqrt{2}} d\theta$

$= \frac{8}{15\pi} \int_0^{2\pi} \left[-\left(1 - \frac{1}{2}\right)^{5/2} - \frac{1}{2^{5/2}} + 1\right] d\theta$

$= \frac{8}{15\pi}(2\pi)\left(1 - \frac{1}{2^{3/2}}\right) = \frac{16}{15}\left(1 - \frac{1}{2\sqrt{2}}\right).$

30 $z = 1 + y = 1 + r\sin\theta$ and the circle in the xy plane with center $(0, 1, 0)$ and radius 1 has

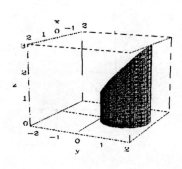

equation $r = 2\sin\theta$. Hence $V = \int_V dV$

$$\int_0^\pi \int_0^{2\sin\theta} \int_0^{r\sin\theta+1} r\, dz\, dr\, d\theta$$

$$= \int_0^\pi \int_0^{2\sin\theta} (r\sin\theta + 1) r\, dr\, d\theta$$

$$= \int_0^\pi \left(\frac{1}{3}r^3 \sin\theta + \frac{1}{2}r^2\right)\Big|_0^{2\sin\theta} d\theta$$

$$= \int_0^\pi \left(\frac{8}{3}\sin^4\theta + 2\sin^2\theta\right) d\theta$$

$$= \frac{16}{3}\int_0^{\pi/2} \sin^4\theta\, d\theta + 4\int_0^{\pi/2} \sin^2\theta\, d\theta$$

$$= \pi + \pi = 2\pi$$

32 $I = \dfrac{M}{\pi a^2 h} \int_0^\pi \int_0^{2a\cos\theta} \int_0^h r^3\, dz\, dr\, d\theta$

$$= \frac{M}{\pi a^2} \int_0^\pi \int_0^{2a\cos\theta} r^3\, dr\, d\theta$$

$$= \frac{M}{4\pi a^2} \int_0^\pi r^4\Big|_0^{2a\cos\theta} d\theta$$

$$= \frac{M}{4\pi a^2} \int_0^\pi 16a^4 \cos^4\theta\, d\theta$$

$$= \frac{8Ma^2}{\pi} \int_0^{\pi/2} \cos^4\theta\, d\theta = \frac{8Ma^2}{\pi} \cdot \frac{3\pi}{16} = \frac{3Ma^2}{2}$$

34

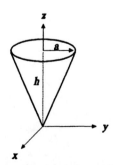

Position the cone as in the figure. Then $I =$

$$\frac{3M}{\pi a^2 h} \int_0^{2\pi} \int_0^a \int_0^{(a-r)h/a} r^3\, dz\, dr\, d\theta$$

$$= \frac{3M}{\pi a^2 h} \int_0^{2\pi} \int_0^a \frac{r^3(a-r)h}{a}\, dr\, d\theta$$

$$= \frac{3M}{\pi a^3} \int_0^{2\pi} \left(\frac{ar^4}{4} - \frac{r^5}{5}\right)\Big|_0^a d\theta = \frac{3Ma^2}{20\pi} \int_0^{2\pi} d\theta$$

$$= \frac{3Ma^2}{10}.$$

36 Let the line through the center of the hemisphere coincide with the z axis. Then $I =$

$$\frac{3M}{2\pi a^3} \int_0^{2\pi} \int_0^a \int_0^{\sqrt{a^2-r^2}} r^3\, dz\, dr\, d\theta$$

$$= \frac{3M}{2\pi a^3} \int_0^{2\pi} \int_0^a r^3 \sqrt{a^2 - r^2}\, dr\, d\theta$$

$$= \frac{3M}{2\pi a^3} \int_0^{2\pi} \frac{2a^5}{15}\, d\theta = \frac{2Ma^2}{5}.$$

38 Place the hemisphere atop the xy plane with its center at the origin. By symmetry, $\bar{x} = \bar{y} = 0$. The moment with respect to the xy plane is

$$\int_V z\, dV = \int_0^{2\pi} \int_0^a \int_0^{\sqrt{a^2-r^2}} zr\, dz\, dr\, d\theta$$

$$= \int_0^{2\pi} \int_0^a \frac{1}{2}(a^2 - r^2) r\, dr\, d\theta$$

$$= \frac{1}{2} \int_0^{2\pi} \int_0^a (a^2 r - r^3)\, dr\, d\theta$$

$$= \frac{1}{2}(2\pi)\left(\frac{1}{2}a^4 - \frac{1}{4}a^4\right) = \frac{1}{4}\pi a^4, \text{ so } \bar{z} = \frac{\pi a^4/4}{2\pi a^3/3}$$

$$= \frac{3a}{8}.$$

40 We are free to place a coordinate system so that R has its center of mass at $(0, 0, 0)$ and $\mathcal{P}$ is the xy

plane. Then $g(P) = z$, so $\int_R g(P)\,\delta(P)\,dV =$

$\int_R z\,\delta(P)\,dV = M\cdot\bar{z} = M\cdot 0 = 0$, as claimed.

15.7 Computing $\int_R f(P)\,dV$ using Spherical Coordinates

2   Spheres and cones

4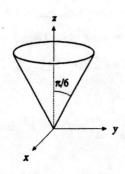

6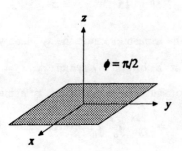

8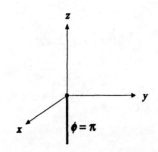

10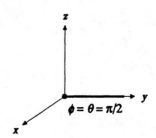

12 $0 \le \theta \le 2\pi,\ 0 \le \phi \le \pi/2,\ 0 \le \rho \le a$

14 $0 \le \theta \le 2\pi,\ 0 \le \phi \le \pi,\ a \le \rho \le b$

16 $0 \le \theta \le 2\pi,\ 0 \le \phi \le \cos^{-1}\dfrac{1}{\sqrt{3}},\ 0 \le \rho \le$

$\dfrac{-\cos\phi + \sqrt{\cos^2\phi + 4\sin^2\phi}}{2\sin^2\phi},\ 0 \le \theta \le 2\pi,$

$\cos^{-1}\dfrac{1}{\sqrt{3}} \le \phi \le \pi/2,\ 0 \le \rho \le \csc\phi\cot\phi$

18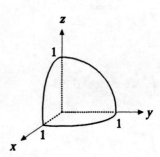

20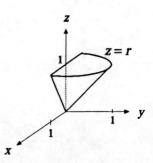

22 First of all, $\rho = \sqrt{x^2 + y^2 + z^2}$. As in polar coordinates, we may take $\tan^{-1}(y/x)$ for (x, y) in the

15.7 Computing $\int_R f(P)\, dV$ using Spherical Coordinates

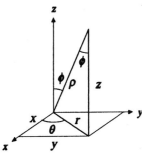

first or fourth quadrants (where θ ranges from $-\pi/2$ to $\pi/2$); for the second and third quadrants we can use $\theta = \tan^{-1}(y/x) + \pi$. With these definitions, θ will vary from $-\pi/2$ up to $3\pi/2$. If it is preferred to have θ between 0 and 2π, redefine θ in the fourth quadrant to be equal to $\tan^{-1}(y/x) + 2\pi$. Recall that $\cos\phi = z/\rho$; therefore $\phi = \cos^{-1}\dfrac{z}{\sqrt{x^2 + y^2 + z^2}}$. Since the inverse cosine gives value from 0 to π, this expression is valid for ϕ regardless of the sign of z.

24 (a) $\rho \sin\phi \cos\theta = 2$

(b) $2\rho \sin\phi \cos\theta + 3\rho \sin\phi \sin\theta + 4\rho \cos\phi = 1$

26 See Figure 14 and Example 4 on page 926. $V =$

$\int_R 1\, dV$

$= \int_0^{2\pi} \int_0^{\tan^{-1} r/h} \int_0^{h/\cos\phi} \rho^2 \sin\phi \, d\rho\, d\phi\, d\theta$

$= \int_0^{2\pi} \int_0^{\tan^{-1} r/h} \dfrac{1}{3}\rho^3 \Big|_0^{h/\cos\phi} \sin\phi\, d\phi\, d\theta$

$= \dfrac{h^3}{3} \int_0^{2\pi} \int_0^{\tan^{-1} r/h} \dfrac{\sin\phi}{\cos^3\phi}\, d\phi\, d\theta$

$= \dfrac{h^3}{6} \int_0^{2\pi} \dfrac{1}{\cos^2\phi} \Big|_0^{\tan^{-1} r/h}\, d\theta = \dfrac{h^3}{6} \int_0^{2\pi} \dfrac{r^2}{h^2}\, d\theta$

$= \dfrac{r^2 h}{6}(2\pi) = \dfrac{\pi r^2 h}{3}$

28 $\int_0^{2\pi} \int_0^{\tan^{-1} a/h} \int_0^{h \sec\phi} (h - \rho\cos\phi)\rho^2 \sin\phi\, d\rho\, d\phi\, d\theta$

$= \dfrac{\pi a^2 h^2}{12}$

30

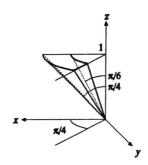

In spherical coordinates $dV = \rho^2 \sin\phi\, d\rho\, d\phi\, d\theta$, so $f(P) = \rho\cos\phi$.

32 Assume that the ball is centered at the origin and that the equatorial plane is the xy plane. Then the average is

$\dfrac{3}{4\pi a^3} \int_0^{2\pi} \int_0^{\pi} \int_0^a \rho^4 \cos^2\phi \sin\phi\, d\rho\, d\phi\, d\theta$

$= a^2/5$.

34 Volume

$= \int_{\theta_0}^{\theta_0 + \Delta\theta_0} \int_{\phi_0}^{\phi_0 + \Delta\phi_0} \int_{\rho_0}^{\rho_0 + \Delta\rho_0} \rho^2 \sin\phi\, d\rho\, d\phi\, d\theta =$

$\dfrac{1}{3}\big[(\rho_0 + \Delta\rho_0)^3 - \rho_0^3\big](\cos\phi_0 - \cos(\phi_0 + \Delta\phi_0))\,\Delta\theta_0$

36 The constant density of the ball is $\delta = \dfrac{3M}{4\pi a^3}$.

Place the ball's center at the origin of a coordinate system and compute I_z. The integrand will be δr^2, which, in spherical coordinates, is equal to $\delta(\rho \sin\phi)^2$. Hence $I_z = \int_R \delta r^2\, dV$

$$= \delta \int_0^{2\pi} \int_0^{\pi} \int_0^a (\rho \sin \phi)^2 \, \rho^2 \sin \phi \, d\rho \, d\phi \, d\theta$$

$$= \delta \int_0^{2\pi} \int_0^{\pi} \int_0^a \rho^4 \sin^3 \phi \, d\rho \, d\phi \, d\theta$$

$$= \delta \int_0^{2\pi} \int_0^{\pi} \frac{a^5}{5} \sin^3 \phi \, d\phi \, d\theta$$

$$= \delta \frac{a^5}{5} \int_0^{2\pi} \left(\frac{\cos^3 \phi}{3} - \cos \phi \right) \Big|_0^{\pi} d\theta$$

$$= \frac{3M}{4\pi a^3} \cdot \frac{a^5}{5} \int_0^{2\pi} \frac{4}{3} \, d\theta = \frac{3Ma^2}{20\pi} \cdot \frac{4}{3} \cdot 2\pi$$

$$= \frac{2Ma^2}{5}.$$

38 $I_z = \int_R \dfrac{3M}{\pi a^2 h} r^2 \, dV =$

$$\frac{3M}{\pi a^2 h} \int_0^{2\pi} \int_0^{\tan^{-1} a/h} \int_0^{h \sec \phi} (\rho \sin \phi)^2 \, \rho^2 \sin \phi \, d\rho \, d\phi \, d\theta$$

$$= \frac{3M}{\pi a^2 h} \int_0^{2\pi} \int_0^{\tan^{-1} a/h} \int_0^{h \sec \phi} \rho^4 \sin^3 \phi \, d\rho \, d\phi \, d\theta$$

$$= \frac{3Ma^2}{10}$$

40 $\int_R \dfrac{g(\rho)}{q} \, dV$

$$= \int_R \frac{g(\rho)}{\sqrt{H^2 + \rho^2 - 2\rho H \cos \phi}} \, dV =$$

$$\int_0^a \int_0^{\pi} \int_0^{2\pi} \frac{g(\rho) \, \rho^2 \sin \phi}{\sqrt{H^2 + \rho^2 - 2\rho H \cos \phi}} \, d\theta \, d\phi \, d\rho$$

$$= \int_0^a \int_0^{\pi} \frac{2\pi g(\rho) \, \rho^2 \sin \phi}{\sqrt{H^2 + \rho^2 - 2\rho H \cos \phi}} \, d\phi \, d\rho =$$

$$\int_0^a \frac{2\pi g(\rho) \rho^2}{\rho H} \sqrt{H^2 + \rho^2 - 2\rho H \cos \phi} \Big|_0^{\pi} d\rho$$

$$= \frac{4\pi}{H} \int_0^a g(\rho) \, \rho^2 \, d\rho = \frac{1}{H} \left(\int_R g(\rho) \, dV \right) = M/H.$$

42 $\int_R x^3 \, dV = 0$ by symmetry over R; similarly for y^3 and z^3.

44 Yes, from physical considerations in analogy with Example 5.

46 (a)

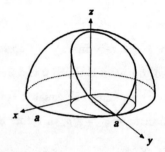

(b) Volume of R

$$= 4 \int_0^{\pi/2} \int_0^{a \sin \theta} \int_0^{\sqrt{a^2 - r^2}} r \, dz \, dr \, d\theta$$

$$= \frac{2a^3}{9} (3\pi - 4).$$

48 Let R_t be the region between concentric spheres of radii a and t. Then $\int_{R_t} f(P) \, dV = \int_{R_t} \dfrac{1}{\rho} \, dV =$

$$\int_0^{2\pi} \int_0^{\pi} \int_t^a \rho \sin \phi \, d\rho \, d\phi \, d\theta = 2\pi(a^2 - t^2).$$

Since $\lim\limits_{t \to 0^+} 2\pi(a^2 - t^2) = 2\pi a^2$, $\int_R f(P) \, dA = 2\pi a^2$. The average of f over R is $\dfrac{\int_R f(P) \, dV}{\text{Volume of } R} =$

$$2\pi a^2 \cdot \frac{3}{4\pi a^3} = \frac{3}{2a}.$$

49 (a) Place the cone with its axis along the z axis. Then by symmetry $\bar{x} = \bar{y} = 0$. By Example

4 of Sec. 15.7, $\bar{z} = \dfrac{h}{4}$. (Note that the cone in this problem is inverted with respect to Example 4.)

(c) The centroid of the generalized cone is 1/4 the distance from the centroid of the base to the vertex.

(d) We already know from Exercise 20 of Sec. 8.4 that the volume of the generalized cone is

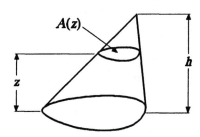

$V = \dfrac{1}{3}Ah$, where A is the area of R. Letting R^* be the solid region occupied by the cone, we have $\bar{z} = \dfrac{3}{Ah}\int_{R^*} z\, dV =$

$\dfrac{3}{Ah}\int_0^h zA(z)\, dz$, where $A(z)$ is the cross-sectional area of the cone at height z. Now $A(z) = \dfrac{A}{h^2}(h-z)^2$, so $\bar{z}$

$= \dfrac{3}{Ah}\int_0^h \dfrac{A}{h^2}z(h-z)^2\, dz$

$= \dfrac{3}{h^3}\int_0^h (h^2 z - 2hz^2 + z^3)\, dz = \dfrac{3}{h^3}\dfrac{h^4}{12}$

$= \dfrac{h}{4}$.

15.S Review Exercises

2 (a) $\int_0^1 x^2 y\, dy = \dfrac{1}{2}x^2 y^2 \Big|_0^1 = x^2/2$

(b) $\int_0^1 x^2 y\, dx = \dfrac{1}{3}x^3 y \Big|_0^1 = y/3$

4 $\int_0^1 \sin(x^2 y)\, dy = -\dfrac{1}{x^2}\cos x^2 y \Big|_0^1$

$= \dfrac{1}{x^2} - \dfrac{1}{x^2}\cos x^2$

6 $\int_0^1 \int_0^x x^2\, dy\, dx = \int_0^{\pi/4}\int_0^{\sec\theta} r^2 \cos^2\theta\, r\, dr\, d\theta$

$= \int_0^{\pi/4} \dfrac{r^4}{4}\cos^2\theta \Big|_0^{\sec\theta} d\theta = \dfrac{1}{4}\int_0^{\pi/4} \sec^2\theta\, d\theta$

$= \dfrac{1}{4}\tan\theta \Big|_0^{\pi/4} = \dfrac{1}{4}$

8 $\int_0^2 \int_0^{\sqrt{2x-x^2}} x\, dy\, dx =$

$\int_0^{\pi/2}\int_0^{2\cos\theta} r\cos\theta\, r\, dr\, d\theta = \dfrac{\pi}{2}$

10 $\int_0^{\pi/4}\int_0^a r^2 \cos\theta\, dr\, d\theta = \int_0^{a/\sqrt{2}}\int_y^{\sqrt{a^2-y^2}} x\, dx\, dy$

$= \dfrac{a^3}{3\sqrt{2}}$

12 $\int_{\pi/4}^{3\pi/4}\int_0^a r^3\, dr\, d\theta$

$= \int_0^{a/\sqrt{2}}\int_x^{\sqrt{a^2-x^2}} (x^2+y^2)\, dy\, dx +$

$\int_{-a/\sqrt{2}}^0 \int_{-x}^{\sqrt{a^2-x^2}} (x^2+y^2)\, dy\, dx = \dfrac{\pi a^4}{8}$

14 $\int_R dA = \int_{-\pi/2}^{\pi/2}\int_1^{1+\cos\theta} r\, dr\, d\theta = 2 + \dfrac{\pi}{4}$ and

$\int_R x\, dA = \int_{-\pi/2}^{\pi/2} \int_1^{1+\cos\theta} r^2 \cos\theta\, dr\, d\theta =$

$\dfrac{15\pi + 32}{24}$, so $\bar{x} = \dfrac{15\pi + 32}{24} \cdot \dfrac{4}{8+\pi} =$

$\dfrac{15\pi + 32}{6(8+\pi)}$. By symmetry, $\bar{y} = 0$.

16. $\int_R f(P)\, dA = \int_0^2 \int_{x^3}^{2x^3} xy\, dy\, dx = 48$

18. $\int_R f(P)\, dA = \int_0^{\pi/4} \int_0^{2\sin\theta} r^3 \cos\theta \sin\theta\, dr\, d\theta$

    $= 1/12$

20. $\int_R dA = \int_0^1 \int_{x^2}^{\sqrt{x}} dy\, dx = 1/3$, $\int_R x\, dA =$

    $\int_0^1 \int_{x^2}^{\sqrt{x}} x\, dy\, dx = 3/20$, and $\int_R y\, dA =$

    $\int_0^1 \int_{x^2}^{\sqrt{x}} y\, dy\, dx = 3/20$. Thus $\bar{x} = \dfrac{3}{20} \cdot \dfrac{3}{1} = \dfrac{9}{20}$

    and $\bar{y} = \dfrac{3}{20} \cdot \dfrac{3}{1} = \dfrac{9}{20}$.

22. $\int_R dA = \int_{-\pi/2}^{\pi/2} \int_0^{\cos x} dy\, dx = 2$

    $\int_R x\, dA = \int_{-\pi/2}^{\pi/2} \int_0^{\cos x} x\, dy\, dx = 0$

    $\int_R y\, dA = \int_{-\pi/2}^{\pi/2} \int_0^{\cos x} y\, dy\, dx = \pi/4$. Thus $\bar{x} =$

    0 and $\bar{y} = \dfrac{\pi}{4} \cdot \dfrac{1}{2} = \dfrac{\pi}{8}$.

24. (a) By symmetry

    (b) $\int_R x^2\, dA + \int_R y^2\, dA = \int_R (x^2 + y^2)\, dA =$

    $\int_R r^2\, dA$

    (c) $\int_R r^2\, dA = \int_0^{2\pi} \int_0^a r^3\, dr\, d\theta = \dfrac{\pi a^4}{2}$

    (d) $\int_R x^2\, dA = \dfrac{1}{2}\left(\int_R x^2\, dA + \int_R x^2\, dA\right)$

$= \dfrac{1}{2} \int_R r^2\, dA = \dfrac{\pi a^4}{4}$

26. $\int_R \ln(x^2 + y^2)\, dA = \int_0^{\pi/4} \int_1^{\sec\theta} r \ln r^2\, dr\, d\theta$

    $= \dfrac{3\pi}{8} + \dfrac{1}{2}\ln 2 - \dfrac{3}{2}$

28. (a) Place the center of the sphere at the origin and take the moment of inertia about the diameter on the z axis. Then $I_z =$

    $\dfrac{3M}{4\pi(b^3 - a^3)} \int_R r^2\, dV =$

    $\dfrac{3M}{4\pi(b^3 - a^3)} \int_0^{2\pi} \int_0^{\pi} \int_a^b (\rho^2 \sin^2\phi) \rho^2 \sin\phi\, d\rho\, d\phi\, d\theta$

    $= \dfrac{3M}{4\pi(b^3 - a^3)} \cdot \dfrac{b^5 - a^5}{5} \cdot \dfrac{4}{3} \cdot 2\pi$

    $= \dfrac{2M(b^5 - a^5)}{5(b^3 - a^3)}$.

    (b) $\lim_{a \to b^-} \dfrac{2M(b^5 - a^5)}{5(b^3 - a^3)} \underset{H}{=} \lim_{a \to b^-} \dfrac{2M(-5a^4)}{-15a^2}$

    $= \dfrac{2Mb^2}{3}$

30. Place the triangle as shown. The area is $\sqrt{3}a^2/4$, so

    $\delta = \dfrac{4M}{\sqrt{3}a^2}$. For convenience, write $b = \sqrt{3}a/2$ so

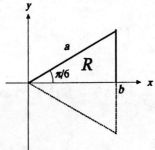

that we may describe R, half the desired triangle, by $0 \le \theta \le \pi/6$, $0 \le r \le b\sec\theta$. Then the moment of inertia is $I_z = 2 \int_R \delta r^2 \, dA$

$= 2\delta \int_0^{\pi/6} \int_0^{b \sec \theta} r^3 \, dr \, d\theta$

$= \dfrac{8M}{\sqrt{3}a^2} \int_0^{\pi/6} \dfrac{1}{4} b^4 \sec^4 \theta \, d\theta$

$= \dfrac{2Mb^4}{\sqrt{3}a^2} \left(\dfrac{1}{3} \tan^3 \theta + \tan \theta \right) \Big|_0^{\pi/6}$

$= \dfrac{2M}{\sqrt{3}a^2} \dfrac{9a^4}{16} \left(\dfrac{1}{9\sqrt{3}} + \dfrac{1}{\sqrt{3}} \right) = \dfrac{5Ma^2}{12}$.

32 For convenience, suppose the diameter of the semidisk lies on the x axis with its center at $(0, 0)$, so that its description is $0 \leq \theta \leq \pi$, $0 \leq r \leq a$.

(a) $M = \int_R 2y \, dA = \int_0^\pi \int_0^a 2r \sin \theta \, r \, dr \, d\theta$

$= \int_0^\pi \int_0^a 2r^2 \sin \theta \, dr \, d\theta = \dfrac{2}{3} a^3 (-\cos \theta) \Big|_0^\pi$

$= \dfrac{4}{3} a^3$

(b) $M_x = \int_R y\delta \, dA = \int_R 2y^2 \, dA$

$= \int_0^\pi \int_0^a 2r^3 \sin^2 \theta \, dr \, d\theta$

$= \dfrac{1}{2} a^4 \left(\dfrac{\theta}{2} - \dfrac{1}{4} \sin 2\theta \right) \Big|_0^\pi = \dfrac{1}{4} \pi a^4$

(c) By symmetry, $\bar{x} = 0$. From (a) and (b), $\bar{y} = \dfrac{M_x}{M}$

$= \dfrac{\pi a^4/4}{4a^3/3} = \dfrac{3\pi a}{16}$.

(d) $I_x = \int_R y^2 \delta \, dA = \int_R 2y^3 \, dA$

$= \int_0^\pi \int_0^a 2r^4 \sin^3 \theta \, dr \, d\theta$

$= \dfrac{2}{5} a^5 \int_0^\pi \sin^3 \theta \, d\theta = \dfrac{8}{15} a^5$.

(e) $I_z = \int_R r^2 \delta \, dA = \int_R 2yr^2 \, dA$

$= \int_0^\pi \int_0^a 2r^4 \sin \theta \, dr \, d\theta = \dfrac{4}{5} a^5$

34 (a) $\displaystyle\lim_{x \to \infty} \int_0^{2\pi} \int_0^s \dfrac{ar}{(a^2 + r^2)^{3/2}} \, dr \, d\theta$

$= \displaystyle\lim_{s \to \infty} 2\pi \left(1 - \dfrac{a}{\sqrt{a^2 + s^2}} \right) = 2\pi$

(b) Yes, because when the point mass is near the plane, the gravitational pull of distant points tends to cancel out (by symmetry). When the point mass is farther from the plane, the gravitational pull tends to combine in the same direction, cancelling out the effect of greater distance.

36 (a) Choose R so that the description is $0 \leq y \leq 1$, $0 \leq x \leq y^2$.

(b) Let R be given by $0 \leq x \leq 1$, $0 \leq y \leq 1$. Either order of the repeated integral requires a nonelementary antiderivative.

38 Volume $= \int_R \left(\sqrt{x^2 + y^2} - xy \right) dA$

$= \int_R (r - r^2 \sin \theta \cos \theta) \, dA$

$= \int_0^{\pi/2} \int_0^1 (r^2 - r^3 \sin \theta \cos \theta) \, dr \, d\theta$

$= \int_0^{\pi/2} \left(\dfrac{1}{3} - \dfrac{1}{4} \sin \theta \cos \theta \right) d\theta = \dfrac{\pi}{6} - \dfrac{1}{8}$

40 (a) $\int_a^b \int_a^x f^{(2)}(t) \, dt \, dx = \int_a^b \int_t^b f^{(2)}(t) \, dx \, dt$

$= \int_a^b f^{(2)}(t)(b - t) \, dt = \int_a^b f^{(2)}(x)(b - x) \, dx$,

giving the desired result.

(b) $\int_a^b f^{(2)}(x)(b-x)\,dx$

$= \int_a^b \left(f^{(2)}(a) + \int_a^x f^{(3)}(t)\,dt\right)(b-x)\,dx$

$= \int_a^b \int_a^x [f^{(2)}(a) + f^{(3)}(t)](b-x)\,dt\,dx$

$= \int_a^b \int_t^b [f^{(2)}(a) + f^{(3)}(t)](b-x)\,dx\,dt$

$= \int_a^b \left[f^{(2)}(a)\frac{(b-t)^2}{2} + f^{(3)}(t)\frac{(b-t)^2}{2}\right]dt$, so

$f(b) = f(a) + f^{(1)}(a)(b-a) + f^{(2)}(a)\frac{(b-a)^2}{2}$

$+ \int_a^b f^{(3)}(t)\frac{(b-t)^2}{2}\,dt.$

(c) $\int_a^x f^{(2)}(t)(b-t)\,dt$

$= -\frac{1}{2}f^{(2)}(t)(b-t)^2\Big|_a^x + \int_a^x \frac{1}{2}(b-t)^2 f^{(3)}(t)\,dt$

$= -\frac{1}{2}f^{(2)}(x)(b-x)^2 + \frac{1}{2}f^{(2)}(a)(b-a)^2 +$

$\frac{1}{2}\int_a^x f^{(3)}(t)(b-t)^2\,dt$. Plugging in b for x,

we have $f(b) = f(a) + f^{(1)}(a)(b-a) +$

$\frac{1}{2}f^{(2)}(a)(b-a)^2 + \frac{1}{2}\int_a^b f^{(3)}(t)(b-t)^2\,dt.$

42 (a) By Exercise 41, $\int_R z\,dV =$

$2\pi \int_0^1 \frac{1}{2}x[g(x)]^2\,dx$, while $\int_R dV = $ Volume

of $R = 2\pi \int_0^1 xg(x)\,dx$. Hence $\bar{z} =$

$\dfrac{\int_R z\,dV}{\int_R dV} = \dfrac{2\pi \int_0^1 \frac{1}{2}x[g(x)]^2\,dx}{2\pi \int_0^1 xg(x)\,dx} =$

$\dfrac{\int_0^1 \frac{1}{2}x[g(x)]^2\,dx}{\int_0^1 xg(x)\,dx}.$

Let A be the plane region that was revolved. Then

$\int_A z\,dA = \int_0^1 \int_0^{g(x)} z\,dz\,dx = \int_0^1 \frac{1}{2}[g(x)]^2\,dx$

and $\int_A dA = \int_0^1 \int_0^{g(x)} dz\,dx = \int_0^1 g(x)\,dx.$

Hence the z coordinate of the centroid is $\dfrac{\int_A z\,dA}{\int_A dA}$

$= \dfrac{\int_0^1 \frac{1}{2}[g(x)]^2\,dx}{\int_0^1 g(x)\,dx}.$

44 We first express the integrand in terms of ρ and ϕ.
Applying the law of cosines twice to the triangle in

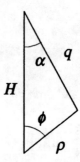

the figure, we obtain $q^2 = H^2 + \rho^2 - 2\rho H \cos\phi$
and $\rho^2 = H^2 + q^2 - 2Hq\cos\alpha$. Hence $q =$

$\sqrt{H^2 + \rho^2 - 2H\rho\cos\phi}$, $\cos\alpha = \dfrac{H^2 + q^2 - \rho^2}{2Hq} =$

$\dfrac{2H^2 - 2H\rho\cos\phi}{2Hq} = \dfrac{H - \rho\cos\phi}{q}$ and $\dfrac{\cos\alpha}{q^2} =$

$\dfrac{H - \rho\cos\phi}{(H^2 + \rho^2 - 2H\rho\cos\phi)^{3/2}}$. Thus $\int_S \dfrac{\cos\alpha}{q^2}\,dV =$

$$\int_0^s \int_0^\pi \int_0^{2\pi} \frac{H - \rho \cos \phi}{(H^2 + \rho^2 - 2H\rho \cos \phi)^{3/2}} \rho^2 \sin \phi \, d\theta \, d\phi \, d\rho$$

$$= 2\pi \int_0^s \int_0^\pi \frac{H - \rho \cos \phi}{(H^2 + \rho^2 - 2H\rho \cos \phi)^{3/2}} \rho^2 \sin \phi \, d\phi \, d\rho.$$

Let $t = -\cos \phi$. Then

$$2\pi \int_0^s \int_{-1}^1 \frac{H + \rho t}{(H^2 + \rho^2 + 2H\rho t)^{3/2}} \rho^2 \, dt \, d\rho.$$ Now let

$u = H^2 + \rho^2 + 2H\rho t$, so $t = \dfrac{u^2 - H^2 - \rho^2}{2H\rho}$ and dt

$= \dfrac{du}{2H\rho}$. The integral is now

$$2\pi \int_0^s \int_{H^2+\rho^2-2H\rho}^{H^2+\rho^2+2H\rho} \frac{H + \dfrac{u - H^2 - \rho^2}{2H}}{u^{3/2}} \rho^2 \frac{du}{2H\rho} \, d\rho$$

$$= \frac{\pi}{2H^2} \int_0^s \int_{(H-\rho)^2}^{(H+\rho)^2} [u^{-1/2} + (H^2 - \rho^2)u^{-3/2}] \rho \, du \, d\rho$$

$$= \frac{\pi}{2H^2} \int_0^s [2u^{1/2} - 2(H^2 - \rho^2)u^{-1/2}]\Big|_{(H-\rho)^2}^{(H+\rho)^2} \rho \, du \, d\rho$$

$$= \frac{\pi}{H^2} \int_0^s \left[(H+\rho) - (H-\rho) - (H^2 - \rho^2)\left(\frac{1}{H+\rho} - \frac{1}{H-\rho}\right)\right] \rho \, d\rho$$

$$= \frac{\pi}{H^2} \int_0^s 4\rho^2 \, d\rho = \frac{4\pi s^3}{3H^2}.$$

46 By Pappus's theorem, the volume of the torus is $2\pi b(\pi a^2) = 2\pi^2 a^2 b$; therefore the density is $\delta =$

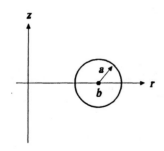

$\dfrac{M}{2\pi^2 a^2 b}$. Recall that a solid of revolution has an equation of the form $z = f(r)$ when expressed in cylindrical coordinates. In the case of the torus, note that every radial cross section is a circle of radius a and center $r = b$, $z = 0$. In the "rz plane" such a circle has the equation $(r - b)^2 + z^2 = a^2$. Using this equation for the torus, we can describe the solid region R it contains by $0 \le \theta \le 2\pi$, $-a \le z \le a$, $b - \sqrt{a^2 - z^2} \le r \le b + \sqrt{a^2 - z^2}$. If we identify L' with the x axis, the desired moment of inertia is I_x. By symmetry, we can compute the result for $z \ge 0$ and double it: $I_x = $

$$\delta \int_R (y^2 + z^2) \, dV =$$

$$2\delta \int_0^a \int_{b-\sqrt{a^2-z^2}}^{b+\sqrt{a^2-z^2}} \int_0^{2\pi} (r^2 \sin^2 \theta + z^2) r \, d\theta \, dr \, dz =$$

$$2\delta \int_0^a \int_{b-\sqrt{a^2-z^2}}^{b+\sqrt{a^2-z^2}} \left[r^3 \left(\frac{\theta}{2} - \frac{\sin 2\theta}{4}\right) + rz^2 \theta\right]\Big|_0^{2\pi} dr \, dz$$

$$= 2\pi\delta \int_0^a \int_{b-\sqrt{a^2-z^2}}^{b+\sqrt{a^2-z^2}} (r^3 + 2rz^2) \, dr \, dz$$

$$= 2\pi\delta \int_0^a \left(\frac{r^4}{4} + r^2 z^2\right)\Big|_{b-\sqrt{a^2-z^2}}^{b+\sqrt{a^2-z^2}} dz$$

$$= 2\pi\delta \int_0^a \left[2b^3\sqrt{a^2 - z^2} + 2b(a^2 - z^2)^{3/2} + 4bz^2\sqrt{a^2 - z^2}\right] dz.$$ Using the trigonometric substitution $z = a \sin \theta$ and Formula 73 from the table of integrals, we have

$$2\pi\delta\left[2b^3 \frac{\pi a^2}{4} + 2b\left(\frac{3\pi a^4}{16}\right) + 4b\left(\frac{\pi a^4}{16}\right)\right]$$

$$= \frac{2\pi M}{2\pi^2 a^2 b}\left(\frac{\pi a^2 b^3}{2} + \frac{5\pi a^4 b}{8}\right) = M\left(\frac{b^2}{2} + \frac{5a^2}{8}\right).$$

48 The area of the region is 5/2, so its density everywhere is $\delta = 2M/5$. The region is described by $0 \leq x \leq 1$, $0 \leq y \leq 2$ and $1 \leq x \leq 2$, $0 \leq y \leq 2 - x$. Hence, the region's moment of inertia about the z axis is $\delta \left[\int_0^1 \int_0^2 (x^2 + y^2)\, dy\, dx \right.$

$\left. + \int_1^2 \int_0^{2-x} (x^2 + y^2)\, dy\, dx \right]$

$= \delta \left[\int_0^1 \left(x^2 y + \frac{1}{3} y^3 \right) \Big|_{y=0}^{y=2} dx \right.$

$\left. + \int_1^2 \left(x^2 y + \frac{1}{3} y^3 \right) \Big|_{y=0}^{y=2-x} dx \right] = \delta \left[\int_0^1 \left(2x^2 + \frac{8}{3} \right) dx \right.$

$\left. + \int_1^2 \left(x^2 (2 - x) + \frac{1}{3}(2 - x)^3 \right) dx \right]$

$= \delta \left[\left(\frac{2}{3} x^3 + \frac{8}{3} x \right) \Big|_0^1 \right.$

$\left. + \frac{1}{3} \int_1^2 (8 - 12x + 12x^2 - 4x^3)\, dx \right]$

$= \frac{2M}{5} \left[\frac{10}{3} + \frac{1}{3}(8x - 6x^2 + 4x^3 - x^4) \Big|_1^2 \right]$

$= \frac{2M}{5} \left[\frac{10}{3} + \frac{1}{3}(8 - 6 \cdot 3 + 4 \cdot 7 - 15) \right] = \frac{26}{15} M$.

50 For the region described by $\alpha \leq \theta \leq \beta$, $0 \leq r \leq f(\theta)$, we have Area $= \int_R 1\, dA$

$= \int_\alpha^\beta \int_0^{f(\theta)} r\, dr\, d\theta = \int_\alpha^\beta \frac{1}{2} r^2 \Big|_0^{f(\theta)} d\theta$

$= \int_\alpha^\beta \frac{1}{2} [f(\theta)]^2\, d\theta.$

52 (a) The area of the region is πr_0^2 and the velocity function is $v(r) = a(1 - r/r_0)^{1/n}$. The average of velocity over the region is $\frac{1}{\pi r_0^2} \int_R v\, dA$

$= \frac{1}{\pi r_0^2} \int_0^{r_0} \int_0^{2\pi} a \left(1 - \frac{r}{r_0} \right)^{1/n} r\, d\theta\, dr =$

$\frac{2\pi}{\pi r_0^2} \int_0^{r_0} \frac{a}{r_0^{1/n}} (r_0 - r)^{1/n} r\, dr.$ Let $s = r_0 - r$,

so the integral becomes

$\frac{2\pi a}{\pi r_0^2 r_0^{1/n}} \int_{r_0}^0 s^{1/n} (r_0 - s)(-ds)$

$= \frac{2a}{r_0^2 r_0^{1/n}} \int_0^{r_0} \left[r_0 s^{1/n} - s^{1 + 1/n} \right] ds$

$= \frac{2a}{r_0^2 r_0^{1/n}} \left[\frac{r_0 s^{1 + 1/n}}{1 + 1/n} - \frac{s^{2 + 1/n}}{2 + 1/n} \right] \Big|_0^{r_0}$

$= 2a \left[\frac{1}{1 + 1/n} - \frac{1}{2 + 1/n} \right]$

$= \frac{2an^2}{(n + 1)(2n + 1)}.$

(b) We average $[v(r)]^2 = a^2 (1 - r/r_0)^{2/n}$ using the same substitution as in (a):

$\frac{1}{\pi r_0^2} \int_0^{r_0} \int_0^{2\pi} a^2 \left(1 - \frac{r}{r_0} \right)^{2/n} r\, d\theta\, dr$

$= \frac{2a^2}{r_0^2 r_0^{2/n}} \int_{r_0}^0 s^{2/n} (r_0 - s)(-ds)$

$= \frac{2a^2}{r_0^{2(1 + 1/n)}} \int_0^{r_0} \left[r_0 s^{2/n} - s^{1 + 2/n} \right] ds$

$= \frac{2a^2}{r_0^{2(1 + 1/n)}} \left[\frac{r_0 s^{1 + 2/n}}{1 + 2/n} - \frac{s^{2 + 2/n}}{2 + 2/n} \right] \Big|_0^{r_0}$

$$= 2a^2 \left[\frac{1}{1 + 2/n} - \frac{1}{2 + 2/n} \right]$$

$$= \frac{a^2 n^2}{(n+1)(n+2)}.$$

54 If $H < a$, then the integral $\int_R \frac{1}{q} \, dV =$

$$\int_0^a \int_0^\pi \int_0^{2\pi} \frac{\rho^2 \sin\phi}{\sqrt{H^2 + \rho^2 - 2\rho H \cos\phi}} \, d\theta \, d\phi \, d\rho$$

$$= \int_0^a \int_0^\pi \frac{2\pi \rho^2 \sin\phi}{\sqrt{H^2 + \rho^2 - 2\rho H \cos\phi}} \, d\phi \, d\rho =$$

$$\int_0^a 2\pi \frac{\rho}{H} \left(\sqrt{H^2 + \rho^2 + 2\rho H} - \sqrt{H^2 + \rho^2 - 2\rho H} \right) d\rho$$

needs to be broken into two parts: one in which $0 \le \rho \le H$ and one in which $H \le \rho \le a$. In the first case, $H - \rho > 0$ as before and we obtain

$$\int_0^H 2\pi \frac{\rho}{H} [(H+\rho) - (H-\rho)] \, d\rho = \int_0^H \frac{4\pi}{H} \rho^2 \, d\rho$$

$$= \frac{4\pi \rho^3}{3H} \bigg|_0^H = \frac{4\pi H^2}{3}.$$ In the second, $H - \rho < 0$,

so we have $\int_H^a 2\pi \frac{\rho}{H} [(H+\rho) - (\rho - H)] \, d\rho =$

$\int_H^a 4\pi \rho \, d\rho = 2\pi \rho^2 \big|_H^a = 2\pi(a^2 - H^2)$. Adding the results together to obtain the total value of the integral, we have $2\pi \left(a^2 - \frac{1}{3} H^2 \right)$. Multiply by the density δ to obtain $\frac{3M}{4\pi a^3} 2\pi \left(a^2 - \frac{1}{3} H^2 \right) =$

$\frac{M}{2a} \left(3 - \frac{H^2}{a^2} \right)$. (Note that this reduces to M/a when $H = a$, as it should.)

16 Green's Theorem

16.1 Vector and Scalar Fields

2 $F = 3y\mathbf{i} + 2x^{10}\mathbf{j}$; $\nabla \cdot F = \frac{\partial}{\partial x}(3y) + \frac{\partial}{\partial y}(2x^{10})$
$= 0 + 0 = 0$

4 $\mathbf{r} = x\mathbf{i} + y\mathbf{j} + z\mathbf{k}$; $\nabla \cdot \mathbf{r} = \frac{\partial}{\partial x}(x) + \frac{\partial}{\partial y}(y) + \frac{\partial}{\partial z}(z)$
$= 1 + 1 + 1 = 3$

6 $F = z(\sec 2xy)\mathbf{i} + \frac{1}{\sqrt{x^2 + y^2}}\mathbf{j} + \ln(\tan^{-1} 2z)\mathbf{k}$;

$\nabla \cdot F = \frac{\partial}{\partial x}(z \sec 2xy) + \frac{\partial}{\partial y}\left(\frac{1}{\sqrt{x^2 + y^2}}\right) +$

$\frac{\partial}{\partial z}[\ln(\tan^{-1} 2z)] = z \sec 2xy \tan 2xy \cdot 2y +$

$\left(-\frac{1}{2}(x^2 + y^2)^{-3/2} \cdot 2y\right) + \frac{1}{\tan^{-1} 2z} \cdot \frac{2}{1 + (2z)^2} = 2yz$

$\sec 2xy \tan 2xy - y(x^2 + y^2)^{-3/2} +$

$\frac{2}{(\tan^{-1} 2z)(1 + 4z^2)}$

10 $F = xy\mathbf{i} + \cos(x + 2y + z)\mathbf{j} + z^2\mathbf{k}$; $\nabla \times F =$

$\begin{vmatrix} \mathbf{i} & \mathbf{j} & \mathbf{k} \\ \frac{\partial}{\partial x} & \frac{\partial}{\partial y} & \frac{\partial}{\partial z} \\ xy & \cos(x+y+z) & z^2 \end{vmatrix} = [0 - (-\sin(x + 2y + z))]\mathbf{i}$

$+ [0 - 0]\mathbf{j} + [-\sin(x + 2y + z) - x]\mathbf{k}$
$= \sin(x + 2y + z)\mathbf{i} - [\sin(x + 2y + z) + x]\mathbf{k}$

12 Let $F = \frac{\hat{\mathbf{r}}}{|\mathbf{r}|^2} = \frac{\mathbf{r}}{|\mathbf{r}|^3}$. Define $r = \|\mathbf{r}\| =$

$\sqrt{x^2 + y^2 + z^2}$. Then $F = \frac{x\mathbf{i} + y\mathbf{j} + z\mathbf{k}}{\left(\sqrt{x^2 + y^2 + z^2}\right)^3} =$

$\frac{x}{r^3}\mathbf{i} + \frac{y}{r^3}\mathbf{j} + \frac{z}{r^3}\mathbf{k}$ and $\nabla \times F =$

$\left[\frac{\partial}{\partial y}\left(\frac{z}{r^3}\right) - \frac{\partial}{\partial z}\left(\frac{y}{r^3}\right)\right]\mathbf{i} + \left[\frac{\partial}{\partial z}\left(\frac{x}{r^3}\right) - \frac{\partial}{\partial x}\left(\frac{z}{r^3}\right)\right]\mathbf{j} +$

$\left[\frac{\partial}{\partial x}\left(\frac{y}{r^3}\right) - \frac{\partial}{\partial y}\left(\frac{x}{r^3}\right)\right]\mathbf{k}$. Now $\frac{\partial}{\partial x}\left(\frac{1}{r^3}\right) =$

$\frac{\partial}{\partial x}[(x^2 + y^2 + z^2)^{-3/2}] = -\frac{3}{2}(x^2 + y^2 + z^2)^{-5/2} \cdot 2x =$

$-\frac{3x}{(x^2 + y^2 + z^2)^{5/2}} = -\frac{3x}{r^5}$. Similarly, $\frac{\partial}{\partial y}\left(\frac{1}{r^3}\right) =$

$-\frac{3y}{r^5}$ and $\frac{\partial}{\partial z}\left(\frac{1}{r^3}\right) = -\frac{3z}{r^5}$. Thus $\nabla \times F =$

$\left[z\left(-\frac{3y}{r^5}\right) - y\left(-\frac{3z}{r^5}\right)\right]\mathbf{i} + \left[x\left(-\frac{3z}{r^5}\right) - z\left(-\frac{3x}{r^5}\right)\right]\mathbf{j} +$

$\left[y\left(-\frac{3x}{r^5}\right) - x\left(-\frac{3y}{r^5}\right)\right]\mathbf{k} = 0.$

14 Let $F = |\mathbf{r}|^2\hat{\mathbf{r}} = \|\mathbf{r}\|\mathbf{r}$. Define $r = \|\mathbf{r}\| =$

$\sqrt{x^2 + y^2 + z^2}$. Then $F = rx\mathbf{i} + ry\mathbf{j} + rz\mathbf{k}$ and

$\nabla \times F = \left[\frac{\partial}{\partial y}(rz) - \frac{\partial}{\partial z}(ry)\right]\mathbf{i} +$

$\left[\frac{\partial}{\partial z}(rx) - \frac{\partial}{\partial x}(rz)\right]\mathbf{j} + \left[\frac{\partial}{\partial x}(ry) - \frac{\partial}{\partial y}(rx)\right]\mathbf{k}$. Now

$\frac{\partial}{\partial x}(r) = \frac{\partial}{\partial x}[(x^2 + y^2 + z^2)^{1/2}] =$

$\frac{1}{2}(x^2 + y^2 + z^2)^{-1/2} \cdot 2x = \frac{x}{r}$. Similarly, $\frac{\partial}{\partial y}(r) =$

$\frac{y}{r}$ and $\frac{\partial}{\partial z}(r) = \frac{z}{r}$. Thus $\nabla \times \mathbf{F} = \left[z \cdot \frac{y}{r} - y \cdot \frac{z}{r}\right]\mathbf{i}$

$+ \left[x \cdot \frac{z}{r} - z \cdot \frac{x}{r}\right]\mathbf{j} + \left[y \cdot \frac{x}{r} - x \cdot \frac{y}{r}\right]\mathbf{k} = 0$.

16 $\mathbf{F} \cdot \mathbf{G}$ is a scalar field.

18 $\nabla \times \mathbf{F} = (R_y - Q_z)\mathbf{i} + (P_z - R_x)\mathbf{j} + (Q_x - P_y)\mathbf{k}$,
so $\nabla \cdot (\nabla \times \mathbf{F})$

$= \frac{\partial}{\partial x}(R_y - Q_z) + \frac{\partial}{\partial y}(P_z - R_x) + \frac{\partial}{\partial z}(Q_x - P_y)$

$= R_{yx} - Q_{zx} + P_{zy} - R_{xy} + Q_{xz} - P_{yz} = 0$.

20 (a), (b), and (c) make sense. (d) and (e) do not.

22 $f(x, y, z) = \ln(x^2 + y^2 + z^2)$

$\nabla^2 f = f_{xx} + f_{yy} + f_{zz} =$

$\frac{2y^2 + 2z^2 - 2x^2}{(x^2 + y^2 + z^2)^2} + \frac{2x^2 + 2z^2 - 2y^2}{(x^2 + y^2 + z^2)^2} + \frac{2x^2 + 2y^2 - 2z^2}{(x^2 + y^2 + z^2)^2}$

$= \frac{2}{x^2 + y^2 + z^2} \neq 0$

24 $f = \frac{1}{\sqrt{x^2 + y^2}} = \frac{1}{r}$; $\nabla^2 f = f_{xx} + f_{yy} =$

$\left(\frac{3x^2}{r^5} - \frac{1}{r^3}\right) + \left(\frac{3y^2}{r^5} - \frac{1}{r^3}\right) = \frac{3r^2}{r^5} - \frac{2}{r^3} = \frac{1}{r^3}$

$\neq 0$.

26 (a) Let $\mathbf{F} = P\mathbf{i} + Q\mathbf{j}$. Then $\nabla \cdot (f\mathbf{F}) =$

$\nabla \cdot (fP\mathbf{i} + fQ\mathbf{j}) = \frac{\partial}{\partial x}(fP) + \frac{\partial}{\partial y}(fQ)$

$= f\frac{\partial P}{\partial x} + P\frac{\partial f}{\partial x} + f\frac{\partial Q}{\partial y} + Q\frac{\partial f}{\partial y}$

$= f\left(\frac{\partial P}{\partial x} + \frac{\partial Q}{\partial y}\right) + \left(P\frac{\partial f}{\partial x} + Q\frac{\partial f}{\partial y}\right)$

$= f\nabla \cdot \mathbf{F} + \nabla f \cdot \mathbf{F}$.

28 $\nabla \times f\mathbf{F} = \left(\frac{\partial}{\partial y}(fR) - \frac{\partial}{\partial z}(fQ)\right)\mathbf{i} +$

$\left(\frac{\partial}{\partial z}(fP) - \frac{\partial}{\partial x}(fR)\right)\mathbf{j} + \left(\frac{\partial}{\partial x}(fQ) - \frac{\partial}{\partial y}(fP)\right)\mathbf{k}$

$= (fR_y + f_y R - fQ_z - f_z Q)\mathbf{i}$
$+ (fP_z + f_z P - fR_x - f_x R)\mathbf{j}$
$+ (fQ_x + f_x Q - fP_y - f_y P)\mathbf{k}$
$= f[(R_y - Q_z)\mathbf{i} + (P_z - R_x)\mathbf{j} + (Q_x - P_y)\mathbf{k}] +$
$[(f_y R - f_z Q)\mathbf{i} + (f_z P - f_x R)\mathbf{j} + (f_x Q - f_y P)\mathbf{k}]$
$= f\nabla \times \mathbf{F} + \nabla f \times \mathbf{F}$

30 $\nabla f \times \nabla g = (f_x\mathbf{i} + f_y\mathbf{j} + f_z\mathbf{k}) \times (g_x\mathbf{i} + g_y\mathbf{j} + g_z\mathbf{k})$
$= (f_y g_z - f_z g_y)\mathbf{i} - (f_x g_z - f_z g_x)\mathbf{j} + (f_x g_y - f_y g_x)\mathbf{k}$.
Thus $\text{div}(\nabla f \times \nabla g) =$

$\frac{\partial}{\partial x}(f_y g_z - f_z g_y) - \frac{\partial}{\partial y}(f_x g_z - f_z g_x) + \frac{\partial}{\partial z}(f_x g_y - f_y g_x)$

$= 0$.

32 $\mathbf{F} = \frac{x}{(x^2 + y^2)^{(k+1)/2}}\mathbf{i} + \frac{y}{(x^2 + y^2)^{(k+1)/2}}\mathbf{j}$. $\nabla \cdot \mathbf{F} =$

$\frac{(x^2 + y^2)^{(k+1)/2} - (k + 1)x^2(x^2 + y^2)^{(k-1)/2}}{(x^2 + y^2)^{k+1}} +$

$\frac{(x^2 + y^2)^{(k+1)/2} - (k + 1)y^2(x^2 + y^2)^{(k-1)/2}}{(x^2 + y^2)^{k+1}}$

$= \frac{2(x^2 + y^2)^{(k+1)/2} - (k + 1)(x^2 + y^2)^{(k+1)/2}}{(x^2 + y^2)^{k+1}}$

$= \frac{2 - (k + 1)}{(x^2 + y^2)^{(k+1)/2}}$.

16.1 Vector and Scalar Fields

(a) If $k = 1$, $k + 1 = 2$ and $\nabla \cdot \mathbf{F} = 0$.

(b) If $k \neq 1$, $k + 1 \neq 2$ and $\nabla \cdot \mathbf{F} \neq 0$.

34 (a) $\mathbf{F} = f(r)\hat{\mathbf{r}} = kr^{-2}\dfrac{\mathbf{r}}{|\mathbf{r}|} = \dfrac{k}{r^3}\mathbf{r}$; $\nabla \cdot \mathbf{F} =$

$\dfrac{\partial}{\partial x}\left(\dfrac{kx}{r^3}\right) + \dfrac{\partial}{\partial y}\left(\dfrac{ky}{r^3}\right) + \dfrac{\partial}{\partial z}\left(\dfrac{kz}{r^3}\right) =$

$k\left(\dfrac{1}{r^3} - \dfrac{3x^2}{r^5}\right) + k\left(\dfrac{1}{r^3} - \dfrac{3y^2}{r^5}\right) + k\left(\dfrac{1}{r^3} - \dfrac{3z^2}{r^5}\right)$

$= 0$

(b) $r = \sqrt{x^2 + y^2 + z^2}$ and $r_x = \dfrac{x}{r}$, so $\left(\dfrac{x}{r}\right)_x =$

$\dfrac{1}{r} - \dfrac{x^2}{r^3}$. Similar results hold with y or z in

place of x. Since $\mathbf{F} = f(r)\hat{\mathbf{r}} = \dfrac{f(r)}{r}\mathbf{r} =$

$f(r)\dfrac{x}{r}\mathbf{i} + f(r)\dfrac{y}{r}\mathbf{j} + f(r)\dfrac{z}{r}\mathbf{k}$, we have $\nabla \cdot \mathbf{F} =$

$\left(\dfrac{f(r)}{r}x\right)_x + \left(\dfrac{f(r)}{r}y\right)_y + \left(\dfrac{f(r)}{r}z\right)_z = f(r)\left(\dfrac{x}{r}\right)_x +$

$\dfrac{x}{r}f'(r)r_x + f(r)\left(\dfrac{y}{r}\right)_y + \dfrac{y}{r}f'(r)r_y + f(r)\left(\dfrac{z}{r}\right)_z$

$+ \dfrac{z}{r}f'(r)r_z = \dfrac{2}{r}f(r) + f'(r)$. We are given

that $\nabla \cdot \mathbf{F} = 0$; so $f'(r) = -\dfrac{2}{r}f(r)$, $\dfrac{f'(r)}{f(r)} =$

$-\dfrac{2}{r}$, $\ln(f(r)) = -2\ln r + C$, $f(r) =$

$e^{-2\ln r + C} = e^C e^{-2\ln r} = kr^{-2}$.

36 We consider only (a); the treatment of (b) is

similar. $\mathbf{curl}\, f(r)\hat{\mathbf{r}} = \mathbf{curl}\, \dfrac{f(r)}{r}\mathbf{r}$

$= \mathbf{curl}\left(f(r)\dfrac{x}{r}\mathbf{i} + f(r)\dfrac{y}{r}\mathbf{j}\right) = \left[\left(f(r)\dfrac{y}{r}\right)_x - \left(f(r)\dfrac{x}{r}\right)_y\right]\mathbf{k}$

$= \left[y \cdot \dfrac{f'(r)x - f(r)\dfrac{x}{r}}{r^2} - x \cdot \dfrac{f'(r)y - f(r)\dfrac{y}{r}}{r^2}\right]\mathbf{k} = \mathbf{0}$.

16.2 Line Integrals

2 The parameterization is $x = 1 - t$, $y = (1 - t)^2$; thus $\mathbf{G}(t)$ describes a portion of the curve $y = x^2$. It starts at $(1, 1)$ when $t = 0$ and ends at $(0, 0)$ when $t = 1$.

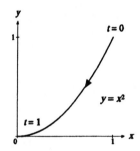

4 The parameterization is $x = 4\cos t$, $y = 5\sin t$. Since $\dfrac{x^2}{16} + \dfrac{y^2}{25} = \cos^2 t + \sin^2 t = 1$, $\mathbf{G}(t)$

describes a portion of the ellipse $\dfrac{x^2}{16} + \dfrac{y^2}{25} = 1$. It

starts at $(4, 0)$ when $t = 0$ and ends at $(0, 5)$ when $t = \pi/2$.

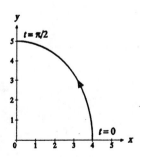

6 The parameterization is $(x, y) = (t + 1, t + 1)$ for $0 \leq t \leq 1$ and $(t + 1, -t + 3)$ for $1 \leq t \leq 2$.

8 A clockwise-oriented parameterization of a circle of radius 2 is $x = 2\cos(\pi - t)$, $y = 2\sin(\pi - t)$. For the given semicircle, the start point $(-2, 0)$ occurs at $t = 0$; the finish point $(2, 0)$ occurs when $t = \pi$. The entire parameterization is $(x, y) = (2\cos(\pi - t), 2\sin(\pi - t))$ for $0 \leq t \leq \pi$.

10 The curve C can be parameterized by $x = 2$ and $y = t$ for $0 \leq t \leq 5$. Thus $\int_C x^2 \, dy =$

$$\int_0^5 x^2 \frac{dy}{dt} \, dt = \int_0^5 2^2 \cdot 1 \, dt = 5t \big|_0^5 = 25.$$

12 The curve C can be parameterized by $x = 1 - t$, $y = t$; $0 \leq t \leq 1$. Hence $\int_C (xy \, dx + x^2 \, dy)$

$$= \int_0^1 \left(xy \frac{dx}{dt} + x^2 \frac{dy}{dt} \right) dt$$

$$= \int_0^1 \left[(1-t)t \cdot (-1) + (1-t)^2 \cdot 1 \right] dt$$

$$= \int_0^1 (2t^2 - 3t + 1) \, dt = \frac{1}{6}.$$

14 (a)

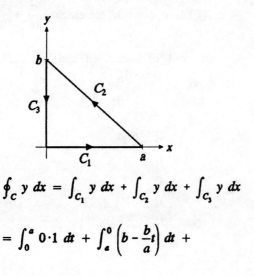

$\oint_C y \, dx = \int_{C_1} y \, dx + \int_{C_2} y \, dx + \int_{C_3} y \, dx$

$= \int_0^a 0 \cdot 1 \, dt + \int_a^0 \left(b - \frac{b}{a} t \right) dt +$

$\int_0^b (b - t) \cdot 0 \, dt = -\frac{ab}{2}$

$= -(\text{Area of triangle})$

(b) $\oint_C y \, dx = \int_0^{2\pi} y \frac{dx}{dt} \, dt$

$= \int_0^{2\pi} (a \sin t)(-a \sin t) \, dt$

$= -a^2 \int_0^{2\pi} \sin^2 t \, dt = -\pi a^2$

$= -(\text{Area of circle})$

16 Divide C into C_1 and C_2 as in the figure. Then C_1 can be parameterized by $x = t$, $y = f(t)$; $a \leq t \leq$

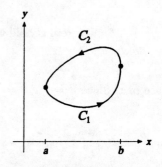

b, and C_2 can be parameterized by $x = b + a - t$, $y = g(t)$; $a \leq t \leq b$. Hence $\oint_C k \, dx$

$= \int_{C_1} k \, dx + \int_{C_2} k \, dx$

$= \int_a^b k \frac{dx}{dt} \, dt + \int_a^b k \frac{dx}{dt} \, dt$

$= \int_a^b k \cdot 1 \, dt + \int_a^b k \cdot (-1) \, dt = 0.$

18 $\oint_C (2y \, dx + 6x \, dy) = 2 \oint_C y \, dx + 6 \oint_C x \, dy$

$= 2(-5) + 6(5) = 20$

20 $\oint_C [(x + 2y + 3) \, dx + (2x - 3y + 4) \, dy] = \oint_C x \, dx$

$+ 2 \oint_C y \, dx + \oint_C 3 \, dx + 2 \oint_C x \, dy -$

16.2 Line Integrals

$3\oint_C y\,dy + \oint_C 4\,dy$

$= 0 + 2(-5) + 0 + 2\cdot 5 - 3\cdot 0 + 0 = 0$

22. Any curve from $(-1, 2)$ to $(3, 5)$ can be parameterized by $x = g(t)$, $y = t$; $2 \leq t \leq 5$ for some $g(t)$, where $g(2) = -1$ and $g(5) = 3$. Then

$\int_C (y + 2)\,dy = \int_2^5 (y + 2)\dfrac{dy}{dt}\,dt$

$= \int_2^5 (t + 2)\cdot 1\,dt = \dfrac{t^2}{2} + 2t\Big|_2^5 = \dfrac{33}{2}.$

24. (a) The line through the points $(0, 0)$ and $(\pi/2, 1)$ is given by $y = \dfrac{2}{\pi}x$. Hence C_1 is parameterized by $x = t$, $y = \dfrac{2}{\pi}t$, $0 \leq t \leq \pi/2$. So $\int_{C_1} x\,dy = \int_0^{\pi/2} x\dfrac{dy}{dt}\,dt =$

$\int_0^{\pi/2} t\dfrac{2}{\pi}\,dt = \dfrac{2}{\pi}\cdot\dfrac{t^2}{2}\Big|_0^{\pi/2} = \dfrac{\pi}{16}.$

(b) Along $y = \sin x$, the path C_2 is parameterized by $x = t$, $y = \sin t$; $0 \leq t \leq \pi/2$. Then

$\int_{C_2} x\,dy = \int_0^{\pi/2} x\dfrac{dy}{dt}\,dt = \int_0^{\pi/2} t\cos t\,dt$

$= t\sin t + \cos t\big|_0^{\pi/2} = \dfrac{\pi}{2} - 1.$

26. $\oint_C f\,dy$

$= -\left(\oint_{C_1} f\,dy + \oint_{C_2} f\,dy + \oint_{C_3} f\,dy + \oint_{C_4} f\,dy\right) =$

$-(1 + 2 + 4 + (-8)) = 1$, by the cancellation principle.

28. $dx = -a\sin t\,dt$, $dy = b\cos t\,dt$, $0 \leq t \leq 2\pi$.

$\oint_C -y\,dx + x\,dy$

$= \int_0^{2\pi} [(-b\sin t)(-a\sin t) + (a\cos t)(b\cos t)]\,dt$

$= \int_0^{2\pi} ab\,dt = 2\pi ab$

30. $dx = -\sin t\,dt$, $dy = \cos t\,dt$, $dz = -\sin t\,dt$,

$\dfrac{\pi}{6} \leq t \leq \dfrac{\pi}{3}$, so $\int_C \dfrac{x\,dx}{yz} + \dfrac{y\,dy}{xz} + \dfrac{8z\,dz}{xy}$

$= \int_{\pi/6}^{\pi/3}\left[\dfrac{\cos t}{\sin t\cos t}(-\sin t) + \dfrac{\sin t}{(\cos t)^2}(\cos t) + \dfrac{8\cos t}{\cos t\sin t}(-\sin t)\right]dt = \int_{\pi/6}^{\pi/3}(\tan t - 9)\,dt$

$= \dfrac{1}{2}(\ln 3 - 3\pi).$

32. The path of integration lies on the cubic $y = x^3$, so $ds = \sqrt{1 + (dy/dx)^2}\,dx = \sqrt{1 + (3x^2)^2}\,dx = \sqrt{1 + 9x^4}\,dx$. The path goes from $(1, 1)$ to $(2, 8)$, so x varies from 1 to 2. The density $f(P)$ equals y, so Mass $= \int_C f(P)\,ds = \int_1^2 x^3\sqrt{1 + 9x^4}\,dx$

$= \dfrac{1}{36}\cdot\dfrac{(1 + 9x^4)^{3/2}}{3/2}\Big|_1^2$

$= \dfrac{1}{54}[(1 + 9\cdot 16)^{3/2} - (1 + 9\cdot 1)^{3/2}]$

$= \dfrac{1}{54}(145^{3/2} - 10^{3/2}).$

34. The path of integration is the cardioid $r = 1 + \cos\theta$ as θ goes from 0 to π. Since $r'(\theta) = -\sin\theta$ and $ds = \sqrt{r^2 + (r')^2}\,d\theta$, we have $ds =$

$\sqrt{1 + 2\cos\theta + \cos^2\theta + \sin^2\theta}\, d\theta =$

$\sqrt{2 + 2\cos\theta}\, d\theta$. The function to be integrated is the height $\sin\theta$, so $\int_C f(P)\, ds$

$= \int_0^\pi \sin\theta \sqrt{2 + 2\cos\theta}\, d\theta$

$= -\dfrac{1}{2} \cdot \dfrac{(2 + 2\cos\theta)^{3/2}}{3/2} \bigg|_0^\pi = -\dfrac{1}{3}(0^{3/2} - 4^{3/2})$

$= -\dfrac{1}{3}(-8) = \dfrac{8}{3}.$

36. $\dfrac{1}{2}\int_C |\mathbf{r} \times \mathbf{T}|\, ds$ is the area swept out by $\mathbf{r}$. (See Exercise 2 of Sec. 13.5.)

38. (a) Label the segments of C as shown in the diagram. On C_1, $y = a - x$ as x goes from a

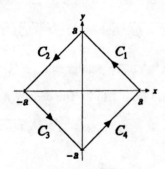

to 0, so $dy = -dx$. On C_2, $y = x + a$ as x goes from 0 to $-a$, so $dy = dx$. On C_3, $y = -(x + a)$ as x goes from $-a$ to 0, so $dy = -dx$. On C_4, $y = x - a$ as x ranges from 0 to a, so $dy = dx$. We therefore have

$\int_{C_1} -y\, dx + x\, dy = \int_a^0 -(a - x)\, dx + x(-dx)$

$= \int_a^0 (-a)\, dx = -ax\big|_a^0 = a^2;$

$\int_{C_2} -y\, dx + x\, dy = \int_0^{-a} -(x + a)\, dx + x\, dx$

$= \int_0^{-a} (-a)\, dx = -ax\big|_0^{-a} = a^2;$

$\int_{C_3} -y\, dx + x\, dy = \int_{-a}^0 (x + a)\, dx + x(-dx)$

$= \int_{-a}^0 a\, dx = ax\big|_{-a}^0 = a^2;\ \int_{C_4} -y\, dx + x\, dy$

$= \int_0^a -(x - a)\, dx + x\, dx = \int_0^a a\, dx$

$= ax\big|_0^a = a^2.$ $\oint_C -y\, dx + x\, dy$ is equal to the sum of the four integrals above, which is $4a^2$. Since the square bounded by C has side of length $\sqrt{2}a$, its area is $2a^2$, which, as claimed, is half the integral.

(b) Label the segments of C as shown in the diagram. On C_1, $y = b\left(1 - \dfrac{x}{a}\right)$ as x goes

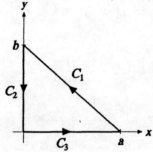

from a to 0, so $dy = -\dfrac{b}{a}\, dx$. On C_2, $x = 0$ as y goes from b to 0, so $dx = 0$. Finally, on C_3, $y = 0$ as x goes from 0 to a, so $dy = 0$. Thus $\int_{C_1} -y\, dx + x\, dy =$

$\int_a^0 -b\left(1 - \dfrac{x}{a}\right) dx + x\left(-\dfrac{b}{a}\, dx\right) =$

$\int_a^0 (-b)\, dx = -bx\big|_a^0 = ab;$

$\int_{C_2} -y\, dx + x\, dy = \int_b^0 0\, dy = 0;$

16.2 Line Integrals

$\int_{C_3} -y \, dx + x \, dy = \int_0^a -0 \, dx = 0.$

$\oint_C -y \, dx + x \, dy$ is equal to the sum of the three integrals above, which is ab. The area of the triangle bounded by C is $\frac{1}{2}ab$, so once again the integral is twice the area.

40 $dV = \frac{\pi}{4}\left(\frac{16-z}{32}\right)^2 ds; \, dM = 487\left(\frac{16-z}{32}\right)^2 \frac{\pi}{4} ds;$

Mass $= \int_C dM = \frac{487\pi}{4} \int_C \left(\frac{16-z}{32}\right)^2 ds =$

$\frac{487\pi}{4} \int_0^{8/3} \left(\frac{16-6t}{32}\right)^2 \sqrt{\left(\frac{dx}{dt}\right)^2 + \left(\frac{dy}{dt}\right)^2 + \left(\frac{dz}{dt}\right)^2} \, dt =$

$\frac{487\pi}{4} \int_0^{8/3} \left(\frac{16-6t}{32}\right)^2 \sqrt{(-8\sin 4t)^2 + (8\cos 4t)^2 + 6^2} \, dt$

$= \frac{2435\pi}{2048} \int_0^{8/3} (256 - 192t + 36t^2) \, dt = \frac{2435\pi}{9}$

≈ 850 lb.

16.3 Four Applications of Line Integrals

2 We have $\int_C \mathbf{F} \cdot d\mathbf{r} = \int_C (x^2\mathbf{i} + 2xy\mathbf{j}) \cdot (dx\mathbf{i} + dy\mathbf{j})$

$= \int_C x^2 \, dx + 2xy \, dy.$ From $x = 2t^2$, $y = 3t^2$ we know $dx = 4t \, dt$, $dy = 9t \, dt$. Hence

$\int_C x^2 \, dx + 2xy \, dy$

$= \int_1^2 (2t^2)^2(4t \, dt) + 2(2t^2)(3t^2)(9t \, dt)$

$= \int_1^2 124t^5 \, dt = \frac{62}{3}t^6 \Big|_1^2 = 1302.$

4 We have $\int_C \mathbf{F} \cdot d\mathbf{r} = \int_C x^2 \, dx + xy \, dy + 3 \, dz$

$= \int_1^2 (2t)^2(2 \, dt) + 2t(3t+1)(3 \, dt) + 3(1 \, dt)$

$= \int_1^2 (26t^2 + 6t + 3) \, dt = \frac{26t^3}{3} + 3t^2 + 3t \Big|_1^2$

$= \frac{218}{3}.$

6 We have $\int_C \mathbf{F} \cdot d\mathbf{r} = \int_C \mathbf{r} \cdot d\mathbf{r}$

$= \int_C x \, dx + y \, dy + z \, dz$

$= \int_0^{2\pi} \cos\theta \, (-\sin\theta \, d\theta) + \sin\theta \, (\cos\theta \, d\theta) = 0.$

8 Work $= \int_C \mathbf{F} \cdot \mathbf{T} \, ds = \int_C x^2y \, dx + y \, dy$

$= \int_C x^2y \, dx + \int_C y \, dy = \int_0^2 2x^3 \, dx + \int_0^4 y \, dy$

$= \frac{x^4}{2}\Big|_0^2 + \frac{y^2}{2}\Big|_0^4 = 8 + 8 = 16$

10 Parameterize the three parts of the path as follows: along C_1, $(x, y) = (t, 0)$ with t in $[0, 2]$; along C_2,

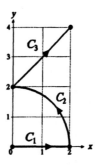

$(x, y) = (2\cos t, 2\sin t)$ with t in $[0, \pi/2]$; and along C_3, $(x, y) = (t, t+2)$ for t in $[0, 2]$. So

$\int_C \mathbf{F} \cdot d\mathbf{r} = \int_C x^2y \, dx + y \, dy =$

$\int_{C_1} x^2y \, dx + y \, dy + \int_{C_2} x^2y \, dx + y \, dy +$

$$\int_{C_3} x^2 y\, dx + y\, dy = 0 +$$

$$\int_0^{\pi/2} 8\cos^2 t \sin t\, (-2\sin t\, dt) + 2\sin t\, (2\cos t\, dt)$$

$$+ \int_0^2 (t^3 + 2t^2)\, dt + (t+2)\, dt$$

$$= \int_0^{\pi/2} (-16\cos^2 t \sin^2 t + 4\cos t \sin t)\, dt +$$

$$\int_0^2 (t^3 + 2t^2 + t + 2)\, dt = (-\pi + 2) + \frac{46}{3}$$

$$= \frac{-3\pi + 52}{3}.$$

12. (a) Let C be the path from $(1, 0)$ to $(b, 0)$ parameterized by $(x, y) = (t, 0)$ for t in $[1, b]$. Then $W(b) = \int_C \mathbf{F}\cdot d\mathbf{r} =$

$$\int_C \frac{-x\, dx - y\, dy}{(x^2 + y^2)^{3/2}} = \int_1^b \frac{-t\, dt}{(t^2)^{3/2}}$$

$$= \int_1^b \left(-\frac{1}{t^2}\right) dt = \frac{1}{t}\bigg|_1^b = \frac{1}{b} - 1.$$

(b) From (a), $\displaystyle\lim_{b\to\infty}\left(\frac{1}{b} - 1\right) = -1$.

14. (a)

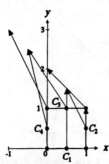

(b) Fluid is neither leaving nor entering; net outward flow is zero.

(c) Let C be parameterized as follows: along C_1, $(x, y) = (t, 0)$ with t in $[0, 1]$; along C_2, $(x, y) = (1, t)$ with t in $[0, 1]$; along C_3, $(x, y) = (1 - t, 1)$ with t in $[0, 1]$; and along C_4, $(x, y) = (0, 1 - t)$ with t in $[0, 1]$. Then

$$\oint_C \mathbf{F}\cdot\mathbf{n}\, ds = \oint_C -y\, dx + (2 - x)\, dy$$

$$= \int_{C_1} -y\, dx + (2 - x)\, dy +$$

$$\int_{C_2} -y\, dx + (2 - x)\, dy +$$

$$\int_{C_3} -y\, dx + (2 - x)\, dy +$$

$$\int_{C_4} -y\, dx + (2 - x)\, dy$$

$$= \int_0^1 0\, dt + \int_0^1 dt + \int_0^1 dt + \int_0^1 (-2\, dt)$$

$$= 0.$$

20. $\oint_C \mathbf{F}\cdot\mathbf{n}\, ds$ represents the net loss of fluid contained in C. Since $\oint_C \mathbf{F}\cdot\mathbf{n}\, ds > 0$, this loss is indeed positive, and fluid is tending to leave the region bounded by C.

22. We see that the angle is $\pi/2$. By (2), $\int_C \dfrac{\mathbf{n}\cdot\hat{\mathbf{r}}}{|\mathbf{r}|}\, ds$

$$= \int_0^1 \frac{\frac{1}{\sqrt{2}}(\mathbf{i} + \mathbf{j})\cdot[(1 - y)\mathbf{i} + y\mathbf{j}]}{(1 - y)^2 + y^2}\, \sqrt{2}\, dy$$

$$= \int_0^1 \frac{dy}{2y^2 - 2y + 1} = \tan^{-1}(2y - 1)\bigg|_0^1 = \frac{\pi}{2},$$ so (2) is verified.

24. Let C be parameterized by $(x, y, z) = (x(t), y(t), z(t))$ for $t_0 \leq t \leq t_1$. So C^* is parameterized by $(x, y) = (x(t), y(t))$ for $t_0 \leq t \leq t_1$. The arc length of C is $\displaystyle\int_C ds = \int_{t_0}^{t_1} \sqrt{\left(\frac{dx}{dt}\right)^2 + \left(\frac{dy}{dt}\right)^2 + \left(\frac{dz}{dt}\right)^2}\, dt$.

16.3 Four Applications of Line Integrals

Letting $ds^* = \sqrt{\left(\dfrac{dx}{dt}\right)^2 + \left(\dfrac{dy}{dt}\right)^2}\, dt$, we see that the arc length of C^* is given by $\int_C ds^*$ (since C^* is simply the projection of C in the $z = 0$ plane).

26 Let C be parameterized by $(x, y) = (x(t), y(t))$ for $t_0 \leq t \leq t_1$ with $x(t_0) = x(t_1)$ and $y(t_0) = y(t_1)$, and assume C is swept out counterclockwise. Let $(x(a), y(a))$ and $(x(b), y(b))$ be the points on C such that a line through 0 is tangent to C (there are exactly two since C is convex). Let θ be the angle between these two lines. As C moves from $(x(a), y(a))$ to $(x(b), y(b))$, the angle subtended is $\pm\theta$, while for the remainder of C the angle subtended is $\mp\theta$. Thus the total angle subtended by C is 0.

28 Label the edges of the rectangle counterclockwise as C_1 through C_4, starting with the right edge. The vectors on C_1 are perpendicular to the edge and of approximate length 5. Similarly, the vectors on C_2 are perpendicular to it and of length 3. On C_3 the vectors are of length 12 and appear to make a 45° angle with the edge. On C_4 the vectors are of length 13 and make an angle of 30°. (All of these numbers are just approximations.)

(a) There is no contribution to the circulation from C_1 and C_2. On C_3 the circulation is approximately $\|\mathbf{F}\| \cos 45°$ (Length of C_3) = $12 \cdot \dfrac{\sqrt{2}}{2} \cdot 10 = 60\sqrt{2}$. Similarly, on C_4 the circulation is $13 \cdot \dfrac{\sqrt{3}}{2} \cdot 20 = 130\sqrt{3}$. Adding these contributions together, we estimate the circulation of $\mathbf{F}$ around C to be about 310.

(b) Using normal components this time, we obtain a flux estimate of $5 \cdot 10 + 3 \cdot 20 + 12 \cdot \dfrac{\sqrt{2}}{2} \cdot 10$

$+ 13 \cdot \dfrac{1}{2} \cdot 20 \approx 325.$

30 (a) $\dfrac{d}{dx}\left(\dfrac{\sin^3 2x}{2x + 1}\right) =$

$\dfrac{(2x + 1)[(3 \sin^2 2x)(\cos 2x)(2)] - (\sin^3 2x)(2)}{(2x + 1)^2}$

$= \dfrac{2 \sin^2 2x \,[(6x + 3)\cos 2x - \sin 2x]}{(2x + 1)^2}$

(b) $\dfrac{d}{dx}(e^{2x} \tan 3x) = e^{2x}[(\sec^2 3x)(3)] +$

$[e^{2x}(2)] \tan 3x = e^{2x}[3 \sec^2 3x + 2 \tan 3x]$

(c) $\dfrac{d}{dx}[\ln(1 + \sec^2 4x)] =$

$\dfrac{(2 \sec 4x)(\sec 4x \tan 4x)(4)}{1 + \sec^2 4x} =$

$\dfrac{8 \sec^2 4x \tan 4x}{1 + \sec^2 4x}$

(d) $\dfrac{d}{dx}[\tan^{-1}(2 \cos 5x)] = \dfrac{2(-\sin 5x)(5)}{1 + (2 \cos 5x)^2}$

$= \dfrac{-10 \sin 5x}{1 + 4 \cos^2 5x}$

(e) $\dfrac{d}{dx}(\sin^{-1} x^2) = \dfrac{2x}{\sqrt{1 - (x^2)^2}} = \dfrac{2x}{\sqrt{1 - x^4}}$

16.4 Green's Theorem

2 $\nabla \cdot \mathbf{F} = 0$ and $\int_A \nabla \cdot \mathbf{F} \, dA = 0$. $\oint_C \mathbf{F} \cdot \mathbf{n} \, ds =$

$\oint_C 5y^3 \, dy + 6x^2 \, dx =$

$\int_0^{2\pi} [5(2 \sin \theta)^3 (2 \cos \theta) + 6(2 \cos \theta)^2 (-2 \sin \theta)] \, d\theta$

$= 0$

4 $\nabla \cdot \mathbf{F} = -\sin(x+y) + \cos(x+y)$, so $\int_A \nabla \cdot \mathbf{F} \, dA$

$= \int_0^a \int_0^{bx/a} [-\sin(x+y) + \cos(x+y)] \, dy \, dx =$

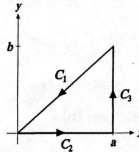

$\int_0^a \left[\cos\left(\frac{a+b}{a}x\right) + \sin\left(\frac{a+b}{a}x\right) - \cos x - \sin x \right] dx$

$= \frac{a}{a+b}[\sin(a+b) - \cos(a+b) + 1] - \sin a +$

$\cos a - 1$. From the figure, $\oint_C \mathbf{F} \cdot \mathbf{n} \, ds =$

$\int_{C_1} \mathbf{F} \cdot \mathbf{n} \, ds + \int_{C_2} \mathbf{F} \cdot \mathbf{n} \, ds + \int_{C_3} \mathbf{F} \cdot \mathbf{n} \, ds =$

$\int_a^0 \frac{-b \cos(x+y) + a \sin(x+y)}{\sqrt{a^2+b^2}} \cdot \frac{\sqrt{a^2+b^2}}{a}(-dx)$

$+ \int_0^a -\sin(x+y) \, dx + \int_0^b \cos(x+y) \, dy$

$= \frac{a}{a+b}[\sin(a+b) - \cos(a+b) + 1] - \sin a +$

$\cos a - 1$.

6 $\oint_C \mathbf{F} \cdot \mathbf{n} \, ds = \int_A \nabla \cdot \mathbf{F} \, dA$

$= \int_A (y \sec^2 x + 2y) \, dA$

$= \int_0^1 \int_0^1 (y \sec^2 x + 2y) \, dy \, dx = \frac{1}{2} \tan 1 + 1$

8 $\oint_C \mathbf{F} \cdot \mathbf{n} \, ds = \int_A \nabla \cdot \mathbf{F} \, dA$

$= \int_A \left(\frac{1}{x^2 y^2} - \frac{1}{x^2 y^2} \right) dA = 0$

10 Since $\mathbf{F} \cdot \mathbf{n} < 0$ on C, $\int_A \nabla \cdot \mathbf{F} \, dA = \oint_C \mathbf{F} \cdot \mathbf{n} \, ds$

< 0.

12 (a) Let $\mathbf{F} = (y^2 + x^3)\mathbf{i} - (x^2 - y^3)\mathbf{j}$. Then

$\oint_C [(x^2 - y^3) \, dx + (y^2 + x^3) \, dy]$

$= \oint_C \mathbf{F} \cdot \mathbf{n} \, ds = \int_A \nabla \cdot \mathbf{F} \, dA$

$= \int_A 3(x^2 + y^2) \, dA.$

(b) $\oint_C [(x^2 - y^3) \, dx + (y^2 + x^3) \, dy]$

$= \int_A 3(x^2 + y^2) \, dA = \frac{3\pi}{2}$

(c) $\oint_C [(x^2 - y^3) \, dx + (y^2 + x^3) \, dy]$

$= \int_0^{2\pi} [(\cos^2 \theta - \sin^3 \theta)(-\sin \theta) +$

$(\sin^2 \theta + \cos^3 \theta)(\cos \theta)] \, d\theta$

$= \int_0^{2\pi} [\cos^2 \theta (-\sin \theta) + \sin^4 \theta +$

$\sin^2 \theta (\cos \theta) + \cos^4 \theta] \, d\theta$

$= 2 \int_0^{2\pi} \sin^4 \theta \, d\theta = \frac{3\pi}{2}$

14 (a) $\mathbf{F}$ is not defined at the origin. Green's theorem does not apply.

16.4 Green's Theorem

(b) $\oint_C \mathbf{F} \cdot \mathbf{n}\, ds = \oint_C \left(\dfrac{x\mathbf{i} + y\mathbf{j}}{a^2}\right) \cdot \left(\dfrac{x\mathbf{i} + y\mathbf{j}}{a}\right) ds$

$= \dfrac{1}{a} \oint_C ds = 2\pi$

(c) From Exercise 13, $\nabla \cdot \mathbf{F} = 0$. Let $\mathcal{A}$ be the disk enclosed by C. Since $\mathbf{F}$ is defined on $\mathcal{A}$, we have $\oint_C \mathbf{F} \cdot \mathbf{n}\, ds = \int_{\mathcal{A}} \nabla \cdot \mathbf{F}\, dA = 0$ by Green's theorem.

16 The remainder of the proof is strictly parallel to that given in the text. Switch P's and Q's, x's and y's, and pay attention to signs.

18 $\nabla \cdot \mathbf{F}(2, 1) \approx \dfrac{\oint_C \mathbf{F} \cdot \mathbf{n}\, ds}{\text{Area of } \mathcal{A}} = \dfrac{0.06}{\pi(0.1)^2} \approx 1.910$

20 $f(x, y)$ is a scalar function, so $\mathbf{F} = \nabla f$ is a vector field. Note that $\mathbf{F} = f_x \mathbf{i} + f_y \mathbf{j}$, so $\nabla \cdot \mathbf{F} = f_{xx} + f_{yy}$. By Green's theorem, we have $\int_{\mathcal{A}} \nabla \cdot \mathbf{F}\, dA =$

$\oint_C \mathbf{F} \cdot \mathbf{n}\, ds = \oint_C -Q\, dx + P\, dy =$

$\oint_C -f_y\, dx + f_x\, dy$; therefore $\int_{\mathcal{A}} (f_{xx} + f_{yy})\, dA =$

$\oint_C f_x\, dy - f_y\, dx$, as claimed.

22 $x = t^6 + t^4$ and $y = t^3 + t$ for t in $[0, 1]$. Observe that (x, y) goes from $(0, 0)$ to $(2, 2)$ as t goes from 0 to 1. The region $\mathcal{A}$ that is bounded by the above curve and the line $y = x$ is shown in the figure. Label the two parts of the boundary C as indicated. The area of $\mathcal{A}$ is equal to $\dfrac{1}{2} \oint_C (-y\, dx + x\, dy)$, which we will compute in two parts—one for C_1 and one for C_2—and add. On C_1, $y = x$ for $0 \leq x \leq 2$, so $dy = dx$. Thus $\dfrac{1}{2} \int_{C_1} (-y\, dx + x\, dy) =$

$\dfrac{1}{2} \int_0^2 (-x\, dx + x\, dx) = 0$. On C_2, the correct orientation requires that t go from 1 to 0; $dx = (6t^5 + 4t^3)\, dt$ and $dy = (3t^2 + 1)\, dt$, so

$\dfrac{1}{2} \int_{C_2} (-y\, dx + x\, dy)$

$= \dfrac{1}{2} \int_1^0 [-(t^3 + t)(6t^5 + 4t^3) + (t^6 + t^4)(3t^2 + 1)]\, dt$

$= -\dfrac{1}{2} \int_0^1 [-3t^8 - 6t^6 - 3t^4]\, dt$

$= \dfrac{3}{2} \int_0^1 [t^8 + 2t^6 + t^4]\, dt$

$= \dfrac{3}{2}\left(\dfrac{t^9}{9} + \dfrac{2t^7}{7} + \dfrac{t^5}{5}\right)\bigg|_0^1 = \dfrac{3}{2}\left(\dfrac{1}{9} + \dfrac{2}{7} + \dfrac{1}{5}\right)$

$= \dfrac{94}{105}$. The area of $\mathcal{A}$ is $\dfrac{94}{105}$ square units.

24 (a) $x = \sin \pi t$ and $y = t - t^2$ as t goes from 0 to

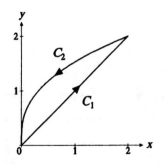

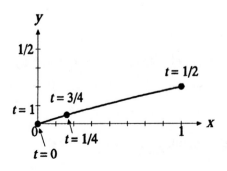

1. We have $(x, y) = (0, 0)$ for $t = 0$ or 1; $(1/\sqrt{2}, 3/16)$ for $t = \frac{1}{4}$ or $\frac{3}{4}$; and $\left(1, \frac{1}{4}\right)$ for $t = \frac{1}{2}$. As the figure shows, the path doubles back on itself and $\mathcal{A}$ is empty. To verify this, note that $\sin \pi(1 - t) = \sin(\pi - \pi t) = \sin \pi t$ and $(1 - t) - (1 - t)^2 = t - t^2$, so the points corresponding to t and $1 - t$ are the same.

(b) No difficulty arises. As noted in (a), the curve doubles back on itself, so all vertical and horizontal cross sections are 0. Hence the area is 0.

(c) As noted above, the area is 0. For your information, if you insist on doing things the hard way, the integral to evaluate is

$$\frac{1}{2} \int_0^1 \left[(t^2 - t)\pi \cos \pi t \, dt + \sin \pi t \, (1 - 2t) \right] dt.$$

26 (a) Let $ABCD$ be a quadrilateral, as shown, with the line AC added. Denote the quadrilateral

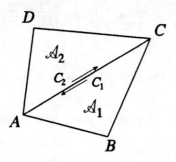

region by $\mathcal{A}$ and its triangular subregions ABC and ACD by $\mathcal{A}_1$ and $\mathcal{A}_2$, respectively. By assumption, Green's theorem holds for $\mathcal{A}_1$ and $\mathcal{A}_2$. Denoting the boundaries by C_1 and C_2, as shown, this means that $\oint_{C_1} \mathbf{F} \cdot \mathbf{n} \, ds +$

$\oint_{C_2} \mathbf{F} \cdot \mathbf{n} \, ds = \int_{\mathcal{A}_1} \nabla \cdot \mathbf{F} \, dA + \int_{\mathcal{A}_2} \nabla \cdot \mathbf{F} \, dA = \int_{\mathcal{A}} \nabla \cdot \mathbf{F} \, dA$. By the cancellation principle, $\oint_{C_1} \mathbf{F} \cdot \mathbf{n} \, ds + \oint_{C_2} \mathbf{F} \cdot \mathbf{n} \, ds = \oint_C \mathbf{F} \cdot \mathbf{n} \, ds$. So $\oint_C \mathbf{F} \cdot \mathbf{n} \, ds = \int_{\mathcal{A}} \nabla \cdot \mathbf{F} \, dA$, and Green's theorem holds for quadrilaterals.

(b) Let P be a polygon with n sides. Consider the diagonals drawn from one of the vertices of P to all other vertices. Then P would be divided into $n - 2$ triangles. So by arguments similar to those given in part (a) and the assumption that Green's theorem holds for triangles, it follows that Green's theorem holds for polygons.

28 Let $\mathbf{F} = \nabla f$. Then $D_\mathbf{n} f = \nabla f \cdot \mathbf{n} = \mathbf{F} \cdot \mathbf{n}$ and $\nabla \cdot \mathbf{F} = \nabla^2 f = f_{xx} + f_{yy}$. From Green's theorem, $\int_{\mathcal{A}} \nabla \cdot \mathbf{F} \, dA = \oint_C \mathbf{F} \cdot \mathbf{n} \, ds$, and it follows that

$$\int_{\mathcal{A}} (f_{xx} + f_{yy}) \, dA = \oint_C D_\mathbf{n} f \, ds.$$

30 (b) and (c) are not divergence-free.

16.5 Applications of Green's Theorem

2 $\mathbf{F} = 3x^2 \mathbf{i} + y \mathbf{j}$. Parameterize the inner boundary, C_1 by $x = \cos \theta, y = \sin \theta$. Then $\mathbf{n} = -x\mathbf{i} - y\mathbf{j}$ and $\oint_{C_1} \mathbf{F} \cdot \mathbf{n} \, ds = \int_0^{2\pi} (-3 \cos^3 \theta - \sin^2 \theta) \, d\theta = -\pi$. Divide the outer boundary, C_2, into four curves and parameterize them as follows: on C_3, $x = 2, -2 \leq y \leq 2$; on C_4, $y = 2, -2 \leq x \leq 2$; on C_5, $x = -2, -2 \leq y \leq 2$; on C_6, $y = -2$,

16.5 Applications of Green's Theorem

$-2 \leq x \leq 2$. Then $\oint_{C_2} \mathbf{F} \cdot \mathbf{n}\, ds = \int_{C_3} \mathbf{F} \cdot \mathbf{n}\, ds +$
$\int_{C_4} \mathbf{F} \cdot \mathbf{n}\, ds + \int_{C_5} \mathbf{F} \cdot \mathbf{n}\, ds + \int_{C_6} \mathbf{F} \cdot \mathbf{n}\, ds = 48 + 8$
$+ (-48) + 8 = 16$. So $\oint_{C_1} \mathbf{F} \cdot \mathbf{n}\, ds + \oint_{C_2} \mathbf{F} \cdot \mathbf{n}\, ds$
$= 16 - \pi = \int_{\mathcal{A}} \nabla \cdot \mathbf{F}\, dA$.

4

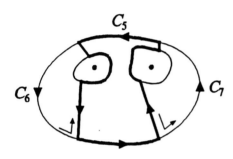

$\oint_{C_3} \mathbf{F} \cdot \mathbf{n}\, ds = \oint_{C_5} \mathbf{F} \cdot \mathbf{n}\, ds + \oint_{C_6} \mathbf{F} \cdot \mathbf{n}\, ds +$
$\oint_{C_7} \mathbf{F} \cdot \mathbf{n}\, ds = \oint_{C_1} \mathbf{F} \cdot \mathbf{n}\, ds + \oint_{C_2} \mathbf{F} \cdot \mathbf{n}\, ds = 7;$
$\oint_{C_4} \mathbf{F} \cdot \mathbf{n}\, ds = \int_{\mathcal{A}} \nabla \cdot \mathbf{F}\, dA = 0$

6 (a) $\nabla \cdot \mathbf{F} = \dfrac{-6xy}{(x^2 + y^2)^4} + \dfrac{6xy}{(x^2 + y^2)^4} = 0$

(b) $\oint_{C_1} \mathbf{F} \cdot \mathbf{n}\, ds =$
$\oint_{C_1} \dfrac{y}{(x^2+y^2)^3}\, dy + \dfrac{x}{(x^2+y^2)^3}\, dx = 0.$
$\oint_{C_2} \mathbf{F} \cdot \mathbf{n}\, ds = \int_{\mathcal{A}_2} \nabla \cdot \mathbf{F}\, dA = 0. \oint_{C_3} \mathbf{F} \cdot \mathbf{n}\, ds$
$= \oint_{C_1} \mathbf{F} \cdot \mathbf{n}\, ds = 0.$

8 $\mathbf{F}(x, y) = x\mathbf{i} + y\mathbf{j}, \nabla \times \mathbf{F} = \mathbf{0}$, and
$\int_{\mathcal{A}} (\nabla \times \mathbf{F}) \cdot \mathbf{k}\, dA = 0$. Also, $\oint_C \mathbf{F} \cdot d\mathbf{r}$

$= \int_0^{2\pi} [a \cos t\, (-a \sin t) + b \sin t\, (b \cos t)]\, dt$
$= (b^2 - a^2) \int_0^{2\pi} \sin t \cos t\, dt = 0.$

10 $\oint_C \mathbf{F} \cdot d\mathbf{r} = \int_{\mathcal{A}} (\nabla \times \mathbf{F}) \cdot \mathbf{k}\, dA = \int_{\mathcal{A}} 3xy\, dA$,
where $\mathcal{A}$ is described by $0 \leq y \leq 1$ and $y \leq x \leq 2 - y$. So $\int_{\mathcal{A}} 3xy\, dA = \int_0^1 \int_y^{2-y} 3xy\, dx\, dy =$
$\int_0^1 (6y - 6y^2)\, dy = 1$, and $\oint_C \mathbf{F} \cdot d\mathbf{r} = 1.$

12 $\oint_C \mathbf{F} \cdot d\mathbf{r} = \int_{\mathcal{A}} (\nabla \times \mathbf{F}) \cdot \mathbf{k}\, dA \approx \int_{\mathcal{A}} 5\mathbf{k} \cdot \mathbf{k}\, dA$
$= 5A$, where A is the area of $\mathcal{A}$.

14 Let $\mathcal{A}$ be a region in the plane bounded by the curves C_1, C_2, and C_3, where C_1 contains C_2 and C_3 and C_2 and C_3 do not overlap. Let $\mathbf{n}^*$ denote the exterior unit normal along the boundary. Then
$\oint_{C_1} \mathbf{F} \cdot \mathbf{n}^*\, ds + \oint_{C_2} \mathbf{F} \cdot \mathbf{n}^*\, ds + \oint_{C_3} \mathbf{F} \cdot \mathbf{n}^*\, ds =$
$\int_{\mathcal{A}} \nabla \cdot \mathbf{F}\, dA$ for any vector field $\mathbf{F}$ defined on $\mathcal{A}$.

18 If $G(x) = \int_c^x f(t)\, dt$, then $G'(x) = f(x)$. By assumption, however, $G(x) = 0$ for all x in (c, d). Since G is constant, its derivative must be 0. Hence $f(x) = 0$ for all x in (c, d).

20 Assume that $f(P_0) = k < 0$. Because f is continuous, there is a disk $\mathcal{D}$ around P_0 such that $f(P) < k/2$ for all points P in $\mathcal{D}$. Then $\int_{\mathcal{D}} f(P)\, dA$
$\leq \int_{\mathcal{D}} \dfrac{k}{2}\, dA = \dfrac{k}{2}(\text{Area of } \mathcal{D}) < 0$, contradicting

our assumption that all such integrals must be 0. Hence f must be 0.

22 Write the integrand of $\int_A (x^2 + y^2)\, dA$ as the divergence of a suitable vector field $\mathbf{F} = P\mathbf{i} + Q\mathbf{j}$. If we let $P_x = x^2$ and $Q_y = y^2$, then $P = \dfrac{x^3}{3}$ and $Q = \dfrac{y^3}{3}$ yields such an $\mathbf{F}$. By Green's theorem,

$$\int_A (x^2 + y^2)\, dA = \int_A \nabla \cdot \mathbf{F}\, dA = \oint_C \mathbf{F} \cdot \mathbf{n}\, ds =$$

$$\oint_C -Q\, dx + P\, dy = \oint_C -\dfrac{y^3}{3}\, dx + \dfrac{x^3}{3}\, dy =$$

$$\dfrac{1}{3}\oint_C (-y^3\, dx + x^3\, dy), \text{ as claimed.}$$

24 (a) Since $\cosh^2 t - \sinh^2 t = 1$, $\mathbf{G}(t) = \cosh t\, \mathbf{i} + \sinh t\, \mathbf{j}$ traces out points on the hyperbola $x^2 - y^2 = 1$. $\mathbf{G}(0) = \mathbf{i}$ and $\mathbf{G}(a) = \cosh a\, \mathbf{i} + \sinh a\, \mathbf{j}$, so $\mathbf{G}(t)$ goes from $(1, 0)$ to $(\cosh a, \sinh a)$ as t goes from 0 to a.

(b) Let C be the closed curve consisting of three segments: C_1, the line segment from O to A; C_2, the hyperbolic curve from A to B; and C_3, the line segment from B to O. Then the area of A is $\dfrac{1}{2}\oint_C (-y\, dx + x\, dy)$. On C_1, $y = 0$ so the line integral over that segment is 0. On C_2, we have $x = \cosh t$, $dx = \sinh t\, dt$, $y = \sinh t$, and $dy = \cosh t\, dt$ as t goes from 0 to a; hence $\dfrac{1}{2}\int_{C_2} (-y\, dx + x\, dy)$

$$= \dfrac{1}{2}\int_0^a (-\sinh t \sinh t\, dt + \cosh t \cosh t\, dt)$$

$$= \dfrac{1}{2}\int_0^a dt = \dfrac{a}{2}.$$ On C_3, we have x going from $\cosh a$ to 0 and $y = x \tanh a$. Hence

$$\dfrac{1}{2}\int_{C_3} (-y\, dx + x\, dy)$$

$$= \dfrac{1}{2}\int_{\cosh a}^0 (-x \tanh a\, dx + x \tanh a\, dx)$$

$= 0$. Hence the area of A is $a/2$.

28 (a) and (c) are not irrotational.

16.6 Conservative Vector Fields

2 Since $\mathbf{F}$ is not defined at P, nothing can be said about $\int_{C_3} \mathbf{F} \cdot d\mathbf{r}$ except that $\int_{C_3} \mathbf{F} \cdot d\mathbf{r} = \oint_C \mathbf{F} \cdot d\mathbf{r} - \int_{C_1} \mathbf{F} \cdot d\mathbf{r} = \oint_C \mathbf{F} \cdot d\mathbf{r} - 5$, where C is the curve consisting of C_1 followed by C_3.

4 From Exercise 2, we see that nothing can be said about $\oint_C \mathbf{F} \cdot d\mathbf{r}$ except that $\oint_C \mathbf{F} \cdot d\mathbf{r} = \int_{C_3} \mathbf{F} \cdot d\mathbf{r} + 5$.

6 $\mathbf{F}(x, y) = \sin y\, \mathbf{i} + (x \cos y + 3)\mathbf{j}$; $P_y = \cos y = Q_x$ and the domain of $\mathbf{F}$ is simply connected, so $\mathbf{F}$ is conservative. Now $f_x = P = \sin y$, and $f = \int f_x\, dx = x \sin y + g(y)$. Hence $Q = f_y = x \cos y + g'(y) = x \cos y + 3$. Thus $g'(y) = 3$ and $g(y) = 3y + C$, where C is constant. Thus $\mathbf{F} = \nabla f$, where $f(x, y) = x \sin y + 3y + C$.

8 $f(x, y) = \sin^3 xy + y + C$

10 If $f(x, y) = \dfrac{1}{2}\ln(x^2 + y^2)$, then $df = \dfrac{x\, dx}{x^2 + y^2} + \dfrac{y\, dy}{x^2 + y^2}$.

12 False, by Example 2.

14 False, by Example 2.

16 From $(0, 0)$ to $(a, 0)$, $\int_{C_1} \mathbf{F} \cdot d\mathbf{r} =$

$\int_{C_1} e^x y\, dx + (e^x + 2y)\, dy =$

$\int_0^a e^x \cdot 0\, dx + (e^x + 2 \cdot 0)\, 0 = 0$; from $(a, 0)$ to

(a, b): $\int_{C_2} \mathbf{F} \cdot d\mathbf{r} = \int_{C_2} e^x y\, dx + (e^x + 2y)\, dy =$

$\int_0^b (e^a + 2y)\, dy = e^a b + b^2 + C$. Thus $f(a, b) =$

$e^a b + b^2 + C$, and $f(x, y) = e^x y + y^2 + C$.

18 Choose a short, closed curve C that contains the origin but does not intersect C_1 or C_2. Since $\nabla \times \mathbf{F} = 0$, we have $\oint_{C_1} \mathbf{F} \cdot d\mathbf{r} = \oint_C \mathbf{F} \cdot d\mathbf{r}$ and $\oint_{C_2} \mathbf{F} \cdot d\mathbf{r}$

$= \oint_C \mathbf{F} \cdot d\mathbf{r}$ by Exercise 17. Thus $\oint_{C_1} \mathbf{F} \cdot d\mathbf{r} =$

$\oint_{C_2} \mathbf{F} \cdot d\mathbf{r}$.

20 (a) Let C_1 be the path from (P_0, T_0) to (P_0, T_1) and let C_2 be the path from (P_0, T_1) to

(P_1, T_1). Then $\int_C \left(\frac{RT\, dP}{P} - R\, dT\right) =$

$\int_{C_1} \left(\frac{RT\, dP}{P} - R\, dT\right) + \int_{C_2} \left(\frac{RT\, dP}{P} - R\, dT\right)$

$= \int_{T_0}^{T_1} (-R)\, dT + \int_{P_0}^{P_1} \frac{RT_1}{P}\, dP$

$= -R\big|_{T_0}^{T_1} + RT_1 \ln P \big|_{P_0}^{P_1}$

$= R(T_0 - T_1) + RT_1 \ln \frac{P_1}{P_0}$.

(b) Similarly, $\int_C \left(\frac{RT\, dP}{P} - R\, dT\right)$

$= RT_0 \ln \frac{P_1}{P_0} + R(T_0 - T_1)$.

(c) Similarly, $\int_C \left(\frac{RT\, dP}{P} - R\, dT\right)$

$= \frac{R(P_1 T_0 - P_0 T_1)}{P_1 - P_0} \ln \frac{P_1}{P_0}$.

22 (a) $r\, dr = \sqrt{x^2 + y^2} \left(\frac{x\, dx}{\sqrt{x^2 + y^2}} + \frac{y\, dy}{\sqrt{x^2 + y^2}}\right)$

$= x\, dx + y\, dy$

(b) $\mathbf{F}(\mathbf{r}) \cdot d\mathbf{r} = f(r) \left(\frac{x\mathbf{i} + y\mathbf{j}}{\sqrt{x^2 + y^2}}\right) \cdot (dx\mathbf{i} + dy\mathbf{j})$

$= f(r) \left(\frac{x\, dx + y\, dy}{r}\right) = f(r)\, dr$

24 Since for any closed curve C, $\oint_C (\mathbf{F} + \mathbf{G}) \cdot d\mathbf{r} =$

$\oint_C \mathbf{F} \cdot d\mathbf{r} + \oint_C \mathbf{G} \cdot d\mathbf{r} = 0$, the answer is yes.

16.S Review Exercises

2 (a) $\oint_C \mathbf{F} \cdot \mathbf{n}\, ds = \int_A \nabla \cdot \mathbf{F}\, dA$

(b) $\oint_C \mathbf{F} \cdot \mathbf{T}\, ds = \int_A (\nabla \times \mathbf{F}) \cdot \mathbf{k}\, dA$

4 If div $\mathbf{F} = 0$ everywhere, $\mathbf{F}$ must also be defined everywhere. So by Green's theorem, $\oint_C \mathbf{F} \cdot \mathbf{n}\, ds$

$= \int_A \nabla \cdot \mathbf{F}\, dA = 0$.

6 div $\mathbf{F} \approx \dfrac{\oint_C \mathbf{F} \cdot \mathbf{n}\, ds}{\text{Area of } A} = \dfrac{-0.003}{\pi (0.01)^2} \approx -9.549$ for

points in the circle.

8 (a) It "sums" the component of force in the direction of travel.

   (b) It "sums" the component of the field perpendicular to the bounding curve.

10 (a) Conservative, because $\nabla \times (3\mathbf{i} + 4\mathbf{j}) = \mathbf{0}$ everywhere.

   (b) Not conservative, because $\nabla \times \mathbf{F} = (-xy \sin xy + \cos xy - xy \cos xy - \sin xy)\mathbf{k} \neq \mathbf{0}$.

   (c) Conservative, because $\nabla \times (x\mathbf{i} + y\mathbf{j}) = \mathbf{0}$ everywhere.

12 $\mathbf{F} = \nabla f = e^x y^2 \mathbf{i} + 2ye^x \mathbf{j}$; $\int_C \mathbf{F} \cdot d\mathbf{r} = \int_C \nabla f \cdot d\mathbf{r} = f(3, 7) - f(1, 1) = 49e^3 - e = e(49e^2 - 1)$.

14 (a) $\mathbf{F} = \mathbf{r}/\|\mathbf{r}\|^{k+1}$, where $\mathbf{r} = x\mathbf{i} + y\mathbf{j} + z\mathbf{k}$ and $r = \sqrt{x^2 + y^2 + z^2}$. Now $\mathbf{F} = P\mathbf{i} + Q\mathbf{j} + R\mathbf{k} = (x/r^{k+1})\mathbf{i} + (y/r^{k+1})\mathbf{j} + (z/r^{k+1})\mathbf{k}$, so $P = x/r^{k+1}$, $Q = y/r^{k+1}$, and $R = z/r^{k+1}$. We will show that the x component of $\nabla \times \mathbf{F}$ is 0 by showing that $R_y - Q_z = 0$. Recall that $r_y = y/r$ and $r_z = z/r$; then $R_y = (zr^{-(k+1)})_y = z((-k+1)r^{-k-2})r_y = -(k+1)zr^{-k-2}(y/r) = -(k+1)yzr^{-k-3}$ and $Q_z = (yr^{-(k+1)})_z = y(-(k+1)r^{-k-2})r_z = -(k+1)yr^{-k-2}(z/r) = -(k+1)yzr^{-k-3} = R_y$. Thus $R_y - Q_z = 0$ and $\nabla \times \mathbf{F}$ has 0 for its x component. Similarly, $R_x - P_z = 0$ and $Q_x - P_y = 0$, so $\nabla \times \mathbf{F} = \mathbf{0}$ for all k.

   (b) $\nabla \cdot \mathbf{F} = 0$ only if $k = 2$. (See Exercise 31 of Sec. 16.1.)

18 $\left| \int_C \mathbf{F} \cdot d\mathbf{r} \right| \leq \int_C |\mathbf{F} \cdot \mathbf{T}| \, ds \leq \int_C M \, ds = M \int_C ds = Ml$

20 (a) ∇f is perpendicular to C. Hence $\nabla f \cdot \mathbf{n} = \mathbf{F} \cdot \mathbf{n} = \pm |\mathbf{F}|$, and $\oint_C \mathbf{F} \cdot \mathbf{n} \, ds = \oint_C \pm |\mathbf{F}| \, ds$. Since $\mathbf{F}$ is undefined at the origin, Green's theorem cannot be applied, and nothing more can be said about $\oint_C \mathbf{F} \cdot \mathbf{n} \, ds$.

   (b) From (a), we know $\nabla f \cdot \mathbf{T} = 0$. Hence $\oint_C \mathbf{F} \cdot \mathbf{T} \, ds = \oint_C \nabla f \cdot \mathbf{T} \, ds = 0$.

22 Let C be any closed curve that does not contain the origin, taken counterclockwise. Let P and Q be the two distinct points on C. Divide C into two curves C_1^* and C_2^* that begin at Q and end at P. Construct a curve C_1, taken counterclockwise, that has C_1^* as a segment and encloses the origin. Let C_2 be the segment C_2^* connected to the segment added to C_1^* to produce C_1. Then $\oint_C \mathbf{F} \cdot \mathbf{n} \, ds = \oint_{C_2} \mathbf{F} \cdot \mathbf{n} \, ds - \oint_{C_1} \mathbf{F} \cdot \mathbf{n} \, ds = 0$. So, by Green's theorem, $\int_\mathcal{A} \nabla \cdot \mathbf{F} \, dA = 0$ for all regions $\mathcal{A}$ that exclude the origin. Hence $\nabla \cdot \mathbf{F} = 0$ at all points in the plane where $\mathbf{F}$ is defined.

24 (a) If a disk is spinning at an angular rate of ω (in radians per unit time), we know that a point r units from the center has a linear velocity of ωr. The vector field $\mathbf{F} = |\mathbf{r}| \hat{\boldsymbol{\theta}} = r\hat{\boldsymbol{\theta}}$ points in the correct direction for circular motion and its magnitude at a point is just the distance of that point from the origin.

(b) Note that on the straight segments C_1 and C_3, F is perpendicular to the path so $\int_{C_1} \mathbf{F} \cdot d\mathbf{r}$ and

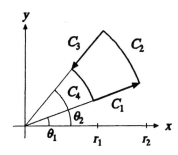

$\int_{C_3} \mathbf{F} \cdot d\mathbf{r}$ are 0. On C_2, we have $\mathbf{F} = r_2 \hat{\boldsymbol{\theta}}$, $\mathbf{T} = \hat{\boldsymbol{\theta}}$ and $ds = r_2 \, d\theta$, so $\int_{C_2} \mathbf{F} \cdot d\mathbf{r} = \int_{C_2} \mathbf{F} \cdot \mathbf{T} \, ds = \int_{\theta_1}^{\theta_2} r_2 \hat{\boldsymbol{\theta}} \cdot \hat{\boldsymbol{\theta}} \, r_2 \, d\theta = \int_{\theta_1}^{\theta_2} r_2^2 \, d\theta$

$= r_2^2(\theta_2 - \theta_1)$. On C_4, we have $\mathbf{F} = r_1 \hat{\boldsymbol{\theta}}$, $\mathbf{T} = -\hat{\boldsymbol{\theta}}$, and $ds = r_1 \, d\theta$, so $\int_{C_4} \mathbf{F} \cdot d\mathbf{r} = \int_{C_4} \mathbf{F} \cdot \mathbf{T} \, ds = \int_{\theta_1}^{\theta_2} r_1 \hat{\boldsymbol{\theta}} \cdot (-\hat{\boldsymbol{\theta}}) r_1 \, d\theta = \int_{\theta_1}^{\theta_2} (-r_1^2) \, d\theta = -r_1^2(\theta_2 - \theta_1)$. Hence $\oint_C \mathbf{F} \cdot d\mathbf{r}$

$= (r_2^2 - r_1^2)(\theta_2 - \theta_1)$.

(c) Counterclockwise; the circulation is positive, so it must agree with the orientation of C.

26 $\mathbf{F} = |\mathbf{r}|^k \hat{\boldsymbol{\theta}} = (x^2 + y^2)^{k/2} \dfrac{-y\mathbf{i} + x\mathbf{j}}{(x^2 + y^2)^{1/2}} =$

$(x^2 + y^2)^{(k-1)/2}(-y\mathbf{i} + x\mathbf{j})$; $(\nabla \times \mathbf{F}) \cdot \mathbf{k} =$

$\dfrac{\partial}{\partial x}(x(x^2 + y^2)^{(k-1)/2}) - \dfrac{\partial}{\partial y}(-y(x^2 + y^2)^{(k-1)/2})$

$= (x^2 + y^2)^{(k-1)/2} + \dfrac{k-1}{2}x(x^2 + y^2)^{(k-3)/2}(2x) -$

$\left(-(x^2 + y^2)^{(k-1)/2} - \dfrac{k-1}{2}y(x^2 + y^2)^{(k-3)/2}(2y)\right)$

$= (k+1)(x^2 + y^2)^{(k-1)/2} = (k+1)r^{k-1}$; $\oint_C \mathbf{F} \cdot d\mathbf{r}$

$= \int_A (\nabla \times \mathbf{F}) \cdot \mathbf{k} \, dA = \int_{\theta_1}^{\theta_2} \int_{r_1}^{r_2} (k+1)r^k \, dr \, d\theta$

$= (\theta_2 - \theta_1)(r_2^{k+1} - r_1^{k+1})$

(a) $(\theta_2 - \theta_1)(r_2^{k+1} - r_1^{k+1}) > 0$ for all $k > 0$, so the paddle wheel must turn counterclockwise for any $k > 0$.

(b) From (a), the paddle wheel will not turn clockwise for all $k > 0$.

(c) The paddle wheel will never be stationary for any $k > 0$.

28 Let w be the component of the water current's velocity in the direction of the boat's trip. In still water, the 2-mile round trip would take time $2/v$. Accounting for the current, the trip would take $\dfrac{1}{v-w} + \dfrac{1}{v+w} = \dfrac{2v}{v^2 - w^2}$. The difference in time for the two trips is $\dfrac{2v}{v^2 - w^2} - \dfrac{2}{v} = \dfrac{2w^2}{v(v^2 - w^2)} > 0$, so the trip in still water takes less time.

30 On C_r we have $ds = r \, d\theta$, so $I(r) =$

$\dfrac{1}{2\pi r} \oint_{C_r} f(P) \, ds = \dfrac{1}{2\pi r} \int_0^{2\pi} f(r, \theta) \, r \, d\theta$

$= \dfrac{1}{2\pi} \int_0^{2\pi} f(r, \theta) \, d\theta$.

32 Let $\mathbf{F} = \nabla f$. Then $\nabla \cdot \mathbf{F} = \dfrac{\partial^2 f}{\partial x^2} + \dfrac{\partial^2 f}{\partial y^2} = 0$. By

Green's theorem, $\int_0^{2\pi} \dfrac{\partial f}{\partial r} \, d\theta = \int_0^{2\pi} \nabla f \cdot \mathbf{n} \, d\theta =$

$$\int_0^{2\pi} \mathbf{F}\cdot\mathbf{n}\, d\theta = \int_A \nabla\cdot\mathbf{F}\, dA = 0. \text{ Thus } \frac{d}{dr}(I(r))$$
$= 0.$

34. (a) $f(x, y) = \ln((x-2)^2 + (y-3)^2);$

$$f_x = \frac{2(x-2)}{(x-2)^2 + (y-3)^2},$$

$$f_y = \frac{2(y-3)}{(x-2)^2 + (y-3)^2},$$

$$f_{xx} = \frac{((x-2)^2 + (y-3)^2)\cdot 2 - 2(x-2)(2(x-2))}{((x-2)^2 + (y-3)^2)^2}$$

$$= \frac{2[(y-3)^2 - (x-2)^2]}{[(x-2)^2 + (y-3)^2]^2}, f_{yy} =$$

$$\frac{[(x-2)^2 + (y-3)^2]\cdot 2 - 2(y-3)[2(y-3)]}{[(x-2)^2 + (y-3)^2]^2}$$

$$= \frac{2[(x-2)^2 - (y-3)^2]}{[(x-2)^2 + (y-3)^2]^2} = -f_{xx}.$$

(b) The domain of f contains all points except $(2, 3)$.

(c) The average of f on the unit circle centered at $(0, 0)$ is $f(0, 0) = \ln 13$.

36. (a) Apply the argument in Exercise 35 to the function $h = f - g$. Then $h = 0$ throughout A, and $f = g$ for all points in A.

(b) Any steady-state temperature distribution is harmonic. So part (a) tells us that if two steady-state temperature distributions are equal on the boundary of some planar region, they must be equal throughout the region.

38. (a) Estimate $\oint_C \mathbf{F}\cdot\mathbf{T}\, ds$ by partitioning the unit circle C into six equal arcs of length $\pi/3$ and taking as sampling points the tail of each

vector in Figure 6. Since the given vectors are all tangent to C, we see that $\mathbf{F}\cdot\mathbf{T} = \|\mathbf{F}\|$. Starting with the vector at $(1, 2)$ and working counterclockwise, we find lengths of approximately 1.2, 1.1, 0.8, 0.9, 1.0, and 1.3. Adding these together and multiplying by $\pi/3$ gives us 6.6 as the approximate value of the circulation.

(b) Since all of the vectors are tangent to C, it appears that the flux is 0.

17 The Divergence Theorem and Stokes' Theorem

17.1 Surface Integrals

2 $dA \approx \cos \gamma \, dS$, where $dA = 0.03$ and $\gamma = 25°$; so $dS \approx \dfrac{0.03}{\cos 25°} \approx 0.033$.

4 (a) S is the portion of the plane cut out by the cardioidal cylinder.

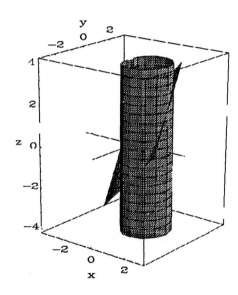

(b) Area of $S = |\sec \gamma|$(Area of A), where γ is the angle between $\mathbf{i} + 3\mathbf{j} - \mathbf{k}$ (a normal to $x + 3y - z = 0$) and $\mathbf{k}$ and Area of $A = \dfrac{1}{2}\int_0^{2\pi} (1 + \cos\theta)^2 \, d\theta = 3\pi/2$. Thus $\cos\gamma = -\dfrac{1}{\sqrt{11}}$, so Area of $S = \dfrac{3}{2}\pi\sqrt{11}$.

6 The surface S is part of a sphere and can be described in spherical coordinates by $0 \le \theta \le 2\pi$, $0 \le \phi \le \pi/4$, and $\rho = a$. Hence Area of $S =$

$\int_S 1 \, dS = \int_0^{2\pi} \int_0^{\pi/4} a^2 \sin\phi \, d\phi \, d\theta$

$= 2\pi a^2(-\cos\phi)\big|_0^{\pi/4} = 2\pi a^2\left(1 - \dfrac{1}{\sqrt{2}}\right)$

$= (2 - \sqrt{2})\pi a^2.$

8 $\mathbf{n} = x\mathbf{i} + y\mathbf{j} + z\mathbf{k}$. So $\mathbf{F}\cdot\mathbf{n} = x^4 + y^3$ and

$\int_S \mathbf{F}\cdot\mathbf{n} \, dS = \int_S (x^4 + y^3) \, dS =$

$\int_0^{2\pi} \int_0^{\pi} \left[(\sin\phi\cos\theta)^4 + (\sin\phi\sin\theta)^3\right] \sin\phi \, d\phi \, d\theta$

$= \int_0^{2\pi} \int_0^{\pi} (\sin^5\phi \cos^4\theta + \sin^4\phi \sin^3\theta) \, d\phi \, d\theta =$

$\int_0^{2\pi} \left(\dfrac{16}{15} \cos^4\theta + \dfrac{3\pi}{16} \sin^3\theta\right) d\theta =$

$\dfrac{16}{15} \cdot \dfrac{3\pi}{4} + \dfrac{3\pi}{16} \cdot 0 = \dfrac{4\pi}{5}.$

10 Using the notation of Example 2, with $R(x,y,z) = z$, we have $\int_S z \cos\gamma \, dS = \int_A (z_2 - z_1) \, dA$, which equals the volume of S because $z_2 - z_1$ is just the cross-sectional length of the solid enclosed by S.

12 $\int_S y \cos\beta \, dS = \int_A (y_2 - y_1) \, dA$ is the volume enclosed by S, where A is the projection of S onto the xz plane.

14 $|\cos\gamma| = \sqrt{1 - x^2 - y^2}$. So the area of the part of the sphere is $\int_S dS = \int_A \dfrac{1}{\sqrt{1 - x^2 - y^2}} \, dA$

$$= \int_{-\pi/2}^{\pi/2} \int_0^{\cos\theta} \frac{r}{\sqrt{1-r^2}}\, dr\, d\theta$$

$$= \int_{-\pi/2}^{\pi/2} \left(-\sqrt{1-r^2}\right)\Big|_0^{\cos\theta} d\theta$$

$$= \int_{-\pi/2}^{\pi/2} (1 - |\sin\theta|)\, d\theta = 2\int_0^{\pi/2} (1 - \sin\theta)\, d\theta$$

$$= \pi - 2.$$

16 (a) Since $z = \sqrt{a^2 - x^2 - y^2}$, $|\cos\gamma| =$

$$\frac{\sqrt{a^2-x^2-y^2}}{a}. \int_S x^2 y\, dS =$$

$$\int_A \frac{ax^2y}{\sqrt{a^2-x^2-y^2}}\, dA$$

$$= \int_0^a \int_0^{\sqrt{a^2-x^2}} \frac{ax^2y}{\sqrt{a^2-x^2-y^2}}\, dy\, dx.$$

(b) $\int_S x^2 y\, dS$

$$= \int_S (\sin\phi\cos\theta)^2 (\sin\phi\sin\theta)\, dS$$

$$= \int_0^{\pi/2} \int_0^{\pi/2} a^5 \sin^4\phi \cos^2\theta \sin\theta\, d\phi\, d\theta$$

(c) From (b), $\int_S x^2 y\, dS$

$$= \int_0^{\pi/2} (a^5 \cos^2\theta \sin\theta) \frac{3\pi}{16}\, d\theta = \frac{\pi a^5}{16}.$$

18 (a) The area of the triangle is half the area of the parallelogram spanned by the vectors $\langle 2-1, 3-1, 4-1\rangle = \langle 1, 2, 3\rangle$ and $\langle 3-1, 4-1, 5-1\rangle = \langle 2, 3, 4\rangle$. By Theorem 3 of Sec. 12.6, the parallelogram's area is $\|\langle 1,2,3\rangle \times \langle 2,3,4\rangle\| = \|\langle -1, 2, -1\rangle\| = \sqrt{6}$, so the triangle's area is $\sqrt{6}/2$.

(b) The surface of the triangle is given by $z =$

$2y - x$. Then $|\cos\gamma| = \dfrac{1}{\sqrt{1+4+1}} = \dfrac{1}{\sqrt{6}}$.

Thus the area is $\int_A \sqrt{6}\, dA = \sqrt{6}(\text{Area of } A)$, where A is the triangle in the xy plane with vertices $(1, 1)$, $(2, 3)$, and $(3, 4)$. Since Area of $A = 1/2$, Area of $S = \sqrt{6}/2$.

20 The surface is given by $z = 3 - 3x - \dfrac{3}{2}y$ and

$|\cos\gamma| = \dfrac{2}{7}$. Thus $\int_S f(P)\, dS$

$$= \int_A \frac{7}{2}(3x + 2y + 2z)\, dA$$

$$= \int_A \frac{7}{2}\left[3x + 2y + 2\left(3 - 3x - \frac{3}{2}y\right)\right] dA$$

$$= \frac{7}{2} \int_0^1 \int_0^{2-2x} (6 - 3x - y)\, dy\, dx$$

$$= \frac{7}{2} \int_0^1 (10 - 14x + 4x^2)\, dx = \frac{91}{6}.$$

22 Since $z = 2 - \dfrac{1}{3}x - \dfrac{2}{3}y$, $|\cos\gamma| =$

$$\frac{1}{\sqrt{\left(\frac{1}{3}\right)^2 + \left(\frac{2}{3}\right)^2 + 1}} = \frac{3}{\sqrt{14}}, \text{ so } \int_S z\, dS =$$

$$\frac{\sqrt{14}}{3} \int_0^4 \int_0^{1-x/4} \left(2 - \frac{1}{3}x - \frac{2}{3}y\right) dy\, dx = \frac{\sqrt{14}}{3} \cdot \frac{8}{3}.$$

Also, Area of $S = \dfrac{\sqrt{14}}{3}(\text{Area of } A) = \dfrac{\sqrt{14}}{3}\cdot 2$.

Thus $\bar{z} = \dfrac{\sqrt{14}}{3} \cdot \dfrac{8}{3} \cdot \dfrac{3}{2\sqrt{14}} = \dfrac{4}{3}$. (Alternatively, we can just observe that the surface is the triangle with

17.1 Surface Integrals

vertices (0, 0, 2), (4, 0, 2/3) and (0, 1, 4/3), so its centroid is (4/3, 1/3, 4/3).)

24 (a) By symmetry.

(b) $\int_S (x^2 + y^2 + z^2)\, dS$

$= \int_0^{2\pi} \int_0^{\pi} (a^2)(a^2 \sin\phi)\, d\phi\, d\theta = 4\pi a^4$

(c) By (a), $\int_S x^2\, dS = \frac{1}{3} \int_S (x^2 + y^2 + z^2)\, dS$

$= \frac{4\pi a^4}{3}$.

(d) $\int_S (2x^2 + 3y^2)\, dS = \frac{20\pi a^4}{3}$

26 $\cos\gamma = \frac{\sqrt{(2a)^2 - x^2 - y^2}}{2a}$, so the area is $\int_S dS$

$= \int_A \frac{2a}{\sqrt{4a^2 - x^2 - y^2}}\, dA$

$= \int_{-\pi/2}^{\pi/2} \int_0^{2a\cos\theta} \frac{2ar}{\sqrt{4a^2 - r^2}}\, dr\, d\theta = 4a^2(\pi - 2)$.

28 $|\cos\gamma| = \frac{1}{\sqrt{c^2/a^2 + c^2/b^2 + 1}}$, so the moment of inertia about the z axis is $I_z =$

$\frac{2M}{ab}\left[\frac{c^2}{a^2} + \frac{c^2}{b^2} + 1\right]^{-1/2} \int_S (x^2 + y^2)\, dS$

$= \frac{2M}{ab} \int_0^a \int_0^{b-bx/a} (x^2 + y^2)\, dy\, dx$

$= \frac{M}{6}(a^2 + b^2)$.

30 S is a sphere of radius a, and A is a point a distance b away from the center of S, where $b < a$. For convenience, place S at the center of a coordinate system, and let A be on the positive z axis. If P is a point on S, then $q = \overline{PA} = \sqrt{a^2 + b^2 - 2ab\cos\phi}$ by the law of cosines. (See p. 927 of the text for a similar derivation.) Now $dS = a^2 \sin\phi\, d\phi\, d\theta$, so the integral over S of $\frac{1}{q}$ is

$\int_S \frac{1}{q}\, dS = \int_0^{2\pi} \int_0^{\pi} \frac{a^2 \sin\phi}{\sqrt{a^2 + b^2 - 2ab\cos\phi}}\, d\phi\, d\theta$

$= 2\pi \cdot \frac{a^2}{2ab} \cdot 2(a^2 + b^2 - 2ab\cos\phi)^{1/2}\Big|_0^{\pi}$

$= \frac{2\pi a}{b}\left(\sqrt{a^2 + b^2 + 2ab} - \sqrt{a^2 + b^2 - 2ab}\right)$

$= \frac{2\pi a}{b}(a + b - (a - b)) = \frac{2\pi a}{b}(2b) = 4\pi a$. The

area of S is $4\pi a^2$, so the average of $\frac{1}{q}$ over S is

$\frac{4\pi a}{4\pi a^2} = \frac{1}{a}$. (We used $b < a$ in simplifying

$\sqrt{a^2 + b^2 - 2ab} = |a - b| = a - b$.)

32 On the top surface, the integrand's constant value is h^2, so the integral over the top is $\pi a^2 h^2$. Let S' be the curved surface and for convenience let $\alpha = \tan^{-1}\frac{a}{h}$. Then $\int_{S'} z^2\, dS =$

$\int_0^{2\pi} \int_0^{\sqrt{a^2+h^2}} (\rho \cos\alpha)^2\, \rho \sin\alpha\, d\rho\, d\theta$

$= \int_0^{2\pi} \int_0^{\sqrt{a^2+h^2}} \rho^3 \cos^2\alpha \sin\alpha\, d\rho\, d\theta$

$= 2\pi \cdot \frac{1}{4}(a^2 + h^2)^2 \cdot \frac{h^2}{a^2 + h^2} \cdot \frac{a}{\sqrt{a^2 + h^2}}$

$= \frac{1}{2}\pi a h^2 \sqrt{a^2 + h^2}$. Therefore the result is

$\pi a^2 h^2 + \frac{1}{2}\pi a h^2 \sqrt{a^2 + h^2}$.

34 The cylinder can be described by $0 \leq \theta \leq 2\pi$, $r = a$, $0 \leq z \leq h$. The integrand is then z^2. The integrals over the top and bottom are $\pi a^2 h^2$ and 0, respectively. The integral over the curved surface is $\int_0^{2\pi} \int_0^h z^2 a \, dz \, d\theta$

$= \frac{2}{3}\pi a h^3$, so $\int_S f(P) \, dS = \pi a^2 h^2 + \frac{2}{3}\pi a h^3$.

36 Let $\mathbf{n}$ be a unit normal to the surface. Let the area of the surface be A and the direction cosines of $\mathbf{n}$ be a, b, and c. Thus $A = \frac{0.01}{a} = \frac{0.02}{b} = \frac{0.03}{c}$

so $1 = a^2 + b^2 + c^2 =$

$\left(\frac{0.01}{A}\right)^2 + \left(\frac{0.02}{A}\right)^2 + \left(\frac{0.03}{A}\right)^2 = \frac{14}{10{,}000 A^2}$. Hence

$A = \frac{\sqrt{14}}{100}$.

38 If we replace $\mathbf{n}$ by its negative, then γ is replaced by $\pi - \gamma$ and we are back to the case discussed in the text. Thus $dA \approx \cos(\pi - \gamma) \, dS = -\cos \gamma \, dS$.

17.2 The Divergence Theorem

4 $\mathbf{F} = z\mathbf{k}$, so $\nabla \cdot \mathbf{F} = 1$ and $\int_V \nabla \cdot \mathbf{F} \, dV =$ Volume of

$V = \frac{2}{3}\pi a^3$. The surface S of V can be broken into

a hemispherical top S_1, whose normal is $\mathbf{n}_1 =$

$\frac{x}{a}\mathbf{i} + \frac{y}{a}\mathbf{j} + \frac{z}{a}\mathbf{k}$ and a bottom disk S_2 whose normal

is $\mathbf{n}_2 = -\mathbf{k}$. Therefore $\int_S \mathbf{F} \cdot \mathbf{n} \, dS =$

$\int_{S_1} \mathbf{F} \cdot \mathbf{n} \, dS + \int_{S_2} \mathbf{F} \cdot \mathbf{n} \, dS = \int_{S_1} \frac{z^2}{a} \, dS +$

$\int_{S_2} (-z) \, dS = \int_A \frac{z^2}{a} \cdot \frac{a}{z} \, dA + \int_{S_2} 0 \, dS$

$= \int_0^{2\pi} \int_0^a \sqrt{a^2 - r^2} \, r \, dr \, d\theta = -\frac{2\pi}{3}(a^2 - r^2)^{3/2} \Big|_0^a$

$= \frac{2}{3}\pi a^3$.

6 $\int_S \mathbf{F} \cdot \mathbf{n} \, dS = \int_V \nabla \cdot \mathbf{F} \, dV = \int_V 3x^2 \, dV =$

$3\int_0^{2\pi} \int_0^\pi \int_a^b (\rho^2 \sin^2 \phi \cos^2 \theta) \rho^2 \sin \phi \, d\rho \, d\phi \, d\theta$

$= \frac{3}{5}(b^5 - a^5) \int_0^{2\pi} \int_0^\pi \sin^3 \phi \cos^2 \theta \, d\phi \, d\theta =$

$\frac{3}{5}(b^5 - a^5) \int_0^{2\pi} \frac{4}{3} \cos^2 \theta \, d\theta = \frac{4\pi}{5}(b^5 - a^5)$.

8 $\int_S \mathbf{F} \cdot \mathbf{n} \, dS = \int_V \nabla \cdot \mathbf{F} \, dV = \int_V 3 \, dV$

$= 12 \int_0^3 \int_0^{\sqrt{(9-x^2)/2}} \int_0^{9-x^2-2y^2} dz \, dy \, dx$

$= 12 \int_0^3 \int_0^{\sqrt{(9-x^2)/2}} (9 - x^2 - 2y^2) \, dy \, dx$

$= 4\sqrt{2} \int_0^3 (9 - x^2)^{3/2} \, dx = \frac{243\pi}{2\sqrt{2}}$

10 $\int_S \mathbf{F} \cdot \mathbf{n} \, dS = \int_V \nabla \cdot \mathbf{F} \, dV = \int_V 3x^2 \, dV =$

$\int_0^{2\pi} \int_0^{\pi/6} \int_0^{\sec \phi} (3\rho^2 \sin^2 \phi \cos^2 \theta)(\rho^2 \sin \phi) \, d\rho \, d\phi \, d\theta$

$= 3 \int_0^{2\pi} \int_0^{\pi/6} \int_0^{\sec \phi} \rho^4 \sin^3 \phi \cos^2 \theta \, d\rho \, d\phi \, d\theta$

$= \frac{3\pi}{5} \int_0^{\pi/6} \sin^3 \phi \sec^5 \phi \, d\phi = \frac{\pi}{60}$

12 $\nabla \cdot \mathbf{F} = 2 + 3 + 5 = 10$, while $\nabla \cdot \mathbf{G} = 3 + 2 + 5$

$= 10$, so $\nabla \cdot \mathbf{F} = \nabla \cdot \mathbf{G}$ and $\int_S \mathbf{F} \cdot \mathbf{n} \, dS =$

17.2 The Divergence Theorem

$$\int_V \nabla \cdot \mathbf{F}\, dV = \int_V \nabla \cdot \mathbf{G}\, dV = \int_S \mathbf{G} \cdot \mathbf{n}\, dS.$$

14 Let S be the boundary of the ball. On S, $\mathbf{F} = 2(x\mathbf{i} + y\mathbf{j} + z\mathbf{k})$, $\mathbf{n} = \frac{1}{2}(x\mathbf{i} + y\mathbf{j} + z\mathbf{k})$, and $\mathbf{F} \cdot \mathbf{n} = x^2 + y^2 + z^2 = 4$. Thus $\int_V \nabla \cdot \mathbf{F}\, dV = \int_S \mathbf{F} \cdot \mathbf{n}\, dS = 4 \int_S dS = 64\pi$.

16 $\nabla \cdot \mathbf{F} = \frac{\partial}{\partial x}(x) + \frac{\partial}{\partial y}(3y + z) + \frac{\partial}{\partial z}(4x + 2z) = 6$, so $\int_S \mathbf{F} \cdot \mathbf{n}\, dS = \int_V \nabla \cdot \mathbf{F}\, dV = \int_V 6\, dV$ is 6 times the volume of the cube. The cube's edges have length 2, so its volume is $2^3 = 8$ and the integral is $6 \cdot 8 = 48$.

18 $\int_S \mathbf{F} \cdot \mathbf{n}\, dS = \int_V \nabla \cdot \mathbf{F}\, dV = \int_V 4\, dV = \frac{4}{3}$

20 $\int_S \mathbf{F} \cdot \mathbf{n}\, dS = \int_V \nabla \cdot \mathbf{F}\, dV = \int_V 3(x^2 + y^2 + z^2)\, dV$
$= 3 \int_0^{2\pi} \int_0^{\pi} \int_0^a \rho^4 \sin\phi\, d\rho\, d\phi\, d\theta = \frac{12\pi a^5}{5}$

22 $\int_S \mathbf{F} \cdot \mathbf{n}\, dS = 4\pi$

24 $\int_S \mathbf{F} \cdot \mathbf{n}\, dS = 4\pi$

26 (a) $\mathbf{E}$ must be perpendicular to the plane. (All parallel components must cancel.)

(b) By symmetry, $\mathbf{E} \cdot \mathbf{n} = 0$ on the curved surface of the cylinder. On the two circular ends, $\mathbf{E} \cdot \mathbf{n} = \|\mathbf{E}\|$, so $\int_S \mathbf{E} \cdot \mathbf{n}\, dS = \|\mathbf{E}\|\pi a^2$ for each one. By Gauss's law, $2\pi a^2 \|\mathbf{E}\| = q\pi a^2$, so $\|\mathbf{E}\| = q/2$. (The field strength does not depend on the distance r from the plane.)

30 (a) div $\mathbf{F}$ must equal 0.

(b) By (a), we must have div $\mathbf{F} = 0$. By Exercise 34(b) of Sec. 16.1, it follows that $f(r) = kr^{-2}$ for some constant k. In other words, $\mathbf{F}$ must obey an inverse-square law.

32 (a) Since div $\mathbf{F} = 0$ for $\mathbf{F} = \frac{\hat{\mathbf{r}}}{\|\mathbf{r}\|^2}$, $\int_S \frac{\hat{\mathbf{r}} \cdot \mathbf{n}}{\|\mathbf{r}\|^2} dS = \int_V 0\, dV = 0$ for any surface S that does not enclose the origin.

(b) As viewed from the origin, the closed surface S consists of two pieces: one part where $\hat{\mathbf{r}} \cdot \mathbf{n} < 0$ and one where $\hat{\mathbf{r}} \cdot \mathbf{n} > 0$. The two parts will have steradian measure that is equal and opposite, cancelling out.

34 Since the identical triangular faces are equidistant from the centroid, each must subtend one quarter of 4π (the measure in all directions): $\frac{1}{4}(4\pi) = \pi$.

36 $\mathcal{V}$ is the ball of center P_0 and radius a. Let $\nabla \cdot \mathbf{F}(P)$ denote the value of $\nabla \cdot \mathbf{F}$ at the point P and V the volume of $\mathcal{V}$. Assuming that $\nabla \cdot \mathbf{F}$ is continuous, we see that $\nabla \cdot \mathbf{F}(P)$ is close to $\nabla \cdot \mathbf{F}(P_0)$ whenever P is near P_0. If the radius a is chosen sufficiently small, therefore, then $\nabla \cdot \mathbf{F}(P) \approx \nabla \cdot \mathbf{F}(P_0)$ for all points P in $\mathcal{V}$. Hence $\int_{\mathcal{V}} \nabla \cdot \mathbf{F}\, dV \approx \nabla \cdot \mathbf{F}(P_0) \int_{\mathcal{V}} dV = (\nabla \cdot \mathbf{F}(P_0))V$. By the divergence theorem, we now have $\frac{1}{V} \int_S \mathbf{F} \cdot \mathbf{n}\, dS = \frac{1}{V} \int_{\mathcal{V}} \nabla \cdot \mathbf{F}\, dV \approx \frac{1}{V}(\nabla \cdot \mathbf{F}(P_0))V = \nabla \cdot \mathbf{F}(P_0)$. In the limit as $a \to 0$, we have equality: $\lim_{a \to 0} \frac{1}{V} \int_S \mathbf{F} \cdot \mathbf{n}\, dS = \nabla \cdot \mathbf{F}(P_0)$.

38 By the divergence theorem, $\int_S (f\mathbf{v} \cdot \mathbf{n})\, dS =$

$\int_V \nabla \cdot f\mathbf{v}\, dV$, so $\int_V \frac{\partial f}{\partial t}\, dV + \int_S (f\mathbf{v} \cdot \mathbf{n})\, dS =$

$\int_V \frac{\partial f}{\partial t}\, dV + \int_V \nabla \cdot f\mathbf{v}\, dV = \int_V \left[\frac{\partial f}{\partial t} + \nabla \cdot f\mathbf{v}\right] dV.$

By the "vanishing-integrals" principle of Sec. 16.5, it follows that $\frac{\partial f}{\partial t} + \nabla \cdot f\mathbf{v} = 0$.

17.3 Stokes' Theorem

2. $\oint_{C_1} \mathbf{F} \cdot d\mathbf{r} = -\oint_{C_2} \mathbf{F} \cdot d\mathbf{r} = \oint_{C_3} \mathbf{F} \cdot d\mathbf{r}$ and $\oint_{C_4} \mathbf{F} \cdot d\mathbf{r} = 0$.

4. $\nabla \times \mathbf{F} = -x\mathbf{i} - 2x\mathbf{j} + (z - 1)\mathbf{k}$, $\mathbf{n} = \frac{1}{\sqrt{3}}(\mathbf{i} + \mathbf{j} + \mathbf{k})$ and $(\nabla \times \mathbf{F}) \cdot \mathbf{n} = \frac{-3x + z - 1}{\sqrt{3}}$, so

$\int_S (\nabla \times \mathbf{F}) \cdot \mathbf{n}\, dS = \frac{1}{\sqrt{3}} \int_S (-3x + z - 1)\, dS =$

$-\frac{5}{6}$. Now $\oint_C \mathbf{F} \cdot \mathbf{T}\, ds = \oint_C y\, dx + xz\, dy + x^2\, dz$.

C consists of three parts: $(1, 0, 0)$ to $(0, 1, 0)$, parameterized by $x = 1 - t$, $y = t$, and $z = 0$; $(0, 1, 0)$ to $(0, 0, 1)$, parameterized by $x = 0$, $y = 1 - t$, and $z = t$; and $(0, 0, 1)$ to $(1, 0, 0)$, parameterized by $x = t$, $y = 0$, and $z = 1 - t$; in each case, $0 \leq t \leq 1$. Thus

$\oint_C y\, dx + xz\, dy + x^2\, dz$

$= \int_0^1 [(-t\, dt) + (0\, dt) + (-t^2)\, dt]$

$= -\frac{1}{2} + 0 - \frac{1}{3} = -\frac{5}{6}.$

6. $\int_S (\nabla \times \mathbf{F}) \cdot \mathbf{n}\, dS = \int_S (2\mathbf{k}) \cdot \frac{(-2x\mathbf{i} + \mathbf{k})}{\sqrt{4x^2 + 1}}\, dS$

$= \int_S \frac{2\, dS}{\sqrt{4x^2 + 1}}$. Since $|\cos \gamma| = \frac{1}{\sqrt{4x^2 + 1}}$ and $\mathcal{A}$, the projection of S on the xy plane, is a disk of radius 2, this integral equals

$\int_\mathcal{A} \frac{2}{\sqrt{4x^2 + 1}} \sqrt{4x^2 + 1}\, dA = 2(\text{Area of } \mathcal{A}) = 8\pi.$

Also, the boundary C is parameterized by $x = 2\cos\theta$, $y = 2\sin\theta$, and $z = 4\cos^2\theta$ for $0 \leq \theta \leq 2\pi$, so $\oint_C \mathbf{F} \cdot \mathbf{T}\, ds = \int_C -y\, dx + x\, dy + z\, dz$

$= \int_0^{2\pi} (4 - 32\cos^3\theta \sin\theta)\, d\theta = 8\pi.$

8. The boundary of S is the circle of radius 2 in the xy plane, centered at the origin. It is parameterized by $x = 2\cos\theta$, $y = 2\sin\theta$ for $0 \leq \theta \leq 2\pi$. Hence

$\int_S (\nabla \times \mathbf{F}) \cdot \mathbf{n}\, dS = \oint_C \mathbf{F} \cdot d\mathbf{r}$

$= \oint_C [(x^2 + y - 4)\, dx + 3xy\, dy]$

$= \int_0^{2\pi} (16 \sin\theta \cos^2\theta - 4\sin^2\theta + 8\sin\theta)\, d\theta$

$= 0 - 4 \int_0^{2\pi} \sin^2\theta\, d\theta + 0 = -4\pi.$

10. Let S be the plane region inside the triangle and let $\mathcal{A}$ be the projection of S onto the xy plane. Then

$\oint_C \mathbf{F} \cdot d\mathbf{r} = \int_S (\nabla \times \mathbf{F}) \cdot \mathbf{n}\, dS$

$= \int_S (e^x \mathbf{k}) \cdot \left(\frac{6\mathbf{i} + 4\mathbf{j} + 3\mathbf{k}}{\sqrt{61}}\right) dS = \int_S \frac{3}{\sqrt{61}} e^x\, dS$

$= \int_\mathcal{A} e^x\, dA = \int_0^2 \int_0^{3-3x/2} e^x\, dy\, dx = \frac{3}{2}(e^2 - 3).$

12. Let S be the plane region inside the rectangle and let $\mathcal{A}$ be the projection of S onto the xy plane. Then

$\oint_C \mathbf{F} \cdot d\mathbf{r} = \int_S (\nabla \times \mathbf{F}) \cdot \mathbf{n}\, dS$

17.3 Stokes' Theorem

$$= \int_S (\sin(x+z)\,\mathbf{i} - \sin(x+z)\,\mathbf{k}) \cdot \left(\frac{-\mathbf{j} + \mathbf{k}}{\sqrt{2}}\right) dS$$

$$= \int_S -\frac{1}{\sqrt{2}} \sin(x+z)\,dS = \int_R -\sin(x+y)\,dA$$

$$= \int_0^1 \int_0^1 -\sin(x+y)\,dy\,dx = \sin 2 - 2\sin 1.$$

14 Since $\mathbf{F} \cdot \mathbf{T} = 0$ on C, $\int_S (\nabla \times \mathbf{F}) \cdot \mathbf{n}\,dS =$

$$\oint_C \mathbf{F} \cdot \mathbf{T}\,ds = 0.$$

16 (a) $\oint_{C_3} \mathbf{F} \cdot d\mathbf{r} = \oint_{C_2} \mathbf{F} \cdot d\mathbf{r} = 3$

(b) $\oint_{C_4} \mathbf{F} \cdot d\mathbf{r} = 0$

18 If $\mathbf{F}$ is a central field, it can be written as $f(r)\mathbf{r}$, so

$$\nabla \times \mathbf{F} = \begin{vmatrix} \mathbf{i} & \mathbf{j} & \mathbf{k} \\ \frac{\partial}{\partial x} & \frac{\partial}{\partial y} & \frac{\partial}{\partial z} \\ f(r)x & f(r)y & f(r)z \end{vmatrix}.$$ The $\mathbf{i}$ component is

$$\frac{\partial}{\partial y}(f(r)z) - \frac{\partial}{\partial z}(f(r)y) = zf'(r)\frac{\partial r}{\partial y} - yf'(r)\frac{\partial r}{\partial z} =$$

$zf'(r)\frac{y}{r} - yf'(r)\frac{z}{r} = 0$. Similarly, the $\mathbf{j}$ and $\mathbf{k}$

components are also 0. Hence all central fields have curl $\mathbf{0}$.

20 S_2 consists of a curved part, S_3, and a circular base, R. On S_3, the exterior unit normal is $\mathbf{n} = x\mathbf{i} + y\mathbf{j}$, while on R it is $\mathbf{n} = -\mathbf{k}$. We have $\nabla \times \mathbf{F} =$

$$\begin{vmatrix} \mathbf{i} & \mathbf{j} & \mathbf{k} \\ \frac{\partial}{\partial x} & \frac{\partial}{\partial y} & \frac{\partial}{\partial z} \\ y & xz & x+2y \end{vmatrix} = (2-x)\mathbf{i} - \mathbf{j} + (z-1)\mathbf{k}, \text{ so}$$

curl $\mathbf{F} \cdot \mathbf{n} = 2x - x^2 - y$ on S_3 and curl $\mathbf{F} \cdot \mathbf{n} = -z + 1 = 1$ on R. As with Exercises 34 and 35 of Sec. 17.1, we can describe S_3 in cylindrical coordinates with $r = 1$, $0 \leq \theta \leq 2\pi$, $0 \leq z \leq x + 2 = \cos\theta + 2$, and $dS = dz\,d\theta$. Hence

$$\int_{S_3} (\nabla \times \mathbf{F}) \cdot \mathbf{n}\,dS =$$

$$\int_0^{2\pi} \int_0^{\cos\theta + 2} (2\cos\theta - \cos^2\theta - \sin\theta)\,dz\,d\theta$$

$$= \int_0^{2\pi} (-\cos^3\theta - \sin\theta\cos\theta + 4\cos\theta - 2\sin\theta)\,d\theta$$

$= 0$. Also, $\int_R (\nabla \times \mathbf{F}) \cdot \mathbf{n}\,dS = \int_R 1\,dS =$ Area

of $R = \pi$. Thus $\int_{S_2} (\nabla \times \mathbf{F}) \cdot \mathbf{n}\,dS = 0 + \pi = \pi$.

Since the orientation of C is opposite to that of S_2,

$$\oint_C \mathbf{F} \cdot d\mathbf{r} = -\pi.$$

22 (a) By the divergence theorem and Exercise 18 of

Sec. 16.1, $\int_S (\nabla \times \mathbf{F}) \cdot \mathbf{n}\,dS =$

$$\int_V \nabla \cdot (\nabla \times \mathbf{F})\,dV = \int_V 0\,dV = 0.$$

(b) A closed curve C divides S into two parts, S_1 and S_2. Because any orientation of C induces opposite orientations on S_1 and S_2, by Stokes' theorem we have $\int_{S_1} (\nabla \times \mathbf{F}) \cdot \mathbf{n}\,dS =$

$-\int_{S_2} (\nabla \times \mathbf{F}) \cdot \mathbf{n}\,dS$. Hence $\int_S (\nabla \times \mathbf{F}) \cdot \mathbf{n}\,dS$

$= \int_{S_1} (\nabla \times \mathbf{F}) \cdot \mathbf{n}\,dS + \int_{S_2} (\nabla \times \mathbf{F}) \cdot \mathbf{n}\,dS$

$= 0$.

24 (a) $\oint_{C_2} \mathbf{F} \cdot \mathbf{T}\,ds = -5$

(b) $\oint_{C_3} \mathbf{F} \cdot \mathbf{T}\,ds = 0$

(c) $\oint_{C_4} \mathbf{F} \cdot \mathbf{T}\, ds = -5$

26 (a) We must have div $\mathbf{F} = 0$.

(b) We must have curl $\mathbf{F} = \mathbf{0}$.

(c) Yes. Since $\oint_C \mathbf{F} \cdot d\mathbf{r} = 0$ for all circles, Stokes' theorem says $\int_S (\nabla \times \mathbf{F}) \cdot \mathbf{n}\, dS = 0$ for every disk S bounded by a circle (no matter how large or small). This can be used to show that $\nabla \times \mathbf{F} = \mathbf{0}$ everywhere, so $\mathbf{F}$ is conservative and $\oint_C \mathbf{F} \cdot d\mathbf{r} = 0$ whether C is a circle or not.

28 Let $\oint_{ABCA} \mathbf{F} \cdot d\mathbf{r}$ be the line integral along the

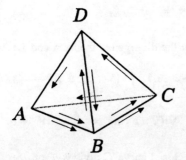

triangle ABC, with the vertices traversed in the order indicated. The orientation of ABC induces a corresponding orientation on the triangles DAB, DCA, and DBC, as indicated in the figure. Now

$\int_S (\nabla \times \mathbf{F}) \cdot \mathbf{n}\, dS = \int_{DAB} (\nabla \times \mathbf{F}) \cdot \mathbf{n}\, dS + \int_{DCA} (\nabla \times \mathbf{F}) \cdot \mathbf{n}\, dS + \int_{DBC} (\nabla \times \mathbf{F}) \cdot \mathbf{n}\, dS$, using the unit normals consistent with the oriented boundaries of each triangle. Since Stokes' theorem is assumed to hold for triangles, we have

$\int_S (\nabla \times \mathbf{F}) \cdot \mathbf{n}\, dS = \oint_{DABD} \mathbf{F} \cdot d\mathbf{r} + \oint_{DCAD} \mathbf{F} \cdot d\mathbf{r} +$

$\oint_{DBCD} \mathbf{F} \cdot d\mathbf{r}$. Using $\int_{DA} \mathbf{F} \cdot d\mathbf{r}$ for the line integral along the straight line from D to A, we break up each of the three closed line integrals into its three edges: $\int_S (\nabla \times \mathbf{F}) \cdot \mathbf{n}\, dS = \int_{DA} \mathbf{F} \cdot d\mathbf{r} + \int_{AB} \mathbf{F} \cdot d\mathbf{r}$

$+ \int_{BD} \mathbf{F} \cdot d\mathbf{r} + \int_{DC} \mathbf{F} \cdot d\mathbf{r} + \int_{CA} \mathbf{F} \cdot d\mathbf{r} +$

$\int_{AD} \mathbf{F} \cdot d\mathbf{r} + \int_{DB} \mathbf{F} \cdot d\mathbf{r} + \int_{BC} \mathbf{F} \cdot d\mathbf{r} + \int_{CD} \mathbf{F} \cdot d\mathbf{r}$

$= \int_{DA} \mathbf{F} \cdot d\mathbf{r} + \int_{AB} \mathbf{F} \cdot d\mathbf{r} + \int_{BD} \mathbf{F} \cdot d\mathbf{r} +$

$\int_{DC} \mathbf{F} \cdot d\mathbf{r} + \int_{CA} \mathbf{F} \cdot d\mathbf{r} - \int_{DA} \mathbf{F} \cdot d\mathbf{r} - \int_{BD} \mathbf{F} \cdot d\mathbf{r}$

$+ \int_{BC} \mathbf{F} \cdot d\mathbf{r} - \int_{DC} \mathbf{F} \cdot d\mathbf{r} = \int_{AB} \mathbf{F} \cdot d\mathbf{r} +$

$\int_{BC} \mathbf{F} \cdot d\mathbf{r} + \int_{CA} \mathbf{F} \cdot d\mathbf{r} = \int_{ABCA} \mathbf{F} \cdot d\mathbf{r}$, showing that Stokes' theorem works for S as well.

30 (a) Since adjacent curves have opposite orientation on their overlapping portions, the right-hand rule gives a consistent choice for the unit normal $\mathbf{n}$, permitting us to orient S.

(b) Yes. If S is orientable, there is a continuous choice of normal $\mathbf{n}$ over the whole surface. If we divide S into smaller regions, the overlapping portions of the boundaries of these regions will be assigned opposite orientations by the right-hand rule.

32 (a) $\nabla \times \mathbf{F} = \nabla \times (x\mathbf{i}) = \mathbf{0}$

(b) For the integral to make sense, we must first define the normal vector $\mathbf{n}$ over the whole surface. We cannot do this because the surface is equivalent to a Möbius band.

(c) Label the segments of C as shown in the figure. Then $\oint_{C_1} \mathbf{F} \cdot d\mathbf{r} =$

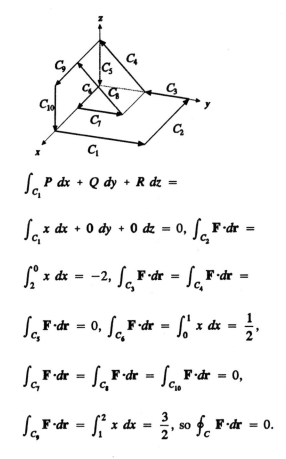

$\int_{C_1} P\,dx + Q\,dy + R\,dz =$

$\int_{C_1} x\,dx + 0\,dy + 0\,dz = 0, \int_{C_2} \mathbf{F}\cdot d\mathbf{r} =$

$\int_2^0 x\,dx = -2, \int_{C_3} \mathbf{F}\cdot d\mathbf{r} = \int_{C_4} \mathbf{F}\cdot d\mathbf{r} =$

$\int_{C_5} \mathbf{F}\cdot d\mathbf{r} = 0, \int_{C_6} \mathbf{F}\cdot d\mathbf{r} = \int_0^1 x\,dx = \frac{1}{2},$

$\int_{C_7} \mathbf{F}\cdot d\mathbf{r} = \int_{C_8} \mathbf{F}\cdot d\mathbf{r} = \int_{C_{10}} \mathbf{F}\cdot d\mathbf{r} = 0,$

$\int_{C_9} \mathbf{F}\cdot d\mathbf{r} = \int_1^2 x\,dx = \frac{3}{2},$ so $\oint_C \mathbf{F}\cdot d\mathbf{r} = 0.$

(d) Nothing. S is not orientable.

17.4 Applications of Stokes' Theorem

2 By Stokes' theorem, $\oint_{C_\mathbf{n}} \mathbf{F}\cdot d\mathbf{r} = \int_{S_\mathbf{n}} (\nabla\times\mathbf{F})\cdot\mathbf{n}\,dS$

$\approx (\nabla\times\mathbf{F})\cdot\mathbf{n}(\text{Area of } S_\mathbf{n}) = 0.0004\pi(\nabla\times\mathbf{F})\cdot\mathbf{n}.$

Hence $\oint_{C_1} \mathbf{F}\cdot d\mathbf{r} = 0.0008$ implies that $(\nabla\times\mathbf{F})\cdot\mathbf{i}$

$\approx \frac{0.0008}{0.0004\pi} = \frac{2}{\pi}$. Similarly, $(\nabla\times\mathbf{F})\cdot\mathbf{j} \approx$

$\frac{0.0012}{0.0004\pi} = \frac{3}{\pi}$ and $(\nabla\times\mathbf{F})\cdot\mathbf{k} \approx \frac{0.0004}{0.0004\pi} =$

$\frac{1}{\pi}$. Thus $\nabla\times\mathbf{F} \approx \langle 2/\pi, 3/\pi, 1/\pi\rangle$.

4 (a) $\nabla f\cdot\mathbf{n}$ gives the rate of change of f in the direction $\mathbf{n}$, the directional derivative.

(b) $(\nabla\times\mathbf{F})\cdot\mathbf{n}$ is a measure of the rotation of the field $\mathbf{F}$ about an axis parallel to $\mathbf{n}$, where $\mathbf{F}$ describes the flow of a fluid.

(c) The turning tendency of $\mathbf{F}$ about a line parallel to $\mathbf{n}$, where $\mathbf{F}$ describes a force field.

6 $\int_S \mathbf{E}\cdot\mathbf{n}\,dS = Q/\epsilon_0,$ so the divergence theorem says

$\int_\mathcal{V} \nabla\cdot\mathbf{E}\,dV = Q/\epsilon_0,$ where $\mathcal{V}$ is the volume

enclosed by S. Let V be the volume of $\mathcal{V}$. For a small volume, $\nabla\cdot\mathbf{E}$ is approximately constant, so

$\int_\mathcal{V} \nabla\cdot\mathbf{E}\,dV \approx (\nabla\cdot\mathbf{E})V$ and $\nabla\cdot\mathbf{E} \approx \dfrac{Q}{\epsilon_0 V} = q/\epsilon_0.$ In

the limit, $\nabla\cdot\mathbf{E} = q/\epsilon_0.$

8 Law 3' states that $\nabla\cdot\mathbf{B} = 0.$ By the divergence theorem, then, we have $\int_S \mathbf{B}\cdot\mathbf{n}\,dS = \int_\mathcal{V} \nabla\cdot\mathbf{B}\,dV$

$= \int_\mathcal{V} 0\,dV = 0,$ which is law 3.

10 Law 2' states that $\nabla\times\mathbf{E} = 0.$ Therefore, by Stokes' theorem, $\oint_C \mathbf{E}\cdot d\mathbf{r} = \int_S (\nabla\times\mathbf{E})\cdot\mathbf{n}\,dS =$

$\int_S 0\,dS = 0,$ which is law 2.

12 Law 4' states that $c^2\nabla\times\mathbf{B} = \mathbf{j}/\epsilon_0.$ By Stokes' theorem, $c^2\oint_C \mathbf{B}\cdot d\mathbf{r} = c^2\int_S (\nabla\times\mathbf{B})\cdot\mathbf{n}\,dS =$

$\int_S (\mathbf{j}/\epsilon_0)\cdot\mathbf{n}\,dS = \dfrac{1}{\epsilon_0}\int_S (\mathbf{j}\cdot\mathbf{n})\,dS,$ which is law 4.

14 (a) Since $\mathbf{F}\cdot\mathbf{k} = 0,$ $\mathbf{F}$ is perpendicular to $\mathbf{k}$, which is the normal vector to the xy plane, and $\mathbf{F}$ is parallel to the xy plane.

(b) $\nabla\times\mathbf{F} = (Q_x - P_y)\mathbf{k},$ so $\nabla\times\mathbf{F}$ is perpendicular to the xy plane and parallel to the z axis.

(c) If **F** represented a water flow, then a paddle wheel would turn most briskly if its axis were parallel to the z axis.

17.S Review Exercises

2. (a) $\mathbf{F} = \dfrac{-y\mathbf{i}}{x^2+y^2} + \dfrac{x\mathbf{j}}{x^2+y^2}$, so div **F**

$$= \dfrac{\partial}{\partial x}\left(\dfrac{-y}{x^2+y^2}\right) + \dfrac{\partial}{\partial y}\left(\dfrac{x}{x^2+y^2}\right)$$

$$= \dfrac{2xy}{(x^2+y^2)^2} + \dfrac{-2xy}{(x^2+y^2)^2} = 0.$$

(b) It is shown in Example 2 of Sec. 16.6 that curl **F** = **0**.

(c) If $f = \tan^{-1}(y/x)$ then $f_x = \dfrac{1}{1+(y/x)^2}\cdot\dfrac{-y}{x^2} = \dfrac{-y}{x^2+y^2}$ and $f_y = \dfrac{1}{1+(y/x)^2}\cdot\dfrac{1}{x} = \dfrac{x}{x^2+y^2}$; thus $\mathbf{F} = \nabla f$ wherever both are defined.

(d) Parameterize the unit circle C by $x = \cos t$, $y = \sin t$. Note that $x^2 + y^2 = 1$, so on C we have $\mathbf{F} = -\sin t\,\mathbf{i} + \cos t\,\mathbf{j}$. Then $\oint_C \mathbf{F}\cdot d\mathbf{r}$

$$= \oint_C P\,dx + Q\,dy$$

$$= \int_0^{2\pi} (-\sin t)(-\sin t\,dt) + \cos t \cos t\,dt$$

$$= \int_0^{2\pi} (\sin^2 t + \cos^2 t)\,dt = \int_0^{2\pi} dt = 2\pi.$$

(e) By (d), the unit circle C is a closed curve such that $\oint_C \mathbf{F}\cdot d\mathbf{r} \neq 0$. Thus **F** is not conservative.

(f) Since the domain of **F** is not simply connected, the fact that curl **F** = **0** does not imply that **F** is conservative. However, in (c) we found that **F** is a gradient, and should therefore be conservative, despite the result of (e). The apparent contradiction arises because of the peculiarities of $\tan^{-1}\dfrac{y}{x}$. It cannot be defined in a continuous manner on any region that contains the origin; it is a discontinuous function and is undefined along the entire y axis.

4. C bounds a square in the yz plane whose sides are of length $2\sqrt{2}$ and whose area is therefore 8. Given the orientation of C, we see that the corresponding unit normal for the enclosed square S is $\mathbf{n} = \mathbf{i}$. To compute $(\nabla \times \mathbf{F})\cdot\mathbf{n}$, all we need then is the $\mathbf{i}$ component of $\nabla \times \mathbf{F}$, which is $R_y - Q_z$. Now $\mathbf{F} = x^2 y e^z \mathbf{i} + (x+y+z)\mathbf{j} + x^2 z\mathbf{k}$, so $Q_z = 1$ and $R_y = 0$; thus $(\nabla \times \mathbf{F})\cdot\mathbf{n} = R_y - Q_z = -1$. By Stokes' theorem, it now follows that $\oint_C \mathbf{F}\cdot d\mathbf{r} =$

$$\int_S (\nabla\times\mathbf{F})\cdot\mathbf{n}\,dS = -\int_S dS = -(\text{Area of }S)$$

$$= -8.$$

6. Let S^* be the disk on the xy plane bounded by $x^2 + y^2 = 4$. Then $\int_S (\nabla\times\mathbf{F})\cdot\mathbf{n}\,dS =$

$$\int_{S^*} (\nabla\times\mathbf{F})\cdot\mathbf{k}\,dS = \int_{S^*} (3y-1)\,dS$$

$$= \int_0^{2\pi}\int_0^2 (3r\sin\theta - 1)\,r\,dr\,d\theta = -4\pi.$$

8. $\int_C x\sin y^2\,dx + \dfrac{x}{y^2+1}\,dy + \tan^{-1} z\,dz$

$$= \int_0^1 t\sin t^2\,dt + \dfrac{t}{t^2+1}\,dt + \tan^{-1} t\,dt =$$

$$\left[-\frac{1}{2}\cos t^2 + \frac{1}{2}\ln(t^2+1) + t\tan^{-1}t - \frac{1}{2}\ln(t^2+1)\right]\Big|_0^1$$

$$= -\frac{1}{2}\cos 1 + \tan^{-1}1 - \left(-\frac{1}{2}\right)$$

$$= \frac{1}{2} - \frac{1}{2}\cos 1 + \frac{\pi}{4}$$

10. $\int_C \frac{y}{x+1}dx + \frac{1}{y^2+1}dy + \frac{1}{z+1}dz$

$$= \int_0^1 \frac{t}{t^2+1} 2t\, dt + \frac{1}{t^2+1}dt + \frac{1}{t^4+1} 4t^3\, dt$$

$$= \int_0^1 \left(2 - \frac{1}{t^2+1} + \frac{4t^3}{t^4+1}\right) dt$$

$$= [2t - \tan^{-1}t + \ln(t^4+1)]\Big|_0^1 = 2 - \frac{\pi}{4} + \ln 2$$

12. (a) It is zero. For, by the divergence theorem,

$$\int_S \mathbf{F}\cdot\mathbf{n}\, dS = \int_V \nabla\cdot\mathbf{F}\, dV = \int_V 0\, dV = 0,$$

where V is the region bounded by S.

(b) Let S_r be the sphere of radius r centered at the origin. If r is large enough so that S and S_r don't intersect, then, by Corollary 1 in Sec. 17.2, $\int_S \mathbf{F}\cdot\mathbf{n}\, dS = \int_{S_r} \mathbf{F}\cdot\mathbf{n}\, dS$. But $|\mathbf{F}\cdot\mathbf{n}|$

$$\leq |\mathbf{F}| \leq \frac{1}{r^3}, \text{ so } \left|\int_{S_r} \mathbf{F}\cdot\mathbf{n}\, dS\right| \leq \int_{S_r} \frac{1}{r^3} dS$$

$$= \frac{1}{r^3}(\text{Area of } S_r) = \frac{1}{r^3}\cdot 4\pi r^2 = \frac{4\pi}{r}. \text{ As}$$

$r \to \infty$, $\frac{4\pi}{r} \to 0$, so $\int_S \mathbf{F}\cdot\mathbf{n}\, dS$ must be zero.

14. (a) $\nabla \times \mathbf{F} = \mathbf{0}$

(b) $\nabla\cdot\mathbf{F} > 0$

16. (a)

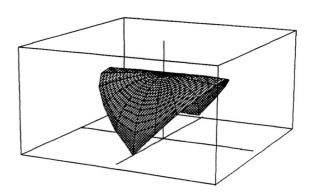

(b) The surface is described by $-a \leq x \leq a$, $-\sqrt{a^2-x^2} \leq y \leq \sqrt{a^2-x^2}$, $z = \sqrt{a^2-x^2}$. Since $\frac{\partial z}{\partial x} = -\frac{x}{\sqrt{a^2-x^2}}$ and $\frac{\partial z}{\partial y}$

$$= 0, |\cos\gamma| = \left[\frac{x^2}{a^2-x^2} + 0 + 1\right]^{-1/2} =$$

$\frac{\sqrt{a^2-x^2}}{a}$. So the area is

$$\int_{-a}^a \int_{-\sqrt{a^2-x^2}}^{\sqrt{a^2-x^2}} \frac{a}{\sqrt{a^2-x^2}}\, dy\, dx$$

$$= \int_{-a}^a \frac{a}{\sqrt{a^2-x^2}} \cdot 2\sqrt{a^2-x^2}\, dx = 2a\int_{-a}^a dx$$

$$= 4a^2.$$

18 (a)

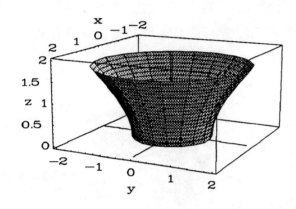

(b) We have $2z \cdot \dfrac{\partial z}{\partial x} = \dfrac{\partial}{\partial x}(z^2) = \dfrac{\partial}{\partial x}(x^2 + y^2 - 1)$

$= 2x$, so $\left(\dfrac{\partial z}{\partial x}\right)^2 = \left(\dfrac{2x}{2z}\right)^2 = \dfrac{x^2}{z^2} =$

$\dfrac{x^2}{x^2 + y^2 - 1}$. Similarly, $\left(\dfrac{\partial z}{\partial y}\right)^2 =$

$\dfrac{y^2}{x^2 + y^2 - 1}$, so $|\cos \gamma| =$

$\left[\dfrac{x^2}{x^2 + y^2 - 1} + \dfrac{y^2}{x^2 + y^2 - 1} - 1\right]^{-1/2} =$

$\sqrt{x^2 + y^2 - 1}$. Hence the area is

$\displaystyle\int_A \dfrac{1}{\sqrt{x^2 + y^2 - 1}}\, dA$, where A is the "washer"

in the xy plane described by $0 \le \theta \le 2\pi$, $1 \le r \le 2$.

20 By Theorem 2 of Sec. 16.6, $\displaystyle\int_C \nabla f \cdot d\mathbf{r}$

$= f(0, 0, 1) - f(1, 0, 0) = 0$.

22 (a)

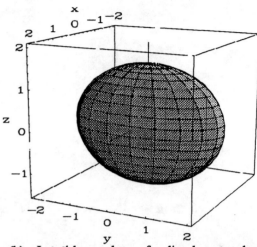

(b) Let S' be a sphere of radius k centered at the origin and small enough to fit entirely within S. Since $\nabla \cdot \mathbf{F} = 0$, Corollary 1 to the two-surface case of the divergence theorem says

$\displaystyle\int_S \mathbf{F} \cdot \mathbf{n}\, dS = \int_{S'} \mathbf{F} \cdot \mathbf{n}\, dS = \int_{S'} \dfrac{\hat{\mathbf{r}}}{r^2} \cdot \hat{\mathbf{r}}\, dS =$

$\dfrac{1}{k^2}(\text{Area of } S') = \dfrac{1}{k^2}(4\pi k^2) = 4\pi$.

(c) Since S is a closed surface, its steradian measure from inside is 4π.

24 (a) $P = f_x$, $Q = f_y$, and $R = f_z$, so $P_y = f_{xy} = f_{yx} = Q_x$. Similarly, $P_z = R_x$ and $Q_z = R_y$.

(b) If $df = P\, dx + Q\, dy + R\, dz$, then $\mathbf{F} = P\mathbf{i} + Q\mathbf{j} + R\mathbf{k} = \nabla f$, so $\mathbf{F}$ is conservative and $\nabla \times \mathbf{F} = \mathbf{0}$. Hence $R_y - Q_z = 0$, $R_x - P_z = 0$, and $Q_x - P_y = 0$.

26 (a) Because $\mathbf{F} = z\mathbf{i}$ is parallel to the circular top and bottom of the cylinder S, the flux through those disks is 0. Now the curved side of S—call it S'—has $\mathbf{n} = \dfrac{x}{a}\mathbf{i} + \dfrac{y}{a}\mathbf{j}$, so $\mathbf{F} \cdot \mathbf{n} =$

$\dfrac{xz}{a}$. By symmetry, $\displaystyle\int_{S'} \dfrac{xz}{a}\, dS = 0$.

Alternatively, parameterize S' by $0 \leq \theta \leq 2\pi$, $0 \leq z \leq h$, and $dS = a\, d\theta\, dz$. Then

$$\int_{S'} \frac{xz}{a}\, dS = \int_0^{2\pi} \int_0^h a \cos\theta\, z\, dz\, d\theta = \frac{1}{2} ah^2 \sin\theta \Big|_0^{2\pi} = 0.$$

(b) By the divergence theorem, $\int_S \mathbf{F} \cdot \mathbf{n}\, dS = \int_V \nabla \cdot \mathbf{F}\, dV = \int_V 0\, dV = 0.$

28 (a) Since $\mathbf{F}$ is defined throughout space and
$\nabla \times \mathbf{F} = (xe^z \cos xy - xe^z \cos xy)\mathbf{i} + (ye^z \cos xy - ye^z \cos xy)\mathbf{j} + (e^z \cos xy - xye^z \sin xy - e^z \cos xy + xye^z \sin xy)\mathbf{k} = \mathbf{0}$,
$\mathbf{F}$ is conservative.

(b) Since $\mathbf{F}$ is defined throughout space and $\nabla \times \mathbf{F} = (0 - 0)\mathbf{k} = \mathbf{0}$, $\mathbf{F}$ is conservative.

30 (a) Let $f(x, y, z) = -\frac{1}{4}(x^2 + y^2 + z^2)^{-2}$. Then

$$\frac{x\mathbf{i} + y\mathbf{j} + z\mathbf{k}}{(x^2 + y^2 + z^2)^3} = \nabla f \text{ and is conservative.}$$

(b) Since $\nabla \times \mathbf{F} \neq \mathbf{0}$, $\mathbf{F}$ is not conservative.

32 Consider the curve C parameterized by $x = 0$, $y = \cos\theta$, $z = \sin\theta$. Then $\oint_C \mathbf{F} \cdot d\mathbf{r} =$

$\oint_C 3\, dx + (-\sin\theta)(-\sin\theta\, d\theta) + (\cos\theta)(\cos\theta\, d\theta)$

$= \int_0^{2\pi} d\theta = 2\pi \neq 0$, and $\mathbf{F}$ is not conservative.

34 All are conservative since each is defined throughout space and the curl is $\mathbf{0}$ in each case.

36 (a) $\oint_{C_4} \mathbf{F} \cdot d\mathbf{r} = \oint_{C_3} \mathbf{F} \cdot d\mathbf{r} = 3$

(b) $\oint_{C_5} \mathbf{F} \cdot d\mathbf{r} = 0$

(c) $\oint_{C_6} \mathbf{F} \cdot d\mathbf{r} = \oint_{C_1} \mathbf{F} \cdot d\mathbf{r} + \oint_{C_2} \mathbf{F} \cdot d\mathbf{r} + \oint_{C_3} \mathbf{F} \cdot d\mathbf{r} = 6$

38 $\nabla \cdot \mathbf{F} = 3$, and $\int_V \nabla \cdot \mathbf{F}\, dV = \int_V 3\, dV = 3 \cdot (\text{Volume of } \mathcal{V}) = \int_S \mathbf{F} \cdot \mathbf{n}\, dS.$

40 $\int_S \mathbf{F} \cdot \mathbf{n}\, dS = \int_V \nabla \cdot \mathbf{F}\, dV = \int_V 12\, dV$
$= 12 \cdot \frac{4}{3}\pi 3^3 = 432\pi$

42 (d) Since $\mathbf{F} = \nabla f$, $\mathbf{F}$ is conservative.

(e) No.

44 All three fields are defined throughout space, so we need check only whether $\nabla \times \mathbf{F} = \mathbf{0}$.

(a) Conservative

(b) Not conservative; $\nabla \times \mathbf{F} = z^2\mathbf{i}$

(c) Conservative

46 (a) $\oint_C \mathbf{F} \cdot \mathbf{T}\, ds = \oint_C 3xy\, dx + 4\, dy + z\, dz$

$= \int_0^1 z\, dz + \int_0^1 4\, dy +$

$\left(\int_1^0 3x\, dx + \int_1^0 z\, dz\right) + \int_1^0 4\, dy = -\frac{3}{2}$

(b) Since $\oint_C \mathbf{F} \cdot \mathbf{T}\, ds \neq 0$, $\mathbf{F}$ is not conservative.

48 Let S^* be the disk on the xy plane bounded by $x^2 + y^2 = 4$. Then $\int_S (\nabla \times \mathbf{F}) \cdot \mathbf{n}\, dS =$

$\int_{S^*} (\nabla \times \mathbf{F}) \cdot \mathbf{k}\, dS = \int_{S^*} (3y - 1)\, dS$

$= \int_0^{2\pi} \int_0^2 (3r \sin\theta - 1)\, r\, dr\, d\theta = -4\pi.$

50 (a) By the divergence theorem, $\int_S (f\nabla g) \cdot \mathbf{n}\, dS =$

$\int_\mathcal{V} \nabla \cdot (f \nabla g) \, dV$. Now $\nabla \cdot (f \nabla g) = f \nabla \cdot (\nabla g) + \nabla f \cdot \nabla g$, by the identity in Exercise 26 of Sec. 16.1, so $\int_\mathcal{V} \nabla \cdot (f \nabla g) \, dV =$

$\int_\mathcal{V} (f \nabla^2 g + \nabla f \cdot \nabla g) \, dV$.

(b) Just switch f and g in (a).

(c) $\int_\mathcal{V} (f \nabla^2 g - g \nabla^2 f) \, dV$

$= \int_\mathcal{V} (f \nabla^2 g + \nabla f \cdot \nabla g) \, dV -$

$\int_\mathcal{V} (g \nabla^2 f + \nabla f \cdot \nabla g) \, dV$

$= \int_S (f \nabla g) \cdot \mathbf{n} \, dS - \int_S (g \nabla f) \cdot \mathbf{n} \, dS$

$= \int_S (f \nabla g - g \nabla f) \cdot \mathbf{n} \, dS$

52 The center of mass is at the origin.

54 Let $\mathbf{c} = \mathbf{A} - \mathbf{B}$. Then $\mathbf{c} \cdot \mathbf{A} = \mathbf{c} \cdot \mathbf{B}$ implies $(\mathbf{A} - \mathbf{B}) \cdot (\mathbf{A} - \mathbf{B}) = 0$. Thus $\|\mathbf{A} - \mathbf{B}\| = 0$, so $\mathbf{A} = \mathbf{B}$.

56 (a) By the divergence theorem, $\int_S \mathbf{c} \cdot \mathbf{n} \, dS =$

$\int_\mathcal{V} \nabla \cdot \mathbf{c} \, dV = \int_\mathcal{V} 0 \, dV = 0$.

(b) Combining (a) with Exercise 53, we have

$\mathbf{c} \cdot \int_S \mathbf{n} \, dS = 0 = \mathbf{c} \cdot \mathbf{0}$ for all $\mathbf{c}$. By Exercise 54, $\int_S \mathbf{n} \, dS = \mathbf{0}$.

58 The parallelepiped $\mathcal{V}$ is spanned by $\mathbf{i} + \mathbf{j} + \mathbf{k}$, $2\mathbf{i} + \mathbf{j} + \mathbf{k}$, and $2\mathbf{i} + 3\mathbf{j} + 4\mathbf{k}$, so by Theorem 4 of Sec. 12.6, its volume V is $\text{abs} \begin{vmatrix} 1 & 1 & 1 \\ 2 & 1 & 1 \\ 2 & 3 & 4 \end{vmatrix} = |-1|$

$= 1$. $\mathbf{F} = 2x\mathbf{i} + 3y\mathbf{j} + 4z\mathbf{k}$, so $\nabla \cdot \mathbf{F} = 9$. By the divergence theorem, $\int_S \mathbf{F} \cdot \mathbf{n} \, dS = \int_\mathcal{V} \nabla \cdot \mathbf{F} \, dV = 9V = 9$.

Appendices

A Real Numbers

2 $a - 2 < b - 2$

4 $(-2)a > (-2)b$

6 $3 - a > 3 - b$

8 $3x - 5 < 7x + 11, -4x < 16, x > 4; (4, \infty)$

10 $2x > 3x + 7, -x > 7, x < -7; (-\infty, -7)$

12 $(-1, 3)$

14 $\emptyset$

16 $(2, 3)$ and $(4, \infty)$

18 $(2, \infty)$

20 $\left(-\frac{1}{2}, 0\right)$ and $\left(\frac{1}{3}, \infty\right)$

22 $|2x - 4| < 1, -1 < 2x - 4 < 1, 3 < 2x < 5,$
$3/2 < x < 5/2$

24 $-2 < 4(x + 2) < 2, -1/2 < x + 2 < 1/2,$
$-5/2 < x < -3/2$

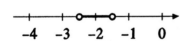

26 None

28 $(1, \infty)$

30 $a = -1, b = 1$

32 $3.1416 = \dfrac{31{,}416}{10{,}000}$

B Graphs and Lines

B.1 Coordinate Systems and Graphs

2

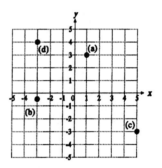

(a) 1st quadrant
(b) 3rd quadrant
(c) 4th quadrant
(d) 2nd quadrant

4 (a) $D = \sqrt{(5-3)^2 + (6-4)^2} = \sqrt{4 + 4}$
$= 2\sqrt{2}$

(b) $D = \sqrt{49 + 0} = 7$

(c) $D = \sqrt{36 + 64} = 10$

6

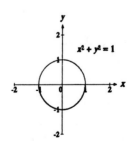

8

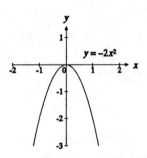

10

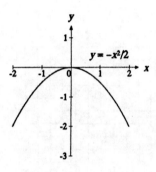

12

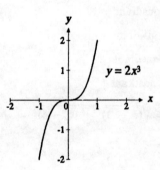

14

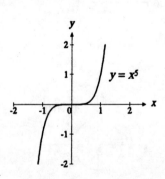

16

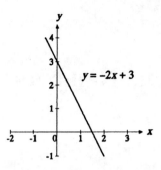

18 $y = 3x - 6$; y intercept, -6; x intercept, 2

20 y intercept, none; x intercepts, $1, -1$

22 y intercept, 4; x intercepts, $-2, 2$

24

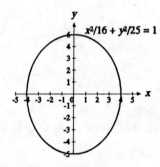

26

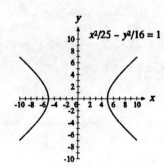

28

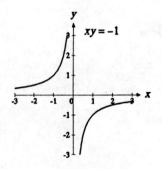

B Graphs and Lines

30

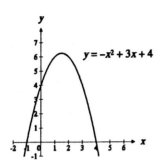

32

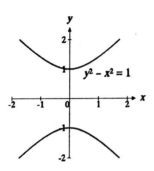

34

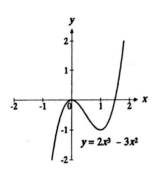

36

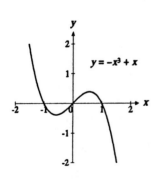

38

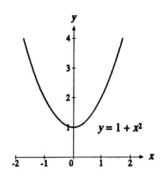

40

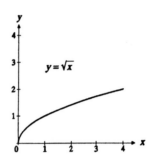

42

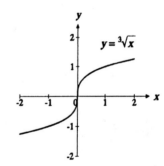

44

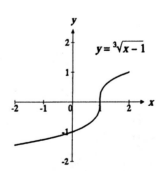

46 $(x + 2)^2 + (y - 3)^2 = 1/4$

B.2 Lines and Their Slopes

2   $m = \dfrac{-2}{-5} = \dfrac{2}{5}$

4   $m = \dfrac{-4}{4} = -1$

6   $m = \dfrac{0}{2} = 0$

8   (a) Negative
    (b) Positive
    (c) Zero
    (d) Positive
    (e) Negative

10 (a) $-\dfrac{1}{2}$

    (b) 2

12 $m_1 = -\dfrac{3}{4}$, $m_2 = -\dfrac{5}{4}$; they are not parallel.

14 $m_1 = \dfrac{4}{7}$, $m_2 = -\dfrac{5}{3}$; they are not perpendicular.

16 (a) $y = -\dfrac{1}{2}x + \dfrac{3}{4}$

    (b) $y = 2x - 2$

    (c) $y = \dfrac{3}{5}x + 3$

18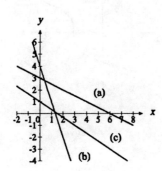

    (a) $m = -\dfrac{1}{2}$, $b = 3$

    (b) $m = -3$, $b = 4$

    (c) $m = -\dfrac{2}{3}$, $b = 1$

20 (a) $y - 0 = \dfrac{4}{3}(x - 0)$

    (b) $y - 5 = 0(x - 2)$, $y - 5 = 0$

22 (a) $m = 2/2 = 1$, $y - 4 = 1(x - 4)$, or $y - 4 = x - 4$

    (b) $m = \dfrac{9}{5}$, $y + 2 = \dfrac{9}{5}(x + 5)$

    (c) $m = \dfrac{5}{3}$, $y - 0 = \dfrac{5}{3}(x - 0)$

24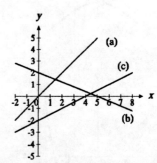

    (a) $x - y = 0$, $y = x$

    (b) $-5y = 2x - 10$, $y = -\dfrac{2}{5}x + 2$

    (c) $x - 2y = 4$, $y = \dfrac{1}{2}x - 2$

26 $m = \dfrac{3}{2}$, so $y = \dfrac{3}{2}x + 2$.

28 $m = 1$, so $y = x + 0$, or $y = x$.

32 (a) $\dfrac{x}{3} + \dfrac{y}{5} = 1$

    (b) $\dfrac{x}{-1} + \dfrac{y}{2} = 1$

    (c) $\dfrac{x}{-1/2} + \dfrac{y}{-3} = 1$

B Graphs and Lines

34 (2, 4)

38 $m_1 = \frac{7}{5}$, $m_2 = \frac{10}{7}$, so the lines are not parallel.

40 $m_1 = -3$, $b = 5$, so $y = -3x + 5$.

C   Topics in Algebra

2 $\dfrac{3}{\sqrt{x}} \cdot \dfrac{\sqrt{x}}{\sqrt{x}} = \dfrac{3\sqrt{x}}{x}$

4 $\dfrac{x^3}{\sqrt{x}} \cdot \dfrac{\sqrt{x}}{\sqrt{x}} = \dfrac{x^3 \sqrt{x}}{x} = x^2 \sqrt{x}$

6 $\dfrac{2}{3 + \sqrt{2}} \cdot \dfrac{3 - \sqrt{2}}{3 - \sqrt{2}} = \dfrac{2(3 - \sqrt{2})}{9 - 2} = \dfrac{6 - 2\sqrt{2}}{7}$

8 $\dfrac{1}{\sqrt{2} - \sqrt{3}} \cdot \dfrac{\sqrt{2} + \sqrt{3}}{\sqrt{2} + \sqrt{3}} = \dfrac{\sqrt{2} + \sqrt{3}}{-1} = -\sqrt{2} - \sqrt{3}$

10 $\dfrac{\sqrt{x + 1} - \sqrt{x}}{x} \cdot \dfrac{\sqrt{x + 1} + \sqrt{x}}{\sqrt{x + 1} + \sqrt{x}} = \dfrac{x + 1 - x}{x(\sqrt{x + 1} + \sqrt{x})}$

$= \dfrac{1}{x(\sqrt{x + 1} + \sqrt{x})}$

12 $\dfrac{\sqrt{u} - \sqrt{v}}{5} \cdot \dfrac{\sqrt{u} + \sqrt{v}}{\sqrt{u} + \sqrt{v}} = \dfrac{u - v}{5(\sqrt{u} + \sqrt{v})}$

14 (a) $(x^2 + 6x + 3^2) + 3 - 3^2 = (x + 3)^2 - 6$

(b) $(x^2 - 6x + 3^2) + 3 - 3^2 = (x - 3)^2 - 6$

(c) $\left(x^2 + 3x + \left(\dfrac{3}{2}\right)^2\right) + 5 - \left(\dfrac{3}{2}\right)^2$

$= \left(x + \dfrac{3}{2}\right)^2 + \dfrac{11}{4}$

16 (a) $\left(x^2 + \dfrac{x}{3} + \left(\dfrac{1}{6}\right)^2\right) + 1 - \left(\dfrac{1}{6}\right)^2$

$= \left(x + \dfrac{1}{6}\right)^2 + \dfrac{35}{36}$

(b) $\left(x^2 + \dfrac{2}{3}x + \left(\dfrac{1}{3}\right)^2\right) - 7 - \left(\dfrac{1}{3}\right)^2$

$= \left(x + \dfrac{1}{3}\right)^2 - \dfrac{64}{9}$

(c) $(x^2 + 10x + 5^2) + 25 - 5^2 = (x + 5)^2$

18 (a) $4(x^2 - 2x + 1) + 5 - 4 \cdot 1^2 = 4(x - 1)^2 + 1$

(b) $3\left(x^2 + \dfrac{2}{3}x + \left(\dfrac{1}{3}\right)^2\right) + 1 - 3\left(\dfrac{1}{3}\right)^2$

$= 3\left(x + \dfrac{1}{3}\right)^2 + \dfrac{2}{3}$

(c) $3\left(x^2 - \dfrac{2}{3}x + \left(\dfrac{1}{3}\right)^2\right) + 1 - 3\left(\dfrac{1}{3}\right)^2$

$= 3\left(x - \dfrac{1}{3}\right)^2 + \dfrac{2}{3}$

20 (a) $x = \dfrac{-5 \pm \sqrt{5^2 - 4 \cdot 2 \cdot (-7)}}{2 \cdot 2} = \dfrac{-5 \pm 9}{4}$, so

$x = -\dfrac{7}{2}$ or 1.

(b) $x = \dfrac{-1 \pm \sqrt{1^2 - 4 \cdot 3 \cdot 5}}{2 \cdot 3} = \dfrac{-1 \pm \sqrt{-59}}{6}$; no

real solutions.

(c) $x = \dfrac{4 \pm \sqrt{4^2 - 4 \cdot 4 \cdot 1}}{2 \cdot 4} = \dfrac{1}{2}$

22 (a) $b^2 - 4ac = 900 - 4 \cdot 9 \cdot 25 = 0$; one.

(b) $b^2 - 4ac = 1 - 4 \cdot 1 \cdot 1 = -3 < 0$; none.

(c) $b^2 - 4ac = 1 - 4 \cdot 1 \cdot (-1) = 5 > 0$; two.

24 (a) $\binom{5}{3} = \dfrac{5 \cdot 4 \cdot 3}{3 \cdot 2 \cdot 1} = 10$

(b) $\binom{6}{3} = \frac{6 \cdot 5 \cdot 4}{3 \cdot 2 \cdot 1} = 20$

(c) $\binom{10}{3} = \frac{10 \cdot 9 \cdot 8}{3 \cdot 2 \cdot 1} = 120$

26 $1 + \binom{12}{1}x + \binom{12}{2}x^2 + \binom{12}{3}x^3 = 1 + 12x + 66x^2 + 220x^3$

28 $\binom{11}{3} = \frac{11 \cdot 10 \cdot 9}{3 \cdot 2 \cdot 1} = 165$

30 (a) $\frac{1 - (-1/2)^6}{1 - (-1/2)} = \frac{21}{32}$

(b) $\frac{8(1 - (1/10)^{10})}{1 - 1/10} = \frac{80}{9}\left(1 - \frac{1}{10^{10}}\right)$

(c) $\frac{2/3(1 - (2/3)^8)}{1 - 2/3} = 2(1 - (2/3)^8) = \frac{12610}{6561}$

32 Note that 1 is a root, so $2x^3 - x^2 - 2x + 1 = (x - 1)(2x^2 + x + 1) = (x - 1)(2x - 1)(x + 1)$.

34 No rational roots

36 $x^4 + x^3 - 3x^2 - 4x - 4$; 2 and -2 are the only rational solutions, so $x - 2$ and $x + 2$ are the desired factors.

38 $3x^2 - 2x - 2 = 3\left(x - \frac{1 + \sqrt{7}}{3}\right)\left(x - \frac{1 - \sqrt{7}}{3}\right)$

40 Assume that $a \neq 0$; then $(a + b)^4 = [a(1 + b/a)]^4 = a^4(1 + b/a)^4$. Let $x = b/a$; we have $(1 + x)^4 = 1 + 4x + 6x^2 + 4x^3 + x^4 = 1 + \frac{4b}{a} + \frac{6b^2}{a^2} + \frac{4b^3}{a^3} + \frac{b^4}{a^4}$, so $(a + b)^4 = a^4[1 + b/a]^4 = a^4 + 4a^3b + 6a^2b^2 + 4ab^3 + b^4$, as claimed.

D Exponents

2 (a) $64^{1/2} = 8$
(b) $64^{-1/2} = 1/8$
(c) $64^{2/3} = 16$
(d) $64^0 = 1$
(e) $64^{5/6} = 32$

4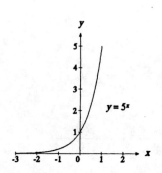

6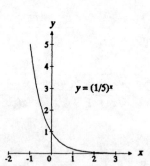

8 (a) $16 = 4^2$
(b) $8 = 4^{3/2}$
(c) $\frac{1}{\sqrt{2}} = 4^{-1/4}$
(d) $4\sqrt{2} = 4^{5/4}$
(e) $16^{35} = 4^{70}$

10 (a) $(\sqrt{a})^2 = a^1$
(b) $\frac{1}{a^2} = a^{-2}$
(c) $1 = a^0$

D Exponents

(d) $\left(\dfrac{a^3}{a^5}\right)^{10} = (a^{-2})^{10} = a^{-20}$

(e) $\dfrac{a}{\sqrt[3]{a}} = a^{1-1/3} = a^{2/3}$

12 (a) Domain: all real numbers; range: all positive real numbers.

(b) Domain: all real numbers; range: all real numbers

(c) Domain: all real numbers; range: all positive real numbers

(d) Domain: all nonnegative real numbers; range: all nonnegative real numbers

14 (a) $(-32)^{4/5} = 16$, $(-1)^{4/5} = 1$, $0^{4/5} = 0$, $1^{4/5} = 1$, $32^{4/5} = 16$

(b)

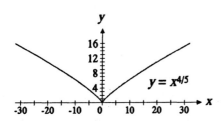

16 2.646 is greater than $\sqrt{7}$.

18 2 and 4

20

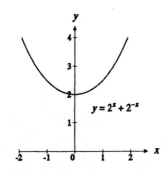

22 (a) $9 \cdot 10^2$

(b) $9.57 \cdot 10^2$

(c) $9.5 \cdot 10^{-2}$

(d) $1.5 \cdot 10^4$

(e) $1.5 \cdot 10^{-3}$

24 (a) Press 5 and $\sqrt{x}$ twice.

(b) Press 5 and $\sqrt{x}$ three times.

(c) Compute 5^3, then press $\sqrt{x}$ three times.

(d) Press 5 and $\sqrt{x}$ twice, then $1/x$.

E Mathematical Induction

2 (a)

k	S_k	Value of S_k
1	1	1
2	1 + 3	4
3	1 + 3 + 5	9
4	1 + 3 + 5 + 7	16
5	1 + 3 + 5 + 7 + 9	25
6	1 + 3 + 5 + 7 + 9 + 11	36
7	1 + 3 + 5 + 7 + 9 + 11 + 13	49
8	1 + 3 + 5 + 7 + 9 + 11 + 13 + 15	64

(b) $S_k = k^2$

(c) Suppose $S_k = k^2$. From (a) we know this is true for $k = 1, 2, \cdots, 8$. Then $S_{k+1} = S_k + (2k + 1) = k^2 + 2k + 1 = (k + 1)^2$. Hence the formula is true for all k.

4 (a) It appears as if we should have $x^k - 1 = (x^{k-1} + x^{k-2} + \cdots + 1)(x - 1)$.

(b) Suppose the formula in (a) is valid for k. Then $x^{k+1} - 1 = x^{k+1} - x + x - 1$

$$= x(x^k - 1) + (x - 1)$$
$$= x[(x^{k-1} + x^{k-2} + \cdots + 1)(x-1)] + (x-1)$$
$$= [x(x^{k-1} + x^{k-2} + \cdots + 1) + 1](x - 1)$$
$$= (x^k + x^{k-1} + \cdots + x + 1)(x - 1),$$

showing that the formula also works for $k + 1$.

5 (a) k must be odd. If $x + 1$ is to be a factor of $x^k + 1$, then -1 must be a root of $x^k + 1$, which is true only when k is odd.

(b) Assume that $x + 1$ divides $x^k + 1$. Then $x^k + 1 = (x + 1)p(x)$, where $p(x)$ is some polynomial. Hence $x^k = (x + 1)p(x) - 1$, so $x^{k+2} = x^2(x + 1)p(x) - x^2$ and $x^{k+2} + 1$
$$= x^2(x + 1)p(x) - x^2 + 1$$
$$= x^2(x + 1)p(x) - (x + 1)(x - 1)$$
$$= (x + 1)[x^2 p(x) - (x - 1)],$$ showing that the result is also true for $k + 2$. Since the $k = 1$ case works, the result is valid for all odd integers.

6 For convenience, write $a = 1 + \sqrt{5}$ and $b = 1 - \sqrt{5}$. The formula $F_k = \dfrac{a^k - b^k}{2^k \sqrt{5}}$ works for F_1 and F_2. We assume that it works for F_{k-1} and F_k and prove that it works for F_{k+1}. Note first that $a^2 = (1 + \sqrt{5})^2 = 6 + 2\sqrt{5} = 4 + 2a$; similarly $b^2 = 4 + 2b$. Then $F_{k+1} = F_{k-1} + F_k =$

$$\dfrac{a^{k-1} - b^{k-1}}{2^{k-1}\sqrt{5}} + \dfrac{a^k - b^k}{2^k \sqrt{5}}$$

$$= \dfrac{2a^{k-1} - 2b^{k-1} + a^k - b^k}{2^k \sqrt{5}}$$

$$= \dfrac{a^{k-1}(2 + a) - b^{k-1}(2 + b)}{2^k \sqrt{5}}$$

$$= \dfrac{a^{k-1}(4 + 2a) - b^{k-1}(4 + 2b)}{2^{k+1}\sqrt{5}} = \dfrac{a^{k+1} - b^{k+1}}{2^{k+1}\sqrt{5}}.$$

8 The product rule says that $D(f_1 f_2) = f_1' f_2 + f_1 f_2'$. Assume that $D(f_1 f_2 f_3 \cdots f_{k-1} f_k) = f_1' f_2 f_3 \cdots f_{k-1} f_k + f_1 f_2' f_3 \cdots f_{k-1} f_k + f_1 f_2 f_3' \cdots f_{k-1} f_k + \cdots + f_1 f_2 f_3 \cdots f_{k-1}' f_k + f_1 f_2 f_3 \cdots f_{k-1} f_k'$. Then
$D(f_1 f_2 f_3 \cdots f_{k-1} f_k f_{k+1}) = D[(f_1 f_2 f_3 \cdots f_{k-1} f_k) f_{k+1}] = (f_1 f_2 f_3 \cdots f_{k-1} f_k) f_{k+1}' + D(f_1 f_2 f_3 \cdots f_{k-1} f_k) f_{k+1} = (f_1' f_2 f_3 \cdots f_{k-1} f_k + f_1 f_2' f_3 \cdots f_{k-1} f_k + f_1 f_2 f_3' \cdots f_{k-1} f_k + \cdots + f_1 f_2 f_3 \cdots f_{k-1}' f_k + f_1 f_2 f_3 \cdots f_{k-1} f_k') f_{k+1} + f_1 f_2 f_3 \cdots f_{k-1} f_k f_{k+1}'$, proving that the rule works in the $k + 1$ case.

10 (a) It equals 0.

(b) We already know that $D^2(P_1) = D^2(ax + b) = 0$. Assume that $D^{k+1}(P_k) = 0$. Write $P_{k+1}' = P_k$ in token of the fact that a polynomial of degree $k + 1$ has derivative of degree k. Then $D^{k+2}(P_{k+1}) = D^{k+1}(D(P_{k+1})) = D^{k+1}(P_k) = 0$, so the rule works for the $k + 1$ case.

F The Converse of a Statement

2 (a) If a^2 is odd, then a is odd. True. If a is even then a^2 is even, so if a^2 is odd, a must be odd.

(b) If $a + a$ is even, then a is odd. False; consider $2 + 2 = 4$.

4 (a) If a^2 is rational, then a is rational. False; consider $a = \sqrt{2}$.

F The Converse of a Statement

(b) If $2a$ is rational, then a is rational. True; if $2a = \frac{p}{q}$, then $a = \frac{p}{2q}$.

6 (a) If a^2 is a multiple of 6, then a is a multiple of 6. True. If a^2 is a multiple of 6, then it must be a multiple of 2 and 3. Hence a must be a multiple of 2 and 3, and hence 6.

   (b) If a^2 is a multiple of 4, then a is a multiple of 4. False; consider $a = 2$.

8 False; consider a rhombus.

10 False; consider the hexagon with vertices $(1, 0)$, $(1, 1)$, $(0, 1)$, $(-1, 0)$, $(-1, -1)$, and $(0, -1)$.

12 True

14 True

16 False; consider $f(x) = x^3$.

18 False; consider $f(x) = |x|$.

20 False; consider $f(x) = \int e^{x^2}\, dx$.

22 False; consider $a_n = (-1)^{n-1}/n$.

24 False; see Example 2 in Sec. 16.6.

G Conic Sections

G.1 Conic Sections

2 $c = 3$ and $2a = 14$, so $a = 7$; then $b^2 = a^2 - c^2 = 49 - 9 = 40$, so $\frac{x^2}{40} + \frac{y^2}{49} = 1$.

4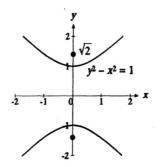

6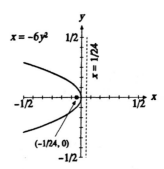

8 $x^2 = -20y$

12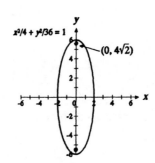

14

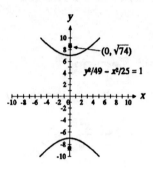

16

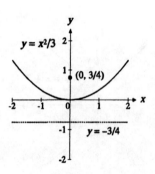

18

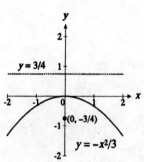

20 Take an xy coordinate system with origin at the garden's center and x axis parallel to the longer side. Drive stakes at $(\pm 3, 0)$, the foci. Tie a 10-foot rope to the stakes and trace out the ellipse.

22 The distance from the focus to the directrix is 7, so the vertex is $(2, 1/2)$ and $c = 7$. We have $(x - 2)^2 = 14\left(y - \dfrac{1}{2}\right)$, $x^2 - 4x + 4 = 14y - 7$, $x^2 - 4x + 11 = 14y$.

24 (b) The piece with A as its focus.

G.2 Translation of Axes and the Graph of $Ax^2 + Cy^2 + Dx + Ey + F = 0$

2 Parabola; focus $(1, 5/4)$

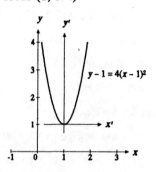

4 Parabola; focus $(-3, 1/4)$

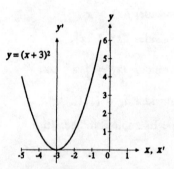

6 Parabola; focus $(3, 17/8)$

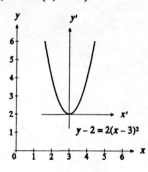

8 Circle; center $(-3, 0)$

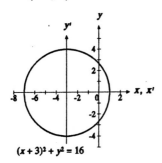

10 Hyperbola; foci $(1 \pm \sqrt{13}, 0)$, asymptotes
$y = \pm \frac{3}{2}(x - 1)$

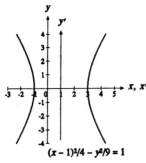

12 Ellipse; foci $(2, \pm\sqrt{3})$

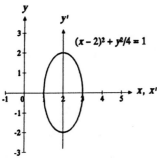

14 $x' = x - 1$, $y' = y - 2$, so $(x')^2 + 4(y')^2 = -4$; the graph is empty.

16 $x' = x - 1$, $y' = y - 3$; $\dfrac{(x')^2}{4} + \dfrac{(y')^2}{13} = 1$

18 $x' = x - 1$, $y' = y - 2$; $(y')^2 - \dfrac{(x')^2}{3} = 1$

20 $x' = x - 5$, $y' = y - 1$, $\dfrac{(x')^2}{25} + \dfrac{(y')^2}{16} = 1$

22 $x' = x - 4$, $y' = y - 1$, $(y')^2 = 4x'$

24 (c) When $A = C$ and $D^2 + E^2 > 4AF$.

G.3 Rotation of Axes and the Graph of $Ax^2 + Bxy + Cy^2 + Dx + Ey + F = 0$

2 $41x^2 + 24xy + 34y^2 - 25 = 0$

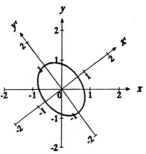

4 $5x^2 + 6xy + 5y^2 - 8 = 0$

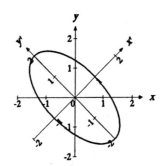

6 $3x^2 + 2\sqrt{3}xy + y^2 + 2x + 2\sqrt{3}y = 0$

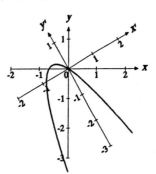

8 $7x^2 - 48xy - 7y^2 - 25 = 0$

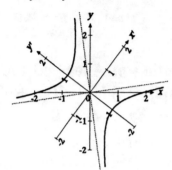

10 $y = \dfrac{x+1}{x+3}$, so $xy + 3y = x + 1$, or $xy - x + 3y - 1 = 0$, where $x \neq -3$. Then $\tan 2\theta = \dfrac{B}{A-C}$

$= \dfrac{1}{0-0}$, undefined; so $2\theta = \pi/2$ and $\theta = \pi/4$.

Then $x = x' \cdot \dfrac{1}{\sqrt{2}} - y' \cdot \dfrac{1}{\sqrt{2}} = \dfrac{1}{\sqrt{2}}(x' - y')$ and $y = \dfrac{1}{\sqrt{2}}(x' + y')$, so the equation becomes

$\dfrac{1}{\sqrt{2}}(x' - y') \cdot \dfrac{1}{\sqrt{2}}(x' + y') - \dfrac{1}{\sqrt{2}}(x' - y') +$

$\dfrac{3}{\sqrt{2}}(x' + y') - 1 = 0,$

$(x')^2 - (y')^2 + 2\sqrt{2}x' + 4\sqrt{2}y' = 2$. By completing the squares, we obtain $\dfrac{(y' - 2\sqrt{2})^2}{4} - \dfrac{(x' + \sqrt{2})^2}{4} = 1$.

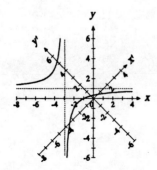

12 $y = \dfrac{x^2 + 2x + 1}{x - 1}$, so $xy - y = x^2 + 2x + 1$ and

$x^2 - xy + 2x + y + 1 = 0$, where $x \neq -1$. Then

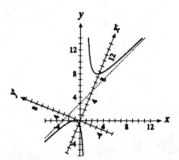

$\tan 2\theta = \dfrac{B}{A-C} = \dfrac{-1}{1-0} = -1$, so $2\theta = \dfrac{3\pi}{4}$

and $\theta = \dfrac{3\pi}{8}$; we have $\sin\theta = \dfrac{1}{2}\sqrt{2 + \sqrt{2}}$ and

$\cos\theta = \dfrac{1}{2}\sqrt{2 - \sqrt{2}}$, by use of the half-angle

formulas. Upon transformation of the equation, we

obtain $\dfrac{1 - \sqrt{2}}{2}(x')^2 + \dfrac{1 + \sqrt{2}}{2}(y')^2 +$

$\dfrac{2\sqrt{2 - \sqrt{2}} + \sqrt{2 + \sqrt{2}}}{2}x'$

$+ \dfrac{\sqrt{2 - \sqrt{2}} - 2\sqrt{2 + \sqrt{2}}}{2}y' + 1 = 0.$

14 $x^2 + xy + y^2 + 3x + 2y = 0$, $A = 1$, $B = 1$, $C = 1$, $D = 3$, $E = 2$, and $F = 0$, so the discriminant $\mathcal{D} = 1^2 - 4 \cdot 1 \cdot 1 = -3 < 0$; an ellipse.

16 $3x^2 - xy - y^2 - 1 = 0$, $A = 3$, $B = -1$, $C = -1$, $D = E = 0$, and $F = -1$, so $\mathcal{D} = (-1)^2 - 4 \cdot 3(-1) = 13 > 0$; a hyperbola.

22 (a) $(3x - y + 1)^2 = 0$, $3x - y + 1 = 0$, $y = 3x + 1$

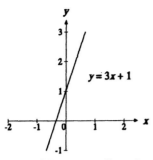

(b) $(x + y + 1)(x - y - 2) = 0$, so $x + y + 1 = 0$ or $x - y - 2 = 0$; these are the lines $y = -x - 1$ and $y = x - 2$.

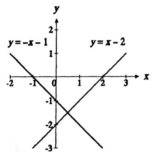

(c) $x^2 - y^2 = 0$, $(x - y)(x + y) = 0$, so $y = x$ or $y = -x$.

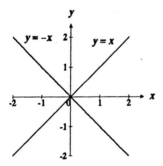

24 $(x', y') = (0, \pm\sqrt{5})$, $(x, y) = (2, -1)$ and $(-2, 1)$

26 $(x', y') = \left(-\dfrac{\sqrt{3}}{4}, \dfrac{1}{4}\right)$, $(x, y) = \left(-\dfrac{1}{2}, 0\right)$

G.4 Conic Sections in Polar Coordinates

2 By Eq. (1), $\dfrac{r}{p - r\cos(\theta - B)} = e$, so $r = ep - er\cos(\theta - B)$; hence $r + er\cos(\theta - B) = ep$ and $r[1 + e\cos(\theta - B)] = ep$, so $r = \dfrac{ep}{1 + e\cos(\theta - B)}$, which is Eq. (2).

4 (b)

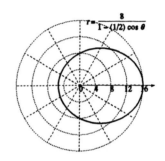

(c) $\dfrac{1}{2}\left(\dfrac{16}{3} + 16\right) = \dfrac{32}{3}$

H Logarithms and Exponentials Defined through Calculus

H.1 The Natural Logarithm Defined as a Definite Integral

2 $L'(x) = 1/x$, so $L''(x) = -1/x^2 < 0$, so the graph of $y = L(x)$ is concave downward.

4 Using four rectangles that overestimate the area for $L(2)$, we have $L(2) < 1 \cdot \dfrac{1}{4} + \dfrac{4}{5} \cdot \dfrac{1}{4} + \dfrac{4}{6} \cdot \dfrac{1}{4} + \dfrac{4}{7} \cdot \dfrac{1}{4} = \dfrac{1}{4} + \dfrac{1}{5} + \dfrac{1}{6} + \dfrac{1}{7} \approx 0.76$. Therefore $L(2) < 1$, so it must be that $e > 2$.

6 (a) Left endpoints give an overestimating sum, so
$$L(2) < 1 \cdot \frac{1}{n} + \frac{n}{n+1} \cdot \frac{1}{n} + \cdots + \frac{n}{2n-1} \cdot \frac{1}{n}$$
$$= \frac{1}{n} + \frac{1}{n+1} + \frac{1}{n+2} + \cdots + \frac{1}{2n-1}.$$ For $n = 10$, we obtain an approximation for $L(2)$ of about 0.72; for $n = 30$, we get about 0.70.

8 (a) If $\sum_{i=1}^{n} \frac{1}{c_i}(x_i - x_{i-1})$ is an approximating sum for $\int_{1}^{b} \frac{1}{x}\,dx$, then $1 = x_0 < x_1 < \cdots < x_n = b$ and $x_{i-1} \le c_i \le x_i$. It follows that $a = ax_0 < ax_1 < \cdots < ax_n = ab$ and $ax_{i-1} < ac_i \le ax_i$, so that $\sum_{i=1}^{n} \frac{1}{ac_i}(ax_i - ax_{i-1})$ is an approximating sum for $\int_{a}^{ab} \frac{1}{x}\,dx$. But
$$\sum_{i=1}^{n} \frac{1}{ac_i}(ax_i - ax_{i-1}) = \sum_{i=1}^{n} \frac{1}{c_i}(x_i - x_{i-1}),$$ so $\int_{1}^{b} \frac{1}{x}\,dx$ and $\int_{a}^{ab} \frac{1}{x}\,dx$ have equal approximating sums, so they must be equal.

(b) By the properties of definite integrals,
$$\int_{1}^{ab} \frac{1}{x}\,dx - \int_{1}^{a} \frac{1}{x}\,dx = \int_{a}^{ab} \frac{1}{x}\,dx$$
$$= \int_{1}^{b} \frac{1}{x}\,dx.$$

(c) Since $L(x) = \int_{1}^{x} \frac{1}{t}\,dt$ the result in (b) implies that $L(ab) - L(a) = L(b)$, so $L(ab) = L(a) + L(b)$.

H.2 Exponential Functions Defined in Terms of Logarithms

2 By Theorem 3, we know that $L(b^x) = xL(b)$, so by implicit differentiation we obtain $\frac{1}{b^x}\frac{d}{dx}(b^x) = L(b)$, so $\frac{d}{dx}(b^x) = b^x L(b)$.

4 By definition, $\frac{L(b^x)}{L(b)} = \frac{L(E(xL(b)))}{L(b)} = \frac{xL(b)}{L(b)} = x$, which shows that $L(x)/L(b)$ is the inverse function of b^x.

I The Taylor Series for $f(x, y)$

2 $f(x, y) = \frac{x}{y}$, $f_x = \frac{1}{y}$, $f_y = -\frac{x}{y^2}$, $f_{xx} = 0$,

$f_{yy} = \frac{2x}{y^3}$, $f_{xy} = -\frac{1}{y^2}$, $f_{xxx} = 0$,

$f_{xxy} = 0$, $f_{yyx} = \frac{2}{y^3}$, $f_{yyy} = -\frac{6x}{y^4}$, $f_{xyx} = 0$,

$f_{xyy} = \frac{2}{y^3}$, $f_{yxx} = 0$, $f_{yxy} = \frac{2}{y^3}$

4 $f(x, y) = \sin(x^2 + y^3)$,
$f_x = 2x\cos(x^2 + y^3)$, $f_y = 3y^2\cos(x^2 + y^3)$
$f_{xx} = -4x^2\sin(x^2 + y^3) + 2\cos(x^2 + y^3)$
$f_{yy} = -9y^4\sin(x^2 + y^3) + 6y\cos(x^2 + y^3)$
$f_{xy} = -6xy^2\sin(x^2 + y^3)$
$f_{xxx} = -8x^3\cos(x^2 + y^3) - 12x\sin(x^2 + y^3)$
$f_{yyy} = (6 - 27y^6)\cos(x^2 + y^3) - 54y^3\sin(x^2 + y^3)$

I The Taylor Series for f(x, y)

$f_{xxy} = f_{xyx} = f_{yxx} = -12x^2y^2\cos(x^2 + y^3) - 6y^2\sin(x^2 + y^3)$

$f_{yyx} = f_{yxy} = f_{xyy} = -18xy^4\cos(x^2 + y^3) - 12xy\sin(x^2 + y^3)$

6 $x^2y = 2 + 4(x - 1) + (y - 2) + 2(x - 1)^2 + 2(x - 1)(y - 2) + (x - 1)^2(y - 2)$

$V_1 - V_2 = 2 + 4(0.1) + (0.1) + 2(0.1)^2 + 2(0.1)(0.1) + (0.1)^2(0.1) - 2 = 0.541$

8 (a) $f(x, y) = \cos(x + y)$

$f_x = -\sin(x + y) = 0$ at $(0, 0)$

$f_y = -\sin(x + y) = 0$ at $(0, 0)$

$f_{xx} = -\cos(x + y) = -1$ at $(0, 0)$

$f_{yy} = -\cos(x + y) = -1$ at $(0, 0)$

$f_{xy} = -\cos(x + y) = -1$ at $(0, 0)$

$f(x, y) = 1 + \frac{-1}{2!}x^2 + \frac{-2}{2!}xy + \frac{-1}{2!}y^2 + \cdots$

$= 1 - \frac{1}{2}x^2 - xy - \frac{1}{2}y^2 + \cdots$

(b) $\cos t = 1 - \frac{t^2}{2!} + \cdots$, so $\cos(x + y)$

$= 1 - \frac{(x+y)^2}{2!} + \cdots = 1 - \frac{1}{2}x^2 - xy - \frac{1}{2}y^2 + \cdots$

10 $f(x, y) = 5 + 6x + 11y - 2x^2 - 3xy + 7y^2 + \cdots$

$f(0, 0) = 5$

$f_x = 6 - 4x - 3y + \cdots$

$f_y = 11 - 3x + 14y + \cdots$

$f_{xx} = -4 + \cdots$

$f_{yy} = 14 + \cdots$

$f_{xy} = -3 + \cdots$

$f_x(0, 0) = 6, f_y(0, 0) = 11, f_{xx}(0, 0) = -4,$
$f_{xy}(0, 0) = -3,$ and $f_{yy}(0, 0) = 14.$

J Theory of Limits

2 We must show that for any $\epsilon > 0$, there is a $\delta > 0$ such that, if $|x - a| < \delta$ and $x \neq a$, then $|f(g(x)) - f(g(a))| < \epsilon$. Since f is continuous at $g(a)$, there is a $\zeta > 0$ such that, if $|y - g(a)| < \zeta$, then $|f(y) - f(g(a))| < \epsilon$. (We can drop the condition $y \neq g(a)$ since if $y = g(a)$ then $|f(y) - f(g(a))| = 0 < \epsilon$.) Similarly, since g is continuous at a, there is a $\delta > 0$ such that, if $|x - a| < \delta$, then $|g(x) - g(a)| < \zeta$. For such x, letting $y = g(x)$ shows that $|f(g(x)) - f(g(a))| < \epsilon$, as required.

4 Let $f(x) = 1/x$. Since $g(a) \neq 0$, f is continuous at $g(a)$ by Exercise 1. Since g is continuous at a, $f \circ g$ is continuous at a. But $(f \circ g)(x) = f(g(x)) = \frac{1}{g(x)}$. Thus $1/g(x)$ is continuous at a.

6 Given $\epsilon > 0$, there is a $\delta > 0$ such that, if $|x - a| < \delta$ and $x \neq a$, then $|g(x) - B| < \beta$, where β is the lesser of $|B|/2$ and $B^2\epsilon/2$. For such x, $|g(x)| > |B|/2$, so $\left|\frac{1}{g(x)} - \frac{1}{B}\right| = \frac{|g(x) - B|}{|B||g(x)|}$

$< \frac{B^2\epsilon/2}{|B||B|/2} = \epsilon$. Hence $\lim_{x \to a} \frac{1}{g(x)} = \frac{1}{B}$.

8 By Theorem 7, all polynomials are continuous. Then, by Exercise 5, any rational function (the ratio of two polynomials) is continuous at all points in its domain.

10 Given E, there is a D_0 such that if $x > D_0$ then $f(x) > E$. There is also a D_1 such that if $x > D_1$ then $g(x) > 0$. Let D be the larger of D_0 and D_1. For $x > D$, $f(x) + g(x) > E + 0 = E$. Thus

$$\lim_{x \to \infty} [f(x) + g(x)] = \infty.$$

12 Given $\epsilon > 0$, there is a D_0 such that if $x > D_0$ then $|f(x) - A| < \epsilon/2$. There is also a D_1 such that if $x > D_1$ then $|g(x) - B| < \epsilon/2$. Let D be the larger of D_0 and D_1. For $x > D$, $|(f(x) + g(x)) - (A + B)| = |(f(x) - A) + (g(x) - B)| \leq |(f(x) - A)| + |(g(x) - B)| < \dfrac{\epsilon}{2} + \dfrac{\epsilon}{2} = \epsilon.$

Hence $\lim_{x \to \infty} (f(x) + g(x)) = A + B$.

14 Given E, there is a D_0 such that if $x > D_0$ then $f(x) > 2E/A$. There is also a D_1 such that if $x > D_1$ then $|g(x) - A| < A/2$. Let D be the larger of D_0 and D_1. For $x > D$, $g(x) > A/2$, so $f(x)g(x) > \dfrac{2E}{A} \dfrac{A}{2} = E$. Thus $\lim_{x \to \infty} f(x)g(x) = \infty$.

K The Interchange of Limits

K.2 The Derivative of $\int_a^b f(x, y)\, dx$ with Respect to y

2 $F(y) = \int_a^b \cos xy\, dx = \dfrac{\sin xy}{y}\bigg|_a^b =$

$\dfrac{1}{y} \sin by - \dfrac{1}{y} \sin ay$, so $F'(y) =$

$\dfrac{1}{y}(b \cos by - a \cos ay) - \dfrac{1}{y^2}(\sin by - \sin ay)$.

Now $\dfrac{\partial}{\partial y}(\cos xy) = -x \sin xy$ and by Formula 57,

$\int_a^b (-x \sin xy)\, dx = -\left(\dfrac{1}{y^2} \sin xy - \dfrac{1}{y} x \cos xy\right)\bigg|_a^b$

$= -\dfrac{1}{y^2} \sin by + \dfrac{1}{y} b \cos by + \dfrac{1}{y^2} \sin ay -$

$\dfrac{1}{y} a \cos ay$, which agrees with the previous result.

4 $G(u, v, w) = \int_u^v f(w, x)\, dx$

(a) $\dfrac{\partial G}{\partial v} = f(w, v)$

(b) $\dfrac{\partial G}{\partial u} = -f(w, u)$

(c) $\dfrac{\partial G}{\partial w} = \int_u^v f_w(w, x)\, dx$

6 $G(u, v) = \int_0^u e^{-vx^2}\, dx$

(a) $\dfrac{\partial G}{\partial u} = e^{-vu^2}$

(b) $\dfrac{\partial G}{\partial v} = -\int_0^u x^2 e^{-vx^2}\, dx$

K.3 The Interchange of Limits

2 $\lim_{x \to 0}\left(\lim_{n \to \infty} \dfrac{nx}{1 + nx}\right) = \lim_{x \to 0}\left(\lim_{n \to \infty} \dfrac{x}{x + 1/n}\right) = \lim_{x \to 0} 1$

$= 1$, while $\lim_{n \to \infty}\left(\lim_{x \to 0} \dfrac{nx}{1 + nx}\right) = \lim_{n \to \infty} \dfrac{0}{1} = 0.$

4 $\lim_{x \to 0}\left(\lim_{y \to 0} \dfrac{x^2}{x^2 + y^2}\right) = \lim_{x \to 0} \dfrac{x^2}{x^2} = 1$, while

$\lim_{y \to 0}\left(\lim_{x \to 0} \dfrac{x^2}{x^2 + y^2}\right) = \lim_{y \to 0} \dfrac{0}{y^2} = 0.$

K The Interchange of Limits

6 $\lim\limits_{x\to\infty}\left(\lim\limits_{y\to\infty}\dfrac{x^2}{x^2+y^2+1}\right) = \lim\limits_{x\to\infty} 0 = 0$, while

$\lim\limits_{y\to\infty}\left(\lim\limits_{x\to\infty}\dfrac{x^2}{x^2+y^2+1}\right) = \lim\limits_{y\to\infty} 1 = 1.$

8 Example 3 shows that f_{xy} and f_{yx} both exist at $(0, 0)$, although their values do not agree. For points apart from the origin, $f(x, y) = xy\cdot\dfrac{x^2-y^2}{x^2+y^2} =$

$\dfrac{x^3y - xy^3}{x^2+y^2}$, which we may differentiate with the

standard rules to compute f_{xy} and f_{yx}. Hence these partial derivatives exist for all (x, y).

10 $\lim\limits_{\Delta t\to 0}\left[\lim\limits_{t\to a}\dfrac{f(t+\Delta t)-f(t)}{g(t+\Delta t)-g(t)}\right] = \lim\limits_{\Delta t\to 0}\dfrac{f(a+\Delta t)-0}{g(a+\Delta t)-0}$

$= \lim\limits_{x\to a}\dfrac{f'(x)}{g'(x)}$, while $\lim\limits_{t\to a}\left[\lim\limits_{\Delta t\to 0}\dfrac{f(t+\Delta t)-f(t)}{g(t+\Delta t)-g(t)}\right]$

$= \lim\limits_{t\to a}\left[\lim\limits_{\Delta t\to 0}\dfrac{\dfrac{f(t+\Delta t)-f(t)}{\Delta t}}{\dfrac{g(t+\Delta t)-g(t)}{\Delta t}}\right] = \lim\limits_{t\to a}\dfrac{f'(t)}{g'(t)}.$

L The Jacobian

L.1 Magnification and the Jacobian

2 (a) The magnifying effect of G is

$\left|\left(\dfrac{d}{ad-bc}\right)\left(\dfrac{a}{ad-bc}\right) - \left(\dfrac{-b}{ad-bc}\right)\left(\dfrac{-c}{ad-bc}\right)\right|$

$= \left|\dfrac{ad-bc}{(ad-bc)^2}\right| = \dfrac{1}{|ad-bc|}.$

(b) Since F and G are inverses, the product of their magnifications is the magnification of the identity map, namely 1.

4 (a) $F(1, 0) = (3, 4)$

$F(0, 1) = (1, 2)$

$F(-1, 0) = (-3, -4)$

$F(0, -1) = (-1, -2)$

(b)

(c) By Exercise 1(a), $u = x - \dfrac{1}{2}y$ and $v =$

$-2x + \dfrac{3}{2}y$, so $\left(x - \dfrac{1}{2}y\right)^2 + \left(-2x + \dfrac{3}{2}y\right)^2 =$

1; simplifying, we obtain $5x^2 - 7xy + \dfrac{5}{2}y^2$

$= 1.$

(d) Since $(-7)^2 - 4\cdot 5\cdot 5/2 < 0$, the image is an ellipse.

(e) The area of the image is $|3\cdot 2 - 1\cdot 4|\pi = 2\pi.$

6 (a) For purposes of parts (b) and (c), let $P_1 = (0, 1)$, $P_2 = (1/2, 1/2)$, and $P_3 = (1, 0)$. Note that all of them lie on the line $u + v = 1$, as required. (Infinitely many other choices are possible.)

(b)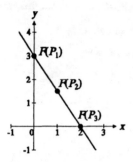

(c) The equation is $\frac{x}{2} + \frac{y}{3} = 1$. See (b).

8 (a)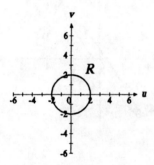

(b)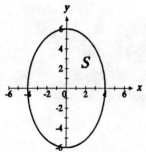

10 (a) We have $x = u + v$ and $y = u - v$. So $u = \frac{1}{2}(x + y)$ and $v = \frac{1}{2}(x - y)$.

(b) We have $\left(\frac{x+y}{2}\right) + 2\left(\frac{x-y}{2}\right) = 1$.

Simplifying, we obtain $3x - y = 2$.

(c) We have $\left(\frac{x+y}{2}\right)^2 + \left(\frac{x-y}{2}\right)^2 = 9$.

Simplifying, we obtain $x^2 + y^2 = 18$.

12 (a)

(u, v)	$(x, y) =$ $(2u + 3v, u + v)$
$(1, 0)$	$(2, 1)$
$(0, 1)$	$(3, 1)$
$(-1, 0)$	$(-2, -1)$
$(0, -1)$	$(-3, -1)$

(b) C is swept out clockwise.

14 (a) $F(3, \pi/6) = (3\sqrt{3}/2, 3/2)$, $F(4, \pi/6) = (2\sqrt{3}, 2)$, $F(5, \pi/6) = (5\sqrt{3}/2, 5/2)$, $F(6, \pi/6) = (3\sqrt{3}, 3)$.

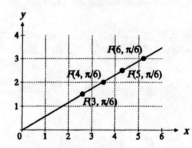

(b) We have $x = u \cos \pi/6$, $y = u \sin \pi/6$, $u > 0$, and the image is $x = \sqrt{3}y$, $y > 0$.

16 (a) $3 = u \cos v$, $3 = u \sin v$, $u > 0$, $0 \le v \le 2\pi$ imply that $\cos v = \sin v$ and thus $v = \pi/4$. Then $3 = u \cdot \frac{1}{\sqrt{2}}$ and $u = 3\sqrt{2}$. Thus

$$(u, v) = \left(3\sqrt{2}, \frac{\pi}{4}\right).$$

(b) $(u, v) = \left(8, \frac{\pi}{3}\right)$

(c) $(u, v) = (6, 0)$

(d) $(u, v) = \left(7, \frac{3\pi}{2}\right)$

18

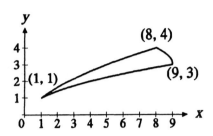

20 The determinant of the matrix $[a_{ij}]$.

L.2 The Jacobian and Change of Coordinates

2 Magnification of F at (u, v) is $\text{abs}\begin{vmatrix} -1/u^2 & 0 \\ 0 & -1/v^2 \end{vmatrix}$

$= \dfrac{1}{u^2 v^2}$.

(a) Magnification at $(2, 3)$ is $\dfrac{1}{2^2 3^2} = \dfrac{1}{36}$.

(b) Magnification at $(1/2, 4)$ is $\dfrac{1}{(1/2)^2 4^2} = \dfrac{1}{4}$.

4 Magnification of F at (u, v) is

$\text{abs}\begin{vmatrix} \dfrac{v^2 - u^2}{(u^2 + v^2)^2} & \dfrac{-2uv}{(u^2 + v^2)^2} \\ \dfrac{-2uv}{(u^2 + v^2)^2} & \dfrac{u^2 - v^2}{(u^2 + v^2)^2} \end{vmatrix} = \dfrac{1}{(u^2 + v^2)^2}$.

(a) Magnification of F at $(3, 1)$ is given by

$\dfrac{1}{(3^2 + 1^2)^2} = \dfrac{1}{100}$.

(b) Magnification of F at $(1, 0)$ is given by

$\dfrac{1}{(1^2 + 0^2)^2} = 1$.

6 Let S be the image of R. Then $\dfrac{\text{Area of } S}{\text{Area of } R} \approx 3$, so

Area of $S \approx 3(0.05) = 0.15$.

8 Let $f(u, v) = (au, bv)$ and let R be the portion of the disk $u^2 + v^2 \le 1$ that lies in the first quadrant. Then S is the image of R under F and $\int_S y\, dA =$

$\int_R (bv) \left| \dfrac{\partial(x, y)}{\partial(u, v)} \right| dA^* = \int_R (bv)(ab)\, dA^*$

$= ab^2 \int_R v\, dA^* = ab^2 \int_0^{\pi/2} \int_0^1 r^2 \sin\theta\, dr\, d\theta$

$= \dfrac{ab^2}{3}$.

10 We first claim F is one-to-one. Let $f(u_1, v_1) =$

$f(u_2, v_2)$. Then $u_1^3 + v_1 = u_2^3 + v_2$ and $\dfrac{v_1^3}{3} = \dfrac{v_2^3}{3}$.

The latter equality implies that $v_1 = v_2$. So $u_1^3 =$

u_2^3 and $u_1 = u_2$. We have $\dfrac{\partial(x, y)}{\partial(u, v)} = \begin{vmatrix} 3u^2 & 1 \\ 0 & v^2 \end{vmatrix} =$

$3u^2 v^2$. Thus Area of $S = \int_S dA = \int_R 3u^2 v^2\, dA^*$

$= \int_1^2 \int_1^2 3u^2 v^2\, dv\, du = \dfrac{49}{3}$.

12 The solution for (a) is provided. (b), (c), and (d) are essentially the same.

(a) $u = 1$ implies that $x = 1 - v^2$ and $y = 2v$, so

$x = 1 - \left(\dfrac{y}{2}\right)^2$. Conversely, if (x, y) satisfies

$x = 1 - \left(\dfrac{y}{2}\right)^2$, then $(u, v) = (1, y/2)$ will be

mapped to (x, y). Hence the image of $u = 1$

is the curve $x = 1 - \left(\dfrac{y}{2}\right)^2$.

(e)

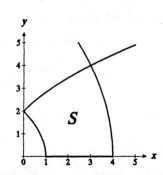

(f) Let F be the map. We claim that F on R is one-to-one. Suppose $f(u_1, v_1) = f(u_2, v_2)$; then $u_1^2 - v_1^2 = u_2^2 - v_2^2$ and $u_1 v_1 = u_2 v_2$, so

$$(v_2^2 - v_1^2)(v_2^2 + u_1^2)$$
$$= v_2^4 + (u_1^2 - v_1^2)v_2^2 - u_1^2 v_1^2$$
$$= v_2^4 + (u_2^2 - v_2^2)v_2^2 - u_2^2 v_2^2 = 0.$$ Since $v_2^2 + u_1^2 > 0$, $v_2^2 - v_1^2 = 0$, and $v_1 = v_2$, from which $u_1 = u_2$ also follows. We have

$$\dfrac{\partial(x,y)}{\partial(u,v)} = \begin{vmatrix} 2u & -2v \\ 2v & 2u \end{vmatrix} = 4u^2 + 4v^2, \text{ so Area}$$

of $S = \int_S dA = \int_R (4u^2 + 4v^2)\, dA^* =$

$\int_1^2 \int_0^1 (4u^2 + 4v^2)\, dv\, du = \dfrac{32}{3}.$

14 (a) First note that the mapping is one-to-one since we can solve the equations uniquely for u and v. The Jacobian is $\dfrac{\partial(x,y)}{\partial(u,v)} =$

$\begin{vmatrix} 1 - 4uv - 4u^3 & -2v - 2u^2 \\ 2u & 1 \end{vmatrix} = 1$, so Area of S

$= \int_S dA = \int_R dA^* = $ Area of R.

(b)

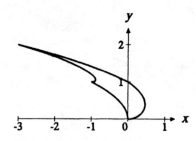

(c) By (a), the area of the image is the same as that of the square, which is 1.

16 (a) Since $\dfrac{\partial(x,y)}{\partial(u,v)} = \begin{vmatrix} 2 & 1 \\ 3 & 2 \end{vmatrix} = 1$, the mapping does not change areas.

(b)

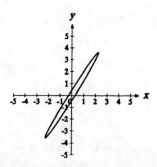

(c) $\int_S x^2\, dA = \int_R (2u + v)^2 \cdot 1\, dA^*$

$= \int_0^{2\pi} \int_0^1 (2r\cos\theta + r\sin\theta)^2\, r\, dr\, d\theta$

$= \dfrac{5\pi}{4}.$

18 $\int_S y^2\, dA = \int_{8/7}^{12/7} \int_{(1-2x)/3}^{(x-2)/2} y^2\, dy\, dx +$

$\int_{12/7}^{17/7} \int_{(1-2x)/3}^{(3-2x)/3} y^2\, dy\, dx + \int_{17/7}^{3} \int_{(x-5)/2}^{(3-2x)/3} y^2\, dy\, dx$

$= \dfrac{170}{343}$

20 $\dfrac{\partial(x,y)}{\partial(u,v)} = \begin{vmatrix} \cos v & -u\sin v \\ \sin v & u\cos v \end{vmatrix} = u.$ By Example 4 of

L The Jacobian

L.1, $\dfrac{\partial(u, v)}{\partial(x, y)} = \begin{vmatrix} \dfrac{x}{\sqrt{x^2+y^2}} & \dfrac{y}{\sqrt{x^2+y^2}} \\ \dfrac{-y}{x^2+y^2} & \dfrac{x}{x^2+y^2} \end{vmatrix}$

$= \dfrac{x^2+y^2}{(x^2+y^2)\sqrt{x^2+y^2}} = \dfrac{1}{\sqrt{x^2+y^2}} = \dfrac{1}{u}$. Thus

$\dfrac{\partial(x, y)}{\partial(u, v)} \cdot \dfrac{\partial(u, v)}{\partial(x, y)} = u \cdot \dfrac{1}{u} = 1.$

22 $\begin{vmatrix} \dfrac{\partial}{\partial z}(r\cos\theta) & \dfrac{\partial}{\partial r}(r\cos\theta) & \dfrac{\partial}{\partial\theta}(r\cos\theta) \\ \dfrac{\partial}{\partial z}(r\sin\theta) & \dfrac{\partial}{\partial r}(r\sin\theta) & \dfrac{\partial}{\partial\theta}(r\sin\theta) \\ \dfrac{\partial z}{\partial z} & \dfrac{\partial z}{\partial r} & \dfrac{\partial z}{\partial\theta} \end{vmatrix}$

$= \begin{vmatrix} 0 & \cos\theta & -r\sin\theta \\ 0 & \sin\theta & r\cos\theta \\ 1 & 0 & 0 \end{vmatrix} = r$

24 Since S has area 6/7, the density is $7M/6$, so I_y

$= \int_S \dfrac{7M}{6} x^2\, dA = \int_R \dfrac{7M}{6}\left(\dfrac{3u+2v}{7}\right)^2\left(\dfrac{1}{7}\right) dA^*$

$= \dfrac{M}{6\cdot 7^2}\int_2^5\int_1^3 (9u^2 + 12uv + 4v^2)\, dv\, du$

$= \dfrac{M}{294}\cdot 1310 = \dfrac{655M}{157}.$

26 (a) Let $u = y/x^2$ and $v = x/y^2$. As (x, y) varies over S, (u, v) varies over the rectangle R described by $1 \le u \le 2,\ 1 \le v \le 2$. Since $y = x^2 u$, $v = \dfrac{x}{(x^2 u)^2} = x^{-3} u^{-2}$. Hence $u^2 v = x^{-3}$, $x = u^{-2/3} v^{-1/3}$, and $y = u^{-1/3} v^{-2/3}$. The Jacobian is $\begin{vmatrix} -\dfrac{2}{3} u^{-5/3} v^{-1/3} & -\dfrac{1}{3} u^{-2/3} v^{-4/3} \\ -\dfrac{1}{3} u^{-4/3} v^{-2/3} & -\dfrac{2}{3} u^{-1/3} v^{-5/3} \end{vmatrix} =$

$\dfrac{1}{3} u^{-2} v^{-2}$. So the area of S is $\int_S dA =$

$\int_R \left|\dfrac{1}{3} u^{-2} v^{-2}\right| dA^* = \dfrac{1}{3}\int_1^2\int_1^2 u^{-2} v^{-2}\, dv\, du$

$= \dfrac{1}{12}.$

(b) Since S is symmetric with respect to the line $y = x$, $\bar{y} = \bar{x} = \dfrac{1}{\text{Area}}\int_S x\, dA$

$= 12 \int_R x \left|\dfrac{1}{3} u^{-2} v^{-2}\right| dA^*$

$= 4 \int_1^2\int_1^2 (u^{-2/3} v^{-1/3}) u^{-2} v^{-2}\, dv\, du$

$= 4 \int_1^2\int_1^2 u^{-8/3} v^{-7/3}\, dv\, du$

$= 4 \int_1^2 u^{-8/3}\left(-\dfrac{3}{4}\right)(2^{-4/3} - 1)\, du$

$= 4\left(-\dfrac{3}{4}\right)(2^{-4/3} - 1)\left(-\dfrac{3}{5} u^{-5/3}\right)\Big|_1^2$

$= \dfrac{9}{40}(9 - 2^{4/3} - 2^{5/3})$. The centroid is

$\left(\dfrac{9}{40}(9 - 2^{4/3} - 2^{5/3}),\ \dfrac{9}{40}(9 - 2^{4/3} - 2^{5/3})\right).$

M Linear Differential Equations with Constant Coefficients

2 Trying $y_p = A\cos x + B\sin x$, we find $A = 2/5$ and $B = 1/5$. The general solution is $y = \frac{2}{5}\cos x + \frac{1}{5}\sin x + Ce^{-2x}$.

4 Trying $y_p = Ae^{2x}$ gives $A = 3/5$; the general solution is $y = \frac{3}{5}e^{2x} + Ce^{x/3}$.

6 Trying $y_p = (Ax + B)e^{2x}$ gives $A = 1/2$ and $B = -1/8$, so the general solution is $y = \frac{1}{8}(4x - 1)e^{2x} + Ce^{-2x}$.

8 The roots of $t^2 + 5t + 6 = 0$ are $t = -2$ and $t = -3$, so $y = C_1 e^{-2x} + C_2 e^{-3x}$.

10 The roots of $2t^2 - t + 3 = 0$ are $t = \frac{1 \pm \sqrt{-23}}{4} = \frac{1}{4} \pm \frac{\sqrt{23}}{4}i$, so $y = \left[C_1 \cos \frac{\sqrt{23}}{4}x + C_2 \sin \frac{\sqrt{23}}{4}x \right] e^{x/4}$.

12 The roots of $4t^2 + 9 = 0$ are $t = \pm\frac{3}{2}i$, so $y = C_1 \cos \frac{3x}{2} + C_2 \sin \frac{3x}{2}$.

14 The roots of $3t^2 - 2t + 3 = 0$ are $t = \frac{1}{3} \pm \frac{\sqrt{8}}{3}i$, so $y = \left[C_1 \cos \frac{\sqrt{8}}{3}x + C_2 \sin \frac{\sqrt{8}}{3}x \right] e^{x/3}$.

16 The roots of $t^2 + t + 1 = 0$ are $t = -\frac{1}{2} \pm \frac{\sqrt{3}}{2}i$, so $y = \left[C_1 \cos \frac{\sqrt{3}}{2}x + C_2 \sin \frac{\sqrt{3}}{2}x \right] e^{-x/2}$.

18 The roots of $t^2 - 3t + 4 = 0$ are $t = \frac{3}{2} \pm \frac{\sqrt{7}}{2}i$, so $y = \left[C_1 \cos \frac{\sqrt{7}}{2}x + C_2 \sin \frac{\sqrt{7}}{2}x \right] e^{3x/2}$.

20 Trying $y_p = Ax^2 + Bx + C$ gives $A = 1$, $B = -2$, and $C = 0$. Combining this with Exercise 16 gives $y = x^2 - 2x + \left[C_1 \cos \frac{\sqrt{3}}{2}x + C_2 \sin \frac{\sqrt{3}}{2}x \right] e^{-x/2}$.

22 Trying $y_p = e^{-2x}(A\cos x + B\sin x) + C\cos 3x + D\sin 3x$, we get $A = 1/2$, $B = -1/2$, $C = -\frac{7}{130}$, and $D = \frac{9}{130}$. The roots of $t^2 + 3t + 2$ are $t = -1$ and $t = -2$, so the general solution is $y = \frac{1}{2}e^{-2x}(\cos x - \sin x) + \frac{1}{130}(9\sin 3x - 7\cos 3x) + C_1 e^{-x} + C_2 e^{-2x}$.

26 The roots of $t^2 - k^2 = 0$ are $t = \pm k$, so $y = C_1 e^{kx} + C_2 e^{-kx}$.

Supplementary Conceptual Problems

These questions are to be answered in full sentences. When appropriate, diagrams may be included. The answers should be in your own words, not just copied from explanations in the text. The number bullet after each question identifies the chapter to which it corresponds.

1. If two positive quantities are both very small, but not zero, what can be said about (a) their product? (b) their quotient? ❷

2. (a) What is meant by the "radian measure" of an angle?
   (b) How is it related to "degrees"? ❷

3. (a) Give an intuitive definition of a continuous function.
   (b) Is $f(x) = 1/x$ continuous?
   (c) What is meant by "$f(x)$ is continuous at a"? ❷

4. (a) What is meant by "the derivative of a function"?
   (b) Give two different applications of the derivative.
   (c) Explain why these applications depend on the idea of the derivative. ❸

5. Why do we use radian measure in calculus instead of degrees? ❸

6. Find the derivative of sin x, using the definition of the derivative. ❸

7. (a) What is meant by "the second derivative of a function"?
   (b) Why is "acceleration" measured by a second derivative? ❹

8. If you were given only the graph of a function $y = f(x)$, how would you go about sketching the graph of its second derivative, $y = f''(x)$? ❹

9. Why is the derivative useful in finding a maximum or minimum of a function defined on a closed interval? ❹

10. (a) What are the assumptions and conclusions in the mean-value theorem?
    (b) Sketching a diagram, explain why the mean-value theorem is plausible.
    (c) What does it tell us about a function defined for all x whose derivative is always positive?
    (d) What does it tell us about a function whose derivative is always 0? ❹

11. What does the first derivative of a function tell us about its graph? ❹

12. What does the second derivative of a function tell us about its graph?

13. (a) What is Newton's method for estimating a root of an equation?
    (b) Using a diagram, derive Newton's formula.

14. Say that you know $f(0.98), f(1)$, and $f(1.02)$. What is the best estimate of $f'(0.98)$? What is the best estimate of $f'(1.02)$? $f''(1)$?

15. (a) Using a diagram, contrast Δf and df.
    (b) Assuming that $f'(a) \neq 0$, explain why df is a good estimate of Δf, in the sense that $\lim_{\Delta x \to 0} \frac{\Delta f}{df} = 1$.

16. (a) If you know $f(a)$ and $f'(a)$, how would you estimate $f(c)$ if c is "near" a?
    (b) Use a diagram to show why your method is plausible.

17. Say that you know that $f(3) = 0, f'(3) = 0$, and $f''(x) \leq 2$ for all x.
    (a) What can you say about $f(5)$?
    (b) Explain your answer in (a) in terms of moving automobiles.

18. Why is the error in using the differential, df, to estimate $\Delta f = f(a + \Delta x) - f(a)$, when Δx is small, approximately $f''(a)(\Delta x)^2/2$?

19. (a) What is the definition of the definite integral?
    (b) State two of its applications.
    (c) Explain why one of these applications is actually an application of the definite integral.

20. (a) Describe three ways to estimate a definite integral.
    (b) Which is the most efficient? The least efficient?
    (c) What formulas justify your answers to (b)?

21. (a) What shortcut is there for computing many definite integrals.
    (b) Does it work for all integrands? Explain.

22. If $f(x) = 0$ except at 1 and $f(1) = 3$, discuss $\int_0^2 f(x)\, dx$. Does it exist? If so, what is it? (Use the definition of definite integral.)

23. (a) What is the geometric meaning of the definite integral of "velocity" from time a to time b? (Do not assume that velocity always has the same sign.)
    (b) Explain.

Supplementary Conceptual Problems

24. (a) What is the link between the derivative and the definite integral?
    (b) Why is that link useful? ❺

25. Why, in calculus, do we prefer to use the base e for exponentials? ❻

26. (a) What is meant by e?
    (b) Using the definition of the derivative, find the derivative of $\log_b x$, where b is a fixed positive number. ❻

27. What is the relation of e to compound interest? ❻

28. (a) What are the assumptions in l'Hôpital's rule?
    (b) What are its conclusions? ❻

29. What is the difference between a "determinate limit" and an "indeterminate limit"? ❻

30. (a) What is meant by an "elementary function"?
    (b) Is the derivative of an elementary function always elementary?
    (c) Is an antiderivative of an elementary function always elementary? ❼

31. (a) What is meant by a convergent improper integral $\int_a^\infty f(x)\,dx$ (where $f(x)$ is continuous)?
    (b) Explain, using a diagram, why the convergence of $\int_a^\infty |f(x)|\,dx$ forces the convergence of $\int_a^\infty f(x)\,dx$. ❽

32. Obtain the formula for finding volume by (a) parallel slabs, (b) concentric shells. (In both cases use both "approximating sums" and also the informal approach.) ❽

33. (a) What is meant by "giving a curve parametrically"?
    (b) Give an example of a curve that is easy to describe parametrically but not explicitly in the form $y = f(x)$. ❾

34. Derive, with the aid of diagrams, the formula for finding area in terms of polar coordinates. ❾

35. Derive, with the aid of diagrams, the formula for finding the arc length of a curve given in the form $y = f(x)$. ❾

36. Derive, with the aid of diagrams, the formula for finding the arc length of a curve given parametrically. ❾

37. Derive, with the aid of diagrams, a formula for finding the area of a surface of revolution if the curve $y = f(x)$ is revolved about the line (a) $y = k$, (b) $x = k$. ⑨

38. (a) What is meant by "curvature"?
    (b) Find the curvature of a circle of radius a.
    (c) Derive the formula for computing curvature for a function $y = f(x)$. ⑨

39. A lawn sprinkler is to be made in the shape of a hemisphere. Small holes all of the same radius will be put in the portion above "latitude" 45°, as shown in the figure. Water issuing out of a hole at angle θ ("latitude θ") travels a horizontal distance $k \sin 2\theta$. (The number k depends only on the water pressure, not on θ.) How should the holes be drilled to ensure that the sprinkler waters the lawn fairly uniformly?

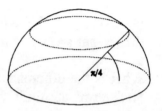

40. Why does the convergence of $\sum |a_n|$ imply the convergence of $\sum a_n$? ⑩

41. Why does the convergence of $\sum a_n c^n$ imply the convergence of $\sum a_n x^n$ for $|x| < |c|$? Hint: $a_n x^n = a_n (x/c)^n c^n$. ⑩

42. The staircase in the figure represents the harmonic series, so its area is $1 + \dfrac{1}{2} + \dfrac{1}{3} + \cdots$. The dashed lines break the area into smaller

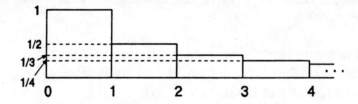

rectangles, one of area 1/2, two of area 1/6, etc. Hence the area is

$$1 + \frac{2}{6} + \frac{3}{12} + \cdots = 1 + \frac{1}{2} + \frac{1}{3} + \cdots. \text{ Thus}$$

$$1 + \frac{1}{2} + \frac{1}{3} + \frac{1}{4} + \cdots = \frac{1}{2} + \frac{1}{3} + \cdots. \text{ Can we conclude that } 1 = 0? \text{ ⑩}$$

Supplementary Conceptual Problems

43. A convergent alternating series has terms that diminish in absolute value. Why does the sum of the series differ from a partial sum by less than the absolute value of the first omitted term? (Include a diagram). ⑩

44. (a) Let $\sum_{n=1}^{\infty} a_n$ be a series of positive terms such that $\lim_{n \to \infty} \frac{a_{n+1}}{a_n} = r <$
    1. Explain why the series converges.

    (b) Let $\sum_{n=1}^{\infty} a_n$ be a series of positive terms such that $\lim_{n \to \infty} \left| \frac{a_{n+1}}{a_n} \right| = r <$
    1. Explain why the series converges. ⑩

45. If we expect a function $f(x)$ to be represented by a series $\sum a_n x^n$, why do we expect a_n to be $f^{(n)}(0)/n!$? ⑪

46. (a) Obtain the Maclaurin series for $\sin x$, $\cos x$, and e^x.
    (b) For which values do they represent the function? ⑪

47. How do complex numbers enable us to relate e^x, $\cos x$, and $\sin x$? ⑪

48. Describe some uses of the Taylor series representation of a function. ⑪

49. Explain why the error in using the Taylor series polynomial $P_3(x; 0)$ to approximate a function $f(x)$ involves the fourth derivative of $f(x)$. ⑪

50. Describe the general idea of "projection," illustrating it for line segments, vectors, and area of planar regions. In each case, explain how to compute the projection. ⑫

51. Explain why any plane has an equation of the form $Ax + By + Cz + D = 0$. ⑫

52. (a) How is the dot product defined?
    (b) How is it computed in terms of components?
    (c) Derive the formula in (b). ⑫

53. (a) How it the cross product defined?
    (b) What is its geometric description?
    (c) Obtain the description in (b) from the definition in (a). ⑫

54. (a) Define the derivative of a vector function $\mathbf{r} = \mathbf{G}(t)$.
    (b) Explain why $\mathbf{r}'$ is called the "velocity vector" if $\mathbf{G}$ describes a moving particle. ⑬

55. (a) Define the vectors **T** and **N** for a plane curve.
    (b) Develop the formula for the acceleration vector in the form $a\mathbf{T} + b\mathbf{N}$. ⑬

56. Sketch how the level curves of $f(x, y)$ might look near point P if (a) f has a local maximum at P, (b) f has a local minimum at P, and (c) f has neither a local maximum nor a local minimum at P. ⑭

57. (a) What is meant by the partial derivative $\partial f / \partial x$, where f is a function of x and y?
    (b) Drawing a graph of $z = f(x, y)$, show the geometric meaning of $\partial f / \partial x$. ⑭

58. (a) What is meant by a directional derivative of $f(x, y)$?
    (b) Show its geometric significance on a graph of $z = f(x, y)$. ⑭

59. Let $u = f(x, y, z)$. Explain why ∇f, evaluated at P_0, is perpendicular to the level surface of f that passes through P_0. ⑭

60. Obtain the formula for the directional derivative in terms of the gradient. ⑭

61. Sketch the key steps that led to the equation
$$\Delta f = \frac{\partial f}{\partial x} \Delta x + \frac{\partial f}{\partial y} \Delta y + \epsilon_1 \Delta x + \epsilon_2 \Delta y.$$ ⑭

62. Let $z = f(x, y)$ and $x_1 = g(y_1, y_2, y_3)$. Obtain the chain rule for $\partial z / \partial y_3$. ⑭

63. Write a short essay concerning the trouble that arises in the chain rule when a "middle" variable is also a "bottom" variable. ⑭

64. Devise a general definition of the definite integral, worded so that
$$\int_a^b f(x) \, dx, \quad \int_R f(P) \, dA, \text{ and } \int_R f(P) \, dV$$
are special cases. ⑮

65. Use some diagrams to explain why we expect $\int_R f(P) \, dA$ to equal a repeated integral of the form $\int_a^b \left(\int_{y_1(x)}^{y_2(x)} f(P) \, dy \right) dx$. ⑮

66. Explain the appearance of the factor r in the integrand when evaluating $\int_R f(P) \, dA$ in polar coordinates. ⑮

67. What does the expression $\rho^2 \sin \phi \, d\rho \, d\theta \, d\phi$ represent? Explain. ⑮

Supplementary Conceptual Problems

68. What theorem about line integrals generalizes the theorem that asserts, if $F' = f$, then $\int_a^b f(x)\,dx = F(b) - F(a)$. ⓖ

69. Prove that the gradient of $f(x, y)$ is perpendicular to the function's level curves. ⓖ

70. Prove that the gradient of $f(x, y, z)$ is perpendicular to the function's level surfaces. ⓖ

71. (a) What is meant by a "central field"?
    (b) Which central fields in the plane are conservative?
    (c) Which central fields in the plane have zero divergence?
    (d) Which central fields in space are conservative?
    (e) Which central fields in space have zero divergence? ⓖ

72. Say that **F** is a conservative vector field in the plane; therefore it is the gradient of a scalar field f.
    (a) Describe two ways of constructing f.
    (b) Is f unique?
    (c) Illustrate the two methods for your own choice of **F**. ⓖ

73. Let **F**(x, y) be defined everywhere except at the point P_0. Assume that it is divergence-free.
    (a) What line integrals are certainly equal to 0?
    (b) What line integrals, not necessarily equal to zero, are equal to each other? ⓖ

74. Explain the cancellation formula for line integrals of the form $\oint_C \mathbf{F}\cdot d\mathbf{r}$. ⓖ

75. Explain the cancellation formula for line integrals of the form $\oint_C \mathbf{F}\cdot \mathbf{n}\,ds$, where **n** is an exterior normal to the region that C bounds. ⓖ

76. (a) What is a "conservative" vector field?
    (b) Give other descriptions of such a field.
    (c) Why is the curl of such a field zero?
    (d) Is the converse of (c) true? ⓖ

77. (a) Explain how the radian measure of the angle subtended at a point by a planar curve can be represented by a line integral.
    (b) Define solid angle and "steradian."
    (c) Explain how the steradian measure of a solid angle is measured by a surface integral. ⓗ

78. Explain why the divergence theorem and Stokes' theorem are both generalizations of Green's theorem.

79. Left $\mathbf{F}(x, y, z) = \hat{\mathbf{r}}/r^2$. Let $\mathcal{S}$ be a surface that encloses the origin. Evaluate $\int_{\mathcal{S}} \mathbf{F} \cdot \mathbf{n}\, dS$ (a) using the divergence theorem, (b) using steradians.